COHERENCE AND QUANTUM OPTICS

COHERENCE AND QUANTUM OPTICS

Proceedings of the Third Rochester Conference on Coherence and Quantum Optics held at the University of Rochester, June 21-23, 1972

Edited by

L. Mandel and E. Wolf

Department of Physics and Astronomy
The University of Rochester
Rochester, New York

PLENUM PRESS · NEW YORK-LONDON · 1973

Library of Congress Catalog Card Number 73-76700
ISBN 0-306-30731-6

A Division of Plenum Publishing Corporation
227 West 17th Street, New York, N. Y. 10011

United Kingdom edition published by Plenum Press, London
A Division of Plenum Publishing Company, Ltd.
Davis House (4th Floor), 8 Scrubs Lane, Harlesden, London, NW10
6 SE, England

Printed in the United States of America

PREFACE

This volume presents the written versions of papers that were delivered at the Third Rochester Conference on Coherence and Quantum Optics, held on the campus of the University of Rochester during the three days of June 21-23, 1972. The Conference was a sequel to two earlier meetings devoted to the same field of modern physics, that were also held in Rochester in 1960 and in 1966.

The scope of the Conference was largely confined to basic problems in the general area of optical coherence and quantum optics, and excluded engineering applications that are well covered by other meetings. Approximately 250 scientists from 9 countries participated, most of whom are active workers in the field. Altogether 72 papers, including 26 invited papers, were presented in 17 sessions. The papers dealt mainly with the subjects of resonant pulse propagation, lasers, quantum electrodynamics and alternative theories, optical coherence, coherence effects in spontaneous emission, light scattering, optical correlation and fluctuation measurements, coherent light interactions and quantum noise. The program was organized by a committee consisting of

N. Bloembergen (Harvard University)
J.H. Eberly (University of Rochester)
E.L. Hahn (University of California at Berkeley)
H. Haken (University of Stuttgart, Germany)
M. Lax (City College of New York)
B.J. Thompson (University of Rochester)
L. Mandel (University of Rochester) }joint secretaries
E. Wolf (University of Rochester)

We are grateful to the Air Force Office of Scientific Research for sponsoring the Conference jointly with the University of Rochester. We are especially indebted to Dr. Marshall C. Harrington of the Air Force Office of Scientific Research for much valuable help and for the personal interest he took in all matters pertaining to the organization of the meeting. We are also appreciative of the help rendered by Mr. Don Parry and his staff of the Conference Office of the University of Rochester.

Finally, we wish to express our thanks to Mrs. Ruth Andrus who took most of the real work connected with the technical preparation of this volume on her shoulders. We are also indebted to her and to Mrs. Margaret Barres for the patience and care with which they typed the major part of the manuscript, and to many others of the staff of our Department for much valuable assistance.

Leonard Mandel Emil Wolf

Department of Physics and Astronomy
University of Rochester
Rochester, New York

January 1973

CONTENTS

W3 COHERENCE EFFECTS IN SPONTANEOUS EMISSION I

Chairman, E.C.G. Sudarshan

W4 COHERENCE I

Chairman, M.H. Nussenzveig

T1 QUANTUM ELECTRODYNAMICS AND ALTERNATIVE THEORIES II

Chairman, R.S. Knox

T6 COHERENT LIGHT INTERACTIONS

Chairman, B.P. Stoicheff

T7 LASERS

Chairman, M. Birnbaum

F1 QUANTUM ELECTRODYNAMICS AND ALTERNATIVE THEORIES III

Chairman, J.R. Klauder

F4 COHERENCE EFFECTS IN SPONTANEOUS EMISSION III

Chairman, S.R. Hartmann

F5 POST DEADLINE SESSION

Chairman, H. Gamo

F6 QUANTUM NOISE

Chairman, W.H. Louisell

SOME NONLINEAR PROPERTIES OF PULSE PROPAGATION

E.L. Hahn

University of California, Berkeley, Calif.

Nonlinear properties of pulse propagation are considered in relation to frequency pulling and pushing of the average carrier frequency of pulses applied off-resonance from the center of absorbing and emitting lines. The average frequency is analyzed as a function of distance of propagation. Some characteristics of propagating pulses in purely homogeneously broadened absorbers are discussed. By means of optical path delay and phase shift, zero-area pulses from a ruby laser are observed to propagate through many absorption lengths in a resonant inhomogeneously broadened ruby sample with anomalously large power outputs.

SELF-INDUCED TRANSPARENCY IN THE BROADLINE AND NARROWLINE LIMITS, AND APPLICATIONS

R.E. Slusher

Bell Laboratories, Murray Hill, N.J.

Recent experimental verification of self-induced transparency theory in the broadline ($T_2^* < \tau_p < T_2', T_1$) and narrowline ($\tau_p < T_2^* \lesssim T_2', T_1$) is described for a Hg laser coincident with a Rb^{87} D-line absorption. Both computer and experimental results show that broadline and narrowline cases give quite similar pulse delays, pulse breakup, etc., if an effective absorption coefficient is used to characterize the narrowline absorber. Effects of incoherent damping (T_2', T_1) are discussed for both cases in terms of a critical absorption length. Applications and extensions of self-induced transparency to pulse shaping of picosecond pulses are discussed using computer simulations.

OBSERVATION OF ZERO-DEGREE PULSE PROPAGATION IN A RESONANT MEDIUM*

H.P. Grieneisen†, J. Goldhar††, and N.A. Kurnit

Massachusetts Institute of Technology, Cambridge, Mass.

Experiments have recently been described[1] in which zero-degree optical pulses[2-5] have been generated and propagated through a resonantly absorbing medium with reduced absorption loss. We briefly review this experiment here and discuss the conditions for low loss propagation. We also describe some initial studies of the evolution of the pulse envelope and discuss the physics underlying this process.

The propagation of a coherent pulse through a resonant medium can be conveniently described by means of the coupled Bloch and Maxwell equations utilized for the discussion of self-induced transparency[6]. For an incident field $\underline{E} = \hat{\varepsilon}\, \mathcal{E}(z,t)\cos(\omega t-kz)$, where $\mathcal{E}(z,t)$ is an envelope which varies slowly on a time scale of $1/\omega$ or a spatial scale of λ, the propagation of the pulse envelope is described, in the case of an absorber which has a symmetric distribution $g(\Delta\omega)$ about the laser frequency, by the reduced wave equation[6]:

---

*Work supported by Air Force Cambridge Research Laboratories, Office of Naval Research and National Aeronautics and Space Administration.

†On leave from Universidade Federal do Rio Grande do Sul, Porto Alegre, Brazil. Work done with partial support from Conselho Nacional de Pesquisas, Brazil.

††Hertz Foundation predoctoral fellow.

$$\frac{\partial \mathcal{E}}{\partial z} + \frac{\eta}{c}\frac{\partial \mathcal{E}}{\partial t} = -\frac{2\pi N\mu\omega}{\hbar\eta c}\int_{-\infty}^{\infty} g(\Delta\omega)\ v(z,t,\Delta\omega)\ d(\Delta\omega) \qquad (1)$$

Here N and η are the number of resonant atoms/unit volume and background refractive index, respectively, and $v(z,t,\Delta\omega)$ is a component of the Bloch vector $(u,v,-w)$ of an atom with resonant frequency $\omega+\Delta\omega$ whose motion in a reference frame rotating about the $\hat{w}$ axis at the laser frequency is given by:

$$\frac{du}{dt} = \Delta\omega v - u/T_2' \qquad (2a)$$

$$\frac{dv}{dt} = -\Delta\omega u - (\mu\mathcal{E}/\hbar)w - v/T_2' \qquad (2b)$$

$$\frac{dw}{dt} = (\mu\mathcal{E}/\hbar)v - (w-w_{eq})/T_1 \qquad (2c)$$

The in-phase and out-of-phase components of the polarization contributed by an atom detuned by frequency $\Delta\omega$ are given by $\mu u(\Delta\omega)$ and $\mu v(\Delta\omega)$, where μ is the component of the transition matrix element along $\hat{\varepsilon}$; $w = \rho_{aa} - \rho_{bb}$ is the population difference between excited and ground state.

If the relaxation times T_1, T_2' are long compared to the pulse duration, the excitation of resonant atoms is determined solely by the integral of the field envelope, since (2b,c) can be integrated to give

$$v(z,t,0) = \sin\phi(z,t) \qquad (3a)$$

$$w(z,t,0) = -\cos\phi(z,t) \qquad (3b)$$

where

$$\phi(z,t) = \frac{\mu}{\hbar}\int_{-\infty}^{t} \mathcal{E}(z,t)dt \qquad (4)$$

The Bloch vector for resonant atoms is turned through the angle ϕ. The pulse "area"[6]

$$\theta(z) = \lim_{t\to\infty} \phi(z,t) \qquad (5)$$

measures the degree of excitation of resonant atoms after passage of the pulse. A pulse whose envelope $\mathcal{E}(z,t)$ has one or more sign changes can have zero area without having zero energy. Such "zero-degree"[2] pulses leave resonant molecules unexcited and thus can propagate with reduced energy loss. It was pointed out in Ref. 2 that this is true even in the presence of level degeneracy, which has an inhibiting influence on the self-induced transparency of $2n\pi$ pulses with $n \geq 1$, since sublevels with different transition matrix elements are then left with different degrees of excitation.

In these experiments[1], the output of a low intensity cw CO_2 or N_2O laser is passed through a GaAs crystal oriented for amplitude modulation. A voltage pulse derived from a charged 50Ω cable and triggered spark gap[7] is propagated over the crystal and produces elliptically polarized light from the linearly polarized laser output. The perpendicular component of this is reflected from a germanium Brewster plate which normally transmits the laser output, and is then passed through an absorption cell. The voltage pulse may be applied to the crystal a second time with either the same or opposite polarity by reflecting it from the open or shorted end of a 50Ω cable which terminates the crystal mount. Reversing the polarity of the pulse produces a 180° phase change in the electric field component reflected by the Brewster plate. This field is proportional to $\sin\Gamma$, where $\pm\Gamma$ ($\simeq\pi/8$ in these experiments) is the phase advance or retardation produced along the fast or slow axis of of the modulator crystal.

Figure 1 shows a CO_2 laser pulse generated with a 2 nsec voltage pulse as observed with a Ge:Cu detector and displayed on a Tektronix oscilloscope with a 1S1 sampling plug-in unit. For the purpose of this figure, a quarter wave plate has been added in front of the modulator, so that the pulses are observed superimposed on a large cw background. In the top trace, two pulses of opposite polarity are applied to the crystal, resulting in opposite changes in the intensity reaching the detector; in the lower trace the second voltage pulse is applied to the crystal with the same polarity as the first. This method gives a reasonable representation of the pulse envelope since the intensity reaching the detector is proportional to $1 + \sin2\Gamma$. For observation of the pulse envelope evolution, it will be desirable to send a cw beam along a separate path and recombine it with the pulse on the detector.

Measurements of the pulse energy transmitted through an absorption cell were performed with both a CO_2 laser and heated CO_2 absorber and an N_2O laser and NH_3 absorber. The latter permitted measurements over many absorption lengths, since the NH_3 ν_2 [as Q(8,7)] transition is coincident to within 10 MHz of the N_2O P (13) transition, and has an absorption coefficient of 0.7 cm^{-1} and Doppler half-width $\Delta\nu$ = 42 MHz[8]. The transmission of 2 nsec and 6 nsec zero-degree and in-phase pulses is shown in Fig. 2 as a function of NH_3 pressure in a 40 cm

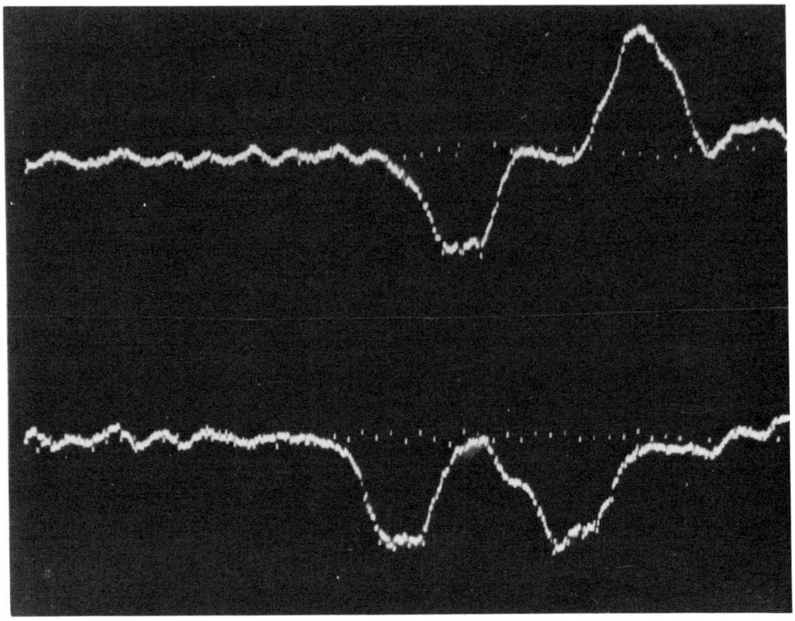

Fig. 1 Pulse observed by Ge:Cu detector with λ/4 plate in front of modulator (see text). Top: Zero-degree pulse obtained by applying out-of-phase pulses to modulator. Bottom: In-phase pulses applied to modulator. Time scale is 2 nsec/div.

absorption cell. The dashed line shows the cw absorption for comparison. Curve A shows the absorption of a zero-degree pulse formed from two out-of-phase 2 nsec pulses, as in the top trace of Fig. 1. At low pressures, the number of absorption lengths is proportional to pressure, and the absorption coefficient may be obtained from the logarithmic slope of the absorption versus pressure. The dashed curve shows the cw absorption for comparison. Curve B shows the absorption of two 2 nsec in-phase pulses as in the bottom trace of Fig. 1. These pulses show an initial absorption coefficient some five times higher than the corresponding zero-degree pulse, but still a factor of five smaller than the cw absorption.

The latter result is of course to be expected from a consideration of the overlap of the Fourier components of the pulse with the resonance line. In the small angle regime in which these experiments have been carried out ($\phi \leq 2^\circ$), the atomic response may be

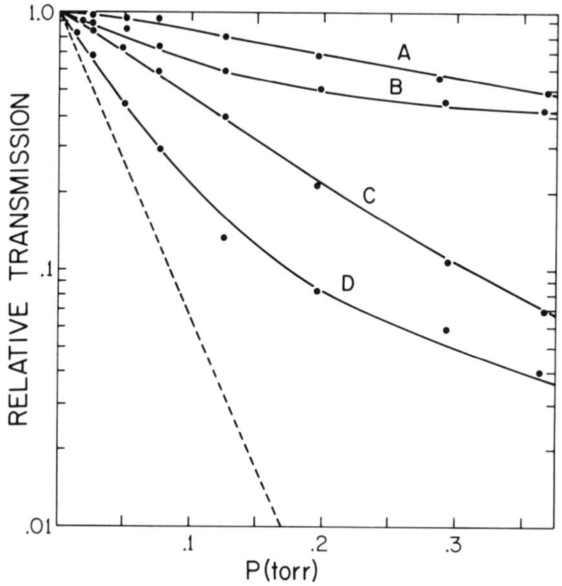

Fig. 2 Transmission of N_2O laser energy through resonant NH_3 absorber as function of pressure, for (A) 2 nsec zero-degree pulses, (B) 2 nsec in-phase pulses, (C) 6 nsec zero-degree pulses, and (D) 6 nsec in-phase pulses. Dashed line gives cw absorption.

approximated by a harmonic oscillator. The pulse evolution may then be described by a linearized theory[5], in which the absorption and dispersion of each Fourier component can be considered independently. A zero-degree pulse has a Fourier spectrum which is zero on resonance (i.e., resonant atoms are left in their initial state after passage of the pulse). Hence, if the resonance line is narrower than the inverse pulse-width, which determines the width of the hole in the Fourier spectrum, very little energy will be extracted from the pulse.

Curves C and D of Fig. 2 show the effect of lengthening the pulse-width to 6 nsec, which is longer than the inhomogeneous relaxation time $T_2^* = 1/2\pi\Delta\nu_D \simeq 4$ nsec. Both zero-degree (C) and in-phase (D) pulses show correspondingly increased absorption, with the zero-degree pulses exhibiting about 1/2 the absorption coefficient of the in-phase pulses. As the pulse lengths are made longer, the absorption for both zero-degree and in-phase pulses approach the cw limit, in this small angle regime. This is not necessarily true in the nonlinear regime, since particular pulse shapes, related to 4π pulses, are predicted to give small attention even for pulses long compared to T_2^* [3,4].

A few general conclusions can be reached regarding the energy absorbed without considering the detailed pulse shape. If one multiplies Eq.(1) on both sides by $\frac{\eta c}{4\pi}\mathcal{E}(z,t)$ and integrates over time from $-\infty$ to $+\infty$, one obtains

$$\frac{ds(z)}{dz} = -\frac{N\mu\omega}{2}\int_{-\infty}^{\infty} g(\Delta\omega)d(\Delta\omega)\int_{-\infty}^{\infty}\mathcal{E}(z,t)v(\Delta\omega,z,t)dt \qquad (6)$$

where

$$s(z) = \frac{\eta c}{8\pi}\int_{-\infty}^{\infty}\mathcal{E}^2(z,t)dt$$

is the pulse energy/unit area. If the pulse duration $\tau = (\int|\mathcal{E}(z,t)|dt)^2/\int\mathcal{E}^2(z,t)dt$ is sufficiently short that the pulse spectrum is broad compared to the resonance line ($\tau \ll T_2', T_2^*$), all of the atoms respond resonantly and obey Eq.(3). From Eq.(4), we have $\mathcal{E}(z,t) = (\hbar/\mu)\,\partial\phi/\partial t$, which, when substituted in Eq.(6) together with (3a), yields

$$\frac{\partial s(z)}{\partial z} = -N\hbar\omega(1-\cos\theta(z))/2 \quad . \qquad (7)$$

The right-hand side is just the energy remaining in the atomic system. The angle θ obeys the area theorem[6]

$$\frac{d\theta(z)}{dz} = -\frac{1}{2}\alpha\sin\theta(z) \qquad (8)$$

where $\alpha = 4\pi N\mu^2\omega T_2^*/\hbar\eta c$ is the small signal linear absorption coefficient of the medium (cf. Eq.(12) below), assuming $T_2^* \ll T_2'$.

This implies that $\theta(z)$ will approach some multiple n=0,1,2... of 2π after a number of absorption lengths α^{-1}. Once this occurs, no further energy will be absorbed in the narrow line limit, according to (7). The initial absorption will itself be very small if $\tau \ll T_2^*$. For example, for a small area square pulse of duration $\tau \ll T_2^*$, Eq.(7) reduces to

$$\frac{ds(z)}{dz} = -\alpha\left(\frac{\tau}{2T_2^*}\right)s(z) \qquad (9)$$

As this pulse propagates, its area decays according to (8), primarily due to the development of a negative lobe which follows the pulse, as will be discussed in more detail below. This negative tail, which can be termed free induction decay[10], superradiance[11], or fluorescent ringing[12], extracts energy from the resonantly excited atoms and grows until it is able to return resonant atoms to the ground state. The reduction in absorption coefficient as this pulse develops into a zero-degree pulse, is evident in curves B and D of Fig. 2. After several absorption lengths, the absorption coefficient begins to approach a value comparable to the initial absorption obtained for zero-degree pulses. Experiments in which the output of one absorption cell was used as the input to another clearly show this behavior[1].

In the small angle regime, this reduction in the absorption coefficient is simply seen to result from the attenuation of resonant Fourier components. It has been shown by Crisp[5,13] that a Fourier domain analysis which accounts for the dispersion as well as absorption of each Fourier component predicts the development of an initially negative tail which oscillates in sign after the pulse propagates many absorption lengths, in agreement with the conclusions of Ref. 12. This analysis also yields the small angle form of the area theorem (Eq.(8)), valid for arbitrary ratio of τ to T_2' and T_2^*.

If the line is not narrow compared to the pulse spectrum, the energy absorbed depends not only on the area, as implied by (7), but on the pulse shape as well. In this more general case, the solution of Eq.(2) for v can be written as integral equation[14,5]:

$$v(z,t,\Delta\omega) = \int_{-\infty}^{t} \exp[(-t+t'')/T_2'] \cos[\Delta\omega(t-t'')] \times \frac{\mu \mathcal{E}(z,t'')}{\hbar} w(z,t'',\Delta\omega)dt'' \quad (10)$$

Substituting this in (6), and letting t' = t-t'', we have

$$\frac{ds(z)}{dz} = -\frac{N\mu\omega}{2}\int_{-\infty}^{\infty} g(\Delta\omega)\, d(\Delta\omega)\int_{-\infty}^{\infty}dt\, \mathcal{E}(z,t)\int_{0}^{\infty}dt'\, \exp(-t'/T_2')\, \cos\Delta\omega t' \times \frac{\mu \mathcal{E}(z,t-t')}{\hbar} w(z,t-t',\Delta\omega) \quad (11)$$

This is not particularly useful unless the solution for w is known; however, a few general conclusions can be drawn. If w is not appreciably altered during the pulse and T_2' or T_2^* is short, the factor of $\exp(-t'/T_2')$ or $\cos\Delta\omega t'$ acts to limit the contribution of the t' integral to small values of t'. The energy absorbed then depends only on values of $\mathcal{E}(z,t)$ at closely spaced times. If $\mathcal{E}(z,t)$ and $\mathcal{E}(z,t-t')$ are different in sign, we have emission rather than absorption of energy. For long relaxation times, this averaging is performed over the whole pulse, and can give as much emission as absorption. For short relaxation times, there is little memory in the system of previous values of the field. If we assume in this limit that $\mathcal{E}(z,t-t') \cong \mathcal{E}(z,t)$, for all t' which contribute significantly to (11), we have by first performing the $\Delta\omega$ integral and then the t' integral

$$\frac{ds(z)}{dz} = -\frac{4\pi N\mu^2\omega T_2}{\hbar\eta c}\, s(z) \equiv -\alpha s(z) \tag{12}$$

where, for convenience, we have taken a Lorentzian frequency distribution,

$$g(\Delta\omega) = \frac{1}{\pi T_2^*[(\Delta\omega)^2 + 1/T_2^{*2}]} \tag{13}$$

and have put $1/T_2 = 1/T_2' + 1/T_2^*$.

In the case of large area pulses, w can deviate significantly from its steady state value, and even change sign, resulting in emission during part of a pulse without the pulse envelope changing sign, as, for example, for 2π pulses. Unlike the linear regime considered above, the effects of homogeneous and inhomogeneous broadening are very different. If $T_2' \ll \tau$, the population will not be appreciably altered except for very intense fields. If, on the other hand, $T_2^* \ll \tau \ll T_2'$, complete inversion of resonant atoms can still be achieved, and frequency groups near resonance can be left highly excited. Figure 3 shows the population $\rho_{aa}(\Delta\omega) = (1+w(0,\tau,\Delta\omega))/2$ excited at the input face of a sample, calculated for a square pulse of duration $\tau \ll T_2'$, with and without a 180° phase change in the middle, for several values of θ. The energy absorbed per unit area in a thin slab is obtained by multiplying this curve by $N\hbar\omega\Delta z g(\Delta\omega)$ and integrating over frequency. Even for pulses as large as $\pi/2$ followed by $\pm\pi/2$, the spectrum of excited atoms strongly resembles the power spectrum of the pulse, which it must reduce to in the small-angle limit. For want of a better term, we shall call this the Bloch spectrum of the pulse. As the pulse intensity is increased further, the nature of the Bloch spectrum changes considerably. Lobes far from resonance become enhanced, and those near resonance can be suppressed. A zero-degree pulse composed of two π pulses can show considerably more absorption than the corresponding 2π

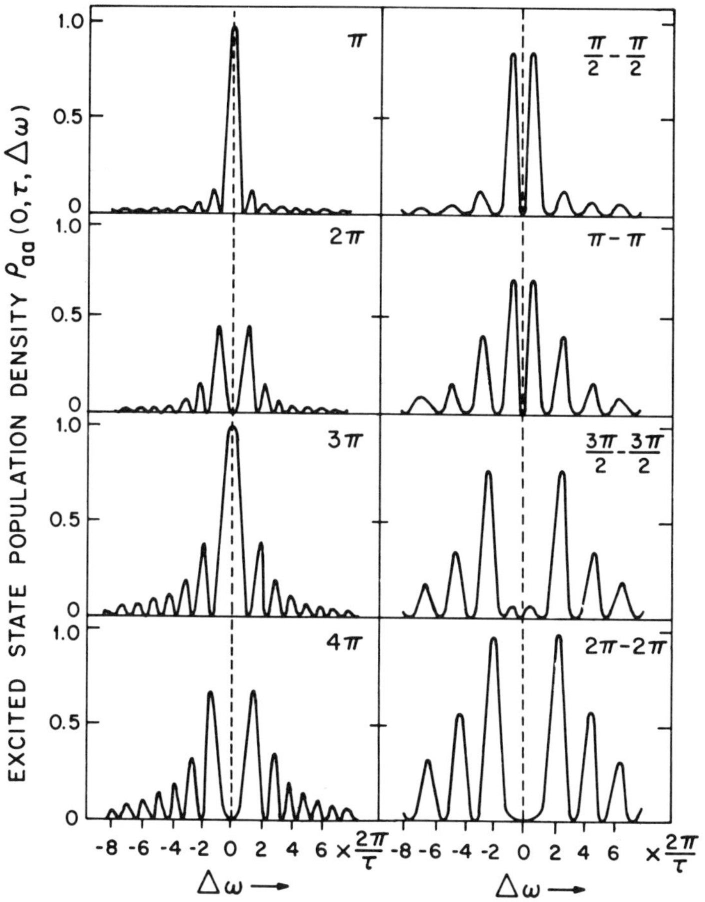

Fig. 3 Excited state population density produced in a thin slab by a short square pulse of duration τ, with and without a 180° phase change at $\tau/2$, as a function of the detuning of the atomic frequency from the frequency of the applied field. Total population change is obtained by multiplying by $g(\Delta\omega)$ and integrating over frequency. Pulse areas are indicated in the figure.

pulse (for a broad line). When the area of each pulse is increased to $3\pi/2$, the zero-degree pulse has a Bloch spectrum which is small over a considerably wider range of frequencies near resonance than the power spectrum of the pulse. Thus, small attenuation can be expected in this case, even for the square pulses considered here, provided the resonance line is narrower than the hole in the Bloch spectrum. More complex pulse shapes have been proposed[3,4] which give better transmission characteristics.

The energy absorbed at distances further into the sample will depend on the modification of the pulse shape by the intervening sample. In general, this is a complicated problem, requiring computer solution of Eqs.(1,2). Analytic solutions have been obtained in the linear regime for a number of pulse shapes[5]. Rather than reviewing such calculations in detail, we shall try to motivate some of our remarks concerning the pulse evolution by considering the effect of a thin slab of sample. The change in pulse shape is due not only to the energy absorbed during the input pulse, but also to the re-radiation by the polarization left in the medium after the pulse passes. The latter is particularly important when $\tau<T_2^*$, since the medium can "ring"[12] for a time T_2^*. In the nonlinear regime, it can be important even if $\tau>T_2^*$, as is evidenced by the phenomenon of edge echoes[15] in NMR. Here, contributions from near-resonant atoms can give rise to a large polarization at a time $\sim\tau$ after a pulse of duration τ.

Consider a pulse incident on a slab of thickness Δz. The field at $z' = z + \Delta z$ and $t' = t + \frac{\eta}{c}\Delta z$ is related to the field at z,t by

$$\mathcal{E}(z',t') = \mathcal{E}(z,t) + \left(\frac{\partial \mathcal{E}(z,t)}{\partial z} + \frac{\eta}{c}\frac{\partial \mathcal{E}(z,t)}{\partial t}\right)\Delta z$$

$$= \mathcal{E}(z,t) - \frac{2\pi N\mu\omega}{\eta c}\Delta z \int_{-\infty}^{\infty} g(\Delta\omega)v(z,t,\Delta\omega)d(\Delta\omega) \qquad (14)$$

according to Eq.(1). The polarization excited by a square input pulse[16] with amplitude $\mathcal{E}_0$ and duration τ applied at t = 0 is, by (2) with $T_2' = \infty$:

$$v(z,t,\Delta\omega) = \frac{\Omega\sin\left[\left(\Omega^2+(\Delta\omega)^2\right)^{\frac{1}{2}}\tau\right]}{\left(\Omega^2+(\Delta\omega)^2\right)^{\frac{1}{2}}}\cos\left(\Delta\omega(t-\tau)\right)$$

$$- \frac{2\Omega\Delta\omega}{\Omega^2+(\Delta\omega)^2}\sin^2\left[\frac{\left(\Omega^2+(\Delta\omega)^2\right)^{\frac{1}{2}}\tau}{2}\right]\sin\Delta\omega(t-\tau) \qquad (15)$$

where $\Omega = \mu\,\mathcal{E}_0/\hbar$. The second term is responsible for the edge echo mentioned above. In the narrow line limit, $\tau<<T_2^*$, (13),(15) yield

$$\langle v(z,t,\Delta\omega)\rangle = \int_{-\infty}^{\infty} g(\Delta\omega)v(z,t,\Delta\omega)d(\Delta\omega) \cong \sin(\Omega\tau)\exp[-(t-\tau/2)/T_2^*] \qquad (16)$$

to first order in τ/T_2^*. The field is then, from (14), for $t>\tau$,

$$\mathcal{E}(z',t') = -\frac{2\pi N\mu\omega}{\eta c}\Delta z\ \sin\theta(z)\ \exp[(-t+\tau/2)/T_2^*] \tag{17}$$

The field radiated after the pulse thus decays exponentially with time constant T_2^* for a sample sufficiently thin that (14) does not have to be iterated. Its sign is negative with respect to the incident field since the radiated field must tend to cancel the incident field during the pulse, for $\theta<\pi$. The area of this tail is obtained by multiplying (17) by $\mu/\hbar$ and integrating from τ to ∞:

$$\theta_{tail}(z') \cong -\frac{2\pi N\mu^2\omega}{\hbar\eta c}\Delta z\ \sin\theta(z)(T_2^*-\tau/2) \tag{18}$$

During the input pulse, the polarization is given approximately by

$$\langle v(z,t)\rangle \cong \sin(\Omega t)\ \exp(-t/2T_2^*) \tag{19}$$

(cf. Eq.(16) with $\tau = t$). The field during the pulse is, from (14) and (19),

$$\mathcal{E}(z',t') = \mathcal{E}_0 - \frac{2\pi N\mu\omega}{\eta c}\Delta z\ \sin(\Omega t)\ \exp(-t/2T_2^*) \tag{20}$$

The area during the time τ when the incident field is present is, for $\tau<<T_2^*$,

$$\theta_{pulse}(z') \cong \theta(z) - \frac{2\pi N\mu^2\omega}{\hbar\eta c}\Delta z\left(\frac{1-\cos\theta(z)}{\Omega}\right) \tag{21}$$

For small θ, this reduces to

$$\theta_{pulse}(z') = \theta(z) - \frac{2\pi N\mu^2\omega}{\hbar\eta c}\left(\frac{\tau}{2}\right)\theta(z)\Delta z \tag{22}$$

Adding this to the small angle form of (18) yields

$$\theta(z') = \theta(z) - \frac{2\pi N\mu^2\omega}{\hbar\eta c}T_2^*\ \theta(z)\Delta z \tag{23}$$

in agreement with the small angle form of (8). We see that, although the area theorem is generally applied in the case of $\tau \gg T_2^*$, it is still well obeyed in the limit $\tau \ll T_2^*$, but the major contribution to the change in area comes from the tail of the pulse.

It might be thought that (16) should contain a term which accounts for radiation damping[17], since, as noted previously, the re-radiation which follows the excitation pulse can be considered to be "superradiant"[11] emission. In order to examine the connection between this pulse-propagation problem and superradiance in more detail, let us calculate the amounts of energy absorbed and re-radiated. The energy incident in our square-input pulse is given by

$$s(z) = \frac{\eta c}{8\pi} \int_0^\infty \mathcal{E}^2(z,t)dt = \frac{\eta c}{8\pi} \mathcal{E}_0^2 \tau \tag{24}$$

while the energy reaching $z' = z + \Delta z$ is obtained by substituting in (24) $\mathcal{E}(z',t')$ from (20), for $t<\tau$, and from (17) for $t>\tau$. Dropping terms second order in τ/T_2^*, this yields

$$s(z') - s(z) \cong - \frac{4\pi N\mu\omega}{\eta c} \Delta z \left(\frac{\eta c}{8\pi}\right) \mathcal{E}_0 \left(\frac{1-\cos\theta(z)}{\Omega}\right)$$

$$+ \frac{4\pi^2 N^2 \mu^2 \omega^2}{\eta^2 c^2} T_2^* (\Delta z)^2 \left(\frac{\eta c}{8\pi}\right) \sin^2\theta(z) \tag{25}$$

For small area pulses, this may be written

$$\Delta s(z) = -\alpha\Delta z\left(\frac{\tau}{2T_2^*}\right) s(z) + (\alpha\Delta z)^2 \frac{\tau}{8T_2^*} s(z) \tag{26}$$

The first term in (25) or (26) represents the energy absorbed from the input pulse; the second term is the energy radiated in the tail. Although the tail accounts for most of the change in area of the pulse, the re-radiated energy is a small fraction of the energy absorbed unless $\alpha\Delta z \gtrsim 1$. But (25) is not valid in this limit since the field in (17) and (20) should properly be modified to account for the changing pulse shape and area[18].

The superradiance approximation[19] treats all of the atoms in the slab of thickness Δz and area A as radiating identically, with an initial radiation rate $I = \frac{\eta c}{8\pi} \mathcal{E}^2 A$, given according to (17), by

$$I = \frac{\eta c}{8\pi} \left(\frac{4\pi^2 N^2 \mu^2 \omega^2}{\eta^2 c^2} \Delta z^2 A\right) \sin^2\theta \tag{27}$$

In terms of the number $\mathfrak{n} = NA\Delta z$ of atoms in the slab, this may be written

$$I = \frac{\mathfrak{n}^2}{4} \sin^2\theta \left(\frac{\hbar\omega}{T_1}\right)\left(\frac{3}{8\pi\eta^2} \frac{\lambda^2}{A}\right)$$

$$= \mathfrak{n}\hbar\omega \sin^2\theta \left(\frac{1}{T_s}\right) \tag{28}$$

where $1/T_1 = 4\eta\mu^2\omega^3/3\hbar c^3$ is the spontaneous emission rate, and

$$\frac{1}{T_s} = \frac{3}{32\pi\eta^2} \mathfrak{n} \frac{\lambda^2}{A} \left(\frac{1}{T_1}\right) \tag{29}$$

is the superradiant decay rate[11,20]. Within the approximation that atoms in the front of the sample do not appreciably alter the field seen by atoms at larger depths in the sample, i.e., $\alpha\Delta z \ll 1$, we thus obtain the enhanced decay rate characteristic of superradiant emission. From the derivation of this result, it is evident that this is only an approximation to the differential equation (1) governing the pulse propagation. As the pulse propagates through the sample, the field radiated by the atoms in the first small thickness produces a negatively phased field which drives atoms further in the sample back toward the ground state. This field grows with increasing z, initially according to Eq.(17), and hence can more and more rapidly extract the energy from the population excited by the first part of the pulse. The time T_s characterizes the re-radiation at a given distance in the medium. Burnham and Chiao[12] have given an analytic expression for the field radiated after a small area δ-function pulse,

$$\theta = \frac{\mu}{\hbar} \mathcal{E}_0 \, \tau\delta(t' - \frac{\eta}{c} z), \quad \tau \ll T_2, T_s$$

which, as modified by Crisp[13] to include relaxation, may be written

$$\mathcal{E}(z,t) = -\mathcal{E}_0 \frac{\tau}{\tau_R} \frac{J_1[z(t/\tau_R)^{\frac{1}{2}}]}{(t/\tau_R)^{\frac{1}{2}}} \exp(-t/T_2) \tag{30}$$

where $t = t' - \frac{\eta}{c} z$ is the time measured from the passage of the input pulse, J_1 is the first order Bessel function, and τ_R is given by[21]

$$\frac{1}{\tau_R} = \frac{2\pi N\mu^2\omega}{\hbar\eta c} z = \frac{\alpha z}{2T_2} = \frac{4}{T_s} \quad . \tag{31}$$

Equation(30) describes an envelope which oscillates in sign as a function of the retarded time t, as indicated in Fig. 4, the first zero occurring at $t = 3.7\tau_R \simeq T_s$. Unless $T_s \lesssim T_2$, which according

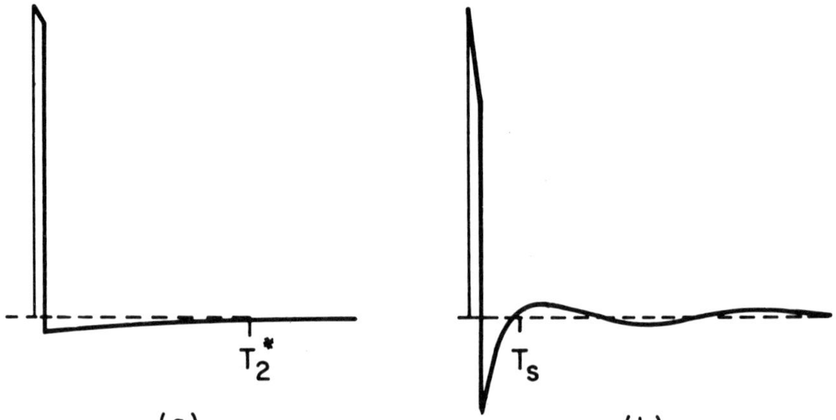

Fig. 4 Evolution of a nearly δ-function input pulse for (a) $\alpha z \ll 1$, (b) $\alpha z \gg 1$.

to (31) implies $\alpha z \gtrsim 8$[22], the factor of $\exp(-t/T_2)$ in (30) will cause the field envelope to decay to zero before it reverses sign. For increasing z, the sign reversal, which arises from an oscillation of the energy between the field and the atomic system, becomes correspondingly faster.

Equation(30) may also be written[13]

$$\mathcal{E}(z,t) = -\mathcal{E}_0\tau\left(\frac{\alpha z}{2tT_2}\right)^{\frac{1}{2}} J_1\left[2\left(\frac{\alpha z t}{2T_2}\right)^{\frac{1}{2}}\right] \exp(-t/T_2) \tag{32}$$

If J_1 is expanded as a power series, the first two terms yield

$$\mathcal{E}(z,t) \simeq -\mathcal{E}_0 \frac{\tau}{2T_2} \exp(-t/T_2)\ \alpha z\left(1-\frac{\alpha z}{4T_2}t\right) \qquad (33)$$

The first term agrees with the small angle form of (17) and predicts a re-radiated field whose peak grows linearly with αz. The second term governs the decrease of the field for small t. The field decays with an initial slope of $(1/T_2)(1+\alpha z/4) = 1/T_2 + 2/T_S$. For pulses not short compared to T_2 and T_S, the distortion within the input pulse must also be considered[5].

A preliminary observation of this pulse envelope distortion is shown in Fig. 5. Unfortunately, some ringing is present due to

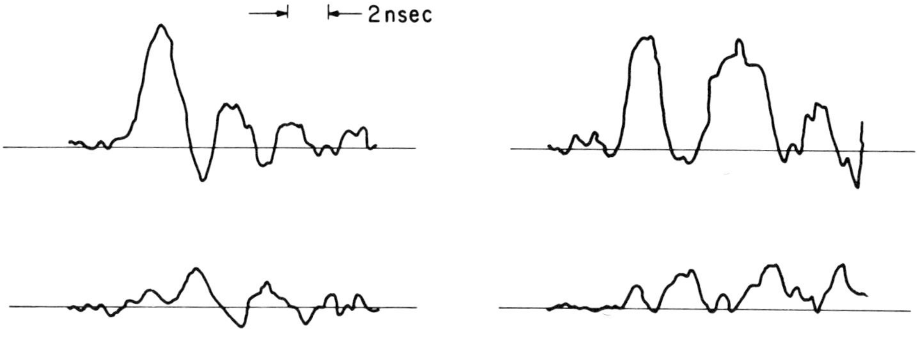

(a) (b)

Fig. 5 (a) Top: 2 nsec N_2O laser input pulse (partially obscured by ringing in amplifier). Bottom: same pulse after traversing 15 absorption lengths of NH_3. (b) Top: zero-degree input pulse formed from two 2 nsec pulses. Bottom: same pulse after traversing 15 absorption lengths of NH_3. A baseline subtraction has been made in order to cancel rf noise from the spark gap.

an amplifier which was needed in order to display the N_2O laser pulses of 10 mw intensity. This spurious electronic ringing attenuates linearly with the signal, however, whereas it can be clearly seen from Fig. 5a that if a single 2 nsec pulse (top trace) is passed through a resonant NH_3 absorber with αz=15, the output of the absorber (bottom trace) shows a second peak which has grown to be larger than the (attenuated) first peak. In Fig. 5b, a zero-degree pulse is shown. The first pulse again has a double peaked structure after passing through the absorber. There is some evidence that a secondary lobe which is forming in the tail of the first pulse partially interferes with the second input pulse, due to a small delay between the two pulses. For a large number of absorption lengths, computer calculations show that this can result in increased absorption for such zero-degree pulses. Some indications of this behavior have been observed[23] and will be discussed elsewhere.

Experiments are in progress to obtain clearer observations of the pulse evolution and to extend these observations to the nonlinear regime where analytic solutions are not available. Also of interest is the propagation behavior in the off-resonance case. These techniques promise to be useful for studying pulse evolution in both absorbing and amplifying media and may find applications in long distance pulse propagation.

Acknowledgments

We wish to thank A. Javan and H.R. Schlossberg for their encouragement and collaboration during the course of these experiments. We also wish to acknowledge numerous stimulating discussions with A. Szöke, M.S. Feld and Y.C. Cho.

References

1. H.P. Grieneisen, J. Goldhar, N.A. Kurnit, A. Javan and H.R. Schlossberg, Bull. Am. Phys. Soc. *17*, 681 (1972), and Appl. Phys. Letters *21*, 559 (1972).
2. C.K. Rhodes, A. Szöke and A. Javan, Phys. Rev. Letters *21*, 1151 (1968).
3. F.A. Hopf, C.K. Rhodes, G.L. Lamb, Jr., and M.O. Scully, Phys. Rev. *A3*, 758 (1971).
4. G.L. Lamb, Jr., *Proceedings of the VII International Quantum Electronics Conference*, to be published; see also G.L. Lamb, Jr., Rev. Mod. Phys. *43*, 99 (1971).
5. M.D. Crisp, Phys. Rev. *A1*, 1604 (1970).
6. S.L. McCall and E.L. Hahn, Phys. Rev. Letters *18*, 908 (1967); Phys. Rev. *183*, 457 (1969); see also H.M. Gibbs and R.E. Slusher, Phys. Rev. *A5*, 1634 (1972).
7. R.C. Fletcher, Rev. Sci. Instr. *20*, 861 (1949).
8. F. Shimizu, J. Chem. Phys. *52*, 3572 (1970); Appl. Phys. Letters *16*, 368 (1970); T. Shimizu and T. Oka, Phys. Rev. *A2*, 1177 (1970).

9. The more general case of level degeneracy can be treated as in Ref. 2.

10. A. Abragam, *Principles of Nuclear Magnetism* (Oxford University Press, London, 1961) Ch. II, III.

11. R.H. Dicke, Phys. Rev. *93*, 99 (1954).

12. D.C. Burnham and R.Y. Chiao, Phys. Rev. *188*, 667 (1969).

13. M.D. Crisp, Opt. Commun. *4*, 199 (1971).

14. F.A. Hopf and M.O. Scully, Phys. Rev. *179*, 399 (1969).

15. A.L. Bloom, Phys. Rev. *98*, 1105 (1955); see also Ref. 14.

16. A square pulse formally violates the slowly varying envelope approximation made for Eq.(1). The pulse can be turned on and off sufficiently slowly to make (1) applicable without appreciably affecting (15).

17. A.M. Ponte Goncalves, A. Tallet and R. Lefebvre, Phys. Rev. *188*, 576 (1969); Phys. Rev. *A1*, 1472 (1970); see also A. Compaan and I.D. Abella, Phys. Rev. Letters *27*, 23 (1971).

18. Similar conclusions have been emphasized by E.L. Hahn, N.S. Shiren and S.L. McCall, Phys. Letters *37A*, 265 (1971); R. Friedberg and S.R. Hartmann, Phys. Letters *37A*, 285 (1971) and *38A*, 227 (1972); R.H. Picard and C.R. Willis, Phys. Letters *37A*, 301 (1971).

19. R.H. Dicke, Ref. 11 and in *Quantum Electronics III*, eds. N. Bloembergen and P. Grivet (Columbia University Press, New York, 1964) Vol. 1, p. 35; N.E. Rehler and J.H. Eberly, Phys. Rev. *A3*, 1735 (1971); R. Bonifacio, P. Schwendimann and F. Haake, Phys. Rev. *A4*, 302 and 854 (1971); see also Y.C. Cho, N.A. Kurnit, and R. Gilmore, in these Proceedings, p. 755.

20. F.T. Arecchi and E. Courtens, Phys. Rev. *A2*, 1730 (1970), as modified by Friedberg and Hartmann, Ref. 18; see also I.D. Abella, N.A. Kurnit and S.R. Hartmann, Phys. Rev. *141*, 391 (1966), Appendix C.

21. In Ref. 12, $1/\tau_R = \omega_p^2 \, z/4c$, where ω_p is the plasma frequency: classically $\omega_p{}^2 = 4\pi f N e^2/m$, where f is the oscillator strength; quantum mechanically, $\omega_p{}^2 = 8\pi N\mu^2\omega/\hbar$.

22. R. Friedberg and S.R. Hartmann, Ref. 18. In the second of these papers and Optics Commun. *2*, 301 (1970), these authors also consider the effect of the backward wave, which gives rise to a frequency shift. This shift should be small in our experiment.

23. H.P. Grieneisen, Ph.D. Thesis, M.I.T., 1972.

HIGHER CONSERVATION LAWS AND COHERENT PULSE PROPAGATION

D.D. Schnack*

Pratt and Whitney Aircraft, East Hartford, Conn.

G.L. Lamb, Jr.

United Aircraft Research Labs., East Hartford, Conn.

1. Introduction

It has been shown that the equations customarily used to describe coherent optical pulse propagation, namely the coupled Maxwell and Bloch equations in the slowly varying envelope approximation, possess conservation laws in addition to the usual one associated with the conservation of energy. These higher conservation laws have been used to determine the amplitude of each of the 2π pulses into which a coherent pulse may decompose as it propagates through an absorbing medium.[1]

Two limitations were found in applying these results: firstly, calculation of the higher conservation laws was somewhat tedious; secondly, the threshold condition for the appearance of an additional pulse as the initial pulse area exceeds 3π, 5π, etc., was not given accurately in previous applications of the method. The present paper provides methods for coping with these restrictions.

In the first case, a relation between the conservation laws of coherent pulse propagation and those for the Korteweg-deVries equation[2,3] is established. Since the first eleven laws for this latter equation have been calculated, corresponding results for coherent pulse propagation are immediately at hand.

---

*Portions of this research are being submitted in partial fulfillment of requirements for the M.S. degree at Trinity College, Hartford, Conn.

Secondly, by a slight modification of a previous usage of the conservation laws to determine pulse amplitudes[1], account may be taken of the optical energy lost to the medium during the initial stages of propagation in which the pulse is reshaped into a sequence of hyperbolic secant pulses. It is shown that when this energy is accounted for, the appearance of each additional pulse in the asymptotic solution as the initial pulse area exceeds 3π, 5π, ..., may be predicted quite accurately by the conservation laws. The technique is applied explicitly to the first three conservation laws.

2. Basic Equations

Propagation of a plane electromagnetic wave in a medium consisting of nondegenerate two-level atoms may be described by a simultaneous solution of the Maxwell equations and the Schrödinger-Bloch equations. If the usual slowly varying envelope approximation is made, the electric field may be written

$$E(x,t) = \frac{\hbar}{\mu} G(x,t) \cos(k_0 x - \omega_0 t) \tag{1}$$

where $\hbar$ is Planck's constant divided by 2π and μ is the dipole matrix element connecting the two resonant atomic levels while k_0 and ω_0 are the wave number and frequency, respectively, of the optical wave. Propagation is assumed to take place in a positive x direction. The envelope function $G(x,t)$ is assumed to vary slowly on the length and time scales of the carrier wave, i.e., $\partial G/\partial x \ll k_0 G$, $\partial G/\partial t \ll \omega_0 G$. For a medium consisting of an assemblage of two-level systems distributed with a uniform density n_0, the macroscopic polarization is $n_0 \int d\Delta\omega g(\Delta\omega) p(\Delta\omega,x,t)$ where $p(\Delta\omega,x,t)$ is the polarization due to an individual atom that is detuned from line center by an amount $\Delta\omega = \omega - \omega_0$ and $g(\Delta\omega)$ is the normalized spectrum of this inhomogeneous broadening. It is assumed that the broadening is symmetric about ω_0. The polarization may be written

$$p(\Delta\omega,x,t) = \mu[P(\Delta\omega,x,t)\sin(k_0 x-\omega_0 t) + Q(\Delta\omega,x,t)\cos(k_0 x-\omega_0 t)]. \tag{2}$$

For pulses much shorter than the atomic dephasing time, T_2, the in-phase and in-quadrature envelope functions $Q(\Delta\omega,x,t)$ and $P(\Delta\omega,x,t)$ respectively, obey the undamped Bloch equations

$$\frac{\partial P}{\partial t} = GN + \Delta\omega Q \tag{3a}$$

$$\frac{\partial Q}{\partial t} = -\Delta\omega P \tag{3b}$$

$$\frac{\partial N}{\partial t} = -GP \tag{3c}$$

where N is the population inversion for a single atom.

The wave equation for the electric field in slowly varying envelope approximation reduces to

$$\frac{\partial G}{\partial t} + c\,\frac{\partial G}{\partial x} = \alpha' c\langle P\rangle \tag{4}$$

where $\alpha' \equiv 2\pi N_0 \omega_0 \mu/\hbar c$ and c is the light velocity in the host medium. The angular brackets signify an average over the inhomogeneously broadened atomic spectrum $g(\Delta\omega)$ as given by

$$\langle\cdots\rangle = \int_{-\infty}^{\infty} d\Delta\omega\, g(\Delta\omega)(\cdots) \quad . \tag{5}$$

As mentioned above, this broadening will be assumed to be symmetric about ω_0.

Equations (3) have exactly the same structure as the Frenet-Serret equations of differential geometry.[4] It is known that the solution to such a set of equations is equivalent to the solution of a Riccati equation.[4] This is seen by first observing that an integral of Eqs.(3) is

$$N^2 + P^2 + Q^2 = 1 \quad . \tag{6}$$

Two new functions are then introduced by writing

$$\frac{N+iP}{1-Q} = \frac{1+Q}{N-iP} = \phi \tag{7a}$$

$$\frac{N-iP}{1-Q} = \frac{1+Q}{N+iP} = -\frac{1}{\psi} = \phi^* \quad . \tag{7b}$$

Equations (7) may be inverted to yield

$$N = \frac{1-\phi\psi}{\phi-\psi} = \frac{2\mathrm{Re}\phi}{|\phi|^2+1} \tag{8a}$$

$$P = i\,\frac{1+\phi\psi}{\phi-\psi} = \frac{2\mathrm{Im}\phi}{|\phi|^2+1} \tag{8b}$$

$$Q = \frac{\phi+\psi}{\phi-\psi} = \frac{|\phi|^2-1}{|\phi|^2+1} \quad . \tag{8c}$$

Equations governing the time dependence of ϕ and ψ are readily deduced by inserting Eqs.(8) into Eqs.(3). It is found that the equations for ϕ and ψ are decoupled; ϕ satisfies the Riccati equation

$$\frac{\partial\phi}{\partial t} = iG\phi + \frac{i}{2}\Delta\omega(\phi^2-1) \tag{9}$$

and ψ satisfies the same equation. One may now employ the usual transformation to convert this Riccati equation to a second-order linear equation. Equation (9) may then be transformed to

$$W_{tt} + \frac{1}{4}[(\Delta\omega)^2 + G^2 + 2iG_t]W = 0 \tag{10}$$

where the subscripts signify differentiation with respect to t. The new dependent variable W is related to ϕ through the transformations

$$W(x,t) = u(x,t)\exp[-\frac{i}{2}\int_{-\infty}^{t} dt'G(x,t')] \tag{11}$$

$$\phi(x,t) = (2i/\Delta\omega)d(\ln u)/dt \quad . \tag{12}$$

The Sturm-Liouville equation given by Eq.(10) provides the contact with the Korteweg-deVries equation to be discussed in the next section.

3. Higher Conservation Laws

It has been shown[2] that there is a close connection between the Korteweg-deVries equation

$$u_t + uu_x + u_{xxx} = 0 \tag{13}$$

and the Sturm-Liouville equation

$$\psi_{xx} + \frac{1}{6}(u-\lambda)\psi = 0 \quad . \tag{14}$$

In particular, if u evolves according to Eq.(13), then

$$\lambda_t = 0 \quad . \tag{15}$$

Equation (13) has the interesting property of possessing an infinite number of conserved quantities, i.e., there exist an infinite sequence of quantities T_n and X_n that are related by equations of the form

$$\frac{\partial T_n}{\partial t} + \frac{\partial X_n}{\partial x} = 0, \qquad n = 1,2,3,... \tag{16}$$

It has been shown that these conservation laws may be used to determine the final pulse amplitudes into which solutions of Eq.(13) may evolve.[5]

It has also been shown[2] that if the evolution of u(x,t) in Eq.(14) is governed by *any* equation such that the eigenvalues in Eq.(14) are invariant, then that equation of evolution possesses all the same conserved densities as does the Korteweg-deVries equation.

Now, with the identification

$$\lambda = -\frac{3}{2}(\Delta\omega)^2 \tag{17}$$

$$u = \frac{3}{2}(G^2 + 2iG_t) \tag{18}$$

it is seen that Eq.(10) is of the form of Eq.(14) except that the space and time variables are interchanged. Hence one may expect that it is expressions involving the function of the electric field envelope appearing in Eq.(18), and time derivatives of this function, that are conserved in a way that is closely related to the way in which functions of u and its space derivatives are conserved for the Korteweg-deVries equation.

The first three conserved densities associated with the Korteweg-deVries equation are[2]

$$T_1 = u \tag{19a}$$

$$T_2 = \frac{1}{2} u^2 \tag{19b}$$

$$T_3 = \frac{1}{3} u^3 - u_x^2 \quad . \tag{19c}$$

Interchanging space and time coordinates and employing Eq.(18), one finds that the corresponding three quantities (fluxes) in optical pulse propagation are

$$F_1 = \frac{1}{2} G^2 \tag{20a}$$

$$F_2 = \frac{1}{4} G^4 - G_t^2 \tag{20b}$$

$$F_3 = \frac{1}{6} G^6 - \frac{10}{3} G^2 G_t^2 + \frac{4}{3} G_{tt}^2 \quad . \tag{20c}$$

Only the real part of the T_n from Eqs.(19) are of concern. The imaginary part yields a perfect time derivative which leads to trivial results that provide no information about the system. (Also, Eq.(20c) merely differs from a previous form given in Ref. 1 by a perfect time derivative.)

The associated conserved densities ρ_n which appear in the conservation laws

$$\frac{\partial \rho_n}{\partial t} + c \frac{\partial F_n}{\partial x} = 0 \quad , \qquad n = 1,2,3,\ldots \tag{21}$$

for coherent optical pulse propagation are now expressed in terms of the polarization and population inversion of the medium by employing Eqs.(3) and (4). It has been shown[1] that

$$\rho_n = T_n + F_n \tag{22}$$

where

$$T_1 = \alpha' c\langle 1+N\rangle \tag{23a}$$

$$T_2 = \alpha' c[G^2\langle N\rangle + 2G\langle \Delta\omega Q\rangle - 2\langle(\Delta\omega)^2(1+N)\rangle] \tag{23b}$$

$$T_3 = \alpha' c[G^4\langle N\rangle - \frac{4}{3} G_t^2\langle N\rangle + \frac{8}{3} G^2 G_t\langle P\rangle + \frac{4}{3} G^3\langle \Delta\omega Q\rangle$$
$$+ \frac{8}{3} G_t\langle(\Delta\omega)^2 P\rangle - \frac{4}{3} G^2\langle(\Delta\omega)^2 N\rangle - \frac{8}{3} G\langle(\Delta\omega)^3 Q\rangle + \frac{8}{3}\langle(\Delta\omega)^4(1+N)\rangle] \tag{23c}$$

and α' is as given below Eqs.(4). (Again, the expression in T_3 differs from that in Ref. 1 because of the above-mentioned change in F_3. Both forms lead to identical results where they are used to analyze pulse propagation.)

Use of these results to determine pulse amplitudes will now be considered.

4. Determination of Pulse Amplitudes

To obtain information about pulse amplitudes from the conservation laws, Eqs.(21) are first converted to integral form by performing integrations over both space and time. Various spatial and temporal intervals may be chosen for these integrations. To motivate the choice made here, we first recall the pulse decomposition process. Basically, an initial pulse profile with envelope area θ_0 evolves into a series of pulses, each of area 2π. As the initial pulse profile enters the resonant medium, pulse reshaping and decomposition begins to take place. During this process, some energy is transferred irreversibly from the pulse to the medium. The region of energy loss may be thought of as extending only for some finite depth L into the resonant medium. Beyond this point, lossless propagation of 2π hyperbolic secant pulses ensues. The number of such pulses is determined by the initial pulse area and the area theorem.[6] The only question left unanswered is the amplitude (and related velocity) of each of these pulses. How the conservation laws may be used to determine these amplitudes will now be outlined.

Integration of Eq.(21) over all time from the remote past up to some current time t and over space from $x = 0$ to $x = x_1$ where $x_1 > L$ (it is assumed that the time t is chosen so that all pulses have passed beyond x_1) yields

$$\int_0^L dx\, \rho_n(x,t) + c \int_{-\infty}^{t} dt'\, F_n(x_1,t') = c \int_{-\infty}^{t} dt'\, F_n(0,t') \qquad (24)$$

where it has been assumed that $\rho_n(x,-\infty) = 0$. Reference to Eqs.(20), (22) and (23) shows that this initial condition for the ρ_n will be satisfied if the resonant medium is being used as an attenuator. The functions $F_n(x_1,t)$ are almost completely determined since beyond $x = L$ the electric field is merely a sum of hyperbolic secant pulses. Only the amplitudes (and hence velocities) of these pulses are unknown. Also, the functions $F_n(0,t)$ are known for a given initial pulse profile.

Since t is chosen such that the pulse has passed the point x_1, the only terms in ρ_n that will contribute to the integral over ρ_n in Eq.(24) are those terms in F_n and T_n which do not contain G or any of its derivatives. From Eqs.(23) it is seen that only the last term in each of the T_n is of this form. It is these terms that are related to the energy lost to the medium. For $N(x,+\infty) = -1$ there is no energy loss, as expected.

It should be noted that as $T_2^* \longrightarrow \infty$, an increasing region of ringing occurs behind the pulse and the assumption of a group of isolated pulses may break down, except for pulses of initial area $2n\pi$.

The terms in the first three T_n that are associated with energy loss are

$$I_1 \equiv \int_0^L dx\, \rho_1(x,t) \longrightarrow \alpha' c \int_0^L dx \langle 1+N \rangle \tag{25a}$$

$$I_2 \equiv \int_0^L dx\, \rho_2(x,t) \longrightarrow -2\alpha' c \int_0^L dx \langle (\Delta\omega)^2 (1+N) \rangle \tag{25b}$$

$$I_3 \equiv \int_0^L dx\, \rho_3(x,t) \longrightarrow \frac{8}{3}\alpha' c \int_0^L dx \langle (\Delta\omega)^4 (1+N) \rangle \, . \tag{25c}$$

Setting

$$\Gamma(\Delta\omega,L,t) \equiv \alpha' c \int_0^L dx (1+N) \tag{26}$$

one may write these integrals as

$$I_n = \kappa_n \int d\Delta\omega\; g(\Delta\omega)\, (\Delta\omega)^{2(n-1)}\, \Gamma(\Delta\omega,x,t) \tag{27}$$

where $\kappa_1 = 1$, $\kappa_2 = -2$, $\kappa_3 = 8/3$. In the absence of population decay constants T_1, the function Γ and hence the I_n will be independent of time.

A thorough numerical analysis of Γ and the I_n would be of considerable interest. For the present it will suffice to show that a very approximate treatment of these terms permits quite accurate determination of pulse amplitudes.

Neglecting the frequency dependence of Γ and using the Gaussian distribution

$$g(\Delta\omega) = \frac{T_2^*}{2\sqrt{\pi}}\, e^{-(T_2^*\Delta\omega/2)^2} \tag{28}$$

one readily obtains

$$
\begin{aligned}
I_1 &= \Gamma \\
I_2 &= -4\Gamma/(T_2^*)^2 \\
I_3 &= 32\Gamma/(T_2^*)^4 \quad .
\end{aligned}
\tag{29}
$$

Henceforth Γ will be treated as a phenomenological constant.

For the initial pulse profile

$$G(0,t) = \frac{\theta_o}{\pi t_o} \operatorname{sech} \frac{t}{t_o} \tag{30}$$

a previously reported[1] calculation has yielded results which may be put in the form

$$\int_{-\infty}^{t} dt\, F_1(0,t) = G(0,t)_{max}\, r \tag{31a}$$

$$\int_{-\infty}^{t} dt\, F_2(0,t) = [G(0,t)_{max}]^3 \cdot \frac{1}{3r} (r^2-2) \tag{31b}$$

$$\int_{-\infty}^{t} dt\, F_3(0,t) = [G(0,t)_{max}]^5 \cdot \frac{8}{45r^3} (r^4-5r^2+7) \tag{31c}$$

where $r = \theta_o/\pi$. (In these integrations the upper limit extends beyond the duration of the initial pulse and may be replaced by ∞.)

After the pulse has evolved into a sequence of hyperbolic secant pulses one may write [6]

$$G(x,t) = \sum_{i=1}^{N} a_i \operatorname{sech}[(t-t_i-x/v_i)/\tau_i] \tag{32}$$

where τ_i and v_i are the half width and velocity of the i'th pulse, respectively, and the t_i are merely delays that serve to separate the various pulses. The number of pulses, N, is determined by the area theorem from the area under the initial pulse envelope.[6] The amplitudes a_i are related to the pulse widths by

$$a_i = 2/\tau_i \tag{33}$$

and the velocities are given by

$$\frac{c}{v_i} = 1 + \alpha' c\tau_i^2 \left< \frac{1}{1+(\tau_i \Delta\omega)^2} \right> . \tag{34}$$

Setting

$$x_i \equiv a_i/G(0,t)_{max} \tag{35a}$$

$$\beta \equiv T_2^*/t_o \tag{35b}$$

$$\gamma \equiv \Gamma t_o/c \tag{35c}$$

one finds that the first three of Eqs.(24) take the form

$$\sum_{i=1}^{N} x_i + \gamma/2r = r/2 \tag{36a}$$

$$\sum_{i=1}^{N} x_i^3 - 12\gamma/r^3\beta^2 = \frac{1}{r}(r^2-2) \tag{36b}$$

$$\sum_{i=1}^{N} x_i^5 + 480\gamma/r^5\beta^4 = \frac{8}{3r^3}(r^4-5r^2+7) . \tag{36c}$$

For $3\pi<\theta_o<5\pi$, two pulses are to be expected and the first two conservation laws with N = 2, and proper choice of β and γ as described below, may be used to determine both x_i and hence the two pulse amplitudes. A similar procedure applies with the three conservation laws for $5\pi<\theta_o<7\pi$.

Setting $\gamma = 0$ yields the equations that were used previously[1] to obtain pulse amplitudes. While the equations with $\gamma = 0$ gave very satisfactory results for the large amplitude pulses, the small pulse that occurs for $\theta_o \sim 3\pi$, 5π was not given satisfactorily. It will now be shown that proper choice of the constants β and γ enables one to obtain correct results in this threshold region as well.

When only two pulses occur, the amplitude of the smaller pulse should vanish at $\theta_o = 3\pi$. This result is assured by determining γ in Eqs.(36a,b) so that $x_3 = 0$ for $\theta_o = 3\pi$. The value of γ obtained in this way will, of course, depend upon the value assumed for $\beta(= T_2^*/t_0)$. For $\beta = 10$ one finds $\gamma = 1.037$ while for $\beta = 1$, the result is $\gamma = 0.673$. A simple test of the validity of such an approach is to use these values of β and γ in the solution of all three equations in the region $5\pi<\theta_0<7\pi$ and observe the value of θ_0 at which the smallest of the three roots vanishes. The result is quite reassuring since the vanishing point is extremely close to 5π (4.99π for $\beta = 10$, 5.05π for $\beta = 1$). The results are shown in Fig.1. The circles are results of numerical solutions of Eqs.(3) and (4) that were performed by M.O. Scully and F.A. Hopf[7].

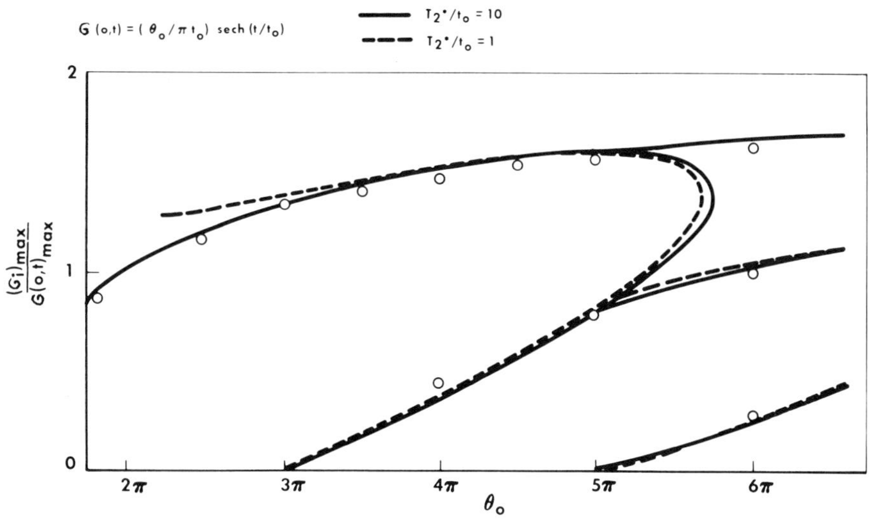

Fig. 1 Pulse amplitudes obtained from conservation laws when energy loss terms are taken into account. Circles represent results of numerical calculations provided by M.O. Scully and F.A. Hopf[7].

References

1. G.L. Lamb, Jr., Rev. Mod. Phys. *43*, 99 (1971).
2. R.M. Miura, C.S. Gardner and M.D. Kruskal, J. Math. Phys. *9*, 1204 (1968).
3. M.D. Kruskal, R.M. Miura, C.S. Gardner and N.J. Zabusky, J. Math. Phys. *11*, 952 (1970).
4. L.P. Eisenhart, *A Treatise on the Differential Geometry of Curves and Surfaces* (Dover Publications, New York, 1960) p. 23.
5. V.I. Karpman and V.P. Sokolov, Soviet Physics JETP *27*, 839 (1968).
6. S.L. McCall and E.L. Hahn, Phys. Rev. *183*, 457 (1969).
7. G.L. Lamb, Jr., M.O. Scully and F.A. Hopf, Appl. Opt. Nov. 1972.

SURVEY OF THE PRESENT STATUS OF NEOCLASSICAL RADIATION THEORY*

E.T. Jaynes

Washington University, St. Louis, Mo.

1. Introduction

Present quantum electrodynamics (QED) contains many very important "elements of truth", but also some clear "elements of nonsense". Because of the divergences and ambiguities, there is general agreement that a rather deep modification of the theory is needed, but in some forty years of theoretical work, nobody has seen how to disentangle the truth from the nonsense. In such a situation, one needs more experimental evidence, but during that same forty years we have found no clues from the laboratory as to what specific features of QED might be modified. Even worse, in the absence of any alternative theory whose predictions differ from those of QED in known ways, we have no criterion telling us *which* experiments would be the relevant ones to try.

It seems useful, then, to examine the various disturbing features of QED, which give rise to mathematical or conceptual difficulties, to ask whether present empirical evidence demands their presence, and to explore the consequences of modified (although

---

*Supported by the Air Force Office of Scientific Research, Contract No. F44620-60-0121.

This talk was scheduled as the first of a long series devoted to Quantum Electrodynamics and alternative theories. Evidently, a fully up-to-date account of the topic could not be given until the other speakers, presenting new evidence bearing on these matters, had been heard. This final version therefore differs from the talk actually given, in deletion of obsolete material and in additions to take account of work reported by others at the Conference.

perhaps rather crude and incomplete) theories in which these features are removed. Any difference between the predictions of QED and some alternative theory, corresponds to an experiment which might distinguish between them; if it appears untried but feasible, then we have the opportunity to subject QED to a new test in which we know just what to look for, and which we would be very unlikely to think of without the alternative theory. For this purpose, the alternative theory need not be worked out as completely as QED; it is sufficient if we know in what way their predictions will differ in the area of interest. Nor does the alternative theory need to be free of defects in all other respects; for if experiment should show that it contains just a single "element of truth" that is *not* in QED, then the alternative theory will have served its purpose; we would have the long-missing clue showing in what way QED must be modified, and electrodynamics (and, I suspect, much more of theoretical physics along with it) could get moving again.

That is, in a nutshell, the program I have visualized for getting my favorite subject, electrodynamics, out of the difficulties that it has been in throughout the adult life of every person here. And up to this point, I think that it is entirely non-controversial; nothing I have said thus far could offend anyone, whatever his personal views about QED. But the trouble starts when we start deciding which specific feature of QED should receive the surgical removal needed to formulate a modified theory. For, I believe our Quantum Electrodynamicists are quite correct in saying that *no* part of QED can be modified without coming into conflict with the basic Copenhagen interpretation of quantum mechanics, as applied to the electrons with which the field interacts. But the Copenhagen interpretation has become something sacred now, no longer a set of hypotheses to be tested by experiment, but an ideology which prescribes the limits of permissible thought in physics. So, no matter what aspect of QED I decide to tamper with, I can expect to be attacked by some fanatic who thinks I am committing blasphemy. And it will do no good to protest that I don't necessarily "believe" the modification; I am merely formulating a tentative hypothesis, to find out what its consequences would be; it is just the inability to comprehend that kind of subtlety that makes a fanatic. So, in spite of the apparently innocent nature of my program when stated in generalities, there is no way to carry it out explicitly without getting into controversy.

This being the case, one might as well be hung for a sheep as a lamb (no pun intended), and so, if I may compound metaphors, I decided to go straight for the jugular vein of QED, and do a surgical removal of field quantization itself. In this, I may be accused of blasphemy, but not of originality. For many physicists (among whom Planck, Schrödinger, and Franken have been quite explicit) have opined that the experimentally observed "quantum effects", such as the $E = h\nu$ relation, should be accounted for by the properties of matter and its interaction with radiation, with no need for any change in the nature of the electromagnetic field itself. From this standpoint, the troubles of QED would be regarded as symptoms that we are trying to take the same thing into account twice, and the theory I want to describe represents merely the working out of some quantitative aspects of an old, but previously not much developed, idea.

Unfortunately, it is necessary to do more than merely describe this theory. Anybody who undertakes to play this game of exploring unconventional ideas will be astonished at the kind of reactions he gets. Not that some applaud your efforts and others deplore them; that you expect, as already noted. What is astonishing is that, after the most carefully written expositions, so many on both sides insist on completely misunderstanding and misrepresenting not only your viewpoint and objectives, but also the plain documentable facts about what you have already accomplished. There is not space here enough to correct all the misinformation about Neoclassical theory to be found in the March 1972 issue of Physics Today. I cringed at the sight of that black box with its sensational tabloid headline: "The Uncertainty Principle Violated!" Let me assure you: the lady's honor is quite intact as far as I am concerned, because she has never set foot in neoclassical theory.

This extraordinary difficulty in communicating unconventional ideas - even when both parties are trying their friendly best to bring it about - means that the person exploring them must be prepared to spend a great deal of time, not on constructive things, but on clearing up past misunderstandings. So, while the primary subject of this article is the status of neoclassical theory, a secondary objective must be to try to correct a long list of almost unbelievably persistent misconceptions about what the theory is, and what we are trying to accomplish with it.

In particular, we must emphasize that (1) while this theory has already made a number of new predictions, it is still in a very incomplete, provisional state with regard to fundamentals, and just for that reason, it is still flexible and can, in many respects, still adapt itself to new facts. (2) As discussions at this conference show, it is hard to keep one's sights on the real issue. I had thought that our motivation and objectives were explained sufficiently clearly two years ago (particularly in Ref. 9, and Ref. 7, Introduction and page 110), but realized too late that I have probably contributed to a new confusion of the issue by the title of this article. So I will emphasize, to the point of belaboring it, that the issue before us is *field quantization*, whether we do or do not need it in order to account for the facts of electrodynamics. The issue is *not* the universal validity of any current form of semiclassical theory, whether concocted by me or anybody else. It is

QED, and not semiclassical theory, which has made pretensions of universal validity.

As the title indicates, we are, of course, interested in the range of validity of semiclassical theories, and in the question whether their present methods can be refined so as to enlarge their domain of validity; and not only for the reasons noted above. Independently of all deep theoretical questions, semiclassical methods have a proven usefulness in current experimental work of quantum optics, such as laser dynamics and coherent pulse propagation, which we would naturally like to increase. Furthermore, even without new experiments, every increase in the domain of applicability of semiclassical methods causes an equal and opposite decrease in the domain where QED can be claimed to be necessary, which sharpens our judgment as to where the faulty feature of QED may lie.

With all this concentration on negative aspects of QED, I will surely be accused of failing to recognize its good features. In defense, let me point out that, to the best of my knowledge, I was the first person to apply QED to problems of coherence in quantum electronics (by this, I mean real QED, and not just "Fermi Golden Rule" type approximations to it). In the middle 1950's, Professor Willis Lamb showed me his then unpublished semiclassical theory of the ammonia maser. But, having exactly the same instincts that I see in young physicists today, I felt very uneasy about the results until I could verify that they followed also from QED. To do this, it was evident that one must go beyond time-dependent perturbation theory and find ways of solving the equations of motion accurately over long times, without losing phase information.

This work, done in 1956, led to a new respect for semiclassical theories and culminated in a report that we finally got out in 1958[1], which included the beginnings of the neoclassical theory, and saw the first appearance of elliptic functions, of which one particular limiting form - the hyperbolic secant pulse - was noted (but without any comprehension of the significance this would have later, thanks to McCall and Hahn). Developments since then[2-9] are found in the Ph.D. theses of many former students, who have contributed a large mass of detailed calculations, including quite a few surprises.

2. Semiclassical Methods

Before we get into its deep philosophical meaning, let's be sure we understand what the term "Neoclassical theory" means pragmatically, by tracing its evolution from older semiclassical methods. What I will call Semiclassical A (SCA) was the original method of incorporating the electromagnetic field into quantum theory, antedating QED. SCA is what we were all taught in our first course in quantum mechanics, defined for our present purposes (which are served adequately by the model of a single nonrelativistic spinless hydrogen atom) by the Schrödinger equation

$$i\hbar\dot{\psi}=\left[\frac{(p-\frac{e}{c}A)^2}{2m}+e\phi\right]\psi \tag{1}$$

in which the electromagnetic potentials A,ϕ are considered given. This equation determines the effect of the field on the atom; from it we obtain the quantum theory of the Zeeman and Stark effects, the Einstein B-coefficients of black-body radiation theory, the Rutherford scattering law, the photoelectric cross-section, and with appropriate generalization, very much more.

SCA is incomplete in that it fails to give the effect of the atom on the field. To supply this, so that one could describe emission and scattering of radiation, there arose the "Klein Vorschrift" described in the Pauli Handbuch article[10], in which one found it necessary to make arbitrary replacements of the form $\langle F^2\rangle \rightarrow 2\langle F^+F^-\rangle$ for field quantities F, in order to obtain sensible results for rate of radiation of energy. Here F^+, F^- are the positive and negative frequency parts of F; thus the Klein Vorschrift was an ancestor of modern normal-ordering methods of QED.

Closely related to this was the "transition current method" (TCM) which is still very much in use today and is described, for example, in Schiff's textbook[11]. In TCM, one specifies initial and final states ψ_i, ψ_f for the electrons, and sandwiches an operator representing current, dipole moment, etc. between them, making the "transition current"

$$J_{fi}(x,t) = \frac{e}{mc}\,\psi_f^*\,(p - \frac{e}{c}A)\,\psi_i \tag{2}$$

or the "transition dipole moment", etc. Then we switch to classical electromagnetic theory, and calculate the fields that would be produced by such a current or dipole moment. In this way, surprisingly, we obtain the correct Einstein A-coefficients for spontaneous emission. TCM also yields many other useful results, such as the Møller e-e scattering formula.

TCM can hardly be considered as a well-motivated physical theory in its own right, because it mixes up the initial and final states in a way that defies any rational physical interpretation. Note, however, that if

$$\Psi = \sum_i a_i \psi_i \tag{3}$$

is a linear combination of stationary states, the quantity

$$J(x,t) = \frac{e}{mc} \operatorname{Re}[\Psi^*(p - \frac{e}{c} A)\Psi] \tag{4}$$

usually called the "probability current" will be interpreted by neoclassical theory as actual current (or, at least, its divergence will equal the divergence of the actual current). Using the expansion (3), we see that the current (4) contains all the transition currents with amplitude factors $a_i^* a_j$:

$$J(x,t) = \sum_{ij} J_{ij}(x,t) \; a_i^* \, a_j \tag{5}$$

Because of the above difficulty of interpretation, and because both the Klein vorschrift and TCM receive an *a posteriori* justification from QED, I would consider that they do not represent parts of any semiclassical theory, but should be regarded as convenient short-cut algorithms contained in QED.

An entirely different way of taking into account the effect of atoms on the field is based on the Ehrenfest theorem. The equations of motion for expectation values

$$\frac{\partial}{\partial t}\langle F\rangle = (i/\hbar)\langle [H,F]\rangle \tag{6}$$

resemble classical deterministic equations, and they reduce to the usual classical equation of motion for the quantity F, as the dispersion

$$(\Delta F)^2 \equiv \langle F^2\rangle - \langle F\rangle^2 \tag{7}$$

tends to zero. This was exploited in the famous 1946 paper of F. Bloch[12] on magnetic rexonance theory, resulting in a theory that I will call Semiclassical B (SCB). Imagine that we have a sample of a few milligrams of some substance containing protons; the number of them is probably of the order of $n \sim 10^{20}$, or more. The operator representing the total magnetic moment of these n protons is the sum of the individual operators:

$$M = M_1 + M_2 + \cdots + M_n \tag{8}$$

and if all spins are related to the sample and to each other in the same way, the expectations of M and M^2 are

$$\langle M\rangle = n\langle M_1\rangle \tag{9}$$

$$\langle M^2\rangle = n\langle M_1^2\rangle + n(n-1)\langle M_1M_2\rangle \tag{10}$$

Defining the mean square fractional fluctuations as $R \equiv (\Delta M)^2/\langle M\rangle^2$, $R_1 \equiv (\Delta M_1)^2/\langle M_1\rangle^2$, we have from (9), (10),

$$R = \frac{\langle M_1M_2\rangle - \langle M_1\rangle^2}{\langle M_1\rangle^2} + \frac{1}{n}\frac{\langle M_1^2\rangle - \langle M_1M_2\rangle}{\langle M_1\rangle^2} \tag{11}$$

The total moment M becomes better defined as $R \to 0$, and we see from(11) that R depends crucially on correlations between spins. We need look only at the two extreme limiting cases of (11). *Case 1*, complete positive correlation: $\langle M_1M_2\rangle = \langle M_1^2\rangle$. Then (11) reduces to $R = R_1$; the total moment of n spins is no better defined than that of a single spin. In other words, there is no "law of large numbers". *Case 2*, no correlation: $\langle M_1M_2\rangle = \langle M_1\rangle\langle M_2\rangle = \langle M_1\rangle^2$. In this case, (11) reduces to

$$R = \frac{1}{n} R_1 \tag{12}$$

and the law of large numbers is resurrected; for large n, the total moment becomes as well-defined as any classical quantity ever was.

Now, the expectation of a single moment obeys the equation of motion

$$\frac{\partial}{\partial t}\langle M_1\rangle = (i/\hbar)\langle [H,M_1]\rangle \tag{13}$$

and, from (9), we need only multiply both sides of this by n to have the equation of motion for $\langle M(t) \rangle$. If the spins are uncorrelated (or more generally, if spin-spin correlations drop off with their separation sufficiently rapidly for an ergodic condition to hold), the relative fluctuation R will be negligible, and we have the deterministic Bloch equations for total magnetic moment of the sample. As I have shown elsewhere[13], it is such considerations that give, in large measure, the explanation for the success of statistical mechanics.

It remains to consider how the total moment $\underline{M}$ affects the radiation field. If the sample is small compared to a wavelength, it seems clear that a well-defined magnetic moment $\underline{M}$ should generate a well-defined classical electromagnetic field via the Maxwell equations

$$\nabla \times \underline{H} - \frac{1}{c}\dot{\underline{E}} = 0 \tag{14}$$

$$\nabla \times \underline{E} + \frac{1}{c}\dot{\underline{H}} = \frac{4\pi}{c}\,\frac{\dot{\underline{M}}}{V} \tag{15}$$

where V is the volume of the sample. In this way, for example, one finds that the open-circuit voltage induced in a coil wound in an arbitrary way about the sample, is given by $v_{oc} = \underline{H}\cdot\dot{\underline{M}}$, where $\underline{H}$ is the magnetic field at the sample due to unit current in the coil. In NMR work, the Q of the receiving circuit is usually so low that radiation damping [i.e., the effect of the field calculated from (14), (15) reacting back on the moment M] can be neglected. In high resolution NMR and ESR, it may be a complicating factor[12].

In the 1946 Bloch paper one finds the sphere representation, fast-passage solutions which include the π-pulse, etc., which prepared the way for the Hahn spin-echo experiment[14]. Further elaborations of the Bloch sphere representation led to a theoretical technique[15] for predicting complicated sequences of spin echoes, which will probably find application soon in the theory of optical pulse echoes.

The SCB method thus initiated, was applied some ten years later in the semiclassical theories of the ammonia maser, by Basov and Prokhorov[16], and by Shimoda, Wang, and Townes[17], who assumed a delta-function velocity distribution. This was generalized to a Maxwellian distribution by Lamb and Helmer[18]. Cummings and I[3], in a work devoted largely to other matters, noted that in an ammonia maser the velocity distribution will not be Maxwellian, because the electrostatic focusser is more efficient for low velocity molecules, and carried out the (by then rather trivial) generalization to an arbitrary velocity distribution. This confirmed that such experimental quantities as the starting current and frequency pulling factor are determined by the slowest few percent of the molecules, being inversely proportional to the mean square and mean cube flight times $<\tau^2>$, $<\tau^3>$, respectively.

From the standpoint of radiation theory, these SCB treatments of the ammonia maser differed from the Bloch magnetic resonance theory mainly in the fact that the field producing the stimulated emission was not "externally applied", but was the field previously radiated by the molecules themselves; in other words, radiation damping was now, due to the high Q of the cavity, an essential part of the theory, with the cavity Q appearing in the expressions for starting current, amplitude of oscillation, and frequency pulling factor. At this microwave frequency (24.8 GHz) ordinary (i.e., cavity-unassisted) spontaneous emission was still completely negligible, corresponding to radiative lifetimes of the order of months, while the cavity-assisted emission took place in less than a millisecond. It is true that the active sample now extended over many wavelengths, but one considered only its interaction with the TM_{01} cavity mode, whose field is constant along the length of the beam, so that again it was the total moment of all the molecules that was considered the source of a classical electromagnetic field.

The fact that the Bloch sphere representation applies equally well to any two-level system was common knowledge at Stanford University when I spent the summer there in 1947. In a sense, it was "obvious" to anyone who knew that (2x2) unitary matrices form a faithful representation of the three-dimensional rotation group; but this does not seem to have been published at the time, and it was left for Feynman, Vernon, and Hellwarth[19] to point it out in 1957, in an article where the hyperbolic secant pulse again puts in a brief appearance [loc. cit., Eq.(19)] under the heading: "Radiation Damping". As that suggests, they were considering only the back half of the pulse, where we move downward to the south pole of the Bloch sphere; but, of course, the analytical solution can be extrapolated backwards past $t = 0$, to give the rising front half of the self-induced transparency pulse. It is incredible, in retrospect, how many times this solution had been found by theoreticians before McCall and Hahn finally realized its significance. One wonders how many other important results are hiding in theoretical papers, unrecognized even by their authors, just waiting for some clever experimenter to show us what they mean.

The first application of SCB to magnetic resonance involved ordinary radio frequencies, $\sim$ 30 MHz, where "quantum effects" had always been considered so negligible as to be impossible even to detect, much less affect any physical phenomena. The step up to the ammonia maser involved a thousand-fold increase in frequency, but here again quantum effects were considered negligible for all ordinary purposes. At room temperature, for example, one had $(\hbar\omega/kT) \sim 3 \times 10^{-3}$, so that thermal noise still predominated over "quantum noise" except at extreme cryogenic temperatures, $T \leq 1°K$. Thus, also here the use of classical electromagnetic theory should not arouse any anxiety even in the most dedicated quantophile. But the next level of application of SCB, Lamb's analysis of the laser[20], represents a further 20,000-fold increase in frequency, into the region $(\hbar\omega/kT) \sim 60$, where some would expect radiation phenomena to be completely dominated by "quantum effects".

The success of Lamb's semiclassical theory in predicting a large mass of experimental facts[21] therefore came as an instructive surprise to some whose education did not include real QED, but only the standard verbal misconceptions of it (i.e., the "buckshot theory" of light, which has propagated through several generations of elementary textbooks) with which we brainwash our undergraduate students. Nearly all of them emerge from this with a mental picture according to which, as the frequency increases, the electromagnetic field gradually acquires some kind of discontinuous, granular structure which wipes out interference effects. Closely related is a persistent literal belief in that over-quoted remark of Dirac, to the effect that a given photon interferes only with itself. From the standpoint of QED such a statement is neither true nor false, but simply meaningless, for "photons" lack the individuality which the statement presupposes; there is just nothing in the mathematical formalism of QED that corresponds to any such notion as "a given photon".

The appearance of the laser as an accomplished fact struck a severe blow to these almost universally held misconceptions about quantum theory. Recall that, as recently as 1963, many physicists thought that, because of Dirac's statement, it was fundamentally impossible to observe interference between independently running lasers. And recall the uproar of 1956, when some of our best known theorists would not believe the Hanbury Brown-Twiss effect, because they thought it violated quantum theory. In both cases the experimental facts were accounted for trivially by classical electromagnetic theory, but to some quantum theorists they appeared as astonishing new phenomena, in need of deep and profound explanation. Such incidents led inevitably to new discussion about the nature of "photons", and the need for QED.

Lamb's SCB theory of the laser differed from SCB treatments of the ammonia maser in several respects. At microwave frequencies, the cavity modes are well-separated in frequency, so that emission into any mode other than the one of interest is negligible. At optical frequencies the cavity "quasimodes" are a discrete set of resonances, superimposed on a continuum of field modes like those of free space. Because of this, and the magnitude of the Einstein

A-coefficient ($\sim 10^8$ sec^{-1}), spontaneous emission into modes other than the ones of interest is hardly ever negligible, and is often the dominant physical process at work. Finally, the optical cavity is of the order of a million wavelengths long, with the normal mode field reversing sign every half-wavelength along the cavity axis. This has two consequences: (1) it is no longer the total moment of the active atoms that is the effective driving force in the classical Maxwell equations; instead, the regions of active atoms is treated as a continuous medium, with an active electric polarization density. (2) Doppler broadening, which could be neglected in the ammonia maser, is now one of the crucial things determining performance; an atom moving at thermal velocities may see up to about fifty phase reversals of the field during a radiative lifetime. By analogy with the effect of flight time in the ammonia maser[3], one would conclude that the "useful" emission must be due mostly to the few percent of active atoms that are moving nearly transversely to the axis of the optical cavity, for a mode tuned exactly on the atom's natural frequency ω_0. For the next optical mode, tuned higher by perhaps $\delta\omega = 5 \times 10^{-7}\ \omega_0$, the "useful" emission will be contributed mostly by another velocity group of atoms, namely those with $v_z \simeq \pm 5 \times 10^{-7}$ c, etc.

In spite of the important role played by spontaneous emission into continuum field modes, Lamb's theory does not consider the actual physical mechanism of this process, but instead invokes a phenomenological damping mechanism which presumably has similar effects. If we define a truncated (2x2) density matrix ρ with rows and columns referring only to the two lasing levels, its equation of motion is taken in the form

$$\dot{\rho} = -i[H,\rho] - \frac{1}{2}(\rho\Gamma + \Gamma\rho) \tag{16}$$

where the damping matrix Γ is diagonal with elements γ_a, γ_b, which are phenomenological constants interpreted as decay rates to unspecified lower levels. Thus ρ, instead of relaxing to a ground state or thermal equilibrium form, damps to zero through a kind of seepage to lower levels which are never brought explicitly into the theory.

The main effect of this is that the time of coherent interaction between any one atom and the field mode of interest is effectively limited to $T_{int} \sim (\gamma_a + \gamma_b)^{-1}$, presumably of the order of a few nanoseconds, regardless of whether the atom is emitting or absorbing. Questions of long-time coherence therefore do not arise, and Lamb takes account of the field interaction by conventional time-dependent perturbation theory carried to third order, making no use of the Bloch sphere representation.

The Lamb SCB laser theory has been described in some detail so that a few of that long list of misconceptions can be pointed out. In a later paper on QED theory of the laser, Scully and Lamb[22] give an Introduction explaining the need for QED here, by pointing out defects in the semiclassical treatment. These points are made: (1) Semiclassical theory "implies that laser radiation in an ideal steady state is absolutely monochromatic". (2) According to semiclassical theory, "oscillations will not grow spontaneously, but require an initial field from which to start". (3) "Still another problem requiring a fully quantum-mechanical theory is to determine the statistical distribution of the energy stored in the laser cavity, i.e., the 'photon' statistics".

To these assertions, we reply as follows: (1) While this criticism may apply to the Lamb Semiclassical theory as actually published, it is not in any way a limitation on that theory; Lamb could easily have calculated the linewidth of an actual laser in the steady state by taking into account the statistical fluctuations in the number of excited atoms. He simply neglected to do so. A semiclassical theory of noise in the ammonia maser (where the main source is now thermal Nyquist noise generated in the cavity walls) has stood for some time[3] as a counter-example to this claim. Perhaps one would reply that the term "ideal steady state" was intended to mean one free of number fluctuations or thermal noise. But even in such a state, so ideal as to be utterly non-physical, it is still true that each excited atom emits a wave train of finite duration, therefore finite spectral width, and therefore the total radiation will have a finite width. Again, Lamb could easily have calculated this in his semiclassical treatment if he had wished to do so. (2) As before, this is not a valid criticism of semiclassical theory *per se;* it describes only the restrictive assumptions that Lamb chose to put into his calculation. He did not get spontaneous buildup of oscillation because he assumed that every atom in the lasing levels was placed by the pumping mechanism into exactly the upper state u_a or the lower one u_b. In reality, of course, the collisional excitation mechanism will place almost every atom (i.e., all except a set of measure zero) in some linear combination $\psi(0) = c_a u_a + c_b u_b$ + (contributions from other non-lasing levels). The excited atom then has, at $t = 0$, a dipole moment proportional to $|c_a^* c_b|$ at the lasing frequency, and oscillations build up spontaneously as well as in QED. From a pragmatic standpoint (i.e., ignoring all philosophical differences, and looking only at the actual calculations done) the main difference between neoclassical theory and other semiclassical methods lies just in the fact that points like this are recognized and taken explicitly into account in neoclassical calculations. (3) By now, our reply can be anticipated; semiclassical theory is quite capable of giving statistical fluctuations in energy. It does this automatically when it is allowed to do so, i.e., when we refrain from putting in restrictive assumptions which amount to denying the possibility of fluctuations. One cannot justify a claim that treatment of energy fluctuations requires a "fully quantum-mechanical theory" until the predictions of both theories have been worked out and compared with experiment. This is, of course, one of the main issues here, since it involves field quantization in a very direct way. Unfortunately, at the present time neither theory has been worked out sufficiently to tell what its predictions are, and the experiments are non-existent.

To those who are surprised by this last remark, being under the impression that the theoretical situation is well understood, having been disposed of, to a large degree, already by Einstein[23] in 1909, we say that we will return to the subject of energy fluctuations in a later section, starting with the recent treatment of Scully and Sargent[24] but carrying the reasoning a few steps further. Be prepared for a bigger surprise.

The final - and by far the greatest - area of applications established for Semiclassical B theory is, of course, to the phenomena of self-induced transparency[25], resonant pulse propagation, chirping, photon echoes, etc. There is no need to go into details here, or even attempt a fair set of references, since this field is developing at such a furious rate. Many other speakers at this Conference will tell us far more about the subject than I can.

It appears that either SCB or QED can be used, almost interchangeably, to discuss the interesting subject of superradiance which, from the Program of this Conference, has now won out over the Schwarz-Hora effect for the honor of being the most discussed and least observed phenomenon in physics.

To turn to the future, many new technological possibilities await the development of reliable and continuously tunable high-power lasers. As one example, I will venture to predict that, by the time another six years have passed, the subject of third-harmonic power generation will be developing as an important side-branch of this field. When the need for it arises, you will find the necessary theory, both QED and SCB, already worked out in the thesis of Duggan[5].

In the area of quantum optics, it is clear that semiclassical theory has led to vastly more real physical predictions than QED. Indeed, the hundreds of existing experiments in this field have, with only two or three exceptions (the experiment of Clauser[26] being outstanding) been predicted and/or explained in terms of SCB theory. It is usually much simpler mathematically than QED (although there is no theorem guaranteeing this for every individual problem), and it gives a simple intuitive picture of what is happening; this is something which is not only pleasing aesthetically, and often necessary for further progress, but which is conspicuously missing in QED. On the other hand, developments in QED inspired by quantum optics (coherent state representation, etc.) have given us a number of elegant theorems. However, they remain sterile, having almost no connection with real experimental facts, and there is a history of frustration[27] in attempts to find experiments which require them. In the face of this, I am glad that I do not have to defend the claim that QED is "the only workable field theory we have".

3. Neoclassical Theory

Throughout the applications of Semiclassical B theory noted above, there was the implicit idea that neglect of field quantization was justified only because of the large number of atoms or molecules involved, and that statistical considerations like Eq.(12) would render the total moment of a sample, or the total polarization of a coherence volume, a well-defined quantity, essentially free of fluctuations, which could then serve as the source of an equally well-defined classical EM field. Although I don't think anyone ever carried out an explicit calculation along the lines of Eq.(11) to verify this, it was always assumed that there was safety in large numbers, so that the SCB calculations were not in conflict with QED, but on the contrary were good approximations to what a (usually far more complicated) QED analysis would have given.

In other words, one had the physical picture of the total moment of a large number of atoms obeying definite, deterministic equations of motion, yet one was not permitted to suppose that the moment of each individual atom behaved in that way. If you asked, "Why not?" you would get different answers from different people, but they would all involve some reference to the uncertainty principle, or complementarity, or the statistical interpretation of quantum mechanics. While no two physicists would agree on just what it was, all felt that, while it was legitimate to talk about the total moment of many atoms, some *Verbot* issued from Copenhagen prevented us from talking about the moment of a single atom in the same way.

Note that it was not just that moments of individual atoms might be unknown to you and me, i.e., that different atoms might have different moments in some statistical distribution as in classical statistical mechanics. In that case, one could say that each atom still has a definite, "objectively real" moment, but that its value was unknown, because it depends not only on the known applied field, but also on unknown microscopic details of the atom's environment. For prediction, one might then have not only a dynamical problem, but also a statistical problem to contend with. The mathematics might get quite involved, but it would remain simple and straightforward conceptually.

If the situation were as just visualized, then consideration of moments of individual atoms would amount to little more than adding statistical considerations to the previous SCB treatments to extend their range of application to the case of a few atoms, where statistical fluctuations must be taken into account, and this should not, after all, present any insurmountable difficulty. But, according to the Copenhagen interpretation, it is far worse than that; not only the numerical value, but also the very concept of the moment of an individual atom, becomes fuzzy in such a way that it is held to be physically meaningless even to ask the question, "How is the moment of an individual atom varying with time?" The reason is connected with the famous von Neumann "hidden variable" theorem[28]. Although quantum theory readily yields certain mathematical quantities $\langle F\rangle = (\psi, F\psi)$ which are called "expectation values", they are not in general expectations over any underlying ensemble, the individual members of which could be identified with the possible "true but unknown" physical situations. In magnetic resonance, for example, one can calculate the expectations $\langle M_x\rangle$, $\langle M_y\rangle$, and expectations of any functions of them: $\langle f_1(M_x,M_y)\rangle$, $\langle f_2(M_x,M_y)\rangle$, $\cdots$, etc. But since M_x and M_y do not commute, there is no underlying joint probability distribution $p(M_x,M_y)$ which yields all those "expectations" by the usual rule of probability theory

$$\langle f(M_x,M_y)\rangle = \iint p(M_x,M_y)\, f(M_x,M_y)\, dM_x\, dM_y \quad . \tag{17}$$

So, having calculated a number of expectation values, if you ask, "What is the ensemble of possible time variations for $M_x(t)$, $M_y(t)$ which would yield my calculated expectation values?", the answer is: "There is *no* such ensemble; your 'expectation values' are expectations over nothing at all. It is not only meaningless to ask what an individual moment is doing, it is even meaningless to ask for an ensemble of *possible* behaviors!"

Now in every other statistical theory ever dreamt of, if such a situation were to arise, one would recognize instantly that a logical contradiction has been found. The obvious, common-sense conclusion would be drawn that our interpretation is in error; a quantity which cannot be written as an expectation, should not be interpreted as an expectation. The mathematical quantities $\langle F\rangle$, whose usefulness is undeniable (they form the source of the radiation field in SCB) ought not to be interpreted physically as mere expectation values; they have a more substantial meaning. As many physicists, including Einstein, Schrödinger, and von Laue[29] have been pointing out for 45 years now, the Copenhagen theory slips here into mysticism; by refusing to recognize this contradiction and clinging to an unjustifiable interpretation, it ends up having to deny the existence of an underlying ensemble, and therefore, of any "objective reality" on the microscopic level.

That this denial is required by the Copenhagen interpretation, has been well recognized by Heisenberg[30], who states it many times. I give three examples: "They (i.e., opponents of the Copenhagen interpretation) would prefer to come back to the idea of an objective real world, whose smallest parts exist objectively in the same sense as stones or trees exist, independently of whether or not we observe them". "The ontology of materialism rested upon the illusion that the kind of existence, the direct 'actuality' of the world around us, can be extrapolated into the atomic range". "An objective description for events in space and time is possible only when we have to deal with objects or processes on a comparatively large scale, •••".

I think most physicists, even though they may profess faithful belief in the Copenhagen interpretation, still share with me a disreputable, materialistic prejudice that stones and trees cannot be either more - or less - real than the atoms of which they are composed. And, if it is meaningless to ask what an individual moment is doing, can it be any more meaningful to ask what their sum is doing?

It seems to me that the proper business of theoretical physics is to recognize these contradictions for what they are, and to try to resolve them. Instead, the Copenhagen school of thought tries to hide them from view, by proclaiming a new philosophy of human knowledge, according to which it is naive even to raise questions about "objective reality", or, for that matter, about anything that the Copenhagen theory cannot answer. Bohm and Bub[31], recognizing this, have rightly emphasized the dangers for the progress of physics in a theory which effectively contains within itself a proclamation of its own infallibility, by the device of declaring to be meaningless any question that the theory is unable to answer. For, if everyone accepted this, then even if the theory were grossly in error, the way to a better theory would be blocked; we would be prohibited from ever raising any question which might permit us to discover the errors.

For these reasons, I think that it is not only desirable, but very likely a prerequisite for any further progress in theoretical physics, that physicists insist on raising, and seeking constructive answers to, physical questions that the Copenhagen interpretation rejects as naive and meaningless, in particular, questions about the detailed mechanism by which an atom interacts with the electromagnetic field. Exactly what is happening within the atom when it is in the process of emitting or absorbing light? How do not only its dipole moment, but the entire underlying charge and current distributions, vary during the interaction? A theory which cannot answer such questions will, I think, be found inadequate to deal with the experimental facts of quantum optics before many more years have passed.

Unlike Bohm and other recent dissenters from the Copenhagen theory, however, I do not think that the way out requires anything so radical as the introduction of new "hidden variables". At least, before going that far out, let's try a more conservative treatment, and retain as much as possible of the present mathematical formalism, which, after all criticism, still contains a very large amount of truth. I have adduced reasons, highly convincing at least to me, indicating that the quantity presently called "expectation of moment" for a single atom, should not be interpreted in that manner. Nevertheless, it clearly has some kind of close connection with the physical notion of dipole moment. So let us give it a new physical interpretation which retains that connection, and see whether we get a more sensible theory which can be interpreted without mysticism. There are many possibilities to be explored here, and of course there is no guarantee that the first one we try out will prove to be the correct one. In other words, we are now at the stage of formulating tentative hypotheses about re-interpretation, not because we believe the new hypothesis is necessarily correct, but rather to find out what its consequences would be. If it proves to be unsatisfactory, then we can try out a different one. I feel strongly that, with enough persistence, this process should lead to the solution of our problems.

That is the philosophical basis for Neoclassical theory, (NCT). Mathematically, the step from SCB to NCT is so trivial as to be hardly noticeable; it consists of nothing more than taking the SCB equations already developed for the total moment of many atoms, and applying them instead to each individual atom. But conceptually, as we have just seen, this amounts to a revolutionary change in viewpoint. The proponents of the Copenhagen interpretation have ignored the dire warnings of Planck, Einstein, Schrödinger, de Broglie, von Laue, about the path they were taking; and so now we are going to ignore the dire warnings of the Copenhagen school, and proceed to do exactly what they have told us cannot be done.

Since, according to this prescription, the "expectation of moment" of an individual atom is now to be used as the source of a classical EM field, the expectation has been re-interpreted as the *actual* value of dipole moment. This amounts to a radical change in the interpretation of the Schrödinger wave function, but, following our conservative plan of action, we will not at this time attempt to postulate exactly what the new interpretation is. Instead we will be guided by the requirement of consistency with the re-interpretation of dipole moment just made, and make other changes in interpretation only to the extent demanded by consistency.

For setting forth the basic properties of NCT, we can still restrict ourselves to the simple model of a non-relativistic, spinless hydrogen atom; it turns out that the needed generalizations all go through effortlessly, in the most obvious way. We will retain all of the Schrödinger mathematical formalism associated with Eq.(1). There is the Hamiltonian

$$H = H_o + V(t) \tag{18}$$

with

$$H_o = \frac{p^2}{2m} - \frac{e^2}{r} \tag{19}$$

$$V(t) = -\frac{e}{mc} \underline{A}\cdot\underline{p} + \frac{e^2}{2mc^2} A^2 \quad , \tag{20}$$

the usual unperturbed stationary states $u_n(x)$:

$$H_o\, u_n(x) = E_n\, u_n = \hbar\omega_n\, u_n(x) \tag{21}$$

and the usual wave function expansion

$$\psi(x,t) = \sum_n a_n(t)\, u_n(x) \quad . \tag{22}$$

Now the net result of the long polemic just concluded is that the quantity

$$\underline{M}(t) = e\int \psi^*(x,t)\underline{r}\, \psi(x,t)\, d^3x = \sum_{n,k} \underline{\mu}_{nk}\, a_n{}^*(t)\, a_k(t) \tag{23}$$

is now taken to represent the dipole moment of the atom, where $\underline{\mu}_{nk}$ are the usual dipole moment matrix elements. From the definition of dipole moment in terms of charge density, $\underline{M} = \int \underline{r}\, \rho(x)\, d^3x$, we conclude that the charge density is given by

$$\rho(x) = e|\psi(x)|^2 \quad , \tag{24}$$

which is exactly Schrödinger's original interpretation of his wave function. According to the well-known conservation theorem which follows from the Schrodinger equation $i\hbar\dot{\psi} = H\psi$, the quantity

$$\underline{J} = \frac{e}{mc} \mathrm{Re}[\psi^* (\underline{p} - \frac{e}{c} \underline{A})\psi] \quad , \tag{25}$$

usually interpreted as (e/c) × (probability current), obeys the equation

$$\nabla \cdot \underline{J} + \frac{1}{c} \dot{\rho} = 0 \quad , \tag{26}$$

and so $\underline{J}$ may be interpreted as electric current density (in emu cm^{-2}), with the proviso that, as far as charge conservation is concerned, any other choice with the same divergence, e.g. $\underline{J}' = \underline{J} + \nabla \times \underline{Q}$, where $\underline{Q}(x)$ is any vector field, will do as well. This is one of the points of "flexibility" of NCT, that I alluded to in the Introduction; different choices of $\underline{Q}$ will alter the radiation from the atom, and at this stage we have only formal simplicity and experimental evidence to help us decide which choice is best. For the time being, we stick to the conventional choice (25).

With charge and current densities identified, we can introduce the radiation field. We use a general modal expansion: define a "cavity" by some volume V bounded by a closed surface S, and let k_λ^2, $\underline{E}_\lambda(x)$ be the eigenvalues and eigenfunctions of the boundary-value problem

$$\nabla \times \nabla \times \underline{E}_\lambda - k_\lambda^2 \underline{E}_\lambda = 0 \quad \text{in } V \quad ,$$

$$\underline{n} \times \underline{E}_\lambda = 0 \quad \text{on } S \quad , \tag{27}$$

where $\underline{n}$ is a unit vector normal to S. The resonant frequencies are $\Omega_\lambda = ck_\lambda$. The vector eigenfunctions $\underline{E}_\lambda(x)$ for which $k_\lambda \neq 0$ form, if V is simply connected, a complete set for expansion of the transverse field; if we use the Coulomb gauge, they will suffice for expansion of the vector potential in the form

$$\underline{A}(x,t) = \sqrt{4\pi}\, c \sum_\lambda Q_\lambda(t)\, \underline{E}_\lambda(x) \quad . \tag{28}$$

The magnetic field is given by

$$\underline{H}(x,t) = \sqrt{4\pi}\, c \sum_\lambda Q_\lambda(t)\, \nabla \times \underline{E}_\lambda(x) \quad , \tag{29}$$

and for the electric field we use expansion coefficients P_λ:

$$\underline{E}(x,t) = -\sqrt{4\pi} \sum_\lambda P_\lambda(t)\, \underline{E}_\lambda(x) \quad . \tag{30}$$

The Maxwell equations

$$\nabla \times \underline{E} + \frac{1}{c}\dot{\underline{H}} = 0 \tag{31a}$$

$$\nabla \times \underline{H} - \frac{1}{c}\dot{\underline{E}} = 4\pi\,\underline{J} \tag{31b}$$

then reduce, using (29), (30), to

$$\dot{Q}_\lambda = P_\lambda \tag{32a}$$

$$\dot{P}_\lambda = -\Omega_\lambda{}^2\, Q_\lambda + \sqrt{4\pi}\, c \int_V \underline{E}_\lambda(x) \cdot \underline{J}(x,t)\, d^3x \tag{32b}$$

respectively. The longitudinal current, being orthogonal to all the $\underline{E}_\lambda(x)$, does not contribute to the integral in (32b).

This formalism has been set up so that the total field energy is

$$H_f = \int_V \frac{E^2 + H^2}{8\pi}\, dV = \sum_\lambda \frac{1}{2}(P_\lambda^2 + \Omega_\lambda^2\, Q_\lambda^2) \quad , \tag{33}$$

and the Hamiltonian equations of motion based on (33) are evidently identical with the free-space Maxwell equations. To write the driven Maxwell equations (31) in Hamiltonian form, we substitute (25) and then (28) into the last term of (32b) and carry out the space integration. The driving force term of (32) then assumes the form

$$\sqrt{4\pi}\, c \int \underline{E}_\lambda \cdot \underline{J}\, d^3x = \sqrt{4\pi}\, \frac{e}{m} \langle \underline{E}_\lambda \cdot \underline{p} \rangle - \frac{4\pi e^2}{m} \sum_\mu \langle \underline{E}_\lambda \cdot \underline{E}_\mu \rangle Q_\mu \quad , \tag{34}$$

in which we have followed the customary notation

$$\langle F \rangle \equiv \int \psi^* F \psi\, d^3x \quad , \tag{35}$$

even though these quantities no longer have the physical meaning of expectation values, but are now to be taken simply as mathematical quantities defined by (35). The "diamagnetic" term of (34), containing Q_μ, has not yet been used in any neoclassical calculation, but we carry it along in order to demonstrate the full consistency of the formalism being developed. From (34), it is evident that if we define an interaction Hamiltonian

$$H_{int} = -\sqrt{4\pi}\, \frac{e}{m} \sum_\lambda \langle \underline{E}_\lambda \cdot \underline{p} \rangle Q_\lambda + \frac{4\pi e^2}{m} \sum_{\lambda\mu} \frac{1}{2} \langle \underline{E}_\lambda \cdot \underline{E}_\mu \rangle Q_\lambda Q_\mu \quad , \tag{36}$$

then the Maxwell equations (31), (32) are identical with the Hamiltonian equations of motion

$$\dot{Q}_\lambda = \frac{\partial H'}{\partial P_\lambda} \tag{37a}$$

$$\dot{P}_\lambda = -\frac{\partial H'}{\partial Q_\lambda} \tag{37b}$$

with $H' = H_f + H_{int}$.

The interaction Hamiltonian (36) was constructed by the sole criterion that its negative derivative should equal (34). It is not obvious, then, whether it bears any relation to the interaction Hamiltonian denoted by V(t) in (20), and chosen by the criterion that it yield the conventional Schrödinger equation of motion for the atom. However, making use of (28), we see that (36) is equal to

$$H_{int} = -\frac{e}{mc}\langle \underline{A}\cdot\underline{p}\rangle + \frac{e^2}{2mc^2}\langle \underline{A}^2\rangle = \langle V(t)\rangle \quad . \tag{38}$$

To make this correspondence, the factor of ½ in the diamagnetic term of (36) was essential.

But (38) now enables us to carry our physical interpretation a step further. For H_{int} , from the relations (31)-(37), clearly has the physical meaning of the interaction energy between atom and field. Therefore, the quantity $\langle V(t)\rangle$, which conventional theory interprets as expectation of interaction energy, must now be re-interpreted as actual interaction energy. From (18), it then appears that the quantity $\langle H_o\rangle$, usually called the expectation of the atom's unperturbed energy, must now be interpreted instead as its actual unperturbed energy. In this manner, the requirement of consistency with our original re-interpretation of dipole moment, leads us to a fairly complete physical interpretation of the whole formalism.

It remains to put the equations of motion for the atom into a form suitable for our purposes. From (22), the Schrödinger equation (1) takes the usual form

$$i\hbar\dot{a}_n = \hbar\omega_n a_n + \sum_k V_{nk}(t)\, a_k \tag{39}$$

which can be written in a "quasi-Hamiltonian" form:

$$i\hbar\, \dot{a}_n = \frac{\partial H''}{\partial a_n^*} \tag{40a}$$

$$i\hbar\, \dot{a}_n^* = -\frac{\partial H''}{\partial a_n} \tag{40b}$$

with

$$H'' = \sum_n \hbar\omega_n\, a_n^*\, a_n + \sum_{nk} V_{nk}\, a_n^*\, a_k$$

$$= \langle H_o \rangle + \langle V(t) \rangle \quad . \tag{41}$$

But we can make Eqs.(40) look much more "classical" by defining real amplitudes $p_n(t)$, $q_n(t)$ as follows:

$$a_n(t) = \frac{p_n - i\omega_n\, q_n}{\sqrt{2\hbar\omega_n}} \tag{42}$$

Rewriting H'' in terms of p_n, q_n, we find that the Schrödinger equation for the atom becomes

$$\dot{q}_n = \frac{\partial H''}{\partial p_n} \tag{43a}$$

$$\dot{p}_n = -\frac{\partial H''}{\partial q_n} \quad , \tag{43b}$$

where now, if we write the Hermitian matrix V_{nk} as

$$V_{nk}(t) = \hbar\sqrt{\omega_n\omega_k}\, (u_{nk} + i\, w_{nk}) \quad , \tag{44}$$

with u real and symmetric, w real and antisymmetric, H'' reduces to

$$H'' = \sum_n \frac{1}{2}(p_n^2 + \omega_n^2\, q_n^2) + \frac{1}{2}\sum_{nk} u_{nk}(p_n p_k + \omega_n\omega_k q_n q_k)$$

$$+ \sum_{nk} w_{nk}\, p_n\, \omega_k\, q_k \quad , \tag{45}$$

and all imaginary quantities have disappeared.

It is only here that the full significance of the term "neoclassical" emerges. For we have created a complete classical

Hamiltonian system which yields equations of motion for both the atom and the field; at this point we can drop the primes in (37) and (43) and define a total Hamiltonian

$$H(p_n,q_n;P_\lambda;Q_\lambda) = H_{at}(p_n,q_n) + H_{int}(p_n,q_n;Q_\lambda) + H_f(P_\lambda,Q_\lambda) \tag{46}$$

where

$$H_{at} = \frac{1}{2}\sum_n (p_n^2 + \omega_n^2\, q_n^2) = \langle H_o \rangle \tag{47}$$

$$H_f = \frac{1}{2}\sum_\lambda (P_\lambda^2 + \Omega_\lambda^2\, Q_\lambda^2) \tag{48}$$

$$H_{int} = (\text{a quadratic form in } p_n,\ q_n, \text{ with coefficients linear and quadratic in } Q_\lambda) = \langle V(t) \rangle\ . \tag{49}$$

The resulting equations of motion (43) are identical with the conventional Schrödinger equation (1) describing how the radiation field affects the state of the atom; they are only transcribed into an unconventional notation. Likewise, the equations of motion (37) are identical with the Maxwell equations for a field driven by the transverse part of the current (25). The fact that they all turn out to be consistent (i.e., the interaction Hamiltonian which gives the correct Schrödinger equation for the atom, also gives the correct Maxwell equations for the field) is perhaps the first indication that there may be some merit in this procedure.

The dynamical variables p_n, q_n, P_λ, Q_λ are, of course, not operators but ordinary numbers as in any classical theory. The "physical quantities" are the atomic wave function and EM field vectors, ψ, E, H. Although we still use the operator $p = -i\hbar\nabla$ in the theory, as in (25), because it is a convenient and familiar notation, it no longer "represents" any particular physical quantity. In neoclassical theory, physical quantities are not represented by operators at all, any more than in classical acoustics. We are, of course, free to use operators whenever this is convenient for mathematical purposes, but whatever commutation relations they may have are simply mathematical relations that carry no physical implications of the "uncertainty principle" type.

A large class of objections to neoclassical theory that have appeared recently[24,32,33] (i.e., that its equations of motion are inconsistent, that it violates energy conservation, that it violates the uncertainty principle, etc.)arises solely from failure to comprehend the points made in the last paragraph. I hope it is now clear from the above derivations, which go into a little more detail than in our previous publications, that (1) the equations of motion for atom and field are completely consistent with each other, down to the diamagnetic term; (2) the total energy H is conserved rigorously, in consequence of the Hamiltonian form of the dynamics; (3) there is no uncertainty principle to violate. In this connection, note that the uncertainty principle that is contained in the Copenhagen theory is not an experimental fact (who has measured the dispersions of two non-commuting quantities in the same state?); it is only a limitation on that theory. Since this is easily the most obscure point in all of physics - surpassing even the second law of thermodynamics in the utter confusion with which it is presented to students - we will return to it in more detail elsewhere.

Now, referring to the program formulated in the Introduction, we have arrived at the point where a fairly definite alternative theory to QED has been constructed (although it would require much generalization before one could think of it as a complete theory of electrodynamics), and the next step is to confront it with experimental facts, to determine: (1) Just what "elements of truth" does it contain? Does it contain any that are not in QED? (2) What "elements of nonsense" are in it? Can the theory be modified to remove them?

As we see from (47), the atom in neoclassical theory is, no less than the EM field, dynamically equivalent to a set of harmonic oscillators, and so at this point you might well say, "Aha - defeat now stares you in the face; for you have dissolved everything away into nothing but classical harmonic oscillators, which will lead inevitably to the Rayleigh-Jeans law, instead of the Planck law, for black-body radiation." Not so! For this model has some tricks in it. There is another uniform integral of the motion in addition to the total energy, of a type that was never dreamt of in classical statistical mechanics, and which completely changes the laws of energy exchange between atom and field. The secret lies in the fact that the interaction Hamiltonian, in its dependence on p_n and q_n, as exhibited in (45), is quadratic rather than linear. In other words, the atom is coupled to the field not directly, but parametrically. Application of a field does not produce any force tending to displace an oscillator coordinate q_n, as we would get from a term in the Hamiltonian linear in q_n; instead, the applied field varies the "masses","spring constants", and "mutual coupling coefficients" of the atomic oscillators (47).

It is clear from (47) that the quantity

$$W_n \equiv \frac{1}{2}(p_n^2 + \omega_n^2 \, q_n^2) \tag{50}$$

is to be interpreted as the energy stored in the n'th vibration mode (formerly n'th stationary state) of the atom. The total energy of the atom, $\sum W_n$, is not a constant of the motion because of the field interaction. But one easily verifies, from (43), (45), that, thanks to the symmetry and anti-symmetry respectively of u_{nk} and w_{nk},

$$\frac{d}{dt}\sum_n \frac{W_n}{\omega_n} = 0 \quad . \tag{51}$$

In other words, we have a law of conservation of *action*. Tracing back, we find that this is the same mathematical relation that the Copenhagen theory interprets as conservation of probability: $\sum|a_n|^2$ = const. (a striking illustration of how much the "natural" physical interpretation of a formalism depends on the particular mathematical form in which it is presented), and that setting the const. equal to unity is equivalent to setting

$$\sum_n \frac{W_n}{\omega_n} = \hbar \quad . \tag{52}$$

Before discussing the physical consequences of this conservation law, let us first put the equations of motion in their most compact, easily surveyed form. Dropping the diamagnetic term, the Schrödinger equation (43) and the Maxwell equations (37) can be written respectively as

$$i\hbar\, \dot{a}_n = \hbar\omega_n\, a_n + \sum_{\lambda k} V_{nk}^{\lambda}\, Q_\lambda\, a_k \tag{53}$$

$$\ddot{Q}_\lambda + \Omega_\lambda^2\, Q_\lambda = -\sum_{nk} V_{nk}^{\lambda}\, a_n^*\, a_k \quad , \tag{54}$$

where

$$V_{nk}^{\lambda} \equiv -\sqrt{4\pi}\,\frac{e}{m}\,(\underline{E}_\lambda \cdot \underline{p})_{nk} \quad . \tag{55}$$

We are here returning to the complex amplitudes $a_n(t)$ because of the familiarity and compactness of the resulting equations. However, we emphasize once again - because it is the most persistently misunderstood point in this theory - that our interpretation of $a_n(t)$ is entirely different from the conventional one. We are regarding the variables p_n, q_n as the fundamental conjugate variables of a set of harmonic oscillators comprising the atom (whose "ultimate physical nature" is a question for the future), and $a_n(t)$ as a complex variable defined by (42), representing amplitude and phase angle in the phase space of the n'th harmonic oscillator. It is so defined that its square magnitude

$$|a_n|^2 = \frac{p_n{}^2 + \omega_n{}^2 q_n{}^2}{2\hbar\omega_n} \tag{56}$$

has now the physical meaning, not of probability, but of energy stored in the n'th mode, in units of $\hbar\omega_n$, or, what is the same thing, as action stored in that mode, in units of $\hbar$.

The action conservation law (52) - an immediate consequence of parametric coupling - has some very obvious, and very important, implications for the laws of energy exchange between field and matter. If the atom is in its n'th oscillation mode (i.e., only p_n, q_n differ from zero), then from (52) its energy is necessarily $\hbar\omega_n$, and the right-hand side of (54) vanishes (in the n'th state, the atom has no permanent dipole moment, and so $V_{nn}^{\lambda} = 0$). Therefore the atom does not excite any field oscillators.

By a suitable external perturbation A(t) it is possible to start with the atom in the n'th mode and to end with it in the m'th. The energy difference $\Delta E = \hbar(\omega_m - \omega_n)$ must then be supplied or absorbed by the radiation field. A possible way of doing this, as we know from conventional solutions of the Schrödinger equation, is to impose a weak field of frequency $(\omega_m - \omega_n)$ for a suitable length of time; this is the phenomenon of stimulated emission, or of absorption.

Suppose now that $\omega_m > \omega_n$. If both the m'th and n'th vibration modes of the atom are excited simultaneously and $V_{mn}^{\lambda} \neq 0$, the right-hand side of (54) oscillates at the difference frequency $\omega_{mn} = \omega_m - \omega_n$ (but, because of the particular way in which $p_m, q_m; p_n, q_n$ are combined in (54), not at the sum frequency). Any field oscillator for which $\Omega_\lambda \approx \omega_{mn}$ is then strongly coupled to the atom, and may be excited to a considerable amplitude. This in turn reacts back on the atom via Eq. (53), causing its state of excitation to change. In passing from mode m to mode n, the atom delivers a total amount of energy $\hbar\omega_{mn}$ to those field oscillators whose frequencies are near to ω_{mn}. Similarly, if the atom is initially excited in the n'th vibration mode, and a field oscillator of frequency $\Omega_\lambda \approx \omega_{mn}$ is originally excited with a greater energy than $\hbar\omega_{mn}$, it can deliver the energy $\hbar\omega_{mn}$ to the atom, leaving it in the m'th vibration mode.

Although the possible energies of field and atom can vary continuously in this theory, if one considers only processes in which the atom changes from one pure oscillation mode to another, the energy of field oscillators will be seen to appear and disappear in units of $\hbar\Omega$. In most experiments, these would be almost the only amounts of energy one could observe to be exchanged, because when a fraction $\alpha\hbar\Omega$ of the energy has been absorbed by an atom, it is left in a state with a large oscillating dipole moment, and continues to interact strongly with the field. Only when it has absorbed the full energy $\hbar\omega_{mn}$, or given the energy $\alpha\hbar\Omega$ back to the field, will it reach a "stationary state", where its dipole moment vanishes and the energy exchange ceases.

In the properties just noted, we see virtually all the "quantum effects" in radiation phenomena on which the early development of quantum theory was based: the Ritz combination principle, the existence of stationary (i.e., nonradiating) states of fixed energy, the interchange of energy in units of $\hbar\Omega$, absorption and induced emission. As we have shown[6,7] before, taking into account the atom's radiation reaction field (which means nothing more than finding the complete solution of (54) for all field modes and putting the result into (53)), leads to prediction of spontaneous emission with the correct Einstein A-coefficients, but a different shape (hyperbolic secant envelope) for the spontaneous emission pulse. Likewise, the beautiful experiments of R.W. Wood[34] on resonance radiation and selective excitation of atoms, are accounted for immediately.

Just before starting the mathematical development of NCT, I mentioned that we would proceed to do exactly what the Copenhagen school has told us cannot be done. That promise has now been fulfilled, for we have all been taught that the aforementioned phenomena cannot be accounted for by classical concepts at all. For example, Bohr[35] has asserted that, "Hence, in the case of atoms, we come upon a particularly glaring failure of the causal mode of description when accounting for the occurrence of radiation processes". Similar assertions are repeated endlessly throughout our textbooks.

We now see just how easy it is to do these "impossible" things; the entire secret lies in the words *parametric coupling*. The Copenhagen interpretation takes the relation $E = \hbar\omega$ as a basic postulate, and never makes any attempt to explain how or why two physical quantities so utterly different as energy and frequency should be so connected. NCT, via (52), explains this as a simple consequence of the dynamics whenever we have parametric coupling to the field, and leaves open the interesting possibility that other systems, with different kinds of field interactions, may not be subject to any such limitation.

This easy initial success of NCT has seemed to me a very powerful argument in favor of the approach used. In the action conservation law (52) we have an "element of truth" which, if not actually missing from QED, is at least present in a more physically appealing form in NCT.

The NCT formalism was set up above in some generality, in a way that emphasizes its consistency and the classical interpretability of (52) and its consequences just noted. For many applications, we can pass to the two-level approximation in which only the amplitudes $a_1(t)$, $a_2(t)$ appear, and by the dipole approximation $(\underline{E}_\lambda \cdot \underline{p})_{nk} \simeq \underline{E}_\lambda \cdot (\underline{p})_{nk}$ and a gauge transformation, the interaction term in the Schrödinger equation (but not in the Maxwell equations - this is the reason why one needs the vector potential in order to write the whole system of equations of motion in Hamiltonian form) takes the form of an electric dipole interaction. Details have been given before [3,6,7], and we recall only the result. If $\omega = \omega_2 - \omega_1 > 0$, the Schrödinger equation (39) or (53) can be written in terms of the dipole moment $M(t)$ defined by (23):

$$\ddot{M} + \omega^2 M = -(2\mu/\hbar)^2 W(t) E(t) \tag{57a}$$

$$\dot{W} = E\dot{M} \tag{57b}$$

where $E(t)$ is the electric field at the atom, and

$$W(t) \equiv \frac{\hbar\omega}{2} \left(|a_2|^2 - |a_1|^2\right) = W_1 + W_2 - \frac{1}{2}\hbar (\omega_1 + \omega_2) \tag{58}$$

is the energy of the atom, referred to a zero lying midway between the levels, W_n being the mode energies(50). μ is the dipole moment matrix element, denoted μ_{12} in (23). Noting the first integral of (57):

$$\dot{M}^2 + \omega^2 M^2 + (2\mu/\hbar)^2 W^2 = \omega^2 \mu^2 \tag{59}$$

which is just the action conservation law (52) in disguised form, and taking the electric field as the sum of external and radiation reaction parts:

$$E(t) = E_{ext}(t) + E_{RR}(t) \quad , \tag{60}$$

with

$$E_{ext}(t) = E_o(t) \cos[\Omega t + \theta_o(t)] \quad , \tag{61}$$

where $E_o(t)$, $\theta_o(t)$ are slowly varying, and

$$E_{RR}(t) = \frac{2}{3c^3}\dddot{M}(t) - \frac{4K}{3\pi c^3}\ddot{M}(t) \quad , \tag{62}$$

(here K is a cutoff described before[7], which is of the order of magnitude $K \sim c/a_0$ with a_0 the Bohr radius; it can be calculated[6] from the detailed current distribution within the atom), we pass to the Bloch sphere representation by introducing dimensionless variables x(t), y(t), z(t):

$$\dot{M} + i\omega M = i\omega\mu(x + iy)\exp[i\Omega t + i\theta_0(t)] \tag{63}$$

$$W = \frac{1}{2}\hbar\omega\, z(t) \quad . \tag{64}$$

Thus, $\mu x(t)$, $\mu y(t)$ are respectively the components of M(t) in phase and 90° ahead of the applied field (61), while z(t) is the atom's energy, in units of $\frac{1}{2}\hbar\omega$. The first integral (59) now reduces to the equation of the unit sphere, $x^2 + y^2 + z^2 = 1$. In this representation, the Schrödinger equation describing slow changes in the energy, and magnitude and phase of the dipole moment, of the atom, has the form[7] of the "secular equations"

$$\dot{x} = \beta zx + (\alpha - \gamma z)y \tag{65a}$$

$$\dot{y} = \beta zy - (\alpha - \gamma z)x + \lambda z \tag{65b}$$

$$\dot{z} = \beta(z^2 - 1) - \lambda y \tag{65c}$$

Here

$$\beta \equiv \frac{2\mu^2\omega^3}{3\hbar c^3} = \frac{1}{2}A \ ; \qquad \gamma \equiv \frac{4K\mu^2\omega^2}{3\pi\hbar c^3} \sim 100\beta \tag{66}$$

are two constants defined by the field interaction (β is half the Einstein A-coefficient for the transition, and γ is the "dynamic Lamb shift" discussed later), while

$$\lambda(t) \equiv (\mu/\hbar)\, E_0(t) \tag{67}$$

$$\alpha(t) \equiv \Omega - \omega + \dot{\theta}_0(t) \quad , \tag{68}$$

are slowly varying measures of the amplitude and momentary frequency of the applied field. Many solutions of these equations have been given[3-7], and Equations (65), although differing in detail in the terms containing γ and β, are essentially equivalent to those of McCall and Hahn[25] in the application to intense, short pulses (i.e., $\lambda >> \beta$, $\gamma\tau << 1$). Equations (65) make a large number of detailed predictions capable of being checked by experiment, and we turn now to one case where new experimental evidence is beginning to appear.

3. Spontaneous Emission

One of the most striking differences between the predictions of QED and NCT concerns the shape of the spontaneous emission pulse from an atom. As discussed in some detail previously[6-9], NCT (via Eqs.(65) with $\lambda=0$), predicts a chirped hyperbolic secant pulse, while QED predicts the usual exponentially damped one, with the tail of the hyperbolic secant pulse $\exp(-\beta t)$ having the same decay constant as the QED pulse. This has not only stimulated some correspondence[8,9] and some experimental efforts[36], but also some more careful thought about quantum theory. This point was first raised at the Coherence Conference here six years ago, but at that time, almost every physicist whom one asked about the shape of a spontaneous emission pulse, would reply that the question was meaningless; the only observable fact is simply whether a photon has or has not been emitted, and any more detailed questions than this are forbidden because they seek to probe below the limits set by the uncertainty principle - or perhaps the Principle of Complementarity - or at least, some prohibition emanating from Copenhagen. So, we are back to that stuff again! There is just no way to avoid it. I am convinced that all fundamental questions in physics today reduce eventually to some question about the Copenhagen interpretation and the need for something better.

Nevertheless, today several laboratories are actively performing experiments to answer such questions, and in theoretical work we see graphs like the one given by Nash and Gordon[33], quite unblushingly comparing the QED and NCT emission rates of an atom as a function of its energy, varying continuously between excited and ground states. No thoughts about "instantaneous quantum jumps", no admonitions from Copenhagen about meaningless questions, impeded that work!

More seriously, let us note what experiment can tell us about the shape of a spontaneous emission pulse, because there is still some confusion about it. Suppose that this is given by the basic function

$$f(t) = \mathrm{Re}[a(t)\, e^{i\omega t}] \quad , \tag{69}$$

where a(t) is a slowly varying complex envelope function. We suppose that the complication of Doppler broadening has been eliminated by placing the emitting atoms in a solid, or by observing the radiation normal to a collimated atomic beam. If the atoms emit independently (i.e., no incipient superradiance or lasing), the total electric field will be a superposition of such pulses occurring at random times t_i: $E(t) = \sum_i f(t-t_i)$. A Michelson interferometer and photocell can then measure the intensity

$$I = \overline{[E(t) + E(t-\tau)]^2} \quad , \tag{70}$$

the bar denoting a time average over a few optical cycles. Now, as we easily verify from the above relations, by observing the maximum and minimum values of I as τ varies over an optical cycle, we can determine some, but not all, details of the function a(t). More specifically, this determines the convolution

$$b(\tau) = \int_{-\infty}^{\infty} a(t)\, a^*(t-\tau)dt \tag{71}$$

as follows. If the interferometer is set at a relative retardation τ, then the Michelson fringe visibility is

$$V(\tau) = \frac{I_{max} - I_{min}}{I_{max} + I_{min}} = \frac{|b(\tau)|}{b(0)} \quad , \tag{72}$$

and a measurement of absolute fringe position (feasible today with a laser-calibrated interferometer) can determine the phase of $b(\tau)$.

From these relations we find the following list of pulse shapes and the corresponding fringe visibility curves:

Damped exponential [$\theta(t)$ = unit step function]:

$$a(t) = A\theta(t)\, e^{-\beta t} \qquad V(\tau) = e^{-\beta|\tau|} \tag{73}$$

Gaussian envelope:

$$a(t) = A\, e^{-qt^2} \qquad V(\tau) = \exp(-\tfrac{1}{2}q\tau^2) \tag{74}$$

Hyperbolic Secant:

$$a(t) = A\, \mathrm{sech}\, \beta t \qquad V(\tau) = \frac{\beta\tau}{\sinh \beta\tau} \tag{75}$$

Chirped Hyperbolic Secant:

$$a(t) = A[\mathrm{sech}\beta t]^{1+i\gamma/\beta} \qquad V(\tau) = \frac{\beta|\sin\gamma\tau|}{\gamma\,\sinh\beta|\tau|} \tag{76}$$

The very different shapes of these visibility curves enable one to distinguish experimentally between different hypotheses concerning the envelope a(t). Measurements of this type are, of course, a long since accomplished fact, and it was only a naive, but astonishingly widespread, misconception of quantum theory that led many to think they were impossible. Were it otherwise, QED would have been disproved by Albert Michelson long before any of us were born.

We leave it as an open question whether other experimental techniques (for example, observing fluctuations and coherence of resonance radiation) might enable us to measure further details of the function a(t) beyond its convolution $b(\tau)$. In any event, it is clear that questions concerning the shape of the spontaneous emission wave train *are* experimentally meaningful, and that a full treatment of the effect of spontaneous emission on noise in laser amplifiers and on stability of laser oscillators will require knowledge of the correct envelope function a(t). It is just a measure of how much progress in understanding we have made that, in this Conference, it never occurred to anyone to raise the kind of Copenhagen NO-NO's that were constantly in the air six years ago.

Today, the measurement, recording, and plotting of fringe visibility curves can be automated to the point where the effect on pulse shape of any change in method of excitation could be determined in a few minutes; experiments of this type could provide a wealth of data checking many details of (65) against the corresponding QED predictions.

A start on experiments of this type has been made in the interesting work of Gibbs[36], reported at this Conference. As suggested before[9], differences between QED and NCT can be seen if one can pump atoms, by laser pulses of controlled amplitude and duration, with an accurate π-pulse, whereupon the entire (or nearly the entire) hyperbolic secant envelope of (75) or (76) can be seen. With the usual inefficient pumping mechanisms the atom is excited, according to NCT, only a small distance from the South pole of the Bloch sphere (i.e., it is left nearly in the ground state, with only a small admixture of the excited state). But then we see only the exponential tail of the hyperbolic secant emission pulse, whose shape (and therefore, spectral distribution and autocorrelation function) are the same as in QED.

Gibbs has attempted more efficient pumping, and instead of analyzing the fluorescence by fringe visibility, has observed the time dependence by direct photoelectric counting. Although he claims to have disproved NCT, we believe that, in the actual conditions of the experiment, the atoms were not pumped far enough from the South pole for differences between the two theories to be observed. According to Eqs.(65), neither the frequency nor the pulse amplitude and shape were correct for pumping to the North pole. If a pulse is sufficiently intense ($\lambda >> \beta$) so that the radiation damping terms in β can be neglected during the pulse, and if the atom starts from the ground state $z = -1$ at the beginning of the pulse, then from (65) we find an integral of the motion

$$(\alpha - \gamma z)^2 = (\alpha + \gamma)^2 + 2\lambda\gamma x \quad . \tag{77}$$

The trajectory during the pulse is therefore the intersection of the parabolic cylinder (77) with the spherical surface $x^2 + y^2 + z^2 = 1$. A family of these trajectories is shown in Ref. 7, Fig. 7, for the case $\alpha = 0$; i.e., the applied field frequency Ω equals the atom's natural frequency ω. Referring to Eqs.(61)-(68), we see that Gibbs' tuning the field to the absorption line (the resonance frequency when $z = -1$), amounts instead to taking $\Omega = \omega - \gamma$, lower by the "dynamic Lamb shift", or $\alpha = -\gamma$. But then, according to (77), the North pole cannot be reached. The maximum attainable value of z is found by setting $x^2 = 1 - z^2$ in (77) and solving the resulting cubic equation. We find

$$1 + z_{max} = \frac{2a}{\sqrt{3}} \sinh[\tfrac{1}{3} \sinh^{-1}(\tfrac{\sqrt{27}}{a})] \ , \tag{78}$$

with

$$a \equiv \frac{2\lambda}{\gamma} = \frac{E}{E_{crit}} \ , \tag{79}$$

where E_{crit} is the critical field at which the trajectory just reaches the equator, $z_{max} = 0$, on the Bloch sphere, before turning downward again. According to (78), in order to pump anywhere near the North pole, requires $a > 10$, which by our estimates is far greater than reached in the Gibbs experiment. Furthermore, (78) is only an approximation which neglects the spontaneous emission terms proportional to β in (65), and which would cause a further southward drift. An estimate based on Eq.(3.28) of Ref. 7 gives for the amount of this drift in latitude, during a pulse of length t,

$$\delta\theta \simeq \beta \int_0^t \sqrt{1-z^2}\, dt \quad , \tag{80}$$

which, for the long-tailed pulse realized in Gibbs' experiment, could amount to about one-quarter radian. Finally, any chirp in the pumping pulse will cause a further decrease in z_{max}.

A full analysis of this experiment will require considerable time and will be reported elsewhere. For the time being, we note that, if the dynamic Lamb shift (i.e., terms proportional to γ in (65), which give rise to the chirping of the pulse (76)) is a real effect, then pumping to the North pole could not have been achieved in the Gibbs experiment unless the laser were retuned to $\Omega=\omega$, the only case where the trajectory (77) can reach the point $z = +1$. In this case, we believe the experiment failed to reach the conditions where differences in the theories could be seen. On the other hand, if the dynamic Lamb shift does not exist, then the Gibbs experiment was, in all probability, a valid disproof of Eqs.(65). One would then investigate whether a different choice of the vector Q discussed following Eq.(26) might give a spontaneous emission law in agreement with the experiment, or whether some other aspect of NCT could be modified.

In any event, two conclusions are: (1) We should have more experiments of this type, with different laser tunings tried out, and with cleaner pulse shapes, in order to make a clear-cut decision. It is, of course, too much to expect that the first experiment tried in a new field will settle all questions; it serves rather to indicate what the real technical problems are in making a decisive experiment. (2) The issue of the reality of the dynamic Lamb shift chirp appears a very crucial one.

In this latter connection, the relation between the two theories is brought out in a beautiful way in the work of Ackerhalt, Eberly, and Knight[37] reported at this Conference. In pseudospin notation, the equations of motion of a two-level atom interacting with its self-made field can be written in a common form. σ_3 is the energy of the atom, in units of $\frac{1}{2}\hbar\omega$, and $\sigma_- = \frac{1}{2}(\sigma_1 - i\sigma_2)$ corresponds to the rotating moment component $\mu^{-1}(M + \dot{M}/i\omega)$. In QED the σ's are operators; in NCT they are numbers, the same numbers that would in QED be called the expectation values of those operators. In either case the equations of motion of a decaying atom take the form

$$\dot{\sigma}_- = [-i(\omega_0+\gamma\sigma_3) + \beta\sigma_3]\sigma_- \tag{81}$$

$$\dot{\sigma}_3 = -4\beta\,\sigma_+\,\sigma_- \tag{82}$$

in which β and γ are the same radiation constants defined in (66), γ is identical with the "Crisp shift" Δ_c of Ackerhalt, Eberly and Knight (AEK). In this form the chirping is evident, since the oscillation frequency is $(\omega+\gamma)$ when $\sigma_3 = 1$ (upper state), and $(\omega-\gamma)$ when $\sigma_3 = -1$ (ground state). Indeed, if we make the substitutions

$z = \sigma_3$, $x - iy = \sigma\_\exp(i\omega t)$, Eqs.(81) and (82) become identical with (65) with $\lambda=\alpha=0$, and have solutions

$$z(t) = -\tanh\beta(t-t_o) = \gamma^{-1}\delta\omega(t) \tag{83}$$

$$[x(t) - iy(t)] = [x(t_o) - iy(t_o)]\operatorname{sech}\beta(t-t_o)\exp[-i\theta(t)] \tag{84}$$

$$\theta(t) = \int_{t_o}^{t}\delta\omega(t)dt = (\gamma/\beta)\log\operatorname{sech}\beta(t-t_o) \quad , \tag{85}$$

where t_o is the time of maximum emission, when the trajectory crosses the equator on the Bloch sphere. The NCT equations thus predict the chirped hyperbolic secant pulse (76).

But now watch closely at how these effects are wiped out by the magic of QED. Starting from the same equations of motion (81), (82), we now have the operator identities

$$\sigma_3\sigma_- = -\sigma_- \tag{86}$$

$$2\sigma_+\sigma_- = 1 + \sigma_3 \quad , \tag{87}$$

as a result of which the coupled nonlinear equations (81), (82) collapse to uncoupled linear ones

$$\dot{\sigma}_- = -[i(\omega_o-\gamma)+\beta]\sigma_- \tag{88}$$

$$\dot{\sigma}_3 = -2\beta(1+\sigma_3) \quad , \tag{89}$$

with the solutions

$$\sigma_-(t) = \sigma_-(0)\, e^{-\beta t}\, e^{-i(\omega_o-\gamma)t} \tag{90}$$

$$[1+\sigma_3(t)] = [1+\sigma_3(0)]\, e^{-2\beta t} \quad , \tag{91}$$

and now we are back to ordinary garden-variety exponential decay with no chirp! This little example is worthy of deep contemplation.

The prediction of chirp in one theory and not in the other,

would appear to be a gross qualitative difference between them, which should be easily accessible to experimental check. One would think that there must surely be some experiment already done, which could settle the question whether this chirp does or does not exist. However, such experiments are surprisingly difficult to find. In the hope of inspiring someone else to invent an experiment, let us note that the easiest thing to see is probably the greater spectral width caused by chirp, if we have efficient, North-pole pumping so that the entire pulse (76) is seen. The exact spectrum of this chirped pulse was given in Ref. 7, Eq.(4.16) with some factors of π omitted. To correct this, and indicate how more general spectra might be calculated, we start from the basic Fourier transform

$$F(\omega) = \int_{-\infty}^{\infty} \operatorname{sech}\beta t \exp\{i[(\omega-\omega_0)t - \theta(t)]\}dt$$

$$= 2^{i(\gamma/\beta)} \frac{\Gamma[\frac{1}{2}+\frac{1}{2}i\beta^{-1}(\omega-\omega_0+\gamma)]\Gamma[\frac{1}{2}-\frac{1}{2}i\beta^{-1}(\omega-\omega_0-\gamma)]}{\beta\Gamma(1+i\beta^{-1}\gamma)} , \qquad (92)$$

for whose evaluation we are indebted to L.P. Benofy. Then, thanks to the identities

$$|\Gamma(1 + iy)|^2 = \pi y \operatorname{cosech}\pi y \quad ,$$

$$|\Gamma(\tfrac{1}{2} + iy)|^2 = \pi \operatorname{sech}\pi y \quad ,$$

we find for the spectral density, normalized to

$$\int_0^{\infty} I(\omega)d\omega = 1 \quad ,$$

$$I(\omega) = \frac{\beta}{4\pi}|F(\omega)|^2 = \frac{1}{2\gamma}\,\frac{\sinh(\pi\gamma/\beta)}{\cosh(\pi\gamma/\beta) + \cosh[\pi(\omega-\omega_0)/\beta]} \cdot \qquad (93)$$

This is essentially flat in the interval $(\omega_0-\gamma)<\omega<(\omega_0+\gamma)$, and zero outside that range, with rapid but smooth transition regions, in which $I(\omega)$ rises from 4% to 96% of its maximum value in an interval $\delta\omega=2\beta$. Thus, the full width of a spontaneous emission line should be about twice the Lamb shift of that line, which is typically of the order of 100 times the "natural line width" of conventional theory. But if we have inefficient excitation, we have an essentially flat-topped spectrum in the narrower width $(\omega_0-\gamma)<\omega<(\omega_0+\gamma z_{max})$, whose upper limit varies with the efficiency of excitation. Hopefully, someone will think of a simple experiment which could confirm or refute this prediction.

5. Field Fluctuations

Another area where it has been claimed that classical electromagnetic theory is inadequate, concerns random fluctuations of fields. These are of two kinds: the thermal fluctuations which we observe in the laboratory as Black Body radiation or Nyquist noise in electrical circuits, and the "zero-point" energy or "vacuum fluctuations" arising from field quantization, whose reality is one of the points at issue here. On the one hand, we have been told here that vacuum fluctuations are "very real things", and that they are the physical cause of spontaneous emission and the Bethe logarithm part of the Lamb shift. Such ideas lead to simple contradictions, which have never been adequately covered up in QED.

For example, if it is true that the Einstein A-coefficients are due physically to zero-point fluctuations of the field, then why is it that the derivation of the Planck law based on the A and B coefficients, leads[33] to a result that does *not* include the zero-point energy? It seems to me that we have here either a flagrant logical contradiction, an error in calculation, or an incorrect interpretation, quite likely all three. For the conventional "derivation" of the Planck law will not bear inspection; we calculate the A-coefficient as if an atom were emitting into field-free space instead of into thermal radiation, and the B-coefficients as if the spontaneous emission were turned off. In reality, these effects interfere with each other[38] in a way that is certainly not negligible. According to either QED or NCT the conventional A - and - B-coefficient argument is just too crude to deal with the problem.

If we were willing to use the same standards of logic as those who accept the conventional derivation, we could claim to have derived the Planck law from NCT; for we too can derive the conventional A and B coefficients. But a valid derivation must obtain the spectral distribution from the full dynamics without making either "Fermi Golden Rule" type approximations or inadmissible physical assumptions about independence of spontaneous and induced processes. At the present time, neither QED nor NCT has produced any respectable derivation of the Planck law. (Of course, its derivation from the canonical ensemble of quantum statistical mechanics is trivial, but the problem here is to produce the detailed physical mechanism by which that distribution is brought about, a problem beyond the scope of equilibrium statistical mechanics).

To those who believe that zero-point fluctuations are the physical cause of the main part of the Lamb shift, they must then be "very real things" at least up to the Compton cutoff frequency, $\hbar\omega = mc^2$, to get the right Bethe logarithm. Please calculate the *numerical value* of the resulting energy density in space, the turbulent power flow from the corresponding Poynting vector, etc., and then tell us whether you still believe the zero-point fluctuations are physically real (for the Poynting vector, we get 6×10^{20} megawatts cm^{-2}; the total power output of the sun is about 2×10^{20} megawatts; real radiation of that intensity would do a little more than just shift the 2S level by 4 microvolts).

Now let us examine a rather milder problem, the energy fluctuations in thermal radiation. According to conventional QED treatments[24], the mean square energy fluctuation of the cavity modes in bandwidth $d\Omega$ is

$$(\Delta E)^2 = (\hbar\Omega)^2[\langle n^2\rangle - \langle n\rangle^2]g(\Omega)\, d\Omega \quad , \tag{94}$$

where $g(\Omega) = \Omega^2 V/\pi^2 c^3$ is the mode density function, and n is the number of photons in a single mode (assumed to have the same probability distribution for all modes in the small frequency interval $d\Omega$). For any field mode in a state describable by the $P(\alpha)$ distribution, we readily find a generalization of Einstein's formula [see Ref. 24, Eq.(21)]:

$$\langle n^2\rangle - \langle n\rangle^2 = [\langle|\alpha|^4\rangle - \langle|\alpha|^2\rangle^2] + \langle n\rangle \quad , \tag{95a}$$

in which the term in square brackets represents the mean-square fluctuations to be expected if α were a classical field variable, while the additive term $\langle n\rangle$ arises solely from field quantization, and was interpreted by Einstein in terms of particles.

If $P(\alpha)$ is Gaussian: $P(\alpha) = (\pi\langle n\rangle)^{-1}\exp(-|\alpha|^2/\langle n\rangle)$, the "classical" contribution reduces simply to $\langle n\rangle^2$, and (95a) becomes

$$\langle n^2\rangle - \langle n\rangle^2 = \langle n\rangle^2 + \langle n\rangle \quad . \tag{95b}$$

In this case, (94) can be written in the suggestive form given by Einstein:

$$(\Delta E)^2 = \frac{1}{N} \langle E\rangle^2 + \hbar\Omega\langle E\rangle \quad , \tag{96}$$

where $\langle E\rangle = \hbar\Omega\langle n\rangle g(\Omega)d\Omega$ is the average energy in the range $d\Omega$, and $N = g(\Omega)d\Omega$ is the number of modes considered. While we have no direct experimental confirmation of this formula, there is at least one case where there are independent theoretical grounds for supposing it is correct. Note that, up to this point, $\langle n\rangle$ can have any frequency dependence. In the case of thermal equilibrium, it is given by the Planck law: $\langle n\rangle = [\exp(\hbar\Omega/kT)-1]^{-1}$, from which we find

$$\frac{\partial\langle n\rangle}{\partial T} = \frac{\hbar\Omega}{kT^2} (\langle n\rangle^2 + \langle n\rangle) \quad , \tag{97}$$

and (96) then reduces to a general theorem of statistical mechanics, relating the energy fluctuations of any thermodynamic system at constant volume to its heat capacity:

$$(\Delta E)^2 = kT^2 C_v \quad . \tag{98}$$

Because of the generality of (98) - it holds equally well in classical or quantum statistical mechanics, whenever we represent thermal equilibrium by the canonical ensemble - there is a strong presumption that (98) is a universally valid relation, quite independently of its above derivation from QED. And in turn, (98) is only a special case of a far more general relation giving fluctuations and covariances of any physical quantities, over any probability distribution derivable from the principle of maximum entropy[39]. So we will accept Einstein's relation (96) for thermal radiation.

How, then, is semiclassical theory to account for the "field quantization" term $\hbar\Omega\langle E\rangle$ of Einstein's relation? To answer this, note that our starting equation (94) presupposed that different mode amplitudes were statistically independent, so that one could simply add up the mean-square fluctuations of the different modes, without any cross-product terms. But a moment's thought about the physical mechanism by which thermal equilibrium is maintained, shows that this cannot be correct. Each elementary emission or absorption process, exchanging an amount of energy $\hbar\Omega$ if it goes to completion, does not do so with just a single mode; it must affect simultaneously the amplitudes of many modes lying in a frequency band $\delta\Omega\sim t^{-1}$, where t is the duration of the process. We have, therefore, a non-zero correlation between the energies stored in two modes, if their frequencies are sufficiently close so that both modes "see" the same elementary emission or absorption process.

Now we have seen already in (11) how much small correlations can affect fluctuations when we are adding up a large number of small terms. The slightest positive correlation in the moments of individual spins was enough to abrogate the usual "law of large numbers". We are now faced with exactly the same phenomenon; although correlations between any two modes are extremely small, and could surely be neglected if we were considering only a few modes, the point is that they are systematic, tending in the same direction for every emission or absorption process, and every pair of modes, and their number grows like N^2, while the number of terms taken into account in (94) is only N. To estimate N: if $V = 10\ cm^3$ and $d\Omega = 10^{-4}\Omega$, then at infrared frequencies where this treatment is relevant, we have $N \simeq 10^7$. Therefore, if intermode correlations $\langle nn' \rangle - \langle n \rangle\langle n' \rangle$ were as large as a millionth of the mode variances $\langle n^2 \rangle - \langle n \rangle^2$, their total contribution might be larger than (94). Obviously, then, before we can make any pretense of having an honest calculation, we must go back to (94) and restore the missing terms.

Let the k'th mode have resonant frequency ω_k. On either classical or quantum electromagnetic theory, we may write the energy stored in this mode as $E_k = \hbar\omega_k n_k$; in classical theory, the number n_k thus defined is a continuously variable positive real number, while in QED it is an operator with non-negative integer eigenvalues. With this notation, the beginnings of the derivation involve the same formal equations in either theory. The full expression for the energy fluctuation is then

$$(\Delta E)^2 = (\hbar\Omega)^2 \sum_{k,r} [\langle n_k n_r \rangle - \langle n_k \rangle\langle n_r \rangle] \quad , \tag{99}$$

in which we sum over all modes whose frequencies are in a narrow range $d\Omega$ about Ω. We choose $d\Omega$ small enough so that we may replace all ω_k by Ω, and large enough so that $N \gg 1$; as noted above, this is no real limitation. We also suppose, as before, that the variation of $\langle n_k \rangle$ over this small frequency interval is negligible, so that we may set all $\langle n_k \rangle = \langle n \rangle$. With these understandings, (94) is seen as an approximation to (99) which retains only the diagonal terms $r = k$.

Einstein's relation (96) may also be written in the form

$$(\Delta E)^2 = (\hbar\Omega)^2 \sum_{k} [\langle n_k \rangle^2 + \langle n_k \rangle] \quad . \tag{100}$$

Now the correlation $\langle n_k n_r \rangle - \langle n_k \rangle\langle n_r \rangle$ will be appreciable when $|\omega_k - \omega_r| \lesssim \delta\Omega \sim t^{-1}$, the spectral width of the emission or absorption process, and negligible otherwise. Therefore, if $d\Omega < \delta\Omega$, (99) and (100) cannot be equal independently of our choice of $d\Omega$; for one is proportional to $d\Omega$, the other to $(d\Omega)^2$. But if $d\Omega >> \delta\Omega$, the general condition for (99), (100) to be equal independently of our choice of $d\Omega$, is

$$\sum_r [\langle n_k n_r\rangle - \langle n_k\rangle\langle n_r\rangle] = \langle n_k\rangle^2 + \langle n_k\rangle \quad , \tag{101}$$

in which we may now sum over all modes r, since contributions outside the range $|\omega_k - \omega_r| \lesssim \delta\Omega$ are negligible.

With the result (101) it is now apparent why neither QED nor semiclassical theory has yet produced any adequate treatment of this problem. For, according to (95b), QED achieves equality in (101) by using only the diagonal term $r = k$, and ignoring the others. Evidently, then, a further calculation is needed, to show that

$$\sum_r{}' [\langle n_k n_r\rangle - \langle n_k\rangle\langle n_r\rangle] = 0 \quad , \tag{102}$$

the prime denoting that the term $r = k$ is deleted. If (102) does not hold, then a correct QED calculation will not lead to the Einstein relation after all. Evidently, a theory which gives a right answer from a demonstrably bad approximation, is thereby in just as much trouble as if it had given a wrong answer from a good calculation.

In classical EM theory, with a gaussian field distribution (or very nearly so, i.e. gaussian but for these small correlations, which are not necessarily described by a multivariate gaussian distribution), the marginal probability distribution of each n_k will still be, to very great accuracy, of the Boltzmann exponential form: $P(n)\,dn = \langle n\rangle^{-1} \exp(-n/\langle n\rangle)dn$. The diagonal term of (101) is then just $\langle n_k\rangle^2$, and so semiclassical theory will have accounted for the Einstein relation if it can be shown that, in contrast to (102),

$$\sum_r{}' [\langle n_k n_r\rangle - \langle n_k\rangle\langle n_r\rangle] = \langle n_k\rangle \quad . \tag{103}$$

Until calculations to check (102), (103) have been carried out, both theories leave us in just the same state of uncertainty, for the same reason. Merely to exhibit the unsolved problems in this symmetrical way, shows how unjustified it is, in our present state of knowledge, to claim that QED is right and classical EM theory wrong, in the matter of energy fluctuations. But let us try to understand the situation a little better, by estimating the magnitude of these correlations.

Consider first semiclassical theory. The following argument makes no pretense of being a rigorous derivation, because it does not go into details of the matter-field interaction; however, it gives us an order-of-magnitude estimate very easily. Suppose that an elementary emission process at frequency Ω produces an increment $\delta n_k \ll 1$, which decays (through absorption processes not analyzed in detail here) with a characteristic lifetime τ, and this occurs, on the average, m times per second. The present value of n_k will be the result of past emissions, and its average is

$$\langle n_k \rangle = m\tau \; \delta n_k \tag{104}$$

since there were, on the average, $m\tau$ such increments during one lifetime τ in the immediate past. Now note that

$$\langle n_k n_r \rangle - \langle n_k \rangle \langle n_r \rangle = \langle (n_k - \langle n_k \rangle)(n_r - \langle n_r \rangle) \rangle \quad . \tag{105}$$

An elementary emission process increases $(n_k - \langle n_k \rangle)(n_r - \langle n_r \rangle)$, on the average, by $\delta n_k \; \delta n_r$; and this also persists for a time of the order τ, so the present value of (105) will be likewise the resultant of emissions over a time τ in the past:

$$\langle n_k n_r \rangle - \langle n_k \rangle \langle n_r \rangle = m\tau \; \delta n_k \; \delta n_r = \langle n_k \rangle \delta n_r \quad . \tag{106}$$

But the total energy emitted in the elementary process is $\sum \hbar\omega_r \; \delta n_r = \hbar\Omega$, or, since all $\omega_r \simeq \Omega$,

$$\sum_r \delta n_r = 1. \tag{107}$$

And so, summing (106) over all $r \neq k$, we have just Eq.(103)!

In QED, the situation is even more interesting. Let $\psi_m (m = 1,2)$ be the ground and excited states of an atom, ϕ_0 the state of the field with n_1 photons in mode 1, n_2 in mode 2, etc., and ϕ_k the field state which differs from ϕ_0 only in that one more photon is in mode k, while ϕ_{-k} is the field state differing from ϕ_0 only in that one photon has been removed from mode k. If the atom is initially in its excited state, $\Psi(0) = \psi_2\phi_0$, then at a later time the state vector will be very accurately (i.e., retaining all terms that can grow secularly in first order),

$$\Psi(t) = a(t)\ \psi_2\phi_0 + \sum_k b_k(t)\ \psi_1\phi_k \quad , \tag{108}$$

but in this state, we find

$$\langle n_r\rangle = n_r + |b_r|^2 \tag{109}$$

$$\langle n_k n_r\rangle = n_k n_r + n_k|b_r|^2 + n_r|b_k|^2 \quad , \tag{110}$$

and the correlation is negative:

$$\langle n_k n_r\rangle - \langle n_k\rangle\langle n_r\rangle = -|b_k|^2\ |b_r|^2 \quad . \tag{111}$$

Evidently, from (109), the probability $|b_k|^2$ now plays the role of δn_r in the classical derivation. When the atom has reached its ground state,

$$\sum_r |b_r|^2 = 1 \quad , \tag{112}$$

which corresponds to (107). If we suppose that this emission process takes place m times per second, and the field relaxes back with a lifetime τ, so that (104) still holds, the equilibrium value of the correlation will be $m\tau$ times (111), and summing over r yields, instead of (102),

$$\sum_r{}' [\langle n_k n_r\rangle - \langle n_k\rangle\langle n_r\rangle] = -\langle n_k\rangle \quad , \tag{113}$$

which just cancels out Einstein's "particle" term! Likewise, we could analyze absorption processes. Starting with the atom in its ground state, $\Psi(0) = \psi_1\phi_0$, we have at time t just (108) with the subscripts 1,2 interchanged, and ϕ_k replaced by ϕ_{-k}. Equations(109) and (110) still hold, but with negative signs for the three terms containing $|b|^2$. This leads back to (111) without modification. Thus, starting from a field state ϕ_0 without correlations, either an emission or absorption leaves it in a new state with negative correlations given by (111).

Evidently, much better calculations to check (102) in QED and (103) in classical theory, are needed before this issue can be finally resolved. We have, however, some grounds for thinking that the situation may be exactly the reverse of what we have all been taught; i.e., the fluctuation term $\hbar\Omega\langle E\rangle$ in (96), which Einstein interpreted as giving the radiation field a "particle" aspect, is accounted for after all by classical EM theory, as the effect of small intermode correlations that Einstein and all subsequent writers except von Laue[40] seem to have neglected. But in QED, the correlations are negative, canceling out the field quantization contribution, and so QED fails to give the presumably correct Einstein fluctuation law.

6. Conclusion

We have not commented on the beautiful experiment reported here by Clauser[26] which opens up an entirely new area of fundamental importance to the issues facing us. The situation is, in fact, so new that it will require much analysis, based on greater knowledge of the exact experimental conditions, before we will be in a position to make any constructive comments beyond the obvious suggestion that the experiment should be repeated with circular polarization. The implications of Bell's theorem[28], as applied to this experiment, are so astonishing that it will require much deep contemplation to digest and understand it.

What it seems to boil down to, is this: a perfectly harmless looking experimental fact (nonoccurrence of coincidences at 90°) which amounts to determining a single experimental point - and with a statistical measurement of unimpressive statistical accuracy - can, at a single stroke, throw out a whole infinite class of alternative theories of electrodynamics, namely all local causal theories. The mind boggles at the thought that any such thing could be possible. I think everybody's first impression is that there must be something wrong in any argument that purports to draw conclusions of such sweeping generality from practically no premises.

At the present time, all I can say is that to date I have not been able to find any flaw in the mathematics or logic, and to the best of my knowledge, nobody else has claimed to do so. Obviously, this argument deserves, and will receive, the closest scrutiny the human mind is capable of bringing to bear on it. If it survives that scrutiny, and if the experimental result is confirmed by others, then this will surely go down as one of the most incredible intellectual achievements in the history of science, and my own work will lie in ruins. I wish John von Neumann were here to see it.

References

1. E.T. Jaynes, Stanford Microwave Laboratory Report #502 (May 1958).
2. E.T. Jaynes in *Quantum Electronics*, ed. C.H. Townes, (Columbia University Press, New York, 1960) p. 287.
3. E.T. Jaynes and F.W. Cummings, Proc. IEEE *51*, 89 (1963).
4. J.H. Eberly, Ph.D. Thesis, Stanford University (1962).
5. M.J. Duggan, Ph.D. Thesis, Stanford University (1963).
6. M.D. Crisp and E.T. Jaynes, Phys. Rev. *179*, 1253 (1969).
7. C.R. Stroud and E.T. Jaynes, Phys. Rev. A *1*, 106 (1970). Note that, in the transition from galley proof to page layout, pages 118 and 119 became scrambled. To make sense, the text should be read in the following sequence:

Page	*Column*	*Lines*
118	1	1 - 3
118	2	4 - end
119	1	1 - 5
118	1	4 - end
118	2	1 - 3
119	1	6 - end

8. D. Leiter, Phys. Rev. A *2*, 259 (1970).
9. E.T. Jaynes, Phys. Rev. A *2*, 260 (1970).
10. W. Pauli, *Die allgemeinen Prinzipien der Wellenmechanik*, Handb. d. Phys. 2. Aufl. Band *24*, 1. Teil (1932), p. 204. Reprinted by Edwards Brothers Inc., Ann Arbor, Mich. (1946).
11. L.I. Schiff, *Quantum Mechanics* (McGraw-Hill Book Co., Inc., New York, 1949) p. 255.
12. F. Bloch, Phys. Rev. *70*, 460 (1946); see also Rabi, Ramsey and Schwinger, Revs. Mod. Phys. *26*, 167 (1954). For a discussion of radiation damping in these experiments, see Bruce, Norberg, and Pake, Phys. Rev. *104*, 419 (1956); S. Bloom, J. Appl. Phys. *28*, 800 (1957).
13. E.T. Jaynes, Phys. Rev. *108*, 171 (1957), particularly Sec. 18.

14. E.L. Hahn, Phys. Rev. *80*, 580 (1950).

15. E.T. Jaynes and A.L. Bloom, Phys. Rev. *98*, 1099, 1104 (1955).

16. N.G. Basov and A.M. Prokhorov, J. Exp. Theor. Phys. USSR, *27*, 431 (1954); *28*, 249 (1955).

17. K. Shimoda, T.C. Wang, and C.H. Townes, Phys. Rev. *102*, 1308 (1956).

18. W.E. Lamb and J.C. Helmer, Stanford Microwave Laboratory Report #311 (1956); J. Appl. Phys. *28*, 212 (1957).

19. R.P. Feynman, F.L. Vernon, and R.W. Hellwarth, J. Appl. Phys. *28*, 49 (1957). Their radiation damping results, with the inevitable hyperbolic secant, had been found also by S. Bloom, J. Appl. Phys. *27*, 785 (1956).

20. W.E. Lamb, Jr., Phys. Rev. *134*, A1429 (1964).

21. A. Szöke and A. Javan, Phys. Rev. Letters *10*, 521 (1963); R.L. Fork and M.A. Pollack, Phys. Rev. *139*, A1408 (1965); B. Pariser and T.C. Marshall, Appl. Phys. Letters *6*, 232 (1965).

22. M.O. Scully and W.E. Lamb, Jr., Phys. Rev. *159*, 208 (1967).

23. A. Einstein, Phys. Zeit. *10*, 185, 323, 817 (1909).

24. M.O. Scully and M. Sargent III, "Physics Today" (March 1972) p. 38. See also the extensive discussion in S. Tomonaga, *Quantum Mechanics* (North-Holland Publ. Co., Amsterdam, 1962) Ch. 2.

25. S.L. McCall and E.L. Hahn, Phys. Rev. Letters *18*, 908 (1967), Phys. Rev. *183*, 457 (1968).

26. J.F. Clauser, "Experimental Limitations to the Validity of Semiclassical Radiation Theories", this volume, p. 111; see also Phys. Rev. Letters *23*, 880 (1969); *28*, 938 (1972).

27. T.L. Paoli, Phys. Rev. *163*, 1348 (1967).

28. J. von Neumann, *Mathematical Foundations of Quantum Mechanics* (Princeton University Press, Princeton, N.J., 1955). For newer developments, see J.S. Bell, Rev. Mod. Phys. *38*, 447 (1966); E.P. Wigner, Am. J. Phys. *38*, 1005 (1970); L.E. Ballentine, Rev. Mod. Phys. *42*, 358 (1970).

29. W. Heisenberg, in *Neils Bohr and the Development of Physics*, ed. W. Pauli (Pergamon Press, New York, 1955) p. 24.

30. W. Heisenberg, *Physics and Philosophy* (Harper and Bros. Publishers, New York, 1958). The above quotations are found on pp. 129, 145, 164.

31. D. Bohm and J. Bub, Rev. Mod. Phys. *38*, 453 (1966).

32. I.R. Senitzky, Phys. Rev. Letters *20*, 1062, 1277 (1968).

33. F.R. Nash and J.P. Gordon, "The Implications of Radiative Equilibrium in Jaynes' Extension of Semiclassical Radiation Theory", presented at this Conference, p. 623.

34. R.W. Wood, *Physical Optics* (Macmillan Co., New York, 1934).

35. N. Bohr, *Atomic Theory and the Description of Nature* (Cambridge University Press, 1934); reprinted in 1961; p. 13. Similar remarks are found on pp. 32, 80, 108.

36. H.M. Gibbs, "A Test of Jaynes' Neoclassical Theory: Incoherent Resonance Fluorescence from a Coherently Excited State", presented at this Conference, p. 83.

37. J.R. Ackerhalt, J.H. Eberly, and P.L. Knight, "A Quantum Electrodynamic Investigation of the Jaynes-Crisp-Stroud Approach to Spontaneous Emission", presented at this Conference, p. 635.

38. M.C. Newstein, Phys. Rev. *167*, 89 (1968); C.R. Stroud, Phys. Rev. A *3*, 1044 (1971).

39. E.T. Jaynes, *Statistical Physics*, Vol. 3, ed. K.W. Ford (W.A. Benjamin Inc., New York, 1963) Ch. 4; see particularly Eq.(21).

40. M. von Laue, Ann. d. Phys. *47*, 853; *48*, 668 (1915).

A TEST OF JAYNES' NEOCLASSICAL THEORY: INCOHERENT RESONANCE FLUORESCENCE FROM A COHERENTLY EXCITED STATE

Hyatt M. Gibbs

Bell Laboratories, Murray Hill, New Jersey

1. Introduction

There were two motivations for observing the incoherent[1] resonance fluorescence from coherently excited Rb atoms. The first was to demonstrate the quantum-electro-dynamic (QED) coherent-optical effect that the fluorescence should have maxima when the atoms are left in a state of maximum excitation and minima when the excitation is minimized[2,3]. This effect in an optically thin sample is analogous to the precession of a permanent magnetic moment driven by an external magnetic field rotating at the Larmor frequency. The second motivation was to test the semiclassical or neoclassical theory (NCT) of Jaynes, Crisp, Stroud, and co-workers[4,5]. NCT assumes that the expectation value of the dipole moment operator is an actual dipole moment which radiates according to classical electrodynamics. Thus NCT predicts a maximum fluorescence for equal admixtures of the ground and excited states and minima when the atom is closest to a pure state whether it is the excited or ground state. Whereas QED predicts maximum fluorescence for a pure excited state, NCT predicts no fluorescence. NCT's electromagnetic field is not quantized so no zero-point fluctuations exist to give rise to spontaneous emission from a pure excited state. The present experiment is sufficiently exacting to distinguish clearly between the predictions of these two theories. Conclusions drawn from computer simulations of the experiment are decisive: the data agree well with QED within experimental errors, but they disagree markedly with NCT.

The plan of this paper is to augment the figures presented at the conference and their detailed captions with a minimum of text material. The experimental details are described in Section 2, the predictions of NCT and QED for a two-level, sharp-line absorber with a very short exciting pulse are summarized in Section 3A, the computer simulation of the actual three-level, almost-sharp-line absorber with the actual excitation pulse is presented in Section 3B, and the experimental observatiuns are compared with the two theories in Section 4. The contradictions of the data and NCT are summarized in Section 5.

2. Experimental Details

Many of the details of the apparatus are already published[6]. The primary alterations were frequency stabilization ($\pm$ 2MHz) of the Hg laser by locking it to the center of a Rb atomic-beam absorption line and detection of the fluorescence by single-photon counting techniques.

The interaction of the laser beam and Rb atomic beam is pictured in Fig. 1 along with details of the atomic beam oven. The atomic beam at the intersection was about .27 mm high and had a density of $\simeq 10^{10}$ atoms/cm^3. The spatial dimensions of the beam oven were dictated by the fact that the oven operated within 3.6 cm bore of a superconducting magnet. A field of about 75 kOe is required to Zeeman tune the Rb D_1 absorption line into coincidence with Hg laser (Fig. 2). A typical magnetic scan of the Rb absorption showing the four ^{87}Rb and six ^{85}Rb resolved components constitutes Fig. 3. Most of the data were taken using the well-resolved lowest-field ^{87}Rb absorption line. Only ^{87}Rb atoms with M_I = 3/2 participated in the experiment; the remaining atoms were useless but harmless. The apparatus is shown in Fig. 4. The efficacy of the laser locking system is displayed in Fig. 5. The $\pm$ 2MHz stability is more than sufficient for the current observations of excitation of a 15-MHz-wide absorber by a short-optical pulse with frequency spread close to 100 MHz.

One of the problems in the experiment is to satisfy the uniform plane-wave condition, i.e., in order that all atoms be excited to the same superposition of ground and excited states, one must make sure that every atom is exposed to the same optical field for the same time. But the frequency-stabilized laser was operated in the lowest-order TEM_{OO} transverse mode which has a Gaussian spatial profile. The uniform plane wave condition was approximated by observing only fluorescence from atoms excited by the uniform central protion of the profile. This was done by imaging the fluorescence upon an output aperture as shown in Fig. 6. The profile of the fluorescence at the output aperture

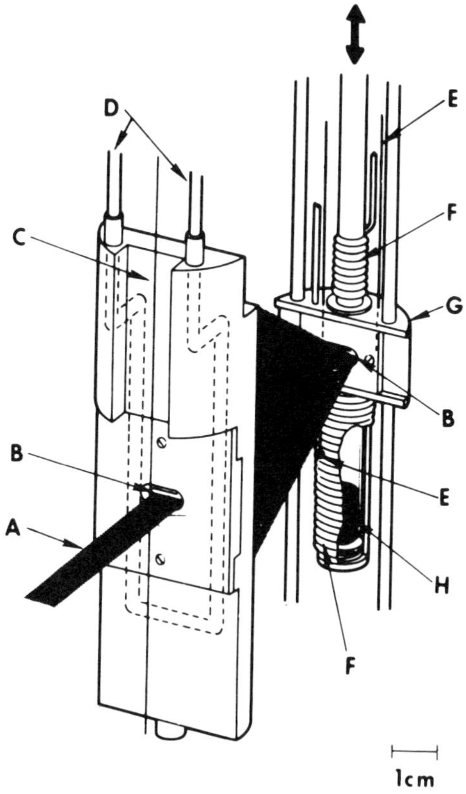

Fig. 1. Sketch of atomic-beam apparatus and intersection of light and atom beams. The 1-cm separation between oven and beam slits is exagerated for clarity. A - Rb Beam; B - Slit; C - Laser Beam; D - Coolant Lines; E - Thermocouple; F - Heater; G - Atomic Beam Oven; H - Rb in Pyrex Tube.

was about the size expected from diffraction and spherical aberrations and the magnification of the lens system. The spatial profile in Fig. 7 is compared with the 100-μm diameter of the output aperture.

Just as important as the spatial profile of the excitation optical pulse is its temporal profile. In the ideal case the optical pulse duration would be negligible compared with the radiative emission times (about 28 nsec here). In the present

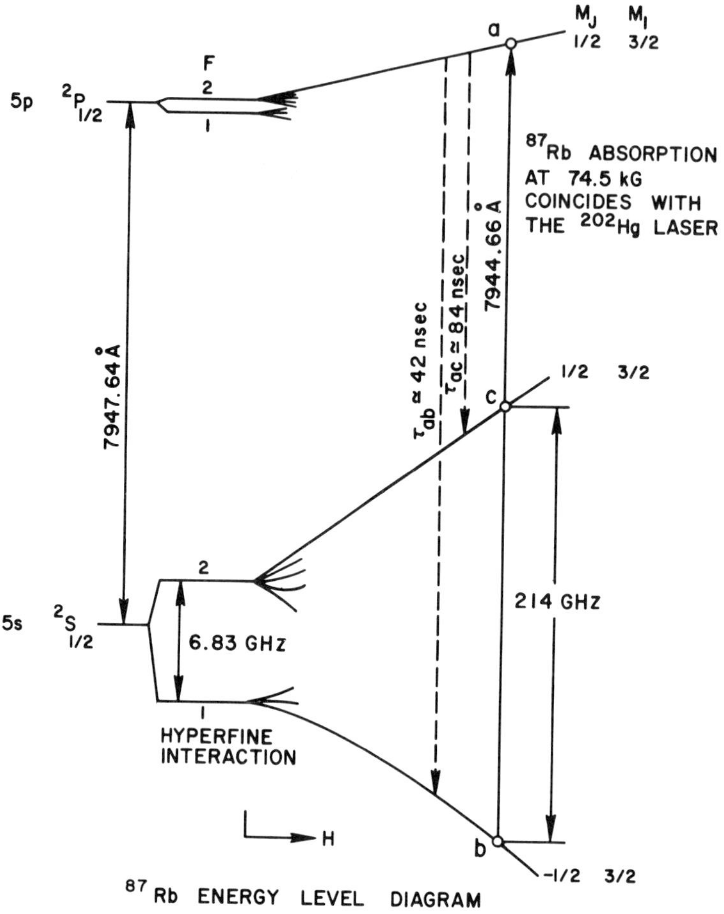

Fig. 2. Diagram of the relevant energy levels of ^{87}Rb as a function of magnetic field strength. The Zeeman interaction at 74.5 kOe lifts the low-field degeneracy and increases the absorption frequency to coincide with the laser emission frequency.

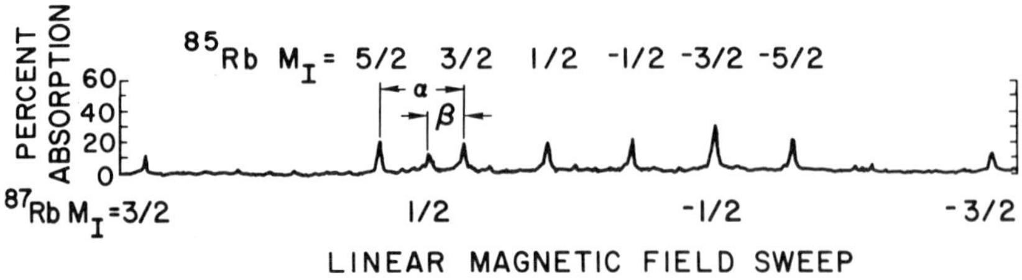

Fig. 3. Absorption of right circularly polarized low-intensity laser light through an atomic beam of natural Rb as a function of magnetic field.

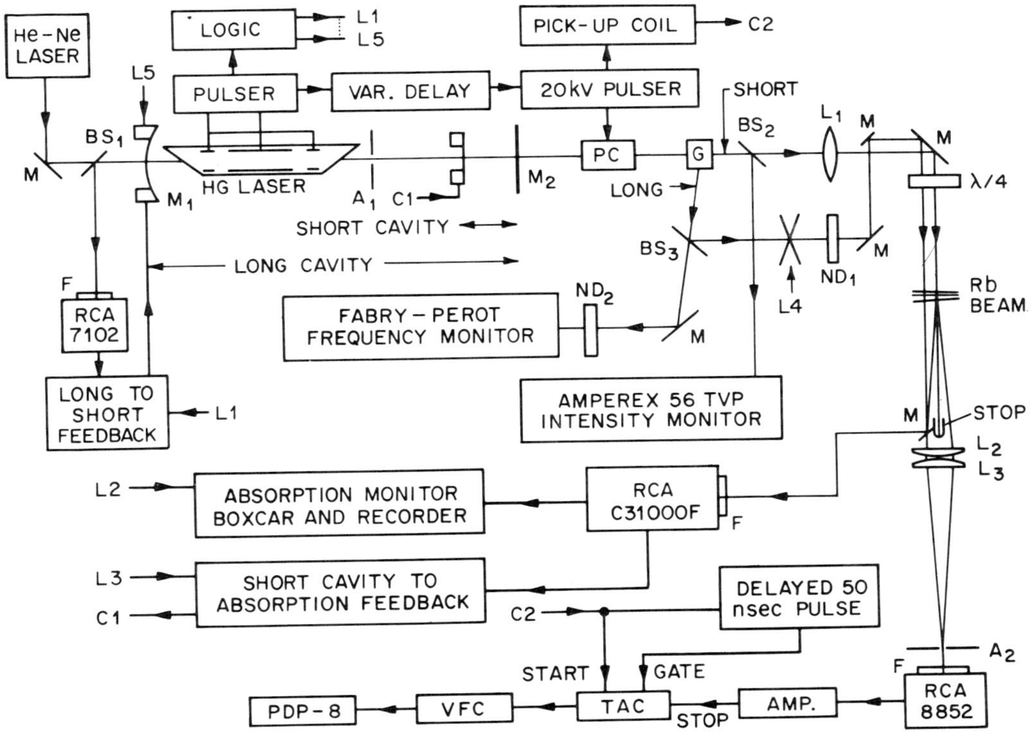

Fig. 4: Block diagram of experimental apparatus. A_1 - aperture to select TEM_{oo} mode; A_2 - aperture to select fluorescence from central, uniform portion of excitation region; BS_i - ith beam splitter; Ci - ith control signal; F - 7945A, 100Å interference filter; G - Glans prism; Li - ith logic pulse; L_i - ith lens; M - mirror; M_1 - 3-m totally reflecting mirror; M_2 - 4% transmission flat output mirror; ND_i - ith neutral density filter; PC - pockels cell; TAC- time-to-amplitude converter; VFC- voltage-to-frequency converter.

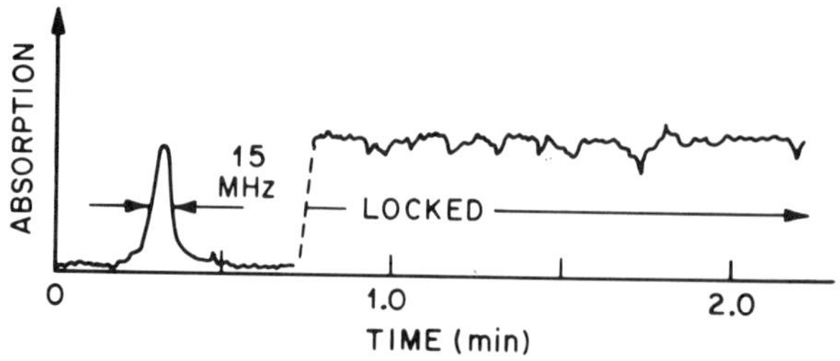

Fig. 5. Laser-absorber frequency locking. The first portion of the scan illustrates the 15 MHz absorption width when the field is scanned with the laser free-running. When the laser frequency is locked to the center of the atomic-beam absorption, the absorption remains close to its maximum value. The frequency difference between the laser and absorber is certainly much less than $\pm$ 7.5 MHz for which the absorption would be one-half its maximum value.

experiment this is not quite the case (see Fig. 8). The FWHM of the intensity of the pulse is only 7 nsec, but there is a weak tail. Also the electric field is proportional to the square root of the intensity, so that its FWHM is closer to 10 nsec. Although it would be desirable to eliminate the tail in an improved experiment, it has been included in the computer simulation. (In the figures shown at the conference the tail of the electric field had the same sign as the main pulse, but it has been found since that the tail had the opposite sign. The differences between the computer simulations here and those presented at the conference are attributable entirely to this discovery. The main difference is the longer predicted decay times, because the negative-tail pulse is able to keep the atoms closer to the excited state (see Figs. 20 and 21). But the data would still be in decisive disagreement with NCT even if the tail were positive.) Improved experiments might also shorten the main pulse. The low power of the laser and diffraction complications with tighter focusing of the laser beam made such a reduction difficult in the present experiment.

The fluorescence detection system utilized standard single-photon counting instrumentation. The pulse induced in a pick-up coil by the firing of the Pockels-cell thyratron started a time-to-amplitude converter (TAC). The arrival of a pulse from the fluorescence 8852 photomultiplier stopped the TAC. A voltage-to-frequency converter and computer-controlled up counter converted the TAC output into a time channel (50, 2 nsec apart).

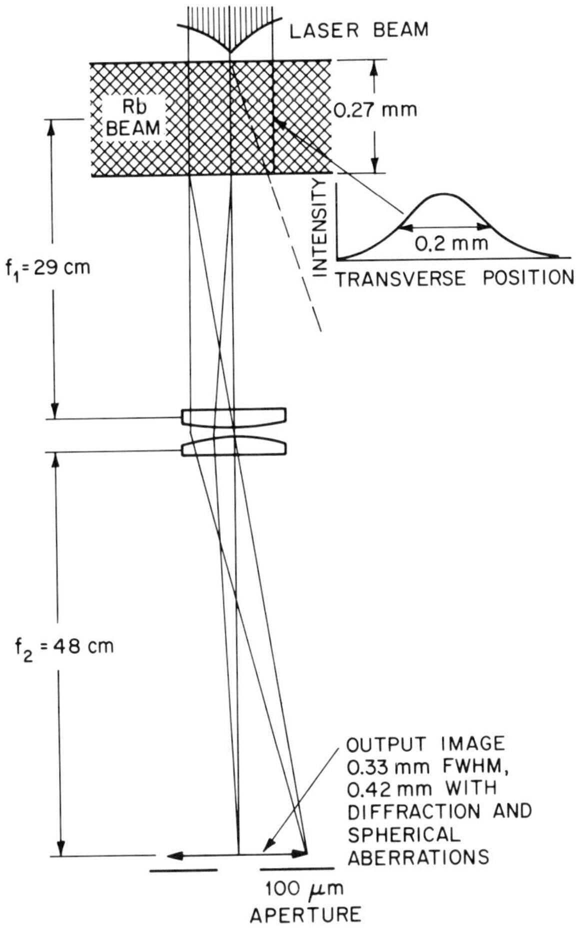

Fig. 6. Uniform plane-wave approximation. The fluorescence emanating from the excited beam atoms was imaged on a 100 μm aperture. The acceptance half angle was always less than 3°, so that the aperture passed fluorescence only from atoms excited by the central, uniform portion of the laser beam. If a large acceptance angle were permitted, such as the dashed line, it would be impossible to distinguish fluorescence originating in the central portion from fluorescence coming from less strongly excited atoms. The small acceptance angle also restricted the fluorescence almost entirely to the 7944.6Å, ΔM = -1 transition since the 7950.7Å, ΔM = 0 radiation is much smaller within that angle. Plano-convex lenses were used with the object and image at the focal lengths to minimize spherical aberrations.

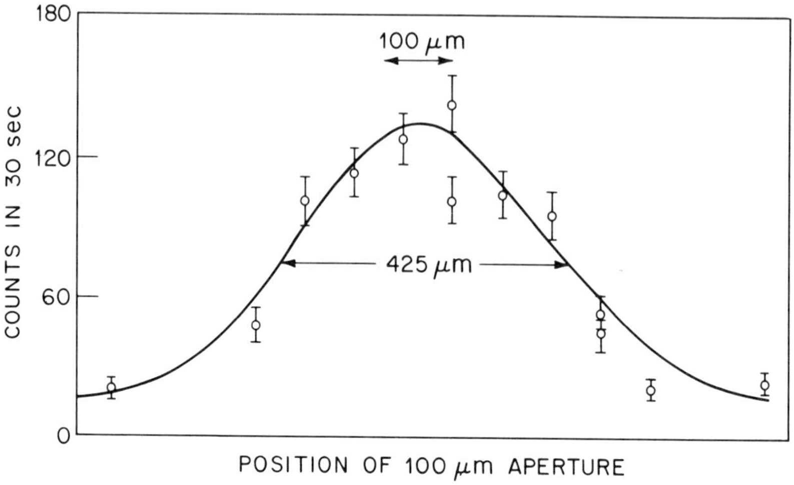

Fig. 7. Experimental transverse spatial profile of the forward fluorescence imaged as in Fig. 6. Vertical and horizontal profiles were obtained at the start of each run to position the 100 μm aperture on the peak of the profile.

One count was then added to that channel. The counting rate was always at least 10 times slower than the 90 counts per second rate of the excitation pulse, preventing pile-up distortions. The computer stored and displayed this fluorescence decay.

Many other aspects of the experiment such as tests for phase shifts in the incident pulse, Fabry-Perot observations of the optical pulse, etc. are discussed in Ref. 6. Other details will be published soon.

3. Predictions of QED and NCT

A. Two-level, Sharp-Line Absorber with Negligibly Short Exciting Pulse

The QED fluorescence rate is given by[4,5] $F_{QED}(t) = (N_2\hbar\omega/\tau_{ab})\rho_{aa}(t)$ where N_2 is the number of 2-level atoms with population $\rho_{aa}(t)$ in the excited state a. Spontaneous emission from state a to state b (with $\hbar\omega$ less energy) is characterized

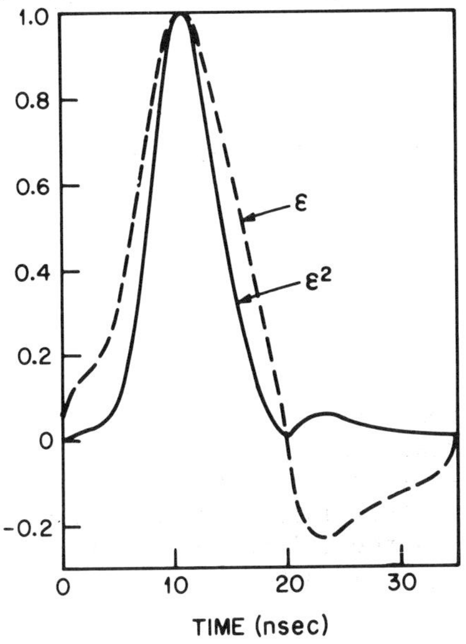

Fig. 8. Input intensity ($\alpha \mathcal{E}^2$) and electric fields ($\mathcal{E}$) used in the computer simulation. The intensity profile was determined by experimental single-photon counting of the input pulse. The change of sign of $\mathcal{E}$, which is reasonable from the shape of the intensity and the expected continuity of $\mathcal{E}$, was verified by detecting the sign reversal of the voltage applied longitudinally to the Pockels cell. For a dipole moment $p = 4.35 \times 10^{-18}$ esu-cm, the intensity units are watts/cm^2 for a pulse of area 1.35 π.

by the spontaneous emission rate $\tau_{ab}^{-1} = \frac{8}{3} (\omega^3/\hbar c^3)\ p^2$; then $\dot{\rho}_{aa}(t) = -\rho_{aa}/\tau_{ab}$, in the absence of external fields. Coherent excitation through a tipping angle θ in a time short compared with relaxation times yields, in the sharpline limit, $\rho_{aa}(t) = (\sin\theta/2)^2 e^{-t/\tau_{ab}}$; $\int_0^T F_{QED}(t)dt \propto (\sin\theta/2)^2$ where the excitation is over by $t = 0$. The fluorescence is then a single exponential decay in time (Fig. 9) and an oscillatory function of pulse area (Fig. 10).

The corresponding NCT equations are:

$$F_{NCT}(t) = \frac{N_2 h\omega}{\tau_{ab}} |\rho_{ab}|^2 = \frac{N_2 \hbar\omega}{\tau_{ab}} \rho_{aa}\rho_{bb},\ \dot{\rho}_{aa} = -\rho_{aa}\rho_{bb}/\tau_{ab},\ \rho_{aa}(t)\rho_{bb}(t) =$$

$$\frac{1}{4}\,\mathrm{sech}^2 \frac{(t-t_m)}{2\tau_{ab}},\ t_m = \tau_{ab}\ \ln \frac{\rho_{aa}(0)}{\rho_{bb}(0)} = 2\tau_{ab}\ \ln\,(\tan \frac{\theta}{2}),$$

$$\int_o^T F_{NCT}(t)dt = \tanh(t_m/2\tau_{ab}) - \tanh[(t_m - T)/2\tau_{ab}] \ .$$

F(T) and $\int F(t)dt$ are shown in Figs. 9 and 10. For weak excitation (small θ) the two theories agree, but for areas between $\pi/4$ and $7\pi/4$ the differences are considerable. An initial pure excited state is unnecessary to detect large discrepancies. The shape of the NCT decay depends upon the initial excitation. For example, in Fig. 9 for $\theta_0 = 90°$ the QED exponential relaxation has a steep initial slope compared with the zero slope of NCT relaxation. For initial tipping angles exceeding 90° the NCT initial slope is positive. So even though the time required to radiate half the excitation energy may not be greatly different in the two theories, the time dependence of that emission may be strikingly different. Also if one restricts the observation time for the integrated fluorescence, one can detect most of the QED emission and very little of the NCT emission as dramatized by Fig. 10.

The results of this analytic solution of the two-level, sharp-line, instantaneous-excitation system illustrate that the differences between QED and NCT predictions are substantial and easily detectable for θ_0 in excess of 90° (equal admixtures of ground and excited states). These solutions include no dynamic shift. NCT predicts that the resonance frequency of an atom shifts to higher frequencies as the atom is excited. This shift is interpreted as a Lamb shift in the weak excitation case where it appears as a static shift. Its magnitude is difficult to calculate with certainty, but it has been estimated as 300 MHz for Na[4]. In the vector model description of an atomic polarization driven by an applied coherent optical field, the polarization could not be driven through the excited state if the laser and absorber frequencies coincided before the pulse was applied. For as the atom is excited its frequency shifts, introducing a time-dependent off-resonance component $\Delta\omega/(2p/\hbar)$ to the effective electric field. The axis of rotation is no longer stationary in time, but rotates from the u axis into the upper portion of the u-W plane. This dynamic shift is included in the computer simulation of the actual system in 3B.

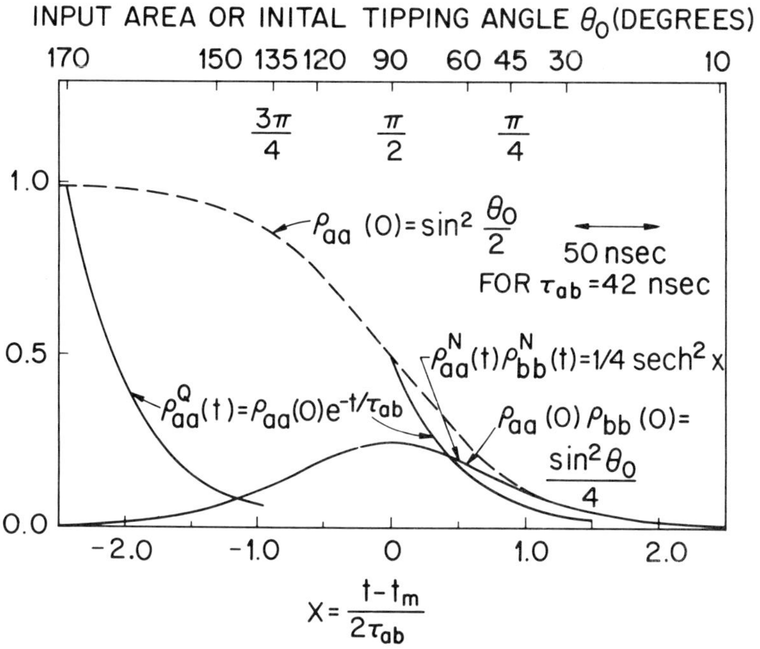

Fig. 9. Comparison of the time dependence of the fluorescence for a quantized (QED) and unquantized (NCT) electro-magnetic field. It is assumed that a two-level system is excited by a pulse of area $\theta_0 = (2p/\hbar)\int_{-\infty}^{\infty}\mathcal{E}(t)dt$ in a time much shorter than the 42 nsec weak-excitation spontaneous emission time. The fluorescences are $F_{QED}(t) = (N\hbar\omega/\tau_{ab}{}^{Q})\rho_{aa}(t)$ and $F_{NCT}(t) = (Nh\omega/\tau_{ab}{}^{N})\rho_{aa}(t)\rho_{bb}(t)$; for a two-level system $\tau_{ab}{}^{Q} = \tau_{ab}{}^{N}$. Then since $F_{NCT} = \rho_{bb}(o)F_{QED}(o)$ and $\rho_{bb} \leq 1$, $F_{NCT}(o) = (N\hbar\omega/\tau_{ab})\sin^2(\theta_0/2)$ is always less than $F_{QED}(o) = (N\hbar\omega/4\tau_{ab})\sin^2\theta_0$ for the same θ_0. But $\rho(t)$ is a different function of time in the two theories so $F_{NCT}(t)$ not only *can* exceed $F_{QED}(t)$ later in the decay, but it *must* since $\int_0^{\infty} F(t)dt = N\hbar\omega\sin^2(\theta_0/2)$ in all theories to conserve energy. By observing fluorescence for only a fixed time interval after the excitation in which most of the QED fluorescence occurs, one excludes that portion ($\pi/2 < \theta < 3\pi/2$) of F_{NCT} which is considerably delayed; see Fig. 2. Two examples of F_{QED} are shown for $\theta_0 = 170°$ and $90°$; the corresponding shapes for F_{NCT} are vastly different. In fact the shapes are easily distinguished for θ_0 above 60°.

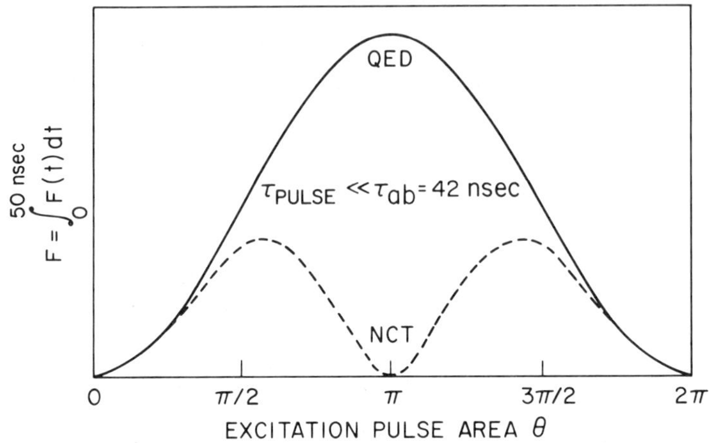

Fig. 10. Comparison of NCT and QED integrated fluorescence of a sharp-line, two-level transition following rapid excitation by a pulse of area θ.

B. Equations for the Three-Level, Almost Sharp-Line Experimental Absorber with Actual Excitation Pulse and Dynamic Shift

The equations used in the QED computer simulation of the experiment are described in detail in Ref. 6. Because the absorption of the pulse was low here, ≈5% total and $\leq$ 30% peak, only two to four integration steps were taken in z, i.e., unlike the self-induced transparency case, the self-consistent solution of Maxwell-Schroedinger coupled equations plays an unimportant role here. That is because the sample contains too few atoms for the field radiated by them to have much effect on the field of the external pulse. But the coupled equations used were the same as in Ref. 6:

$$\dot{u} = v\Delta\omega - u/T_2' \tag{1a}$$

$$\dot{v} = -u\Delta\omega - \kappa^2 \mathcal{E} W/\omega - v/T_2' \tag{1b}$$

$$\dot{W} = v\mathcal{E}\omega - (X + W)/T_1 \tag{1c}$$

$$\dot{X} = -(2/T_2' - 1/T_1)(X + W) \tag{1d}$$

$$\frac{\partial \mathcal{E}}{\partial z} + \frac{1}{c}\frac{\partial \mathcal{E}}{\partial t} = -\frac{2\pi\omega}{c} \int_{-\infty}^{\infty} g(\Delta\omega) v d(\Delta\omega) \; . \tag{2}$$

The pseudopolarization vector $\underline{P} = u\hat{u}_o + v\hat{v}_o - \frac{W\kappa}{\omega}\hat{w}_o$ has components

$$u + iv = \frac{Np}{2} \langle \sigma_x + i\sigma_y \rangle \tag{3}$$

and
$$W = \frac{N\hbar\omega_o}{2} \langle \sigma_z \rangle . \tag{4}$$

Also
$$X = \frac{N\hbar\omega_o}{2} \langle I \rangle , \tag{5}$$

$$\frac{1}{T_1} = \frac{1}{2\tau_{ac}} + \frac{1}{\tau_{ab}} , \tag{6}$$

$$\frac{1}{T_2'} = \frac{1}{2\tau_{ac}} + \frac{1}{2\tau_{ab}} , \tag{7}$$

$$\kappa = 2p/\hbar \; . \tag{8}$$

$\mathcal{E}$ is the slowly varying envelope of the electric field of carrier frequency ω incident upon absorbers with frequencies $g(\omega)$ centered about ω_0. The radiative emission terms in Eqs. (1) were obtained in Ref. 6 in the Weisskopf-Wigner approximation following Mollow and Miller.

In NCT Crisp and Jaynes have derived explicit formulae (Eqs. 30) for a multi-level system in which only the energy difference between two levels is resonant with the optical frequency. With the following identifications $x = -u/N_3p$, $y = v/N_3p$, $z = -W/W_3{}^o$, $\varepsilon_1 = \kappa\mathcal{E}$, and $\varepsilon_2 = 0$ their Eqs. (30) become

$$\dot{u} = v\Delta\omega - \left(\frac{W}{2W_3{}^o\tau_{ab}}\right) u - \left(\frac{X + W_3{}^o}{2W_3{}^o\tau_{ac}}\right) u - \Gamma_{ab}\frac{W}{W_3{}^o} v - \Gamma_{ac}\frac{X + W_3{}^o}{W_3{}^o} v \tag{9a}$$

$$\dot{v} = -u\Delta\omega - \frac{\kappa^2 \mathcal{E} W}{\omega} - \left(\frac{W}{2W_3{}^o\tau_{ab}}\right)v - \left(\frac{X + W_3{}^o}{2W_3{}^o\tau_{ac}}\right)v + \Gamma_{ab}\frac{W}{W_3{}^o}u + \Gamma_{ac}\frac{X + W_3{}^o}{W_3{}^o}u \tag{9b}$$

$$\dot{W} = v\mathcal{E}\omega + \frac{(X + W)(X - W)}{2W_3{}^o\tau_{ab}} - \frac{(X + W)(X + W_3{}^o)}{2W_3{}^o\tau_{ac}} \tag{9c}$$

$$\dot{X} = -(X + W)(X + W_3{}^o)/2W_3{}^o\tau ac \quad , \tag{9d}$$

with $W_3{}^o = -N_3\hbar\omega_0/2$ where N_3 is the number of Rb atoms in the three-level system. These equations differ from those of QED by the addition of the dynamic shift Γ terms and by the fact that in NCT the decay of an excited state is affected by the population of the lower state in the transition:

$$\dot{\rho}_{aa}^{Q} = -\rho_{aa}^{Q}/\tau_{ab} - \rho_{aa}^{Q}/\tau_{ac} \quad ; \tag{10a}$$

$$\dot{\rho}_{aa}^{N} = -\rho_{aa}^{N}\rho_{bb}^{N}/\tau_{ab} - \rho_{aa}^{N}\rho_{cc}^{N}/\tau_{ac} \quad . \tag{10b}$$

$$\dot{\rho}_{ab}^{Q} = -\rho_{ab}^{Q}/2\tau_{ab} - \rho_{ab}^{Q}/2\tau_{ac} \quad ; \tag{11a}$$

$$\dot{\rho}_{ab}^{N} = (\rho_{aa}^{N} - \rho_{bb}^{N})\,\rho_{ab}^{N}/2\tau_{ab} - \rho_{ab}^{N}\rho_{cc}^{N}/2\tau_{ac} \quad . \tag{11b}$$

The dynamic shift terms appear as time-dependent frequency shifts since in the $\dot{u}$ equation they contain a v factor as does the $v\Delta\omega$ frequency-shift term. It can be shown that $\Gamma_{ab} = 2\Gamma_{ac}$ for the s-p transition in Rb using the definition of the Γ's and the angular momentum properties of the Rb three-level system.

The results of the computer simulations will be presented with the data in the next section.

4. Comparison of Experimental Data and Computer Simulations

A. Time Dependence of Fluorescence

The number of fluorescence counts following within each 2-nsec decay channel are shown in Fig. 11 for an excitation area,

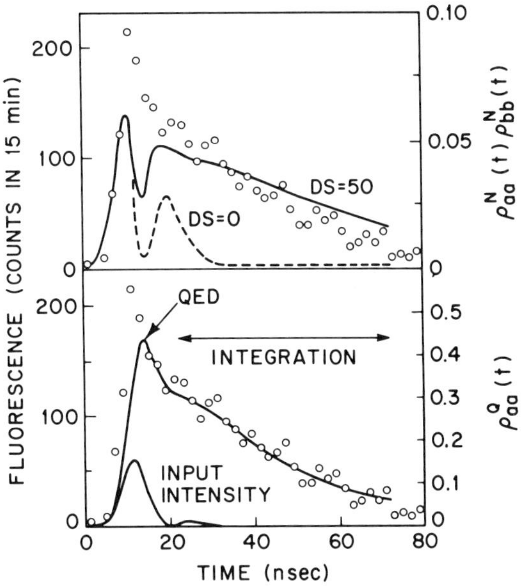

Fig. 11. Comparison of $F_{NCT} \propto \rho^N_{aa}(t)\rho^N_{bb}(t)$ and $F_{QED} \propto \rho^Q_{aa}(t)$ with experimental data for an input pulse area of π. The input pulse is shown in more detail in Fig. 8. Reflections of the excitation pulse into the detector added counts to the fluorescence signal in the 5 to 15 nsec region. The theoretical curves are computer simulations for the actual three-level system including 15 MHz absorption width, finite absorption, and the actual input pulse. The QED and NCT (with 50-MHz dynamic shift) curves are normalized to equal the data at 35 nsec after which the external field vanishes. The NCT curve with no dynamic shift is clearly a ridiculous description of the observations; it is shown on the same scale as the DS = 50 curve. The output aperture diameter was 100 μm. A comparison of the experimental and theoretical best-fit single-exponential decay times is made in Fig. 16.

$A = \theta_0 = (2p/\hbar) \int_{-\infty}^{\infty} \mathcal{E}(t)dt$, of about π. The results are in excellent agreement with QED except in the 6-14 nsec interval where light scattered from the incident pulse adds counts which are not fluorescence. That this was the case was verified by its presence with the atomic beam blocked. Also it increased linearly with the incident pulse intensity unlike the resonance fluorescence. NCT is shown for no dynamic shift (DS = 0), which is in clear disagreement, and for a dynamic shift of 50. The latter value was chosen because it gave the best fit of the NCT to the integrated fluorescence data of Fig. 12. But the DS = 50 prediction of the lifetime is too long as evidenced by the slope of the NCT decay curve beyond 30 nsec; this is further substantiated in Fig. 16 to be discussed below.

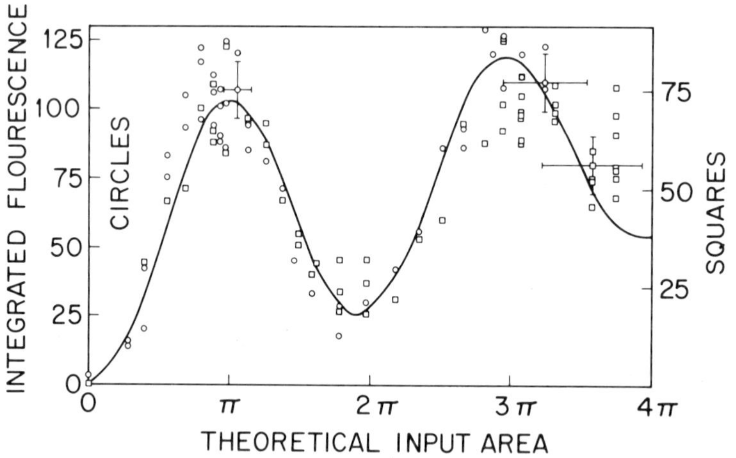

Fig. 12. Fluorescence integrated from 22 to 72 nsec (see Fig. 11) as a function of input pulse area. The squares and circles are counts in 100 and 80 sec intervals for peak cw absorption coefficients of .21 and .32, respectively. The QED curve is normalized to yield a minimum weighted variance with the circled points of 2.2. The areas assigned to the experimental points in order to agree with the shape of the QED theoretical curve are in good agreement with the areas deduced from observations of self-induced transparency; see Fig. 17.

B. Integrated Fluorescence

The fluorescence was electronically integrated from 22 to 72 nsec and recorded as a function of input pulse intensity or area as in Fig. 12. Much more integrated fluorescence data were taken than decay curves. The averaging effects of only approximating the uniform plane wave condition and of amplitude fluctuations of the laser are conservatively indicated by the horizontal error bars in Fig. 12. This averaging is more detrimental the larger the input area since the absolute uncertainty in angle becomes a larger fraction of π. The data above 3π are also slightly less certain because in order to obtain sufficient power it was necessary to increase the Pockels-cell voltage and to remove a linear polarizer used to improve the purity of the laser output polarization. Also at higher pulse intensities the details of the pulse shape become more important. The QED simulation is seen to be in good agreement with the experimental data.

The NCT simulations are in poor agreement with the data as seen by comparing Fig. 13 with Fig. 12. This was done numerically for the dynamic shifts indicated by calculating the weighted variance (WVAR) defined as

$$\mathrm{WVAR} = \sum_{i=1}^{M} W_i (E_i - T_i)^2/(M-p).$$

E_i and T_i are the observed and theoretical values of one of the M data points and p is the number of parameters to be determined (only one, the normalization, here). W_i, the weight of the ith data point, is the inverse of the variance of the ith data point. For E_i counts the uncertainty in E_i is $E_i^{1/2}$ and the variance is just E_i. Then

$$\mathrm{WVAR} = \sum_{i=1}^{M} (E_i - T_i)^2/E_i(M - p)$$

which is equal to one if E differs from T only statistically, i.e., only because of errors in measurements. Then the square root of the uncertainty is roughly the ratio of the average difference between E and T to the average uncertainty in the value of E. For QED in Fig. 12 this ratio is about 1.4, whereas for NCT it is 4 even for a dynamic shift of 50 MHz for which the WVAR is minimized. Although this disagreement is substantial, greater statistical evidence (ratio of 18) is provided in Fig. 14 by combining data from several runs. Also data taken 20 MHz off resonance were in

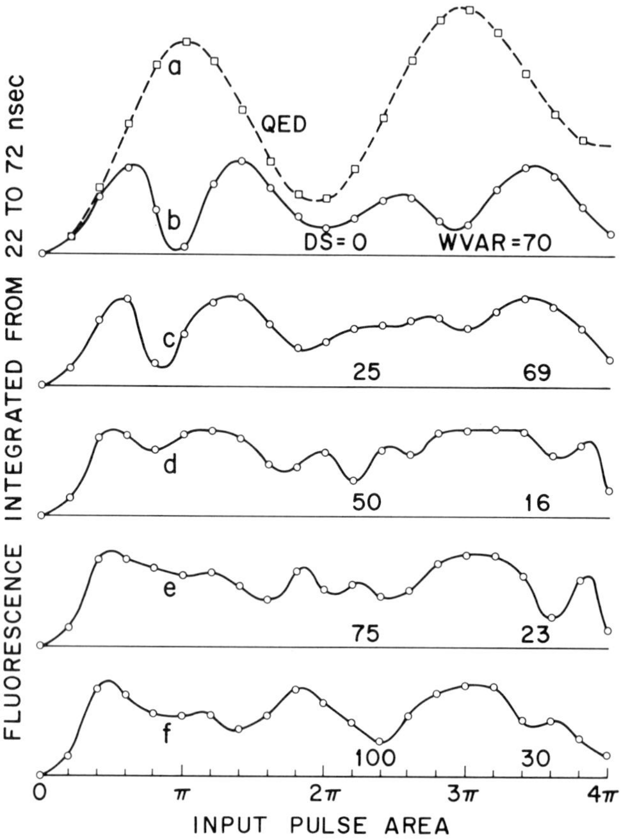

Fig. 13. Comparison of QED and NCT integrated fluorescence as a function of input pulse area for various dynamic shifts DS in MHz. The circles and squares are the calculated values, and the solid curves were drawn visually to connect those values. The WVAR's are the weighted variances of the theoretical curves as a fit to the circle data in Fig. 12. The fluorescences are on the same scale emphasizing that the NCT fluorescence is a factor of two or more lower than the QED fluorescence for $.7\pi < \theta < 1.3\pi$ and during the time interval 22 to 72 nsec. The fluorescences are equal for weak excitation ($\theta \lesssim .3\pi$). From visual observation as well as the WVAR's it is clear that QED is in much better agreement with the data than NCT.

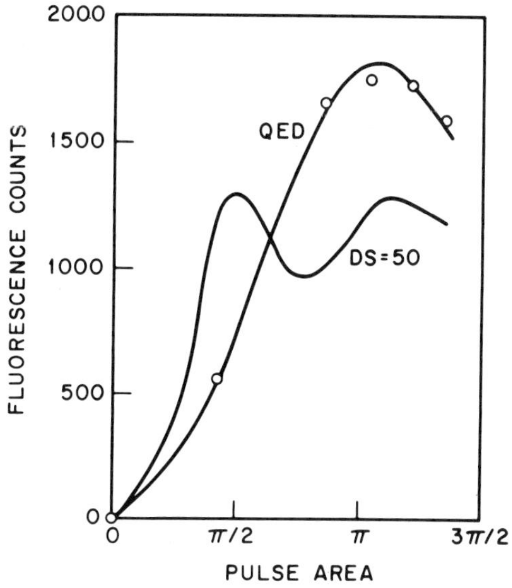

Fig. 14. Comparison of QED and NCT integrated fluorescences. The combined data from several runs have small uncertainties (square root of the number of counts) yielding a weighted variance of 1.1 for QED and 347 for NCT with a 50-MHz dynamic shift. Other dynamic shifts result in even larger variances. Thus on the average the difference between the theoretical and experimental values is 18 times larger for NCT (DS = 50) than QED.

reasonable agreement with QED but not with NCT (see Fig. 15). Since the input pulse is not hyperbolic secant in its time dependence, it is unable to return the polarization as close to its initial value when it is off resonance, i.e., the minimum in Fig. 12 is deeper than in Fig. 15. For the same reason there is an off-resonance contribution to the effective field resulting in one complete rotation of the polarization vector before the area of the input pulse is 2π, i.e., the minumum in Fig. 15 occurs at about 1.6π compared with $\approx 2\pi$ in Fig. 12.

C. Apparent Exponential Lifetimes

Several decay curves for various input areas have been least-squares fitted to a single exponential. The best-fit lifetimes are consistent with the prediction of QED that the decay should

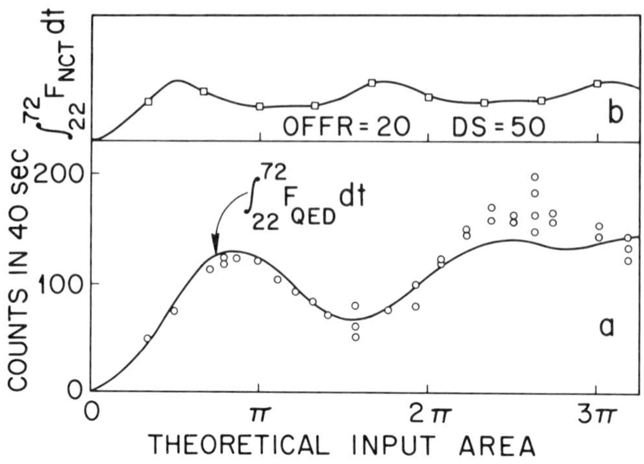

Fig. 15. Comparison of QED and NCT integrated fluorescences with the laser frequency 20 MHz below the center of the absorber frequency. The experimental points (circles) in (a) are in reasonable agreement with the QED prediction (solid curve) shown also in (a), but they are in disagreement with the NCT (DS = 50) prediction shown in (b). The squares in (b) are the calculated NCT points.

be an exponential with 28-nsec decay time dependent of the input areas (see Fig. 16). The shape of a NCT decay curve is a function of the input pulse area, or more accurately where one begins on the decay curve depends upon the initial area. But even though almost complete inversion of the a-b system is achieved in the NCT simulation, the decay is always at least crudely represented by a single exponential over the 40-to-70 nsec interval. This is because the a-c system is never inverted and a-c relaxation is always at work (but its decay time is 42 nsec or longer). By fitting the NCT decay curve to an exponential over this interval, one can compare the average decay rates with observations as in Fig. 16. Again the NCT simulation times are several standard deviations longer than the observed values.

D. Self-Induced Transparency

In the discussion so far no calibration of the input area

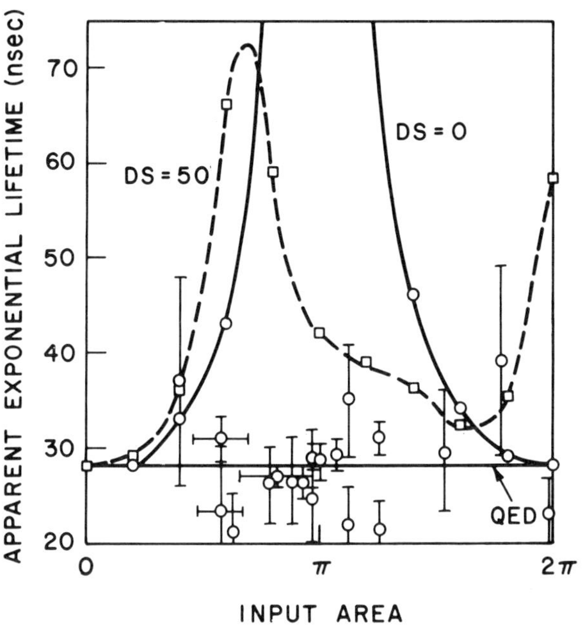

Fig. 16. Comparison of observed apparent exponential decay times with QED and NCT predictions. The experimental uncertainty in area is no more than ± 10% unless shown otherwise. The data and theoretical predictions were fitted to a single exponential over the range 40 to 70 nsec. Although the NCT lifetimes are often several times longer than the 28-nsec QED prediction, the decay curves are approximated well by a single exponential yielding a standard deviation of external consistency of 2% or less for the theoretical lifetimes. The measured decay times are in excellent agreement with previous determinations and the QED prediction and many standard deviations in disagreement with NCT (DS = 0 or 50).

has been introduced; rather the first minimum in the experimental fluorescence was identified as the first minimum in the QED simulation. The same input area normalization was used in the NCT simulation. Perhaps a better, but still poor, agreement with the data would be reached with zero dynamic shift if the π minimum in Fig. 13b were identified with the 2π dip in the data. That possibility is ruled out by self-induced transparency observations reported in Fig. 17. The input areas as labeled agree with the self-induced transparency calibration.

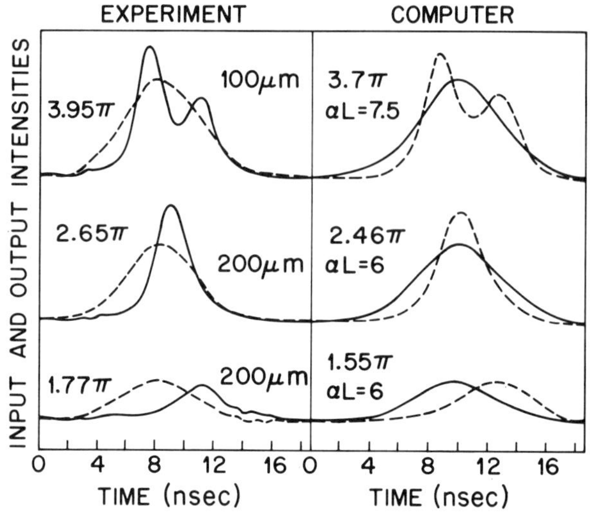

Fig. 17. Self-induced transparency in an optically thick beam verifying that the integrated fluorescence minimum occurs at an area of 2π. The larger slit used to increase the absorption resulted in a 45-MHz absorption width. The absorption coefficients αL used in the computer simulation were experimental values which could be slightly low if unabsorbable left circularly polarized light were present in the incident beam. Good agreement between the observations and the simulation is obtained for an input area slightly smaller than the experimental identification from Fig. 12 and a 20 or 30% higher αL. But these differences are well within the uncertainties. It is certainly clear that the integrated fluorescence minimum labeled 2π in Fig. 12 could not be in fact a π minimum. During the short time of the SIT pulse relaxation plays such an insignificant role that one can hardly distinguish between QED and NCT (DS = 0) computer simulations.

E. Simulated State-Population Evolutions

It is interesting to compare theoretical simulations under various conditions to see how closely complete inversion of the a-b system is approached and to compare absolute fluorescences. Figure 18 illustrates the degradation of the minima in the NCT fluorescence at π and 3π as a static (OFFR) or dynamic (DS) shift is introduced. It also demonstrates the larger magnitude of the QED fluorescence and the non-exponential field-free (t > 35 nsec)

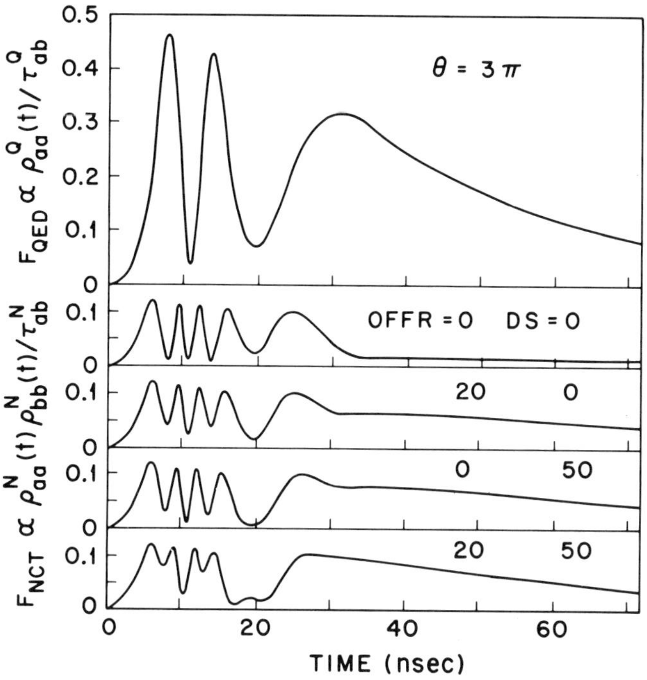

Fig. 18. Time dependence of the NCT and QED fluorescences for an input pulse of area 3π. A value of 0.5 on the F_{QED} curve corresponds to $\rho_{aa}{}^{Q} = 1/2$ and $\rho_{bb}{}^{Q} = 0$ with $\rho_{cc}{}^{Q} = 1/2$. The inversion of the a - b system is then almost complete at 8 nsec, but $\rho_{aa}{}^{Q} \approx 0.3$ or 60% of its maximum value at the end of the pulse. Because the 3π pulse has a negative tail the area of the positive portion is almost 4π, i.e., F_{QED} almost reaches its 4π minimum at 20 nsec and then increases back to a 3π maximum as the negative tail reverses the direction of the rotation of the polarization. F_{QED} and F_{NCT} are on the same scale which illustrates the lower fluorescence predicted by NCT. The maxima in F_{NCT} occur when $\rho_{aa} = \rho_{bb}$, i.e., when the dipole moment of the ab system is a maximum, rather than when ρ_{aa} is a maximum as in the QED case. Minima in F_{NCT} occur when ρ_{aa} or ρ_{bb} is a minimum. The addition of a static shift (OFFR) of 20 MHz or a dynamic shift (DS) of 50 MHz reduces the π and 3π minima corresponding to the inability of the applied field to invert the a - b system off resonance. The OFFR = 20, DS = 50 decay curve beyond 35 nsec is a good example of the non-exponential shape of NCT decay.

decays in NCT. Figures 19 to 21 show the time evolution of ρ_{aa} and ρ_{bb} ($\rho_{aa} + \rho_{bb} + \rho_{cc} = 1$). The most striking conclusion is that the shape of the input pulse and the fact that NCT relaxation vanishes as inversion is approached combine to produce almost complete inversion of the a-b system in the absence of a dynamic shift (Fig. 20). Even with a dynamic shift of 50 MHz, Fig. 21, the a-b system is inverted for most of the integration interval.

5. Discrepancies between the Data and NCT

A. Integrated Fluorescence

It was shown in Section 4B and Figs. 12 to 15 that the QED

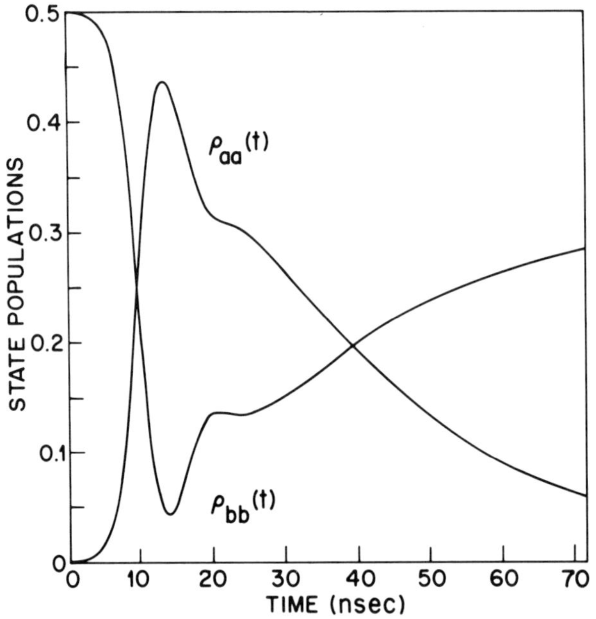

Fig. 19. State-population time dependence in QED for an input pulse area of π. Since ρ_{cc} is 0.5 or greater, ρ_{aa} must be no larger than 0.5. Because of the negative tail the polarization is rotated by about 1.25π and then by -0.25π. The first time the pulse area passes through π, $\rho_{aa} \approx 0.435$ at $t \approx 13.5$ nsec. At the end of the pulse ($t = 35$ nsec) when the area is again π, ρ_{aa} is only 0.225 because of spontaneous emission from a to b and c during the pulse. Nonetheless, the a - b system is left slightly inverted. A ρ_{aa} as high as .37 (out of 0.5) is predicted by the computer simulation at the end of the positive portion of $\mathcal{E}(t)$.

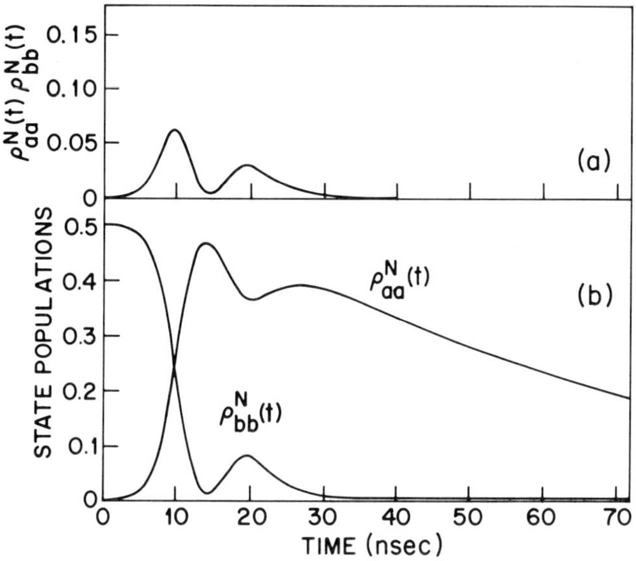

Fig. 20. State-population time dependence in NCT on resonance and with no dynamic shift. $F_{NCT} \propto \rho_{aa}{}^N(t)\rho_{bb}{}^N(t)$ for an input area of π is shown in (a) corresponding to the small value of the integrated fluorescence in Fig. 12. Part (b) shows that this results from the smallness of ρ_{bb}. The input pulse of Fig. 8 is very effective in inverting the a - b system according to NCT. The system is rapidly inverted by the first π portion of the pulse. As the rotation slows down in rotating from π to 1.25π and then back to π, ρ_{bb} is always small and the NCT relaxation to b is weak. Relaxation to c with $\rho_{cc} = 1/2$ is still appreciable, accounting for the decay of ρ_{aa}. But F_{NCT}, defined as the a to b fluorescence, is almost zero after the pulse ends at 35 nsec.

simulation agrees with the data within the experimental error both on resonance and 20 MHz off resonance. The NCT simulation fails decisively in both cases with or without an arbitrary dynamic shift.

B. Decay Time

It was shown in Section 4A and C and Figs. 11 and 16 that there is no experimental evidence for any lengthening of the decay time as complete inversion of the a-b system is approached. The

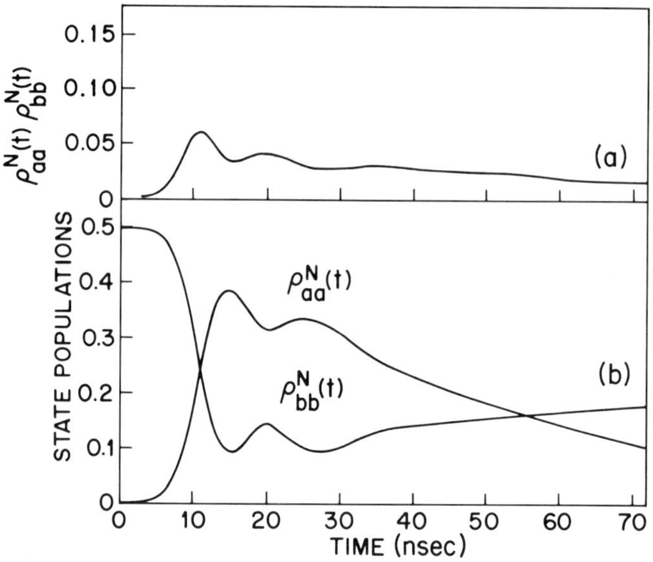

Fig. 21. State-population time dependence in NCT on resonance and with a 50-MHz dynamic shift for an input area of 0.8π. Even with a dynamic shift of 50 MHz, the NCT computer simulation predicts appreciable inversion of the a - b system, ie., $\rho_{aa} > \rho_{bb}$, for most of the 22 to 72 nsec integration period. This gives rise to the minimum at 0.8π in Fig. 13d.

NCT simulation predicts sufficiently large inversion such that the predicted lifetime lengthening would be many times the experimental uncertainties.

C. Dipole Moment

One must insist that the fluorescence be the same in the NCT and QED simulations for weak excitation. Suppose initially $\rho_{aa}(o) = \varepsilon << 1$, $\rho_{bb}(o) = 1/2 - \varepsilon$, $\rho_{cc}(o) = 1/2$. Then $F_{QED} \equiv N_3\hbar\omega\rho_{aa}{}^Q/\tau_{ab}{}^Q = N_3\hbar\omega\varepsilon/\tau_{ab}{}^Q$ and $F_{NCT} \equiv N_3\hbar\omega\rho_{aa}{}^N\rho_{bb}{}^N/\tau_{ab}{}^N = N_3\hbar\omega\varepsilon(1/2 - \varepsilon)/\tau_{ab}{}^N \approx N_3\hbar\omega\varepsilon/2\tau_{ab}{}^N$. In order that these two fluorescences be equal one must have $\tau_{ab}{}^N = \tau_{ab}{}^Q/2$. Since $\tau_{ac} = 2\tau_{ab}$ by the angular momenta properties of the transition and the dipole operator, $\tau_{ac}{}^N = \tau_{ac}{}^Q/2$ also. This factor of two difference in the lifetimes in the two theories was inserted in the computer simulations. Without it the NCT would be even more inconsistent with the data here since the decay times would be doubled. But a factor of two shorter lifetime means a $\sqrt{2}$ larger dipole moment, inconsistent with the value calculated from the wave functions of the Rb atom[7]. It would also imply a factor of two lower power requirement for a 2π pulse in contradiction with a previous measurement of that power[6]. It should be noted that the factor of two in lifetime discussed here arises from (a) the fact that there are two equally populated ground states (b) the QED decay rate is multipled by the lower state population to obtain the NCT decay rate.

This contradiction between the theoretical lifetime and the observed lifetime is independent of this experiment.

References

1. Coherent resonance fluorescence is calculated to be $\lesssim$ 10% of the incoherent fluorescence even at $\theta = \pi/2$.
2. I. D. Abella, N. A. Kurnit, and S. R. Hartmann, Phys. Rev. *141*, 391 (1966). S. L. McCall and E. L. Hahn, Phys. Rev. *183*, 457 (1969). The value of the dipole moment is derived in Ref. 6.
3. H. P. Grieneisen, N. A. Kurnit, and A. Szöke, Optics Comm. *3*, 259 (1971). In this reference is described a similar experiment in which the F versus θ oscillations are almost averaged out by level degeneracies and broadline absorption.
4. M. D. Crisp and E. T. Jaynes, Phys. Rev. *179*, 1253 (1969) particularly Eqs. (14) and (30). C. R. Stroud, Jr. and E. T. Jaynes, Phys. Rev. A *1*, 106 (1970). D. Leiter, Phys. Rev. A *2*, 259 (1970). E. T. Jaynes, Phys. Rev. A *2*, 260 (1970).
5. Several articles claiming to disprove the NCT have appeared during this experiment: R. K. Nesbet, Phys. Rev. Letters *27*, 553 (1971); R. K. Nesbet, Phys. Rev. A *4*, 259 (1971); J. F. Clauser, Phys. Rev. A, to be published; F. R. Nash and J. P. Gordon, to be published. These papers reanalyze previous experiments using NCT. The present experiment has the advantage of being the experiment suggested by Jaynes; it also demonstrates the validity of QED in a new regime.
6. Many details are contained in H. M. Gibbs and R. E. Slusher, Phys. Rev. Letters *24*, 638 (1970); R. E. Slusher and H. M. Gibbs, Phys. Rev. A *5*, 1634 (1972); H. M. Gibbs and R. E. Slusher, "Sharp-Line Self-Induced Transparency," to be published.
7. O. S. Heavens, J. Opt. Soc. Am. *51*, 1058 (1961).

EXPERIMENTAL LIMITATIONS TO THE VALIDITY OF SEMICLASSICAL RADIATION THEORIES *

John F. Clauser

University of California, Berkeley, Calif.

Recently there has been speculation that quantization of the electromagnetic field is unnecessary. In support of this viewpoint, a large number of effects - long thought to require such a quantization - have successfully been rederived semiclassically. It has been hoped that in a suitably treated semiclassical theory one could account for all of the present data of atomic physics as well as eliminate the problem of divergences in quantum electrodynamics.

It is the purpose of this paper to show that there is at least one first-order effect that cannot be predicted by a semiclassical theory. Further, it will be shown that currently existing experimental data exclude semiclassical theories in general.

The pertinent experiment is a measurement of the polarization correlation of photons emitted successively in a $J = 0 \to J = 1 \to J = 0$ atomic cascade. Such an experiment has been performed by Kocher and Commins[1], and more recently by Freedman and Clauser[2]. In their experiments, the emitted cascade photons were selected by interference filters, analyzed with linear polarizers, and detected individually with photomultiplier tubes in coincidence. A quantum mechanical description of the photons required that the coincidence rate vanish when the polarizers are crossed. This prediction is a direct consequence of angular momentum and parity conservation.

Since isolated coincidences are observed, a semiclassical description of photons must describe them as short pulses of electromagnetic radiation. Two fundamental assumptions will be made for these pulses:

(1) For a classical electromagnetic wave of any incident polarization, transmission of a linear polarizer varies as a + b $\cos^2(\theta-\phi)$ where θ is the orientation of the linear polarization of the pulse, and ϕ is the orientation of the polarizer axis.

(2) The probability of electron emission at a photomultiplier cathode is proportional to the incident intensity.

Both of these assumptions are rather well tested experimentally for classical electromagnetic radiation and evidently cannot be modified within the framework of Maxwell's equations.

We consider the situation when the two polarizers are crossed. For the coincidence rate to vanish in this configuration, a semi-classical theory requires that the linear polarization direction of at least one of the pulses must be orthogonal to the axis of the associated polarizer. This clearly cannot happen, if the orientation of the pair of polarizers is averaged over all directions for a given ensemble of photons. Thus any semiclassical theory must predict a nonvanishing of the coincidence rate with polarizers crossed. Consistent with the efficiencies of the polarizers, both experiments observed this rate to vanish; thus these results exclude semiclassical theories in general. A rigorous mathematical treatment of the above discussion has been performed, and the results rule out semiclassical theories with high statistical accuracy.[3] In addition, the Jaynes-Crisp-Stroud predictions for this situation have been calculated and are in direct disagreement with experiment.

The difference in the predictions of the two theories arises from a neglect in the semiclassical calculation of interference terms of one photon with the other. It is one of the most curious predictions by the quantum theory that this nonlocal polarization interference persists, even when the photons are remote from each other, and have different frequencies.

---

* Work supported by the U.S. Atomic Energy Commission.

1. C.A. Kocher and E.D. Commins, Phys. Rev. Lett. *18*, 575 (1967).

2. S.J. Freedman and J.F. Clauser, Phys. Rev. Lett. *28*, 938 (1972).

3. J.F. Clauser, Phys. Rev. *A6*, 49 (1972).

TIME DELAY STATISTICS OF PHOTOELECTRIC EMISSIONS: AN EXPERIMENTAL TEST OF CLASSICAL RADIATION THEORY*

W. Davis and L. Mandel

University of Rochester, Rochester, N.Y.

1. Introduction

For many years the photoelectric effect has been taken to provide evidence for the quantum nature of electromagnetic radiation. However, as has been pointed out from time to time[1-3], if the electromagnetic field is treated as a classical, c-number perturbation acting on the bound electron, essentially the same conclusions can be reached as by quantum electrodynamics. For example, photoemission takes place only when the frequency ω is high enough for $\hbar\omega$ to exceed the work function W of the electrons,

$$\hbar\omega > W , \tag{1}$$

and the probability of photoemission at time t is proportional to the expectation value of the light intensity I(t) at time t,

$$\begin{array}{l}\text{photoemission probability} \\ \text{at time t within } \Delta t\end{array} = \text{constant} \times \langle I(t)\rangle \Delta t . \tag{2}$$

The latter conclusion implies that photoemission can commence, with non-zero probability, as soon as the photodetector is exposed to the electromagnetic field. However, as both conclusions follow from the quantum mechanics of the electron treated as an open system, energy conservation is not automatically taken into account.

If we require that energy remains conserved over all times

* This work was supported by the National Science Foundation and by the Air Force Office of Scientific Research.

long compared with an optical period, then classical electrodynamics leads to the additional condition for photoemission,

$$\int_{t}^{t+T} dt' \int_{S} \underline{P}(\underline{r},t') \cdot \underline{dS} \geqslant W \quad , \tag{3}$$

where $\underline{P}(\underline{r},t)$ is the Poynting vector of the electromagnetic field, the surface integral is taken over the sensitive, exposed area of the photocathode, and t to t+T is the time interval of observation. According to this condition, when a photodetector is first exposed to the electromagnetic field, photoemission cannot commence until a critical energy accumulation time τ_c has elapsed such that

$$\int_{t}^{t+\tau_c} dt' \int_{S} \underline{P}(\underline{r},t') \cdot \underline{dS} = W \quad . \tag{4}$$

To a certain extent condition (3) is inconsistent with Eq. (2). However condition (3) appears to be inescapable if we wish to describe the field as a classical wave that obeys the energy conservation principle.

A number of experiments have been performed that throw some light on the question whether there exists some delay in photoemission following the turn-on of the light beam.[4-7]. The three oldest experiments provide some upper limit on a possible delay time, but throw no light on the question of the validity of Eq. (3), because the power or some other key parameter was not measured. The best experiment is that of Tyler[7], who demonstrated clearly that photoemission can commence long before the critical accumulation time τ_c is reached. Nevertheless, a certain amount of ambiguity in interpretation remains. If the potential well in which the electron is trapped is broad, as it usually is in practice, the electron has a wide spectrum of possible energy levels [8], and the initial electron state may not be the state of lowest energy. Secondly, if the light intensity at the photocathode is not zero before the light beam is nominally turned on, energy can in principle be accumulated beforehand. In Tyler's experments the beam was turned on and off with the help of a Kerr cell, and this had an on-off transmission ratio of about 250:1, which is not large. Finally, Tyler's experiments were concentrated in the region $\tau<\tau_c$ and could not therefore yield information about possible changes of emission probability as τ passes through the value τ_c.

For these reasons we set out to study the statistics of photoelectric emission under conditions in which there is very high discrimination between the on and off phases of the light beam, and at various times following the turn-on, before, at, and after the critical accumulation time τ_c. If there is always enough energy available for photoemission after the time τ_c, and not always enough before time τ_c, we might expect to see this reflected not only in the variation of the mean number of photoemissions, but also in the fluctuations of the numbers. Measurements of the probability distribution at different times therefore seemed to be desirable.

2. The Experiment

The set-up for the experiment is illustrated in Fig. 1. The

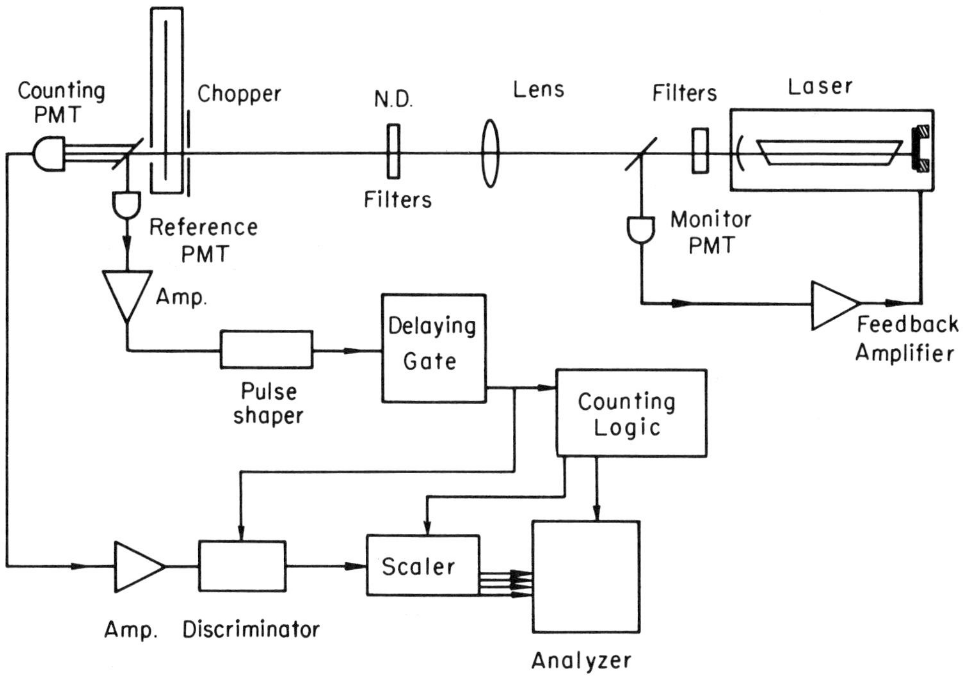

Fig. 1 The Experimental Set-up

light source was a single-mode He:Ne laser, that was operated far above threshold, and was controlled by a feedback arrangement to maintain its intensity constant. The beam was focused by a long focus lens to a focal spot of about 10^{-2} cm diameter on the knife-edge of a rotating disc. This disc served as shutter for turning the beam on and off. It had 24 apertures near the rim, each of which carried a knife edge, and was rotated at 60 revs./sec by a synchronous motor. It permitted the beam to be turned on and off in times of order 5 μsec, 1440 times per second. Although the switching time was very long compared with possible electronic switching times, it was still short compared with τ_c, which was arranged to be about 20 μsec. However, this method of switching permits very high discrimination between the nominal on and off intensities of the light, which were in the ratio of at least 10^4:1.

A small reference phototube located immediately behind the shutter served to determine the turn-on time. The main light beam was strongly attenuated and then fell on a counting phototube, whose pulses, after amplification and pulse shaping, were sent via a gate to a scaler, where they were registered. The gate controlled the counting interval T, and, with the help of a delay circuit, it could be centered at various times τ from 1 or 2 μsec to 100 μsec following the turn-on. The number registered in the scaler at the end of the counting interval T was used as an address in a multichannel analyzer, to which the information was transferred and where it was accumulated. After many counting intervals, the histogram of numbers stored in the memory of the analyzer provided a direct measure of the probability distribution $p(n,\tau)$ of the number of counts registered in the time interval T after a delay τ.

According to quantum electrodynamics, for a single-mode laser this probability should be given by the Poisson distribution [9]

$$p(n,\tau) = \frac{(RT)^n e^{-RT}}{n!} , \tag{5}$$

where R is the mean counting rate due to the light beam, and should be independent of τ. According to Eq. (3) on the other hand, the probability for a count should be larger for $\tau>\tau_c$ than for $\tau<\tau_c$. $p(1,\tau)$ should be appreciably smaller for $\tau<<\tau_c$ than for $\tau>\tau_c$, and for the same reason $p(2,\tau)$ should be much smaller for $\tau<<2\tau_c$ than for $\tau>2\tau_c$, etc. We might also expect to see an appreciable change in the variance of n at times of order $\tau\simeq\tau_c$, $\tau\simeq2\tau_c$, etc.

The power of the laser beam was determined with a calibrated bolometer, and this allowed the quantum efficiency of the photodetector to be calibrated in turn. The light intensity was adjusted to be $\sim 10^{-14}$ watts after attenuation, such that the critical accumulation time τ_c was 20 ± 1.8 μsec. The measurement interval T was fixed at 10 μsec, and the delay was varied from 10 μsec to 60 μsec, as measured to the center of the interval T. The quantum efficiency was only about 4% for the red laser light, so that the mean number of counts RT per measurement interval was only about 0.02.

Due to the background or dark counting rate of the photodetector, some correction to the measured values of $p_M(n,\tau)$ was necessary. If $p_B(n)$ is the probability distribution of counts due to the background alone, we have the relationship

$$p_M(n,\tau) = \sum_{m=0}^{n} p(n-m,\tau)\,p_B(m) \quad . \tag{6}$$

Separate measurements of $p_M(n,\tau)$ and $p_B(n)$ then allow $p(n,\tau)$ to be extracted by deconvolution from Eq. (6).

3. Results

The results of 6 different measurements of the probability distribution $p(n,\tau)$ are shown in Fig. 2, together with the theoretical Poisson distributions based on Eq. (5). It will be seen that there is no evidence of any variation of $p(n,\tau)$ with τ, and that moreover, $p(n,\tau)$ is nearly Poissonian. There is a small departure from the Poisson form, that is independent of τ, and is probably due to acoustic modulation of the laser by the rotating shutter. Such a modulation, of order 1% in depth was visible at the outputs of the reference and monitor phototubes.

Table 1 shows the first 4 moments of the measured probability and of the corresponding Poisson distribution, for a typical run taken at $\tau = 30$ μsec. The departure from the ideal Poisson form is small but evident.

	Experimental	Poissonian
$\langle n \rangle$	0.02208 ± .00015	0.02208
$\langle (\Delta n)^2 \rangle$	0.02223 ± .00016	0.02208
$\langle (\Delta n)^3 \rangle$	0.02253 ± .00029	0.02208
$\langle (\Delta n)^4 \rangle$	0.02459 ± .00035	0.02355

Table 1 Counting Moments for a Typical Run

($\tau = 30$ μsec. Total no. of samples $\sim 10^6$)

The results are clearly in agreement with the predictions of quantum electrodynamics and of the semiclassical Eq. (2), and show no evidence for the existence of any critical energy accumulation

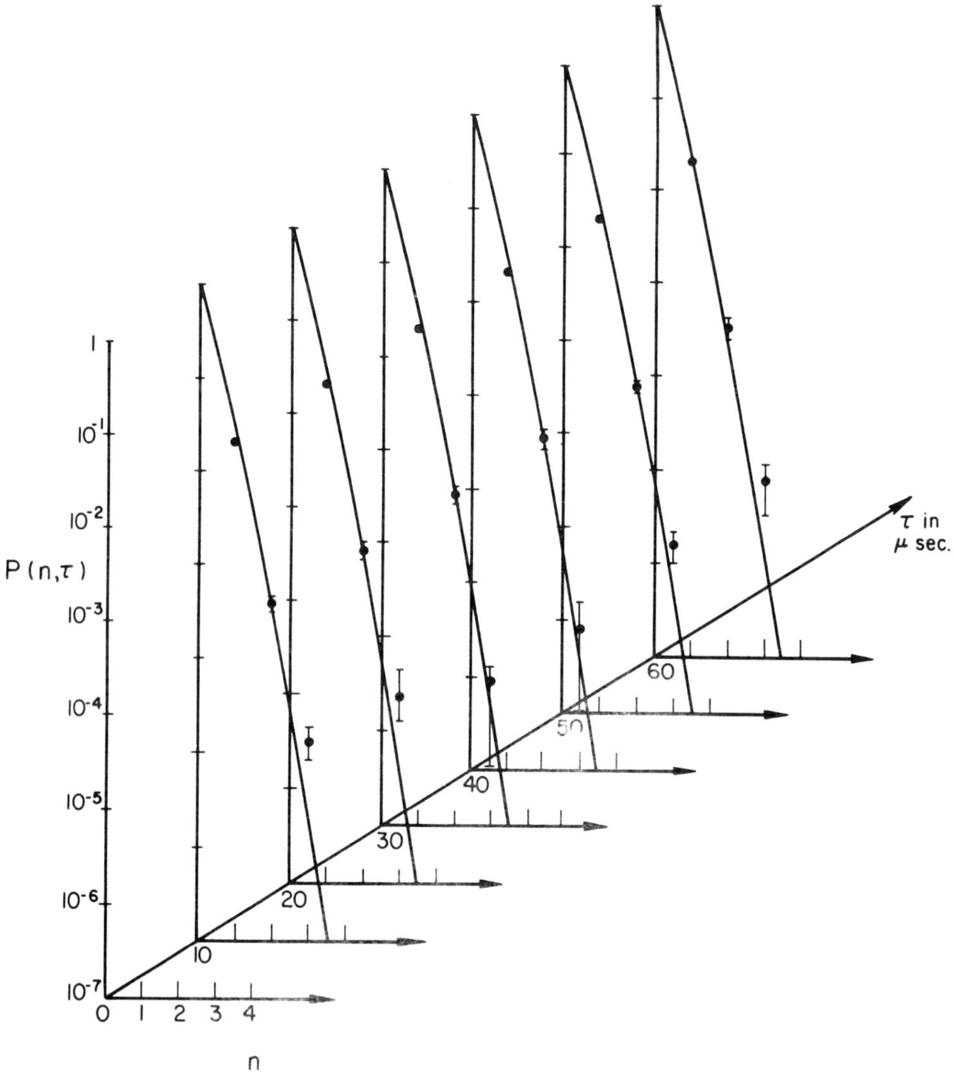

Fig. 2 Experimental values for the counting probability $p(n,\tau)$ for 6 different values of delay τ. The full curves are Poisson distributions of parameter $\langle n \rangle = RT$. The critical energy accumulation time τ_c was 20 ± 1.8 μsec.

time. Of course, it is always possible to argue that we have not adequately accounted for all changes of electron energy in treating the transition from the initial to the final electron state. Our experimental results evidently have no bearing on this question. But, within the context of the usual descriptions of the electron and of the field, it appears that the classical energy flow condition (3) is violated.

References

1. See for example L.I. Schiff, *Quantum Mechanics* 3rd Ed. (McGraw-Hill, New York, N.Y., 1955) p. 397.

2. L. Mandel, E.C.G. Sudarshan and E. Wolf, Proc. Phys. Soc. (London) *84*, 435 (1964).

3. W.E. Lamb, Jr. and M.O. Scully, *Polarization, Matter and Radiation. Jubilee Volume in honor of Alfred Kastler* (Presses Universitaires de France, Paris, France, 1969) p. 363.

4. E. Marx and K. Lichtenecker, Ann. d. Phys. *41*, 124 (1913).

5. E. Meyer and W. Gerlach, Ann. d. Phys. *45*, 177 (1928).

6. E.O. Lawrence and J.W. Beams, Phys. Rev. *32*, 478 (1928).

7. C.E. Tyler, *The Element of Time in the Photoelectric Effect*, Ph.D. Thesis, Washington University (1969).

8. See for example, Landau and Lifshitz, *Non-Relativistic Quantum Mechanics* (Pergamon Press, London, 1958) 1st ed., Section 21, p. 69.

9. See for example, L. Mandel and E. Wolf , Rev. Mod. Phys. *37*, 231 (1965).

NON-LINEAR RADIATION REACTION

R. K. Bullough

University of Manchester Institute of Science and Technology, England

1. Introduction: operator equations of motion

In a series of papers Jaynes and co-workers (Jaynes and Cummings, 1963; Crisp and Jaynes, 1969; Stroud and Jaynes, 1970) and others (e.g. Stroud, Eberly, Lama and Mandel, 1972) have looked at the effect of a radiation reaction field of the type first considered by Lorentz

$$\underline{E}_{\text{self}}(t) = \frac{2}{3c^3}\dddot{\underline{P}}(t) \tag{1.1}$$

reacting on an oscillating dipole $\underline{P}(t)$. Rather less work has been done on the *operator* self-field

$$\underline{e}_{\text{self}}(t) = \frac{2}{3}\frac{e}{c^3}x_{os}\,\hat{\underline{u}}\,\dddot{\sigma}_x(t) \tag{1.2}$$

although Series (1969) for example has investigated some consequences of this. We shall be very much concerned with this operator field in this paper and the notation is already in the form we shall use: ex_{os} is the matrix element of the dipole operator ex between two atomic states $|s\rangle$ and $|0\rangle$ so that $ex_{os} = e\langle 0|x|s\rangle$: $\hat{\underline{u}}$ is a unit vector in the direction of x. We consider spinless 1-electron atoms so that $x_{os} = x_{so}$. As usual e is the charge on the electron: c is the velocity of light *in vacuo*. The operator $\sigma_x(t)$ is the operator part of the dipole operator in second quantized notation and Heisenberg representation. We confine attention to 2-level atoms so that $\sigma_x(t)$ proves to be a Pauli spin operator. This paper may, nevertheless, be said to be in small part an examination of the role of the photon propagator

$$\underline{\underline{F}}(\underline{x},\underline{x}';\ t-t') = i\hbar^{-1}\ [\underline{e}(\underline{x},t),\underline{e}(\underline{x};t')] \tag{1.3}$$

in non-linear optical theory. The quantities $\underline{e}(\underline{x},t)$ are electric field operators, which here evolve as free field operators, but the commutator is a c-number because

$$\underline{e}(\underline{x},t) = i\ \sqrt{\frac{2\pi\hbar}{V}} \sum_{\underline{k},\lambda} \hat{\underline{\epsilon}}_{\underline{k},\lambda}\omega_k^{\frac{1}{2}}\ \{a_{\underline{k},\lambda}(t)e^{i\underline{k}.\underline{x}} - a^{+}_{\underline{k},\lambda}(t)e^{-i\underline{k}.\underline{x}}\ \}; \tag{1.4}$$

$\lambda = 1$ or 2 is a polarization index; $\hat{\underline{\epsilon}}_{\underline{k},\lambda}$ is a unit polarization vector orthogonal to $\underline{k}$; $\omega_k \equiv ck$ and V is the quantizing volume. This field operator is a linear combination of boson annihilation and creation operators and the commutator $\underline{\underline{F}}$ must be a c-number. It is obviously a second rank tensor.

In fact (Jordan and Pauli, 1928)

$$\underline{\underline{F}}(\underline{x},\underline{x}';t-t') = (\nabla\nabla - c^{-2}\underline{\underline{U}}\partial^2/\partial t^2)\{\delta(t-t'-rc^{-1})r^{-1} - \delta(t-t'+rc^{-1})r^{-1}\ \} \tag{1.5}$$

in which $\underline{\underline{U}}$ is the unit tensor and $r = |\underline{x}-\underline{x}'|$. If (1.3) is to be a 'causal' propagator it must be multiplied by $\theta(t-t'):\theta(x)=1, x>0;\ = 0,\ x<0$. In this case (1.3) is the 'retarded commutator' of $\underline{e}(\underline{x},t)$ and only the term in $\delta(t-t'-rc^{-1})r^{-1}$ is retained in the curly bracket in (1.5).

We look at the operator self-field with this c-number propagator in mind. Evidently we are concerned with all modes of the field although a single mode can still dominate a particular situation: the all mode limit nevertheless raises profound problems in the interpretation of self-fields like (1.1) or (1.2). We are also concerned with the limit $V\to\infty$, but this does not eliminate the possibility of treating a merely finite region of space containing matter interacting with ambient radiation or the vacuum.

This paper is concerned with non-linear radiation reaction in a comprehensive sense. It implicitly embraces the 'neo-classical' one-atom spontaneous emission theory of Jaynes *et al.* since the expectation value of (1.2) is the semiclassical (or classical) self-field (1.1). It includes Dicke's (1954) theory of superradiance. It embraces recent extensions by Rehler and Eberly (1971) and Agarwal (1970, 1971a, 1971b) of the superradiant all mode problem as well as the more recent work (Stroud, Eberly, Lama and Mandel, 1972). In a different way Bonifacio and Preparata (1970) treat the many-atom one-mode superradiant problem to which (1.2) does not apply. However a field like (1.2) is inapplicable more because the system studied is a closed system rather than because the theory is restricted to a single mode. It is also interesting to note that the master equation we shall obtain within self-field theory (see equation (2.29b) below) is identical with that obtained by Bonifacio, Schwendimann and Haake (1971a,b) (see also Bonifacio and Schwendimann 1970) but the reasons for this must be discussed elsewhere. All of this work is implicitly or explicitly in rotating wave approximation (r.w.a.). Exact solutions of the r.w.a. Hamiltonian by Jaynes and Cummings (1963), Tavis and Cummings(1968), Walls (1970) for a single mode and by Walls and Barakat (1970), Swain (1972) and by my own group are therefore also relevant.

The essence of this work is that N atoms (supposedly in a region small compared with a cubic wavelength) are effectively placed on a single site. An important implicit feature of this paper is that whilst superradiant effects (essentially the enhancement of radiation rates by a factor N) occur for isolated regions of these dimensions they do not occur in extended dielectrics. This is already apparent in linear theory (Bullough,1970) where we become concerned with the refractive index: in this theory the damping from radiation reaction is necessarily linear. In non-linear theory we become concerned with phenomena like self-induced transparency (McCall and Hahn, 1967, 1969; Bullough 1971; Bullough and Ahmad, 1971): if radiation reaction is included in this theory the damping becomes essentially non-linear.

We start first from a Hamiltonian for N 2-level atoms labelled by i on an arbitrary but fixed configuration of sites $\underline{x}_i$. We work in a pseudo-spin description with spin densities

$$\sigma_x(\underline{x}) \equiv \sum_{i=1}^{N} \sigma_1^{(i)} \delta(\underline{x}-\underline{x}_i) , \qquad \sigma_y(\underline{x}) \equiv \sum_{i=1}^{N} \sigma_2^{(i)} \delta(\underline{x}-\underline{x}_i) ,$$

$$\sigma_z(\underline{x}) \equiv \sum_{i=1}^{N} \sigma_3^{(i)} \delta(\underline{x}-\underline{x}_i)$$ in Schrödinger representation. The

Pauli matrices $\sigma_\alpha^{(i)}$ (α = 1,2,3) satisfy commutation relations

$\underline{\sigma}^{(i)} \times \underline{\sigma}^{(j)} = 2i\, \delta_{ij} \underline{\sigma}^{(j)}$: the spin densities satisfy

$\underline{\sigma}(x) \times \underline{\sigma}(\underline{x}') = 2i\, \delta(\underline{x}-\underline{x}')\, \underline{\sigma}(\underline{x})$. In dipole approximation the

Hamiltonian is

$$H = \hbar\sum_{\underline{k},\lambda}\omega_k(a^{\dagger}_{\underline{k},\lambda}a_{\underline{k},\lambda} + \tfrac{1}{2}) + \tfrac{1}{2}\hbar\omega_s\sum_{\underline{x}}\sigma_z(\underline{x})$$

$$-i\left(\frac{2\pi\hbar}{V}\right)^{\frac{1}{2}}\sum_{\underline{x},\underline{k},\lambda}\omega_k^{\frac{1}{2}}ex_{os}\,\underline{u}.\hat{\underline{\varepsilon}}_{\underline{k},\lambda}\sigma_x(\underline{x})(a_{\underline{k},\lambda}e^{i\underline{k}.\underline{x}} - a^{\dagger}_{\underline{k},\lambda}e^{-i\underline{k}.\underline{x}})$$

(1.6)

The summation over $\underline{x}$ is a formal one to be interpreted as an integration. It is important to note that

$$[a_{\underline{k},\lambda},\sigma_{x,y,z}(\underline{x})] = [a^{+}_{\underline{k},\lambda},\sigma_{x,y,z}(\underline{x})] = 0 \quad (1.7a)$$

so that $\underline{e}$ commutes with all matter operators $\underline{\sigma}$:

$$[\underline{\sigma},\underline{e}] = \underline{\underline{0}} \quad (1.7b)$$

It is convenient to move into Heisenberg representation in which it follows immediately that

$$\dot{a}_{\underline{k},\lambda}(t) = +i\hbar^{-1}[H,a_{\underline{k},\lambda}] = -i\omega_k a_{\underline{k},\lambda}(t) + \hbar^{-1}\sum_{\underline{x}}C(\underline{k},\lambda)\sigma_x(\underline{x},t)e^{-i\underline{k}.\underline{x}}$$

$$\dot{a}^{+}_{\underline{k},\lambda}(t) = +i\hbar^{-1}[H,a^{\dagger}_{\underline{k},\lambda}] = +i\omega_k\, a^{\dagger}_{\underline{k},\lambda}(t) + \hbar^{-1}\sum_{\underline{x}}C(\underline{k},\lambda)\sigma_x(\underline{x},t)e^{+i\underline{k}.\underline{x}}$$

(1.8)

in which $C(\underline{k},\lambda) \equiv \sqrt{\left(\frac{2\pi\hbar}{V}\right)}ex_{os}\,\hat{\underline{u}}.\hat{\underline{\varepsilon}}_{\underline{k},\lambda}\sqrt{\omega_k}$. A Fourier transformation on time t to frequency ω (e.g. $a_{\underline{k},\lambda}(t) \rightarrow \bar{a}_{\underline{k},\lambda}(\omega)$) yields

$$\bar{a}_{\underline{k},\lambda}(\omega) = i \sum_{\underline{x}} \frac{C(\underline{k},\lambda)\sigma_x(\underline{x},\omega)e^{-i\underline{k}.\underline{x}}}{\hbar(\omega - \omega_k)}$$

$$\bar{a}^{+}_{\underline{k},\lambda}(\omega) = i \sum_{\underline{x}} \frac{C(\underline{k},\lambda)\sigma_x(\underline{x},\omega)e^{+i\underline{k}.\underline{x}}}{\hbar(\omega + \omega_k)} \tag{1.9}$$

and (1.4) becomes

$$\underline{e}(\underline{x},\omega) = -\sum_{\underline{k},\lambda,\underline{x}'} \frac{2\pi\omega_k}{V} ex_{os}\ \hat{\underline{u}}.\hat{\underline{\varepsilon}}_{\underline{k},\lambda}\hat{\varepsilon}_{\underline{k},\lambda}\left[\frac{e^{+i\underline{k}.(\underline{x}-\underline{x}')}}{\omega - \omega_k} - \frac{e^{-i\underline{k}.(\underline{x}-\underline{x}')}}{\omega+\omega_k}\right]\sigma_x(\underline{x}',\omega). \tag{1.10}$$

This inverts to

$$\underline{e}(\underline{x},t) = \underline{e}_o(\underline{x},t) + \int_{-\infty}^{t}\int \tilde{\underline{\underline{F}}}(\underline{x},\underline{x}';t-t').\hat{\underline{u}}\ ex_{os}\sigma_x(\underline{x}',t')d\underline{x}'dt' \tag{1.11}$$

in which $\underline{e}_o(\underline{x},t)$ is a free field. It can only be given by (1.4) in which $a_{\underline{k},\lambda}(t)$ and $a^{\dagger}_{\underline{k},\lambda}(t)$ continue to evolve as free field operators so that $a_{\underline{k},\lambda}(t) = a_{\underline{k},\lambda}(0)e^{-i\omega_k t}$ with the adjoint relation for $a^{\dagger}_{\underline{k},\lambda}(t)$.

The propagator $\tilde{\underline{\underline{F}}}$ proves to be

$$\tilde{\underline{\underline{F}}}(\underline{x},\underline{x}';t-t') \equiv (\nabla\nabla - \nabla^2\underline{\underline{U}})\ \delta(t-t'-rc^{-1})\ r^{-1} \quad . \tag{1.12}$$

We assume a causal condition by interpreting ω as $\omega+i\delta$ ($\delta>0$). This propagator is transverse (div $\tilde{\underline{\underline{F}}} = 0$). This is because the Hamiltonian (1.6) omits the dipole-dipole Coulomb interaction. We shall include this correctly by replacing (1.12) by the causal part of (1.5) namely

$$\underline{\underline{F}}(\underline{x},\underline{x}';t-t') \equiv (\nabla\nabla - c^{-2}\underline{\underline{U}}\ \partial^2/\partial t^2)\ \delta(t-t'-rc^{-1})r^{-1} \tag{1.13}$$

This is not the place to consider this step in more detail. Certainly $\underline{e}(\underline{x},t)$ now satisfies Maxwell's wave equation in the operator form

$$\text{curl curl } \underline{e}(\underline{x},t) - c^{-2}\partial^2/\partial t^2 \underline{e}(\underline{x},t) = 4\pi e x_{os}\underline{\hat{u}}\partial^2/\partial t^2 \sigma_x(\underline{x},t). \quad (1.14)$$

There is no problem in obtaining a Bloch equation from the Hamiltonian (1.6). As long as $\underline{e}(\underline{x},t)$ commutes with the matter operators $\sigma_x(\underline{x},t)$, Heisenberg's equations of motion are equivalent to

$$\dot{\underline{\sigma}}(\underline{x},t) = \underline{\omega}(\underline{x},t) \times \underline{\sigma}(\underline{x},t) \quad (1.15a)$$

or

$$\dot{\sigma}_x(\underline{x},t) = - \omega_s \sigma_y(\underline{x},t)$$

$$\dot{\sigma}_y(\underline{x},t) = + \omega_s \sigma_x(\underline{x},t) - \omega_1(\underline{x},t) \sigma_z \underline{x},t)$$

$$\dot{\sigma}_z(\underline{x},t) = + \omega_1(\underline{x},t) \sigma_y(\underline{x},t) \quad (1.15b)$$

The vector operator density $\underline{\omega}(\underline{x},t)$ is

$$\underline{\omega}(\underline{x},t) \equiv (\omega_1(\underline{x},t), 0, \omega_s)$$

$$\omega_1(\underline{x},t) \equiv -2ex_{os} \hbar^{-1} \underline{\hat{u}}.\underline{e}(\underline{x},t) \quad . \quad (1.15c)$$

Evidently only the component ω_1 depends on $(\underline{x},t)$.

From (1.14) we have

$$\underline{e}(\underline{x},t) = \underline{e}_0(\underline{x},t) + \int_{-\infty}^{t}\int_V \underline{F}(\underline{x},\underline{x}');t-t').\underline{\hat{u}}\ ex_{os}\sigma_x(\underline{x}',t')d\underline{x}'dt'. \quad (1.16)$$

As long as the free field operator is ineffective, as it can be if it only operates on an initial state which is the photon vacuum state, (1.16) replaces the effective field operator by a linear functional of a matter operator: to this extent equations (1.15) are now non-linear integro-differential operator equations involving only the matter operators. The operators $\underline{\sigma}$ and the total field operator $\underline{e}$ act in a Hilbert space which is the direct product of two spaces spanned respectively by free matter states alone and free field states alone. The result (1.16) is equivalent, as we show later, to projection from the product space to the sub-space spanned by free matter states. This projection carries with it difficulties which, however, are partly consequent on the all mode limit which is embodied in the self-fields (1.2) and (1.1).

The result (1.16) appears to be unsatisfactory since $\underline{e}(\underline{x},t)$ and $\underline{\sigma}(\underline{x}',t')$ commute at t=t'=0 and hence at all equal times t=t'. However, this is not necessarily incompatible with (1.16) since $\sigma_{\underline{x}}(\underline{x}',t')$ is evaluated at all retarded times t=t'. Nevertheless, we now show that there is a very real problem associated with the self-field in the all mode limit: the commutation relations (1.7) are broken and the theory is inconsistent. Nesbet (1971) has realized that some difficulty surrounds a result like (1.16); but he has failed to understand the precise character of the difficulty and his 'semi-quantized' theory is mistaken (whilst his expression analogous to (1.16) is surely incorrectly stated)†.

The result (1.16) applies in the all mode limit: at the site $\underline{x}_i$ of the ith atom the field is

$$\underline{e}(\underline{x}_i t) = \underline{e}_o(\underline{x}_i,t) + \underline{e}_{int}(\underline{x}_i,t) + \int_{-\infty}^{t}\int_{V} \underline{\underline{F}}(\underline{x}_i,\underline{x}';t-t')).\hat{\underline{u}}\ ex_{os}\sigma_1^{(i)}(t')\delta(\underline{x}'-\underline{x}_i)d\underline{x}'dt'. \tag{1.17a}$$

The inter-atomic field $\underline{e}_{int}$ at $\underline{x}_i$ is just

$$\underline{e}_{int}(\underline{x}_i,t) = \int_{-\infty}^{t}\int_{V} \underline{\underline{F}}(\underline{x}_i,\underline{x}';t-t').\hat{\underline{u}}\ ex_{os}\sum_{j\neq i}\sigma_1^{(j)}(t')\delta(\underline{x}'-\underline{x}_j)d\underline{x}'dt' \tag{1.17b}$$

† Compare the papers by Lama and Mandel at this Conference.

which causes no difficulties as long as no $\underline{x}_j$ coincides with $\underline{x}_i$. The self-field of atom i is the last term in (1.17a). It evidently involves the integral

$$\int_V \underline{\underline{F}}(\underline{x},\underline{x}' ; t-t')\delta(\underline{x}-\underline{x}')d\underline{x}' \tag{1.18}$$

the convergent part of which is arguably $\frac{2}{3c^3}\delta'''(t-t')\,\underline{\underline{U}}$. Another, possibly clearer, argument for this self-field starts from (1.10), integrates this on $\delta(\underline{x}-\underline{x}')$ and discards divergent integrals $\int_0^\infty k^2dk$ and $\int_0^\infty dk$. The first integral is a self-dipole-dipole interaction and is spurious. The second integral is associated with the kinetic electron mass renormalization of Bethe (1947) and with the Lamb shift of Stroud and Jaynes (1970) and will be looked at again. Either way we now find

$$\underline{e}_{self}(\underline{x}_i,t) = \frac{2}{3c^3}\,\hat{\underline{u}}\;ex_{os}\,\dddot{\sigma}_1^{(i)}(t) \tag{1.19}$$

which (with $\sigma_1^{(i)} \rightarrow \sigma_x$) is the operator self-field (1.2).

The result (1.19) breaks the commutation relation (1.7). For (1.7) is to apply for any number of atoms N and so applies also for N=1 with a single atom placed at $\underline{x}_i$. The interatomic field (1.17b) vanishes and the self-field (1.19) together with the free field $\underline{e}_o(\underline{x},t)$ constitute the total field. The free field commutes with the matter operators at t=0 (and only then) but the self-field does not vanish there. This evolves as a Heisenberg operator with the full Hamiltonian H. We can evaluate commutators $[\underline{e}_{self},\sigma_2^{(i)}(t)]$ and $[\underline{e}_{self},\sigma_3^{(i)}(t)]$ from Heisenberg's equations of motion for the third derivative $\dddot{\sigma}_1^{(i)}(t)$ using the 1-atom Hamiltonian H assuming the commutation relations (1.7). We find the commutator with $\sigma_2^{(i)}(t)$ vanishes but that with $\sigma_3^{(i)}(t)$ does not. This is inconsistent and the commutation relations (1.7) cannot apply. The all mode limit which leads to (1.19) therefore breaks a constant of the motion.

The causal condition assumed at (1.12) is equivalent to radiating boundary conditions and the choice of an open system. The inconsistency problem is not avoided, however, by rejecting causality and closing the system. For a closed system in the all mode limit we find the propagator

$$\underline{\underline{F}}_p(\underline{x},\underline{x}';t-t') \equiv (\nabla\nabla - c^{-2}\underline{\underline{U}}\partial^2/\partial t^2)\tfrac{1}{2}[\delta(\tau-rc^{-1})r^{-1}+\delta(\tau+rc^{-1})r^{-1}] \tag{1.20}$$

with $\tau \equiv t - t'$. The self-field (1.19) certainly vanishes for all t but it can be formally extended to include a field which is also implicit in (1.20) namely

$$\underline{e}_{self}(\underline{x}_i,t) = \frac{2}{3}\frac{\hat{\underline{u}}}{c^3} ex_{os}\ddot{\sigma}_1^{(i)}(t)\{\frac{2}{\pi}\int_0^\infty dk\} . \tag{1.21}$$

This is the operator form of the dipolar self-field of Stroud and Jaynes (1970). The integral is convergent for an extended atom, but this field also breaks the commutation relations. These constants of the motion are therefore broken by the all mode limit. We examine yet another operator self-field in the appendix (Appendix 2) to this paper. This field correctly describes the conventional Bethe (1947) form of the non-relativistic Lamb shift and is analogous to (1.21). It does not avoid the problem created by the breakdown of the commutation relations. This problem is certainly avoided by considering a finite number of modes, but solutions are now oscillatory for a closed system and it is not clear whether or how to break the constants of the motion to create an open system. The step from (1.18) to (1.19) has been used successfully in optical scattering theory (Bullough et al. 1968). We shall therefore investigate some of the consequences of operator self-fields like (1.19) and (1.21) in this paper.

2. Generalized Non-linear Radiation Reaction and Superradiance

The enhanced radiation rate which is the characteristic feature of superradiance relies on all N atoms occupying a region small compared with a wavelength. We therefore consider the case of N atoms occupying the same single site $\underline{x}_o$ but work with the non-linear operator equations (1.15). In this case $\sigma_x(\underline{x},t) = \{\sum_{i=1}^N \sigma_1^{(i)}(t)\}\,\delta(\underline{x}-\underline{x}_o)$ for example. Both atomic self-fields and interatomic fields are evaluated at this single point $\underline{x}_o$ so that the *total* field is by (1.19) just

$$\underline{e}(t) = \frac{2}{3}\frac{\hat{\underline{u}}}{c^3} ex_{os}\sum_{i=1}^N \dddot{\sigma}_1^{(i)}(t) + \underline{e}_o(t)$$

$$\equiv \frac{2}{3}\frac{\hat{\underline{u}}}{c^3} ex_{os}\dddot{\sigma}_x(t) + \underline{e}_o(t) . \tag{2.1}$$

We shall assume for the moment that it is possible simply to omit the free field. In any case the density function $\delta(\underline{x}-\underline{x}_o)$

factors throughout the Bloch equations (1.15) which are equivalent without free field to

$$\ddot{\sigma}_x(t) + \Gamma_o \omega_s^{-2} \dddot{\sigma}_x(t)\, \sigma_z(t) + \omega_s^{2} \sigma_x(t) = 0$$

$$\sigma_z(t) = \sigma_z(t_o) + \Gamma_o \omega_s^{-4} \int_{t_o}^{t} \dddot{\sigma}_x(t')\dot{\sigma}_x(t')\,dt'$$

$$\sigma_y(t) = - \omega_s^{-1} \dot{\sigma}_x(t).$$

(2.2)

This system of equations is a non-linear operator equation generalization of Lorentz's damped oscillator equation. The quantity Γ_o is the Einstein A-coefficient: $\Gamma_o \equiv \frac{4}{3} \frac{e^2 x_{os}^2 \omega_s^3}{\hbar c^3}$

Because this is a system of operator equations and the field (2.1) breaks the commutation relations (1.7), there is a difficulty in placing the field operator e, that is the operator $\omega_1(t) \equiv - 2ex_{ox}\hbar^{-1} e(t)$ in its proper order in the two operator products $\omega_1(t)\, \sigma_z(t)$ and $\omega_1(t)\, \sigma_y(t)$ in (1.15). We have noted that this difficulty is peculiar to the case of an infinity of field modes: a functional of $\sigma_x(t)$ is replaced by the field (2.1) which depends on $\dddot{\sigma}_x(t)$. We first place the field operator (2.1) before the matter operators σ_z and σ_y but we shall be obliged to review this prescription shortly.

Equations (2.2) linearize by replacing $\sigma_z(t)$ by its initial value $\sigma_z(t_o)$. They then have expectation values

$$\langle\ddot{\sigma}_x\rangle + \Gamma_o \omega_s^{-2} \langle\dddot{\sigma}_x\rangle\, (N_+ - N_-) + \omega_s^{2} \langle\sigma_x\rangle = 0$$

$$\langle\sigma_z(t)\rangle = \langle\sigma_z(t_o)\rangle = N_+ - N_-$$

$$\langle\sigma_y(t)\rangle = - \omega_s^{-1} \langle\dot{\sigma}_x(t)\rangle \quad . \qquad (2.3)$$

These equations assume the system starts in (or arrives at) a number state with N_+ atoms in the upper state and N_- in the lower at an initial (or final) time $t = t_o$. Therefore both $\langle\sigma_x(t_o)\rangle = 0$ and $\langle\sigma_y(t_o)\rangle = 0$ and these initial conditions mean $\langle\sigma_x(t)\rangle = 0$ throughout the motion (at least in r.w.a.) unless t_o is a singular point of the equations. Singular points occur at $t = \pm\infty$ and by the device of setting $t_o = \mp\infty$ we find within r.w.a. the alternative solution that

$$\langle\sigma_x(t)\rangle = Ae^{m\Gamma_o t} \cos(\omega_s t + \delta) \tag{2.4}$$

where $m \equiv \frac{1}{2}(N_+ - N_-)$ and A, δ are *arbitrary* parameters. The solution grows from $t = -\infty$ if $m > 0$ and damps toward $t = +\infty$ if $m < 0$. This singular type of solution underlies the semi-classical limit of the superradiance problem and the spontaneous emission theory of Jaynes et al.

The choice $\sigma_z(t) = \sigma_z(t_o)$ linearizes the operator equations and converts the theory to the theory of superradiant oscillators. The operator character of $\sigma_z(t_o)$ still changes the solutions in a significant way however. For example, with $t_o \equiv 0$ and certainly not a singular point, actual operator solutions within r.w.a. are

$$\sigma_+(t) = \sigma_+(o)\, e^{\frac{1}{2}\sigma_z(o)}\, e^{i(\omega_s t + \delta)}$$

$$\sigma_-(t) = e^{\frac{1}{2}\sigma_z(o)}\, \sigma_-(o)\, e^{-i(\omega_s t + \delta)} \tag{2.5}$$

providing we are careful to preserve the adjoint character of the equations for $\sigma_\pm \equiv \frac{1}{2}(\sigma_x \pm i\sigma_y)$. These equations are

$$\sigma_+ + \Gamma_o \omega_s^{-2}\, \dddot{\sigma}_+\, \sigma_z(0) + \omega_s^2\, \sigma_+ = 0$$

$$\ddot{\sigma}_- + \Gamma_o \omega_s^{-2}\, \sigma_z(o)\, \dddot{\sigma}_- + \omega_s^2\, \sigma_- = 0 \tag{2.6}$$

which means placing the negative and positive frequency parts of (2.1) in different orders with respect to $\sigma_z(o)$ in the two equations. This is not the prescription we first adopted of placing e before σ_z but there is one good reason for it. The free field was eliminated from the operator equations (2.2) simply by dropping it. However, providing the initial state is always a product state of matter state and *the photon vacuum state* it may be eliminated in the following way: the free field is split into positive and negative frequency parts $e_o^{\pm}(t)$: within the adjoint preserving prescription $e_o^{(+)}$ and $e_o^{(-)}$ must be placed in an adjoint relationship in the equations (2.2) but there are two ways of doing this: but if in addition $e_o^{(-)}$ is placed to the left of matter operators and $e_o^{(+)}$ to the right these free field parts act respectively on the bra and ket vacuum states and annihilate. This prescription not only preserves adjoint relationships but it also places operators in normal order: we refer to it as the normal ordering prescription henceforth. The consequences of the other prescription in which e precedes both σ_y and σ_z in equations (1.15) are analyzed in an appendix (Appendix 1) to this paper. This other prescription does not have the nice feature of automatically eliminating the free field.

The same situation obtains in the non-linear theory. We shall now get exact solutions† of the non-linear N atom all mode superradiant problem. We first look for solutions of the Bloch equation (1.15) coupled by the field (2.1) in the form

$$\sigma_{\pm}(t) \equiv \tfrac{1}{2}(\sigma_x(t) \pm i\,\sigma_y(t)) \equiv R_{\pm}(t)\, e^{\pm i\omega_s t}$$

$$\sigma_z(t) \equiv R_3(t) \quad . \tag{2.7}$$

We assume $R_{\pm}(t)$ and $R_3(t)$ are slowly varying operators and work in r.w.a. which rejects terms $e^{\pm 2i\omega_s t}$. Providing we preserve the adjoint relations between operators throughout the equations and assume that the initial state contains the photon vacuum state so that the free field is annihilated by full implementation of the normal ordering prescription we can reach

† This general solution which is 'exact' with r.w.a. is due to my student R. Saunders. It was found independently of Agarwal's (1970) solution with which however it coincides. The argument uses only Heisenberg's equations of motion and the field (2.1) and is very different from Agarwal's.

$$\dot{R}_+ = \tfrac{1}{2}\,\Gamma_o\, R_+ R_3 , \qquad \dot{R}_- = \tfrac{1}{2}\,\Gamma_o\, R_3\, R_-$$

$$\dot{R}_3 = -\ 2\,\Gamma_o R_+\, R_- \quad . \tag{2.8}$$

These are exactly the results we reach within rotating wave approximation by taking the Hamiltonian (1.6) in that approximation and placing all atoms at $\underline{x}_o$. This r.w.a. Hamiltonian is

$$H = \hbar \sum_{\underline{k},\lambda} \omega_k (a^+_{\underline{k},\lambda}\, a_{\underline{k},\lambda} + \tfrac{1}{2}) + \tfrac{1}{2}\hbar\omega_s\, \sigma_z$$

$$-i \sum_{\underline{k},\lambda} C\ (\underline{k},\lambda)\,[a_{\underline{k},\lambda}\, \sigma_+ - a_{\underline{k},\lambda}{}^+\, \sigma_-\,] \tag{2.9}$$

with the definitions (2.7). The coupling constant $C(\underline{k},\lambda)$ is defined below (1.8). The total spin operator

$$\hat{J}^2 \equiv \sigma_z{}^2 + 2\ (\sigma_+\sigma_- \ + \sigma_-\ \sigma_+\ )$$

$$= R_3^2 + 2\ (R_+R_- \ + R_-R_+\ ) \tag{2.10}$$

now commutes with H,so total spin is a constant of the motion providing this is not broken by the effective field e. This field again proves to be given by (2.1): it has the negative and positive frequency parts

$$\underline{e}^{(\mp)}(t) = \frac{2}{3}\, \frac{e x_{os}}{c^3}\ \hat{\underline{u}}\ \dddot{\sigma}_\pm(t) \ + \underline{e}_o{}^{(\mp)}(t) , \tag{2.11a}$$

$$= \mp\, i\, \frac{2}{3}\, \frac{e x_{os}}{c^3}\ \omega_s{}^3\, \hat{\underline{u}}\, \sigma_\pm\ (t) + \underline{e}_o{}^{(\mp)}\ (t) \quad . \tag{2.11b}$$

The step from (2.11a) to (2.11b) is both consistent and necessary within r.w.a.

We now find easily that for *any* matter operator A(t)

$$\dot{A}(t) = -\tfrac{1}{2}i\omega_s[A(t),R_3(t)] - \tfrac{1}{2}\Gamma_0\{[A(t),R_+(t)]R_-(t)$$

$$-R_+(t)[A(t),R_-(t)]\} \quad . \qquad (2.12)$$

The field terms (2.11b) are placed, by prescription, in an adjoint normally ordered relationship in the two terms within the curly bracket. Equations (2.8) now follow immediately by choosing A(t) successively as $R_+(t)e^{i\omega_s t}$, $R_-(t)e^{-i\omega_s t}$ and $R_3(t)$.

The important equation in $\dot{R}_3$ in (2.8) is equivalent to

$$\dot{R}_3 = -\tfrac{1}{2}\Gamma_0\,[\hat{J}^2 - R_3^{\,2} + 2R_3\,] \qquad (2.13)$$

providing the commutation relations for $R_\pm$ and R_3 hold: these are derived from those for $\sigma_\pm$ and $\sigma_z \equiv R_3$ and are

$$[R_\pm,\ R_3] = \pm 2\,R_\pm\ , \quad [R_+,\ R_-\,] = R_3 \qquad (2.14)$$

as usual. Total spin $\hat{J}^2$ is not a constant of the motion described by (2.8) so that the question of failure of the commutation relations is non-trivial. Haken (1970), for example, has considered this problem in essentially the same context.

For one atom (2.13) is easily solved: we can observe that $\hat{J}^2 = 3I$ and $R_3^{\,2} = I$ (in which I is the unit operator) are constants of the motion if such constants hold. (These results apply at t=0 and hence for all time; but the difficulty of course is that (2.11) means that the motion is not governed by (2.9). Then we can set

$$\hat{N} \equiv \tfrac{1}{2}(R_3 + I) \qquad (2.15)$$

to get

$$\hat{N}(t) = \hat{N}(0)\,e^{-\Gamma_0 t} \qquad (2.16)$$

The operator $\hat{N}$ is the number operator for the 1 atom upper state: a variation substitutes for R_3 in (2.13) and uses $\hat{N}^2 = \hat{N}$ which is a constant of the motion. Then

$$\frac{d\hat{N}}{dt} = -\tfrac{1}{4}\Gamma_o\,[\hat{J}^2 - 3\,I - 4\,\hat{N}^2 + 8\,\hat{N}\,]$$

$$= -\Gamma_o\,\hat{N} \quad . \tag{2.17}$$

This variation generalizes to the Dicke problem of N 2-level atoms at the same site $\underline{x}_o$. The non-linear operator equation (2.13) applies here just as well as it did for 1 atom. By choosing number operators like $\hat{N}$ the operator equation is linearized exactly. For two atoms we can find the diagonal elements of the density matrix

$$|C_{1,1}|^2 = e^{-3/2\Gamma_o t}\,\{\cos\tfrac{1}{2}\sqrt{7}\Gamma_o t - {}^{3}/\sqrt{7}\sin\tfrac{1}{2}\sqrt{7}\Gamma_o t\ \}$$

$$|C_{1,o}|^2 = e^{-3/2\Gamma_o t}\,\{\ 8/\sqrt{7}\ \sin\tfrac{1}{2}\sqrt{7}\Gamma_o t\ \}$$

$$|C_{1,-1}|^2 = 1 - e^{-3/2\,\Gamma_o t}\,\{\cos\tfrac{1}{2}\sqrt{7}\Gamma_o t + 5/\sqrt{7}\ \sin\tfrac{1}{2}\sqrt{7}\Gamma_o t\ \}\ ;$$

$$|C_{o,o}|^2 = 1\ . \tag{2.18}$$

The density matrix is calculated with basis states $|r,m\rangle$ which are Dicke's (1954) states: $\tfrac{1}{4}\hat{J}^2|r,m\rangle = r(r+1)\,|r,m\rangle$, $\tfrac{1}{2}R_3(o)|r,m\rangle = m|r,m\rangle$: r is Dicke's co-operation number, $m \equiv \tfrac{1}{2}(N_+ - N_-)$ where $N_\pm$ refer to numbers of atoms in their upper (lower) states at $t=0$. The initial state at $t=0$ is $|r,m\rangle$ and the diagonal elements are labelled $|C_{r,m}|^2$.

These elements are not non-negative. We have checked the method of calculation for N atoms on the same site coupled with constant C to a single resonant mode. The equation of motion for R_3 proves to be

$$\ddot{R}_3 + C^2\hbar^{-2}[\hat{J}^2 - 3R_3^2 + 2(2\hat{M} + I)R_3] = 0 \tag{2.19}$$

in which $\hat{J}^2$ is total spin and $\hat{M}$ is a second constant of the motion. (This constant appears explicitly in (2.34) below). The equation has a simple oscillatory solution for N=1 which agrees with the eigenvalues for this problem obtained by Jaynes and Cummings (1963). It is intrinsically non-linear for N=2 but may be exactly linearized by substituting for number operators appropriate to the 2-atom problem as in the variant leading to (2.17) above. The solutions are again oscillatory and agree with the exact solution of the N-atom-Q-mode Dicke problem we report briefly below. Thus the peculiar results (2.18) again illustrate the problem surrounding the field operators (2.1) and (2.11) which depend on the all mode limit.

For N=1 the operator solution (2.16) agrees with Weisskopf-Wigner (1930) theory. This theory is equivalent to assuming the r.w.a. Hamiltonian (2.9) together with an outgoing condition as we show below. From the operator equation for R_+ in (2.8) we find

$$\dot{R}_+ = \tfrac{1}{2}\Gamma_o R_+ \{2\hat{N}(0)e^{-\Gamma_o t} - I\} \tag{2.20}$$

and with $R_+(0)\hat{N}(0) = 0$, $R_+(t) = R_+(0)\, e^{-\frac{1}{2}\Gamma_o t}$. If the atom starts in its upper state at t=0 the density matrix is diagonal throughout the motion: it is

$$\begin{bmatrix} e^{-\Gamma_o t} & 0 \\ 0 & 1-e^{-\Gamma_o t} \end{bmatrix} \tag{2.21}$$

This is incoherent in the sense that there is no phase information: phase information is carried by off-diagonal elements of the density matrix. The result (2.21) agrees exactly with Weisskopf-Wigner theory providing that theory is assigned a normalized spectrum. The Weisskopf-Wigner spectrum (essentially $\frac{\Gamma_o}{2\pi}\frac{\omega^3}{\omega_s^3}\frac{1}{(\omega_s-\omega)^2+\frac{1}{4}\Gamma_o^2}$) is normalizable if but only if $\omega^3\omega_s^{-3}$ is set equal to unity.

This one atom theory can be extended to include the Lamb shift.

In rotating wave approximation the self-field operator can be expressed as

$$-i\,\frac{2}{3}\,\frac{ex_{os}}{c^3}\,[\omega_s{}^3(1 + \tfrac{1}{2}\,i\,K)\,R_+(t)\,e^{+i\omega_s t}] + \text{c.c.} \tag{2.22}$$

where

$$K \equiv [\frac{2c}{\pi\omega_s}\int_0^\infty dk + \frac{2}{\pi}\log_e(\omega_c/\omega_s)] \tag{2.23}$$

The factor one half multiplying K in (2.22) is a feature of r.w.a. An interesting situation surrounds this factor and this is analyzed in the appendix (Appendix 2). Here we note only that the integral $\frac{2}{\pi}\frac{c}{\omega_s}\int_0^\infty dk \equiv K_o$ is the Stroud-Jaynes (1970) integral whilst the logarithm is the source of Bethe's (1947) form of the non-relativistic Lamb shift. In order to obtain either term in convergent form we need to use a cut-off: the logarithm in (2.23) is already cut off at the Compton wavelength of the electron which is equivalent to cut-off at $\omega = \omega_c \equiv ck_c \equiv m_e c^2\hbar^{-1}$ where m_e is the electron mass. This is Bethe's (1947) cut-off and the integral K_o must then be identified as his kinetic electron mass renormalization term†. However the factor one half in (2.22) now means that the energy *spacing* is shifted by only one half of the total Bethe shift. The point here is that both the atomic ground state of energy E_o and the excited state energy E_s should receive Bethe shifts identical in magnitude but opposite in sign. In the appendix it is shown how this situation can be described within the present theory: in essence if the complete rotating wave approximation is not made two integrals K_o cancel and two logarithmic terms add. On the other hand within semi-classical (or neo-classical) theory everything is based directly on the c-number propagator $\underline{F}$ of (1.13) and the consequence is that two integrals K_o add whilst two logarithmic terms subtract. This is the origin of the Jaynes-Stroud (1970) neo-classical Lamb shift; the theory is equivalent to using the operator self-field (1.2) extended by(1.21).

For the present we remain in r.w.a. and examine the Lamb shift in this case. With the self-field (2.22) the equations (2.8) become

---

† Integrals $\frac{2}{\pi}\,\frac{c^3}{\omega_s{}^3}\int_0^\infty k^2dk$ and $\frac{2}{\pi}\,\frac{c^2}{\omega_s{}^2}\int_0^\infty kdk$ also contribute to K: the first integral is part of the dipolar Coulomb interaction and is spurious; the second is the 'quadratic divergence' assignable to the square of the vector potential concealed in H.

$$\dot{R}_{\pm} = \tfrac{1}{2}(\Gamma_o \pm i\,B_o)\,R_{\pm}\,R_3$$

$$\dot{R}_3 = -\,2\,\Gamma_o\,R_+\,R_- \;;\quad B_o \equiv \tfrac{1}{2}K\,\Gamma_o \qquad (2.24)$$

and K is defined by (2.23). The r.w.a. to $\sigma_{\pm}(t)$ is therefore

$$\sigma_{\pm}(t) = R_{\pm}(0)\; e^{-\frac{1}{2}\Gamma_o t}\; e^{+i(\omega_s-\Delta\omega_s)t} \qquad (2.25)$$

where $\Delta\omega_s$ is exactly the Bethe level shift for a single level

$$\Delta E_B = \hbar\,\Delta\omega_s = \frac{2}{3\pi}\;\frac{e^2 x_{os}{}^2}{c^3}\,\text{-}\;\omega_s{}^3\,\log_e\left(\frac{\omega_c}{\omega_s}\right) \qquad (2.26)$$

as long as the integral K_o is dropped. This result can be obtained from Louisell's (1964) formulation of Weisskopf-Wigner theory: it is precisely one half of the shift expected for the energy spacing as we have noted and the error is a consequence of the r.w.a. The appendix exhibits this explicitly but cannot, of course, eliminate inconsistencies from the argument.

The key point, however, is that the solution of the non-linear operator equations is the 'linear' Weisskopf-Wigner result. On the other hand equations (2.24) decorrelate to yield the Crisp-Jaynes-Stroud non-linear neo-classical result

$$\langle R_{\pm}(t)\rangle = \tfrac{1}{2}\text{sech}\,\tfrac{1}{2}\Gamma_o t\; e^{\pm i(\omega_s t+\xi(t))}$$

$$\langle R_3(t)\rangle = -\tanh\tfrac{1}{2}\,\Gamma_o\,t \qquad (2.27)$$

where $\xi(t) = \frac{1}{2}K\log_e\cosh\frac{1}{2}\Gamma_o t$. The result is identically theirs if the integral K_o is retained, the logarithm discarded and $\frac{1}{2}K(=\frac{1}{2}K_o)$ then replaced by K $(=K_o)$. The factor two comes in from the neoclassical photon propagator $\underline{\underline{F}}$ as we have explained. The dynamical aspect of the Lamb shift comes in from the decorrelation, the characteristic feature of neo-classical theory, however.

The N atom problem preserves all the features of the one atom problem and we now look at this. Since (2.12) is true for all matter operators A(t) the equation of motion for the density operator is as follows: Set $A(t) = \rho_{(m,n)}(t)$ the Heisenberg operator which has matrix elements

$$\langle r,m' | \rho_{(m,n)}(o) | r,n' \rangle = \delta_{mm'} \delta_{nn'} \tag{2.28}$$

at t=0. The total spin r is supposed a constant of the motion. Since $[\rho_{(m,n)}, R_3] = 2(m-n)\, \rho_{(m,n)}$ at $t = 0$ and we suppose this commutation relation a constant for $t > 0$ we find the Heisenberg operator equation

$$\dot{\rho}_{(m,n)} = -\tfrac{1}{2}\Gamma_o \{\rho_{(m,n)} R_+ R_- - 2R_+ \rho_{(m,n)} R_- + R_+ R_- \rho_{(m,n)}\} . \tag{2.29a}$$

Since the expectation value of $\rho_{(m,n)}$ is the (n,m) -matrix element of the density matrix it then follows that

$$\dot{\rho} = -\tfrac{1}{2}\Gamma_o \{R_+ R_- \rho - 2R_- \rho R_+ + \rho R_+ R_-\} \tag{2.29b}$$

is the equation of motion for the density operator in Schrödinger representation.

Equation (2.29b) is just the master equation for the normally ordered density operator obtained by Agarwal (1970). He used normal ordering projection techniques on the Liouville equation, projecting from the matter-photon Hilbert space to a matter Hilbert sub-space. This projection, equivalent to tracing over all photon states, is here actually executed by the step in which $\underline{e}$ is related to σ_x by (1.11) or (1.16) providing the free field plays no role. This eliminates the total field operator from the theory whether all atoms occupy the same site or not (whilst in the case of an extended dielectric there is the additional feature of an operator optical 'Extinction Theorem' to eliminate the free field operator). Furthermore Agarwal's normal ordering condition is equivalent to the normal ordering prescription we adopted after (2.6). It is interesting to note in addition that (2.29b) is the master equation obtained by Bonifacio et al. (Bonifacio, Schwendimann and Haake 1971 a,b) and used by Haake and Glauber (1972) providing the characteristic time Γ_o^{-1} and the explicit form of $R_\pm$ are re-interpreted. The reasons for this curious situation must be considered elsewhere however.

The master equation (2.29b) includes the N-atom generalized Lamb shift within r.w.a. in the form

$$\dot{\rho} = -\tfrac{1}{2}\{(\Gamma_0 + i\,B_0)\rho\, R_+R_- + (\Gamma_0 - i\,B_0)\, R_+R_-\rho - 2\, R_-\rho\, R_+\} \quad . \tag{2.30}$$

The density matrix with basis states $|r,m>$, $|r,n>$ satisfies

$$\dot{\rho}_{mn} = \rho_{m+1,n+1}\Gamma_0 \{f(n)\ f(m)\}^{\frac{1}{2}}$$

$$- \rho_{m,n}\ \tfrac{1}{2}\{(\Gamma_0 - i\,B_0)\ f(n-1) + (\Gamma_0 + i\,B_0)\ f(m-1)\} \tag{2.31}$$

in which

$$f(m) \equiv (r-m)\ (r+m+1)$$

$$f(m-1) \equiv (r+m)\ (r-m+1) \quad . \tag{2.32}$$

Notice that this equation only connects elements on the same diagonal stripe. This is consistent with the fact that if the system starts in a number state the density matrix remains diagonal throughout the motion.

These equations have an exact analytic solution. Our solution is that obtained by my student R. Saunders: it apparently agrees with that found by Agarwal (1970) but is perhaps more compact. Simple examples of the case in which the N-atom system starts in the state $|r,r>$ where $r = \frac{1}{2}N$ are (with the notation $A_j(t) \equiv \rho_{r-j\ r-j}(t)$):

2 atoms, $r = 1$: $A_0(t) = e^{-2\Gamma_0 t}$

$$A_1(t) = 2\Gamma_0 t\, e^{-2\Gamma_0 t}$$

$$A_3(t) = 1 - e^{-2\Gamma_0 t} - 2\,\Gamma_0 t\ e^{-2\Gamma_0 t}$$

3 atoms, $r = \frac{3}{2}$: $A_0(t) = e^{-3\Gamma_o t}$

$$A_1(t) = 3\,[e^{-3\Gamma_o t} - e^{-4\Gamma_o t}]$$

$$A_2(t) = 12\, e^{-3\Gamma_o t}[\Gamma_o t - 1 + e^{-\Gamma_o t}]$$

$$A_3(t) = 1 + e^{-3\Gamma_o t}\,[8 - 12\,\Gamma_o t - 9e^{-\Gamma_o t}]$$

4 atoms, $r = 2$: $A_0(t) = e^{-4\Gamma_o t}$

$$A_1(t) = 2\,[e^{-4\Gamma_o t} - e^{-6\Gamma_o t}]$$

$$A_2(t) = 6\, e^{-4\Gamma_o t}\,[1 - e^{-2\Gamma_o t}(2\,\Gamma_o t + 1)\,]$$

$$A_3(t) = 36e^{-4\Gamma_o t}\,[e^{-2\Gamma_o t}\,(1 + \Gamma_o t) + \Gamma_o t - 1]$$

$$A_4(t) = 1 - \sum_{j=0}^{3} A_j(t)$$

5 atoms, $r = \frac{5}{2}$: $A_0(t) = e^{-5\,\Gamma_o t}$

$$A_1(t) = \frac{5}{3}\,[e^{-5\Gamma_o t} - e^{-8\Gamma_o t}\,]$$

$$A_2(t) = \frac{1}{60}\,\left[\,\frac{e^{-5\Gamma_o t}}{12} - \frac{e^{-8\Gamma_o t}}{3} + \frac{e^{-9\Gamma_o t}}{4}\right]$$

$$A_3(t) = \;.\;.\;.$$

etc.

$$(2.32)$$

Whilst this general solution includes partial solutions reported by others (e.g. by Dillard and Robl, 1969) and, in contrast with (2.18) is certainly positive indefinite on the main diagonal, it may still not be a totally correct interpretation of the equations (note again that since the equations reject divergent terms in the self-fields, and since the constants of the motion and commutation relations break down there can be no correct answer to the superradiance problem). Another 'exact' solution can be obtained as follows.

We work in r.w.a. with N atoms on the same site and the number of field modes restricted to Q only. In principle we can take the limit $Q \to \infty$ if this limit exists; but in practice it cannot do so because of the divergence problem when both $Q \to \infty$ and the wave number k is unbounded. Nevertheless a physically meaningful limiting procedure may exist. The Hamiltonian is (2.9): it is cleaner now to label modes $(\underline{k},\lambda)$ by the index j : $1 \leq j \leq Q$. The key observation is that there are two constants of the motion: total spin $\hat{J}^2$ as in (2.10) and the 'total number' operator

$$\sum_{j=1}^{Q} a_j^+ a_j + \tfrac{1}{2}\sigma_z \equiv \hat{M} . \tag{2.34}$$

If both $\frac{1}{4}\hat{J}^2$ and $\hat{M}$ are definite with eigenvalues $r(r+1)$ and M, any state vector is

$$|\psi(t)\rangle = \sum_{m=-r}^{+r} \sum_{\{n_i\}} C_{r,m\,\{n_i\}}(t)\, |r,m\rangle\, |\{n_i\}\rangle \tag{2.35}$$

where the set of integers $\{n_i\}$ including $n_i = 0$ label the photon states and the condition $\sum_{i=1}^{Q} n_i + m = M$ on the double sum applies.

The numbers r,M are constants of the motion which define a Hilbert sub-space to which the motion is necessarily confined. Since M bounds the photon numbers n_i, the sub-space is of finite dimension and even small if there are few atoms and Q, the number of modes is small. The equations of motion for the $C_{r,m,\{n_i\}}(t)$ are soluble by Laplace transforms and the results involve the eigenvalues λ_i of H. Indeed the $C_{r,m,\{n_i\}}(t)$ are linear combinations of $e^{-i\lambda_i \hbar^{-1} t}$ as we can expect. Since H is Hermitian the λ_i are real and the matter-photon system oscillates. If the λ_i are irrational the common period may be large but for spontaneous emission and true superradiance we need a dense set of eigenvalues and therefore an infinite number of (plane wave) modes.

Figures 1, 2, 3 and 4 illustrate results for two simple cases and are self-explanatory. We are here primarily interested in the all-mode limit: this limit is easy for the 1 atom problem.

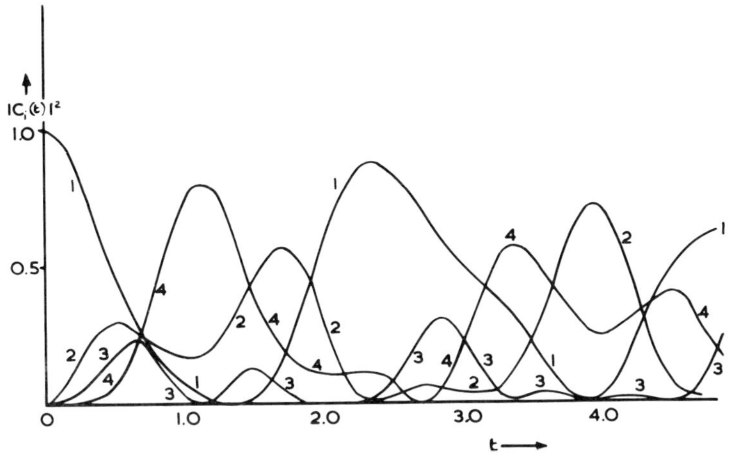

Fig. 1 Dicke problem: method based on equations (2.34) and (2.35). Occupation numbers $|C_{r,m}(t)|^2 \equiv |C_{r,m\ \{r-m\}}(t)|^2$ for the case of 3 atoms, one resonant mode, no initial photons (N=3, r = 3/2, M = 3/2). Key: $|C_{3/2,3/2}|^2$=1, $|C_{3/2,1/2}|^2$ =2, $|C_{3/2,-1/2}|^2$=3, $|C_{3/2,-3/2}|^2$ =4.

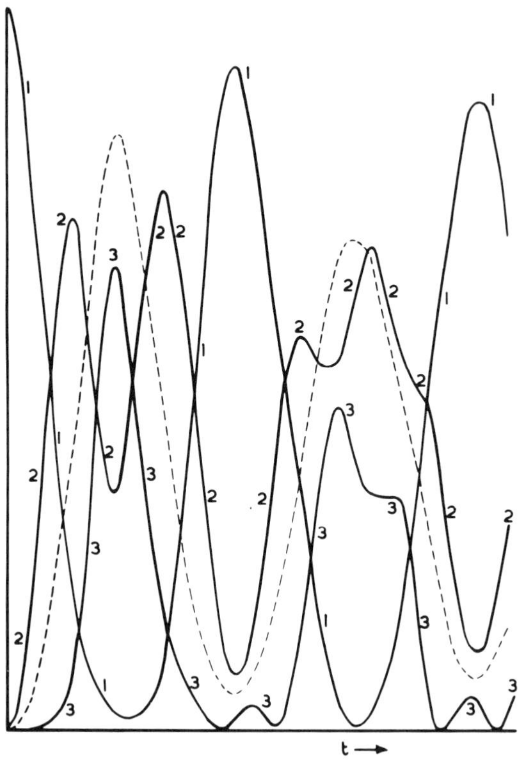

Fig. 2 Dicke problem: method based on equations (2.34) and (2.35). Occupation numbers for highest and lowest states with total occupation number for the intermediate states in the case of 6 atoms, one resonant mode, no initial photons (N = 6, r = 3, M = 3). Key: $|C_{3,3}|^2 = 1$, $\sum_{m=-2}^{+2} |C_{3,m}|^2 = 2$, $|C_{3,-3}|^2 = 3$, mean photon no./max. photon no.=4= ----- .

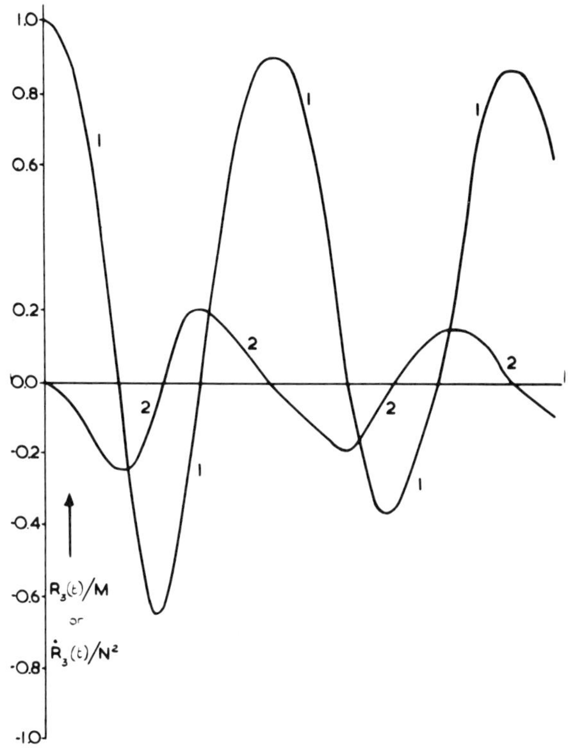

Fig. 3 Dicke problem as in Fig. 2 R_3 (curve 1) and $\dot{R}_3$ (curve 2) scaled by M and N^2 respectively (N=6, r=3, $M=3^3$).

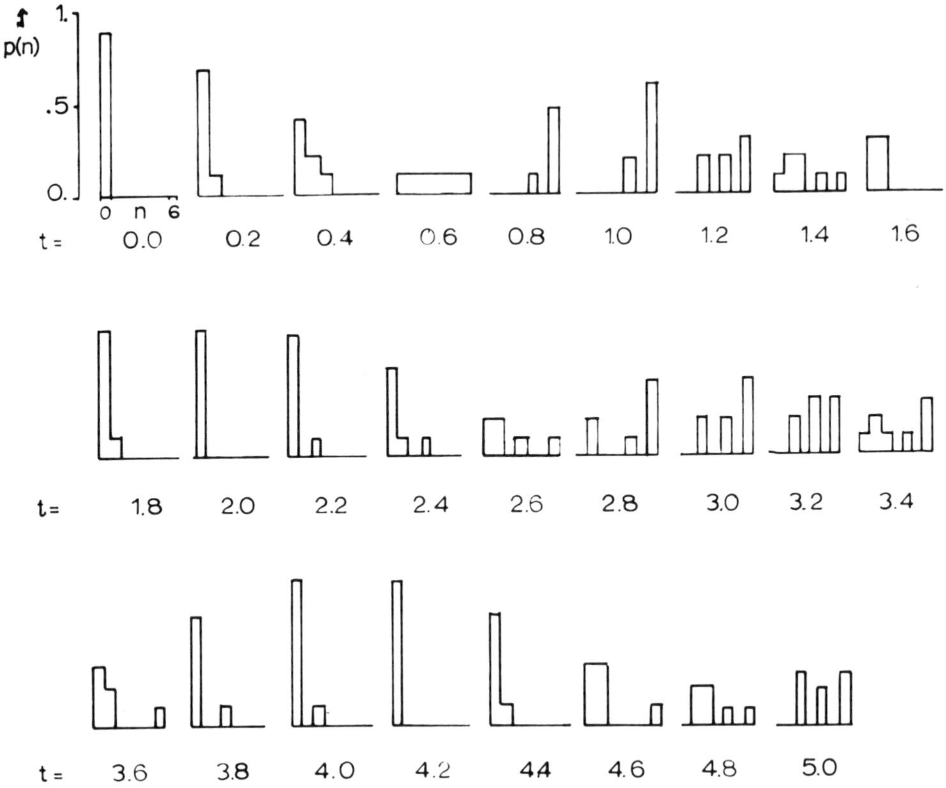

Fig. 4 The probability p(n) of finding n photons at time t for N =6, r = 3, M = 3.

By usual methods we can pick up a complex eigenvalue from

$$\tfrac{1}{2}\hbar\omega_s - \lambda = \frac{2}{3\pi} e^2 x_{os}^2 \, c \int_0^\infty \frac{k^3 dk}{\omega_k - \omega_s - i\gamma} = \hbar\, \Delta\omega_s + \tfrac{1}{2} i\hbar\, \Gamma_0 \quad . \qquad (2.36)$$

The frequency shift $\Delta\omega_s$ is exactly that in (2.26). Indeed it includes the Coulomb interaction and can be formally reduced to contain the integral K of (2.23) where the leading term retained is $\frac{\omega_s}{c} K_0 \equiv \frac{2}{\pi} \int_0^\infty dk$. Unfortunately we have so far found no simple limiting procedure for the 2 or more atom problems. We have noticed that the solutions in (2.38) can be found by but only by *approximation* to these all mode limits; but it seems probable that the approximation is no worse than dropping terms $\Gamma_0 \omega_s^{-1}$ which appear in the self-fields: these appear because the self-field (2.11a) must be converted to the self-field (2.11b) within r.w.a. and this approximation is additional to restricting the Hamiltonian (1.6) to the r.w.a. form (2.9). The case N=1 is a special case: the relation between the 2-atom 'exact' solution (2.18) and the 1-atom solution contained in (2.17) shows this again in a different way.

A problem of some interest surrounds the constant of the motion (2.34) (this constant also applies to the r.w.a. Hamiltonian with distinct atomic sites). The elements of the matter density matrix are (with $|\psi(t)\rangle$ given by (2.35)) just the trace over photon states

$$\sum_{\{n_i\}} \langle \{n_i\} | \langle r,m|\psi(t)\rangle\langle\psi(t)|r,n\rangle | \{n_i\} \rangle \qquad (2.37)$$

Since $\sum_{i=1}^{Q} n_i + m = M$, this trace exists only if m = n. The matter density matrix is therefore diagonal for all motions and all times for every N and Q and this result is consistent with the earlier results, like (2.21) for 1 atom, in the all mode limit. Evidently the density matrix governed by (2.31) will have off-diagonal elements if, but only if, the system starts at the initial time with off-diagonal elements; this is so because (2.31) only connects elements on the same diagonal. But an off-diagonal element of the density matrix at the initial time is excluded by the condition (2.34) since (2.37) still applies.

All the problems considered in this section assume an initial state which is a product state of a matter number state and a photon number state. This makes M definite. The operators $\sigma_\pm$

which appear in the Hamiltonian (2.9) have no matrix representations for *any* Hilbert sub-space which has M definite: thus there is no motion of the type considered where $\sigma_\pm$ have non-zero matrix representations and these operators are arguably zero. Of course σ_z, $\sigma_+\sigma_-$, $\sigma_+ a_j$, $a_j^+\sigma_-$ and $a_j^+ a_j$ (j labels the modes) have matrix representations and are well defined. A correct solution of the operator equations confined to these motions therefore appears to be

$$R_+ = R_- = 0, \qquad R_3 \neq 0. \tag{2.38}$$

This result raises a problem in the semi-classical limit.

The Dicke (1954) 'semi-classical limit' is obtained as the 'neo-classical' result of decorrelating equations (2.8). A solution is then

$$\langle R_\pm^{(i)}(t)\rangle \equiv \overline{R}_\pm(t), \qquad \langle R_3^{(i)}(t)\rangle \equiv \overline{R}_3(t)$$

independent of i with

$$\overline{R}_\pm(t) = \tfrac{1}{2}\,\mathrm{sech}\,\tfrac{1}{2}N\,\Gamma_o t\; e^{\pm i(\omega_s t+\delta)}$$

$$\overline{R}_3(t) = -\ \tanh \tfrac{1}{2}N\,\Gamma_o t \quad . \tag{2.39}$$

With Dicke basis and in Schrödinger representation.

$$|\psi(t)\rangle = \sum_{m=-r}^{+r} C_{r,m}(t)\,|\,r,m\rangle \quad . \tag{2.40}$$

Not all the off-diagonal elements of the density matrix $C^*_{r,m\pm1}(t)\,C_{r,m}(t)$ can vanish for all times t if $\overline{R}_\pm(t)$ are given by (2.39).

The 'semi-classical limit' (2.39) apparently breaks the constant of the motion (2.34), makes the operator $\hat{M}$ indefinite, and samples a dense set of its eigenvalues. Thus this 'limit' is not the mathematical limit of any sequence of solutions of the Dicke problem which has r.w.a. Hamiltonian (2.19) and in which the initial states are number states. If the semi-classical limit state is a limit of a sequence of coherent states as suggested by Bonifacio, Kim and Scully (1969), for example, the semi-classical limit (2.39) is not the limit as $N \to \infty$ of N atom Dicke problems and it is not the limit as $N \to \infty$ of any problems studied in this section. One conclusion is that superradiance must evidently be incoherent unless the system is coherently 'prepared' by the choice of initial conditions: another is that both the nature of semi-classical limits (which are coherent) and the emergence of coherent optical processes themselves need further work for their complete understanding.

3. Conclusion

The main results of the theory are: (i) the operator equations of motion for N coupled 2-level atoms are non-linear: in particular matter operators satisfy a non-linear operator Bloch equation but the field and dipole operators couple the linear operator Maxwell equations; (ii) for single atoms, or for N atoms on the same site, the operator equations become equivalent to three coupled non-linear operator equations one of which is a direct generalization of Lorentz's equation of motion for an oscillator damped by its own reaction field; (iii) the solution of these equations is nevertheless a linear type solution: in particular for one atom, and within the conventional robust attitude to the divergence problem, the Weisskopf-Wigner solution to the spontaneous emission problem is the correct one whilst the level shift is the Bethe shift; (iv) nevertheless decorrelation of the operator equations yields solutions of the Jaynes-Crisp-Stroud type for 1 atom and Dicke's semi-classical superradiant solution with $N\Gamma_o$ instead of Γ_o for N atoms.

The paper contains a number of exact solutions of superradiant problems the majority worked within the terms of operator self-field theory: it solves the problem of N atoms on the same site and Q modes and the problem of N atoms on the same site and all modes and presents several alternative solutions of these. It provides a comprehensive theory of N atom Lamb type level shifts both within second quantization and within the 'neo-classical' theory. It provides a field operator approach to the several master equations which have been reported by others.

The paper does not explicitly treat the physically important case of an extended dielectric. However, here we have found, in every case we have examined, that coherence narrowing eliminates the superradiant feature $\Gamma_o \to N\Gamma_o$ in agreement with linear theory. Superradiance of this type is *not* a consequence of the non-linearity of the operator equations but is a consequence solely of superposing all atoms on or close to the same site. Consequently in a dielectric of dimensions small compared with a wavelength $\lambda_s \equiv 2\pi c \omega_s^{-1}$ some of the observed 'superradiant' effects can be classified as diffraction effects whilst in an extended dielectric incoherent superradiance is non-linear homogeneous broadening now reduced by coherence narrowing. The remaining part of the phenomenon is very much a problem in non-linear dielectric theory: in particular if the initial inversion is prepared by an incident pulse, this part of the problem is precisely a problem in non-linear pulse propagation whilst the non-linear homogeneous broadening can exhibit power narrowing in the amplifier with an opportunity for self-sustained oscillations.

A conclusion therefore is this: certainly the naive description of superradiance in which neighboring atoms cooperatively enhance the radiation rate from the system as a whole is relevant to small samples; but it has little relevance to macroscopically extended dielectrics.

Appendix 1: the normal ordering prescription

Another operator ordering prescription places e (or ω_1) before σ_z and σ_y in those two places in equations (1.15) where it appears. The equations analogous to (2.8) are

$$\dot{R}_\pm = \tfrac{1}{2}\Gamma_0 R_\pm R_3, \quad \dot{R}_3 = -\Gamma_0(R_+R_- + R_-R_+) \ . \tag{A.1}$$

The equations for $R_\pm$ are not adjoint and the solution is $R_+ = R_- = 0$ for all times $t \geqslant t_0$ where t_0 is the initial time. This is comparable with the result in r.w.a. that $R_\pm$ have no matrix representations (see p.28).

Although causality is already built into the propagator $\underline{\underline{F}}$ the equation for R_3 is not outgoing. An additional outgoing condition takes the negative frequency part (only) of the self-fields in conformity with Feynman's rules: the self-field on atom i becomes $\frac{4}{3}\frac{ex_{os}}{c^3}\dddot{\sigma}_+^{(i)}(t)$ by choosing the 'outgoing' propagator $\underline{\underline{F}}^{(-)}(\underline{x},\underline{x}';t-t')$ instead of $\underline{\underline{F}}$. This is related to $\underline{\underline{F}}$ defined in (1.13) by replacing $\delta(t-t'-rc^{-1})r^{-1}$ there by $\delta_-(t-t'-rc^{-1})r^{-1}$. This involves the negative frequency part of the δ-function in the definition

$$\delta_-(x) = \frac{1}{\pi}\int_{-\infty}^{0} e^{-i\omega x}\, d\omega \ . \tag{A.2}$$

The total field on atom i is now

$$\underline{e}(t) = \frac{2}{3c^3}\,\hat{\underline{u}}\; ex_{os}\; [2\,\dddot{\sigma}_+^{(i)}(t) + \sum_{j\neq i} \sigma_1^{(j)}(t)] \qquad (A.3)$$

Note that in the case of a single atom this is precisely *twice* the negative frequency part of the self-field in r.w.a. exhibited in (2.11). The equations (A.1) now become

$$\dot{R}_{\pm} = \pm \tfrac{1}{2}\, \Gamma_o \sum_i R_{\pm}^{(i)}\, R_3^{(i)} + \tfrac{1}{2}\Gamma_o\, R_{\pm} R_3$$

$$\dot{R}_3 = -\,2\,\Gamma_o\, R_+ R_- \qquad (A.4)$$

for N atoms, and these reduce to

$$\dot{R}_{\pm} = \Gamma_o\, R_+ R_3,\; \dot{R}_- = 0,\; \dot{R}_3 = -\,2\,\Gamma_o\, R_+ R_- \qquad (A.5)$$

for 1 atom. The solution $R_+ = R_- = 0$ remains unchanged but the equation for $\dot{R}_3$ is identical with that in (2.8) and all the solutions in §2 which occupy a number state in the course of their motion are still valid.

It was this prescription and solution which was described in the pre-Conference summary to this paper.

Appendix 2: the Lamb shift

The final form of this paper has benefited from discussions with J.H. Eberly before the Conference. Ackerhalt, Eberly and Knight (AEK) present an interesting paper at the Conference which argues in terms of self-fields outside r.w.a. and finally obtains the total Lamb shift for the energy spacing of the 2-level atom as twice the Bethe single level Lamb shift. This result agrees with that obtained by Bullough and Caudrey (1971) by related but certainly different methods but it does not agree with the results (2.25) and (2.26) obtained above which are in r.w.a. Nor does it agree with the result of using the operator self-field (1.21) which is outside r.w.a. This operator self-field yields the Jaynes-Stroud shift but there is no dynamical contribution unless the operator equations are decorrelated. In this comment I show the relationship between these two possible results outside r.w.a. by deriving both of them. The reason why two different results can be obtained for the same quantity is because the self-field theory of this paper is not internally consistent. However, the AEK self-field theory is not internally consistent either and I know of meaningful one which is.

In order to examine the situation in dipole approximation but outside r.w.a. we return to (1.9) with all atoms on the same site $\underline{x} = \underline{0}$. Then (1.9) reads

$$\bar{a}_{\underline{k},\lambda}(\omega) = \frac{i\, C(\underline{k},\lambda)\, \sigma_x(\omega)}{\hbar(\omega - \omega_k)} \tag{B.1a}$$

$$\bar{a}^{+}_{\underline{k},\lambda}(\omega) = \frac{i\, C(\underline{k},\lambda)\, \sigma_x(\omega)}{\hbar(\omega + \omega_k)} \tag{B.1b}$$

Fourier transforms of adjoint operators are not adjoint. In r.w.a. $\sigma_x(\omega)$ becomes $\sigma_-(\omega)$ in (B.1a) and $\sigma_+(\omega)$ in (B.1b). Equations (B.1) themselves are not in r.w.a., however, and we can proceed from them in at least two different ways. The two illustrated are these:

First: calculate the Fourier transform of the field (1.4) at $\underline{x} = \underline{0}$ by

$$\bar{\underline{e}}(\underline{0},\omega) = i\sqrt{\left(\frac{2\pi\hbar}{V}\right)} \sum_{\underline{k},\lambda} \hat{\varepsilon}_{\underline{k},\lambda}\omega_k^{1/2} \; \{ \bar{a}_{\underline{k},\lambda}(\omega) - \bar{a}^{+}_{\underline{k},\lambda}(\omega) \} \, . \tag{B.2}$$

The field is interpreted as a coupled Heisenberg operator and the result can only be written as

$$\bar{\underline{e}}(\underline{0},\omega) = \{\frac{4\pi}{V} \sum_{\underline{k},\lambda} \frac{k^2(\underline{\underline{U}} - \hat{\underline{k}}\hat{\underline{k}})}{k^2 - \omega^2 c^{-2}} \} \cdot \hat{\underline{u}}\; ex_{os}\; \sigma_x(\omega) \tag{B.3}$$

in which the curly bracket is written as the formal integral

$$\frac{4}{3\pi} \int_0^\infty k^4 dk \;\; [k^2 - \omega^2 c^{-2}]\; \underline{\underline{U}} \quad . \tag{B.4}$$

The scalar integral diverges as $\int^\infty k^2 dk$, a relic of the Coulomb interaction in dipole interaction. The remaining part is $-\frac{2}{3}\frac{\omega^2}{c^2} k_s K_o$ in which $k_s \equiv \omega_s c^{-1}$ since the logarithms from P-integrals at the next order cancel. It follows that the self-field causing the Lamb shift is given by (1.21).

Since $\ddot{\sigma}_x(t) = \ddot{\sigma}_+(t) + \ddot{\sigma}_-(t)$, there are two adjoint self-fields $\frac{2}{3c^2}{}^3k_s K_o\, ex_{os}\, {}^x\ddot{\sigma}_\pm(t)$. The shift is determined by the number K_0 defined below (2.23) and the commutation relations are broken whether the free field is included or not. In particular the free field commutes with the matter operators at t = 0 but the self-field which does not commute does not vanish there. It is only because the total field does not commute that we need two adjoint self-fields. The choice of these introduces an ambiguity.

Second: we calculate $\underline{e}^{(-)}(t)$ and $\underline{e}^{(+)}(t)$ as the Fourier transforms of $\overline{\underline{e}}^{(\mp)}(\omega)$ where from (B.1)

$$\overline{\underline{e}}^{(+)}(\omega) \equiv i\sqrt{\left(\frac{2\pi\hbar}{V}\right)}\sum_{\underline{k},\lambda} \hat{\underline{\varepsilon}}_{\underline{k},\lambda}\omega_k^{\frac{1}{2}}\, \overline{a}_{k,\lambda}(\omega) \qquad \text{(B.5a)}$$

$$\overline{\underline{e}}^{(-)}(\omega) \equiv -i\sqrt{\left(\frac{2\pi\hbar}{V}\right)}\sum_{\underline{k},\lambda} \hat{\underline{\varepsilon}}_{\underline{k},\lambda}\omega_k^{\frac{1}{2}}\, \overline{a}^{+}_{\underline{k},\lambda}(\omega)\ . \qquad \text{(B.5b)}$$

These two fields reduce to

$$\overline{e}^{(\mp)}(\omega) = \frac{2}{3}\ ex_{os}\ \left\{\frac{1}{\pi}\int_0^\infty \frac{k^3 dk}{k \pm \omega c^{-1}}\right\}\sigma_x(\omega)\ . \qquad \text{(B.6)}$$

If the leading term retained is in K_0 we get from the P-integral

$$e^{(\mp)}(t) \simeq \frac{2}{3}\, ex_{os}k_s^3\ \{\tfrac{1}{2} K_0\, \sigma_x(t) \pm i\, \tfrac{1}{2} K_1\, \sigma_y(t)\} \qquad \text{(B.7)}$$

where $K_1 \equiv \frac{2}{\pi}\log_e \frac{\omega_c}{\omega_s}$. We have cut-off at the Compton frequency. These fields with free field and damping fields added do not appear quite to coincide with the AEK fields. Nevertheless they yield the doubled Bethe shift.

Instead of (2.8) we get by the normal ordering prescription for placing the *fields* that

$$\dot{R}_+ + \tfrac{1}{2}\,\Gamma_o\,[(1 + \tfrac{1}{2}iK)\,R_+R_3 + \tfrac{1}{2}i\,(K_o - K_1)R_3R_+]$$

$$\dot{R}_- = \tfrac{1}{2}\,\Gamma_o\,[(1 - \tfrac{1}{2}iK)\,R_3R_- - \tfrac{1}{2}i\,(K_o - K_1)R_-R_3] \quad . \qquad (B.8)$$

For one atom $R_\pm R_3 = \mp R_\pm$ and $R_3\,R_\mp = \mp R_\mp$ so that

$$\dot{R}_\pm = -\tfrac{1}{2}\,\Gamma_o\,(1 \pm i\,K_1)\,R_\pm \qquad (B.9)$$

Then

$$\sigma_\pm\,(t) = R_\pm(0)\,e^{-\frac{1}{2}\Gamma_o t}\,e^{\pm i(\omega_s - 2\Delta\omega_s)t} \qquad (B.10)$$

in comparison with (2.25). Equation (2.26) still defines $\Delta\omega_s$ but now there is no need to drop the integral K_o which cancels exactly. The same line of argument is applicable to the equation of motion (2.12) and hence to the N-atom master equation (2.30) and I am indebted to a fruitful remark by G.S. Agarwal in this connection. It can also be checked that with the particular field which contains (B.7) the equation for $\dot{R}_3$ is still precisely $\dot{R}_3 = -2\,\Gamma_o R_+ R_-$.

We have also explored direct integration of the t-dependent first order differential equations equivalent to (B.1). The self-field so obtained demonstrably breaks the commutation relations (1.7) in but only in the all mode limit.

References

Agarwal, G.S., 1970, Phys. Rev. *A2*,2038.
Agarwal, G.S., 1971a, Phys. Rev. *A3*, 1783.
Agarwal, G.S., 1971b, Phys. Rev. *A4*, 1791.
Bethe, H.A., 1947, Phys. Rev. *72*, 339.
Bonifacio, R., and Preparata, G., 1970, Phys. Rev. *A2*, 336.
Bonifacio, R., Kim, D.M., and Scully, M.O., 1969, Phys.Rev. *187*,441.
Bonifacio, R., and Schwendimann, P., 1970, Nuovo Cimento Lett. *2*, 509, 512.
Bonifacio, R., Schwendimann, P., and Haake, F., 1971a, Phys.Rev. *A4*,302.
Bonifacio, R., Schwendimann, P., and Haake, F., 1971b, Phys. Rev. *A4*,454.
Born, M., and Wolf, E., 1959, *Principles of Optics* (London:Pergamon).
Bullough, R.K., 1970, *Physics of Quantum Electronics*, 1969, Optical Sciences Center, University of Arizona, Tech. Report No. 45.
Bullough, R.K., 1971, *Physics of Quantum Electronics*,1970, Optical Sciences Center, University of Arizona, Tech. Report No. 66.
Bullough, R.K., and Ahmad, F., 1971, Phys. Rev. Lett. *27*,330.
Bullough, R.K., and Caudrey, P.J., 1971, J. Phys. A: Gen. Phys.*4*, L 41.
Crisp, M.D., and Jaynes, E.T., 1969, Phys. Rev. *179*, 1253.
Dicke, R.H., 1954, Phys. Rev. *93*,99.
Dillard, M., and Robl, H.R., 1969, Phys. Rev. *184*,312.
Haake, F., and Glauber, R.J., 1972, Phys. Rev. *5*, 1457.
Haken, H., 1970, in *Quantum Optics* (S.M. Kay and A. Maitland, Eds.) (Academic Press: London) p.249 and references there.
Jaynes, E.T., and Cummings, F.W., 1963, Proc. I.E.E.E. *51*, 89.
Jordan, P., and Pauli, W., 1928, Zeit. Phys. *47*, 151.
Louisell, W.H., 1964, *Radiation and Noise in Quantum Electronics*, (McGraw Hill: New York) pp. 191-197.
McCall, S.L., and Hahn, E.L., 1967, Phys.Rev. Lett. *18*,908.
McCall, S.L., and Hahn, E.L., 1969, Phys. Rev. *183*, 457.
Nesbet, R.K., 1971, Phys. Rev. Lett. *27*, 553.
Rehler, N.E., and Eberly, J.H. 1971, Phys. Rev. *A3*, 1735.
Series, G.W., 1969, in *Optical Pumping and Atomic Line Shape*, ed. T. Skalinski, Panstwowe Wydawnictwo Namkowe, Warszwa pp. 25-41.
Stroud, Jr., C.R., and Jaynes, E.T., 1970, Phys. Rev. *A1*, 106.
Stroud, Jr., C.R., Eberly, J.H., Lama, W.L., and Mandel, L., 1972, Phys. Rev. *A3*, 1094.
Swain, S., 1972, J. Phys. A: Gen. Phys. *5*, L3.
Tavis, M., and Cummings, F.W., 1968, Phys. Rev. *170*, 379.

Walls, D.F., 1970, in *Quantum Optics* (S.M. Kay and A. Maitland,eds.) (Academic Press: London) p. 501.
Walls, D.F., and Barakat, R., 1970, Phys. Rev. *A1*, 446.
Weisskopf, V., and Wigner, E.P., 1930, Z. Phys. *63*, 59.

MASTER EQUATIONS IN THE THEORY OF INCOHERENT AND COHERENT SPONTANEOUS EMISSION [+]

G.S. Agarwal [++]

Universität Stuttgart, Stuttgart, Germany

The subject of spontaneous emission is a very old one. Recently this has received a whole new momentum because it has become possible to observe some of the peculiar phenomena associated with it. Of particular interest is the phenomenon of superradiance which was discovered by Dicke [1] in 1954. He found that the radiation rate from a collection of identical two-level atoms or molecules is, under certain circumstances depending on the excitation of the system, proportional to the square of the number of atoms. Dicke employed the second order perturbation theory to calculate the characteristics of the emitted radiation. We would, of course, like a description which enables us to calculate time dependent statistical properties.

The first treatment of the time dependent properties of the spontaneously emitted radiation was given by Weisskopf and Wigner [2]. It was then applied to a wide variety of problems by Weisskopf [3]. Since then the method has been improved upon and reformulated in modern language. The methods currently in use to study the spontaneous emission may be divided into the following categories:

(a) Quantum Electrodynamic Methods

[+] Invited paper given at "Third Rochester Conference on Coherence and Quantum Optics", June 21-23, 1972 Rochester, New York.

[++] On leave of absence from the Department of Physics and Astronomy University of Rochester, Rochester, New York 14627

(1) Weisskopf - Wigner [2]
(2) Heitler - Ma [4]
(3) Goldberger - Watson [5]
(4) Low [6]

(b) Neo-classical Method due to Jaynes and Coworkers [7]. The methods of Goldberger - Watson and of Low are also known as the "resolvant operator" and Green's function method respectively and are rather closely related to each other. The methods of Weisskopf-Wigner and Heitler - Ma have also much in common. The method of Low has been recently explored in great detail by Chang and Stehle [8]. All the above methods appear to have the drawback that they work very well if the number of atoms taking part in spontaneous emission is very few, say one or two. In fact very little has been done using such methods for the case of more than two atoms [9].

We have developed a method which allows us to calculate all the characteristics of the emitted radiation [10]. There is no restriction, in our treatment (which is fully quantum mechanical) on either the size of the system or the number of atoms. We have made use of the master equation techniques of non-equilibrium statistical mechanics [11], [12]. Methods similar in spirit, to that which I would describe now have been developed by Lehmberg [13] and by Bonifacio et al. [14].

The Hamiltonian for a collection of identical N two-level atoms interacting with a quantized radiation field is given by [15].

$$H = \omega\Sigma_1^N s_j^z + \Sigma_{\underline{ks}}\omega_{ks} a^+_{\underline{ks}} a_{\underline{ks}} + \Sigma_{i\neq j} V_{ij} s_i^+ s_j^-$$

$$+ \Sigma_{\underline{ks}j} \{g_{j\underline{ks}} a_{\underline{ks}} (s_j^+ - s_j^-) + \text{H.C.}\} \qquad (\hbar = 1), \tag{1}$$

where ω is the energy separation between two atomic levels, $s_j^{\pm z}$ are the components of the spin angular momentum operator (spin -½ value) and the coupling coefficient is given by

$$g_{j\underline{ks}} = - (\frac{i\omega}{c})(\frac{2\pi c}{L^3})^{\frac{1}{2}} \frac{1}{\sqrt{k}} (\underline{\varepsilon}_{ks} \cdot \underline{d} \quad e^{i\underline{k}\cdot\underline{R}_j} \tag{2}$$

Here $\underline{R}_j$ is the position of the jth atom and all other symbols have the usual meaning. In (1) V_{ij} is the dipole-dipole interaction and is given by

$$V_{ij} = \left\{ \underline{d}_i \cdot \underline{d}_j - \frac{3(\underline{d}_i \cdot \underline{R}_{ij})(\underline{d}_j \cdot \underline{R}_{ij})}{R_{ij}^2} \right\} \frac{1}{R_{ij}^3} \; . \tag{3}$$

Let $\rho_{A+R}(t)$ be the density operator for the combined atoms-field system and let $\rho_A(t)$ be the reduced density operator corresponding to the atomic system alone. $\rho_A(t)$ and $\rho_{A+R}(t)$ are related by

$$\rho_A(t) = Tr_R \rho_{A+R}(t) \; , \tag{4}$$

where Tr_R denotes the trace over the radiation field variables. The initial state of the system is given by

$$\rho_{A+R}(0) = \rho_A(0)\rho_R(0) \quad , \; \rho_R(0) = |\{0\}\rangle\langle\{0\}| \; , \tag{5}$$

where we leave the state of the atomic system quite arbitrary and where the field is in vacuum state. It is clear that the radiation field in vacuum state interacting with the atomic system behaves like a reservoir at zero temperature. The problem of spontaneous emission thus reduces to that of relaxation at zero temperature. We can now use the master equation techniques to obtain the master equation for $\rho_A(t)$. We introduce the projection operator P defined by

$$P \ldots = \rho_R(0) \; Tr_R \; \ldots \; , \tag{6}$$

and then

$$P\rho_{A+R}(t) = \rho_R(0) \; \rho_A(t) . \tag{7}$$

It is easy to show that $P\rho_{A+R}(t)$ satisfies the equation [11], [12]

$$\frac{\partial}{\partial t} [P\rho_{A+R}(t)] + iP\hat{L}(t)P\rho_{A+R}(t) + \int_0^t d\tau P\hat{L}(t)U(t,\tau)(1-P)\hat{L}(\tau)P\rho_{A+R}(\tau)$$

$$= 0 \quad , \tag{8}$$

where

$$U(t,\tau) = T\exp\{-i\int_{\tau}^{t} dt'(1-P)\hat{L}(t')(1-P)\},\quad \hat{L}(\tau) \equiv [H(\tau),\quad].$$

On substituting (1)-(3), (7) in (8) and on making the Born approximation, the Markovian approximation and the rotating wave approximation, we obtain the master equation

$$\frac{\partial\rho}{\partial t} = -i\omega_{\gamma}\Sigma_i [s_i^+ s_i^-, \rho] - i\,\Sigma_{i\neq j}(\Omega_{ij} + V_{ij})[s_i^+ s_j^-, \rho]$$

$$- \Sigma_{i,j}\,\gamma_{ij}(s_i^+ s_j^- \rho - 2s_j^- \rho s_i^+ + \rho s_i^+ s_j^-), \tag{9}$$

where we have suppressed the subscript A from ρ. In Eq. (9) γ_{ii} is equal to ½ (natural life time)$^{-1}$ and is given by

$$\gamma_{ii} = \gamma = \frac{2\omega^3|d|^2}{3c^3}, \tag{10a}$$

and ω_γ is the renormalized frequency given by

$$\omega_\gamma = \omega + \Omega\,,\quad \Omega = \frac{-\gamma}{\pi}\ln\{(\frac{\omega_c}{\omega} + 1)|\frac{\omega_c}{\omega} - 1|\}\,, \tag{10b}$$

where ω_c is the cut-off frequency.

The parameters γ_{ij} and Ω_{ij} are found to be given by

$$\gamma_{ij} = \gamma\Lambda_{ij}(\frac{\omega}{c}\, r_{ij})\,,\quad \underline{r}_{ij} = \underline{R}_i - \underline{R}_j\,, \tag{11}$$

$$\Lambda_{ij}(x) = \frac{3}{2}\{[1 - (\hat{d}\cdot\hat{r}_{ij})^2]\frac{\sin x}{x} + [1-3(\hat{d}\cdot\hat{r}_{ij})^2][\frac{\cos x}{x^2} - \frac{\sin x}{x^3}]\}, \tag{12}$$

and

$$\Omega_{ij} = -(\frac{\gamma c}{\omega})\frac{1}{\pi}\int_0^\infty k^2 dk\Lambda_{ij}(k\mathbf{r}_{ij})\{k^2 - \frac{\omega^2}{c^2}\}^{-1}. \tag{13}$$

It should be noted that the integrand in (13) has poles at $k = \pm\omega/c$, 0. The pole at k=0 comes from the singularity of Λ_{ij}. The integral in (13) is evaluated in the standard manner [16] and one finds that

$$\Omega_{ij} = -V_{ij} + \Omega_{ij}^{(2)}, \tag{14}$$

where

$$\Omega_{ij}^{(2)} = \gamma\Delta_{ij}\left(\frac{\omega}{c} r_{ij}\right),$$

$$\Delta_{ij}(x) = \frac{3}{2}\left\{[1-3(\hat{d}\cdot\hat{r}_{ij})^2]\left[\frac{\sin x}{x^2} + \frac{\cos x}{x^3}\right] - \frac{\cos x}{x}[1-(\hat{d}\cdot\hat{r}_{ij})^2]\right\}. \tag{15}$$

The terms $\Omega_{ij}^{(2)}$ and V_{ij} stem from the poles at $k = \omega/c$ and at k=0 respectively. On substituting (14) into (9) we obtain

$$\frac{\partial\rho}{\partial t} = -i\omega_\gamma \Sigma_i [s_i^+ s_i^-, \rho] - i\,\Sigma_{i\neq j}\Omega_{ij}^{(2)}\,[s_i^+ s_j^-, \rho]$$

$$+ \Sigma_{ij}\gamma_{ij}\,\{s_i^+ s_j^- \rho - 2\, s_j^- \rho s_i^+ + \rho\, s_i^+ s_j^-\}\,. \tag{16}$$

We thus see that the dipole-dipole interaction term, which is static in nature, cancels with the contribution from the k=0 pole in (13) and we may refer to $\Omega_{ij}^{(2)}$ as the interaction energy between i^{th} and j^{th} atom [16],[17]. This term may also be obtained by considering the radiation field, at the point $\underline{R}_j$, due to a dipole at $\underline{R}_i$.

The master equation (16) enables us to calculate the properties of the atomic system during spontaneous emission. In order to calculate the coherence properties of the radiation field, we supplement the master equation (16) with equations which relate the field operators to atomic operators. One such equation is obtained by direct integration of the Heisenberg equation for $a_{\underline{k}s}(t)$

$$a_{\underline{k}s}(t) = a_{\underline{k}s}(0)\, e^{-i\omega_{ks}t} - i\,\Sigma_j g_{j\underline{k}s} \int_0^t s_j^-(\tau)\, e^{-i\omega_{\underline{k}s}(t-\tau)}\, d\tau. \tag{17}$$

Since the radiation field is in vacuum state, it is clear that the first term in (17) does not contribute as far as the normally ordered correlations are concerned. For instance, the mean number of photons in the mode $\underline{k}$,s is given by

$$\langle a^+_{\underline{ks}} a^-_{\underline{ks}} \rangle_t = \Sigma_{jl} g^*_{j\underline{ks}} g_{l\underline{ks}} \int_0^t \int \langle s^+_j(\tau_1) s^-_l(\tau_2) \rangle \, e^{i\omega_{ks}(\tau_2-\tau_1)} \, d\tau_1 d\tau_2. \tag{18}$$

The two-time correlation function $\langle s^+_j(\tau_1) s^-_1(\tau_2)\rangle$ appearing in (18) can be obtained from the solution of the master eq. (16) and the quantum regression theorem [18]. We define the radiation rate I(t) as the rate of change of the energy of the atomic system [19], i.e.

$$I(t) = -\omega \frac{d}{dt} \Sigma_j \, s^z_j(t) . \tag{19}$$

One may further show by using (17) and the expansion for the electric field operator $\underline{E}(\underline{r},t)$ that its positive frequency part in the radiation zone is given by [20]

$$\underline{E}^{(+)}(\underline{r},t) \sim \left(\frac{\omega^2}{c^2}\right) \Sigma_i \{\underline{d} - \hat{r}_i(\hat{r}_i \cdot \underline{d})\} \, r_i^{-1} \exp\{\frac{i\omega}{c}(r_i - r\} \, s_i^{(-)}(t - \frac{r_i}{c}),$$

$$\underline{r}_i = \underline{r} - \underline{R}_i . \tag{20}$$

We have again ignored the free field contribution to $\underline{E}^{(+)}$, as it does not contribute to normally ordered field correlations. Equation (16)-(20) are the basic relations of the master equation approach to spontaneous emission. It is easily shown from (16) that $\langle s^z_i \rangle$ satisfies the equation

$$\frac{d}{dt} \langle s^z_i \rangle + \Sigma_j \, \gamma_{ij} \{\langle s^+_i s^-_j \rangle + \langle s^+_j s^-_i \rangle\} = 0 \tag{21}$$

and hence the radition rate is given by

$$I(t) = 2\omega \, \Sigma_{ij} \gamma_{ij} \, \langle s^+_i(t) s^-_j(t) \rangle . \tag{22}$$

We first discuss [21] time independent characteristics of the emitted radiation and establish connection with some of the work of Dicke. It is clear that the perturbation theoretic result for the radiation rate is

$$I_o = 2\omega \Sigma_{ij} \gamma_{ij} \langle s^+_i s^-_j \rangle_o , \tag{23}$$

where $\langle\ \rangle_o$ refers to the expectation value with respect to the initial state. For small systems, i.e, systems whose linear dimensions are small compared to the wave length of the emitted radiation, we have $\gamma_{ij} \simeq \gamma$ and we may write (23) as

$$I_o = 2\omega\gamma\, \Sigma_{ij} \langle s_i^+ s_j^- \rangle_o \tag{24}$$

$$= 2\omega\gamma \langle s^+ s^- \rangle_o \,. \tag{25}$$

In going from (24) to (25) we have introduced the collective operators $s^{\pm,z}$ defined by

$$s^{\pm,z} = \Sigma_i\, s_i^{\pm,z} \,. \tag{26}$$

We calculate I_o when the system was initially excited to a state

$$\rho(0) = \Pi_i |\theta_o,\phi_o\rangle_i\ {}_i\langle\theta_o,\phi_o| \,, \tag{27}$$

where $|\theta_o,\phi_o\rangle_i$ [Θ- state] is given by

$$|\theta_o,\phi_o\rangle_i = \cos\frac{\theta_o}{2}\, e^{i\phi_o/2} |-\rangle_i + \sin\frac{\theta_o}{2}\, e^{-i\phi_o/2} |+\rangle_i \,, \tag{28}$$

and where $|+\rangle_i$ and $|-\rangle_i$ are the excited and the ground states of the ith atom respectively. From (24) and (27) we have

$$I_o = 2\gamma\omega N \sin^2(\tfrac{\theta_o}{2}) \{1 + (N-1)\cos^2(\tfrac{\theta_o}{2})\} \,. \tag{29}$$

From (29) we see that for large N

$$I_o \propto N^2 \qquad \text{for } \theta_o = \pi/2,$$

which shows that a system excited to a state of the form (27) gives rise to superradiant emission if $\theta_o = \pi/2$. It is important to note that (27)(with $\theta_o = \pi/2$) is a state in which there are no atomic correlations but the dipole moment is maximum. We may refer to this type of emission as the superradiance of first kind and it is wholly due to the fact that the dipole moment of the system has a finite value. Next if the system was excited to a Dicke state $|N/2,m\rangle$ then

$$I_o = 2\gamma\omega\{\frac{N}{2}(\frac{N}{2}+1) - m^2 + m\} \quad , \tag{30}$$

which shows that

$$I_o \propto N^2 \quad \text{if } m = 0 \quad .$$

Therefore a system excited to a state $|N/2,0\rangle$ leads to the superradiant emission which may be referred to as the superradiance of second kind. This kind of superradiance is due to the presence of atomic correlations (dipole moment of the system being zero), for instance we have

$$\langle S_i^+ S_j^- \rangle - \langle S_i^+ \rangle \langle S_j^- \rangle \simeq \frac{1}{4} \quad .$$

The energy-energy correlations are, of course, not very important

$$\langle S_i^z S_j^z \rangle - \langle S_i^z \rangle \langle S_j^z \rangle = -\frac{1}{4(N-1)} \quad . \tag{32}$$

We now discuss the time dependent properties of the radiation. We first consider a few special cases and establish connection with the work of Weisskopf and Wigner.

(a) One Two-Level Atom Problem:

In this case the master equation (16) leads to the equations of motion

$$\frac{\partial}{\partial t}\langle S^{\pm}\rangle = \pm i\omega_\gamma \langle S^{\pm}\rangle - \gamma\langle S^{\pm}\rangle, \quad \frac{\partial}{\partial t}\langle S^{\pm}\rangle = -2\gamma\langle S^{\pm}\rangle \quad . \tag{33}$$

These equations are easily integrated and we find that

$$\langle S^{\pm}\rangle_t = e^{(\pm i\omega_\gamma - \gamma)(t-t')}\langle S^{\pm}\rangle_{t'}, \quad \langle S^+S^-\rangle_t = e^{-2\gamma t}\langle S^+S^-\rangle_o \quad , \tag{34}$$

which, of course, imply the familiar exponential decay. On using quantum regression theorem and (34) we also obtain

$$\langle s^+(t)s^-(t')\rangle = \exp\{i\omega_\gamma(t-t')-\gamma|t+t'|\}\langle s^+s^-\rangle_0. \tag{35}$$

On using (18) and (35) we find that the mean number of photons in the mode $\underline{k}$,s is given by

$$\langle a^+_{\underline{k}s}a_{\underline{k}s}\rangle_t = \frac{|g_{\underline{k}s}|^2}{(\gamma^2+x^2)}\{1+e^{-2\gamma t}-2e^{-\gamma t}\cos xt\}\langle s^+s^-\rangle_0,$$

$$x = (\omega_\gamma - kc). \tag{36}$$

Let $P_{\underline{k}s}(t)$ be the probability that there is a photon of momentum $\underline{k}$ and polarization s. We now show [22] that $P_{\underline{k}s}(t)$ is also given by Eq. (36). It is clear that

$$P_{\underline{k}s}(t) = \langle|\underline{k}s\rangle\langle\underline{k}s|\rangle_t,$$

where $|\underline{k},s\rangle$ is a state in which there is a photon in the mode $\underline{k}$,s. $P_{\underline{k}s}(t)$ may also be written as

$$P_{\underline{k}s}(t) = \langle a^\dagger_{\underline{k}s}|\{0\}\rangle\langle\{0\}|a_{\underline{k}s}\rangle_t = \langle a^\dagger_{\underline{k}s}\,{}^o_o\prod_{\underline{k}s} e^{-a^\dagger_{\underline{k}s}a_{\underline{k}s}}\,{}^o_o\, a_{\underline{k}s}\rangle_t, \tag{37}$$

where ${}^o_o\;{}^o_o$ denote the normal ordering. Since in our problem there is at most only one photon, it is clear that only one term in (37) contributes and then we have

$$P_{\underline{k}s}(t) = \langle a^+_{\underline{k}s}(t)\; a_{\underline{k}s}(t)\rangle\;. \tag{38}$$

On combining (36) and (38) we obtain the classic result of Weisskopf-Wigner theory.

(b) Two Atom case

For the sake of the simplicity of calculations we assume that the two atoms are within a wavelength of each other and we will also ignore the effect of the terms involving $\Omega_{ij}^{(2)}$. Here we just give the time dependence of the total energy and the distribution of photons. It is found that

$$W(t) \equiv \Sigma_i \langle s_i^z(t)\rangle = \{W(0) + \frac{3}{4} + \langle \underline{s}_1 \cdot \underline{s}_2\rangle_o + 4\gamma t \langle s_1^+ s_1^- s_2^+ s_2^-\rangle_o$$

$$\times\, e^{-4\gamma t} - \{\frac{3}{4} + \langle \underline{s}_1 \cdot \underline{s}_2\rangle_o\} \quad , \tag{39}$$

which implies that

$$W(\infty) = -\{\frac{3}{4} + \langle \underline{s}_1 \cdot \underline{s}_2\rangle_o\} \; , \tag{40}$$

From (40) we conclude that each atom is not necessarily left in its ground state after the emission process has been completed. A similar result has been found by Stroud et al. [23] who used neo-classical theory. We next assume that each atom was initially in its excited state, then we find that the two time correlation $\langle s^+(t)s^-(t')\rangle$ is given by

$$\langle s^+(t)s^-(t')\rangle = \{8\gamma t' \, e^{-2\gamma(t+t')} + 4\, e^{-2\gamma(t+t')} - 2e^{-4\gamma t}\}$$

$$e^{i\omega(t-t')}, \quad t > t'. \tag{41}$$

On combining (41) and (18) we obtain

$$\langle a_{\underline{k}s}^+ a_{\underline{k}s}\rangle \xrightarrow{t\to\infty} \frac{2|g_{\underline{k}s}|^2(x^2+40\gamma^2)}{(x^2+16\gamma^2)(x^2+4\gamma^2)} \quad \alpha \quad P_{\underline{k}s}(\infty), \tag{42}$$

which gives the distribution of photons in the mode $\underline{k}$,s and is a sum of two Lorentzian lines. A similar result has been derived recently by Arecchi et al. [24] who made use of the resolvant operator method [5].

(c) Multi-atom case

We have already seen that the radiation rate can be computed if the two particle mean value $\langle s_i^+(t)s_j^-(t)\rangle$ is known. On using eq. (16) we find that $\langle s_i^+ s_j^-\rangle$ satisfies the equation

$$\frac{\partial}{\partial t} \langle s_i^+ s_j^- \rangle + 2\gamma \langle s_i^+ s_j^- \rangle = \gamma_{ij}\{\langle (1+2s_i^z)\, s_j^z \rangle + \langle (1+2s_j^z) s_i^z \rangle\}$$
$$+ 2\, \Sigma_{l\neq i\neq j}\{\gamma_{jl}\langle s_i^+ s_j^z s_l^- \rangle + \gamma_{il}\langle s_l^+ s_i^z s_j^- \rangle\} \ . \tag{43}$$

It is seen that the equation of motion for the two-particle mean value contains the three particle mean values and one thus obtains the whole hierarchy of equations. Hence in order to obtain the radiation rate, we must resort to approximate methods. Before we discuss various decoupling procedures, we discuss an exactly soluble model. The results from this model will be very useful in making appropriate approximations for the case of two-level atoms.

This model is obtained if we replace each of the two-level atoms by a harmonic oscillator. The details of the emission from a system of harmonic oscillator are described in refs. [10a], [10b]. The master equation describing spontaneous emission from such a system is

$$\frac{\partial \rho}{\partial t} = - i\omega_\gamma \Sigma_i [a_i^+ a_i, \rho] - i\, \Sigma_{i\neq j}\, \Omega_{ij}^{(2)}\, [a_i^+ a_j, \rho]$$
$$- \Sigma_{i,j}\, \gamma_{ij}[a_i^+ a_j \rho - 2a_j \rho a_i^+ + \rho a_i^+ a_j), \tag{44}$$

where a_i and a_i^+ are the boson annihilation and creation operators satisfying the usual commutation rules. The numerical value of ω_γ now differs from the one given by (10b). It is convenient to transform (44) to coherent state representation[25]

$$\rho = \int \Phi(\{z\}) \,|\{z\}\rangle \langle\{z\}|\, d^2(\{z\}). \tag{45}$$

From (44) and (45) we obtain

$$\frac{\partial \Phi}{\partial t} = \Sigma_i \{i\omega_\gamma \frac{\partial}{\partial z_i} (z_i \Phi) + \text{c.c.}\} + \Sigma_{i\neq j}\, \Omega_{ij}^{(2)}\{ i\frac{\partial}{\partial z_i} (z_j \Phi) + \text{c.c}\}$$
$$+ \Sigma_{ij}\gamma_{ij}\{ \frac{\partial}{\partial z_i} (z_j \Phi) + \text{c.c.}\} \ . \tag{46}$$

The corresponding Langevin equations are

$$\dot{z}_i = - i\omega_\gamma z_i - i \Sigma_{i\neq j} \Omega_{ij}^{(2)} z_j - \Sigma_j \gamma_{ij} z_j. \qquad (47)$$

The beauty of eqs. (47) is that they are linear and therefore can be solved exactly. We consider again for the sake of simplicity small systems and ignore the effect of the terms $\Omega_{ij}^{(2)}$. In such a case the solution of (46) subject to the initial condition

$$\Phi(0) = \Pi_i \, \delta^{(2)} (z_i - z_i^o),$$

is given by

$$\Phi(t) = \Pi_i \, \delta^{(2)} (z_i - \bar{z}_i(t)).$$

where

$$\bar{z}_i(t) = z_i^o - \frac{1}{N} (1 - e^{-\gamma t N}) \Sigma_j z_j^o .$$

The radiation rate from this system is

$$I(t) = 2\gamma\omega \, e^{-2\gamma t N} \, \Sigma_{ij} \langle a_i^+ a_j \rangle_o .$$

The steady state solution is

$$\rho(\infty) = |\{z_i^o - \frac{1}{N} \Sigma z_j^o\}\rangle \langle\{z_i^o - \frac{1}{N} \Sigma z_j^o\}|$$

which is identical to the ground state of the system only if $z_i^o = z_o$ for all i. We consider the case of following initial excitations.

(a) Initial coherent excitation:

We assume that each of the oscillator was initially excited to a coherent state $|z_o\rangle$. One may then show the following:

I) The radiation rate is given by

$$I_{coh}(t) = 2\gamma\omega \, N^2 \, e^{-2\gamma N t} |z_o|^2$$

which is proportional to the square of the number of particles and the decay constant is N times than that due to a single particle. The coherent state thus constitutes a superradiant state of the system.

II) The state of the system at time t is

$$\rho(t) = \Pi_i |z_o\, e^{-N\gamma t}>_i\; _i<z_o\, e^{-N\gamma t}| \; ,$$

showing that the system remains in a superradiant state.

III) No correlations are induced among any two oscillators. It should be noted that in our previous terminology these are the features of the superradiance of first kind.

(b) Initial Fock State Excitation (Inocherent Excitation):

If each of the oscillator was initially excited to a Fock state, then the radiation rate is

$$I_F(t) = 2\gamma\omega\, e^{-2\gamma Nt}\, \Sigma_i <a_i^+ a_i>_o$$

which is, of course, the normal incoherent emission. One also finds that the correlations are induced among different oscillators as a result of emission, for example, one has

$$<a_i^+(t)\; a_j(t)> - <a_i^+(t)><a_j(t)> = -\frac{2n}{N}\; \sinh(n\gamma t)\; e^{-N\gamma t} \quad (i\neq j).$$

where n indicates the initial excitation of the system.

One may do similar calculations for large systems as the eq. (47) are linear. We quote here only the result for N=2. We find the following results for the radiation rate for two cases discussed above:

$$I_F(t) = 2\omega\{\gamma\, \cosh\,(2\gamma_{12}t) - \gamma_{12}\, \sinh(2\gamma_{12}t)\}\; e^{-2\gamma t}\Sigma_i <a_i^+(0)a_i(0)> \tag{48}$$

$$I_{coh}(t) = 2\omega|z_o|^2 e^{-2\gamma t}\{(\gamma-\gamma_{12})e^{2\gamma_{12}t}(1-\cos\phi_o)+(\gamma+\gamma_{12})\, e^{-2\gamma_{12}t}(1+\cos\phi_o)\} \; . \tag{49}$$

In obtaining (49) we assumed a phase difference ϕ_0 between the initial amplitudes of the two oscillators. It is also interesting that the terms $\Omega_{12}^{(2)}$ do not contribute to (48) and (49). The changes in the effective decay constants have been observed recently in a beautiful experiment by Lama et al [26].

We now return to the analysis of multi-atom case. We confine our attention only to small systems and ignore the effect of terms involving $\Omega_{ij}^{(2)}$. The master equation (16) in terms of collective variables becomes [27]

$$\frac{\partial \rho}{\partial \tau} = -\frac{1}{2}(s^+s^-\rho - 2s^-\rho s^+ + \rho s^+s^-) \quad , \quad \tau = 2\gamma t \, , \tag{50}$$

where ρ now refers to the density operator in the interaction picture. On taking the matrix element of both sides of (50) with respect to Dicke states $|s,m\rangle$ where s is Dicke's cooperation number, we obtain

$$\frac{\partial \rho_{mn}}{\partial \tau} = (\nu_{m+1}\nu_{n+1})^{1/2} \rho_{m+1,n+1} - \frac{1}{2}(\nu_m+\nu_n) \rho_{m,n} ; \rho_{mn} = \langle sm|\rho|s\,n\rangle \tag{51}$$

where

$$\nu_m = (s - m + 1)(s + m) . \tag{52}$$

It is seen that $2\gamma\nu_m$ is the probability that the system makes the transition from the state $|s,m\rangle$ to $|s,m-1\rangle$. For diagonal elements we find that

$$\frac{\partial \rho_{mn}}{\partial t} = \nu_{m+1} \rho_{m+1,m+1} - \nu_m \rho_{m,m} , \tag{53}$$

which is of the form of Pauli master equation. It should be noted that one could have obtained eq. (53) just from the arguments of detailed balance. However, we would not have had any information about the off-diagonal elements which are, of course, needed in the evaluation of two time correlation functions of the form $\langle s^+(t)s^-(t')\rangle$.

Since the eq. (51) is in the form of a difference differential equation it is easily solved by taking its Laplace transform and then by iteration. We find that the Laplace transform

$$\tilde{\rho}_{m,n}(\beta) = \int_0^\infty d\tau e^{-\tau\beta}\rho_{m,n}(\tau) \ , \quad \mathrm{Re}\ \beta > 0,$$

of

$\rho_{m,n}(\tau)$ is given by[27a]

$$\tilde{\rho}_{m,n}(\beta) = \Sigma_{\ell\geq 0}(\Pi_1^\ell \nu_{m+k}\nu_{n+k})^{1/2} \ (\Pi_0^\ell[\beta+ \tfrac{1}{2}(\nu_{m+k}\nu_{n+k})]^{-1})\rho_{m+\ell,n+\ell}(0). \tag{54}$$

We first note that the steady state solution is given by

$$\rho_{m,n}(\infty) = \delta_{m,n}\delta_{n,-s} \ ,$$

which implies that the steady state value of the energy is

$$W(\infty) = -\ s.$$

A special case of the result is given by eq. (40). It should also be noted that in the quantum theory, the cooperation number takes only the integer or half integer values.

As a second consequence of the exact solution, we note

$$\langle s^{\pm}(t)\rangle = 0 \quad \text{if} \quad \rho(0) = |s,m\rangle\langle s,m| \ ,$$

i.e., a system excited to a state with zero dipole moment will remain in a state with zero dipole moment and therefore such a system will not show superradiance of first kind.

It is also to be noted that once eq. (51) has been solved in $|s,m\rangle$representation, the solution may be used to calculate the properties of the system for any initial excitation. For example, we may express the Θ - state as

$$\Pi_i|\theta_o\phi_o\rangle_i = \Sigma_{-N/2}^{+N/2}\binom{N}{N/2\ -m} \quad (\cos\frac{\theta_o}{2}\ e^{-i\phi_o/2})^{N/2-m}$$

$$\times(\sin\frac{\theta_o}{2}\ e^{i\phi_o/2})^{N/2+m}\ |N/2,m\rangle\ , \tag{55}$$

and thus the initial condition for Θ - excitation is

$$\rho_{m,n}(0) = \binom{N}{N/2-m}^{1/2} \binom{N}{N/2-n}^{1/2} e^{i\phi(m-n)} (\cos\frac{\theta_o}{2})^{N-m-n} (\sin\frac{\theta_o}{2})^{N+M+n} . \tag{56}$$

Equation (51) is in a form which is very attractive for numerical computations. Numerical work on this master equation has been done by Bonifacio et al.[14] and by Ponte Goncalves-Tallet [28]. Figures 1 and 2 give the radiation rate as a function of time for the case of initial excitations $|N/2,N/2>$ and $|\theta_o,\phi_o>$ ($\theta_o=\pi/2$) respectively.

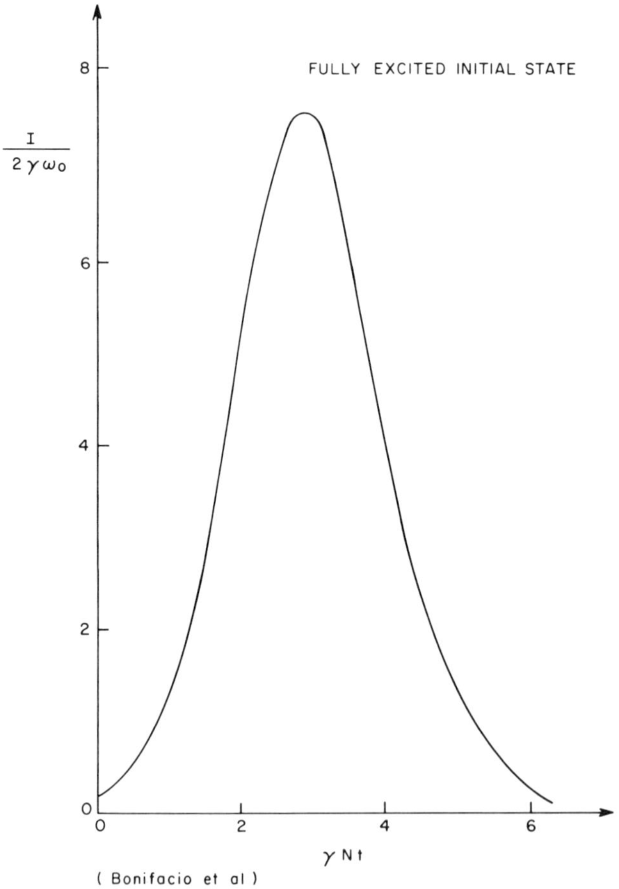

Fig. 1 The radiation rate as a function of time for the initial excitation $|N/2,N/2>$ for the case of 200 atoms.

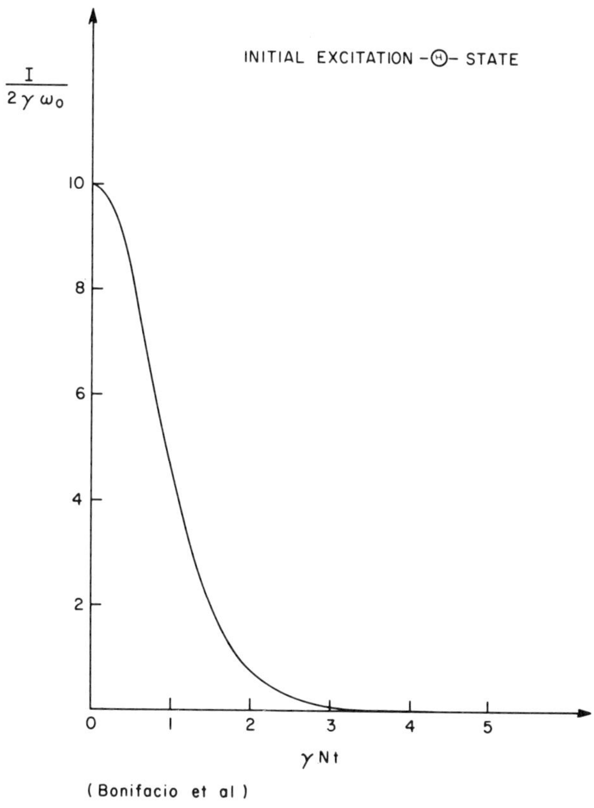

Fig. 2 The radiation rate as a function of time for the initial excitation $|\pi/2,\phi_0>$ for the case of 200 atoms

Figure 3 gives the behaviour of the intensity fluctuation

$$\Gamma^{(2)} \equiv <s^+s^+s^-s^-> - <s^+s^->^2 = \Sigma \nu_m \nu_{m-1} \rho_{mm} - (\Sigma \nu_m \rho_{mm})^2 .$$

for the case when the system was initially excited to a completely inverted state.

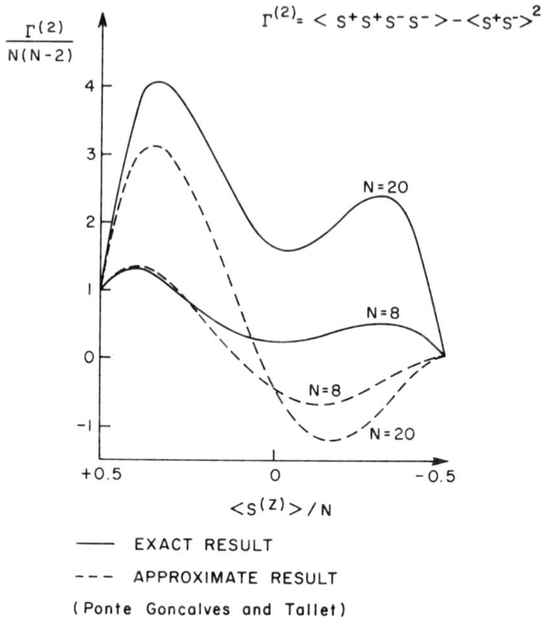

Fig. 3 The intensity fluctuation $\Gamma^{(2)}$ as a function of the energy for the initial excitation $|N/2,N/2\rangle$ (full curve). The dotted curve represents the intensity fluctuation if the assumption of Θ - state is made for all times.

It is also possible to obtain some analytic results for the radiation rate by decoupling the hierarchy of equations which we discussed earlier. We would now discuss such decoupling schemes [10a,c,d]. For small systems there is permutation symmetry which implies that

$$\langle \underline{S}_i \cdot \underline{S}_j \rangle = \frac{1}{4} ,$$

if all the atoms are initially excited to a permutationally symmetric state. From (21) and this result we obtain

$$\frac{\partial W}{\partial t} + 2\gamma(W + \frac{N}{2}) + 2\gamma(N-1)\{W + \frac{N}{2} - N\langle S_i^+ S_j^+ S_i^- S_j^- \rangle\} = 0 \, . \qquad (57)$$

The natural question is - how can we express the two-particle correlation in terms of one-particle mean values. We may do the following [29]

$$\text{(I)} \quad \langle s_i^+ s_j^+ s_i^- s_j^- \rangle \approx \langle s_i^+ s_i^- \rangle \langle s_j^+ s_j^- \rangle \qquad (i \neq j)$$

$$\text{(II)} \approx \langle s_i^+ s_i^- \rangle \langle s_j^+ s_j^- \rangle + \langle s_i^+ s_j^- \rangle \langle s_j^+ s_i^- \rangle .$$

The choice of the approximation (I) or (II) will depend on the initial excitation of the systems as the approximation has to be consistent with it.

I) Initial Θ - Excitation ($\theta < \pi$)

In this case it is easily seen that the approximation (II) is not consistent with the initial condition and hence we proceed by making the approximation (I) i.e.

$$\langle s_i^+ s_j^+ s_i^- s_j^- \rangle \approx \langle s_i^+ s_i^- \rangle \langle s_j^+ s_j^- \rangle = \frac{1}{N^2} \left(W + \frac{N}{2}\right)^2 . \tag{58}$$

On combining (57) and (58), we find that

$$\frac{\partial W}{\partial t} = 2\gamma \left(W + \frac{N}{2}\right) \left\{ \left(1 - \frac{1}{N}\right) \left(W' - \frac{N}{2}\right) - 1 \right\} . \tag{59}$$

This equation is easily integrated and one finds that

$$I(t) = \frac{\omega \gamma N^3}{2(N-1)} \ \text{sech}^2 \, N\gamma(t-\tau) \quad , \tau = \frac{1}{2N\gamma} \ \ell n \left\{ \frac{(N-1)}{1 + N \cot^2(\theta_0/2)} \right\} . \tag{60}$$

We thus obtain the well known [19] "Sech" behaviour for the radiation rate. We have also studied the accuracy of the approximation (58). This is done by working with equations of motion for two-particle mean values and making approximation on three-particle mean values. We find that $\langle s_i^+ s_j^+ s_i^- s_j^- \rangle$ obeys the equation

$$\frac{\partial}{\partial t}\langle s_i^+ s_j^+ s_i^- s_j^-\rangle + 4\gamma\langle s_i^+ s_j^+ s_i^- s_j^-\rangle + \gamma\Sigma_{\ell\neq i\neq j}\{\langle s_i^+ s_j^+ s_i^- s_j^-\rangle$$

$$+ \langle s_i^+ s_j^+ s_i^- s_\ell^-\rangle + c.c.\} = 0 . \qquad (61)$$

We now make the approximation of type (I) on three particle mean value,

$$\langle s_i^+ s_j^- s_k^+ s_k^-\rangle \approx \langle s_i^+ s_j^-\rangle \langle s_k^+ s_k^-\rangle \qquad (i \neq j \neq k). \qquad (62)$$

We then find that

$$\langle s_i^+ s_j^+ s_i^- s_j^-\rangle - \langle s_i^+ s_i^-\rangle\langle s_j^+ s_j^-\rangle \sim 0(\frac{1}{N})$$

i.e. the correction terms to (58) are of the order 1/N. The details of the method are given in ref. [10d]. These results are supported by the numerical computations of the master equation.

(2) Initial Excitation to Dicke Superradiant State $|N/2,0\rangle$

In this case because of relation (32) it is not possible to make either of the approximations (I) or (II) on the two-particle mean value. We use eq. (61) in conjunction with the approximation (62) to calculate the two-particle mean value $\langle s_i^+ s_j^+ s_i^- s_j^-\rangle$ which then we substitute in (57). We obtain an equation which is very similar to (59). The radiation rate is found to be given by

$$I(t) = 2\gamma\omega\,\frac{N^2}{4}\,(1 - \frac{2}{N})^{-1}\,\mathrm{sech}^2[N\gamma(t+\tau)], \tau = \frac{1}{2N\gamma}\ \ell n(\frac{N+2}{N-2}). \qquad (63)$$

The striking similarity between (60) [with $\theta_0 = \pi/2$] and (63) is to be noticed. This is not very surprising, even though the origin of the two types of superradiance is different, if we recall that for Θ - state, we had

$$\rho_{mm}(0) = (\frac{1}{2})^N \frac{N!}{(N/2-m)!\,(N/2+m)!}$$

and it is, of course, well known that this binomial distribution is sharply peaked near m = 0 for large N.

(3) Initial Excitation to Dicke state $|N/2,N/2\rangle$

We finally consider the case when the system was initially excited to Dicke state $|N/2,N/2\rangle$ (completely inverted system). In this case both the approximations I and II are consistent with the initial condition. However the approximation I would be a poor choice for the simple reason that even for harmonic oscillator model, which is described by linear equations, we found that the correlations were induced among different oscillators when the system was initially excited to a Fock state. We are therefore motivated to adopt (II) which in conjunction with $\langle \underline{s}_i \cdot \underline{s}_j \rangle = 1/4$ leads to the relation

$$\langle s_i^+ s_j^+ s_i^- s_j^- \rangle \approx \langle s_i^+ s_i^- \rangle + \frac{1}{2} - \{\frac{1}{4} + \langle s_i^+ s_i^- \rangle - \langle s_i^+ s_i^- \rangle^2\}^{1/2} .$$

On substituting this in (57) we obtain

$$\frac{\partial W}{\partial t} + 2\gamma\,(W + \frac{N}{2}) - \gamma N(N-1) + 2\gamma N(N-1)\{\frac{1}{2} - \frac{W^2}{N^2}\}^{1/2} = 0 \tag{64}$$

Equation (64) is easily integrated by quadratures and we show the result for the radiation rate in fig. 4. We also show the result which one would have obtained by making the approximation (I). Our results are again in agreement with the direct numerical computations of the master equation. It should also be noted that near the point of superradiant emission

$$\langle s_i^z s_j^z \rangle - \langle s_i^z \rangle \langle s_j^z \rangle \approx .05$$

A similar result has been found in ref. [30].

We will conclude this article by giving a few generalizations. We have so far considered atoms or molecules with only two levels. We have also studied the collective emission from three level systems and this will be discussed elsewhere. Here we only quote some results for the case of a single atom with 3 "unequidistant" levels. The master equation for this system is

$$\frac{\partial \rho_{ij}}{\partial t} = \delta_{ij}\,\Sigma_{k\neq i}\,\gamma_{ik}\,\rho_{kk} - \frac{1}{2}\,(\Gamma_i + \Gamma_j)\,\rho_{ij}, \tag{65}$$

where γ_{ik} is the transition probability from the state $|k\rangle$ to $|i\rangle$ and Γ_i is the transition probability from the state $|i\rangle$ to all other states. In deriving (65) we have ignored the frequency shift terms. Suppose we are interested in the distribution of photon which is emitted in the transition from $|3\rangle$ to $|2\rangle$ (it is assumed that

$E_3>E_2>E_1$; E = energy). Then we need to calcualte the correlation function $\langle P_{32}(t_1)P_{23}(t_2)\rangle$, where $P_{ij} = |i\rangle\langle j|$. We find that the distribution photon in this transition is given by

$$P_{\underline{k}s}(\infty) = |g_{\underline{k}s32}|^2 \Gamma_3^{-1} (\Gamma_3 + \Gamma_2)\{(\omega_{ks}-\omega_{32})^2 + \frac{1}{4}(\Gamma_3+\Gamma_2)^2\}^{-1},$$

$$\omega_{ij} = E_i - E_j. \tag{66}$$

This result agrees with the result obtained by using Weisskopf-Wigner theory [2] and Goldberger-Watson theory [31].

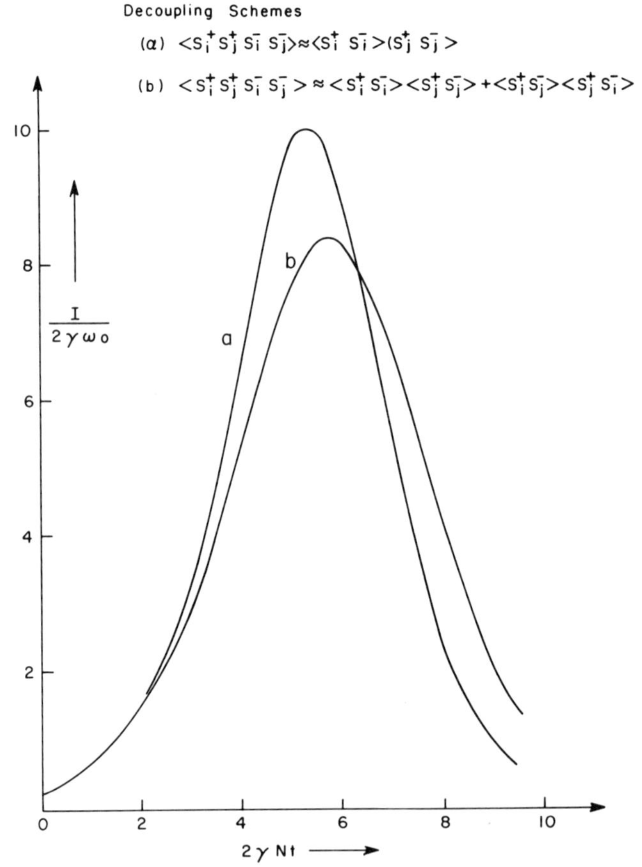

Fig. 4 The radiation rate as a function of time for the initial excitation $|N/2,N/2\rangle$ for the case of 200 atoms according to
(a) the decoupling approximation (I)
(b) the approximation (II).

Recently there also has been some interest in the study of spontaneous emission in presence of black body radiation [32]. The inclusion of a black body radiation field is trivial in our theory. The appropriate master equation is

$$\frac{\partial\rho}{\partial t} = - i\omega[s^z,\rho] -\gamma\ (1+k)(s^+s^-\rho-2s^-\rho s^++\rho s^+s)$$

$$- \gamma k(s^-s^+\rho-2s^+\rho s^- + \rho s^-s^+), \qquad (67)$$

where

$$k = [\exp\ (\frac{\omega}{k_B T}) - 1]^{-1} ,$$

where k_B is the Boltzmann constant and T is the temperature of the field. In deriving (67) we have ignored temperature dependent frequency shift terms. This master equation can be handled in the same manner as equation (50). We obtain, for example

$$\frac{\partial}{\partial t} \langle s^z\rangle = -2\gamma\langle s^+s^-\rangle - 4\gamma k\langle s^z\rangle \qquad (68)$$

which shows that there is an enhancement in the decay rate.

We may also generalize our theory to include the effect of external fields. We may, for instance, have a three-level atom whose levels $|3\rangle$ and $|2\rangle$ may be coupled by a potential and there is no radiative transition between the levels $|3\rangle$ and $|2\rangle$, but there is a radiative transition between the levels $|2\rangle$ and $|1\rangle$. Or we may have strongly driven two-level atoms. If the Hamiltonian of the total system is expressed as

$$H = H_A + H_R + H_{AR} + H_{ext}(t),$$

then the appropriate master equation in Born and Markovian approximations is

$$\frac{\partial}{\partial t} [P\rho(t)] + i\hat{L}_{ext}(t)P\rho(t) + \int_0^\infty d\tau\rho_R(0)\ Tr_R\{[H_{AR}(t),$$

$$[V(t,t-\tau)H_{AR}(t-\tau)\ V^+(t,t-\tau),\ P\rho(t)]]\} = 0, \qquad (69)$$

where

$$V(t,\tau) = T \exp\{-i \int_{\tau}^{t} H_{ext}(t')dt'\} . \tag{70}$$

Using this master equation we may study interference effects. For the case of three-level atomic system, which we mentioned above, we find that the photon distribution is given by

$$P_{\underline{ks}}(\infty) = |H_{AR,0}|^2 |V_0|^2 \{[(\omega_{ks}-\omega_{21})(\omega_{ks}-\omega_{31}) - V]^2 + \frac{1}{4}\Gamma_2^2 (\omega_{ks} - \omega_{31})^2\}^{-1} \tag{71}$$

which is in agreement with the result found in ref. [34], by using Heitler-Ma method.

Finally our theory is also easily generalized to include the effects of inhomogeneous broadening and of counter rotating terms. Such generalizations are discussed in refs. [10d, 10e].

The author would like to thank Deutsche Forschungsgemeinschaft for a grant which enabled him to attend the Conference.

Footnotes

1. R.H. Dicke, Phys. Rev. *93*, 99 (1954).
2. V. Weisskopf and E. Wigner, Z. Physik *63*, 54 (1930), ibid.,*65* 18 (1931).
3. V. Weisskopf, Ann. Physik *9*, 23 (1931), Z. Physik *85*, 451 (1933).
4. W. Heitler and S.T. Ma, Proc. Roy. Ir. Ac. *52*, 109 (1949); W. Heitler *Quantum Theory of Radiation*, (Oxford University Press, 3rd edition) § 16.
5. M.L. Goldberger and K.M. Watson *Collision Theory* (John Wiley New York, 1964), Chapter 8: Generalizations of this method are given in A.S. Goldhaber and K.M. Watson, Phys. Rev.*160*, 1151 (1967); L. Mower, Phys. Rev. *165*, 145 (1968).
6. F.E. Low, Phys. Rev.*88*, 53 (1952).
7. E.T. Jaynes and F.W. Cummings, Proc.IEEE*51*, 89 (1963); M.D. Crisp and E.T. Jaynes, Phys. Rev.*179* , 1253 (1969); C.R. Stroud and E.T. Jaynes, Phys. Rev. *A1*, 106 (1970)
8. C.S. Chang and P. Stehle, Phys. Rev. *A4* , 641 (1971).
9. M.Dillard and H.R. Robl, Phys. Rev.*184*, 312 (1969).
10. G.S. Agarwal (a) Phys. Rev. *A2*, 2038 (1970); (b) Phys. Rev. *A3*, 1783 (1971); (c) Nuovo Cim. Lett. *2* , 49 (1971); (d) Phys. Rev. *A4*, 1791 (1971); (e) Phys. Rev. *A4*, 1778 (1971); (f) Phys.Rev. *A7*

(1973)[in press]. Many of the results, which I would present here, have already appeared in these references.
11. R.W. Zwanzig in *Lectures in Theoretical Physics*, ed. W.E. Brittin et al (Wiley, New York 1961) Vol. III; Physica *33*, 119 (1964); see also G.S. Agarwal, Phys. Rev. *178*, 2025 (1969).
12. For the use of such methods to other problems in quantum optics, we refer to G.S. Agarwal in *Progress in Optics*, ed. E. Wolf (North-Holland Publ. Co., Amsterdam) Vol. XI (in press).
13. R.H. Lehmberg, Phys. Rev. *A2*, 883 (1970); ibid *A2*, 889 (1970).
14. R. Bonifacio, P. Schwendimann and F. Haake, Phys. Rev. *A4*, 302 (1971); ibid *A4*, 854 (1971); R. Bonifacio and P. Schwendimann, Nuovo Cim. Letters *3*, 512 (1970).
15. See e.g. H. Haken in *Handbuch der Physik* ed. S. Flugge (Springer, Berlin 1970), Vol. XXV/2c, p. 28.
16. See e.g. E.A. Power and S. Zienau, Phil. Trans. Roy. Soc. *251*, 427 (1959).
17. E.A. Power, Nuovo Cim *6*, 7 (1957); M.J. Stephen, J. Chem. Phys. *40*, 669 (1964).
18. M. Lax, Phys. Rev. *172*, 350 (1968),
19. J.H. Eberly and N.E. Rehler, Phys. Lett. *29A*, 142 (1969); N.E. Rehler and J.H. Eberly, Phys. Rev. *A3*, 1735 (1971).
20. cf. ref. 13.
21. This part of our presentation follows closely ref. 10c.
22. The author would like to thank Dr. F. Haake for a discussion in which this point came up.
23. C.R. Stroud, J.H. Eberly, W.L. Lama and L. Mandel, Phys. Rev. *A5* , 1094 (1971).
24. F.T. Arecchi, G.P. Banfi and V. Fassato-Bellani, Nuovo Cim. (in print).
25. E.C.G. Sudarshan, Phys. Rev. Letters *10*, 277 (1965); R.J. Glauber, Phys. Rev. *131*, 2766 (1963). For the general transformation of the operator equations into c-number equations, see G.S. Agarwal and E. Wolf, Phys. Rev. Lett. *21*, 180 (1968); Phys. Rev. *D2*, 2187 (1970).
26. W.L. Lama, R. Jodoin and L. Mandel, Am. J. Phys. *40* , 32 (1972).
27. Starting from a different viewpoint this master equation has also been obtained by Bonifacio et al, ref. 14.

(a) $\Omega_{ij}^{(2)}$, in principle, depends on the positions of ith and jth atom even for small systems. The permutation symmetry is also violated and it does not appear anymore convenient to work with the collective operators. However, there may be geometries for which $\Omega_{ij}^{(2)} \simeq \Omega$, i.e. it is on the average independent of i,j (cf. ref. 23). In such cases the solution (54) is modified to

$$\tilde{\rho}_{m,n}(\beta) = \Sigma_{\ell\geq 0}(\Pi_1^\ell \nu_{m+k}\nu_{n+k})^{1/2}(\Pi_0^\ell\ [\beta + \frac{1}{2}\{(1+\frac{i\Omega}{\gamma})\nu_{m+k} + (1-\frac{i\Omega}{\gamma})\nu_{m-k}\}]^{-1})\rho_{m+\ell,n+\ell}(0).$$

28. A.M. Ponte Goncalves and A. Tallet, Phys. Rev. *A4* , 1319 (1971).
29. Such decoupling procedures are, of course, very familiar in non-equilibrium statistical mechanics, see e.g. D.N. Zubarev, Sov. Phys. Uspekhi *3*, 320 (1960); see also L.P. Kadanoff and G. Baym *Quantum Statistical Mechanics* (Benjamin Inc., New York 1962).
30. R. Bonifacio and M. Gronchi, Nuovo Cim. Lett.*1*, 1105 (1971); V. DeGiorgio, Opt. Commun. *2*, 362 (1971).
31. N.M. Kroll in *Quantum Optics and Quantum Electronics* ed. Cohen Tannoudji et al. (Gordon and Breach N.Y. 1965) p. 47.
32. cf. ref. (8); J.E. Walsch, Phys. Rev. Lett. *27*, 208 (1971); G. Barton, Phys. Rev.*A5*, 468 (1972).
33. see e.g. ref. 12.
34. P.R. Fontana and D.J. Lynch, Phys. Rev.*A2* , 347 (1970).

INFLUENCE OF RESONANT FREQUENCY SHIFTS ON SUPERRADIANT DAMPING *

R. Friedberg and S. R. Hartmann

Columbia University, New York, New York

and

Jamal T. Manassah

Institute for Advanced Study, Princeton, New Jersey and Columbia University, New York, New York

The interaction between two electric dipoles $\vec{p}_1$ and $\vec{p}_2$ at position $\vec{r}_1$ and $\vec{r}_2$, analyzed at frequency kc, is [1]

$$v = - e^{ikr}[(\vec{p}_1 \cdot \vec{p}_2 - \hat{r} \cdot \vec{p}_1\, \hat{r} \cdot \vec{p}_2)\frac{k^2}{r}$$

$$+ (\vec{p}_1 \cdot \vec{p}_2 - 3\hat{r} \cdot \vec{p}_1\, \hat{r} \cdot \vec{p}_2)(\frac{ik}{r^2} - \frac{1}{r^3})\quad]$$

$$= \frac{1}{r^3}(\vec{p}_1 \cdot \vec{p}_2 - 3\hat{r} \cdot \vec{p}_1\, \hat{r} \cdot \vec{p}_2) - \frac{k2}{2r}(\vec{p}_1 \cdot \vec{p}_2 + \hat{r} \cdot \vec{p}_1\, \hat{r} \cdot \vec{p}_2)$$

$$- \frac{2}{3}\, i\, k^3\, \vec{p}_1 \cdot \vec{p}_2 + O[r]$$

where $\vec{r}_1 - \vec{r}_2 = r\hat{r}$.

---

* This work was supported in part by the Joint Services Electronics Program (U.S. Army, U.S. Navy and U.S. Air Force) under Contract DAAB07-69-C-0383, in part by the Atomic Energy Commission, in part by the Air Force Office of Scientific Research, OAR, under AFOSR Grant 70-1866.

The first term in the series expansion is the electrostatic potential; the second is the leading nonstatic correction. Both these terms, being real, lead to shifts in the resonance frequency of two-level atoms, whereas the third term, being imaginary, gives rise to damping.

The frequency shift arising in a system of many atoms, from the electrostatic potential, was investigated by Lorentz [2] for an infinite medium and by Kittel [3] for one of finite extent. (Kittel actually considered the magnetostatic interaction, for which the numerical results are different but the main features similar). Both authors considered the medium as linear.

Fain [4] introduced nonlinearity *via* the two-level atom model, which he studied in the formalism of Dicke.[5] He showed that the frequency shift is linear in the degree of excitation and changes sign when the population is inverted. This result has been criticized by Arecchi and Kim,[6] on the ground that the transition between two unstable levels is not a completed physical process and that frequency can be defined only with respect to all the light emitted until the system reaches its ground state. Without denying the importance of studying this whole process, we wish to point out that in a sample small compared to the wavelength, the first two (real) terms in the potential expansion far exceed the third (imaginary) term, so that the resulting frequency shifts are much greater than the inverse time for superradiant decay. Hence the uncertainty principle does *not* preclude measuring these shifts at different times as the excited population changes during the radiation process.

Fain [4] also considered for the first time the shift due to the second (nonstatic) term, but his treatment of the sample geometry was crude and yielded only an order-of-magnitude result. Correct numerical values of this shift were obtained for a small sphere by Plumier [7] and for a variety of other cases by ourselves. [8,9] We have also emphasized that cooperative frequency shifts in a system of identical atoms are a phenomenon not peculiar to superradiant emission.[8] Shifts of the *same* size are predicted in the absorption line and in ordinary incoherent emission.[8]

Recently, Stroud *et al.*[10] have attempted to follow the course of superradiant emission from a small sample by studying the "super Bloch vector" which represents the sum of all atomic moments. Their treatment depends on the assumption that the system wavefunction, if initially symmetric in all the atoms, will remain so throughout the radiative decay. Actually, this assumption is quite far from the truth. For example, in a small sphere (radius $\ll k_{res}^{-1}$ where k_{res} is the wave vector of the resonance radiation) of uniform density, the leading nonstatic contribution to the frequency shift varies (r.m.s. deviation ~ 12%) significantly through the sample, so that the shifted resonance frequency of atoms near the center differs from that of atoms near the surface by much more than the inverse superradiant decay time. Therefore, the different parts of the sphere lose coherence *very quickly* - before any significant fraction of the excitation energy has been radiated. [11] This conclusion is even stronger in a small sample of nonellipsoidal shape, since there the much larger static contribution also varies with position.

We must therefore, modify the picture, current since Dicke's work, of the approach from a state with $m > - r$ to a state with $m = - r$. It is often assumed that r remains nearly constant while m decreases through coherent emission to the value -r, after which $|m|$ and r rise together, by incoherent emission to the ground-state value (1/2)N. Instead, we find that r decreases very quickly to the value $|m|$ after which incoherent emission takes over. The decay $r \to |m|$ happens so fast that the *fractional* change in m during that time is negligible, although the absolute *number* of photons coherently emitted may be macroscopic. [11] This situation is already recognized in the case of strong inhomogeneous broadening; we are only pointing out that even if such broadening is absent locally, a global (i.e. positioning correlated) broadening is produced by the radiative fields themselves.

The "super Bloch vector" approach also runs into difficulty in large samples (>> wavelength) excited for coherent radiation in one direction. Here the frequency shifts are unimportant, but the radiative damping rate varies strongly with position. Atoms at the exit face will have radiated most of their energy before those at the entrance face have lost any significant fraction of theirs. (This effect has nothing to do with "cooperation length" [12] and persists if the speed of light is taken as infinite).

It follows from the above that the treatments of Rehler and Eberly [13], Bonifacio *et al.* [14] and numerous others [15] will need revision.

Note: After this paper was delivered, Professors Mandel, Freedhoff, and Senitzky all separately pointed out to us that if N identical atoms are arranged equidistantly on a ring, with axial polarization, they will all feel the same field. The same is true if there are two rings, one obtained from the other by an arbitrary axial displacement and an arbitrary azimuthal rotation. We believe no further elaboration is possible. The rings must not be of macroscopic thickness, but must contain atoms in single file.

References

1. John David Jackson, *Classical Electrodynamics* (John Wiley and Sons, Inc., New York, London (1962)), p.271 Eq. (9.18).
2. H.A. Lorentz, *Theory of Electrons* (Dover, New York, 2nd Ed., (1952)).
3. C. Kittel, Phys. Rev. *73*, 155 (1948)
4. V.M. Fain, Soviet Phys. JETP *36*, 798 (1959).
5. R.H. Dicke, Phys. Rev. *93*, 99 (1954).
6. F.T. Arecchi and D.M. Kim, Opt. Comm. *2*, 324 (1970).
7. R. Plumier, Physica *28*, 423 (1962).
8. R. Friedberg, S.R. Hartmann and J.T. Manassah, Phys. Letters *35A*, 161 (1971).
9. R. Friedberg and S.R. Hartmann, Optics Comm. *2*, 301 (1970).
10. C.R. Stroud, J.H. Eberly, W.L. Lama and L. Mandel, Phys. Rev. *A5*, 1094 (1972).
11. An explicit formula for the dephasing time can be found in R. Friedberg, S.R. Hartmann and Jamal T. Manassah, Phys. Letters *40A*, 365 (1972).
12. F.T. Arecchi and Eric Courtens, Phys. Rev. *A2*, 1730 (1971).
13. N.E. Rehler and J.H. Eberly, Phys. Rev. *A3*, 1735 (1971).
14. R. Bonifacio, P. Schwendimann and Fritz Haake, Phys. Rev. *A4*, 854 (1971).
15. For examples from this conference we cite (a) Kenneth G. Whitney: "*A Quantum Electrodynamic View of Superradiance as a Competition between Stimulated and Spontaneous Atomic Decay*" (b) G.S. Agarwal: "*Master Equations in the Theory of Incoherent and Coherent Spontaneous Emission*" (c) R.K. Bullough: "*Non-linear Radiation Reaction*", (a) p. 767, (b) p. 157, (c) p. 121, this volume.

SPONTANEOUS EMISSION RATE IN THE PRESENCE OF A MIRROR

Karl H. Drexhage

Eastman Kodak Company, Rochester, N.Y.

Recently it was shown experimentally that the decay time of fluorescence is influenced by a mirror[1-3]. It was decreased or increased depending on the distance from the mirror. Owing to the very low density of excited molecules in these experiments, the observed phenomenon must be ascribed to an influence of the mirror on the rate of spontaneous emission of each excited molecule alone. A theoretical treatment is given which is based on the radiation of a classical oscillator, and it is applied to the actual experimental situations of a metal mirror, a boundary between media of different refractive indices, and a system of two or more parallel mirrors. The strong birefringence of the environment of the fluorescing molecules and competing nonradiative processes is taken into account in order to permit a quantitative comparison with the experimental data.

An oscillating electric dipole of dipole moment $\mu = \mu_0 \cos\omega t$ creates an electric field whose amplitude at a distance $r >> \lambda$ ($\lambda = 2\pi c/\omega$) is given by $E_0 = (4\pi^2\mu_0/\lambda^2 r)\sin\alpha$, where α is the angle between the direction of r and the axis of the dipole. The total energy L_f, emitted per unit time, is obtained through integration as $L_f = 16\pi^4 c n_1 \mu_0^2/3\lambda^4$ (n_1=refractive index of the isotropic environment). We assume that the rate $1/\tau$ of spontaneous emission of the molecule is proportional to L_f. If we now place the dipole in front of a mirror, the electric field in the far zone can be obtained as a superposition of the direct and reflected waves, and the angular distribution of radiation is strongly dependent on the distance d from the mirror and on the orientation of the dipole. For instance, in the case of the dipole axis being parallel to the mirror normal the electric-field amplitude is given by

$$E^{\parallel}_{o,M} = (4\pi^2\mu_o/\lambda^2 r)\sin\alpha[1+\rho_\parallel^2-2\rho_\parallel \cos(4\pi n_1 d\cos\alpha/\lambda-\delta_\parallel)]^{\frac{1}{2}} ,$$

where $\rho_\parallel$ and $\delta_\parallel$ are the reflection coefficient and the phase shift at the mirror, respectively. If the reflectivity of the mirror is less than unity ($\rho_\parallel<1$) because of transmission, there will also be a wave beyond the mirror with the amplitude

$$\tilde{E}^{\parallel}_{o,M} = (4\pi^2\mu_o/\lambda^2 r)\sin\alpha(1-\rho_\parallel^2)^{\frac{1}{2}} .$$

It is assumed that a mirror whose reflectivity is reduced owing to absorption (as is the case with metals) has an identical effect upon the oscillator. The total power emitted by the dipole is again found by integration and is assumed to be proportional to the rate $1/\tau^{\parallel}_M$ of spontaneous emission in the presence of the mirror. Thus one obtains[2,3], with the abbreviations $z = 4\pi n_1 d/\lambda$ and $u = \cos\alpha$:

$$\tau/\tau^{\parallel}_M = 1 - \frac{3}{2}\int_0^1 \rho_\parallel(1-u^2)\cos(zu - \delta_\parallel)du ,$$

and in the case where the dipole axis is perpendicular to the mirror normal, correspondingly:

$$\tau/\tau^{\perp}_M = 1 + \frac{3}{4}\int_0^1 [\rho_\perp\cos(zu - \delta_\perp) + \rho_\parallel u^2\cos(zu - \delta_\parallel)]du .$$

The same expressions are obtained if the emitting dipole is replaced by an absorbing one and the energy is calculated that is absorbed per unit time from light sources that are uniformly distributed on a shell with radius $r \gg \lambda$(reciprocity theorem). Because all experimental data were obtained on a europium chelate[1-3], where the emitting dipole does not have a preferred orientation, the expected emission rate $1/\tau^a_M$ is obtained as the average:

$$1/\tau^a_M = \frac{1}{3}(1/\tau^{\parallel}_M) + \frac{2}{3}(1/\tau^{\perp}_M) .$$

In the case of a metal mirror the phase shift $\delta_{\parallel}$ is strongly dependent on the angle α in a way which gives rise to an expected increase of the decay time $\tau_M^{\parallel}$ by about two orders of magnitude for $z \approx 3.5$. The decay time $\tau_M^{\perp}$ is not expected to vary by more than a factor of 2. If the boundary between two media of refractive index n_1 and n_2 constitutes the reflector, the reflection coefficients and phase shifts are given by Fresnel's formulae. Here the strongest effect is expected at d = 0[3]. For this distance the above integrals can be solved, and it is found with $n = n_1/n_2$ that:

$$\tau/\tau_M^{\parallel} = \langle n^3+2n^2-2n-1-3n^2(n^2+1)^{-\frac{1}{2}}\ell n[n-1+n^{-1} + (n^{-1}-1)\ (n^2+1)^{\frac{1}{2}}]\rangle/(n^7+n^5-n^3-n)$$

and

$$\tau/\tau_M^{\perp} = \langle n^7 + \frac{3}{2}\,n^5-2n^4+ 2n^3- \frac{3}{2}\,n^2-1 + \frac{3}{2}\,n^4(n^2+1)^{-\frac{1}{2}}\ell n[n-1+n^{-1} + (n^{-1}-1)\ (n^2+1)^{\frac{1}{2}}]\rangle/(n^7+ n^5- n^3- n) \quad .$$

For instance, one obtains with n = 1.5 $\tau_M^{\parallel}/\tau = 3.30$ and $\tau_M^{\perp}/\tau = 1.08$ and with n = 2.0 $\tau_M^{\parallel}/\tau = 9.28$ and $\tau_M^{\perp}/\tau = 1.05$. These equations are of practical importance, as they relate to the case of molecules adsorbed at an interface.

The above considerations are extended to the case of a dipole between two (or more) parallel mirrors, for which the most detailed experimental data are available[2,3]. If nonradiative decay processes compete with the emission (quantum yield $q < 1$), the ratio of the real decay times τ' and τ_M' is given by

$$\tau'/\tau_M' = 1 + q[(\tau/\tau_M)-1].$$

Thus a value $q < 1$ gives rise to a less pronounced effect of the mirror on the decay time. Assuming a value $q = 0.7$ the experimental data agree very well with the theory. A remarkable difference between a silver and gold mirror, observed experimentally[2], can be explained in terms of the different optical constants of these materials.

References

1. K.H. Drexhage, M. Fleck, H. Kuhn, F.P. Schäfer, and W. Sperling, Ber. Bunsenges. Phys. Chem. *70*, 1179 (1966).

2. K.H. Drexhage, *Optische Untersuchungen an neuartigen monomolekularen Farbstoffschichten* (Habilitations-Schrift, University of Marburg, Germany, 1966).

3. K.H. Drexhage, J. Luminesc. *1,2*, 693 (1970).

ATOMIC COHERENT STATES IN QUANTUM OPTICS

F. T. Arecchi

Università di Pavia & Laboratori C.I.S.E., Milano, Italy

E. Courtens

IBM Research Laboratory, Zürich, Switzerland

R. Gilmore*

Massachusetts Institute of Technology, Cambridge, Mass.

H. Thomas

J. W. Goethe Universität, Frankfurt, Germany

1. Introduction+

The central problem of Quantum Optics (laser theory, superradiance, resonant propagation, etc.) is the description of the interaction between N atoms and an electromagnetic field confined in a cavity of finite volume. A suitable model Hamiltonian for this problem is the following one ($\hbar = 1$)

$$H = \sum_k \omega_k a_k^+ a_k + \frac{\omega_o}{2} \sum_{i=1}^{N} S_{3(i)} + \sum_{k,i} g_k (a_k S_i^+ e^{i\underline{k}\cdot\underline{x}_i} + a_k^+ S_i^- e^{-i\underline{k}\cdot\underline{x}_i}) ,$$

*Present address: Physics Department, University of South Florida, Tampa, Florida 33620

+This paper is a shortened version of a longer contribution submitted to the Physical Review.

where a_k, a_k^+ are Bose operators describing the k-th field mode and $S_i^{\pm}$, S_{3i} are Pauli operators describing the atom located at position $\underline{x}_i$ as a two-level system.

We can introduce the collective operators

$$J_k^{\pm} = \sum_i S_i^{\pm} e^{\pm i\underline{k}\cdot\underline{x}_i}$$

$$J_z = \sum_i S_{3i} .$$

They obey the commutation rule:

$$[J_k^+, J_{k'}^-] = \sum_i S_{3i} e^{i(\underline{k}-\underline{k}')\cdot\underline{x}_i}$$

This reduces to $\delta_{kk'}J_z$ in the following particular cases which are of extreme physical importance:

i) point laser (cavity volume $<<\lambda^3$)
ii) single mode laser (traveling wave)
iii) traveling wave field in an amplifying or absorbing medium.

In such cases the above operators obey standard angular momentum commutation rules. The associated Heisenberg equations of motion become (leaving out for simplicity the k index)

$$\dot{a} = -i\omega a - i g J^-$$

$$\dot{J}^- = -i\omega_o J^- + igaJ_z$$

$$\dot{J}_z = -ig(aJ^+ - a^+J^-) ,$$

plus similar equations for a^+ and J^+. This set of five equations is *not* closed. For instance, to solve the second equation, we must know the evolution of the binary operator aJ_z , whose equation of motion will imply ternary operators, and so on.

In the self-consistent approximation (SCA), or semi-classical approach, we introduce the approximation

$$<a\, J^+> \approx <a>\, <J^+> ,$$

that is, we consider only the interaction among the mean fields. The three equations for $J^{\pm}$, J_z can be summarized in the vector equation

$$\langle\dot{\underline{J}}\rangle = \underline{\Omega} \times \langle\underline{J}\rangle \ ,$$

where $\underline{\Omega} \equiv (g\langle a\rangle, 0, \omega_0)$. This is a Bloch [1] equation for an angular momentum $\underline{J}$ precessing around a classical field $\langle a\rangle$.

The field equation becomes

$$\langle\dot{a}\rangle = -\, i\omega\langle a\rangle - ig\langle J^-\rangle \quad .$$

This is the equation for a field acted upon by a classical current $\langle J\rangle$. Starting with a field in the vacuum state, it leads to a particular field state with $\langle a\rangle \neq 0$ which was introduced by Bloch and Nordsieck to deal with the "infrared catastrophe", [2] then formalized by Schwinger [3], and later used by Glauber in Quantum Optics, [4-5] under the name of coherent state.

We will similarly call a coherent atomic state, or Bloch state, that state with zero induced dipole ($\langle J^-\rangle \neq 0$) generated from the ground state by a classical field.

We want to show that these states, besides being generated by classical sources, give expectation values of quantum operators whose limits, for large excitations, are the classical value (see Table I).

	Atomic states		Field states
Classical states	Classical current	↘ ↙	Classical field
	↕	╳	↕
Quantum states	Coherent atomic state		Coherent field state

Table I. Classical excitation and coherent states. The single arrows indicate the direction of production of coherent states starting from classical states. The double arrows indicate states connected by the correspondence principle.

2. Properties of the Field Coherent States

We begin by summarizing the properties of the field coherent states. [3,4,5]

1. Model Hamiltonian: a useful model for the interaction of a single mode of the electromagnetic field (neglecting polarization) with a classical driving current is:

$$H = H_o + H_{pert}$$
$$H_o = \hbar\omega a^\dagger a + E_o I = \hbar\omega n + E_o I$$
$$H_{pert} = \lambda(t) a^\dagger + \lambda^*(t) a \quad .$$

Here $a^\dagger a$ is the number operator, while $a^\dagger$ and a are the creation and annihilation operators, respectively.

2. Commutation relations: the field operators obey the commutation relations

$$[n , a^\dagger] = + a^\dagger \qquad [n , I] = 0$$
$$[n , a] = - a \qquad [a , I] = 0$$
$$[a^\dagger , a] = - I \qquad [a^\dagger , I] = 0 \quad .$$

The operators n, $a^\dagger$, a, I span a Lie algebra h_4 , called the "harmonic oscillator algebra."

3. Diagonal states: eigenstates of the Hamiltonian H_o are states $|n>$ of the photon field containing a fixed number n of photons

$$H_o|n> = (n\hbar\omega + E_o)|n>$$
$$|n> = \frac{(a^\dagger)^n}{\sqrt{n!}}|0> \quad .$$

These states are called "Fock" states.

4. Ground state: The ground state $|0>$ is defined up to a complex phase factor by

$$a|o> = 0 \quad \text{or} \quad e^a|0> = |0> \quad .$$

This state also obeys

$$n|0> = a^\dagger a|0> = 0 \quad .$$

5. Unitary translation operator: under the influence of a classical current, the ground state $|0\rangle$ will evolve under the influence of a unitary operator

$$U(\alpha) = e^{\alpha a^\dagger - \alpha^* a}$$

$$U(\alpha)|0\rangle = |\alpha\rangle \quad .$$

In general, $\alpha(t)$ is a time-dependent complex number, and $\alpha(t) \neq \lambda(t)$. These states are called "Coherent" states (Glauber states).

6. Eigenvalue equation of the coherent state: the coherent state $|\alpha\rangle$ obeys an eigenvalue equation which may be derived from the equation defining the ground state $|0\rangle$, which is also a coherent state ($\alpha = 0$):

$$\{U(\alpha) a U^{-1}(\alpha)\} \{U(\alpha)|0\rangle\} = (a - \alpha)|\alpha\rangle \quad .$$

7. Baker-Campbell-Hausdorff formulas: these formulas involve the rearrangement of the orders of exponential operator products. They are extremely useful for facilitating computations. For the Lie algebra h_4 a useful BCH relation is

$$e^{(\alpha a^\dagger + \beta a)} = e^{-\frac{1}{2}\alpha\beta} e^{\beta a} e^{\alpha a^\dagger} = e^{+\frac{1}{2}\alpha\beta} e^{\alpha a^\dagger} e^{\beta a} \quad .$$

8. Expansion of $|\alpha\rangle$ in terms of $|n\rangle$: such an expansion is greatly facilitated by the application of a BCH relation:

$$|\alpha\rangle = U(\alpha)|0\rangle = e^{-\frac{1}{2}\alpha^*\alpha} e^{\alpha a^\dagger} e^{-\alpha^* a}|0\rangle$$

$$= e^{-\frac{1}{2}\alpha^*\alpha} \sum_0^\infty \frac{(\alpha a^\dagger)^n}{n!} |0\rangle$$

$$= e^{-\frac{1}{2}\alpha^*\alpha} \sum_0^\infty \frac{(\alpha)^n}{\sqrt{n!}} |n\rangle \quad .$$

9. Non-orthogonality: there is a non-denumerable number of coherent states $|\alpha\rangle$, but a denumerable number of basis vectors $|n\rangle$ spanning the Hilbert space of possible states of the harmonic oscillator. We should expect the states $|\alpha\rangle$ to have a non-zero overlap. This inner product may be computed by means of the eigenstate expansion above. It may also be computed using a BCH relation:

$$<\alpha|\beta> = <0|U^\dagger(\alpha)U(\beta)|0>$$

$$= e^{+\alpha^*\beta - \frac{1}{2}[\alpha^*\alpha+\beta^*\beta]}$$

$$|<\alpha|\beta>|^2 = e^{-|\alpha-\beta|^2} \quad .$$

10. Over-completeness: the identity operator I may be resolved either in terms of the diagonal states or the coherent states:

$$\int \frac{d^2\alpha}{\pi} |\alpha> <\alpha| = I = \sum_{n=0}^{\infty} |n> <n| \quad .$$

11. Uncertainty relations: the canonically conjugate operators p, q are linear combinations of the operators $a^\dagger$, a, which are hermitian:

$$p = \frac{a-a^\dagger}{i\sqrt{2}}$$

$$[p, q] = -i \; .$$

$$q = \frac{a+a^\dagger}{\sqrt{2}}$$

These non-commuting hermitian operators have minimum uncertainty within a coherent state

$$(\Delta p)^2 \; (\Delta q)^2 = (\tfrac{1}{2})^2$$

$$(\Delta p)^2 = <\alpha|(p - \bar{p})^2|\alpha>$$

$$\bar{p} = <\alpha|p|\alpha> \quad .$$

12. Generating functions: such functions greatly facilitate the computation of matrix elements of ordered products of creation and annihilation operators within coherent states. The ordering may be:

normal $(<\alpha|(a^\dagger)^m (a)^n|\alpha> = (\alpha^*)^m (\alpha)^n)$,

anti-normal $(<\alpha|(a)^n(a^\dagger)^m|\alpha>$, or

fully symmetrized $(<\alpha|S\{(a)^n(a^\dagger)^m\}|\alpha>)$.

It is clear that

$$\langle\alpha|(a)^n(a^\dagger)^m|\alpha\rangle = \left(\frac{\partial}{\partial\gamma}\right)^n\left(\frac{\partial}{\partial\delta}\right)^m \langle\alpha|e^{\gamma a}\, e^{\delta a^\dagger}\,|\alpha\rangle\Big|_{\gamma=\delta=0} \quad .$$

The matrix element on the right is a generating function whose value is easily computed:

$$\langle\alpha|e^{\gamma a}\, e^{\delta a^\dagger}\,|\alpha\rangle = \langle 0|e^{-\frac{1}{2}\alpha^*\alpha}\, e^{\alpha^* a}\, e^{\gamma a}\, e^{\delta a^\dagger}\, e^{\alpha a^\dagger}\, e^{-\frac{1}{2}\alpha^*\alpha}|0\rangle$$

$$= e^{-\alpha^*\alpha}\, \langle 0|e^{(\alpha^*+\gamma)a}\, e^{(\alpha+\delta)a^\dagger}\,|0\rangle \ .$$

The application of a BCH relation considerably simplifies this matrix element:

$$= e^{-\alpha^*\alpha}\langle 0|e^{(\alpha+\delta)a^\dagger}\, e^{(\alpha^*+\gamma)(\alpha+\delta)}\, e^{(\alpha^*+\delta)a}|0\rangle$$

$$= e^{-\alpha^*\alpha} e^{(\alpha^*+\gamma)(\alpha+\delta)} \quad .$$

At this time we stress that the most important reason for studying the coherent states is the semi-classical theorem: An electromagnetic field mode, originally in a coherent state $|\alpha\rangle$, or in particular in its ground state $|0\rangle$, will evolve into a coherent state under the influence of a classical external driving current when the radiation reaction can be neglected.

3. Properties of a Single Two-Level System

We turn our attention now to a system with two internal degrees of freedom, such as an electron or a two-level atom. We ignore all external degrees of freedom, such as translational, etc. Under these conditions, the treatment of the two-level system is remarkably similar to the treatment of the oscillator. We summarize the properties of a two-level system:

1. Model Hamiltonian: a useful model for the interaction of the two-level system with a classical driving field is

$$H = H_o + H_{pert}$$

$$H_o = \hbar\omega S_z + E_o I_2$$

$$H_{pert} = \lambda(t)S_+ + \lambda^*(t)S_- \quad .$$

Here $S_\pm$, S_z are the usual angular momentum operators for $j = \frac{1}{2}$.

2. Commutation relations: the operators $\underline{S}$ are proportional to the Pauli spin operators

$$\underline{S} = \tfrac{1}{2}\underline{\sigma} \ .$$

They obey the commutation relations

$$\begin{array}{ll} [S_z, S_+] = +S_+ & [S_z, I_2] = 0 \\ [S_z, S_-] = -S_- & [S_+, I_2] = 0 \\ [S_+, S_-] = 2S_z & [S_-, I_2] = 0 \end{array} \quad .$$

The operators S_z, S_+, S_-, I_2 span the Lie algebra u(2).

3. Diagonal states: eigenstates of the Hamiltonian H_o are the familiar angular momentum eigenstates $|{}^j_{m_j}>$, $j = \frac{1}{2}$, $m_j = \pm\frac{1}{2}$:

$$H_o|{}^{\frac{1}{2}}_m> = (\hbar\omega m + E_o)|{}^{\frac{1}{2}}_m>$$

$$|{}^{\frac{1}{2}}_{\frac{1}{2}}> = \frac{S_+^1}{\sqrt{1!}}|{}^{\frac{1}{2}}_{-\frac{1}{2}}> \quad .$$

These states are called "Dicke" states.

4. Ground state: the ground state $|{}^{\frac{1}{2}}_{-\frac{1}{2}}>$ is defined up to a complex phase factor by

$$S_-|{}^{\frac{1}{2}}_{-\frac{1}{2}}> = 0 \ , \text{ or } \quad e^{S_-}|{}^{\frac{1}{2}}_{-\frac{1}{2}}> = |{}^{\frac{1}{2}}_{-\frac{1}{2}}> \quad .$$

This state also obeys

$$(S_z + \tfrac{1}{2})|{}^{\frac{1}{2}}_{-\frac{1}{2}}> = 0 \quad .$$

5. Unitary translation operator: under the influence of a classical driving field, an arbitrary state will evolve under the influence of a unitary 2×2 transformation matrix:

$$|\psi(t)\rangle = U(t,t_o)\,|\psi(t_o)\rangle$$

$$U(t,\, t_o) \,\varepsilon\, U(2) \quad .$$

A subset of operations in U(2) will leave the ground state essentially invariant:

$$U(t,t_o)\,|g\rangle = \left(\begin{array}{c|c} U_2(1) & \\ \hline & U_1(1) \end{array}\right)\begin{pmatrix} 0 \\ 1 \end{pmatrix} = \begin{pmatrix} 0 \\ e^{i\phi} \end{pmatrix} = e^{i\phi}\,|g\rangle \; .$$

The phase factor may be absorbed into the definition of the ground state. The subgroup $U_2(1)$ is the stability subgroup of the ground state $|g\rangle$. Under an arbitrary perturbation the ground state can evolve into states existing in 1-1 correspondence with the coset representatives

$$\frac{U(2)}{U_2(1)U_1(1)} = \frac{SU(2)}{U_2(1)} \cong S^2 .$$

This space is a sphere. This sphere is called the "Bloch sphere." The corresponding states are called "Bloch" states. Such states exist in one-to-one correspondence with points on the surface of a sphere. Each may be obtained by applying a coset representative to the south polar state (Fig. 1). The coset representative is a rotation through an angle $\theta(t)$ about an axis in the x-y plane making an angle $\phi - \frac{\pi}{2}$ with the positive x axis:

$$U(\theta,\phi) = e^{(\zeta J_+ - \zeta^* J_-)}$$

$$\zeta = \frac{\theta}{2}\, e^{-i\phi}$$

$$U(\theta,\phi)\,|{}^{\frac{1}{2}}_{-\frac{1}{2}}\rangle = |{}^{\frac{1}{2}}_{\theta\phi}\rangle = |{}^{\frac{1}{2}}_{m}\rangle\, \mathcal{D}^{\frac{1}{2}}_{m,-\frac{1}{2}}(\theta\phi)$$

$$= |{}^{\frac{1}{2}}_{-\frac{1}{2}}\rangle \cos\frac{\theta}{2} + |{}^{\frac{1}{2}}_{\frac{1}{2}}\rangle\, e^{-i\phi} \sin\frac{\theta}{2} \quad .$$

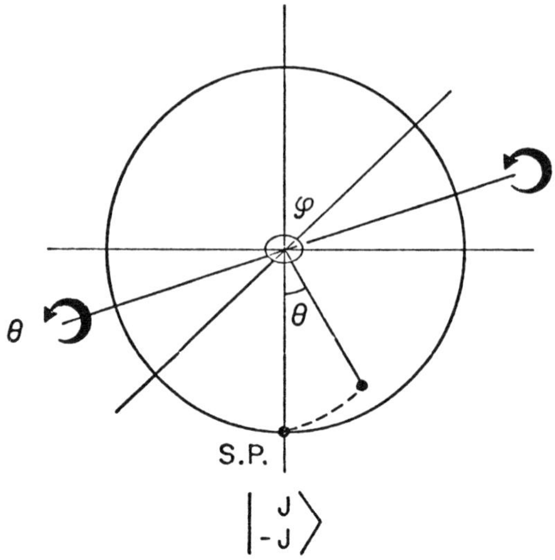

Fig. 1. Definition of the rotation $R_{\theta,\phi}$ in angular momentum space.

7. BCH formulas: reparameterizations of the group SU(2) exist and are useful for computational purposes. These formulas are computed by mapping various sets of parameters onto a particular SU(2) group element $\begin{pmatrix} x & iy \\ iy^* & x \end{pmatrix}$ in various different ways. For example,

$$e^{w_+J_+ + w_-J_- + w_zJ_z} = \begin{pmatrix} \cosh K + \frac{w_z}{2}\frac{\sinh K}{K} & w_+\frac{\sinh K}{K} \\ w_-\frac{\sinh K}{K} & \cosh K - \frac{w_z}{2}\frac{\sinh K}{K} \end{pmatrix}$$

with $K = \sqrt{w_+w_- + w^2/4}$.

Straightforward expansion of the exponential and matrix multiplication also gives

$$e^{x_+J_+}e^{(\ln x_z)\,J_z}e^{x_-J_-} = \begin{pmatrix} \sqrt{x_z} + \frac{x_+x_-}{\sqrt{x_z}} & \frac{x_+}{\sqrt{x_z}} \\ \frac{x_-}{\sqrt{x_z}} & \frac{1}{\sqrt{x_z}} \end{pmatrix} ,$$

$$e^{y_-J_-} e^{(\ln y_z) J_z} e^{y_+J_+} = \begin{pmatrix} \sqrt{y_z} & y_+\sqrt{y_z} \\ y_-\sqrt{y_z} & \frac{1}{\sqrt{y_z}} + y_+y_-\sqrt{y_z} \end{pmatrix} .$$

Equating each of these expressions for the same group element with each other provides an analytic relationship among the sets of parameters w, x, and y.

$$e^{w_+J_+ + w_-J_- + w_zJ_z} = e^{x_+J_+} e^{(\ln x_z)J_z} e^{x_-J_-} = e^{y_-J_-} e^{(\ln y_z)J_z} e^{y_+J_+} .$$

Many other BCH relations can be obtained in this way for the group SU(2). As an example

$$e^{(\zeta J_+ - \zeta^* J_-)} = e^{\tau J_+} e^{\ln(1 + \tau^*\tau)J_z} e^{-\tau^* J_-}$$

$$= e^{-\tau^* J_-} e^{-\ln(1 + \tau^*\tau)J_z} e^{\tau J_+}$$

$$\zeta = e^{-i\phi} \frac{\theta}{2}$$

$$\tau = e^{-i\phi} \tan \frac{\theta}{2} .$$

ζ and τ are related by a projective transformation. Since the BCH relation is valid in the defining representation of the Lie group SU(2), it is valid in all representations of this group. For this reason we have replaced S by J in the expression above.

Additional properties of the Bloch states for a two-level system could be stated now, but will be deferred until later.

4. Symmetrized States

The Hamiltonian which describes the interaction of N identical two-level systems with a classical internal driving field is

$$H_{TOT} = \Sigma H_i + (\text{particle-particle}) + \ldots$$

We assume all interactions among the two-level systems can be neglected, and that

$$H_i = \hbar\omega_i (S_z)_i + \lambda_i(t)(S_+)_i + \lambda_i^*(t)(S_-)_i \quad .$$

The time evolution operator for each individual particle is a unitary 2×2 matrix. If the system is initially in a direct product state, it will evolve in time into a direct product state

$$|\psi(t)\rangle = \prod_{i=1}^{N} |\phi_i(t)\rangle = \prod_{i=1}^{N} U_i(t,t_0)|\phi_i(t_0)\rangle$$

$$= U(t,t_0)|\psi(t_0)\rangle$$

$$U(t,t_0) \ \varepsilon \ SU(2) \times SU(2) \times \ldots \times SU(2) = [SU(2)]^N$$

$$U_i(t,t_0) \ \varepsilon \ \{SU(2)\}_i \ .$$

If the atoms are sufficiently close together so that they all see the same field dependence, or if for some other reason they each obey the same Hamiltonian, then

$$\omega_i = \omega$$

$$\lambda_i(t) = \lambda(t) \quad .$$

If each atom is initially in the same state, say the ground state, then each will evolve under exactly the same unitary 2×2 matrix. The state at any time t will be a direct product of identical single-particle Bloch states.

The entire system will evolve under a transformation from the group $SU(2) \otimes P_N$. The $(2)^N$ basis vectors of the form

$$\left|{}^{\frac{1}{2}}_{m_1}\right\rangle \times \left|{}^{\frac{1}{2}}_{m_2}\right\rangle \times \ldots \times \left|{}^{\frac{1}{2}}_{m_N}\right\rangle \quad , \qquad m_i = \pm\frac{1}{2}$$

which carry an irreducible representation of $\{SU(2)\}^N$, carry a reducible representation of the subgroup $SU(2) \otimes P_N$. Linear combinations of these bases may be constructed which carry irreducible representations [6-8] of the direct product subgroup (Fig. 2). These representations are described by a Young partition (frame) containing no more than 2 rows:

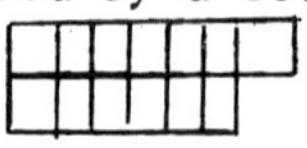

λ_1 λ_2 , with (contd.p14)

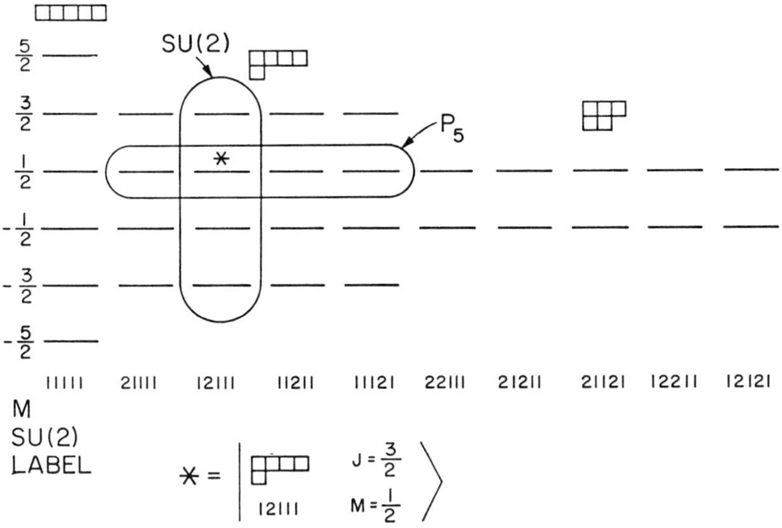

Fig. 2. An arbitrary state in the symmetry-adapted basis is labeled by: (i) The U(2) invariant subspace J in which it lies, and its position M within that space. (ii) The P_N invariant subspace $\underline{\lambda}$ in which it lies, and its position i within that space. Moreover, the intersection of any P_N invariant subspace with any U(2) invariant subspace is at most one-dimensional, so the quantum numbers J, M and $\underline{\lambda}$, i are sufficient for a complete labeling of states. The intersection is exactly one-dimensional when the partition describing the U(2) and P_N invariant subspace are identical: $2J=\lambda_1-\lambda_2$. The state marked * in the figure is labeled $\left|\begin{matrix} J = 3/2 \\ M = 1/2 \end{matrix} ; \begin{matrix} \underline{\lambda} = (4,1) \\ \overline{i} = 12111 \end{matrix}\right\rangle$.

Under a U(2) operation R, this state is mapped into a linear combination of states within the same vertical box

$$R\left|\begin{matrix} 3/2 \\ 1/2 \end{matrix} ; \begin{matrix} (4,1) \\ 12111 \end{matrix}\right\rangle = \sum_M \left|\begin{matrix} 3/2 \\ M \end{matrix} ; \begin{matrix} (4,1) \\ 12111 \end{matrix}\right\rangle \mathcal{D}^{3/2}_{M,1/2}(R) \quad .$$

Under a P_4 operation p, it is mapped into a linear combination of states within the horizontal box

$$p\left|\begin{matrix} 3/2 \\ 1/2 \end{matrix} ; \begin{matrix} (4,1) \\ 12111 \end{matrix}\right\rangle = \sum_j \left|\begin{matrix} 3/2 \\ 1/2 \end{matrix} ; \begin{matrix} (4,1) \\ j \end{matrix}\right\rangle \Gamma^{(4,1)}_{j,12111}(p) \quad .$$

In other words, group operations do not affect the labels classifying the invariant subspace (upper line); they only affect the appropriate *internal* state labels (lower line). Note that states belonging to the same M but different $\underline{\lambda}$ values are not necessarily degenerated.

$$\lambda_1 + \lambda_2 = N$$

$$\mathrm{Dim}^{\lambda_1,\lambda_2}\{SU(2) \otimes P_N\} = \mathrm{Dim}^{\lambda_1,\lambda_2}\{SU(2)\} \times \mathrm{Dim}^{\lambda_1,\lambda_2}\{P_N\} \quad .$$

The direct product group may be further restricted to either of its subgroups, SU(2) or P_N. Then the irreducible representation (λ_1,λ_2) of SU(2) $\otimes$ P_N is further reducible. Under SU(2) $\otimes$ $P_N \downarrow$ SU(2), (λ_1,λ_2) of the direct product group splits into $\mathrm{Dim}^{\lambda_1,\lambda_2}(P_N)$ copies of the irreducible representation (λ_1,λ_2) of SU(2) with $2J = \lambda_1 - \lambda_2$. Under the restriction SU(2) $\otimes$ $P_N \downarrow P_N$, (λ_1,λ_2) of SU(2) $\otimes$ P_N splits into $\mathrm{Dim}^{\lambda_1,\lambda_2}$[SU(2)] copies of the irreducible representation (λ_1,λ_2) of P_N.

$$\mathrm{Dim}^{(\lambda_1,\lambda_2)}\{SU(2)\} = \lambda_1 - \lambda_2 + 1$$

$$\mathrm{Dim}^{(\lambda_1,\lambda_2)}\{P_N\} = \begin{pmatrix} \lambda_1 + \lambda_2 \\ \lambda_2 \end{pmatrix} - \begin{pmatrix} \lambda_1 + \lambda_2 \\ \lambda_2 - 1 \end{pmatrix}$$

Part of the subgroup restricting process

$$[SU(2)]^N \downarrow SU(2) \otimes P_N \begin{matrix} \nearrow SU(2) \\ \searrow P_N \end{matrix}$$

is indicated in Fig. 2. It should be noted that the group SU(2) acts "vertically" and maps states in a vertical invariant subspace (fixed J or Young frame) into a linear combination of states in the *same* subspace. The group P_N acts "horizontally" and maps states in a horizontal invariant subspace (same Young frame) into a linear combination of states in the *same* subspace.

A theorem due to Weyl [6,7] states that the intersection of any SU(r) invariant subspace with any P_N invariant subspace is at most one-dimensional, and that it is exactly one-dimensional only when the Young partition labeling a representation of SU(r) is exactly the same as the Young partition describing a representation of P_N.

Then each symmetrized atomic state is uniquely labeled by:

1. specifying a Young partition labeling a representation of SU(r) $\otimes$ P_N;

2. specifying a label describing the position of the basis within an SU(r) invariant subspace. For SU(2), this is the M_J quantum number. For SU(r), this is the Gel'fand-Tsetlein pattern; [9,10,11]

3. specifying a label describing the position of the basis within a P_N invariant subspace. This is a Young Tableau [12] or a Yamanouchi symbol - they are both equivalent [11] to Gel'fand-Tsetlein patterns for the permutation groups.

When each of the N atoms evolves under the same field dependence,

$$H_{TOT} = \sum_{i=1}^{N} \hbar\omega(S_z)_i + \sum_{i=1}^{N} \lambda(t)(S_+)_i + \lambda^*(t)(S_-)_i$$
$$= \hbar\omega \sum_1^N (S_z)_i + \lambda(t) \sum_1^N (S_+)_i + \lambda^*(t)\sum_1^N(S_-)_i \quad .$$

If $p \varepsilon P_N$ is an arbitrary permutation operation, then

$$p\, H_{TOT}\, p^{-1} = H_{TOT} \, .$$

This result can be written symbolically

$$[H_{TOT}, P_N] = 0 \quad .$$

By Schur's Lemma, H_{TOT} is a multiple of the identity within any irreducible representation of P_N. As a result, not only are the labels $\bar{\lambda}$ describing irreducible representations of P_N good quantum numbers, but so also are the Yamanouchi symbols describing the bases within the representation $\bar{\lambda}$ of P_N.

We conclude that if such a system is initially within a single vertical column (Fig. 2 - single Yamanouchi label), it must remain within the same SU(2) {SU(r)} invariant subspace. This selection rule can lead to enhanced transition matrix elements, and has been called superradiance.[13] It arises for roughly the following reason: There is a kind of conservation law for transition probabilities within the kinds of systems described here. When some decay channels are closed, those remaining open must have an enhanced amplitude.

5. Properties of the Atomic Coherent States

Under the influence of a classical external field, the system of N identical atoms described in the previous section will evolve from the ground state $|{}^J_{-J}>$ (J = N/2) into an atomic coherent state. The properties of the atomic coherent states greatly resemble the properties of a single two-level system. These properties can be written down in complete analogy with the properties of the field coherent states.

1. Model Hamiltonian:

$$H_{TOT} = H_o + H_{pert}$$

$$H_o = \hbar\omega\Sigma(S_z)_i + E_o\Sigma(I)_i = \hbar\omega J_z + E_o J_o$$

$$H_{pert} = \lambda(t)\sum_1^N (S_+)_i + \lambda^*(t)\sum_1^N (S_-)_i = \lambda(t)J_+ + \lambda^*(t)J_-$$

2. Commutation relations: the commutation relations of the operators $\bar{J} = (\bar{S}_i)$ can be determined immediately from those of the individual particle operators

$$[(S_z)_i\,,\,(S_\pm)_j] = \pm(S_\pm)_j\,\delta_{ij}\,,\ \text{etc.}$$

In particular, we find the operators J_z, $J_\pm$, J_o span the Lie algebra U(2)

$$[J_z, J_+] = +J_+ \qquad [J_z, J_o] = 0$$

$$[J_z, J_-] = -J_- \qquad [J_+, J_o] = 0$$

$$[J_+, J_-] = 2J_z \qquad [J_-, J_o] = 0 \quad .$$

3. Diagonal states: these are eigenstates of J^2 and J_z:

$$J^2\left|\begin{matrix}J\\M\end{matrix}\right\rangle = J(J+1)\left|\begin{matrix}J\\M\end{matrix}\right\rangle$$

$$J_z\left|\begin{matrix}J\\M\end{matrix}\right\rangle = M\left|\begin{matrix}J\\M\end{matrix}\right\rangle \qquad J = \frac{N}{2}\ .$$

4. Ground state: this is defined by

$$J_-\left|\begin{matrix}J\\-J\end{matrix}\right\rangle = 0$$

up to a phase factor, which is the $(2J)^{th}$ power of the phase factor for a single two-level system.

5. Unitary translation operator: the ground state evolves into a coherent state under

$$\left|\begin{matrix} J \\ \theta\phi \end{matrix}\right\rangle = e^{(\zeta J_+ - \zeta^* J_-)} \left|\begin{matrix} J \\ -J \end{matrix}\right\rangle$$

$$= \left|\begin{matrix} J \\ M \end{matrix}\right\rangle \left\langle\begin{matrix} J \\ M \end{matrix}\right| e^{(\zeta J_+ - \zeta^* J_-)} \left|\begin{matrix} J \\ -J \end{matrix}\right\rangle .$$

The rotation matrix elements are

$$\mathcal{D}^J_{M,-J}\{e^{(\zeta J_+ - \zeta^* J_-)}\} = \binom{2J}{J \pm M}^{\frac{1}{2}} (\cos \tfrac{\theta}{2})^{J-M} (e^{-i\phi} \sin \tfrac{\theta}{2})^{J+M} .$$

6. Eigenvalue equations: an arbitrary state in the J-invariant subspace obeys

$$J^2 \left|\begin{matrix} J \\ \cdot \end{matrix}\right\rangle = J(J+1) \left|\begin{matrix} J \\ \cdot \end{matrix}\right\rangle .$$

A diagonal state obeys in addition

$$J_z \left|\begin{matrix} J \\ M \end{matrix}\right\rangle = M \left|\begin{matrix} J \\ M \end{matrix}\right\rangle .$$

The extremal south-polar state also obeys

$$J_- \left|\begin{matrix} J \\ -J \end{matrix}\right\rangle = 0 .$$

These properties of the ground state can be transferred to properties of the coherent states:

$$\{U(\theta\phi) J^2 U^{-1}(\theta\phi)\} \left|\begin{matrix} J \\ \theta\phi \end{matrix}\right\rangle = J(J+1) \left|\begin{matrix} J \\ \theta\phi \end{matrix}\right\rangle$$

$$\{U(\theta\phi)J_z U^{-1}(\theta\phi)\}\left|\begin{matrix}J\\ \theta\phi\end{matrix}\right\rangle = -J\left|\begin{matrix}J\\ \theta\phi\end{matrix}\right\rangle$$

$$\{U(\theta\phi)J_- U^{-1}(\theta\phi)\}\left|\begin{matrix}J\\ \theta\phi\end{matrix}\right\rangle = 0 \quad .$$

7. BCH relations: the analytic transformations of parameters obtained for two-level systems for the group SU(2) are valid within an arbitrary representation of this group.

8. Expansion of the coherent states: these states may be expanded directly in terms of the SU(2) rotation matrices, or in terms of BCH relations. The expansions are identical:

$$\left|\begin{matrix}J\\ \theta\phi\end{matrix}\right\rangle = \left|\begin{matrix}J\\ M\end{matrix}\right\rangle \mathcal{D}^J_{M,-J}\{e^{(e^{-i\phi}\frac{\theta}{2}J_+ - e^{i\phi}\frac{\theta}{2}J_-)}\}$$

$$= [\cos\frac{\theta}{2}]^{2J} e^{(e^{-i\phi}\tan\frac{\theta}{2}J_+}\left|\begin{matrix}J\\ -J\end{matrix}\right\rangle \quad .$$

9. Non-orthogonality: the Bloch states have non-zero overlap

$$\left\langle\begin{matrix}J\\ \theta\phi\end{matrix}\middle|\begin{matrix}J\\ \theta'\phi'\end{matrix}\right\rangle = [\cos\frac{\theta}{2}\cos\frac{\theta'}{2} + e^{i(\phi-\phi')}\sin\frac{\theta}{2}\sin\frac{\theta}{2}]^{2J}$$

$$\left|\left\langle\begin{matrix}J\\ \theta\phi\end{matrix}\middle|\begin{matrix}J\\ \theta'\phi'\end{matrix}\right\rangle\right|^2 = \left(\frac{1 + \hat{n}(\Omega)\cdot\hat{n}(\Omega')}{2}\right)^J \quad .$$

Here $\hat{n}(\Omega)$ is the unit vector drawn from the center to the point $\Omega = (\theta,\phi)$ on the surface of the Bloch unit sphere.

10. Over-completeness: the identity may be resolved in terms of either Bloch or Dicke states:

$$\frac{2J+1}{4\pi}\int\left|\begin{matrix}J\\ \theta\phi\end{matrix}\right\rangle d\mu(\theta\phi)\left\langle\begin{matrix}J\\ \theta\phi\end{matrix}\right| = I_{2J+1} = \sum_{M=-J}^{+J}\left|\begin{matrix}J\\ M\end{matrix}\right\rangle\left\langle\begin{matrix}J\\ M\end{matrix}\right|$$

$$d\mu(\theta\phi) = \sin\theta d\theta d\phi \ .$$

11. Uncertainty relations: the relations

$$\Delta J_x^2 \, \Delta J_y^2 \geq (\tfrac{1}{2})^2 \, \Delta J_z^2$$

become, after rotation through $(\theta\phi)$:

$$U(\theta\phi)(J_x, J_y, J_z)U^{-1}(\theta\phi) = (J_\xi, J_\eta, J_\zeta)$$

$$\Delta J_\xi^2 \quad \Delta J_\eta^2 \quad \geq (\tfrac{1}{2})^2 \, \Delta J_\zeta^2 \quad .$$

This assumes the minimum value within the coherent atomic state.

12. Generating functions: these can be constructed for computing expectation values of ordered products of the angular momentum operators. The generating function for the anti-normal ordering

$$\left\langle \begin{matrix} J \\ \theta\phi \end{matrix} \right| (J_-)^r (J_z)^s (J_+)^t \left| \begin{matrix} J \\ \theta\phi \end{matrix} \right\rangle$$

is simply computed:

$$\left\langle \begin{matrix} J \\ \theta\phi \end{matrix} \right| e^{a_- J_-} \, e^{a_3 J_3} \, e^{a_+ J_+} \left| \begin{matrix} J \\ \theta\phi \end{matrix} \right\rangle$$

$$= (\cos^2 \tfrac{\theta}{2})^{2J} \left\langle \begin{matrix} J \\ -J \end{matrix} \right| e^{(e^{i\phi}\tan \frac{\theta}{2} + a_-)J_-} \, e^{a_3 J_3} \, e^{(e^{-i\phi}\tan \frac{\theta}{2} + a_+)J_+} \left| \begin{matrix} J \\ -J \end{matrix} \right\rangle$$

$$= (\cos^2 \tfrac{\theta}{2})^{2J} \left\langle \begin{matrix} J \\ -J \end{matrix} \right| e^{(\ )' J_+} \, e^{a_3' J_3} \, e^{(\ )' J_-} \left| \begin{matrix} J \\ -J \end{matrix} \right\rangle$$

$$= (\cos^2 \tfrac{\theta}{2})^{2J} \left[e^{-a_3/2} + e^{a_3/2} (a_- + e^{i\phi}\tan \tfrac{\theta}{2})(a_+ + e^{-i\phi}\tan \tfrac{\theta}{2}) \right]^{2J} \quad .$$

Additional generating functions for normal and symmetrized orders have been obtained simply from this generating function using BCH relations.

6. Contraction and Coherent State Properties

The properties of the atomic coherent states bear a striking resemblance to the properties of the field coherent states. This correspondence is made manifest by contracting [14-18] the algebra u(2) to the algebra h_4, and by contracting the unitary irreducible representations of the group U(2) to the UIR of H_4.

The coherent state properties 1, 2, and 7 depend on the algebra. The remainder of the properties depend on the *representation* of the algebra or associated group.

We first contract the algebra by performing a change of basis which becomes singular in a certain limit.

$$\begin{pmatrix} h_+ \\ h_- \\ h_z \\ h_o \end{pmatrix} = \begin{pmatrix} c & \cdot & \cdot & \cdot \\ \cdot & c & \cdot & \cdot \\ \cdot & \cdot & 1 & \frac{1}{2c^2} \\ \cdot & \cdot & \cdot & 1 \end{pmatrix} \begin{pmatrix} J_+ \\ J_- \\ J_z \\ J_o \end{pmatrix} .$$

The new basis vectors obey the following commutation relations

$$[h_z, h_\pm] = \pm h_\pm$$
$$[h_+, h_-] = 2c^2 h_z - I$$
$$[\underline{h}, \ I\ ] = 0 \qquad .$$

Although the transformation is singular in the limit $c \to 0$ and its inverse fails to exist, the commutation relations above have a well-defined limit. They are isomorphic with the C.R. of the algebra h_4.

This contraction $u(2) \to h_4$ (property #2) allows the contraction of the atomic Hamiltonian to the oscillator Hamiltonian (property #1) and of the U(2) BCH relations to the H_4 BCH relations (property #7). We illustrate the latter contraction by an example:

$$e^{(\zeta J_+ - \zeta^* J_-)} = e^{\tau J_+} e^{\ln(1+\tau^*\tau) J_z} e^{-\tau^* J_-} = e^{-\tau^* J_-} e^{-\ln(1+\tau^*\tau) J_z} e^{\tau J_+}$$

$$\Big\downarrow \quad \begin{array}{l} \zeta^{(*)}/c \to \alpha^{(*)} \\ cJ_\pm \to \begin{cases} a^\dagger \\ a \end{cases} \\ c^2 J_z \to c^2 a^\dagger a - \tfrac{1}{2} I \end{array}$$

$$e^{(\alpha a^\dagger - \alpha^* a)} = e^{\alpha a^\dagger} e^{-\frac{1}{2}\alpha^*\alpha} e^{-\alpha^* a} = e^{-\alpha^* a} e^{\frac{1}{2}\alpha^*\alpha} e^{\alpha a^\dagger} .$$

The remaining coherent state properties 3-6, 8-12 are obtained by contracting the representations of U(2) to those of H_4. Since U(2) is compact, its UIR are finite-dimensional. Since H_4 is non-compact, it has no faithful finite-dimensional UIR. Therefore, we must let $J \uparrow \infty$ as $c \downarrow 0$ to construct UIR of H_4 from those of U(2). Since energies are measured from the ground state, we demand

$$\lim_{c\to 0} h_3 \left| \begin{matrix} J \\ -J \end{matrix} \right\rangle = \text{Lim}\,(J_3 + \frac{1}{2c^2}) \left| \begin{matrix} J \\ -J \end{matrix} \right\rangle = 0 \left| \begin{matrix} J \\ -J \end{matrix} \right\rangle .$$

Thus we demand $2Jc^2 \to 1$ as $c \downarrow 0$. Under this condition, the Dicke states contract to the Fock states:

$$\text{Lim}\, h_3 \left| \begin{matrix} J \\ M \end{matrix} \right\rangle = \text{Lim}\,(J_3 + \frac{1}{2c^2}) \left| \begin{matrix} J \\ M \end{matrix} \right\rangle$$

$J \to +\infty$

$M \to -\infty$

$J + M = n$ fixed

$$= \text{Lim}\,(M + J) \left| \begin{matrix} J \\ M \end{matrix} \right\rangle \to n\,\text{Lim} \left| \begin{matrix} \infty \\ J+M \end{matrix} \right\rangle$$

$$\text{Lim}\, h_+ \left| \begin{matrix} J \\ M \end{matrix} \right\rangle = \text{Lim}\, cJ_+ \left| \begin{matrix} J \\ M \end{matrix} \right\rangle$$

$$= \text{Lim} \left| \begin{matrix} J \\ M+1 \end{matrix} \right\rangle \sqrt{[2Jc^2 + (J+M)c^2][(J+M)+1]}$$

$$a^\dagger \left| \begin{matrix} \infty \\ n \end{matrix} \right\rangle = \left| \begin{matrix} \infty \\ n+1 \end{matrix} \right\rangle \sqrt{n+1} .$$

In addition, under the limit

$$\lim_{c\downarrow 0} \frac{e^{-i\phi}}{2c} \to \alpha$$

the Bloch states contract to Glauber states. It can easily be verified that all properties of the coherent atomic states (Section 5) contract to the corresponding property of the coherent field states (Section 2). For example

$$|\alpha\rangle = \lim_{c\to 0} \left| \begin{matrix} J \\ \theta\phi \end{matrix} \right\rangle = \lim_{c\to 0} (1 + \tau^*\tau)^{-J} e^{\tau J_+} \left| \begin{matrix} J \\ -J \end{matrix} \right\rangle$$

$$= \lim_{c\to 0} \left(1 - 2c^2 \frac{\alpha^*\alpha}{2}\right)^{-\frac{1}{2}c^2} e^{(\tau/c)(cJ_+)} \left| \begin{matrix} J \\ -J \end{matrix} \right\rangle$$

$$= e^{-\frac{1}{2}\alpha^*\alpha} e^{\alpha a^\dagger} \left| \begin{matrix} \infty \\ 0 \end{matrix} \right\rangle .$$

7. Comments

Not only do all properties associated with the field coherent states follow from the corresponding properties of the atomic coherent states by contraction, but also all properties associated with H_4 flow from the corresponding property in U(2). Thus the harmonic oscillator eigenfunctions are contractions of the spherical harmonics. Shift operators, Casimir invariants, special functions, eigenvalue equations, generating functions, orthogonality and completeness relations, (etc.) in H_4 are contracted directly from the corresponding property in U(2).

There is a rich variety of BCH analytic reparameterizations for SU(2). Each contracts to its counterpart in H_4. Several distinct SU(2) BCH relations may contract to the same H_4 BCH relation. The SU(2) BCH relations can be used to construct H_4 BCH relations which are difficult to obtain by other means. For instance

$$e^{\lambda J_z} e^{(\alpha_+ J_+ + \alpha_- J_-)} =$$

$$\exp\left[2 \cosh\theta \sinh\frac{\lambda}{2} J_z + \alpha_+ \frac{\sinh\theta}{\theta} e^{\lambda/2} J_+ + \alpha_- \frac{\sinh\theta}{\theta} e^{-\lambda/2} J_-\right] \frac{\Omega/2}{\sinh\Omega/2} ,$$

where $\theta^2 = \alpha_+\alpha_-$, $\cosh\Omega/2 = \cosh\theta \cosh \lambda/2$

contracts to (using $\text{Lim} \frac{\alpha_\pm}{c} = \gamma_\pm$)

$$e^{\lambda a^\dagger a} e^{(\gamma_+ a^\dagger + \gamma_- a)} =$$

$$e^{-\frac{1}{4}\lambda\gamma_+\gamma_-[1+\frac{2}{\lambda}\coth\frac{\lambda}{2} - \coth^2\frac{\lambda}{2}]} \exp[\lambda a^\dagger a + \frac{\lambda/2}{\sinh\lambda/2}(\gamma_+ e^{\lambda/2} a^\dagger + \gamma_- e^{-\lambda/2} a)] .$$

This result has already been derived [19] in a more complicated way to compute certain thermal averages.

8. Conclusion

The construction and labeling of fully symmetrized states describing N identical two-level systems has been described and illustrated. We have shown how selection rules related to permutation group labels can lead to the existence of coherent states.

When each of N identical two-level atoms evolves from the ground state under the same classical field dependence, the total system state will be a coherent atomic state. The properties of the Bloch states have been enumerated and compared with the corresponding properties of the field coherent states.

The strong similarities between the Bloch and Glauber states are made manifest by a group contraction process. All properties of the field coherent states are obtained by contraction of the corresponding property of the atomic coherent states.

The atomic and field coherent states are "dual" to each other and of physical importance because of the semi-classical duality theorem:

When radiation reaction and spontaneous decay can be neglected, then

1. under the influence of a classical current, an electromagnetic field initially in a coherent state, or in particular its ground state, will evolve into a coherent state;
2. under the influence of a classical electromagnetic field, an atomic system initially in a coherent state, or in particular, its ground state, will evolve into a coherent state.

	Dicke state $\|M = -J\cos\theta\rangle$	Bloch state $\|\theta,\phi\rangle$
Spontaneous emission	$[J^2\sin^2\theta + 2J\sin^2\frac{\theta}{2}]I_0$	$[J^2\sin^2\theta + 2J\sin^4\frac{\theta}{2}]I_0$
Stimulated emission	$-2J\cos\theta$	$-2J\cos\theta$
Dipole moment	0	$J\sin\theta(\underline{p}\, e^{i\omega t + i\phi} + c.c.)$
Classical emission	0	$[J^2\sin^2\theta]I_0$

Table II. Comparison of emission rates and dipole moments of Dicke and Bloch states of the same energy expectation value.

	Operators	Coordinates	Eigenvalues	Eigenstates	Coherent states
Angular momentum	$cJ_+, cJ_-, J_z + \frac{1}{2c^2}$	$\frac{\theta}{2c}\, e^{-i\phi}$	$2c^2 J, J+M$	$\|J,M\rangle$ (Dicke)	$\|\theta,\phi\rangle$ (Bloch)
Harmonic oscillator	$a^\dagger, a, a^\dagger a$	a	$1, n$	$\|\infty,n\rangle$ (Fock)	$\|\alpha\rangle$ (Glauber)

Table III. Rules for the contraction of the angular momentum algebra to the harmonic oscillator algebra. The limit of the angular momentum quantities (1st line) for $c \to 0$ are the corresponding harmonic oscillator quantities (2nd line).

References

1. F. Bloch, Phys. Rev. *70*, 460 (1946).
2. F. Bloch and A. Nordsieck, Phys. Rev. *52*, 54 (1937).
3. J. Schwinger, Phys. Rev. *91*, 728 (1953).

4. R. J. Glauber, Phys. Rev. *130*, 2529 (1963).
5. R. J. Glauber, Phys. Rev. *131*, 2766 (1963).
6. H. Weyl, *The Theory of Groups and Quantum Mechanics* (Dover Publications, New York, 1931).
7. H. Weyl, *The Classical Groups* (Princeton University Press, Princeton, N. J., 1946).
8. H. Boerner, *Representations of Groups* (North Holland, Amsterdam, 1963).
9. I. M. Gel'fand, M. L. Tsetlein, Doklady Akad. Nauk. S.S.S.R. *71*, 825 (1950).
10. I. M. Gel'fand, M. L. Tsetlein, Doklady Akad. Nauk. S.S.S.R. *71*, 1017 (1950).
11. R. Gilmore, Jl. Math. Phys. *11*, 3420 (1970).
12. M. Hammermesh, *Group Theory* (Addison Wesley, Reading, Mass. 1962).
13. R. H. Dicke, Phys. Rev. *93*, 99 (1954).
14. I. E. Segel, Duke Math. Jl. *18*, 221 (1951).
15. E. Inönü, E. P. Wigner, Proc. Nat'l. Acad. Sci. (U.S.A) *39*, 510 (1953).
16. E. J. Saletan, Jl. Math. Phys. *2*, 1 (1961).
17. W. Miller, *Lie Theory and Special Functions* (Academic Press, New York, 1968).
18. A. O. Barut, L. Girardello, Commun. Math. Phys. *21*, 41 (1971).
19. G. H. Weiss, A. A. Maradudin, Jl. Math. Phys. *3*, 771 (1962).

COHERENT STATES IN MODERN PHYSICS

Robert Gilmore

University of South Florida, Tampa, Florida

1. Introduction

In the preceding work [1] a number of algebraic techniques have been used to construct and label the symmetrized states describing an ensemble of N identical two-level atoms. Under certain physically attainable circumstances, such states evolve into "coherent atomic states" under a classical driving field. The properties of the atomic coherent states were stated and compared with the properties of the field coherent states. The atomic and field coherent states were found to be related by a group contraction process.

The procedure used in the previous work is algebraic in nature and is valid for other systems besides the electromagnetic field or two-level atoms. The central feature in this work was to make a model of a physical Hamiltonian in which the mathematical operators involved are elements of a Lie algebra. Under these conditions the system state will evolve in time within a vector space which carries a unitary irreducible representation of the Lie group associated with the Lie algebra.

In particular, under the influence of a classical driving field and neglecting radiation reaction, the ground state will evolve into a "coherent" state. There is a one-to-one correspondence between coherent states and points in a particular coset space. This is the quotient of the Lie group G under consideration, by the largest subgroup H leaving the ground state essentially invariant. We obtain coherent states by applying elements from the coset G/H to the ground state.

The properties of such coherent states may be enumerated in detail. Some of their properties depend on the Lie algebra (derived from the Hamiltonian) alone. Other properties depend on the particular representation space to which the ground state belongs. All properties 1 → 12 can be stated explicitly for any system amenable to the treatment outlined above.

We present here the application of this technique to the following four systems:

1. r-level atoms
2. superconductors
3. superfluids
4. electrons in metals in the presence of electric and magnetic fields.

2. r-level Atoms

In this section we treat r-level atoms in exactly the same way that two-level atoms were treated previously.

1a. Model Hamiltonian for a single r-level atom: if the algebra u(2) is useful for describing a two-level system, then the algebra u(r) is useful for describing an r-level system.

$$\mathcal{H} = \mathcal{H}_o + \mathcal{H}_{pert}$$

$$\mathcal{H}_o = \kappa \underline{v}\cdot\underline{H}$$

$$\mathcal{H}_{pert} = \sum_{\underline{\alpha}} \lambda^{\underline{\alpha}}(t) E_{\underline{\alpha}} \quad .$$

The generators $\underline{H} = (H_1, H_2, \ldots H_r)$ belong to the Cartan subalgebra [2] of u(r), $\underline{v}$ is a vector in the r-dimensional root space, and κ is a constant with the dimensions of energy. The constants $\kappa, \underline{v}$ are chosen so as to reproduce the static energy level splitting for the unperturbed atom (Fig. 1).

A transition from the jth to the ith level is effected by the operator

$$E_{\underline{\alpha}} = E_{\underline{e}_i - \underline{e}_j} = U^i{}_j \quad .$$

The time-dependent functions $\lambda^{\underline{\alpha}}(t)$ describe the magnitude with which a field mode couples to an atomic resonance frequency. If a transition is forbidden, $\lambda^{\underline{\alpha}}(t) = 0$. The $\underline{\alpha}$ sum extends over all

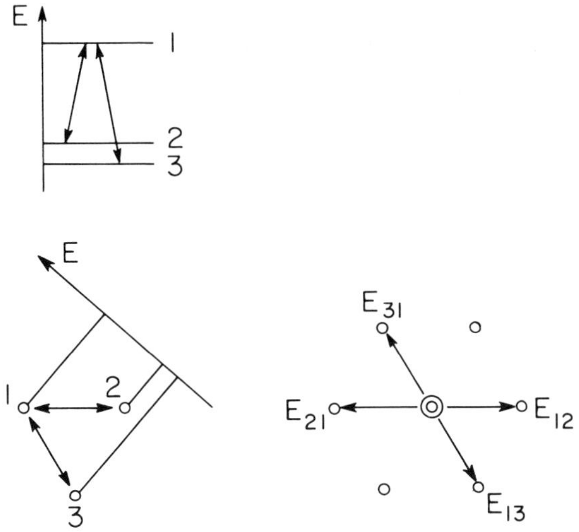

Fig. 1. The energy level spectrum of a three-level atom can be reproduced, up to a scaling factor κ whose dimensions are energy, by projecting the weights $\underline{m}$ in the first fundamental representation $\lambda = \square$ onto an axis $\underline{v}$. The operators which cause transitions between the states of a three-level atom are the shift operators $\underline{\alpha}$ in the root space diagram for SU(3). In this figure the transition between the states two and three is forbidden, so the shift operators E_{23}, E_{32} will not occur in the perturbing Hamiltonian.

non-zero roots in the root space of u(r).

In order for the Hamiltonian to be hermitian, $\kappa\underline{v}$ must be real and $\lambda^{\underline{\alpha}}(t)^{\dagger} = \lambda^{-\underline{\alpha}}(t)$. When the driving field is classical, $\lambda^{\underline{\alpha}}(t)$ is a c-number.

1b. Model Hamiltonian for N identical r-level atoms: when particle-particle interactions can be neglected, then

$$\mathcal{H}_{TOT} = \sum_{s=1}^{N} \mathcal{H}_s$$

$$\mathcal{H}_s = \kappa_s \underline{v}_s \cdot (\underline{H})_s + \sum_{\alpha} \lambda^{\underline{\alpha}}_s(t)(E_{\underline{\alpha}})_s \ .$$

If all particles interact with the same external classical field, then $\mathcal{H}_{TOT}$ may be simplified

$$\mathcal{H}_{TOT} = \kappa \underline{v} \cdot \sum_{s=1}^{N} (\underline{H})_s + \sum_{\alpha} \lambda^{\underline{\alpha}}(t) \sum_{s=1}^{N} (E_{\underline{\alpha}})_s \quad .$$

2a. Single particle commutation relations:

$$[U^i{}_j, U^r{}_s] = U^i{}_s \delta_j{}^r - U^r{}_j \delta_i{}^s$$

$$H_i = U^i{}_i$$

$$E_{\underline{e}_i - \underline{e}_j} = U^i{}_j \quad .$$

These are the commutation relations for u(r).

2b. Many particle commutation relations: the operators for the individual atoms t,t' are kinematically independent. As a result

$$[(U^i{}_j)_t, (U^r{}_s)_{t'}] = [U^i{}_j, U^r{}_s]_t \, \delta_{tt'}$$

In particular, the operators

$$\tilde{U}^i{}_j = \sum_{t=1}^{N} (U^i{}_j)_t$$

simply obey the u(r) C.R:

$$[\tilde{U}^i{}_j, \tilde{U}^r{}_s] = \tilde{U}^i{}_s \delta_j{}^r - \tilde{U}^r{}_j \delta_s{}^i \quad .$$

3a. Eigenstates of the unperturbed Hamiltonian: the single-atom eigenstates of the Hamiltonian $\mathcal{H}_o$ are the eigenstates of the generators H_i in the Cartan subalgebra of u(r). These states are uniquely described by a weight $\underline{m}$ belonging to the weight space diagram of the fundamental representation of u(r). Moreover

$$\mathcal{H}_o |\underline{m}\rangle = \kappa \underline{v} \cdot \underline{H} |\underline{m}\rangle = \kappa \underline{v} \cdot \underline{m} |\underline{m}\rangle \quad .$$

3b. Symmetrized states for N identical r-level atoms: such states are constructed and labeled following exactly the same procedure used in the two-level case. When each atom evolves under its own individual field dependence, the dynamical transformation group for the system is

$$\prod_{t=1}^{N} \{U(r)\}_t \simeq \{U(r)\}^N \quad .$$

The $(r)^N$ direct product bases carry an irreducible representation of this group.

When each atom sees the same field dependence, the dynamical transformation group reduces to $U(r) \otimes P_N$. The $(r)^N$ dimensional irreducible representation of the original group breaks up into a number of irreducible representations of the direct product group $U(r) \otimes P_N$. Each such representation is characterized by a Young frame or a Young partition of N. The original representation is simply reducible: Each representation $(\bar{\lambda}; U(r) \otimes P_N)$ occurs exactly once in the reduction of the direct product representation, when $\bar{\lambda}$ is a proper partition of N into no more than r rows, and not at all for $\bar{\lambda}$ otherwise. Finally, the direct product group $U(r) \otimes P_N$ may be restricted to either of its factors $U(r)$ or P_N, in which case $(\bar{\lambda}; U(r) \otimes P_N)$ is further reducible. This reduction scheme is illustrated in Fig. 2.

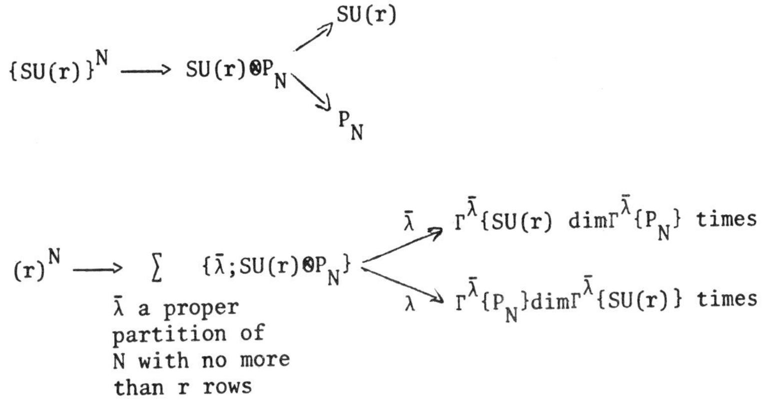

Fig. 2. Top: the dynamical transformation group for N independent systems, each with r internal degrees of freedom, is $\{SU(r)\}^N$. When each atom obeys the same Hamiltonian, the dynamical transformation group is reduced to $SU(r) \otimes P_N$. We can further look at the action of each component separately. SU(r) acts "vertically" while P_N acts "horizontally." Bottom: the $(r)^N$ direct product states for N independent r-level atoms carry an irreducible representation of $\{SU(r)\}^N$. This representation is simply reducible to a direct sum of irreducible representations of the subgroup under the restriction $\{SU(r)\}^N \downarrow SU(r) \otimes P_N$. Only those representations occur for which $\underline{\lambda}$ is a proper partition of N containing no more than r rows. Under the further reduction of $SU(r) \otimes P_N$ to either of its subgroups, an $SU(r) \otimes P_N$ irreducible representation $\underline{\lambda}$ is reducible to a direct sum of representations of type $\underline{\lambda}$ of the subgroup. The number of copies of this representation is equal to the dimension of the representation of type $\underline{\lambda}$ of the other group.

Once again, the Weyl [3,4] intersection theorem is applicable. The U(r) invariant subspaces (vertical) are characterized by a partition $\bar{\lambda}$ of N into no more than r rows. The P_N invariant subspaces (horizontal) are characterized by a partition $\bar{\lambda}'$ of N. The intersection of a U(r) invariant subspace with a P_N invariant subspace is at most one-dimensional. It is exactly one-dimensional when the $\bar{\lambda}$ characterizing representations of U(r), and the $\bar{\lambda}'$ describing representations of P_N are identical:

$$\bar{\lambda} = \bar{\lambda}' \quad .$$

As a result, any symmetrized state is uniquely described by specifying its location within a P_N invariant subspace and a U(r) invariant subspace. Bases within U(r) invariant subspaces are labeled by Gel'fand-Tsetlein patterns. [5,6] Bases within P_N invariant subspaces are Yamanouchi symbols, or Young tableaux, both of which are equivalent to each other and which are essentially the Gel'fand-Tsetlein patterns for the symmetric group. [7]

These labels provide a mechanism for uniquely labeling the fully symmetrized states of N identical r-level atoms. [8] Such states are shown schematically below, together with a postal analogy:

$$\left| \begin{matrix} \text{Young frame} \\ \text{Gel'fand-Tsetlein pattern} \; ; \; \text{Yamanouchi symbol} \end{matrix} \right\rangle \cong \left| \begin{matrix} \text{Manhattan} \\ \text{2nd. Ave; 49th St.} \end{matrix} \right\rangle$$

These symmetrized eigenstates of the unperturbed Hamiltonian have energy eigenvalues determined from

$$\mathcal{H}_{0\,TOT} \left| \begin{matrix} \bar{\lambda} \\ \underline{M} \end{matrix} \right\rangle = \kappa \underline{v} \cdot \underline{M} \left| \begin{matrix} \bar{\lambda} \\ \underline{M} \end{matrix} \right\rangle$$

Several Gel'fand-Tsetlein patterns may give rise to the same weight $\underline{M}$. This accounts for the multiple degeneracy of some of the energies indicated in Figs. 3,4.

4. Ground state: a large number of shift operators annihilate the ground state of the single r-level atom in the following sense

$$E_{\underline{\alpha}} |g\rangle = 0$$

Not all of these operators are independent. In fact, only r-1 of the operators suffice to define the ground state up to a phase factor. These are

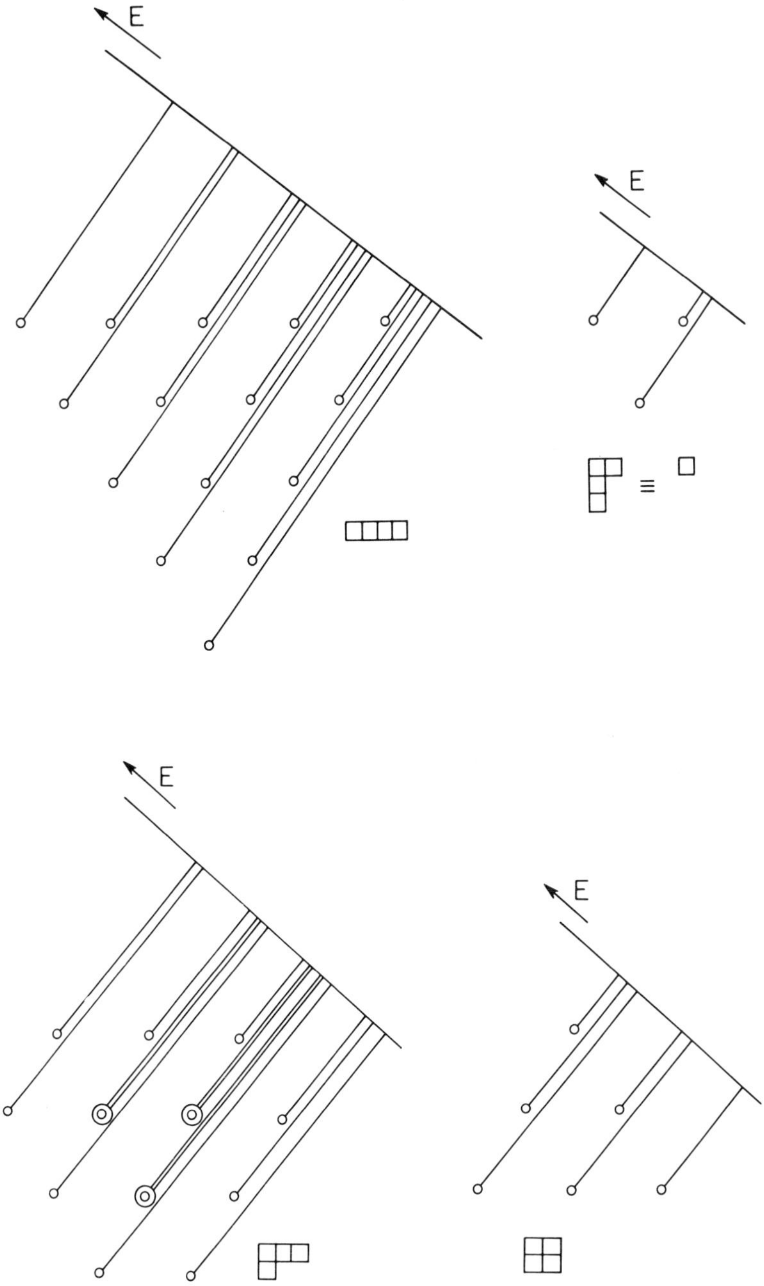

Fig. 3 (For caption, see following page).

Fig. 3. When four identical three-level atoms evolve under identical Hamiltonians, the dynamical transformation group is $SU(3)\otimes P_4$. The energy level spectrum for this system is simply obtained by projecting the appropriate weight space diagram $\bar{\lambda}$ of SU(3) onto the axis $\underline{v}$. The four representations of SU(3) involved in the reduction described in Fig. 2 are shown here.

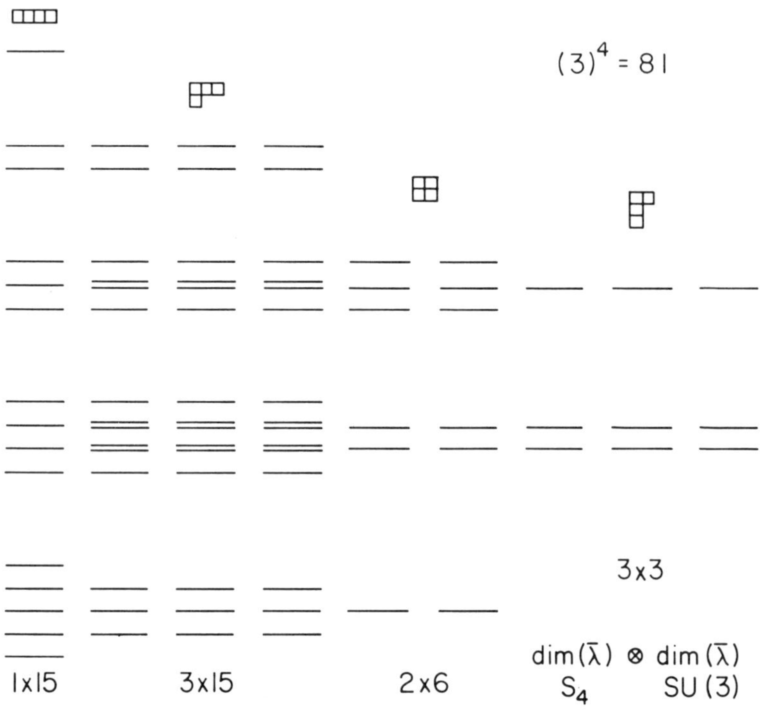

Fig. 4. The energy level spectrum for an ensemble of four identical three-level atoms can be determined directly from the projections illustrated in Fig. 3. The double degeneracies shown arise from the weight degeneracy in the representations (3,1) of SU(3). The representation $(\underline{\lambda};SU(3))$ occurs dim $(\underline{\lambda},P_4)$ times in this energy level scheme. The group SU(3) acts "vertically" within the SU(3) invariant subspace $(\underline{\lambda},SU(3))$, while P_N acts horizontally within a P_N invariant subspace $(\underline{\lambda}',P_N)$. By the Weyl intersection theorem, these invariant subspaces have an intersection which is at most one-dimensional. The intersection is non-zero *only* when $\underline{\lambda}=\underline{\lambda}'$. Therefore, Young partitions which label the representation of $SU(r)\otimes P_N$, and Gel'fand-Tsetlein patterns (vertical labels) and Yamanouchi symbols (horizontal labels) serve to label symmetrized atomic states completely.

$$U^1{}_2\ ,\ U^2{}_3\ ,\ U^3{}_4\ ,\ \ldots\ ,\ U^{r-1}{}_r\ .$$

Within *any* representation of the algebra or group, the ground state is defined by

$$\Gamma^{\bar{\lambda}}(U^i{}_{i+1})\left|\begin{matrix}\bar{\lambda}\\ \text{gnd}\end{matrix}\right\rangle = 0$$

$$\Gamma^{\bar{\lambda}}(e^{U^i{}_{i+1}})\left|\begin{matrix}\bar{\lambda}\\ \text{gnd}\end{matrix}\right\rangle = e^{\Gamma^{\bar{\lambda}}(U^i{}_{i+1})}\left|\begin{matrix}\bar{\lambda}\\ \text{gnd}\end{matrix}\right\rangle = \left|\begin{matrix}\bar{\lambda}\\ \text{gnd}\end{matrix}\right\rangle .$$

In particular, for a single particle $\bar{\lambda} = |\square|$. We will associate the ground state $\left|\begin{matrix}\bar{\lambda}\\ \text{gnd}\end{matrix}\right\rangle$ with the lowest weight state $\left|\begin{matrix}\bar{\lambda}\\ M^{\ell}\end{matrix}\right\rangle$, in some particular ordering of the weights.

5. Unitary translation operator: assume that a single r-level atom evolves from the ground state under an arbitrary Hamiltonian. Then it will evolve according to

$$|\psi(t)\rangle = U(t,t_o)\,|\text{gnd}\rangle$$

$$U(t,t_o) \in U(r) \quad .$$

The operations in the subgroup $U(r-1) \otimes U(1)$ leave the ground state essentially invariant:

$$\left(\begin{array}{c|c} U(r-1) & 0 \\ \hline 0 & U(1) \end{array}\right)\left(\begin{array}{c} 0 \\ \hline 1 \end{array}\right) = \left(\begin{array}{c} 0 \\ \hline e^{i\phi} \end{array}\right) .$$

The ground state can always be redefined by absorbing the phase factor. Therefore, there is a one-to-one correspondence between coherent states - the states into which the ground state can evolve under a classical driving field - and the coset representatives in

$$\frac{U(r)}{U(r-1)\otimes U(1)} = \frac{SU(r)}{U(r-1)} = S^{2(r-1)} \quad .$$

This coset space is topologically equivalent to the surface of a unit sphere of dimensionality 2(r-1). This is most easily seen as follows:

$$\left(\begin{array}{c|c} & c_r \\ & \cdot \\ & \cdot \\ & c_2 \\ \hline -c_r^* \cdots -c_2^* & r_1 \end{array}\right) \left(\begin{array}{c} 0 \\ \cdot \\ \cdot \\ 0 \\ \hline 1 \end{array}\right) \longrightarrow \left(\begin{array}{c} c_r \\ \cdot \\ \cdot \\ c_2 \\ \hline r_1 \end{array}\right) .$$

Since the ground state can be redefined by absorbing a phase factor, r_1 can always be taken real. Therefore, the coefficients describing the coherent state obey the normalization condition

$$r_1^2 + c_2^* c_2 + \ldots + c_r^* c_r = 1 \quad .$$

Thus the single atom coherent states exist in one-to-one correspondence with the points in $S^{2(r-1)}$.

The coset representatives can be expressed as exponentials of elements in the Lie algebra:

$$\mathrm{EXP}\left(\begin{array}{c|c} 0 & A \\ \hline -A^{\dagger} & 0 \end{array}\right) = \left(\begin{array}{c|c} \cos\sqrt{(AA^{\dagger})} & A\,\dfrac{\sin\sqrt{(A^{\dagger}A)}}{\sqrt{(A^{\dagger}A)}} \\ \hline -A^{\dagger}\,\dfrac{\sin\sqrt{(AA^{\dagger})}}{\sqrt{(AA^{\dagger})}} & \cos\sqrt{(A^{\dagger}A)} \end{array}\right)$$

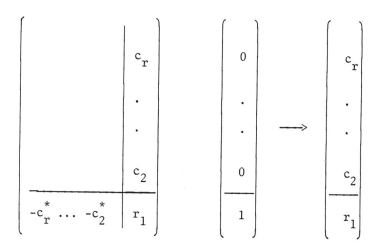

$$\longrightarrow \left(\begin{array}{c|c} [I_{r-1} - XX^{\dagger}]^{\frac{1}{2}} & X \\ \hline -X^{\dagger} & [1-X^{\dagger}X]^{\frac{1}{2}} \end{array}\right)$$

$$X = \begin{pmatrix} c_r \\ \cdot \\ \cdot \\ \cdot \\ c_2 \end{pmatrix} , \qquad c_i \text{ complex} .$$

Since the (r-1)x1 submatrices A(t) in the Lie algebra, and X(t) in the coset, serve to define the coherent state, the coherent states will be labeled either by A or X.

The ground state of N identical r-level systems evolves along the same path in $S^{2(r-1)}$ as does the single atom:

$$\left| \begin{matrix} \{N,\dot{0}\} \\ A \end{matrix} \right\rangle = \text{EXP} \left(\begin{array}{c|c} 0 & A \\ \hline -A^{\dagger} & 0 \end{array}\right) \left| \begin{matrix} \{N,\dot{0}\} \\ \text{gnd} \end{matrix} \right\rangle$$

$$= \left| \begin{matrix} \{N,\dot{0}\} \\ \underline{M} \end{matrix} \right\rangle \left\langle \begin{matrix} \{N,\dot{0}\} \\ \underline{M} \end{matrix} \right| \text{EXP} \left(\begin{array}{c|c} 0 & A \\ \hline -A^{\dagger} & 0 \end{array}\right) \left| \begin{matrix} \{N,\dot{0}\} \\ \text{gnd} \end{matrix} \right\rangle .$$

The coefficients on the right-hand side may be computed in detail. Let n_j be the length of the single row Young partition describing U(j) in the Gel'fand-Tsetlein pattern above. Then

$$M_j = n_j - n_{j-1} - \frac{N}{r}$$

$$\left| \begin{matrix} \{N,\dot{0}\} \\ A \end{matrix} \right\rangle = \sum \left| \begin{matrix} \{N,\dot{0}\} \\ M \end{matrix} \right\rangle \left(\frac{N!}{n_1!(n_2-n_1)! \ \ldots (N-n_{r-1})!} \right)^{\frac{1}{2}}$$

$$\otimes\ (r_1)^{N-n_{r-1}} (c_2)^{n_{r-1}-n_{r-2}} \cdots (c_{r-1})^{n_2-n_1} (c_r)^{n_1} .$$

6. Eigenvalue equations: these are simply determined by operating on the eigenvalue equations obeyed by the ground state. Three types of eigenvalue equations arise:

1. from Casimir operators
2. from diagonal operators
3. from shift down operators.

The results are expressed in terms of the coset operations (X):

$$(X)(\text{Casimir operator})(X)^{-1} \left| \begin{matrix} \bar{\lambda} \\ X \end{matrix} \right\rangle = (\text{Casimir eigenvalue}) \left| \begin{matrix} \bar{\lambda} \\ X \end{matrix} \right\rangle$$

$$(X)(\text{diagonal operator})(X)^{-1} \left| \begin{matrix} \bar{\lambda} \\ X \end{matrix} \right\rangle = (\text{lowest weight}) \left| \begin{matrix} \bar{\lambda} \\ X \end{matrix} \right\rangle$$

$$(X)(\text{annihilation operators})(X)^{-1} \left| \begin{matrix} \bar{\lambda} \\ X \end{matrix} \right\rangle = 0 \left| \begin{matrix} \bar{\lambda} \\ X \end{matrix} \right\rangle .$$

These equations have been computed explicitly.[8]

7. BCH formulas: large numbers of BCH analytic reparameterizations can be constructed for any Lie group with a faithful matrix representation. For the purposes at hand, the BCH formulas of greatest use involve a decomposition of the Lie algebra of the form

$$\Delta = \Delta^{+} + \Delta^{d} + \Delta^{-} .$$

In the faithful $r \times r$ matrix representation of SU(r), Δ^{+}, Δ^{-}, Δ^{d} consist of all upper triangular, lower triangular, and diagonal $r \times r$ matrices. The triangular matrices have only 0 on the major diagonal. Then BCH formulas can be constructed explicitly by finding vectors in these three subspaces which map onto a particular SU(r) group element under these three parameterizations:

$$\text{EXP}\ (\Delta_1^{+}+\Delta_1^{d}+\Delta_1^{-}) \begin{matrix} = \text{EXP}\ (\Delta_2^{+})\ \text{EXP}\ (\Delta_2^{d})\ \text{EXP}\ (\Delta_2^{-}) \\ \| \\ = \text{EXP}\ (\Delta_3^{-})\ \text{EXP}\ (\Delta_3^{d})\ \text{EXP}\ (\Delta_3^{+}) . \end{matrix}$$

8. Coherent state expansions: These BCH formulas can be used to construct explicit expansions for the coherent states in terms of the states diagonal with respect to $\underline{H}$:

$$\text{"A"}\left|\begin{matrix}\bar{\lambda}\\ \text{gnd}\end{matrix}\right\rangle = \left|\begin{matrix}\bar{\lambda}\\ \underline{M}\end{matrix}\right\rangle\left\langle\begin{matrix}\bar{\lambda}\\ \underline{M}\end{matrix}\right| \text{EXP}\left(\begin{array}{c|c} 0 & A \\ \hline -A^{\dagger} & 0\end{array}\right)\left|\begin{matrix}\bar{\lambda}\\ \text{gnd}\end{matrix}\right\rangle .$$

The coefficients in this expansion are determined by

$$\left\langle\begin{matrix}\bar{\lambda}\\ M\end{matrix}\right|\text{EXP}\left(\begin{array}{c|c} 0 & A \\ \hline -A^{\dagger} & 0\end{array}\right)\left|\begin{matrix}\bar{\lambda}\\ \text{gnd}\end{matrix}\right\rangle = \left\langle\begin{matrix}\bar{\lambda}\\ \underline{M}\end{matrix}\right|\text{EXP}\left(\begin{array}{c|c} 0 & A' \\ \hline 0 & 0\end{array}\right)\text{EXP}_i\left(\begin{array}{c|c} \diagup\!\diagup\!\diagup & 0 \\ \hline 0 & d_r\end{array}\right)$$

$$\text{EXP}\left(\begin{array}{c|c} 0 & 0 \\ \hline -A'^{\dagger} & 0\end{array}\right)\left|\begin{matrix}\bar{\lambda}\\ \text{gnd}\end{matrix}\right\rangle .$$

On the right-hand side, the first exponential of an annihilation operator simply gives +1 times the ground state. The next operator also gives back the ground state times the complex phase factor of modulus unity $\exp(iM^{\ell}_{r}d_r)$. Here M^{ℓ}_{r} is the rth component of the lowest weight $\underline{M}^{\ell}$ corresponding to the ground state. The coefficient of $\left|\begin{matrix}\bar{\lambda}\\ M\end{matrix}\right\rangle$ in the expansion of $\left|\begin{matrix}\bar{\lambda}\\ A\end{matrix}\right\rangle$ is then seen to be the matrix element of a shift up operator between the ground state and the diagonal state $\left|\begin{matrix}\bar{\lambda}\\ M\end{matrix}\right\rangle$.

These matrix elements can all be constructed explicitly [5,7] within an arbitrary representation λ of SU(r). Such representations arise when the system evolves from the ground state of a $\bar{\lambda}$ SU(r) invariant subspace. Such a state is a "local" ground state. When the system evolves from the total system ground state, $\bar{\lambda} = \{N,0\}$ and the matrix elements are multinomial coefficients.

9. Non-orthogonality: the coherent states are non-orthogonal with non-zero overlap given by

$$\left\langle\begin{matrix}\bar{\lambda}'\\ B\end{matrix}\middle|\begin{matrix}\bar{\lambda}\\ A\end{matrix}\right\rangle = \left\langle\begin{matrix}\bar{\lambda}'\\ \text{gnd}\end{matrix}\right|(B)^{\dagger}(A)\left|\begin{matrix}\bar{\lambda}\\ \text{gnd}\end{matrix}\right\rangle$$

$$= \left\langle \begin{matrix} \bar{\lambda}' \\ \text{gnd} \end{matrix} \middle| (B)^{-1}(A) \middle| \begin{matrix} \bar{\lambda} \\ \text{gnd} \end{matrix} \right\rangle$$

$$= \left\langle \begin{matrix} \bar{\lambda}' \\ \text{gnd} \end{matrix} \middle| \begin{matrix} \bar{\lambda} \\ \text{"-B+A"} \end{matrix} \right\rangle \quad \delta^{\bar{\lambda}',\bar{\lambda}} \quad .$$

Here "-B+A" are the coordinates of the element in the Lie algebra which maps onto the product of coset representatives $(B)^{-1}(A)$. Such a product is not in general a coset representative. This product can always be written as the product of an element in the stability subgroup, followed by a coset representative. When $\bar{\lambda}' = \bar{\lambda} = \{N,\dot{0}\}$

$$\left\langle \begin{matrix} \{N,\dot{0}\} \\ A \end{matrix} \middle| \begin{matrix} \{N,\dot{0}\} \\ A' \end{matrix} \right\rangle = (r_1 r_1' + c_2^* c_2' + \ldots + c_r^* c_r')^N \ .$$

10. Over-completeness: within any $(\bar{\lambda};\ SU(r))$ invariant subspace, the coherent states $\left| \begin{matrix} \bar{\lambda} \\ X \end{matrix} \right\rangle$ form an overcomplete system in terms of which the identity operator may be resolved. We find

$$\frac{\dim(\bar{\lambda})}{\mathrm{Vol}(G/H)} \int \left| \begin{matrix} \bar{\lambda} \\ X \end{matrix} \right\rangle d\mu(X) \left\langle \begin{matrix} \bar{\lambda} \\ X \end{matrix} \right| = I_{\dim(\bar{\lambda})} = \sum_{\bar{M}\epsilon\bar{\lambda}} \left| \begin{matrix} \bar{\lambda} \\ \bar{M} \end{matrix} \right\rangle \left\langle \begin{matrix} \bar{\lambda} \\ \bar{M} \end{matrix} \right| \ .$$

Here $d\mu(X)$ is the invariant measure induced on the coset $G/H=SU(r)/U(r-1)\simeq S^{2(r-1)}$ from the Haar measure on $SU(r)$. It is

$$d\mu(X) = \delta([r_1^2 + c_2^* c_2 + \ldots + c_r^* c_r]^{\frac{1}{2}} - 1)\, dr_1 d^2c_2 \ldots d^2c_r \ .$$

Then

$$\mathrm{Vol}(G/H) = \mathrm{Vol}\left(\frac{SU(r)}{U(r-1)}\right) = \mathrm{Vol}\left(S^{2(r-1)}\right) = \frac{2\pi^{r-\frac{1}{2}}}{\Gamma(r-\frac{1}{2})} \ .$$

11. Uncertainty relations: expectation values and uncertainty relations for any of the shift operators $U^i{}_j$ can be computed. Within the representation $\lambda = \{N,\dot{0}\}$

$$\left\langle \begin{matrix} \{N,\dot{0}\} \\ X \end{matrix} \middle| U^i_j \middle| \begin{matrix} \{N,\dot{0}\} \\ X \end{matrix} \right\rangle = N c_i^* c_j$$

$$\left\langle \begin{matrix}\{N,\dot{0}\}\\ X\end{matrix} \right| U^i_j\, U^i_j \left| \begin{matrix}\{N,\dot{0}\}\\ X\end{matrix} \right\rangle = N(N-1)(c_i^* c_j)^2$$

$$\left\langle \begin{matrix}\{N,\dot{0}\}\\ X\end{matrix} \right| U^i_j\, U^j_i \left| \begin{matrix}\{N,\dot{0}\}\\ X\end{matrix} \right\rangle = N(N-1)(c_i^* c_i)(c_j^* c_j) + Nc_i^* c_i \quad .$$

No summations (over i,j) are involved in any of the expectation values above. Uncertainty relations for the hermitian combinations

$$\mathrm{Re}\; U^i_j = \tfrac{1}{2}(U^i_j + U^j_i)$$

$$\mathrm{Im}\; U^i_j = \frac{1}{2i}(U^i_j - U^j_i)$$

can be computed directly from the expectation values given above.

12. Generating functions: once again, generating functions for expectation values of products of the U(r) operators $U^i{}_j$ within any coherent state can be constructed. These all involve computing matrix elements of exponential operators, viz.

$$\left\langle \begin{matrix}\bar{\lambda}\\ X\end{matrix} \right| \mathrm{EXP}\ (\Delta^-)\ \mathrm{EXP}\ (\Delta^d)\ \mathrm{EXP}\ (\Delta^+) \left| \begin{matrix}\bar{\lambda}\\ X\end{matrix} \right\rangle \quad .$$

The explicit computation of such matrix elements is greatly facilitated by application of the appropriate BCH relation. A number of useful BCH relations can be constructed for any desired operator ordering.

We close this section on the properties of r-level atoms, and their coherent states, with four comments:

1. All results discussed above hold for the particular case r=2. In particular, this discussion reduces to the discussion of the properties of atomic coherent states on the Bloch sphere.

2. We reemphasize that properties 1, 2, 7 are properties of the Lie *group* or *algebra*, while properties 3-6, 8-12 depend on the unitary irreducible *representations* of the group. This distinction may not be apparent in the work above. This is because SU(r) is compact and has only finite-dimensional UIR. When the group described in 1, 2, 7 is non-compact, all its non-trivial UIR are infinite-dimensional. Then, although finite-dimensional (non-unitary) representations occur in 1, 2, 7, only infinite-dimensional representations will occur in the discussions of the remaining points 3-6, 8-12.

3. It is also possible to discuss the effects of noise in the classical driving fields $\lambda^{\underline{\alpha}}(t)$ in terms of the theory of Brownian motion [9,10] on symmetric spaces. The results are expressible in terms of convolution integrals over the space G/H. These integrals converge quickly and beautifully when expressed in terms of the special functions $\Gamma^{\lambda}{}_{0,\underline{M}}(G/H)$. These results are derived and presented in Ref. 8.

4. Finally, we stress again the importance of these coherent states: The ground state (or some other coherent state) of a system will evolve into a coherent state under the influence of a classical driving field when radiation reaction and spontaneous emission can be neglected.

3. Superconducting Systems

We consider now a system of electrons in a crystal lattice. The electrons interact with the lattice through the lattice vibrations, and with each other through the Coulomb interaction. An approximate Hamiltonian for this system is [11-13]

$$\mathcal{H} = \mathcal{H}_{KE} + \mathcal{H}_{X'tal} + \mathcal{H}_{C}$$

$$\mathcal{H}_{KE} = \sum_{\underline{k},\sigma} \varepsilon_{\underline{k}}\, f^{\dagger}_{\underline{k},\sigma} f_{\underline{k},\sigma}$$

$$\mathcal{H}_{X'tal} = -\tfrac{1}{2} \sum_{\substack{\underline{k}\neq\underline{k}' \\ \underline{q}}} \sum_{\sigma',\sigma} \frac{\omega_{\underline{k}-\underline{k}'}\, M^2_{\underline{k}-\underline{k}'}}{\omega^2_{\underline{k}-\underline{k}'} - (\varepsilon_{\underline{k}} - \varepsilon_{\underline{k}'})^2} \otimes f^{\dagger}_{\underline{k}',\sigma'} f^{\dagger}_{\underline{q}-\underline{k}',\sigma} f_{\underline{q}-\underline{k},\sigma} f_{\underline{k},\sigma'}$$

$$\mathcal{H}_{C} = \sum_{\substack{\underline{k}\neq\underline{k}' \\ \underline{q}}} \sum_{\sigma',\sigma} \frac{2\pi (e^*)^2}{|\underline{k}-\underline{k}'|^2} \otimes f^{\dagger}_{\underline{k}',\sigma'} f^{\dagger}_{\underline{q}-\underline{k}',\sigma} f_{\underline{q}-\underline{k},\sigma} f_{\underline{k},\sigma'}$$

In these expressions

1. $\varepsilon_{\underline{k}}$ is the energy of an electron with wave vector $\underline{k}$;
2. $\omega_{\underline{k}}$ is the energy of a phonon with wave-vector $\underline{k}$;
3. $M_{\underline{k}-\underline{k}'}$ is a matrix element describing phonon-induced scattering of electrons from state $\underline{k}'$ into state $\underline{k}$;
4. e^* is the electron charge. We include the * to emphasize that screening and exchange should be taken into account in this term;

5. σ is a physical spin index which can assume only the values "up" and "down": ↑,↓;
6. $f^{\dagger}_{\underline{k},\sigma}$, $f_{\underline{k},\sigma}$ are electron creation and annihilation operators for the state $(\underline{k},\sigma)$.

Since electrons are Fermions, the operators $f^{\dagger}$, f obey *anti-*commutation relations rather than commutation relations:

$$[f^{\dagger},f]_{+} = \{f^{\dagger},f\} = f^{\dagger}f + f\,f^{\dagger} = 1$$
$$f\,f = f^{\dagger}f^{\dagger} = 0 \quad .$$

Operators describing different modes *anti-commute,*

$$[f^{\dagger}_{\underline{k}},f_{\underline{k}'}]_{+} = 0 \qquad \text{etc.}$$

The terms $\mathcal{H}_{X'tal}$, $\mathcal{H}_{C}$ describe the scattering of an electron pair $(\underline{k},\underline{q}-\underline{k})$ of total momentum $\underline{q}$ into another electron pair $(\underline{k}',\underline{q}-\underline{k}')$ with the same total momentum $\underline{q}$ and same spins, and with $\underline{k}' \neq \underline{k}$.

In superconducting systems the most important [14] electron pairs are those with opposite momenta and spins. For this reason, we linearize the Hamiltonian above by dropping all terms with either $\underline{q} \neq 0$ or $\sigma + \sigma' \neq 0$. The result is the reduced Hamiltonian

$$\mathcal{H}_{RED} = \sum_{\underline{k}} \varepsilon_{\underline{k}}(f^{\dagger}_{\underline{k}}f_{\underline{k}} + f^{\dagger}_{-\underline{k}}f_{-\underline{k}}) - \sum_{\underline{k}}{}' \, f^{\dagger}_{\underline{k}}f^{\dagger}_{-\underline{k}} \sum_{\underline{k}'} V_{\underline{k},\underline{k}'}f_{-\underline{k}'}f_{\underline{k}'}$$
$$- \sum_{k} f_{-\underline{k}}f_{\underline{k}} \sum_{\underline{k}'} V_{\underline{k}\underline{k}'} \, f^{\dagger}_{\underline{k}'} \, f^{\dagger}_{-\underline{k}'} \quad .$$

In this expression the signs (+,-) imply spin (up, down) states. The interaction potential is

$$V_{\underline{k},\underline{k}'} = V_{\underline{k}'\underline{k}} = \frac{\omega_{\underline{k}-\underline{k}'} \; M^2_{\underline{k}-\underline{k}'}}{\omega^2_{\underline{k}-\underline{k}'} - (\varepsilon_{\underline{k}} - \varepsilon_{\underline{k}'})^2} - \frac{4\pi(e^*)^2}{|\underline{k}-\underline{k}'|^2} \quad .$$

This Hamiltonian can be further simplified by introducing the operators

$$(J_+)_{\underline{k}} = f^{\dagger}_{\underline{k}}f^{\dagger}_{-\underline{k}}$$
$$(J_-)_{\underline{k}} = f_{-\underline{k}}f_{\underline{k}} \quad .$$

Although the fermion operators obey *anti*-commutation relations, bilinear combinations of these operators obey *commutation* relations:

$$[(J_+)_{\underline{k}}\ ,\ (J_-)_{\underline{k}'}] = 2(J_3)_{\underline{k}}\ \delta_{\underline{kk}'}$$

$$2(J_3)_{\underline{k}} = f^{\dagger}_{\underline{k}} f_{\underline{k}} + f^{\dagger}_{-\underline{k}} f_{-\underline{k}} - 1 \quad .$$

The operators $\underline{J}$ for a single mode obey the commutation relations of the algebra $SU(2)$:

$$[J_3, J_{\pm}] = \pm J_{\pm}$$

$$[J_+, J_-] = +\ 2J_3 \quad .$$

Operators describing different modes commute:

$$[(\underline{J})_{\underline{k}}\ ,\ (\underline{J})_{\underline{k}'}] = [\underline{J},\underline{J}]_{\underline{k}}\ \delta_{\underline{kk}'}$$

The $\underline{J}$'s act *like* spin operators but they are not *physical* spin operators. They may be called "pseudo-spin" operators. Within a single mode we find

$$J_3 \left| \begin{matrix} n_{\underline{k}} = 1 \\ n_{-\underline{k}} = 1 \end{matrix} \right\rangle = +\tfrac{1}{2}$$

$$J_3 \left| \begin{matrix} n_{\underline{k}} = 0 \\ n_{-\underline{k}} = 0 \end{matrix} \right\rangle = -\tfrac{1}{2} \quad .$$

The $\underline{J}$'s thus describe the occupation number space for electrons with opposite momenta and opposite spin.

The Hamiltonian $\mathcal{H}_{RED}$ can be expressed in terms of the operators $\underline{J}$:

$$\mathcal{H}_{ALG} = \sum_{\underline{k}} \varepsilon_{\underline{k}} (2J_3 + 1)_{\underline{k}} - \sum_{\underline{k}} (J_+)_{\underline{k}} \lambda_{\underline{k}} + (J_-)_{\underline{k}} \lambda^*_{\underline{k}}$$

$$\lambda_{\underline{k}} = \tfrac{1}{2} {\sum_{\underline{k}'}}' V_{\underline{kk}'}\ (J_-)_{\underline{k}'} \quad .$$

In the remainder of this section we will treat $\lambda(\underline{k})$ as a $\underline{k}$ dependent interaction potential rather than an interaction dependent on other modes $\underline{k}'$.

In the absence of an interaction ($V \to 0$), all electron states below the Fermi level will be occupied ($J_3 \to +\frac{1}{2}$) and those with $k > k_F$ will be empty ($2J_3 + 1 \to 0$). This discontinuity at the Fermi level can be emphasized by adding a constant term to the Hamiltonian. Then $\mathcal{H}_{ALG}$ is the direct sum of single-mode contributions, one for each $\underline{k}$:

$$\mathcal{H}(\underline{k})_{ALG} = 2(\varepsilon_{\underline{k}} - \varepsilon_F)(J_3) - \{J_+\lambda(\underline{k}) + J_-\lambda^*(\underline{k})\} \quad .$$

It is understood that in this expression the operators $\underline{J}$ refer specifically to the $\underline{k}$ mode.

This algebraic Hamiltonian can be considerably simplified by means of a Bogoliubov transformation:[15]

$$\mathcal{H}_{ALG} = e^{i\underline{\theta}\cdot\underline{J}}\,|E|J_3\,e^{-i\underline{\theta}\cdot\underline{J}}$$

$$\underline{\theta} = \theta(i_x \cos\phi,\ i_y \sin\phi,\ 0) \quad .$$

The parameters $\theta_{\underline{k}}$, $\phi_{\underline{k}}$ are given explicitly by

$$\cos\theta_{\underline{k}} = \frac{\varepsilon_{\underline{k}} - \varepsilon_F}{[(\varepsilon_{\underline{k}} - \varepsilon_F)^2 + \lambda^*(\underline{k})\lambda(\underline{k})]^{\frac{1}{2}}}$$

$$e^{i(\phi_{\underline{k}} - \frac{\pi}{2})} = [\lambda^*(\underline{k})/\lambda(\underline{k})]^{\frac{1}{2}}$$

$$|E|^2 = 2^2[(\varepsilon_{\underline{k}} - \varepsilon_F)^2 + \lambda^*(\underline{k})\lambda(\underline{k})]$$

From the transformed structure of $\mathcal{H}_{ALG}$, it is clear that the ground state of the $\underline{k}^{th}$ mode of $\mathcal{H}_{ALG}(\underline{k})$ is

$$|gnd\rangle = e^{i\underline{\theta}\cdot\underline{J}}\left|\begin{matrix}\frac{1}{2}\\ -\frac{1}{2}\end{matrix}\right\rangle \quad .$$

As usual, this rotation operator can be considerably simplified using a BCH relation:

$$|\mathrm{gnd}\rangle = [\cos\frac{\theta}{2}]^{2J}\, e^{\{e^{-i(\phi-\frac{\pi}{2})}\tan\frac{\theta}{2}\}J_+}\left|\begin{matrix}\frac{1}{2}\\ -\frac{1}{2}\end{matrix}\right\rangle$$

$$|J| = \tfrac{1}{2} \quad .$$

From these results, it can be seen directly that, for $\varepsilon_{\underline{k}} < \varepsilon_F$, $\theta \simeq \pi$ and the ground state has $n_{\underline{k}} = n_{-\underline{k}} \simeq 1$, while for $\varepsilon_{\underline{k}} > \varepsilon_F$, $\theta \simeq 0$ and the ground state has $n_{\underline{k}} = n_{-\underline{k}} \simeq 0$.

The ground state wave function is a direct product of single-mode ground state functions:

$$\left|\begin{matrix}\mathrm{TOTAL}\\ \mathrm{GROUND}\\ \mathrm{STATE}\end{matrix}\right\rangle = \prod_{\mathrm{all}\ \underline{k}} [\cos\frac{\theta_{\underline{k}}}{2}]\, e^{\{e^{-i(\phi_{\underline{k}}-\frac{\pi}{2})}\tan\frac{\theta_{\underline{k}}}{2}\}f^{\dagger}_{\underline{k}}f^{\dagger}_{-\underline{k}}}\,|\mathrm{VACUUM}\rangle$$

$$|\mathrm{VACUUM}\rangle = \prod_{\mathrm{all}\ \underline{k}}\left|\underline{k}\ ;\ \begin{matrix}\frac{1}{2}\\ -\frac{1}{2}\end{matrix}\right\rangle \quad .$$

We conclude with the comment that when $V_{\underline{k}\underline{k}'}$ is positive (attractive) on the average when $\varepsilon_{\underline{k}} - \varepsilon_{\underline{k}'}$ is small and $\underline{k},\underline{k}'$ are near the Fermi surface, then it is energetically desirable for many $\underline{k}$ modes near the Fermi surface to be in states not "parallel to the z axis in pseudo spin (occupation number) space." As a result, an energy gap can exist between the ground and first excited state, and the system can become superconducting.

4. Superfluid Systems

Superfluid systems can be treated in an analogous way. A useful model Hamiltonian proposed by Bassichis and Foldy [16] has recently been treated algebraically by Solomon.[17] Space limitations prevent us from summarizing his elegant treatment here. It is sufficient to note that the superfluid ground state is obtained by applying a "Bloch rotation" (ie. A Bogoliubov transformation) to an extremal state in a semi-bounded [18,19] unitary irreducible representation of the non-compact group SU(1,1):

$$\left|\begin{matrix}\text{superfluid}\\ \text{ground}\\ \text{state}\end{matrix}\right\rangle = e^{-\frac{1}{2}(J_+ - J_-)\theta}\left|\begin{matrix}j\\ n=0\end{matrix}\right\rangle \quad .$$

The ground state is most easily obtained explicitly by applying an appropriate SU(1,1) BCH formula. The result is simply

$$\left|\begin{matrix}\text{superfluid}\\ \text{ground}\\ \text{state}\end{matrix}\right\rangle = \prod_{\underline{k}\neq 0} \{\cosh \frac{\theta(\underline{k})}{2}\}^{2j} e^{-\tanh \frac{\theta(\underline{k})}{2} b^{\dagger}_{+\underline{k}} b^{\dagger}_{-\underline{k}}} \left|\begin{matrix} j \\ 0 \end{matrix}\right\rangle .$$

All terms are defined in Ref. 17. We simply note that this result holds for an arbitrary j describing a UIR of SU(1,1), and that use of the BCH formula has obviated the cumbersome summations appearing in Ref. 17.

5. Charged Particles in External Field

A three-dimensional harmonic oscillator can be simply treated using algebraic techniques. This treatment remains valid if the oscillator consists of a charged particle moving in an external uniform [20-22] magnetic field. The treatment can be further extended to cover the motion of a charged particle in external time-dependent electric and magnetic fields which are not necessarily crossed. [23-26] In fact, any polynomial Hamiltonian containing terms no higher than second order is amenable to such treatment. In particular, the Hamiltonian for a relativistic spin ½ particle can be so treated, once it has been transformed to a quadratic form by eliminating the small components in first order. Space limitations once again prevent us from summarizing these treatments. We merely note in closing that the Hamiltonian for such a system is essentially a linear superposition of generators contracted from U(4). The general $U(N+1) \to \sum_{i=1}^{N} (h_4)_i$ contraction is given by (Lim $d \downarrow 0$):

$$\begin{aligned}
&\text{Lim } U_{i,i} + \frac{1}{(N+1)d^2}\xi && \to n_i \\
&\text{Lim } dU_{i,N+1} && \to a_i^{\dagger} \\
&\text{Lim } dU_{N+1,i} && \to a_i \\
&\text{Lim } \sum_{i=1}^{N+1} U_{i,i} = \xi && \to I \quad .
\end{aligned}$$

The generators u_{ij} of u(N+1) obey

$$[U_{ij}, U_{rs}] = U_{is}\delta_{jr} - U_{rj}\delta_{is}$$

from which the commutation relations of the contracted operators can be constructed, viz:

$$[a_i^\dagger, a_i] = \mathrm{Lim}\ [dU_{i,N+1}, dU_{N+1,i}]$$

$$= \mathrm{Lim}\ d^2 (n_i - \frac{1}{(N+1)d^2} \xi)$$

$$- \mathrm{Lim}\ d^2 (U_{N+1,N+1} = \xi - \sum_{i=1}^{N} (N_i - \frac{1}{(N+1)d^2} \xi)) = - I \ .$$

It should therefore come as no surprise that the energy eigenvalue spectrum for such problems has a semi-bounded discrete [20] lattice structure. Such a spectrum is simply the weight space diagram for a contracted representation of the form ($\lambda_1 \to \infty$, $\lambda_2 = 0$) of u(4) or u(3,1). In such a representation all weights (energy eigenvalues) are non-degenerate. Moreover, the coherent states are contractions of the coherent states on SU(3, ±1)/U(3), or in the general case, SU(N,±1)/U(N).

6. Conclusion

We have extended our previous discussion of two-level atoms to the case of r-level atoms. We have shown how the symmetrized states are labeled by the properties of the representations of SU(r) and P_N. We described how coherent states in this case arise both physically (evolution from a ground state) and mathematically (application of a coset operation to an extremal state). The properties of these coherent states were enumerated. All properties reduce, in the case r = 2, to the corresponding properties of the atomic coherent states.[1] These in turn reduce to the field coherent states under the contraction u(2) → h_4. An explicit distinction was made between the group or algebra describing a system, and the representation for the group, which must be unitary.

These methods were then applied successively to three additional systems:

1. superconductors
2. superfluids
3. charged particles in external fields.

In each discussion, we showed how the system under discussion could be represented, after suitable simplifying assumptions, by a model Hamiltonian which was amenable to algebraic treatment.

Once the Hamiltonian has been shown to be an element is some identifiable Lie algebra ("algebraized"), the essential work has been done. The coherent states are constructed and discussed as described in para. 2 and Ref. 1. We have not discussed the properties of the coherent states of these three systems in any detail, since such a discussion has been carried out thoroughly in para. 2. We only mention that in all these cases, the system ground state is a coherent state.

References

1. F. T. Arecchi, E. Courtens, R. Gilmore, H. Thomas, this volume p. 191. See also the paper by the same authors to be published.
2. S. Helgason, *Differential Geometry and Symmetric Spaces* (Academic Press, New York, 1962). N. Jacobson, *Lie Algebras* (Wiley-Interscience, New York, 1962).
3. H. Weyl, *The Theory of Groups and Quantum Mechanics* (Dover Publications, New York, 1931).
4. H. Weyl, *The Classical Groups* (Princeton University Press, Princeton, New Jersey, 1946).
5. I. M. Gel'fand, M. L. Tsetlein, Doklady Akad. Nauk SSSR *71*, 825-828 (1950).
6. I. M. Gel'fand, M. L. Tsetlein, Doklady Akad. Nauk SSSR *71*, 1017-1020 (1950).
7. R. Gilmore, J. Math. Phys. *11*, 3420-3427 (1970).
8. R. Gilmore, Ann. Phys. New York (to be published).
9. M. F. Perrin, Ann. Scient. L'Ecole Normale Sup. *45*, 1-51 (1928).
10. E. B. Dynkin, Am. Math. Soc. Transl. (2) *72*, 203-228 (1968).
11. J. Bardeen, D. Pines, Phys. Rev. *99*, 1140-1150 (1955).
12. D. Pines, Phys. Rev. *109*, 280-287 (1958).
13. P. W. Anderson, Phys. Rev. *112*, 1900-1916 (1958).
14. J. Bardeen, L. N. Cooper, J. R. Schrieffer, Phys. Rev. *106*, 162-164 (1957); J. Bardeen, L. N. Cooper, J. R. Schrieffer, Phys. Rev. *108*, 1175-1204 (1957).
15. N. N. Bogoliubov, I. Sov. Phys. JETP *34(7)*, 41-46 (1958); II. Sov. Phys. JETP *34(7)*, 51-55 (1958).
16. W. H. Bassichis, L. L. Foldy, Phys. Rev. *133A*, 935-943 (1964).
17. A. J. Solomon, J. Math. Phys. *12*, 390-394 (1971).
18. W. Miller, On Lie Algebras and Some Special Functions of Mathematical Physics, Memoirs of the American Mathematical Society #50 (Am. Math. Soc., Providence, R. I. 1964).
19. W. Miller, *Lie Theory and Special Functions* (Academic Press, New York, 1968).
20. A. Feldman, A. H. Kahn, Phys. Rev. (3) *B1*, 4584-4589 (1970).
21. I. A. Malkin, V. I. Man'ko, Sov. Phys. JETP *28*, 527-532 (1969).
22. I. A. Malkin, V. I. Man'ko, Sov. J. Nucl. Phys. *8*, 731-735 (1969).

23. I. A. Malkin, V. I. Man'ko, D. A. Trifonov, Sov. Phys. JETP *31*, 386-390 (1970).
24. I. A. Malkin, V. I. Man'ko, D. A. Trifonov, Phys. Lett. *30A*, 414 (1969).
25. I. A. Malkin, V. I. Man'ko, Sov. Phys. JETP *32*, 949-953 (1971).
26. I. A. Malkin, V. I. Man'ko, Phys. Lett. *32A*, 243-244 (1970).

DEBYE REPRESENTATION AND MULTIPOLE EXPANSION OF THE QUANTIZED FREE ELECTROMAGNETIC FIELD *

A. J. Devaney

University of Rochester, Rochester, New York

Introduction

Quite often in classical electrodynamics, a problem is simplified, considerably, by choosing an appropriate representation for the field. Well-known examples are the Cauchy initial-value problem and the Dirichlet or Neumann boundary-value problem. For the Cauchy problem an ideal representation for the field is a plane-wave expansion [1]; this being so because the plane-wave fields form a complete set of solutions to the wave equation into which the Cauchy data can be easily expanded. For the case of Dirichlet or Neumann boundary conditions given on an infinite plane surface, together with Sommerfeld's radiation condition at infinity, a plane-wave expansion is again appropriate, although in this case it is necessary to include evanescent (inhomogeneous) plane waves in the expansion [2]. Plane-wave expansions are not always the most appropriate representation to use, however. For example, with Dirichlet or Neumann conditions prescribed on the surface of a sphere, it is desirable to expand the field into a multipole expansion since the multipole fields form a complete, orthogonal set of functions on such surfaces [3].

In the case of the quantized electromagnetic field a similar situation is encountered. Rather than speaking of representations for the field, one now speaks of representations for the field operators. The most commonly used representation is a plane-wave expansion, identical in form to that used for the classical field, where, however, the plane-wave amplitude is given in terms of

* Research supported by the U.S. Air Force Office of Scientific Research.

creation and annihilation operators [4]. These operators are associated with Fock states of well-defined energy, linear momentum and spin. In some applications, it is desirable to use a basis of Fock states of well-defined energy, total angular momentum (i.e., orbital angular momentum plus spin), and parity (i.e., the operator that causes reflection through the origin). The appropriate representation for the field operators is, in this case, in terms of a multipole expansion [5] which is again identical in form to the multipole expansion of the classical field where, however, the expansion coefficients (multipole moments) are now creation and annihilation operators associated with Fock states of fixed energy, total angular momentum, and parity.

In the present paper, a procedure is developed for obtaining the multipole expansion for the quantized electromagnetic field, complete with the commutation relations for the multipole moment operators, from the plane-wave representation for this field. Moreover, we also derive the closely related Debye representation [6] for the quantized field and determine the commutation relations for the Debye potential operators.

The theory presented here has several advantages over the usual treatments. First, the *validity* of the multipole expansion is shown to follow at once from the validity of the plane-wave expansion. This advantage is quite important since the usual procedures for establishing multipole expansions for even the classical electromagnetic field are quite involved, generally requiring elaborate proofs to establish completeness of the multipole fields. Secondly, the commutation relations for the multipole moments are easily obtained from the well-known commutation relations for the creation and annihilation operators occuring in the plane-wave expansion for the field operators; there is no need to go through the laborious procedure of obtaining the field Hamiltonian in terms of the multipole moments and then identifying the canonically conjugate "position" and "momentum" variables [7]. Third, the procedure leads directly to the Debye representation, together with commutation relations for the Debye potential operators. This representation and the associated commutation relations appears to be new. Finally, the entire theory can also be used to obtain the Debye representation and multipole expansion for the classical electromagnetic field - even for the case when sources are present [8]. Indeed, the technique presented here is essentially "classical" in nature and herein lies one of its principle advantages.

Whittaker's Technique for Obtaining Multipole Expansions

Whittaker [9] appears to be the first to obtain a multipole expansion directly from a plane-wave representation. Restricting his attention to scalar fields satisfying the homogeneous wave equation, he was able to represent the fields in plane-wave expansions of the following form:

$$\psi(\underline{r},t) = \frac{1}{2\pi}\int_{-\infty}^{\infty} d\omega e^{-i\omega t}\int_{-\pi}^{\pi} d\beta\int_{0}^{\pi} d\alpha\sin\alpha\, A(\frac{\omega}{c}\underline{s})e^{i\frac{\omega}{c}\underline{s}.\underline{r}} \tag{2.1}$$

where the unit propagation vector $\underline{s}$ is given by

$$\underline{s} = \sin\alpha\cos\beta\,\underline{u}_x + \sin\alpha\sin\beta\,\underline{u}_y + \cos\alpha\,\underline{u}_z\ , \tag{2.2}$$

with $\underline{u}_x$, $\underline{u}_y$, $\underline{u}_z$ being unit Cartesian vectors along the x,y,z axes respectively. By taking the plane-wave amplitude $A(\omega/c\ \underline{s})$ to be proportional to a spherical harmonic $Y_\ell^m(\alpha,\beta)$, Whittaker was able to construct a multipole field of degree ℓ and order m; i.e., he showed that [10]

$$j_\ell(\frac{\omega}{c} r)\, Y_\ell^m(\theta,\phi) = \frac{(-i)^\ell}{4\pi}\int_{-\pi}^{\pi} d\beta\int_{0}^{\pi} d\alpha\sin\alpha\, Y_\ell^m(\alpha,\beta)\, e^{i\frac{\omega}{c}\underline{s}.\underline{r}}\ . \tag{2.3}$$

where $j_\ell(\omega/c\ r)$ denotes the spherical Bessel function of the first kind. Making use of (2.3), it is then an easy matter to construct the multipole expansion for the field $\psi(\underline{r},t)$ from the plane-wave representation (2.1). In particular, one finds that

$$\psi(\underline{r},t) = \frac{1}{2\pi}\int_{-\infty}^{\infty} d\omega\, e^{-i\omega t}\sum_{\ell=0}^{\infty}\sum_{m=-\ell}^{\ell} a_\ell^m(\omega) j_\ell(\frac{\omega}{c} r) Y_\ell^m(\theta,\phi)\ , \tag{2.4}$$

where the multipole moments $a_\ell^m(\omega)$ are determined from the plane-wave amplitude $A(\omega/c\ \underline{s})$ as follows:

$$a_\ell^m(\omega) = 4\pi i^\ell\int_{-\pi}^{\pi} d\beta\int_{0}^{\pi} d\alpha\sin\alpha\, A(\frac{\omega}{c}\underline{s}) Y_\ell^{m*}(\alpha,\beta)\ . \tag{2.5}$$

Although Whittaker dealt with classical (c number) fields, it is clear that the above procedure is equally valid when $\psi(\underline{r},t)$ is an operator field; in this case the plane-wave amplitude is an operator and, consequently, so are the multipole moments defined in (2.5).

For the case of the (free) electromagnetic field it is again possible to represent the field in terms of a plane-wave expansion. In particular, the electric and magnetic fields $\underline{E}(\underline{r},t)$ and $\underline{B}(\underline{r},t)$ can be expressed as follows: [11]

$$\text{(2.6a)}\quad \underline{E}(\underline{r},t) = \frac{1}{2\pi}\int_0^\infty d\omega e^{-i\omega t}\int_{\pi}^{\pi} d\beta \int_0^\pi d\alpha \sin\alpha \tilde{\underline{E}}(\frac{\omega}{c}\,\underline{s}) e^{i\frac{\omega}{c}\underline{s}\cdot\underline{r}} + \text{h.c.},$$

$$\text{(2.6b)}\quad \underline{B}(\underline{r},t) = \frac{1}{2\pi}\int_0^\infty d\omega e^{-i\omega t}\int_{-\pi}^{\pi} d\beta \int_0^\pi d\alpha \sin\alpha \tilde{\underline{B}}(\frac{\omega}{c}\,\underline{s}) e^{i\frac{\omega}{c}\underline{s}\cdot\underline{r}} + \text{h.c.},$$

where h.c. stands for the Hermitian conjugate of the first term. For the case of the classical electromagnetic field, the plane-wave amplitudes $\tilde{\underline{E}}(\omega/c\ \underline{s})$ and $\tilde{\underline{B}}(\omega/c\ \underline{s})$ are given in terms of the values of $\underline{E}(\underline{r}.t)$ and $\underline{B}(\underline{r},t)$ at $t = 0$ (i.e., in terms of the Cauchy data) while in the case of the quantized field these plane-wave amplitudes are operators that can be expressed as follows:

$$\text{(2.7a)}\quad \tilde{\underline{E}}(\frac{\omega}{c}\,\underline{s}) = i(\hbar\omega)^{\frac{1}{2}}\,\frac{\omega^2}{c^3}\,\hat{\underline{a}}(\frac{\omega}{c}\,\underline{s}) \quad ,$$

$$\text{(2.7b)}\quad \tilde{\underline{B}}(\frac{\omega}{c}\,\underline{s}) = i(\hbar\omega)^{\frac{1}{2}}\,\frac{\omega^2}{c^3}\,\underline{s}\times\hat{\underline{a}}(\frac{\omega}{c}\,\underline{s}).$$

As usual $\hbar = h/2\pi$, with h being Planck's constant. The operator $\hat{\underline{a}}(\omega/c\ \underline{s})$ is given in terms of annihilation operators

$$\text{(2.8)}\quad \hat{\underline{a}}(\frac{\omega}{c}\,\underline{s}) = \sum_{j=1,2}\hat{a}_j(\frac{\omega}{c}\,\underline{s})\,\underline{\varepsilon}_j(\underline{s}) \ ,$$

where $\underline{\varepsilon}_j(\underline{s})$ $(j = 1,2)$ are unit polarization vectors which are perpendicular to the propagation vector $\underline{s}$. The annihilation operators $\hat{a}_j(\omega/c\ \underline{s})$ $(j = 1,2)$ satisfy the commutation relations

$$\text{(2.9a)}\quad [\hat{a}_j(\frac{\omega}{c}\,\underline{s}),\ \hat{a}_{j'}(\frac{\omega'}{c}\,\underline{s}')] = 0$$

(2.9b) $[\hat{a}_j^+ (\frac{\omega}{c}\underline{s}), \hat{a}_{j'}^+ (\frac{\omega'}{c}\underline{s}')] = 0$

(2.9c) $[\hat{a}_j (\frac{\omega}{c}\underline{s}), \hat{a}_{j'}^+ (\frac{\omega'}{c}\underline{s}')] = \delta(\frac{\omega}{c}\underline{s} - \frac{\omega'}{c}\underline{s}')\,\delta_{j,j'}$,

where $\hat{a}_{j'}^+ (\omega'/c\ \underline{s}')$ denotes the creation operator (i.e., the adjoint of the annihilation operator $\hat{a}_{j'}(\omega'/c\ \underline{s}')$, and $\delta(\omega/c\ \underline{s} - \omega'/c\ \underline{s}')$ is Dirac's delta function and $\delta_{j,j'}$ is the Kronecker delta function. From this point on we shall restrict our attention to the quantized field and, consequently, take the plane-wave amplitudes to be the operators defined in (2.7a) and (2.7b).

Following Whittaker's procedure, it is possible to expand the plane-wave amplitudes $\tilde{\underline{E}}(\omega/c\ \underline{s})$ and $\tilde{\underline{B}}(\omega/c\ \underline{s})$ into a series of spherical harmonics. What results is a series expansion of the electric and magnetic field operators into the *scalar* multipole fields $j_\ell(\omega/c\ r)\ Y_\ell^m(\theta,\phi)$ with vector valued operator coefficients. Unfortunately, this series expansion is *not* the multipole expansion for these field operators. The reasons for this are that in the first place, any given term in the expansion need not have zero divergence. This means that the series is not a mode expansion of the Maxwell equations (or, equivalently, of the Heisenberg equations of motion for the field). Secondly, the fields comprising this expansion do not possess the proper transformation properties under coordinate system rotation required of a multipole field; each term in a multipole expansion should have rotational properties of spherical tensors. An expansion of the form (2.4) has this property since the expansion coefficients are *scalars*. However, the expansion loses this property when vector-valued expansion coefficients are employed.

Multipole Expansion for the Electromagnetic Field Operators

From the results of the previous section, we see that the spherical harmonics $Y_\ell^m(\alpha,\beta)$ play a central role in the theory of scalar multipole expansions: they occur both as the angular functions in the multipole fields and as the plane-wave amplitudes of these fields. Moreover, the transformation properties of the multipole fields under coordinate system rotation are precisely the same as the transformation properties of the spherical harmonics: both possess rotational properties of spherical tensors. The above observations suggest that electromagnetic multipole fields can be constructed by taking plane-wave amplitudes that are "vector field equivalents" of the spherical harmonics. One such field is the *vector spherical harmonic* [12] $\underline{Y}_\ell^m(\alpha,\beta)$ of degree ℓ and order m.

This field is a spherical tensor of rank ℓ and is related to the spherical harmonic as follows:

$$(3.1) \qquad \underline{Y}_\ell^m(\alpha,\beta) = \underline{L}\, Y_\ell^m(\alpha,\beta),$$

where $\underline{L}$ is the orbital angular momentum operator

$$(3.2) \qquad \underline{L} = -i\frac{\omega}{c}\,\underline{s} \times \nabla_{\underline{s}} = -i\left[\underline{u}_\beta \frac{\partial}{\partial\alpha} - \frac{1}{\sin\alpha}\,\underline{u}_\alpha \frac{\partial}{\partial\beta}\right].$$

$\underline{s}$ is the unit propagation vector defined in (2.2) and $\underline{u}_\alpha$ and $\underline{u}_\beta$ are unit vectors in the α,β directions respectively. $\nabla_{\underline{s}}$ denotes the gradient operator in the $\omega/c,\alpha,\beta$ space. These fields can also be expressed as linear combinations of spherical harmonics of different order. In particular, we have that

$$(3.3) \qquad \underline{Y}_\ell^m(\alpha,\beta) = [(\ell-m)(\ell+m+1)]^{\frac{1}{2}}\underline{\varepsilon}_-\, Y_\ell^{m+1}(\alpha,\beta) +$$

$$[(\ell+m)(\ell-m+1)]^{\frac{1}{2}}\underline{\varepsilon}_+\, Y_\ell^{m-1}(\alpha,\beta) + m\,\underline{u}_z\, Y_\ell^m(\alpha,\beta),$$

where

$$(3.4) \qquad \underline{\varepsilon}_+ = \frac{1}{2}(\underline{u}_x + i\,\underline{u}_y)$$

$$\underline{\varepsilon}_- = \frac{1}{2}(\underline{u}_x - i\,\underline{u}_y).$$

Eqs. (3.3) is readily obtained from Eqs. (3.1) by expressing $\underline{L}$ in terms of the raising and lowering operators L^+ and L^- and its z component L_z

By making use of (3.3) and Whittaker's plane-wave expansion of the scalar multipole fields given in (2.3), we can construct "vector field equivalents" of the scalar multipole fields. We find that

$$(3.5)\quad \frac{(-i)^{\ell}}{4\pi}\int_{-\pi}^{\pi} d\beta \int_{0}^{\pi} d\alpha \sin\alpha\, \underline{Y}_{\ell}^{m}(\alpha,\beta)\, e^{i\frac{\omega}{c}\underline{s}\cdot\underline{r}} =$$

$$j_{\ell}\left(\frac{\omega}{c} r\right)\underline{L}_{r}\, Y_{\ell}^{m}(\theta,\phi) = i\nabla \times \underline{r}\, j_{\ell}\left(\frac{\omega}{c} r\right) Y_{\ell}^{m}(\theta,\phi) \quad ,$$

where $\underline{L}_r$ denotes the angular momentum operator in $\underline{r}$ space; i.e.

$$(3.6)\quad \underline{L}_r = - i\, \underline{r} \times \nabla \equiv i\nabla \times \underline{r} \quad .$$

The fields $i\nabla \times \underline{r}\, j_{\ell}(\omega/c\, r)\, Y_{\ell}^{m}(\theta,\phi)$ possess the properties required of electromagnetic multipole fields: they are solutions to the reduced wave equation, have zero divergence, and are spherical tensors of rank ℓ. These fields are not, however, sufficient by themselves to represent the electric and magnetic field operators; they do not form a complete set of modes for expanding these field operators. The reason for this is that the vector spherical harmonics $\underline{Y}_{\ell}^{m}(\alpha,\beta)$ do not form a complete set for expanding the plane-wave amplitudes $\underline{\tilde{E}}(\omega/c\ \underline{s})$ and $\underline{\tilde{B}}(\omega/c\ \underline{s})$. A complete set is obtained, however, by including the vector fields $\underline{s} \times \underline{Y}_{\ell}^{m}(\alpha,\beta)$. The set $\underline{Y}_{\ell}^{m}(\alpha,\beta)$ and $\underline{s} \times \underline{Y}_{\ell}^{m}(\alpha,\beta)$ form a complete set for expanding transverse vector fields defined on the surface of a sphere. (The completeness of these fields is a consequence of Hodge's Decomposition Theorem (see next section)). Consequently, we can write

$$(3.7a)\quad \underline{\tilde{E}}\left(\frac{\omega}{c}\underline{s}\right) = \sum_{\ell=1}^{\infty}\sum_{m=-\ell}^{\ell} \frac{(-i)^{\ell}}{4\pi}\left(\frac{\omega}{c}\right)\{a_{\ell,1}^{m}(\omega)\underline{s} \times \underline{Y}_{\ell}^{m}(\alpha,\beta) + a_{\ell,2}^{m}(\omega)\underline{Y}_{\ell}^{m}(\alpha,\beta)\}\ ,$$

$$(3.7b)\quad \underline{\tilde{B}}\left(\frac{\omega}{c}\underline{s}\right) = \sum_{\ell=1}^{\infty}\sum_{m=-\ell}^{\ell} \frac{(-i)^{\ell}}{4\pi}\left(\frac{\omega}{c}\right)\{-a_{\ell 1}^{m}(\omega)\underline{Y}_{\ell}^{m}(\alpha,\beta) + a_{\ell,2}^{m}(\omega)\underline{s} \times \underline{Y}_{\ell}^{m}(\alpha,\beta)\}\ ,$$

where we have used the fact that (cf. eqs. (2.7a) and (2.7b)) $\underline{\tilde{B}}\,(\omega/c\ \underline{s}) = \underline{s} \times \underline{\tilde{E}}\,(\omega/c\ \underline{s})$. Note that there are no vector spherical harmonics of zero degree and order. Because of this, there are no terms corresponding to $\ell = 0$ in the expansions (3.7a) and (3.7b) and, consequently, in the multipole expansion of the field. This means that there are no (transverse) states of the field having zero total angular momentum. Substituting (3.7a) and (3.7b) into

(2.6a) and (2.6b) respectively, we obtain the multipole expansion for the electric and magnetic field operators. We find that

$$\text{(3.8a)}\quad \underline{E}(\underline{r},t) = \frac{1}{2\pi}\int_0^\infty d\omega\, e^{-i\omega t}\sum_{\ell=1}^{\infty}\sum_{m=-\ell}^{\ell}\{a_{\ell,1}^m(\omega)\,{}^e\underline{E}_\ell^m(\underline{r},\omega)$$

$$+\, a_{\ell,2}^m(\omega)\,{}^h\underline{E}_\ell^m(\underline{r},\omega)\} + \text{h.c.} \quad ,$$

$$\text{(3.8b)}\quad \underline{B}(\underline{r},t) = \frac{1}{2\pi}\int_0^\infty d\omega\, e^{-i\omega t}\sum_{\ell=1}^{\infty}\sum_{m=-\ell}^{\ell}\{a_{\ell,1}^m(\omega)\,{}^e\underline{B}_\ell^m(\underline{r},\omega)$$

$$+\, a_{\ell,2}^m(\omega)\,{}^h\underline{B}_\ell^m(\underline{r},\omega)\} + \text{h.c.} \quad ,$$

where we have used (3.5) and the related expansion

$$\text{(3.5')}\quad \nabla \times \nabla\times\underline{r}\, j_\ell(\tfrac{\omega}{c} r)\, Y_\ell^m(\theta,\phi)$$

$$= \frac{(-i)^\ell}{4\pi}\left(\frac{\omega}{c}\right)\int_{-\pi}^{\pi} d\beta \int_0^{\pi} d\alpha \sin\alpha\, \underline{s} \times \underline{Y}_\ell^m(\alpha,\beta)\, e^{i\frac{\omega}{c}\underline{s}\cdot\underline{r}} ,$$

and where the *electric* and *magnetic multipole fields* of degree ℓ and order m are defined as follows:

Electric Multipole Field of Degree ℓ *and order m*

$$\text{(3.9a)}\quad {}^e\underline{E}_\ell^m(\underline{r},\omega) = \nabla \times \nabla\times \underline{r}\, j_\ell(\tfrac{\omega}{c} r)\, Y_\ell^m(\theta,\phi) \quad ,$$

$$\text{(3.9b)}\quad {}^e\underline{B}_\ell^m(\underline{r},\omega) = -\, i\,\frac{\omega}{c}\, \nabla \times \underline{r}\, j_\ell(\tfrac{\omega}{c} r)\, Y_\ell^m(\theta,\phi) \; ,$$

Magnetic Multipole Field of Degree ℓ *and order m*

$$\text{(3.9c)}\quad {}^h\underline{B}_\ell^m(\underline{r},\omega) = \nabla \times \nabla \times \underline{r}\, j_\ell(\tfrac{\omega}{c} r)\, Y_\ell^m(\theta,\phi) \; ,$$

(3.9d) $$ {}^{h}\underline{E}_{\ell}^{m}(\underline{r},\omega) = i\frac{\omega}{c}\,\nabla \times \underline{r}\; j_{\ell}(\frac{\omega}{c} r)\, Y_{\ell}^{m}(\theta,\phi) \quad . $$

${}^{e}\underline{E}_{\ell}^{m}(\underline{r},\omega)$ and ${}^{e}\underline{B}_{\ell}^{m}(\underline{r},\omega)$ are called the *electric* and *magnetic parts*, respectively, of the electric multipole field of degree ℓ and order m. Similarly, ${}^{h}\underline{E}_{\ell}^{m}(\underline{r},\omega)$ and ${}^{h}\underline{B}_{\ell}^{m}(\underline{r},\omega)$ are called the *electric* and *magnetic parts* of the magnetic multipole field of degree ℓ and order m. It is easy to verify from the above definitions that the magnetic part of the electric multipole field, and the electric part of the magnetic multipole field are everywhere perpendicular to the radius vector $\underline{r}$, i.e.,

(3.10) $$ \underline{r}\cdot {}^{e}\underline{B}_{\ell}^{m}(\underline{r},\omega) \equiv \underline{r}\cdot {}^{h}\underline{E}_{\ell}^{m}(\underline{r},\omega) \equiv 0. $$

For this reason, the electric multipole field is sometimes referred to as a *transverse magnetic* or *TM field*, and the magnetic multipole field as a *transverse electric* or *TE field*.

The multipole moment operators $a_{\ell,1}^{m}(\omega)$ and $a_{\ell,2}^{m}(\omega)$ (termed the electric and magnetic multipole moments, respectively) are easily determined from (3.7a) and (3.7b). In particular, by making use of the orthogonality relations

(3.11a) $$ \int_{-\pi}^{\pi} d\beta \int_{0}^{\pi} d\alpha \sin\alpha\; \underline{Y}_{\ell}^{*m}(\alpha,\beta)\cdot\underline{Y}_{\ell'}^{m'}(\alpha,\beta) = \ell(\ell+1)\,\delta_{\ell,\ell'}\,\delta_{m,m'} \quad , $$

(3.11b) $$ \int_{-\pi}^{\pi} d\beta \int_{0}^{\pi} d\alpha \sin\alpha\; [\underline{s} \times \underline{Y}_{\ell}^{*m}(\alpha,\beta)\,]\cdot[\underline{s} \times \underline{Y}_{\ell'}^{m'}(\alpha,\beta)\,] $$

$$ = \ell(\ell+1)\,\delta_{\ell,\ell'}\,\delta_{m,m'} \quad , $$

(3.11c) $$ \int_{-\pi}^{\pi} d\beta \int_{0}^{\pi} d\alpha \sin\alpha\; [\underline{s} \times \underline{Y}_{\ell}^{*m}(\alpha,\beta)\,]\cdot\underline{Y}_{\ell'}^{m'}(\alpha,\beta) = 0, $$

we have that

$$(3.12a)\quad a^{m}_{\ell,1}(\omega) = \frac{4\pi i^{\ell+1}}{\ell(\ell+1)}\frac{\omega}{c^2}(\hbar\omega)^{\frac{1}{2}}\int_{-\pi}^{\pi}d\beta\int_{0}^{\pi}d\alpha\sin\alpha\ \underline{\hat{a}}(\frac{\omega}{c}\underline{s}).[\underline{s}\times\underline{Y}^{*m}_{\ell}(\alpha,\beta)],$$

$$(3.12b)\quad a^{m}_{\ell,2}(\omega) = \frac{4\pi i^{\ell+1}}{\ell(\ell+1)}\frac{\omega}{c^2}(\hbar\omega)^{\frac{1}{2}}\int_{-\pi}^{\pi}d\beta\int_{0}^{\pi}d\alpha\sin\alpha\ \underline{\hat{a}}(\frac{\omega}{c}\underline{s}).\ \underline{Y}^{*m}_{\ell}(\alpha,\beta)\ ,$$

where we have used (2.7a). The relations (3.11a)-(3.11c) are easily established by making use of (3.3) and the fact that the spherical harmonics are orthonormal on the unit sphere.

The multipole moment operators are completely defined in terms of the annihilation operator $\underline{\hat{a}}$ $(\omega/c\ \underline{s})$ through eqs.(3.12a) and (3.12b). These defining relations are rather complicated, however, and consequently, it is desirable to define the multipole moment operators directly via their commutation relationships. Making use of (3.12a) and (3.12b) and the commutation relationships of the annihilation operators $\hat{a}_j(\omega/c\ \underline{s})$ given in the previous section, it is not difficult to show that

$$(3.13a)\quad [a^{m}_{\ell,j}(\omega),\ a^{m'}_{\ell',j'}(\omega')\ ] = 0,$$

$$(3.13b)\quad [a^{+m}_{\ell,j}(\omega),\ a^{+m'}_{\ell',j'}(\omega')\ ] = 0,$$

$$(3.13c)\quad [a^{m}_{\ell,j}(\omega),\ a^{+m'}_{\ell',j'}(\omega')\ ] = \frac{16\pi^2\hbar\omega}{c\ell(\ell+1)}\,\delta(\omega-\omega')\delta_{\ell,\ell'}\delta_{m,m'}\delta_{j,j'}\ ,$$

with $\ell = 1,2,\ldots;\ m = -\ell,-\ell+1,\ldots,\ell-1,\ell;\ j = 1,2$. For a derivation of the above commutators the reader is referred to reference 8, App. VIII.

The above commutation relations establish the multipole moment operators $a^{m}_{\ell,j}(\omega)$ and their adjoints $a^{+m}_{\ell,j}(\omega)$ as being annihilation and creation operators, respectively. By defining a number operator in the usual way, i.e.,

$$(3.14)\quad N^{m}_{\ell,j}(\omega) = a^{+m}_{\ell,j}(\omega)\,a^{m}_{\ell,j}(\omega)\quad ,$$

it is possible to construct a set of Fock states in entirely the same manner as is done for the number operator $N_j(\omega/c\ \underline{s}) = \hat{a}_j^+(\omega/c\ \underline{s})\ \hat{a}_j(\omega/c\ \underline{s})$. It is not difficult to show that the set of Fock states so constructed define photon states of well defined energy (labeled by the quantum number ω), total angular momentum (ℓ, m), and parity $(-1)^{\ell+j}$. An important application for such states (and for the multipole expansion) is in calculations of spontaneous and induced emission by excited atoms [13].

The Debye Representation for the Field Operators

In obtaining the multipole expansion of the field operators in the previous section, we made use of the fact that the vector spherical harmonics $\underline{Y}_\ell^m(\alpha,\beta)$ and the related fields $\underline{s} \times \underline{Y}_\ell^m(\alpha,\beta)$ together constitute a complete set for expanding (well-behaved) transverse vector fields defined on the surface of a sphere. We shall now show that this result, and indeed the entire theory of the Debye representation and multipole expansion, follows from a theorem due to Hodge [14]. Hodge's theorem deals with vector fields $\underline{A}(\underline{k})$ that are defined on and everywhere tangent to the surface of the sphere k = constant. Putting $\underline{k} = \omega/c\ \underline{s}$, and denoting the surface of the sphere $k = \omega/c$ by Ω, Hodge's theorem reads as follows:

Hodge's Decomposition Theorem

Given a transverse vector field $\underline{A}(\omega/c\ \underline{s})$ *(i.e.* $\underline{s}\cdot\underline{A}(\omega/c\ \underline{s}) = 0$) *defined and continuous, with continuous partial derivatives up to second order on* Ω. *Then there exist unique functions* $\alpha_1(\omega/c\ \underline{s})$ *and* $\alpha_2(\omega/c\ \underline{s})$ *defined and continuous, with continuous first partial derivatives on* Ω *and such that*

$$\int_{-\pi}^{\pi} d\beta \int_0^{\pi} d\alpha \sin\alpha \quad \alpha_j(\frac{\omega}{c}\ \underline{s}) = 0 \ , \qquad (j = 1,2) \qquad , \tag{4.1a}$$

$$\underline{A}(\frac{\omega}{c}\ \underline{s}) = \underline{L}\ \alpha_1(\frac{\omega}{c}\ \underline{s}) + \underline{s} \times \underline{L}\ \alpha_2(\frac{\omega}{c}\ \underline{s}) \ , \tag{4.1b}$$

where $\underline{L}$ *is the angular momentum operator defined in Eqs. (3.2).*

For a proof of Hodge's theorem the reader is referred to the paper by Wilcox, or to the author's thesis.

Although Hodge's theorem was originally established for classical (c number) fields it is equally valid for field operators.

In particular, since the plane-wave amplitudes defined in Eqs. (2.7a) and in Eqs. (2.7b) are transverse vector field operators defined on Ω, we can write [15]

$$(4.2a)\quad i(\hbar\omega)^{\frac{1}{2}} \frac{\omega^2}{c^3}\, \underline{\hat{a}}(\tfrac{\omega}{c}\underline{s}) = \frac{\omega}{c}\,[\underline{s} \times \underline{L}\,\tilde{\Pi}_1(\tfrac{\omega}{c}\underline{s}) + \underline{L}\,\tilde{\Pi}_2(\tfrac{\omega}{c}\underline{s})\,] ,$$

$$(4.2b)\quad i(\hbar\omega)^{\frac{1}{2}} \frac{\omega^2}{c^2}\, \underline{s} \times \underline{\hat{a}}(\tfrac{\omega}{c}\underline{s}) = \frac{\omega}{c}\,[- \underline{L}\tilde{\Pi}_1(\tfrac{\omega}{c}\underline{s}) + \underline{s} \times \underline{L}\tilde{\Pi}_2(\tfrac{\omega}{c}\underline{s})] ,$$

where the factor ω/c has been inserted for later convenience. (4.2b) is obtained from (4.2a) upon making use of the operator identity $\underline{s} \times \underline{s} \times \underline{L} \equiv - \underline{L}$. Substituting the above representations for the plane-wave amplitudes into the plane-wave expansions (2.6a) and (2.6b) and integrating by parts we obtain

$$(4.3a)\quad \underline{E}(\underline{r},t) = \nabla \times \nabla \times \underline{r}\,\Pi_1(\underline{r},t) - \frac{1}{c}\nabla \times \underline{r}\,\frac{\partial}{\partial t}\,\Pi_2(\underline{r},t),$$

$$(4.3b)\quad \underline{B}(\underline{r},t) = \frac{1}{c}\nabla \times \underline{r}\,\frac{\partial}{\partial t}\,\Pi_1(\underline{r},t) + \nabla \times \nabla \times \underline{r}\,\Pi_2(\underline{r},t) ,$$

where the scalar field operators $\Pi_j(r,t)$ $(j = 1,2)$ are given by

$$(4.3c)\quad \Pi_j(\underline{r},t) = \frac{1}{2\pi}\int_0^\infty d\omega e^{-i\omega t}\int_{-\pi}^{\pi} d\beta \int_0^{\pi} d\alpha \sin\alpha\, \tilde{\Pi}_j(\tfrac{\omega}{c}\underline{s}) e^{i\frac{\omega}{c}\underline{s}\cdot\underline{r}} + \text{h.c.} \quad .$$

The calculation leading to (4.3a) and (4.3b) is straightforward and is performed in reference 8, App. III.

The representation for the electromagnetic field operators given in (4.3a) and (4.3b) is called the Debye representation, and the scalar field operators $\Pi_j(\underline{r},t)$ $(j = 1,2)$ are called Debye potentials. This representation has a long and rich history in classical electromagnetic theory but, except in the form of a multipole expansion, has not been used in quantum electrodynamics. We see that it results when the plane-wave amplitudes of the electromagnetic field operators are represented according to Hodge's Decomposition Theorem.

The plane-wave amplitudes $\tilde{\Pi}_j(\frac{\omega}{c}\underline{s})$ (j = 1,2) of the Debye potentials can be determined in terms of the annihilation operator $\hat{\underline{a}}(\frac{\omega}{c}\underline{s})$. To do this we make use of the operator identities

(4.4a) $$\underline{L} \cdot (\underline{s} \times \underline{L}) \equiv (\underline{s} \times \underline{L}) \cdot \underline{L} \equiv 0 ,$$

(4.4b) $$(\underline{s} \times \underline{L}) \cdot (\underline{s} \times \underline{L}) \equiv \underline{L}.\underline{L} = L^2 = -\left(\frac{1}{\sin\alpha}\frac{\partial}{\partial\alpha} \sin\alpha \frac{\partial}{\partial\alpha} + \frac{1}{\sin^2\alpha}\frac{\partial^2}{\partial\beta^2}\right),$$

to find from (4.2a) that these plane-wave amplitudes satisfy the following linear differential equations:

(4.5a) $$L^2 \tilde{\Pi}_1 (\frac{\omega}{c}\underline{s}) = i(\hbar\omega)^{\frac{1}{2}} \frac{\omega}{c^2} (\underline{s} \times \underline{L}) \cdot \hat{\underline{a}} (\frac{\omega}{c}\underline{s}),$$

(4.5b) $$L^2 \tilde{\Pi}_2 (\frac{\omega}{c}\underline{s}) = - i(\hbar\omega)^{\frac{1}{2}} \frac{\omega}{c^2} \underline{L}. \hat{\underline{a}} (\frac{\omega}{c}\underline{s}) .$$

We also require, on account of Hodge's theorem, that

(4.5c) $$\int_{-\pi}^{\pi} d\beta \int_0^{\pi} d\alpha \sin\alpha \, \tilde{\Pi}_j(\frac{\omega}{c}\underline{s}) = 0 , \quad (j = 1,2).$$

The above differential equations, together with the condition (4.5c) uniquely specify the two plane-wave amplitudes $\tilde{\Pi}_j(\omega/c\,\underline{s})$ (j = 1,2). Indeed, it can be easily verified that unique solutions to (4.5a) and (4.5b) are readily obtained in the form of a series of spherical harmonics $Y_\ell^m(\alpha,\beta)$. Moreover, it is clear that these series solutions, when substituted into the expressions (4.2a) and (4.2b), simply yield the decomposition of the plane-wave amplitudes that lead to the multipole expansion of the field operators (cf. Eqs. (3.7a) and Eqs. (3.7b). Since this expansion was obtained in the previous section we shall not pursue this technique for solving the above differential equations and consider instead an integral form for the solutions in terms of the Green function of the operator L^2. This Green function $G(\underline{s},\underline{s}')$ was found by Zermelo [16] and is given by

$$(4.6)\quad G(\underline{s},\underline{s}') = \sum_{\ell=1}^{\infty}\sum_{m=-\ell}^{\ell} \frac{Y_\ell^{m*}(\alpha,\beta)Y_\ell^m(\alpha,\beta\ )}{\ell\ (\ell+1)} = -\ \frac{1}{2\pi}\log(\sin\ \frac{\gamma(\underline{s},\underline{s}')}{2}),$$

where $\gamma(\underline{s},\underline{s}')$ is the geodesic distance from $\underline{s}$ to $\underline{s}'$ on the surface of the unit sphere. In terms of $G(\underline{s},\underline{s}')$ we have

$$(4.7a)\quad \tilde{\Pi}_1(\frac{\omega}{c}\underline{s}) = \int_{-\pi}^{\pi} d\beta' \int_0^{\pi} d\alpha'\sin\alpha'\ [G(\underline{s},\underline{s}')][i(\hbar\omega)\ \frac{\omega}{c^2}(\underline{s}'x\underline{L}').\hat{a}(\frac{\omega}{c}\underline{s}')],$$

$$(4.7b)\quad \tilde{\Pi}_2(\frac{\omega}{c}\underline{s}) = \int_{-\pi}^{\pi} d\beta' \int_0^{\pi} d\alpha'\sin\alpha'\ [G(\underline{s},\underline{s}')][i(\hbar\omega)\ \frac{\omega}{c^2}\ \underline{L}'.\hat{\underline{a}}\ (\frac{\omega}{c}\ \underline{s}')],$$

where $\underline{L}'$ denotes the angular momentum operator in the primed coordinates.

Although the above expressions for the plane-wave amplitudes of the Debye potentials are somewhat complicated, these amplitudes have rather simple commutation relations. In particular, one finds that [17].

$$(4.8)\quad [\tilde{\Pi}_j\ (\frac{\omega}{c}\ \underline{s})\ \tilde{\Pi}_j^+\ (\frac{\omega'}{c}\ \underline{s}')] = \frac{\hbar\omega}{c}\ \delta_{j,j'}\ \delta(\omega-\omega')\ \ G(\underline{s},\underline{s}')\qquad ,$$

with all other commutators vanishing. Finally, upon making use of (4.8), we find that the Debye potential operators obey the following commutation relations

$$(4.9)\quad [\Pi_j(\underline{r},t),\ \Pi_{j'}(\underline{r}',t')] = i\hbar\delta_{j,j'}\int_{-\pi}^{\pi} d\beta \int_0^{\pi} d\alpha\sin\alpha\ G(\underline{u}_r,\underline{s})\ \cdot$$

$$\cdot\ D^+(\underline{r}-r'\underline{s},t-t') - i\hbar\delta_{j,j'}\int_{-\pi}^{\pi} d\beta \int_0^{\pi} d\alpha\sin\alpha[G(\underline{u}_{r'},\underline{s})]D^+(\underline{r}'-r\underline{s},t'-t),$$

where $\underline{u}_r$ and $\underline{u}_{r'}$ denote unit vectors in the $\underline{r}$ and $\underline{r}'$ directions, respectively. Here $D^+(R,\tau)$ denotes the positive frequency part of the Schwinger function, i.e.,

$$(4.10)\quad D^+(R,\tau) = \frac{1}{2\pi}\int_0^\infty d\omega\, e^{-i\omega\tau}\left[\frac{e^{i\frac{\omega}{c}R}}{R} - \frac{e^{-i\frac{\omega}{c}R}}{R}\right] .$$

It is interesting to note that the two Debye potential operators $\Pi_1(\underline{r},t)$ and $\Pi_2(\underline{r}',t')$ commute at *all* pairs of space-time points.[1] This result is in sharp contrast to the commutator for the Cartesian vector components of the electric or magnetic field operators which vanishes only off the light cone joining the two space-time points. On the other hand, the commutator of either Debye potential with itself at two space-time points is much more complicated than that encountered with the Cartesian vector components of the electromagnetic field operators. Indeed, it is seen that this commutator does *not* necessarily vanish off the light cone joining the two space-time points.

Acknowledgement

The author wishes to gratefully acknowledge the contribution of Professor Emil Wolf, who first suggested the procedure used in this paper, and who was advisor on the thesis from which this paper was abstracted.

References

1. R. Courant and D. Hilbert, *Methods of Mathematical Physics*, (Interscience Publishers, New York, 1962) Vol.II.,Chap.III,§5.

2. J.A. Stratton, *Electromagnetic Theory*,(McGraw-Hill, New York, 1941) §6.7.

3. See, for example, reference 2, §7.11.

4. W. Heitler, *The Quantum Theory of Radiation*,(Oxford University Press, London, 1954) Chap. II.

5. See, for example, J.M. Blatt and V.F. Weisskopf, *Theoretical Nuclear Physics*, John Wiley and Sons, New York, 1952, App. B.See also, A.M. Messiah, *Quantum Mechanics*, (John Wiley and Sons,

1966) Vol. II, Chap. XXI.

6. For a treatment of the Debye representation of the classical field see: T.J. I'a. Bromwich, Phil. Mag., Ser 6, *38* (1919) 143; C.H. Wilcox,J. Math. Mech., *6*,(1957), 167. The relationship of this representation to that of the radial Hertz vector representation has been established by: A. Sommerfeld, contribution in P. Frank and R.V. Mises,*Riemann-Weber's Differentialgleichungen der Mathematischer Physik*,(Dover Publications, 1961); A. Nisbet, Physica, *21*(1955), 799. See also, C.J. Bouwkamp and H.B.G. Casimir,Physica, *20*(1954) 539.

7. As, for example, done in B.W. Shore and D.H. Menzel, *Principles of Atomic Spectra*, (John Wiley and Sons, New York, 1968) Chap.X.

8. A.J. Devaney, *A New Theory of the Debye Representation of Classical and Quantized Electromagnetic Fields*, Ph.D. Thesis, University of Rochester, 1971. The classical theory will also appear in a future paper by Devaney and E.Wolf.

9. E.T. Whittaker, Math. Ann., *57*(1902), 333. See also, F. Rohrlich, *Classical Charged Particles*, (Addison-Wesley, Reading, Mass., 1965) §4.3.

10. We use the definitions of $Y_\ell^m(\alpha,\beta)$ and $j_\ell(\omega/c\ r)$ that are given in A. Messiah, *Quantum Mechanics*, (North-Holland Publishing Co., Amsterdam) Vol.I. App. B II and B.IV, (third printing 1965).

11. See reference 8, §3.3.

12. Definitions of the vector spherical harmonics $\underline{Y}_\ell^m(\alpha,\beta)$ vary from author to author, but are, essentially, equivalent to the one used here. For a treatment of vector spherical harmonics see: E.L. Hill, Amer. Jour. Phys., *22*(1954) 211.

13. See, for example, reference 7.

14. C. Wilcox (see reference 6) also used Hodge's theorem in a study of the Debye representation. His use of the theorem differs completely from that given here, however.

15. Here we assume that the operator $\underline{a}(\omega/c\ \underline{s})$ possesses the necessary continuity properties required in Hodge's theorem. This assumption seems justifiable since, in a coherent state representation, the (diagonal) matrix elements of this operator are of the form of a classical plane-wave amplitude and, in the classical case, the plane-wave amplitudes are generally entire analytic functions of $\omega/c\ \underline{s}$ (cf. reference 8).

16. See, for example, the paper by Wilcox. The Green function is derived also in R. Courant and D. Hilbert, *Methods of Mathematical Physics*, (Interscience Publishers, New York, 1953) Vol. I., 378.

17. For a derivation of the commutators given in (4.8) and (4.9) the reader is referred to reference 8, App. VIII.

VOLUME OF COHERENCE

Andrzej Zardecki*, Claude Delisle and Jacques Bures

Université Laval, Quebec, Canada

1. Introduction

The volume of coherence has been usually defined as the volume of a right-angled cylinder whose base is the area of coherence and whose height is the coherence length.[1] It is also the volume corresponding to one cell of phase space of photons.[2,3] We shall show in this paper that the former definition can be retained only in some particular cases whereas the latter one describes a greater variety of cases provided the notion of a cell of phase space is adequately specified.

A convenient approach is based on considering the photocount statistics. The counting probability in a thermal optical field is described by the Bose-Einstein distribution if the observation time and the size of the photodetector are both sufficiently small. In a general case no closed form expression is known for the probability $p(n,T)$ that n photoelectrons will be registered in a time interval T. Mandel[4] found an approximate formula valid for an arbitrary time interval T but restricted to the case of a point-like detector. This formula corresponds to the distribution of n bosons over a number of cells s of phase space. As the counting interval T becomes very large, s tends to the ratio of T/T_c, T_c being the coherence time. By plotting s as function of T, the coherence time can be determined. In previous papers[5-7] we introduced similarly the concept of the coherence area by deriving the approximate photocount probability distribution for the case when light is detected by an extended photocathode during a short time T. The distribution is formally

*On leave of absence from Institute of Physics, The Warsaw Technical University, Warsaw, Poland.

identical with that of Mandel; the number of degrees of freedom depends, however, on the spatial coherence characteristics rather than the spectral density of light which characterizes temporal coherence properties.

If we remain within the limits of the second moment approximation, the arguments can be extended further, i.e. to include both the temporal and spatial coherence characteristics. The basic new result is a general definition of the volume of coherence which seems to be appropriate for any incoherent thermal source. It should be evident that the factorization of the volume of coherence into the product of the coherence area and coherence time depends essentially on the factorization of the mutual coherence function at the detector surface. This property is related in turn to the spectral purity of light.

The volume of coherence is also defined when the light is emitted from a partially coherent source. An interesting feature which appears is that this case cannot be formally reduced to the former one. Finally, using a pseudothermal source we present some experimental results supporting our considerations.

2. Distribution of Photons

We represent the electromagnetic field contained within a finite volume of space as a superposition of normal modes. The annihilation operator a_k of the k'th mode has eigenvectors $|\alpha_k\rangle$ of the form

$$a_k|\alpha_k\rangle = \alpha_k|\alpha_k\rangle \tag{1}$$

with α_k playing the role of eigenvalues. In the diagonal Glauber-Sudarshan representation of the density operator of the field, the probability distribution p(m) of the number of photons m within the given volume of space at time t can be expressed as[1,8],

$$p(m) = \left\langle \frac{U^m}{m!} e^{-U} \right\rangle$$

or

$$p(m) = \int \Phi(\{\alpha_k\}) \frac{U^m}{m!} e^{-U} d^2\{\alpha_k\} \quad , \tag{2}$$

where

$$U = \sum_m |\alpha_k|^2 \tag{3}$$

and $\Phi(\{\alpha_k\})$ is the function specifying the density operator. The factorial moment generating function of the distribution p(m), defined as

$$D(s) = \langle(1 - s)^m\rangle \quad , \tag{4}$$

by virtue of (2) becomes

$$D(s) = \langle\exp\{-sU\}\rangle$$

$$D(s) = \int\Phi(\{\alpha_k\})\, e^{-sU}\, d^2\{\alpha_k\} \quad . \tag{5}$$

The weight function $\Phi(\{\alpha_k\})$ corresponding to the thermal light is[9]

$$\Phi(\{\alpha_k\}) = \sum_k \frac{1}{\pi\langle m_k\rangle}\, e^{-|\alpha_k|^2/\langle m_k\rangle} \quad . \tag{6}$$

From (5) and (6) it follows that

$$D(s) = \prod_k (1 + s\langle m_k\rangle)^{-1} \quad . \tag{7}$$

If the field is excited in M modes only and if the average number of photons in each of these modes is the same, i.e., $\langle m_k\rangle = \langle m\rangle/M$ where $\langle m\rangle$ denotes the total number of photons in the field, then (7) becomes

$$D(s) = \left(1 + \frac{s\langle m\rangle}{M}\right)^{-M} \quad . \tag{8}$$

The probability distribution p(m), which can be derived from (8) with the help of the relation

$$p(m) = \frac{1}{m!}\,(-1)^m\,\frac{d^m}{ds^m}\,D(s)\Big|_{s=1} \tag{9}$$

is

$$p(m) = \frac{\Gamma(m+M)}{m!\,\Gamma(M)}\left[1 + \frac{\langle m\rangle}{M}\right]^{-M}\left[1 + \frac{M}{\langle m\rangle}\right]^{-m} \quad . \tag{10}$$

This formula, as was recognized by Mandel[4], describes the probability of finding m bosons distributed among M equal cells of phase space. It is worth mentioning that the number of modes M is not necessarily an integer. It corresponds rather to the number of distinct statistical states; for instance the number of states radiated within a given solid angle within a certain frequency range.[2] The derivation of the formula (10) presented here bears no restrictions concerning the allowed values of M inherent in the derivation based on the combinatorial analysis. From the point of view of photoelectric detection, the importance of Eq.(10) lies in the fact that the photoelectron distribution is approximately described by a formula formally identical to (10). When the appropriate identification of variables is made, the parameter M will here, and also in the case of photoelectric detection, be said to determine the number of degrees of freedom of light.

3. Photoelectric Distribution in Partially Coherent Light

Let the light of instantaneous strength $V(r,t)$, where $V(\underline{r},t)$ is the fluctuating field value in the complex analytic signal representation, fall on an extended photodetector during a time interval T. We assume that the statistical properties of the light are those of stationary thermal beam. The probability distribution of photoelectrons ejected from the photodetector surface is given by the formula formally equivalent to (10)

$$p(n,T) = \frac{\Gamma(n+N)}{n!\Gamma(N)} \left[1 + \frac{\langle n \rangle}{N}\right]^{-N} \left[1 + \frac{N}{\langle n \rangle}\right]^{-n} , \tag{11}$$

where $\langle n \rangle$ is the average number of photoelectrons.[6] The parameter N is chosen in such a way that the second moment of the distribution is exact. In other words Eq.(11) gives the distribution of photoelectrons correctly up to the second moment. This requirement permits us to express N in the following form:

$$N = TS_d/\omega(T,S_d) , \tag{12}$$

where S_d is the area of the detector surface and the function $\omega(T,S_d)$ is determined both by the quantities related to the experiment, i.e. T,S_d and by the behaviour of the mutual coherence function $\Gamma(\underline{r}_1,\underline{r}_2,t_1-t_2)$ across the detector surface and within the detection time. Specifically we have

$$\omega(T,S_d) = \frac{1}{TS_d i^2} \int_{S_d} d^2\underline{r}_1 \int_{S_d} d^2\underline{r}_2 \int_{-T/2}^{T/2} \int_{-T/2}^{T/2} |\Gamma(\underline{r}_1,\underline{r}_2,t_1-t_2)|^2 dt_1 dt_2 \tag{13}$$

the average intensity i per unit surface of the photodetector being defined as

$$i = \frac{1}{S_d} \int_{S_d} \Gamma(\underline{r},\underline{r},0)\, d^2\underline{r} \quad . \tag{14}$$

The mutual coherence function, which obeys the wave equation, propagates from the source of light to the detector surface in the form of a wave. Consequently the parameter N can be related to the properties of the source. To this end we make use of the generalized van Cittert-Zernike theorem when a wave field is produced by an extended, not necessarily quasi-monochromatic, primary source.[10]

Let (ξ,η) be the coordinates of a typical source point and let $\underline{r}_1(x_1,y_1)$ and $\underline{r}_2(x_2,y_2)$ be the coordinates of two running points of the detector. If the linear dimensions of the source and the detector are small compared to the distance D from the detector to the source, the mutual coherence function can be expressed as

$$\Gamma(\underline{r}_1,\underline{r}_2,t_1-t_2) = \frac{1}{D^2} \int_0^\infty \exp\{i[\psi(\nu)-2\pi\nu(t_1-t_2)]\}d\nu$$

$$\times \int_{S_s} I(\xi,\eta,\nu)\, \exp\{-i\, \frac{k}{D}\, [(x_1-x_2)\xi + (y_1-y_2)\eta]\}d\xi d\eta \tag{15}$$

with

$$\psi(\nu) = \frac{2\pi\nu}{c}\, \frac{[(x_1^2+y_1^2) - (x_2^2+y_2^2)]}{2D} \quad . \tag{16}$$

We assume that the light propagates in vacuum, so that $k = 2\pi\nu/c$, c being the velocity of light. $I(\xi,\eta,\nu)$ in (15) denotes the intensity per unit area of the source, per unit frequency range. The second integral in (15) extends over the source surface. It follows from (15) that the intensity distribution across the detector surface is uniform in this approximation and is given by

$$\Gamma(\underline{r},\underline{r},o) = \frac{1}{D^2} \int_0^\infty d\nu \int_{S_s} I(\xi,\eta,\nu)\, d\xi d\eta \quad . \tag{17}$$

We shall calculate now the function $\omega(T,S_d)$ formally in the limit when both T and S_d tend to infinity. Note that although we assumed in (15) that the detector surface is small, the mutual coherence function becomes effectively zero outside a region in the vicinity of the origin of the (x,y) plane; it does not thus contribute appreciably to the value of the integral in (13). The integration over time in (13) yields for large T

$$\int_{-\frac{1}{2}T}^{\frac{1}{2}T} \int_{-\frac{1}{2}T}^{\frac{1}{2}T} dt_1\, dt_2\, \exp\{-2\pi i(t_1-t_2)(\nu_1-\nu_2)\} \int_{-T}^{T} (t-|t|)\exp\{-2\pi i(\nu_1-\nu_2)dt\}$$

$$= T\delta(\nu_1-\nu_2) \quad . \tag{18}$$

Hence by virtue of (13)-(18) we are left with

$$\omega = \frac{F\int_0^\infty d\nu \int_{S_s} I(\xi_1,\eta_1,\nu)d\xi_1\, d\xi_2 \int_{S_s} I(\xi_2,\eta_2,\nu)d\xi_2\, d\xi_2}{[\int_0^\infty d\nu \int_{S_s} I(\xi,\eta,\nu)d\xi\, d\eta]^2} \tag{19}$$

where we have included the integration over the detector variables in the factor F defined as

$$F = \frac{1}{S_d} \iint_{S_d} dx_1\, dx_2\, \exp\{-i\,\frac{2\pi\nu}{cD}(x_1-x_2)(\xi_1-\xi_2)\}$$

$$\iint_{S_d} dy_1\, dy_2\, \exp\{-i\,\frac{2\pi\nu}{cD}(y_1-y_2)(\eta_1-\eta_2)\} \quad . \tag{20}$$

In the large detector limit the integral in (20) becomes independent of the detector shape. If we choose a detector of square shape with sides parallel to the x-, y-axes of the coordinate system whose origin coincides with the center of the square, Eq.(20), analogously to (18), can be replaced by a delta function,

$$F = (\frac{cD}{\nu})^2\ \delta(\xi_1-\xi_2)\ \delta(\eta_1-\eta_2) \quad . \tag{21}$$

From (19) and (21) we get

$$\omega(T = \infty,\ S_d = \infty) = \frac{(cD)^2 \int_0^\infty \frac{d\nu}{\nu^2} \iint_{S_d} [I(\xi,\eta,\nu)]^2\, d\xi d\eta}{[\int_0^\infty d\nu \iint_{S_s} I(\xi,\eta,\nu)\, d\xi d\eta]^2} \tag{22}$$

and the parameter N as given by (12), becomes

$$N^{-1} = \frac{c^2D^2}{S_dT} \frac{\int_0^\infty \frac{d\nu}{\nu^2} \int_{S_s} [I(\xi,\eta,\nu)]^2 \, d\xi d\eta}{[\int_0^\infty d\nu \iint_{S_s} I(\xi,\eta,\nu) d\xi d\eta]^2} \quad . \tag{23}$$

Equation(23) constitutes the desired result, as it gives the parameter N of the distribution (19) in terms of the spectral properties of the source. It is seen that N is proportional to the volume of detection, $V_d = S_dT$, which is assumed to be very large. The parameter $\omega(T = \infty, S_d = \infty)$, appears as a natural measure of volume of coherence since it determines the volume of detection per one statistical degree of freedom, provided the volume of detection is large. We thus arrive at the following definition of the volume of coherence

$$V_c = \frac{S_dT}{N} = \omega(T = \infty, S_d = \infty) \quad . \tag{24}$$

It should be noted that the volume of coherence defined by (24) with ω given by (22), does not reduce, in general, to the product of the area of coherence and coherence time. An adequate description of the photocount experiments requires the more general definition (24) and (22).

4. Coherence Time and Area

In order to find under what conditions the volume of coherence factorizes, it will be useful to recall the concepts of the coherence time and the coherence area. Suppose the photodetection is performed by a point detector, i.e. a detector whose radius is much smaller than the radius of coherence. If we define the time of coherence as the time of detection per degree of freedom, the time of detection being sufficiently large, we obtain[4]

$$T_c = \int_0^\infty I^2(\nu) \, d\nu / [\int_0^\infty I(\nu) \, d\nu]^2 \tag{25}$$

where $I(\nu)$ is the spectral density of light. It may be observed that within the accuracy of the approximation involved in (15) and (17)

$$I(\nu) = \frac{1}{D^2} \int_{S_s} I(\xi,\eta,\nu)\,d\xi d\eta \quad . \tag{26}$$

In a similar manner, when the counting time interval is much smaller than the coherence time, the area of coherence has been defined[6] as the ratio of the area of detection in the large detector limit to the number of degrees of freedom. For some simple geometrical configurations, under the assumption that the light is quasi-monochromatic and the source intensity distribution is uniform, the area of coherence was shown to be

$$S_c = \frac{\lambda_o^2 D^2}{S_s} \tag{27}$$

where λ_o is the mean wavelength. This simple result was obtained earlier and with different justification by Forrester et al[11], Brown and Twiss[3], Hauss[12], Cummins and Swinney[13]. Now, in a general case of arbitrary source distribution $I(\xi,\eta)$ we can calculate N along similar lines as we did in the previous section, the only difference being that for $T<<T_c$, we can replace (13) by

$$\omega(T<<T_c,S_d) = \frac{T}{S_d i^2} \int_{S_d} d^2\underline{r}_1 \int_{S_d} d^2\underline{r}_2 \; |\Gamma(\underline{r}_1, \underline{r}_2,0)|^2 \quad . \tag{28}$$

For the area of coherence we obtain

$$S_c = \lambda_o^2 D^2 \frac{\int_{S_s} I^2(\xi,\eta)\, d\xi d\eta}{[\int_{S_s} I(\xi,\eta)\, d\xi d\eta]^2} \tag{29}$$

which becomes (27) if the intensity per unit area of the source, $I(\xi,\eta)$, is uniform.

The results contained in (25) and (29) can be summarized as

$$T_c = \omega(T>>T_c,\; S_d<<S_c) \tag{30}$$

$$S_c = \omega(T<<T_c,\; S_d>>S_c) \quad . \tag{31}$$

Going back to Eq.(22), we see that if the spectral density of the source can be varied independently of the position, i.e. if

$$I(\xi,\eta,\nu) = I_1(\xi,\eta)\ I_2(\nu) \tag{32}$$

and if in addition the light is quasimonochromatic, then

$$\omega(T = \infty,\ S_d = \infty) = T_c S_c \tag{33}$$

with T_c and S_c given by (25) and (29) respectively. The integrand $I(\nu)$ in (25) can be replaced by $I_2(\nu)$ since due to the assumption (32) and the defining equation (26), the integrals containing $I_1(\xi,\eta)$ will cancel.

The situation described by (32) embraces in particular the case of a source having a uniform normalized spectral density. All such sources lead for sufficiently small path differences to spectrally pure optical fields[14]. More generally, if we assume that the light across the detector surface is cross-spectrally pure in the sense that the complex degree of coherence factorizes

$$\gamma(\underline{r}_1,\underline{r}_2,t_1-t_2) = \gamma(\underline{r}_1,\underline{r}_2,0)\ \gamma(t_1-t_2) \tag{34}$$

Then, starting from (13), it is readily shown that $\omega(T,s)$ can be represented as a product of two factors

$$\omega(T,S_d) = \frac{1}{S_d}\int_{S_d} d^2\underline{r}_1 \int_{S_d} d^2\underline{r}_1\ |\gamma(\underline{r}_1,\underline{r}_2,0)|^2\ \frac{1}{T}\iint_{-\frac{1}{2}T}^{\frac{1}{2}T} |\gamma(t_1-t_2)|^2 dt_1\,dt_2 \tag{35}$$

In the limit $T \to \infty$ and $S_d \to \infty$ we obtain once again the expression for the volume of coherence in the factorized form. This splitting of the function $\omega(T,S_d)$ into the product of two factors related uniquely to spatial and temporal aspects of coherence was discussed in a previous paper[15] in connection with the photocount distribution of partially coherent Gaussian light detected at L discrete space points.

5. Generalization for the Case of a Partially Coherent Source

Light emanating from an incoherent source becomes partially

coherent in the process of propagation. A secondary source located at a certain distance from a primary incoherent source will be thus in general partially coherent. By making use of the van Cittert-Zernike theorem, the volume of coherence has been expressed in terms of the spectral density of the source. In order to deal with a more general case use will be made of the propagation laws of the mutual coherence. We start again with formulas (12) and (13) which determine the parameter N of the distribution (11). The mutual coherence function $\Gamma(\underline{r}_1,\underline{r}_2, t_1-t_2)$ at the surface of the detector can be expressed in terms of the mutual coherence at a fictitious surface which will be regarded as the surface of the source and denoted S in the following. If the detector is situated sufficiently far away from the source, we have, neglecting the inclination factors[10],

$$\Gamma(P_1,P_2,t_1-t_2) = \int_{S_s}\int_{S_s} \frac{\Gamma(Q_1,Q_2,t_1-t_2 - \frac{D_1-D_2}{c})}{D_1 D_2}\, dS_1\, dS_2 \tag{36}$$

In (36) dS_1 and dS_2 denote elements of the surface of the source centered on the points S_1 and S_2, and D_1 and D_2 are the distances $S_1 P_1$ and $S_2 P_2$ respectively. Now, the mutual coherence of the source $\Gamma(P_1,P_2,t_1-t_2)$ is a Fourier transform of the mutual source intensity $\hat{\Gamma}(Q_1,Q_2,\nu)$

$$\Gamma(Q_1,Q_2,t_1-t_2) = \int_0^\infty \hat{\Gamma}(Q_1,Q_2,\nu)\, \exp\{-2\pi i\nu(t_1-t_2)\}d\nu\ . \tag{37}$$

In the Fraunhofer approximation Eqs.(15) and (17) are now replaced by

$$\Gamma(\underline{r}_1,\underline{r}_2,t_1-t_2) = \frac{1}{D^2}\int_0^\infty \exp\{i[\psi(\nu) - 2\pi\nu(t_1-t_2)]\}d\nu$$

$$\times \int_{S_s} \hat{\Gamma}(\xi_1,\eta_1,\xi_2,\eta_2,\nu)\, \exp\{-i\,\frac{k}{D}[(x_1\xi_1-x_2\xi_2)$$

$$+ (y_1\eta_1-y_2\eta_2)]\}d\xi_1 d\eta_1 d\xi_2 d\eta_2 \tag{38}$$

$$\Gamma(\underline{r},\underline{r},0) = \frac{1}{D^2} \int_0^\infty d\nu \int_{S_s} \hat{\Gamma}(\xi_1,\eta_1,\xi_2,\eta_2,\nu)$$

$$\exp\{-i \frac{k}{D}[x (\xi_1-\xi_2) + y (\eta_1-\eta_2)]\} d\xi_1 d\xi_2 d\eta_1 d\eta_2 \tag{39}$$

where $\psi(\nu)$ is given by (16). $\underline{r}_1 = (x_1,y_1)$, $\underline{r}_2 = (x_2,y_2)$ denote the position of the points P_1 and P_2 of the detector, and (ξ_1,η_1), (ξ_2,η_2) denote the position of the points Q_1 and Q_2 of the source.

After substitution of (38) and (39) into (13), the integration over time can be disposed of just as we did in the previous section. The integration over the detector variables which appear only in the exponential factors yields delta functions in the large detector limit. The final result of these straightforward but tedious manipulations is

$$\omega(T = \infty, S_d = \infty) = S_d \frac{\int_0^\infty \frac{d\nu}{\nu^4} \int_{S_s} \int_{S_s} |\hat{\Gamma}(Q_1,Q_2,\nu)|^2 \, dS_1, dS_2}{[\int_0^\infty \frac{d\nu}{\nu^2} \int_{S_s} \hat{\Gamma}(Q,Q,\nu) \, dS]^2} . \tag{40}$$

If an incoherent source is represented by a delta function, i.e.

$$\hat{\Gamma}(Q_1,Q_2,\nu) = \hat{I}(Q,\nu) \, \delta(Q_1-Q_2) , \tag{41}$$

it becomes obvious from (40) that the two distinct limiting processes of going to an infinite detection volume, i.e. $S_d \to \infty$ and $T \to \infty$, and passing from a partially coherent to an incoherent source cannot be interchanged.

Indeed, putting (41) into (40) leads to a divergent expression. That means that in order to treat the limiting case of an incoherent source, we have to go back to the considerations of Section 3.

6. Experimental Results

The experimental set-up used to verify the theory has been described elsewhere[16]. A schematic diagram of the principle of the experiment is shown on Fig. 1. Both the source and the detector

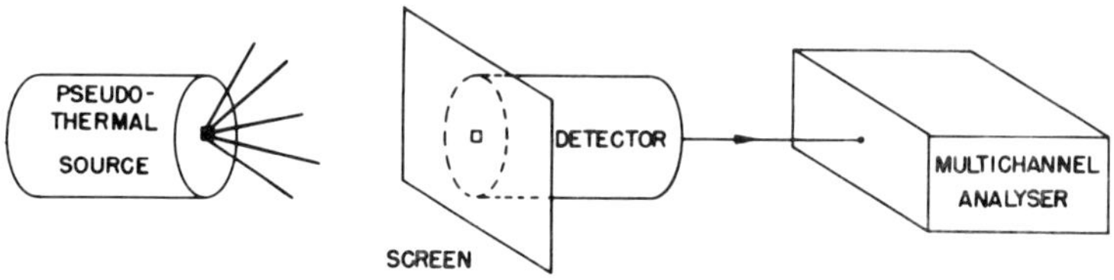

Fig. 1

are diaphragmed with a circular aperture. In order to get different ratios of T/T_c, or more precisely T/ξ, $\xi = \omega(T, S_d \ll S_c)$, the ground glass was mounted in such a way that it could be rotated at four different speeds[17]. The structure of the ground glass of the pseudo-thermal source being uniform, it is reasonable to assume a uniform spectral density over the entire surface of the source. It is then possible to factorize the function $I(Q,\nu)$ of the Eq.(22) such that,

$$I(Q,\nu) = I_1(Q)\ I_2(\nu) \quad .$$

We also assume that the intensity $I_1(Q)$ is uniform over the entire surface of the source while the spectral density $I_2(\nu)$ is arbitrary. Under these conditions, the volume of coherence reduces to the product of coherence time and coherence surface.

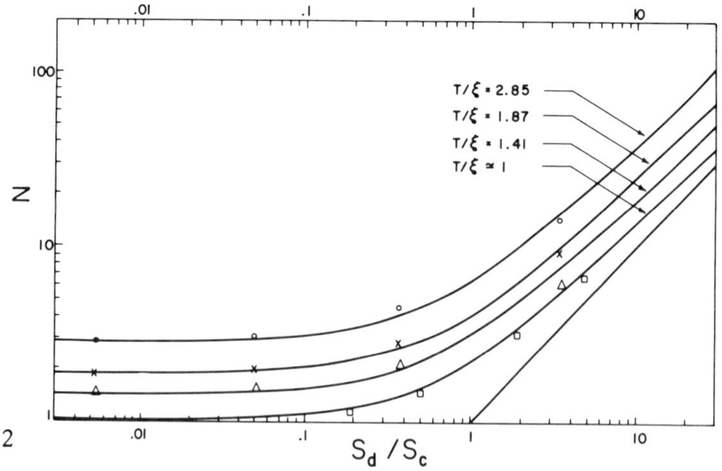

Fig. 2

Figure 2 shows N vs. S_d/S_c for four different speeds of rotation of the ground glass. On this log-log graph, the factorization of the volume of coherence appears simply as an appropriate displacement of the lower curve ($T \ll T_c$) towards higher values of N when the ratio T/ξ increases. One may then use any of the asymptotes of the different curves to determine the area of coherence. The agreement between the experimental data and theoretical curves confirms the hypothesis concerning the factorization of the volume of coherence in the present experimental set-up.

References

1. L. Mandel and E. Wolf, Rev. Mod. Phys. *37*, 231 (1965).
2. A. Kastler, *Quantum Electronics, Proceedings of the Third International Congress*, ed. N. Bloembergen and P. Grivet (Dunod et Cie., Paris, Columbia University Press, New York, 1964) p. 3.
3. R.H. Brown and R.Q. Twiss, Proc. Roy. Soc. (London) *243A*, 291 (1958).
4. L. Mandel, Proc. Phys. Soc. *74*, 113 (1959).
5. C. Delisle, J. Bures and A. Zardecki, J. Opt. Soc. Am. *61*, 1583 (1971).
6. J. Bures, C. Delisle and A. Zardecki, Can. J. Phys. *50*, 760 (1972).
7. A. Zardecki, C. Delisle and J. Bures, Opt. Comm. (in press).
8. F. Ghielmetti, Phys. Letters *12*, 210 (1964).
9. R.J. Glauber, Phys. Rev. *131*, 2766 (1963).
10. M. Born and E. Wolf, *Principles of Optics*, 4th ed. (Pergamon Press, New York, 1970) Ch. X.
11. A.T. Forrester, R.A. Gudmundsen and P.O. Johnson, Phys. Rev. *99*, 1691 (1955).
12. H.A. Haus, in *Proceedings of the International School of Physics "Enrico Fermi" XLII Course*, ed. R.J. Glauber (Academic Press, New York, 1969) p. 111.
13. H.Z. Cummins and H.L. Swinney, in *Progress in Optics*, ed. E. Wolf (North-Holland Publishing Co., Amsterdam, 1970) *VIII*.
14. L. Mandel, J. Opt. Soc. Am. *51*, 1342 (1961).
15. J. Bures, C. Delisle and A. Zardecki, Can. J. Phys. *49*, 3064 (1971).
16. J. Bures and C. Delisle, Can. J. Phys. *49*, 1255 (1971).
17. L.E. Estes, L.M. Narducci and R.A. Tuft, J. Opt. Soc. Am. *61*, 1301 (1971).

A QUANTUM TREATMENT OF SPONTANEOUS EMISSION WITHOUT PHOTONS*

W.L. Lama and L. Mandel

University of Rochester, Rochester, N.Y.

1. Introduction

The question whether it is possible to account for spontaneous emission of electromagnetic radiation from atoms without quantum electrodynamics has lately been the subject of further discussion.[1-5] In a recent article Nesbet[5] has considered an approach to the problem that differs substantially from the neoclassical approach of Jaynes and his co-workers.[1-3] In this theory the electromagnetic field is expressed explicitly in terms of its sources, which are quantized, and it obeys the algebra of the sources, while the concept of the free boson field is discarded altogether. Although he has referred to it as 'semi-quantized radiation theory', the theory is actually a fully quantized one, in the sense that no c-number currents or fields appear. When the rate of energy flow into the far electromagnetic field written in normal order is equated to the rate of energy loss of a two-level atom, Nesbet's theory apparently leads to exponential decay of the atomic energy.[5,6]

Such a source field approach is close in spirit to one advocated by Schwinger[7] and has a certain philosophical appeal. The concept of the free field is after all an abstraction. Fields are produced by sources, and can no more be divorced from their sources than the sources can be divorced from their fields. It might well be argued that the field is just an aspect of the composite quantum system 'source with its field', rather than one of two coupled quantum systems. Moreover, for a two-level atom, such a system is one of

*This work was supported by the Air Force Office of Scientific Research and by the National Science Foundation.

the simplest imaginable, that spans a two-dimensional Hilbert space. We have recently treated the closed system consisting of a two-level atom with its associated field[8] in some closed cavity, and have shown that it evolves in time very much as predicted by quantum electrodynamics.

However, as we show below, Nesbet's argument[5] for accounting for exponential atomic decay is not consistent with quantum mechanics. In order to save the situation, we are led to make several additional assumptions and a re-interpretation of the nature of the quantum system. It is no longer possible to ask meaningful questions about the nature of the field or of the atom, for the two-level system is irreducible and cannot be decomposed into parts. As a result, there are severe limitations of this approach when attempts are made to go beyond the immediate domain of the simple quantum system, for example to describe the far field of the atom. However, within these limitations we are able to account for the exponential decay of the atom with the same lifetime as is given by quantum electrodynamics.

2. The Naive Energy Balance Argument

Let us first reproduce the essential steps of the simple energy balance argument given by Nesbet.[5] We suppose that the bare atom has only two energy eigenstates $|1\rangle$ and $|2\rangle$, forming a complete set, to which we assign energy eigenvalues $e_0-\frac{1}{2}\hbar\omega$ and $e_0+\frac{1}{2}\hbar\omega$, respectively. By the well known analogy between this system and a spin $\frac{1}{2}$ in a magnetic field, we obtain the appropriate dynamical variables by introducing two anti-commuting raising and lowering operators $\hat{R}^{(+)}$ and $\hat{R}^{(-)}$, defined by[9]

$$\left.\begin{aligned} \hat{R}^{(+)}|1\rangle &= |2\rangle & \hat{R}^{(-)}|1\rangle &= 0 \\ \hat{R}^{(+)}|2\rangle &= 0 \quad , & \hat{R}^{(-)}|2\rangle &= |1\rangle \quad , \end{aligned}\right\} \quad (1)$$

together with three Hermitian operators

$$\left.\begin{aligned} \hat{R}_1 &\equiv \tfrac{1}{2}(\hat{R}^{(+)} + \hat{R}^{(-)}) \\ \hat{R}_2 &\equiv \frac{1}{2i}(\hat{R}^{(+)} - \hat{R}^{(-)}) \\ \hat{R}_3 &\equiv \frac{1}{2}[\hat{R}^{(+)}, \hat{R}^{(-)}] \quad , \end{aligned}\right\} \quad (2)$$

which obey angular momentum commutation rules

$$[\hat{R}_\ell, \hat{R}_m] = i\, \varepsilon_{\ell mn} \hat{R}_n \quad , \tag{3a}$$

and the anti-commutation rules

$$\{\hat{R}_\ell, \hat{R}_m\} = \frac{1}{4} \delta_{\ell m} \quad . \tag{3b}$$

Then the energy of the bare atom is $e_\circ + \hbar\omega\hat{R}_3$, and the transition electric dipole moment $\hat{\underline{\mu}}$ (when there is no permanent dipole moment) may be written as

$$\hat{\underline{\mu}} = 2\langle 1|\hat{\underline{\mu}}|2\rangle\, \hat{R}_1 \equiv 2\underline{\mu}\, \hat{R}_1 \quad . \tag{4}$$

From classical electromagnetic theory, the rate of energy flow in the far field at a distance r from the dipole, at time $t+r/c$, is given by $(8\pi/3)\ddot{\mu}^2(t)$. If we neglect retardations, and write this energy flow as a normally ordered operator in the two-dimensional Hilbert space of the atom, we have an expression of the form $C\,\hat{R}^{(+)}\hat{R}^{(-)}$, where C is some c-number. The expectation value of this operator is then put equal to the rate of loss of the energy expectation of the atom, so that we write

$$C\langle\psi(t)|\hat{R}^{(+)}\hat{R}^{(-)}|\psi(t)\rangle = -\frac{d}{dt}\langle\psi(t)|\hbar\omega\hat{R}_3|\psi(t)\rangle \quad ,$$

where $|\psi(t)\rangle$ is the state at time t in the Schrodinger picture, or since

$$\hat{R}^{(+)}\hat{R}^{(-)} = \hat{R}_3 - \frac{1}{2} \quad ,$$

we have

$$\left(\frac{d}{dt} + \frac{C}{\hbar\omega}\right)\langle\psi(t)|(\hat{R}_3 - \tfrac{1}{2})|\psi(t)\rangle = 0 \quad . \tag{5}$$

This leads to the exponentially decaying solution for the mean atomic energy

$$\langle\psi(t)|(\hat{R}_3 - \tfrac{1}{2})|\psi(t)\rangle = \langle\psi(0)|(\hat{R}_3 - \tfrac{1}{2})|\psi(0)\rangle\; e^{-Ct/\hbar\omega} \quad , \tag{6}$$

for any initial state $|\psi(0)\rangle$ of the atomic system, as claimed.[5]

However, if the result (6) is to be consistent with quantum mechanics, there must exist some unitary evolution operator $\hat{U}(t)$, operating on the state vectors in the two-dimensional Hilbert space spanned by the states $|1\rangle$ and $|2\rangle$, such that

$$|\psi(t)\rangle = \hat{U}(t)|\psi(0)\rangle \quad , \tag{7}$$

in Eq.(6). But since Eq.(6) is meant to hold for any initial state $|\psi(0)\rangle$, it implies the operator equality

$$\hat{U}^\dagger(t)(\hat{R}_3 - \tfrac{1}{2})\,\hat{U}(t) = (\hat{R}_3 - \tfrac{1}{2}) \quad e^{-Ct/\hbar\omega} \quad . \tag{8}$$

The impossibility of satisfying this equation is already apparent if we compare the c-number terms on the left and on the right. It is shown explicitly in Appendix A that there exists no unitary evolution operator $\hat{U}(t)$ within the two-dimensional Hilbert space that satisfies Eq.(8). Eq.(6) is therefore not consistent with quantum mechanics.

3. Energy of the Quantum System

In the foregoing energy balance argument, the energy was supposed to reside entirely in the excitation of the 'bare' atom. However, this is not in keeping with our attempt to treat the atom with its field as a unified quantum system. There is more to the system than the bare atom. Indeed, we know from classical considerations that the energy of a point dipole resides largely in the near field of the dipole.[10] It might seem, therefore, that contributions from the electromagnetic energy of the near field, of the form

$$\frac{1}{8\pi}\int_V (\underline{\hat{E}}^2 + \underline{\hat{B}}^2)\, d^3x \quad ,$$

where V is some volume surrounding the atom, of dimensions of order a wavelength, should be included in the total energy of the system. We assume that the electric and magnetic fields $\underline{\hat{E}}(\underline{r},t)$ and $\underline{\hat{B}}(\underline{r},t)$ are given by the usual dipole formulae of classical electrodynamics[11]

$$\underline{\hat{E}}(\underline{r},t) = \left[\frac{3}{r^5}\,\hat{\mu}(t-\frac{r}{c}) + \frac{3}{cr^4}\,\dot{\hat{\mu}}(t-\frac{r}{c}) + \frac{1}{c^2r^3}\,\ddot{\hat{\mu}}(t-\frac{r}{c})\right](\underline{n}\cdot\underline{r})\underline{r}$$

$$- \left[\frac{1}{r^3}\,\hat{\mu}(t-\frac{r}{c}) + \frac{1}{cr^2}\,\dot{\hat{\mu}}(t-\frac{r}{c}) + \frac{1}{c^2r}\,\ddot{\hat{\mu}}(t-\frac{r}{c})\right]\underline{n} \tag{9}$$

$$\hat{\underline{B}}(\underline{r},t) = \left[\frac{1}{cr^3}\,\dot{\hat{\mu}}(t-\frac{r}{c}) + \frac{1}{c^2r^2}\ddot{\hat{\mu}}(t-\frac{r}{c})\right](\underline{n}\times\underline{r}) \quad , \tag{10}$$

where $\underline{n}$ is a unit vector in the direction of the dipole, except that, since $\hat{\mu}$ is a Hilbert space operator, $\hat{\underline{E}}(\underline{r},t)$ and $\hat{\underline{B}}(\underline{r},t)$ are operators also. Like $\hat{\mu}$, they operate within the two-dimensional Hilbert space of the atom. Our theory of the field is therefore not at all classical. Moreover, we find that for states for which the expectation values of $\hat{\underline{E}}(\underline{r},t)$ and $\hat{\underline{B}}(\underline{r},t)$ vanish, their variances do not generally vanish, so that a phenomenon analogous to vacuum fluctuations is implied by Eqs.(9) and (10).

In addition to the contribution from the field energy, it might seem that the total energy of the system needs to be supplemented by the potential - or interaction - energy of the atomic dipole in its own field. The classical form of this contribution is $-\underline{\mu}\cdot\underline{E}$, where $\underline{E}$ is the radiation reaction field.[1-3] In order to form a Hermitian operator, we may symmetrize this by writing $-\frac{1}{2}(\hat{\underline{\mu}}\cdot\hat{\underline{E}} + \hat{\underline{E}}\cdot\hat{\underline{\mu}})$.

If we sum all three contributions to the energy, and equate the negative derivative of the result to the integral of the (symmetrized) Poynting vector over the surface S enclosing V, we obtain the equation

$$-\frac{d}{dt}\left[\hbar\omega\hat{R}_3 + \frac{1}{8\pi}\int_V(\hat{\underline{E}}^2 + \hat{\underline{B}}^2)\,d^3x - \frac{1}{2}(\hat{\underline{\mu}}\cdot\hat{\underline{E}}+\hat{\underline{E}}\cdot\hat{\underline{\mu}})\right]$$

$$= \frac{c}{4\pi}\int_S \frac{1}{2}(\hat{\underline{E}}\times\hat{\underline{B}} - \hat{\underline{B}}\times\hat{\underline{E}})\cdot\underline{dS} \quad . \tag{11}$$

Unfortunately, this equation is no more acceptable as a quantum mechanical equation than Eq.(8). One reason is that, in the two-dimensional Hilbert space in which we are working, all symmetrized operator products are c-numbers, and both the additional contributions to the energy and the Poynting vector as written in Eq.(11) fall into this category.

We shall adopt two fairly drastic changes of approach in order to arrive at meaningful results. First of all, in order to form Hermitian operators corresponding to classical variables, we take the anti-symmetrized operator product where necessary. For fermion operators whose anti-commutators are c-numbers, this would seem to be the more natural rule. For the energy flow operator $\hat{F}(t)$ we therefore write

$$\hat{F}(t) = \frac{c}{4\pi}\int_S -\frac{1}{2i}\,(\hat{\underline{E}}\times\hat{\underline{B}} + \hat{\underline{B}}\times\hat{\underline{E}})\cdot\underline{dS} + \eta(t) \quad , \tag{12}$$

where $\eta(t)$ is some c-number function that has to be included since the process of anti-symmetrization leaves the operator undefined up to a c-number. Secondly, we abandon all attempts to separate the total energy into contributions from the bare atom, the field, and the interaction between atom and field. Such a procedure is suitable for two coupled quantum systems, but not for a single irreducible quantum system. To be sure, the field and the atom are the essence of our quantum system, but they will henceforth be treated as having lost their separate identities, for there can be no structure in a two-dimensional Hilbert space. Instead, we shall use the energy balance equation

$$-\frac{d}{dt}\hat{H}(t) = \hat{F}(t) \quad , \tag{13}$$

with $\hat{F}(t)$ given by Eq.(12) and $\underline{\hat{E}}$ and $\underline{\hat{B}}$ given by Eqs.(9) and (10) in the Heisenberg picture, to determine the energy of the system, with the assumption that $\hat{H}(t)$ is also the time-dependent Hamiltonian. For the radius r of the spherical surface S over which we integrate in Eq.(12) we shall choose a radius of order an optical wavelength, in accordance with our scheme of treating the near electromagnetic field as an inseparable part of the quantum system. Needless to say, the energy balance Eq.(13) does not necessarily imply that the energy of the system is well defined at all times. If the state is not an eigenstate of $\hat{H}(t)$ there is a dispersion of the energy, and energy is balanced only on the average.

4. Solution of the Energy Balance Equation

In order to apply Eqs.(9) and (10), we need to evaluate the derivatives of μ in the Heisenberg picture. In general $\hat{\mu}(t)$ is given by the unitary transformation,

$$\hat{\mu}(t) = \hat{U}^{\dagger}(t)\hat{\mu}\,\hat{U}(t) \quad , \tag{14}$$

where $\hat{U}(t)$ is the time evolution operator

$$\hat{U}(t) = T \quad e^{-\frac{i}{\hbar}\int_0^t \hat{H}(t')\,dt'} \quad , \tag{15}$$

and T is the time-ordering operator. Thus the explicit time dependence of the Hamiltonian $\hat{H}(t)$ is required, which is presently unknown. However, for an infinitesimal time displacement δt, the evolution operator becomes

$$\hat{U}(\delta t) = 1 - \frac{i}{\hbar}\hat{H}(0)\;\delta t \quad . \tag{16}$$

If we adopt for H(t) the most general possible form for the two-dimensional Hilbert space,

$$\hat{H}(t) = \alpha(t)\ \hat{R}_1 + \beta(t)\ \hat{R}_2 + \gamma(t)\ \hat{R}_3 + \xi(t) \quad , \tag{17}$$

where $\alpha(t)$, $\beta(t)$, $\gamma(t)$, $\xi(t)$ are c-numbers, then the dipole moment at time δt, to the first order in δt, is given by

$$\hat{\mu}(\delta t) = \hat{U}^{\dagger}(\delta t)\ 2\mu\hat{R}_1\ \hat{U}(\delta t)$$

$$= 2\mu[\hat{R}_1 + \frac{\delta t}{\hbar}(\beta\hat{R}_3 - \gamma\hat{R}_2)] \quad , \tag{18}$$

where $\beta \equiv \beta(0)$, $\gamma \equiv \gamma(0)$, etc. Because $\hat{H}(t)$ is explicitly time dependent, $\dot{\hat{\mu}}(t)$ and $\ddot{\hat{\mu}}(t)$ are given by

$$\dot{\hat{\mu}}(t) = \frac{1}{i\hbar}\hat{U}^{\dagger}(t)[\hat{\mu}(0),\hat{H}(t)]\ \hat{U}(t) \quad , \tag{19}$$

$$\ddot{\hat{\mu}}(t) = \frac{1}{(i\hbar)^2}\hat{U}^{\dagger}(t)[[\hat{\mu}(0),\hat{H}(t)],\ \hat{H}(t)]\ \hat{U}(t)$$

$$+ \frac{1}{i\hbar}\hat{U}^{\dagger}(t)[\hat{\mu}(0),\dot{\hat{H}}(t)]\ \hat{U}(t) \quad . \tag{20}$$

For the short time δt, Eqs.(19) and (20) become

$$\dot{\hat{\mu}}(\delta t) = \frac{2\mu}{\hbar}\{(\beta\hat{R}_3-\gamma\hat{R}_2) + \delta t[\dot{\beta}\hat{R}_3-\dot{\gamma}\hat{R}_2 + \frac{\alpha}{\hbar}(\beta\hat{R}_2+\gamma\hat{R}_3) - \frac{1}{\hbar}(\beta^2+\gamma^2)\hat{R}_1]\} \tag{21}$$

$$\ddot{\hat{\mu}}(\delta t) = \frac{2\mu}{\hbar^2}\{\alpha(\beta\hat{R}_2+\gamma\hat{R}_3) - (\beta^2+\gamma^2)\hat{R}_1 + \hbar(\dot{\beta}\hat{R}_3-\dot{\gamma}\hat{R}_2)$$

$$+ \delta t[\hbar(\ddot{\beta}\hat{R}_3-\ddot{\gamma}\hat{R}_2) + \dot{\alpha}(\beta\hat{R}_2+\gamma\hat{R}_3) - 3(\beta\dot{\beta}+\gamma\dot{\gamma})\hat{R}_1$$

$$+ 2\alpha(\dot{\beta}\hat{R}_2+\dot{\gamma}\hat{R}_3) - (1/\hbar)(\alpha^2+\beta^2+\gamma^2)(\beta\hat{R}_3-\gamma\hat{R}_2)]\} \quad . \tag{22}$$

Anti-symmetrization of the Poynting vector in Eq.(12) leads to

$$\hat{F}(t) = -\frac{1}{3ir^3}\{[\hat{\mu}(t),\dot{\hat{\mu}}(t)] + \frac{r}{c}[\hat{\mu}(t),\ddot{\hat{\mu}}(t)]\} + \eta(t) \quad ,$$

so that, to the first order in δt, we have the equation

$$\hat{F}(\delta t) = \frac{4\mu^2}{3\hbar r^3}\{(\beta\hat{R}_2+\gamma\hat{R}_3) + \frac{r}{\hbar c}(\alpha\gamma+\hbar\dot{\beta})\hat{R}_2 - \frac{r}{\hbar c}(\alpha\beta-\hbar\dot{\gamma})\hat{R}_3$$

$$+ \delta t[\dot{\beta}\hat{R}_2+\dot{\gamma}\hat{R}_3+\frac{\alpha\gamma}{\hbar}\hat{R}_2 - \frac{\alpha\beta}{\hbar}\hat{R}_3+\frac{r}{c}(\ddot{\beta}\hat{R}_2+\ddot{\gamma}\hat{R}_3)$$

$$- \frac{r\alpha}{c\hbar^2}(\alpha\beta\hat{R}_2+\alpha\gamma\hat{R}_3-(\beta^2+\gamma^2)\hat{R}_1) - \frac{r}{\hbar c}(\beta\dot{\alpha}+2\dot{\beta}\alpha)\hat{R}_3$$

$$+ \frac{r}{\hbar c}(\gamma\dot{\alpha}+2\dot{\gamma}\alpha)\hat{R}_2 - \frac{r}{\hbar c}(\beta\dot{\gamma}-\gamma\dot{\beta})\hat{R}_1]\} + \eta + \dot{\eta}\delta t \ . \qquad (24)$$

Finally, we insert Eq.(24) in the energy balance equation (13), with

$$\frac{d}{dt}\hat{H}(\delta t) = \dot{\alpha}\hat{R}_1 + \dot{\beta}\hat{R}_2 + \dot{\gamma}\hat{R}_3 + \dot{\xi}$$

$$+ \delta t[\ddot{\alpha}\hat{R}_1 + \ddot{\beta}\hat{R}_2 + \ddot{\gamma}\hat{R}_3 + \ddot{\xi}] \ . \qquad (25)$$

Equating coefficients of like powers of δt yields the equations

$$-\dot{\alpha}\hat{R}_1-\dot{\beta}\hat{R}_2-\dot{\gamma}\hat{R}_3-\dot{\xi} = \left(\frac{4\mu^2}{3\hbar r^3}\right)[\beta\hat{R}_2+\gamma\hat{R}_3+\frac{r}{c}\left(\frac{\alpha\gamma}{\hbar}+\dot{\beta}\right)\hat{R}_2 - \frac{r}{c}\left(\frac{\alpha\beta}{\hbar}-\dot{\gamma}\right)\hat{R}_3] + \eta \ , \qquad (26)$$

and

$$-\ddot{\alpha}\hat{R}_1-\ddot{\beta}\hat{R}_2-\ddot{\gamma}\hat{R}_3-\ddot{\xi} = \left(\frac{4\mu^2}{3\hbar r^3}\right)[\dot{\beta}\hat{R}_2+\dot{\gamma}\hat{R}_3+\frac{\alpha}{\hbar}\left(\gamma\hat{R}_2 - \beta\hat{R}_3\right)$$

$$+ \frac{r}{c}\left(\ddot{\beta}\hat{R}_2+\ddot{\gamma}\hat{R}_3\right) - \frac{r\alpha}{c\hbar^2}\left(\alpha\beta\hat{R}_2+\alpha\gamma\hat{R}_3 - (\beta^2+\gamma^2)\hat{R}_1\right)$$

$$- \frac{r}{c\hbar}\left(\beta\dot{\alpha} + 2\dot{\beta}\alpha\right)\hat{R}_3 + \frac{r}{c\hbar}\left(\gamma\dot{\alpha} + 2\dot{\gamma}\alpha\right)\hat{R}_2$$

$$+ \frac{r}{c\hbar}\left(\gamma\dot{\beta} - \beta\dot{\gamma}\right)\hat{R}_1] + \dot{\eta} \ . \qquad (27)$$

Since $\hat{R}_1$, $\hat{R}_2$, $\hat{R}_3$ are linearly independent operators, the coefficients of these operators in Eqs.(26) and (27) must be equal. Thus we find from Eq.(26)

$$\dot{\alpha} = 0 \tag{28a}$$

$$\dot{\beta} = -\frac{1}{\tau_o}\left[\beta + \frac{r}{c}\left(\frac{\alpha\gamma}{\hbar} + \dot{\beta}\right)\right] \tag{28b}$$

$$\dot{\gamma} = -\frac{1}{\tau_o}\left[\gamma + \frac{r}{c}\left(-\frac{\alpha\beta}{\hbar} + \dot{\gamma}\right)\right] \tag{28c}$$

$$\dot{\xi} = -\eta \quad ; \tag{28d}$$

where $\tau_o \equiv 3hr^3/4\mu^2$, and another set of four equations follows from Eq.(27), of which the first is

$$\ddot{\alpha} = \frac{r}{c\hbar\tau_o}\left[-\frac{\alpha}{\hbar}(\beta^2+\gamma^2) - (\gamma\dot{\beta}-\beta\dot{\gamma})\right] . \tag{29}$$

If we make the assumption that these equations hold at all times, not just at t = 0, the equations are readily solved. On multiplying Eq.(28b) by β and Eq.(28c) by γ and adding, we obtain

$$\frac{1}{2}\frac{d}{dt}(\beta^2+\gamma^2) + \frac{1}{\tau}(\beta^2+\gamma^2) = 0 \quad , \tag{30}$$

where we have written

$$\tau \equiv \tau_o + r/c \quad , \tag{31}$$

and the solution is

$$\beta^2(t) + \gamma^2(t) = \left(\beta^2(0)+\gamma^2(0)\right) \exp(-2t/\tau) \quad . \tag{32}$$

From Eq.(28a),

$$\alpha(t) = \text{constant} \quad , \tag{33}$$

so that $\ddot{\alpha} = 0$. On the other hand, with the help of Eqs.(28b), (28c), and (31), we have from Eq.(29)

$$0 = -\frac{\alpha r}{c\hbar^2\tau}[\beta^2(0)+\gamma^2(0)] \exp(-\frac{2t}{\tau}) \quad , \tag{34}$$

which requires that

$$\alpha = 0 \quad . \tag{35}$$

Equations(28b) and (28c) then yield immediately,

$$\beta(t) = \beta(0)\ \exp(-t/\tau) \quad , \tag{36}$$

and

$$\gamma(t) = \gamma(0)\ \exp(-t/\tau) \quad , \tag{37}$$

and the other equations derivable from Eq.(27) are all consistent with these solutions.

The constants $\beta(0)$ and $\gamma(0)$ and the function $\xi(t)$ can be determined from physical considerations. It is convenient to write the density operator of the two-level system in the Heisenberg picture in the form

$$\hat{\rho} = \left(\sin\tfrac{1}{2}\theta\ \exp(\tfrac{1}{2}i\phi)\,|1\rangle + \cos\tfrac{1}{2}\theta\ \exp(-\tfrac{1}{2}i\phi)\,|2\rangle\right)$$
$$\left(\sin\tfrac{1}{2}\theta\ \exp(-\tfrac{1}{2}i\phi)\langle 1| + \cos\tfrac{1}{2}\theta\ \exp(\tfrac{1}{2}i\phi)\langle 2|\right) \quad , \tag{38}$$

where θ and ϕ are parameters corresponding to the polar and azimuthal angles of the Bloch vector. The upper state corresponds to $\theta = 0$, and the lower state to $\theta = \pi$. Then the expectation values of the energy and energy flow are given by

$$\langle\hat{H}(t)\rangle = \mathrm{Tr}[\hat{\rho}\hat{H}(t)]$$
$$= \frac{1}{2}\exp(-t/\tau)[\beta(0)\ \sin\theta\ \sin\phi + \gamma(0)\ \cos\theta] + \xi(t) \quad , \tag{39}$$

and

$$\langle\hat{F}(t)\rangle = \mathrm{Tr}[\hat{\rho}\hat{F}(t)]$$
$$=(1/2\tau)\exp(-t/\tau)\ [\beta(0)\ \sin\theta\ \sin\phi + \gamma(0)\ \cos\theta] + \eta(t) \ . \tag{40}$$

From Eqs.(39) and (40) it follows that

$$\langle\hat{H}\rangle - \tau\langle\hat{F}\rangle = \xi(t) + \tau\,\dot{\xi}(t) \quad .$$

If we adopt the convention that $\langle\hat{H}\rangle$ vanishes in the state in which the energy flow $\langle\hat{F}\rangle$ vanishes, then in this state

$$\xi(t) + \tau\,\dot{\xi}(t) = 0 \ ,$$

or

$$\xi(t) = \xi(0)\,\exp(-t/\tau) \ . \tag{41}$$

Moreover, if we also adopt the convention that at $t = 0$

$$\langle 2|\hat{H}|2\rangle - \langle 1|\hat{H}|1\rangle = \hbar\omega \quad ,$$

then

$$\gamma(0) = \hbar\omega \quad , \tag{42}$$

and Eqs.(39) and (40) reduce to

$$\langle\hat{H}\rangle = \exp(-t/\tau)\ [\tfrac{1}{2}\beta(0)\ \sin\theta\ \sin\phi + \tfrac{1}{2}\hbar\omega\ \cos\theta + \xi(0)] \tag{43}$$

and

$$\langle\hat{F}\rangle = \frac{1}{\tau}\,\exp(-t/\tau)\ [\tfrac{1}{2}\beta(0)\ \sin\theta\ \sin\phi + \tfrac{1}{2}\hbar\omega\ \cos\theta + \xi(0)] \quad . \tag{44}$$

This represents exponential decay of the system with time constant τ given by Eq.(31). We have already argued that r should be made of order the optical wavelength, in order that the whole near field of the atom is incorporated in the quantum system. If we choose

$$r = c/\omega \quad ,$$

then

$$\tau = \frac{3\hbar c^3}{4\mu^2\omega^3} + \frac{1}{\omega} = \tau_0 + \frac{1}{\omega} \quad , \tag{45}$$

which differs from the usual lifetime given by quantum electrodynamics (τ_0) only by the optical period $1/\omega$, and is therefore indistinguishable from it in practice.

The parameter $\xi(0)$ is undetermined, but does not appear to have a great deal of physical significance in that the contributions to $\langle \hat{H} \rangle$ and $\langle \hat{F} \rangle$ involving $\xi(0)$ are independent of the state of the system. $\xi(0)$ appears to play a role somewhat similar to a reference level.

If we put $\xi(0) = \frac{1}{2}\hbar\omega$, we can readily show that $\beta(0)$ should be zero also. For, from physical consideration, $\langle \hat{F} \rangle$ should be non-negative no matter what the initial state of the system may be. Equation (44) for $\langle \hat{F} \rangle$ then requires that

$$\hbar\omega(1 + \cos\theta) \geq \beta(0) \sin\theta$$

or

$$\beta(0)/\hbar\omega \leq \cot\frac{\theta}{2} \quad ,$$

for all θ in the range 0 to π. This condition can only be satisfied if $\beta(0) = 0$.

The expressions for the energy and for the energy flow rate then become

$$\hat{H}(t) = \hbar\omega(\hat{R}_3 + \frac{1}{2}) \exp(-t/\tau) \tag{46}$$

and

$$\hat{F}(t) = \frac{\hbar\omega}{\tau} (\hat{R}_3 + \frac{1}{2}) \exp(-t/\tau) \quad , \tag{47}$$

while the corresponding expectation values are

$$\langle \theta,\phi|\hat{H}(t)|\theta,\phi \rangle = \frac{\hbar\omega}{2} (1 + \cos\theta) \exp(-t/\tau) \quad , \tag{48}$$

$$\langle \theta,\phi|\hat{F}(t)|\theta,\phi \rangle = \frac{\hbar\omega}{2\tau} (1 + \cos\theta) \exp(-t/\tau) \quad . \tag{49}$$

We have therefore arrived at a Hamiltonian of an exceptionally simple kind. If $\xi(0)$ is not chosen to be $\frac{1}{2}\hbar\omega$, the Hamiltonian is a little less simple, but it still satisfies the commutation relation

$$[\hat{H}(t), \hat{H}(t')] = 0 \quad .$$

5. Discussion

Having found the time-dependent Hamiltonian for the system, we can, in principle, find the expectation value of any dynamical variable of interest. Indeed, since $\hat{H}(t)$ and $\hat{H}(t')$ commute at all times t and t', the unitary time evolution operator $\hat{U}(t)$ has the simple form

$$\hat{U}(t) = \exp\left[-i\omega\tau\left(1-\exp(-t/\tau)\right)\hat{R}_3\right] \quad . \tag{50}$$

In practice, however, it is difficult to attach physical significance to most of the dynamical variables of the quantum system that we have been considering. For example, although we might like to think of $\hat{U}^{\dagger}(t)\hbar\omega(\hat{R}_3+\frac{1}{2})\ \hat{U}(t)$ as the energy of the bare atom, or of $\mu(t)$ as the transition dipole moment of the bare atom, these variables have no physical significance for the composite quantum system of atom plus field. In a two-dimensional Hilbert space there can be no internal structure, and it makes no sense to ask questions about the quantum system that imply that such structure exists. The source and its field are inseparable, and form an irreducible system[12].

By way of illustration, we note that the expectation value of what we might like to regard as the 'energy of the bare atom' $\hat{U}^{\dagger}(t)\hbar\omega(\hat{R}_3+\frac{1}{2})\ \hat{U}(t)$ is given by

$$\langle\theta,\phi|\hat{U}^{\dagger}(t)\hbar\omega(\hat{R}_3+\tfrac{1}{2})\ \hat{U}(t)|\theta,\phi\rangle = \tfrac{1}{2}\hbar\omega(1+\cos\theta) \quad , \tag{51}$$

and remains constant in time as the system is losing energy. Evidently

$\hbar\omega(\hat{R}_3+\frac{1}{2})$ is not a physically meaningful variable for our quantum system. Similarly, if we took the far field of the atomic dipole to be given by the familiar formula[11]

$$\hat{\underline{E}}(\underline{r},t) = \frac{\hat{\mu}(t-r/c)}{c^2 r}\left[\frac{(\underline{n}\cdot\underline{r})\underline{r}}{r^2} - \underline{n}\right] , \tag{52}$$

then, with the help of the Heisenberg equations of motion, we would find that

$$\langle\theta,\phi|\hat{\underline{E}}(\underline{r},t)|\theta,\phi\rangle = \frac{\mu\omega\sin\theta}{c^2 r\tau}\left[\frac{(\underline{n}\cdot\underline{r})\underline{r}}{r^2} - \underline{n}\right]\Big\{-\omega\tau\exp[-\tfrac{2}{\tau}(t-\tfrac{r}{c})]$$

$$\times\cos[\omega\tau(1-\exp[-\tfrac{1}{\tau}(t-\tfrac{r}{c})]) + \phi]$$

$$+\exp[-\tfrac{1}{\tau}(t-\tfrac{r}{c})]\sin[\omega\tau(1-\exp[-\tfrac{1}{\tau}(t-\tfrac{r}{c})]) + \phi]\Big\} , \tag{53}$$

which represents an oscillation of instantaneous frequency $\omega\exp[-(t-r/c)/\tau]$. But once again this is not a physically meaningful quantity. Indeed, in a sense the far field should not be expected to be a part of our quantum system, for it can be measured without appreciably disturbing the system.

It is clear that the number of physically meaningful questions that can be answered by using the present approach, within the confines of a two-dimensional Hilbert space, is very limited. Nevertheless, for those features, like the energy flow and the lifetime, for which results can be obtained by this theory, they are found to coincide with the corresponding results given by quantum electrodynamics. At least within this limited sense, the approach seems to be justified.

Appendix A: The Impossibility of Satisfying Eq.(8) within a Two-Dimensional Hilbert Space

In order to show that there exists no unitary evolution operator $\hat{U}(t)$ satisfying Eq.(8), we consider the most general possible form of $\hat{U}(t)$, given by

$$\hat{U}(t) = \exp -i[\alpha(t)\hat{R}_1 + \beta(t)\hat{R}_2 + \gamma(t)\hat{R}_3 + \xi(t)] . \tag{A1}$$

Since $\hat{R}_1$, $\hat{R}_2$, $\hat{R}_3$, together with the unit operator, form a complete linearly independent set of operators for the two-dimensional Hilbert space, the exponent in Eq.(A1) is the most general possible Hermitian form. With the help of the operator expansion theorem we find

$$\hat{U}^{\dagger}(t)\hat{R}_3\hat{U}(t) = (1/\eta^2)\Big[\big(\gamma\alpha(1-\cos\eta) - \beta\eta\sin\eta\big)\hat{R}_1 + \big(\gamma\beta(1-\cos\eta) + \alpha\eta\sin\eta\big)\hat{R}_2 + \big(\gamma^2 + (\alpha^2+\beta^2)\cos\eta\big)\hat{R}_3\Big] \quad , \tag{A2}$$

where

$$\eta^2 \equiv \alpha^2 + \beta^2 + \gamma^2 \quad . \tag{A3}$$

Substitution of Eq.(A2) in Eq.(8) and comparison of coefficients of $\hat{R}_1$, $\hat{R}_2$, $\hat{R}_3$ leads to the relations

$$\gamma\alpha(1-\cos\eta) - \beta\eta\sin\eta = 0 \quad , \tag{A4}$$

$$\gamma\beta(1-\cos\eta) + \alpha\eta\sin\eta = 0 \quad , \tag{A5}$$

$$\gamma^2 + (\alpha^2+\beta^2)\cos\eta = \eta^2\exp(-Ct/h\omega) \quad . \tag{A6}$$

From (A4) and (A5)

$$\gamma(\alpha^2+\beta^2)(1-\cos\eta) = 0 \quad ,$$

and

$$(\alpha^2+\beta^2)\sin\eta = 0 \quad ,$$

which requires that

$$\text{either } \alpha = 0,\ \beta = 0 \ ,$$

$$\text{or} \quad \eta = 2\pi n \quad ,$$

where n is an integer. In both cases Eq.(A6) then reduces to the impossible condition

$$1 = \exp(-Ct/h\omega) \quad ,$$

which is obtained by comparing c-number terms in Eq.(8) directly.

References

1. E.T. Jaynes and F.W. Cummings, Proc. IEEE *51*, 89 (1963).
2. M.D. Crisp and E.T. Jaynes, Phys. Rev. *179*, 1253 (1969).
3. C.R. Stroud, Jr. and E.T. Jaynes, Phys. Rev. *A1*, 106 (1970).
4. R.K. Nesbet, Phys. Rev. *A4*, 259 (1971).
5. R.K. Nesbet, Phys. Rev. Letters *27*, 553 (1971).
6. A somewhat similar argument was implicit also in the work of N.E. Rehler and J.H. Eberly, Phys. Rev. *A2*, 1735 (1971).
7. J.S. Schwinger, *Proceedings of the 1967 International Conference on Particles and Fields*, ed. C.R. Hagen, G. Guralnik and V.A. Mathur (Interscience Publishers, John Wiley, New York, 1967) p. 128.
8. W.L. Lama and L. Mandel, Phys. Rev. *A6*, 2247 (1972).
9. We label all Hilbert space operators by the caret ^.
10. L. Mandel, J. Opt. Soc. Am. *62*, 1011 (1972).
11. See for example M. Born and E. Wolf, *Principles of Optics, 4th Ed.* (Pergamon Press, Oxford, 1970) Section 2.2.3.
12. We have recently been made aware of a treatment of the atomic decay problem by G.W. Series, *Proceedings of the International Conference on Optical Pumping and Atomic Lineshape*, Warsaw, 1968, p. 25, in which the electromagnetic field is also expressed explicitly in terms of source operators, and the free field plays no role. By working with a larger Hilbert space, Series is able to derive both the decay rate and the Lamb shift.

FINITE QUANTUM AND CLASSICAL ELECTRODYNAMICS

E. C. G. Sudarshan

University of Texas at Austin, Austin, Texas

In this paper I wish to report on a continuing investigation of relativistic quantum theory which is free from mathematical absurdities, in agreement with physical results and is conceptually satisfactory. [1] While the results so far obtained in the theory are encouraging and are of general applicability to strong, electromagnetic and weak interactions, for a variety of reasons I will confine my attention to finite quantum electrodynamics. Even apart from the primary concern of this Conference, electrodynamics is the prototype of all field theories. It is universal and it has been the beginning of our understanding of the quantum nature of the primary constituents of the universe. Moreover, quantum electrodynamics has made many remarkable quantitative predictions like the anomalous magnetic moment of the electron and the Lamb shift. At a conceptual level quantum electrodynamics reconciles the principles of statistical mechanics with the existence of systems with infinite number of degrees of freedom; and demonstrates that virtual quantum field theoretic processes are real enough to produce observed quantitative physical effects.

Troubles in Electrodynamics

In spite of all these satisfactory features quantum electrodynamics is a sick theory. Many straightforward questions get meaningless answers. The mathematical formalism itself contains at best ambiguous and at worst meaningless expressions. The appearance of divergent quantities makes the calculations more of a magical ritual than a mathematical derivation. The divergence

* Work supported by the U.S. Atomic Energy Commission.

problems eliminated from the Rayleigh-Jeans paradox for the black body reappear in the case of interacting systems as divergences in virtual processes.

In the myth of creation light was the first to be created. Harmonies abounded in the successive stages of creation and disharmony came only with the coming together of the serpent, the woman, the man and the apple. But quantum field theory would rewrite the story and have it that the disharmony has its roots in creating anything other than light which could emit or absorb it! There were no divergent radiative corrections until the sun and the moon were created. The coupling constant is the root of all disharmony.

Yet it is said that the Creator found what he had created on the six days to be "Good". All the disharmony and error crept in when the snake persuaded the woman to take an arbitrary independent point of view about the nature of things. Much laborious work by the good Lord was necessary to correct these errors.

Let us draw a moral from this myth and assert that the interacting radiation and matter system is not inconsistent either physically or mathematically. But this consistency is for the complete system of interacting constituents. The traditional version of quantum electrodynamics is not consistent in this sense. We must start afresh, but the theory must not only describe the usual kind of electrons and photons but be in quantitative agreement with the observed interactions.

Some of the divergences of electrodynamics are an inheritance from classical theory. [2] Classical radiation theory in conjunction with the principles of statistical mechanics give rise to the Rayleigh-Jeans paradox. The classical electron is inconsistent. If it were not for nonelectromagnetic forces the electron should have exploded. Shall we follow Poincaré and postulate such nonelectromagnetic forces of electromagnetic strength? [3] What could these nonelectromagnetic forces be? Should we now invent other forces like surface tension or incompressible matter, and then give up the connection between forces and coupled fields? But if we invent new fields to produce these forces would they not have their own quanta? Where are they?

Classical Electrodynamics with Nonelectromagnetic Fields

Regarding these nonelectromagnetic stabilizing forces we note that for relativistic invariance of the theory this new field must behave in the same manner as the electromagnetic field and yet give an attractive force between like charges. In a classical Lagrangian theory such a result can be obtained from a Lagrangian of the form

$$L = -\sum_r \tfrac{1}{2} m_r \dot{x}^{\mu} \dot{x}_{r\mu} - \int d^3x \tfrac{1}{4} F_{\mu\nu}(x) F^{\mu\nu}(x)$$

$$+ \sum_r e_r \dot{x}_r^{\mu} A_{\mu}(x_r) - \int d^3x \tfrac{1}{2} M^2 B^{\mu}(x) B_{\mu}(x)$$

$$+ \int d^3x \tfrac{1}{4} G_{\mu\nu}(x) G^{\mu\nu}(x) + \sum_r e_r \dot{x}_r^{\mu} B_{\mu}(x_r),$$

where

$$F_{\mu\nu} = \partial_{\nu} A_{\mu} - \partial_{\mu} A_{\nu}$$

$$G_{\mu\nu} = \partial_{\nu} B_{\mu} - \partial_{\mu} B_{\nu}$$

Note that the coupling of A_{μ} and B_{μ} are through the *same* coupling constants so that only the linear combination $A_{\mu} + B_{\mu}$ enters the interaction. The sign of the *free* energy term of the satellite field B is of the sign *opposite* of the usual one.

The equation of motion satisfied by the particles and fields can now be worked out by standard methods. We see that there is a relative sign change between the A_{μ} and B_{μ} fields in field equations. (This sign could be removed by replacing B_{μ} by $-B_{\mu}$ but then it will show up in the Lorentz force equation.) Precisely because of this sign change the *mutual* effect of two charged particles by virtue of their coupling to these two fields are of *opposite* signs. We may say that while the charged particles interact with both the electromagnetic field $F_{\mu\nu}$ and the nonelectromagnetic field $G_{\mu\nu}$ in the same way, they act as sources for these two fields in opposite ways. The "absorption" is the same for both fields but the "emission" is opposite. Hence the difference.[4]

If we wish to develop this theory further and formulate its Hamiltonian equations we would conclude that it is necessary for the $B_{\mu}, G_{\mu\nu}$ fields to have a Poisson bracket with the sign opposite that of the $A_{\mu}, F_{\mu\nu}$ fields. With this minor change the equations of motion of the system can be cast in a Hamiltonian form. We can verify that the self interaction between two charged particles is finite. In the specially simple case of static force the potential has the form

$$V(r) = \frac{e^2}{r} \left(1 - e^{-mr}\right)$$

which no longer exhibits any explosive tendencies for $r \to 0$. We could, at the same time verify that the infinite self stress of the classical electron has also disappeared. There is a finite self stress which does not lead to any difficulties.

While this pleasant result has been achieved we must remember that we had to pay a price: we do have the field B_μ for which the free field energy is negative definite. We would find it embarrassing to have such negative energy fields being really absorbed or emitted. Yet how can we eliminate them?

This paper is devoted to quantum electrodynamics and so I shall not answer this question except to say that there is a method of devising such a theory. Instead we shall now jump to quantum electrodynamics.

Quantum Electrodynamics with Photon Satellites

Since we have to deal with the quantum theory of radiation and matter we shall take as our model the quantum electrodynamics of electrons and photons, the theory of coupled Maxwell-Dirac fields. One starts with the Lagrangian density:

$$L(x) = \overline{\psi}(x)(\gamma^\mu i\partial_\mu - m)\,\psi(x) - \tfrac{1}{4} F^{\mu\nu}F_{\mu\nu}$$

$$+ \; e\,\overline{\psi}(x)\,\gamma^\mu\psi(x)\,A_\mu(x)$$

Quantizing this system for $e = 0$ is quite straightforward and it leads to photons obeying Bose statistics and electrons and positrons obeying Fermi statistics. But when $e \neq 0$ and there is genuine interaction, the theory apparently breaks down. If we attempt to calculate the self energy of the electron we get an infinite result. This result was to be expected since we inherit it from the classical theory. (The modification of quantum theory is to reduce the divergence from "quadratic" to "logarithmic", but still it is infinite.) But in addition to this new infinities arise. The photon can virtually dissociate into an electron-positron pair and the modification of the propagation properties of the photon is again infinite: this is the so-called "vacuum polarization". Still another kind of infinity appears in the modification of the basic photon electron interaction. All these effects enter into the various higher order effects, the so-called "radiative corrections". But "corrections" which are infinite are no corrections but catastrophes.

The ingenuity of the theorist has found ways of getting around in this absurd world and extract some meaningful predictions out of an apparently meaningless theory. The story of renormalized quantum electrodynamics is a heroic one and an intriguing one. It predicted correctly the $\frac{\alpha}{2\pi}$ Kusch-Schwinger anomalous moment of the electron; and still higher order corrections. None the less the theory is absurd: renormalizations, yes, but infinite renormalizations, no.

Let us make a finite quantum electrodynamics using the lessons learnt in classical electrodynamics with the satellite field B_u. With its inclusion the Lagrangian density is amended to read

$$L(x) = \overline{\psi}(x)\,(\gamma^{\mu}_{1}\,\partial_{\mu} - m)\,\psi(x) - \tfrac{1}{4} F^{\mu\nu} F_{\mu\nu}$$

$$+ \quad \tfrac{1}{4} G^{\mu} G_{\mu} - \tfrac{1}{2} M^2 B^{\mu} B_{\mu}$$

$$+ \quad e\,\overline{\psi}\,(x)\,\gamma^{\mu}\overline{\psi}(x)\,(A_{\mu}(x) + B_{\mu}(x)\,)\;.$$

For e = 0 we can again quantize this system as before; the only modification is that the B_{μ}, $G_{\mu\nu}$ field is now quantized with the sign of the commutator brackets opposite the usual one. Its quanta are particles of mass M, but because of the peculiar sign of the commutator we get a strange kind of state space. The vector space now turns out to be an innerproduct space with an indefinite scalar product. We obtain a quantum mechanical system with an indefinite metric. [1,5] But choosing this system all the equations of motion are satisfied as in the classical theory. Since a quantized field is equivalent to a collection of oscillators, we may recognize the appearance of an indefinite metric in the case of a single oscillator degree of freedom.

Let us, accordingly choose a single degree of freedom with variables q and p satisfying

$$[q,p] \;=\; -\,i\;.$$

If we now write

$$a = \frac{\omega q + ip}{\sqrt{2}\omega^{\frac{1}{2}}}\;, \qquad a^{+} = \frac{\omega q - ip}{\sqrt{2}\omega^{\frac{1}{2}}}\;,$$

then

$$[a,a^{+}\,] = -\;1.$$

If we demand that a ground state exists such that it is annihilated by the destruction operator a, then

$$a\,|\,0\,\rangle = \;0,$$

and we obtain the sequence of states

$$|n\rangle = \frac{(a^+)^n}{\sqrt{n!}} |0\rangle$$

are pseudoorthonormal in the sense

$$\langle n' | n\rangle = (-1)^n \delta_{nn'} .$$

The state $|1\rangle$ has the square -1 since

$$\langle 1|1\rangle = \langle 0|aa^+|0\rangle = \langle 0|[a,a^+] + a^+a|0\rangle$$

$$= -1 + \langle 0|a^+a|0\rangle = -1 .$$

Thus the opposite sign of the commutator automatically leads to an indefinite metric theory. But the opposite sign of the commutator came from the need to have the correct equations of motion, which in turn is necessitated by having to have the nonelectromagnetic force cancel the singularity of the electromagnetic force at short distances.

Since in every interaction the A_μ and B_μ fields enter with the same coupling we may, if we so choose, insert an effective propagation function which is the algebraic sum of the propagators for the electromagnetic and the nonelectromagnetic fields. Now, the photon propagator has $\delta(x^2)$ and $(x^2)^{-1}$ singularities and these are precisely cancelled by the corresponding singularities of the satellite propagator. Equivalently, in momentum space the standard photon propagation function had an inverse p^2 dependence; so does the satellite propagator but the leading term has an opposite sign. Consequently the effective propagator will fall off as the inverse p^4; it is "superconvergent".

Quantum Electrodynamics without Divergences

This newly obtained superconvergence makes the electron self energy finite and generally improves the convergence of all quantities. But it is not sufficient. In quantum electrodynamics we have also the self energy of the photon ("vacuum polarization"), which is also infinite and not made any less divergent because of the satellite photon since the simplest process involves only an electron-positron pair and no photon propagator. Some people like to make it disappear by appeal to a magical incantation called gauge invariance, but I have not found any way to get convinced by this exorcism ritual.

We must get rid of this infinity. We can do this only by use of electron satellites; each electron must have at least two satellites so that the propagators achieve the necessary degree of superconvergence.[1]

Having obtained such a superconvergent electron propagator it follows that all the diagrams of quantum electrodynamics which have at least one pair of external lines now converge. There is no need to have the photon satellite, though there is no reason not to have it. Having it does not remove any more divergences.

There are still the vacuum processes which correspond to the spontaneous creation and subsequent total annihilation of electrons, positrons and photons which lead from vacuum to vacuum. Due to the infinite volume of space time the amplitude for this process must diverge and cannot be made finite. However, in the observed transition matrices these infinities automatically cancel out and we may hence ignore them altogether.

My own personal preference is to seek nonelectromagnetic stabilizing influences in the presence of satellite electrons. I must admit that there is no reason to forbid photon satellites except that the simplest possibility is to have as few satellites as needed for convergence; but perhaps we have both kinds of satellites. Let experiment decide.

Theory of Shadow States

Divergence troubles have been eliminated, but what about the appearance of negative probabilities? After all, the probabilistic interpretation of quantum mechanics would breakdown if negative *probabilities* are introduced. We must therefore prevent these states from appearing in the initial or final states despite the virtual processes involving such. Without such a device our theory is not physically useful. We already saw a similar requirement in classical theory where the nonelectromagnetic field had negative field energy.

This is accomplished by the discovery of the theory of "shadow states".[1] Shadow states are states that enter the mathematical description of the dynamical variables and the interaction but do not contribute to incoming or outgoing probability fluxes. The probabilities are conserved entirely amongst the physical states only. A proper presentation of the theory of shadow states will be beyond the scope of this paper but it amounts to having standing rather than running waves in the shadow channels; running waves are allowed only in physical channels. The difference in behavior of these channels is entirely due to the choice of the appropriate boundary conditions, a freedom not constrained by quantum field theory. One can demonstrate these results for models which can be solved in closed form but they could be verified for field theories in general.

Consider, for example, a two-channel quantum mechanical model with a free Hamiltonian

$$H_o = \begin{bmatrix} H_o^{(1)} & 0 \\ 0 & H_o^{(2)} \end{bmatrix}$$

and an interaction

$$V = \begin{bmatrix} V_{11} & V_{12} \\ V_{21} & V_{22} \end{bmatrix}$$

If ϕ_λ is an eigenfunction H_0 with eigenvalue λ the usual perturbation result for the corresponding eigenfunction for the total Hamiltonian

$$H = H_o + V = \begin{bmatrix} H_o^{(1)} + V_{11} & V_{12} \\ V_{21} & H_o^{(2)} + V_{22} \end{bmatrix}$$

is given by the familiar formula

$$\psi_\lambda = \phi_\lambda + G_o(\lambda) V \phi_\lambda + G_o(\lambda) V G_o(\lambda) V \phi_\lambda + \dots$$

$$= (1 - G_o(\lambda) V)^{-1} \phi_\lambda$$

where

$$G_o(\lambda) = (\lambda - H_o + i\varepsilon)^{-1}$$

is the (retarded) Green's function, the boundary conditions associated with the $i\varepsilon$ ensuring that there are "plane waves" plus "outgoing spherical waves". The corresponding scattering amplitude matrix is given by the coefficient of the diverging spherical wave

$$T(\lambda) = \underset{H_0 \to \lambda}{\text{Lt}} \{G_0(\lambda)\ (\psi_\lambda - \phi_\lambda)\}$$

$$= V + V\,G(\lambda)\,V + \dots$$

$$= V\,(1 - G_0(\lambda)\,V)^{-1}$$

This expression satisfies the unitarity relation

$$T - T^\dagger = T\,(G_0(\lambda) - G_0^\dagger(\lambda))\,T^\dagger$$

assuring ourselves of conservation of probability. The probability is conserved amongst *both* the channels; of course transitions between channels occur only because the interaction V has cross matrix elements between the two channels.

We now alter the boundary conditions and choose the new Green's function

$$G(\lambda) = \begin{bmatrix} (\lambda - H_0^{(1)} + i\,\varepsilon)^{-1} & 0 \\ 0 & P\left(\dfrac{1}{\lambda - H_0^{(1)}}\right) \end{bmatrix}$$

$$= (\lambda - H_0^{(1)} + i\,\varepsilon\,\pi^{(1)})^{-1}$$

where $\pi^{(1)}$ is the projection operator to the first channel. In the second channel we have standing waves. Given this choice of Green's function we can compute a new scattering amplitude

$$T = V + V\,G\,V + \dots$$

$$= V\,(1 - G(\lambda)\,V)^{-1}$$

This expression does not quite satisfy the unitarity relation but the truncated quantity

$$\tau = \pi^{(1)}\,T\,\pi^{(1)}$$

does satisfy a unitarity relation of the form:

$$\tau - \tau^* = \tau\,\{(\lambda - H_0^{(1)} + i\varepsilon)^{-1} - (\lambda - H_0^{(1)} - i\varepsilon)^{-1}\}\,\tau^*.$$

This assures us that there are no transitions between the two channels even though there is an interaction coupling them. The probability is conserved in channel 1 by itself. We call states in channel 2 as shadow states. Shadow states contribute to the dynamics but do not enter into probability conservation. Like the wind in Camelot which blows during the night to sweep away the fallen leaves, the shadow states in an indefinite metric theory blow away the infinities and absurdities but yet never see the light of day.

The theory is given more detailed exposition elsewhere [1] and shown to be capable of giving finite "unitary" transition amplitudes between physical states which obey conservation of probability.[7] Since we may choose as physical states only states not containing any satellite quanta we shall have no problem with negative probabilities. All probabilities are positive. Causality problems do not seem to provide any terrors either. [8]

Confrontation with Experiments

Thus we have developed a conceptually satisfactory framework for finite quantum electrodynamics. We have convergence due to a suitable use of an indefinite metric theory, yet the probability interpretation is not upset. But what about the quantitative aspects of the successful predictions of quantum electrodynamics? Surprisingly enough, it turns out that both the anomalous moment of the electron and the Lamb shift are quite insensitive to these modifications. So is much of the data on Compton scattering. [9] We should look for possible departures in large angle muon and electron pair processes at very high energy.

What kind of qualitatively new predictions does one expect? We can immediately say that the satellites should exhibit themselves as resonances in the respective channels. [1] If photon satellites are there we must see photonic resonances.[5] We must also distinguish them from the observed vector meson-associated resonances. From what has been said above we know that electron satellites are expected. These satellites are more striking in the sense that we have no other mechanism for their occurrence.[10]

There are other technical features which would be crucial: many of these resonances would have their phases rotate opposite to the usual case, i.e., the complex amplitudes will rotate clockwise. At the opening of the pseudothresholds associated with continuum shadow states the amplitude will have a point of nonanalyticity: The amplitude will be changing from one analytic function into a distinct one at this point. This piecewise analyticity would be a trademark of a shadow state theory.[11]

Return to Harmony

In conclusion we remark that harmony can be restored in creation provided we use a consistent formulation without mathematical absurdities from the beginning. This involves a generalization involving satellite fields and an indefinite metric and the theory of shadow states. It is a new theory. The standing waves in the shadow channel are the mathematical counterpart of the physical requirement that stability requires the presence of nonelectromagnetic forces at all times.

I have not talked about the mathematical problems of taking products of field operators at the same point. But let me add here that once the indefinite metric is properly used only suitable combinations of the fields and their satellites appear in the interaction; and the mathematical ambiguities disappear. [1]

At the turn of the century, Planck showed us that basic divergences of field theory required the bold experiment of a new conceptual structure. Should we not continue his work?

References

1. For systematic expositions see,
   E.C.G. Sudarshan, Proc. XIX Solvay Congress (1967) "*Fundamental Problems in Elementary Particle Physics*", Interscience, New York (1968).
   E.C.G. Sudarshan, Fields & Quanta *2*, 175 (1972).
   E.C.G. Sudarshan, Lecture at the Latin American Summer School, University of Mexico (1971).
2. For an account of classical electron theory, see
   F. Rohrlich, "*The Classical Electron*", (Addison-Wesley, New York)
3. H. Poincaré, Comptes Rend. *140*, 1504 (1905) and Rend.Circl.Mat. Palermo *21*, 129 (1906).
4. Imaginary coupling constants are an unnecessary complication without any physical significance.
5. T.D. Lee and G.C. Wick, Phys. Rev. *D2*, 1033 (1970).
6. See, for example, J.M. Jauch & F. Rohrlich, "*Theory of Photons and Electrons*", (Addison-Wesley, New York) (1954).
7. C.A. Nelson & E.C.G. Sudarshan, paper submitted to Physical Review (in press).
   J.L. Richard, Comm. Math. Phys. (in press).
   C.A. Nelson, LSU preprint (1972).
8. H. Rechenberg & E.C.G. Sudarshan, Max-Planck preprint (1972).
9. M.E. Arons, M.Y. Han & E.C.G. Sudarshan, Phys. Rev. *137*, B1085 (1965).
10. E.C.G. Sudarshan, Proc. Muon Conference, University of Colorado (in press).
11. M.G. Gundzik & E.C.G. Sudarshan, Phys. Rev.*D6*, 796 (1972).

THE USE OF CLASSICAL SOLUTIONS IN QUANTUM OPTICS

I. R. Senitzky*

U.S. Army Electronics Command, Fort Monmouth, N.J.

The problem I propose to consider is that of a number of identical two-level-systems coupled to the entire radiation field, which is represented by a large number of modes. I will discuss both the phenomenon of free decay of the systems from initial excited states, and the forced oscillation by the systems under an external resonant driving field. It will be seen that the classical results play an important role in supplementing the quantum mechanical analysis.

We will idealize the situation with respect to the two-level systems by the assumption that these systems are identically coupled to each mode under consideration; in other words, they see the same field. Under such idealized circumstances, the effect of the two-level systems on the field is that of a number of identical spin 1/2 systems that can be combined into a single angular momentum system of total quantum number L_o, L_o having any value up to one-half the number of two-level systems. The problem thus becomes one of the interaction between a single angular momentum system - of possibly large quantum number L_o - and the radiation field. The analysis will be presented both classically and quantum mechanically, in a parallel fashion.

The angular momentum system, to which I will refer as the atomic system, can be described by the Hamiltonian $H = \hbar\, \omega\, \ell_3$ where ω is the natural resonance frequency of the system (if the system were a magnetic dipole, ω would be the angular precession

*Present address: Department of Physics, Technion Israel Institute of Technology, Haifa, Israel.

frequency in an external magnetic field). ℓ_1, ℓ_2, and ℓ_3 are dimensionless angular momentum components. It will be more convenient to use the variables ℓ_+ and ℓ_-,

$$\ell_{\pm} = 2^{-\frac{1}{2}}(\ell_1 \pm i\,\ell_2) \equiv L_{\pm}\, e^{\pm i\omega t} ,$$

and furthermore, to introduce the reduced variables L_+ and L_-, which, for the free atomic system, are constants, and have a slow time variation caused by coupling to other systems.

Equations of motion, both classical and quantum mechanical, can be written down for the interaction between the atomic system and a large number of denumerable field modes with close frequency spacing. One can then solve for the field variables in terms of the angular momentum variables, substitute back into the equations of motion, and obtain a set of equations that contain only the angular momentum variables (which are unknown), and the initial - or free - field variables (which are known).[1] These equations will also contain, as parameters, the constants specifying the coupling between the atomic system and each mode, as well as the density of modes. For free decay, the classical form of these equations can be solved easily,

$$L_3(t) = -L_0 \tanh \alpha L_0(t - t_0),$$

where

$$\alpha \equiv \tfrac{1}{4}\pi\, |\gamma(\omega)|^2 \rho(\omega), \qquad t_0 = (\alpha L_0)^{-1} \tanh^{-1}[L_3(0)/L_0],$$

$|\alpha(\omega)|$ being an average over the absolute value of the coupling constants referring to the modes of frequency in the neighborhood of ω, and $\rho(\omega)$ being the density of modes in the same neighborhood. The classical solution for L_3 is well known. We note that for $L_3(0) = L_0$, the classical solution is $L_3(t) = L_0$, and we have a condition of unstable equilibrium. We will refer to the initial condition $L_3(0) = L_0$, in which the atomic system is fully excited, as the unstable equilibrium initial condition even in the quantum mechanical discussion, later. The amplitudes of the oscillating components of the angular momentum are given by

$$L_{\pm}(t) = \{\tfrac{1}{2}[L_0^2 - L_3^2(t)]\}^{\frac{1}{2}} \exp\{\pm i[\theta_0 + \tilde{\alpha}\textstyle\int_0^t dt_1 L_3(t_1)]\},$$

where $\tilde{\alpha}$ will be defined shortly. If $L_3(t)$ changes slowly, $\tilde{\alpha}\, L_3(t)$ is approximately the radiative frequency shift of the atomic system,

$$\Delta\omega \simeq \tilde{\alpha}\, L_3(t) .$$

$\tilde{\alpha}$ is a constant that consists of three terms,

$$\tilde{\alpha} \equiv \alpha_2 + \alpha_3 - 2\alpha_4 \equiv \int_0^\infty d\nu |\gamma(\nu)|^2 \rho(\nu) \left[\frac{P}{\nu-\omega} + \frac{1}{\nu+\omega} - \frac{2}{\nu} \right].$$

The interesting aspect of this decomposition is that α_3 and α_4 would be missing if the rotating wave approximation were used. Thus, the frequency shift is qualitatively affected - even to the possible extent of having a divergent integral rather than a convergent integral for $\tilde{\alpha}$ - when the rotating wave approximation is used. The same result applies to a quantum mechanical calculation of the frequency shift. If we consider an initial energy state described by the quantum number m,

$$L_3(o)|m\rangle = m|m\rangle,$$

then for t much less than the decay time, the matrix elements of $L_\pm$ between two consecutive energy states yield the frequency shift

$$\Delta\omega \simeq \tilde{\alpha}m - (\alpha_2 - \alpha_4).$$

For a two-level system, it becomes

$$\Delta\omega = -\tfrac{1}{2}(\alpha_2 - \alpha_3),$$

and for m large (positively or negatively), it approaches the classical frequency shift. It is seen, therefore, that for obtaining frequency shifts, the rotating-wave approximation should not be used. In energy considerations, however, it is a good approximation.

We consider the free decay, or the spontaneous emission, of the atomic system energy for large L_o. From uncertainty-principle considerations, as well as from a comparison of the classical and quantum mechanical equations of motion,[1] it can be concluded that a system of large angular momentum behaves like a classical system except when its energy is near maximum *and* when the external field acting on it is weak. In fact, it is necessary to treat the system quantum mechanically only for

$$L_o - L_3 \lesssim 1 .$$

We make this a formal quantum mechanical condition by putting expectation value brackets around L_3. The most interesting spontaneous emission problem, and the one for which the solution differs most from the classical solution, is that for which the atomic system is initially in its maximum energy state. We therefore consider the problem (with the unstable equilibrium initial condition)

$$L_3(o) \mid > \;=\; L_o \mid > .$$

Quantum mechanically, the atomic system does emit in this state, and the expectation value of the energy goes down. Let t_1 be the time necessary for the energy expectation value to decay (by spontaneous emission) sufficiently from its maximum value so that the atomic system may be treated classically. We then solve the problem quantum mechanically for $t < t_1$, and join the solution to the classical solution for $t > t_1$. Since the quantum mechanical solution is a statistical solution, we must express the initial classical value in statistical terms. $L_3(t)$, for $t > t_1$, thus becomes a classical random variable, because of its statistical initial value.

Using a dimensionless time,

$$\tau = \alpha t,$$

we expand $<L_3(\tau)>$ in a Taylor's series,

$$< L_3(\tau) > \;=\; L_o + \sum_{n=1}^{\infty} < L_3^{(n)}(o) > \frac{\tau^n}{n!} .$$

The initial derivatives can be obtained for arbitrary L_o by an iterative scheme (based on the equations of motion),[1]

$$< L_3^{(1)}(o)> \;=\; -\; 2L_o, \qquad <L_3^{(2)}(o)> \;=\; -\; 2^2(L_o^2 - L_o),$$

$$< L_3^{(3)}(o) > \;=\; -\; 2^3\,(L_o^3 - 4L_o^2 + 2L_o),$$

$$< L_3^{(4)}(o)> \;=\; -\; 2^4\,(L_o^4 - 11\,L_o^3 + 19\,L_o^2 - 7\,L_o),$$

- -

$$< L_3^{(n)}(o) > \;=\; -\; 2^n\,[L_o^n - (2^n - n - 1)\,L_o^{n-1} + - - -].$$

It is seen that the derivatives can be expressed as polynomials in L_o. The Taylor's series can now be rearranged in decending powers of L_o, yielding

$$<L_3(\tau)> \;=\; L_o\;\{1 - (e^{2L_o\tau}/L_o)\,(1 - e^{-2L_o\tau}) + (e^{2L_o\tau}/L_o)^2\,[1 - e^{-2L_o\tau}(1 + 2\,L_o\tau)] - \ldots\} .$$

If we choose τ_1 (which is αt_1) as the dimensionless time at which we change over from the quantum mechanical to the classical solution, then τ_1 has to satisfy the inequalities

$$L_o >> e^{2L_o\tau_1} >> 1 .$$

The first inequality implies that a few terms in the Taylor's series will be a good approximation, and the second inequality implies that the changeover to a classical solution is permissable. A convenient choice for τ_1 is $1/L_o$.

If $\langle L_3(\tau)\rangle$ is known in the quantum mechanical region, one can obtain the higher moments of L_3 by an iterative procedure (again based on the equations of motion), [1]

$$\langle L_3^{(1)}\rangle = \langle L_3^2\rangle - \langle L_3\rangle - L^2 ,$$

$$\langle L_3^{(2)}\rangle = 2\,[\langle L_3^3\rangle - 2\langle L_3^2\rangle + (1 - L^2)\langle L_3\rangle + L^2],$$

and so on, with $L^2 = L_o(L_o + 1)$. The classical solution for $\tau > L_o^{-1}$ is now given by

$$L_3(\tau) = - L_o \tanh\{L_o\tau - \tfrac{1}{2}\ln 2L_o + \ln[1 - (1 + D)e^{-2}]\} ,$$

where D is a random variable, some of its properties being [1]

$$\langle D\rangle = 0, \qquad \langle D^2\rangle^{\frac{1}{2}} = e^2(1 - e^{-2})^{\frac{1}{2}}$$

$$D_{max} = e^2(1 - e^{-2}), \qquad \langle D^3\rangle < 0 .$$

It is worth noting that the expression for $L_3(\tau)$ *without the random term* is often given as an approximate expression for $\langle L_3(\tau)\rangle$, obtained by an approximation which neglects the statistical fact that $\langle L_3^2\rangle \neq \langle L_3\rangle^2$, in general.

As an illustration of the use of this solution, we obtain, the average time of decay of the energy to zero (that is, to half the initial value) or, to put it another way, the average time for which the radiated power reaches a maximum. This time is obtained by setting the argument of the hyperbolic tangent equal to zero, and then averaging. Expanding the logarithm about zero, retaining only the first two terms, and utilizing the r.m.s. value of D, we obtain

$$\langle \tau_{decay} \rangle \simeq \frac{1}{2L_o} (\ln 2 L_o + 1).$$

Note that the second term in the parenthesis is the contribution of the statistical term.

Another simple illustration of the use of the statistical expression for $L_3(\tau)$ arises when one asks "what is the maximum power radiated by the atomic system?", a question which has produced some disagreement in the literature. In any one experiment, D is a constant, and is random with respect to an ensemble of experiments. Thus, in any one experiment, the power radiated, given by $-\dot{L}_3$, is a maximum when the argument of the hyperbolic tangent vanishes, the maximum being L_o^2 . In other words, the *average of the maximum power* radiated is L_o^2. This does not mean, however, that the *maximum of the average power* radiated is L_o^2 . The maximum of the average will obviously be less, since L_o has to be multiplied by the average of the square of the hyperbolic secant over a range of values of the argument. For instance, if we take $L_o \tau = \frac{1}{2} \ln 2L_o$, and average the radiated power over the three values $D = 0, \pm\frac{1}{2}\sqrt{\langle D^2 \rangle}$, we obtain $0.83\, L_o^2$.

In the analysis of the free decay of the atomic system, the classical solution was used *to supplement* the quantum mechanical solution, or, one can say, *as an aid* in the quantum mechanical solution. In the case of an external field, it will be seen that the classical solution is needed in order to *interpret* the quantum mechanical solution. Here, for large L_o and a strong external driving field, we have, from general considerations, an essentially classical system under all conditions, and naturally, expect that there should be only slight differences between the classical and quantum mechanical solutions. The result turns out to be rather surprising.

We consider, for simplicity, the atomic system to be initially in the ground state. Classically, the equation of motion for L_3 can be written

$$\ddot{L}_3 - 3\alpha L_3 \dot{L}_3 + [\Omega^2 - \alpha^2 (L_o^2 - L_3^2)]\, L_3 = 0,$$

where

$$\Omega^2 = 2\,|\,A\,|^2 ,$$

and $|A|$ is the strength of the driving field. We consider the case of a strong driving field,

$$\Omega >> \alpha\, L_o .$$

Then the classical equation for the energy becomes

$$\ddot{L}_3 - 3\,\alpha\, L_3\dot{L}_3 + \Omega^2 L_3 = 0 \ .$$

If there were no coupling to the radiation field, α would be zero and we would have the equation for a periodic sinusoidal oscillation of period $2\pi/\Omega$. The α term makes this a nonlinear second order differential equation, which can be analyzed in the L_3, $\dot{L}_3$ plane. Such an analysis shows that L_3 *still* oscillates periodically, but not longer sinusoidally. Moreover, to first order in α, the period is the same, namely, $2\pi/\Omega$.

The quantum mechanical equation for $\langle L_3\rangle$, also for a strong driving field, is

$$\langle\ddot{L}_3\rangle + \frac{3}{2}\,\alpha\langle\dot{L}^3\rangle + \Omega^2\langle L_3\rangle - \frac{3}{2}\,\alpha\,\langle\dot{L}_3L_3 + L_3\dot{L}_3\rangle = 0 \ .$$

If the expectation value brackets are discarded and the noncommutativity of L_3 and $\dot{L}_3$ is ignored, this equation is the same as the classical equation *except* for the $\dot{L}_3$ term. Now, this is a very important term and produces a damping of the oscillation of $\langle L_3\rangle$. Thus, the $\dot{L}_3$ term alters the solution qualitatively. A particularly simple equation for $\langle L_3\rangle$ is obtained when $L_o = \frac{1}{2}$, in which case the symmetrized term in the equation vanishes, and the equation becomes

$$\langle\ddot{L}_3\rangle + \frac{3}{2}\,\alpha\,\langle\dot{L}_3\rangle + \Omega^2\langle L_3\rangle = 0.$$

This equation has the approximate solution

$$\langle L_3\rangle = -\tfrac{1}{2}\exp\left(-\frac{3}{4}\alpha t\right)\cos\Omega t \quad ,$$

an exponentially damped sinusoidal oscillation.

We therefore have an apparent paradox: On the one hand, a strongly driven system of large L_o should behave almost classically, and yet there is a qualitative difference between the time-dependence of the classical energy and that of the quantum mechanical expectation value of the energy. The explanation must be sought in the statistical aspects of the quantum mechanical result. The expectation value denotes an ensemble average, each member of the ensemble being a large angular momentum system coupled to a radiation field and driven by an external field. Although it is customary to predict the result of an experiment involving large (or macroscopic) systems by an expectation value, it would be incorrect to do so in the present case. The expectation value must be interpreted as being the average over an ensemble of undamped oscillations, each oscillation differing from the classical energy oscillation only slightly by a small random variation in frequency. As is well known, the average over undamped oscillations with a spread in frequency, or with random fluctuation in frequency, will be a damped oscillation.

A comparison of $\langle L_3\rangle^2$ and $\langle L_3{}^2\rangle$ confirms this interpretation. If we consider a time short compared to the damping time α^{-1} but long compared to a period of oscillation $2\pi/\Omega$, then

$$\langle L_3\rangle^2 \simeq L_o{}^2 \left(1 - \frac{3}{2}\alpha t\right) \cos^2 \Omega t,$$

which is consistent with exponential damping of $\langle L_3\rangle$. However, we get [1]

$$\langle L_3{}^2\rangle \simeq \left[L_o{}^2 - \frac{3}{2}\alpha t L_o(L_o - \tfrac{1}{2})\right] \cos^2 \Omega t$$

$$+ \left[\tfrac{1}{2} L_o + \alpha t L_o (L_o - \tfrac{1}{2})\right] \sin^2 \Omega t,$$

which is consistent with an ensemble of *undamped* oscillations that have a variation in frequency. An ensemble of damped oscillations, where each member would resemble the expectation value, is inconsistent with the existence of the term $\alpha t L_o{}^2 \sin^2 \Omega t$. In this instance, the classical solution is more descriptive of the result of a single experiment than the quantum mechanical expectation value, and motivates the correct interpretation of the quantum mechanics.

References

1. I.R. Senitzky, Phys. Rev. *A6*, 1175 (1972).

RAMAN EFFECT IN SEMICLASSICAL THEORY

M. D. Crisp

Owens-Illinois Technical Center, Toledo, Ohio

Semiclassical radiation theory describes the interaction of light with matter in terms of a classical electromagnetic field perturbing a quantum mechanical atom. In a recent paper[1], Nesbet claims to have demonstrated a clear disagreement between the semiclassical theory of the Raman effect and experimental data on Raman scattering. It will be shown here that, when properly applied, semiclassical theory does provide a completely adequate description of the Raman effect. This result eliminates the necessity of forming a "semiquantized theory" as described in reference 1.

A formulation of semiclassical theory which is appropriate for the Raman effect may be found in reference 2. The analysis presented in reference 2 included the effect of an atom's radiation field reacting back upon the atom. This radiation reaction field resulted in a nonexponential spontaneous decay of the atom and radiative frequency shifts of the light emitted by the atom. It will be shown here that a description of the Raman effect does not require the inclusion of this radiation reaction field. Nesbet's semiclassical analysis appears to have overlooked this point.

In order to describe the Raman effect semiclassically it is necessary to consider the time dependent expectation of the dipole moment operator $\langle\mu\rangle$. In semiclassical theory it is assumed that $\langle\mu\rangle$ can be treated as a real dipole moment which is capable of radiating electromagnetic fields in accordance with Maxwell's equations. The Raman scattered light will be shown to appear as a component of this radiated field.

For the sake of simplicity the analysis presented below

will be done for a one-electron atom. The calculation is further simplified by considering the interaction of a left circularly polarized light wave with an atom in a magnetic field whose direction is along the direction of propagation of the light beam. If the atom is located at the origin then it is perturbed by a field of the form

$$\underline{A}\ (0,t) = \frac{c\mathcal{E}}{\omega} [\hat{\underline{e}}_x \cos\omega t + \hat{\underline{e}}_y \sin \omega t] \tag{1}$$

in the dipole approximation.

The evolution of the atom under the influence of this monochromatic plane wave is described by the Hamiltonian

$$H = H_a + V_D + V, \tag{2}$$

where the atomic Hamiltonian H_a has a complete set of eigenstates $|\ell\rangle$ such that

$$H_a\ |\ell\rangle = E_\ell\ |\ell\rangle\ .$$

The diamagnetic term in the Hamiltonian $(e^2/2mc^2)\underline{A}^2$ is given by

$$V_D = e^2 \mathcal{E}^2/2m\omega^2 \tag{3}$$

for the circularly polarized plane wave. This term is diagonal in the basis $|\ell\rangle$. The off-diagonal part of the perturbation is $-(e/mc)\underline{A}\cdot\underline{p}$ in the Coulomb gauge and its matrix elements are

$$V_{\ell n} = \frac{-i\Omega_{\ell n}\mu_{\ell n}}{\omega}\ e^{\pm i\omega t}\ , \tag{4}$$

in the basis $|\ell\rangle$. The atomic transition frequencies are

$$\Omega_{\ell n} = (E_\ell - E_n)/\hbar\ ,$$

and the dipole moment matrix element is

$$\mu_{\ell n} = \langle\ell|ex|n\rangle$$

where x is the electron's position coordinate along the x-axis.

The choice of the sign of the exponential in Eq.(4) depends on the change of the magnetic quantum number M. If $M_\ell = M_n + 1$ then the - sign is used; if $M_\ell = M_n - 1$ then the + sign applies.

When second order time dependent perturbation theory as outlined in reference 3 is applied to the Hamiltonian of Eq.(2), the resulting time evolution operator can be written as

$$U^I_{\ell n} = \delta_{\ell n} + \frac{i\Omega_{\ell n}\mu_{\ell n}}{\hbar\omega}\,\mathcal{E}\left[\frac{e^{i(\Omega_{\ell n}\pm\omega)t}-1}{\Omega_{\ell n}\pm\omega}\right] + \sum_\gamma \frac{\Omega_{\ell\gamma}\Omega_{\gamma n}\mu_{\ell\gamma}\mu_{\gamma n}}{(\Omega_{\gamma n}\pm\omega)\hbar^2\omega^2}\,\mathcal{E}^2 \tag{5}$$

$$\left[\frac{e^{i(\Omega_{\ell n}\pm\omega\pm\omega)t}-1}{(\Omega_{\ell n}\pm\omega\pm\omega)} - \frac{e^{i(\Omega_{\ell\gamma}\pm\omega)t}-1}{\Omega_{\ell\gamma}\pm\omega}\right],$$

in the interaction picture. The second term in the brackets of Eq.(5) results from the sudden turning on the perturbation and it is usually ignored in the quantum electrodynamic derivation of the Kramer-Heisenberg formula[5]. The plus and minus signs in the term $\exp i(\Omega_{\ell n}\pm\omega\pm\omega)t$ are determined by the selection rules relating the states ℓ, γ, and n.

In order to understand the implications of Eq. (5), it is useful to apply it to a specific atomic level scheme for which a Raman scattering experiment can be done. A very clean experiment involving Raman scattering in atomic thallium has recently been done by Weingarten *et al* [6]. The actual experiment involves several different competing Raman transitions; a typical one is illustrated in Fig. 1. Quantitative agreement with the experiment would require inclusion of more than one intermediate state, but the level scheme of Fig.1 is adequate to illustrate the semiclassical description of Raman scattering.

For the three states chosen, the selection rules and values of the magnetic quantum number require that the perturbation V has the form

$$V = -\hbar\begin{bmatrix} 0 & 0 & \alpha_{13}e^{i\omega t} \\ 0 & 0 & \alpha_{23}e^{i\omega t} \\ \alpha_{31}e^{-i\omega t} & \alpha_{32}e^{-i\omega t} & 0 \end{bmatrix} \tag{6}$$

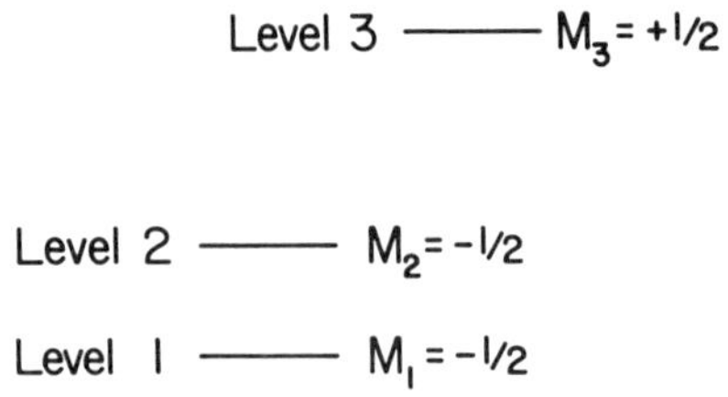

Fig.1. Raman Effect in Semiclassical Theory
This figure shows three levels that might be involved in a typical Raman scattering experiment. For comparison with the experiment of reference 6, the levels could be considered to be sublevels of the $6^2P_{1/2}$, $6^2P_{3/2}$, and $7^2S_{1/2}$ levels of atomic thallium.

where the coefficient $\alpha_{\ell n}$ is given by

$$\alpha_{\ell n} = i(\mu_{\ell n} \mathcal{E}/\hbar\omega)\Omega_{\ell n} \quad . \tag{7}$$

The expression for the time evolution operator that corresponds to the perturbation of Eq.(6) is

$$U^I(t) = \begin{bmatrix} 1 + \dfrac{|\alpha_{13}|^2}{(\Omega_{31}-\omega)} \left[it + \dfrac{e^{-i(\Omega_{31}-\omega)t}-1}{\Omega_{31}-\omega} \right] \\ \dfrac{\alpha_{23}\alpha_{31}}{(\Omega_{31}-\omega)} \left[\dfrac{e^{i\Omega_{21}t}-1}{\Omega_{21}} + \dfrac{e^{-i(\Omega_{32}-\omega)t}-1}{\Omega_{32}-\omega} \right] \\ \alpha_{31} \dfrac{e^{i(\Omega_{31}-\omega)t}-1}{\Omega_{31}-\omega} \end{bmatrix}$$

$$\frac{\alpha_{13}\alpha_{32}}{(\Omega_{32}-\omega)}\left[-\frac{e^{-i\Omega_{21}t}-1}{\Omega_{21}}+\frac{e^{-i(\Omega_{31}-\omega)t}-1}{\Omega_{31}-\omega}\right]$$

$$1+\frac{|\alpha_{23}|^2}{(\Omega_{32}-\omega)}\left[it+\frac{e^{-i(\Omega_{32}-\omega)t}-1}{\Omega_{32}-\omega}\right]$$

$$\alpha_{32}\,\frac{e^{i(\Omega_{32}-\omega)t}-1}{\Omega_{32}-\omega}$$

$$(8)$$

$$-\alpha_{13}\,\frac{e^{-i(\Omega_{31}-\omega)t}-1}{\Omega_{31}-\omega}$$

$$-\alpha_{23}\,\frac{e^{-i(\Omega_{32}-\omega)t}-1}{\Omega_{32}-\omega}$$

$$1-\frac{|\alpha_{31}|^2}{(\Omega_{31}-\omega)}\left[it-\frac{e^{i(\Omega_{31}-\omega)t}-1}{\Omega_{31}-\omega}\right]-\frac{|\alpha_{32}|^2}{(\Omega_{32}-\omega)}\left[it-\frac{e^{i(\Omega_{32}-\omega)t}-1}{\Omega_{32}-\omega}\right]$$

This expression enables one to follow the time development of an atomic system under the influence of the incident light wave.

The evolution of a particular atomic system can be described by specifying the density matrix for its pure state. The density matrix evolves according to

$$\rho(t) = U(t)\rho(0)U^{\dagger}(t) \quad , \tag{9}$$

where $\rho(0)$ describes the initial state of the atom. In the discussion that follows it will be assumed for the sake of simplicity that the initial state of the atom is represented by a diagonal density matrix. The time evolution operator in the Schrödinger picture $U(t)$ is related to $U^{I}(t)$ by

$$U(t) = e^{-iH_a t/\hbar} U^{(I)}(t) \quad . \tag{10}$$

The transition rate to level 1 of the three level atom is equal to $d\rho_{11}(t)/dt$ and the transition rate to level 2 is equal to $d\rho_{22}(t)/dt$. These rates can be obtained through the use of Eq.(9) and

$$i\hbar\dot{\rho} = H\rho - \rho H. \tag{11}$$

When applied to the three level system under consideration, Eq.(11) implies

$$\dot{\rho}_{11} = i\left[\alpha_{13}\rho_{31}e^{+i\omega t} - \rho_{13}e^{-i\omega t}\alpha_{31}\right] \tag{12}$$

and

$$\dot{\rho}_{22} = i\left[\alpha_{23}\rho_{32}e^{+i\omega t} - \rho_{23}\alpha_{32}e^{-i\omega t}\right] \quad . \tag{13}$$

Expressions for ρ_{31}, ρ_{13}, ρ_{32} and ρ_{23} can now be found through the use of Eq.(9). When this is done the transition rates are found to be

$$\dot{\rho}_{11} = i\alpha_{13}\left[U^{*}_{11}U_{31}\rho_{11}(0) + U^{*}_{12}U_{32}\rho_{22}(0) + U^{*}_{13}U_{33}\rho_{33}(0)\right] \tag{14}$$

$$e^{+i\omega t} + \text{complex conjugate},$$

and

$$\dot{\rho}_{22} = i\alpha_{23}\left[U^*_{21}U_{31}\rho_{11}(0) + U^*_{22}U_{32}\rho_{22}(0) + U^*_{23}U_{33}\rho_{33}(0)\right] \tag{15}$$

$$e^{+i\omega t} + \text{complex conjugate.}$$

From Eq.(14) it is seen that the probability per unit time of the atom making a transition from level 2 to level 1 is equal to

$$W_{\text{anti-Stokes}} = \frac{2|\mu_{13}|^2|\mu_{32}|^2\mathcal{E}^4\Omega_{31}{}^2\Omega_{32}{}^2}{\hbar^4\omega^4(\Omega_{31}-\omega)(\Omega_{32}-\omega)}\,\rho_{22}(0)\,\frac{\sin\Omega_{21}t}{\Omega_{21}}, \tag{16}$$

and the transition probability from level 1 to 2 is equal to

$$W_{\text{Stokes}} = \frac{2|\mu_{13}|^2|\mu_{32}|^2\mathcal{E}^4\Omega_{31}{}^2\Omega_{32}{}^2}{\hbar^4\omega^4(\Omega_{31}-\omega)(\Omega_{32}-\omega)}\,\rho_{11}(0)\frac{\sin\Omega_{21}t}{\Omega_{21}}\;. \tag{17}$$

Equations (16) and (17) show that the anti-Stokes and Stokes transition probabilities are proportional to the initial population of level 2 and level 1, respectively. Thus the ratio of the two transition probabilities depends on the initial preparation of the atomic system and is generally not equal to unity. This result is in contradiction to the prediction of Nesbit's semiclassical analysis.

Raman scattering experiments are usually done with collections of atoms that are initially in thermodynamic equilibrium. The ensemble average (which is denoted by a bar) of the density matrix for such an initial state is

$$\overline{\rho_{\ell n}(0)} = \frac{\delta_{\ell n}e^{-E_\ell/kT}}{\sum_j e^{-Ej/kT}}\;. \tag{18}$$

When an ensemble average of the transition probabilities shown in

Eqs.(16) and (17) is formed using Eq.(18) it is predicted that the observed ratio of Stokes to anti-Stokes transition probabilities is equal to

$$\frac{\overline{W}_{\text{Stokes}}}{\overline{W}_{\text{anti-Stokes}}} = \frac{\overline{\rho_{11}(0)}}{\overline{\rho_{22}(0)}} = \frac{e^{-E_1/kT}}{e^{-E_2/kT}} \quad . \tag{19}$$

This is the experimentally verified result which Nesbet claimed could not be obtained semiclassically.

A complete semiclassical description of the Raman effect requires a calculation of the dipole moment that radiates the scattered radiation. The time dependent expectation of the dipole moment operator $\underline{\mu}$ is given by

$$\langle\underline{\mu}\rangle = \text{Tr}\,[\underline{\mu}\rho(t)] = \text{Tr}\,[\,\underline{\mu}U\rho(0)U^{\dagger}\,] \quad . \tag{20}$$

In addition to terms oscillating at the Stokes and anti-Stokes frequencies, Eq. (20) contains terms which oscillate at the incident frequency and correspond to Rayleigh scattering and terms that are due to the abrupt turning on of the perturbation. The components of the expectation of the atom's dipole moment that give rise to the Stokes and anti-Stokes radiation are

$$\langle\underline{\mu}\rangle_{\text{Stokes}} = -\frac{|\mu_{32}|^2|\mu_{31}|^2\mathcal{E}^3\Omega_{31}^{\,2}\,\Omega_{32}}{(\Omega_{31}-\omega)\,(\Omega_{32}-\omega)\Omega_{21}\,\hbar^3\omega^3}\,\rho_{11}(0)$$

$$\times \quad (\hat{\underline{e}}_x + i\hat{\underline{e}}_y)e^{-i(\omega-\Omega_{21})t} + \text{complex conjugate} \tag{21}$$

$$\langle\underline{\mu}\rangle_{\text{anti-Stokes}} = -\frac{|\mu_{32}|^2|\mu_{31}|^2\mathcal{E}^3\Omega_{32}^{\,2}\,\Omega_{31}}{(\Omega_{31}-\omega)(\Omega_{32}-\omega)\Omega_{21}\,\hbar^3\omega^3}\rho_{22}(0)\,(\hat{\underline{e}}_x+i\hat{\underline{e}}_y)$$

$$\times \quad e^{-i(\omega+\Omega_{21})t} + \text{complex conjugate} \quad . \tag{22}$$

The semiclassical picture of the Raman effect treats these expressions as components of an actual dipole moment which radiates electromagnetic fields according to classical electrodynamics.

REFERENCES

1. R. K. Nesbet, Phys. Rev. Letters *27*, 553 (1971).
2. M. D. Crisp and E. T. Jaynes, Phys. Rev. *179*,1253 (1969).
3. P. A. M. Dirac, *The Principles of Quantum Mechanics* (Oxford, 1938), Fourth Edition, Sec. 44, p. 172.
4. L. D. Landau and E. M. Lifshitz, *The Classical Theory of Fields* (Addison-Wesley, Massachusetts).
5. J. J. Sakurai, *Advanced Quantum Mechanics* (Addison-Wesley, Massachusetts, 1967), Sec. 2-5.
6. R. A. Weingarten, L. Levin, A. Flusberg and S. R. Hartmann, Phys. Letters *39A*, 38 (1972).

THEORY OF RESONANT LIGHT SCATTERING PROCESSES IN SOLIDS

Joseph L. Birman*

New York University, New York, N.Y.

A survey is presented of some recent work on the theory of resonant Brillouin and Raman scattering in insulating solids. The major themes which are examined are the effect of intermediate state interactions, and effects of spatial dispersion.

1. Introduction

Light scattering studies of solids have recently experienced a remarkable upsurge, both experimentally and theoretically. [1] In large part this dramatic increase in activity over the last 5-6 years can be attributed to sophisticated laser spectroscopy which now makes possible refined and quantitative studies of the intensity of scattered light as a function of various crystal and laser parameters. Most of this experimental work has been done using laser sources with discrete exciting frequencies. In the near future lasers with continuously variable (tunable) source frequencies are expected to become increasingly available, thus enlarging still further the scope of possible experiments.

In particular, many studies have been made of light scattering processes which occur when the incident, or scattered light frequency lies near an absorption level, or band. These are generally denoted as resonance light scattering processes, and the intensity (cross section) of such processes is generally (not always) strongly frequency dependent. Major interest attaches to these types of

* Supported in part by NSF and AROD.

processes owing to their enhanced intensity (resonant scattering processes can be 10-100 times more intense than non-resonant processes), and to the belief that resonant processes provide a royal road to the unravelling of all the microscopic mechanisms which contribute to light scattering.

Because of the rather intense activity in this area, certain aspects of the observed phenomena are still not completely understood, certain theoretical ideas are still in process of elaboration. Hence in this paper I shall review briefly some of the present activity and themes in this field to serve in part as an introduction to current work.

Some of my own recent work on light scattering has been done in collaboration with Dr. A.K. Ganguly, Dr. B. Bendow, Dr. R. Zeyher, Dr. C.S. Ting and Prof. W. Brenig, and I am glad to acknowledge their participation in our mutual work.

2. The Hamiltonian and Related Matters

The microscopic theory of Raman scattering concerns itself with obtaining the scattering cross-section from an assumed Hamiltonian, and comparing the prediction with experiment. Thereby it is hoped to simultaneously test the validity of the assumed Hamiltonian, and determine the major specific contributing parts to the particular scattering processes under study.

In this paper we restrict attention to Raman scattering by phonons in transparent insulators. The major physical mechanism often assumed for the scattering process can easily be put into words: the incident photon of energy $\hbar\omega_1$ is removed from the radiation field causing a virtual electronic excitation in the insulator; in the virtual excited state an interaction between phonon and electron occurs producing a real phonon of energy $\hbar\omega_0$ and causing a transition to a different virtual excited electronic state; when the excited electronic state returns to the ground state the scattered photon of energy $\hbar\omega_2$ is produced. This picture invokes the electronic (intermediate) manifold, and therefore resonance processes occur when incident or scattered photon is close to an absorption corresponding to an electronic transition.

The Hamiltonian for Raman scattering in this Bloch picture can be written in the schematic form [2]

$$H^{BL} = H_o + H_{ER} + H_{EL} \tag{2.1}$$

where

$$H_o = H_{RAD} + H_{LATT} + H_{ELEC} \tag{2.2}$$

with the terms in H_o representing the uncoupled photon, lattice, and electron fields respectively. The interaction terms are electron radiation coupling

$$H_{ER} = \sum \{f\alpha_{k_h} \alpha^+_{k_e} A_{k_1} + \text{c.c.}\} \quad \delta(k_1 - (k_h + k_e)) \tag{2.3}$$

and electron-lattice coupling terms

$$H_{EL} = \sum \{g\alpha^+_{k-q} \alpha_k b^+_q + \text{c.c.}\} \quad . \tag{2.4}$$

In these expressions only the lowest order terms for the coupling are given; the "bare" coupling coefficients f and g depend on the electron and radiation, and the electron and lattice, quantum numbers, respectively; the electron, photon and phonon creation operators are denoted α^+_k, A^+_k, b^+_q, with $k_{e,h}$ the electron(hole) wave vector; k_1 the photon, and q the phonon, wave vector. The system described by this Hamiltonian is physically a finite crystal; ergo one needs to deal properly with all the quantized fields inside the finite crystal as well as match the interior fields properly to their external extensions. We shall make some remarks on this later.

The form of H given above already involves some physical assumption regarding the detailed mechanism of Raman scattering and it is worthwhile to call attention to this by denoting this as a Hamiltonian in the Bloch picture, or non-interacting picture for the electrons. Alternate pictures will be introduced below.

Finally we observe that very many important physical details are concealed in the abbreviated form of H. Thus for specific sets of quantum numbers (initial and final electronic states) an electronic transition can be allowed, or forbidden; in the former case the corresponding coupling function f will be independent of $\kappa = k_e - k_h$ while in the latter case the first non-zero contribution to f will be a linear in κ. Likewise, depending on specifics, g may be constant or may depend linearly on q, etc. We will often be concerned with allowed one-phonon Raman scattering in some particular medium, in which case both f and g may be taken as constant.

3. First Theme: Intermediate State Interactions

In this Section we shall discuss the theoretical predictions for Raman scattering based on various "pictures". Our intention here is to examine the theory in the successive stages of elaboration which occur when "intermediate state interactions" are turned on. The three pictures considered are: the Bloch or non-interacting picture; the exciton picture; and the polariton picture. In the next section we shall discuss matters related to spatial dispersion, which is assumed absent here.

The simplest way to deal with Raman scattering on the basis of the Hamiltonian (2.1)-(2.4) is to use conventional perturbation theory. Consider the complete set of eigenstates of the system to be product eigenstates of H_o; so for example the initial state would be

$$|i\rangle = |n\rangle_R \, |m\rangle_L \, |\lambda\rangle_E , \qquad (3.1)$$

where n, m, λ are a complete set of quantum numbers for the radiation field (photons), lattice field (phonons) and electron field respectively. Then λ includes band indices as well as wave vectors of electron and hole. For Raman scattering by a transparent insulator, for incident frequencies in the frequency region near the gap, the electronic manifold λ, can be restricted to a given pair of bands (valence and conduction) which can, furthermore, be taken as simple spherical bands with reduced effective mass μ. Using third order perturbation theory the Raman scattering probability is given as

$$I(\omega_1) \sim \sum_f \left| \sum_{a,b} \frac{\langle f|H_{ER}|b\rangle\langle b|H_{EL}|a\rangle\langle a|H_{ER}|0\rangle}{(\omega_b - \omega_1)(\omega_a - \omega_1)} \right.$$

$$\left. + (\circlearrowleft) \right|^2 \delta(\omega_1 - \omega_f) \qquad (3.2)$$

[($\circlearrowleft$) means a suitable permutation of operators],

where initial and final states are denoted $|i\rangle$, $|f\rangle$ respectively, and intermediate states are $|a\rangle$, $|b\rangle$. If now attention is restricted to the given pair of bands, and the intermediate state energies are given as: $\varepsilon_a = \varepsilon_g + h\,\bar{K}^2$ where ε_g is the energy gap and where $\bar{K}$ is the *relative* wave vector, and $\varepsilon_b = \varepsilon_a + \hbar\omega_o$ then the sum over intermediate states becomes an integral over the relative wave vector $\bar{K}$. Further assuming direct, allowed transitions, the coupling coefficients f and g are constant, so that the integral can be carried out. For ω_1 near ω_g we find as the principle contribution $(\omega_1 < \omega_g)$ [17]:

$$I(\omega_1) \sim (f^2 g)^2 \left| \int_0^{k_{MAX} \to \infty} d^3k \; [\omega_b - \omega_1)(\omega_a - \omega_1)]^{-1} \right|^2$$

$$\sim (f^2 g)^2 \; [(\omega_g + \omega_o - \omega_1)^{\frac{1}{2}} - (\omega_g - \omega_1)^{\frac{1}{2}}] \;|^2$$

$$= (f^2 g)^2 \; [h(\omega_1)]^2 \quad . \tag{3.3}$$

Two points are memorable about this: first, the expression (3.3) is the contribution from continuum states (electron and hole) in their separate uncorrelated Bloch states); and secondly the expression (3.3) is non-divergent at $\omega_1 = \omega_g$. Far from resonance (e.g. $\omega_1 << \omega_g$) the expression for $I(\omega_1)$ becomes constant

$$I(\omega_1) \sim A^2 \qquad\qquad \omega_1 << \omega_g \tag{3.4}$$

There is some evidence [3] that phonon Raman scattering (from TO branch) in some II-VI compounds follows an expression of the form

$$I(\omega_1) \sim |A - Bh(\omega_1)|^2 \tag{3.5}$$

where A is a constant, $B \sim f^2 g$ and $h(\omega_1)$ is the frequency dependent term in square brackets in (3.3). This has been interpreted to indicate that for this process the (saturated) contributions from other bands far from ω_g gives rise to A while the closer band pair states gives the second term. These interesting cancellation effects are still under active scrutiny.

A next stage in elaboration of the model [4] occurs if the electrons and holes are allowed to interact, via the residual Coulomb interaction, to produce excitons. This interaction can be understood to occur in the intermediate excited state, and results in the replacement of electron operators α^+ by exciton operators a^+. Under this transcription $\alpha^+ \to a^+$, and the unperturbed part H_E of the Hamiltonian H_o, now appears in the exciton operators.

Consequently in the eigenstates such as (3.1) the electronic factor is now an exciton eigenfunction. In the interaction terms we now have:

$$\bar{H}_{ER} \sim \sum f' \, (a^{+}_{k_1} A_{k_1} + c.c.) \tag{3.6}$$

$$\bar{H}_{EL} \sim \sum g' \, (a^{+}_{k-q} a_k b^{+}_q + c.c.) \tag{3.7}$$

where a^{+}_k creates an exciton with total (center of mass) momentum k_1, etc. Note that $\bar{H}_{ER}$ is now the exciton-radiation interaction which is *bilinear* in operators and $\bar{H}_{EL}$ is the exciton lattice interaction, which is trilinear in operators. The new coupling coefficients are now f' and g'.

Again, using perturbation theory a simple expression can be obtained for the Raman scattering intensity near resonance. The third order perturbation theory result is formally the same as given in (3.2). Only now, the intermediate state sum is, even in case of simple bands, to be transformed into a *sum* over discrete bound exciton states plus an integral over the exciton continuum states. Recall that the exciton energy spectrum is $\varepsilon_a = \varepsilon_g - R'/n^2$ for discrete, $\varepsilon_a = \varepsilon_g + \hbar \bar{K}^2/2\mu$ for continuum states where R' is the exciton Rydberg, and n an integer and again $\varepsilon_b = \varepsilon_a + \hbar\omega_o$. At present take the exciton c.m. stationary. Then for ω_1 very near the 1s exciton energy: $\omega_1 \sim \omega_{ex} = \omega_g - R'$ the term n=1 dominates the sum and one finds the intensity of phonon Raman scattering to behave like

$$I^{EX}(\omega_1) \sim \left| (\omega_{ex} - \omega_1)(\omega_{ex} + \omega_o - \omega_1) \right|^{-2} \quad . \tag{3.8}$$

Note the predicted "in" and "out" resonances which seem to be verified by experiment for one phonon scattering (LO branch) in II-VI insulators.[5] The actual expression for one phonon Raman scattering in the exciton picture, at *any* frequency requires inclusion of the contribution from both discrete and continuum states, and cannot be put in closed analytic form like (3.8) except very close to the poles. Actual Raman intensities predicted by the exciton theory always involve an interplay between both discrete and continuum contributions: in certain frequency regimes (for $\omega_1 > \omega_{ex}$) cancellations can occur owing to different signs of the discrete and continuum parts. [6] As earlier mentioned, there are indications that LO phonon Raman scattering in II-VI compounds, for $\omega_1 < \omega_g$ is well described [7] by the exciton intermediate state picture. But, for $\omega_1 > \omega_g$ the sparse experimental data is in disagreement with theory,[7] even including a more refined bare exciton theory.[8]

Before turning to the next stage of elaboration of the theory it appears appropriate to make a remark regarding higher order (multiphonon) Raman scattering. In the naive exciton picture multiphonon Raman scattering arises by an iteration of the term $\bar{H}_{EL}$ (repeated exciton scattering by phonons) in the virtual state. This produces a Raman intensity which is predicted to be [4]

$$I^{EX}_{nth\ order} \sim \sum_f \left| \sum_{a_1, a_2, \ldots a_n} \frac{\langle f|\bar{H}_{ER}|a_n\rangle\langle a_n|\bar{H}_{EL}|a_{n-1}\rangle \ldots \langle a_1|\bar{H}_{ER}|0\rangle}{(\varepsilon_n - \varepsilon_1)(\varepsilon_{n-1} - \varepsilon_1)\ldots(\varepsilon_{a_1} - \varepsilon_1)} \right|^2$$

$$\times\ \delta\ (\omega_f - \omega_i)$$

$$\sim (f^2 g^n)^2 \left[(\omega_{ex} + n\omega_o - \omega_1)(\omega_{ex} + (n-1)\omega_o - \omega_1)\ldots(\omega_{ex} - \omega_1)\right]^{-2} \qquad (3.9)$$

where a few obvious changes in notation were made and only the most divergent term was retained in (3.9). In some very recent experimental studies on two photon scattering in II-VI compounds, carried out with a tunable dye laser, it has been reported [9] that (3.9) gives for (n=2) excellent agreement with experiment, in regard to the frequency dependence of $I^{EX}(\omega_1)$ for $\omega_1 \to \omega_{ex}$. Note that the nth order Raman scattering intensity is according to (3.9) predicted to have (n+1) resonances, namely the "in" and "out" plus those at "intermediate" energies. The predicted occurance of "intermediate" resonances has not been verified by experiments to date. Of course in all the expressions for the Raman intensity, complex (intermediate state) energies should occur ($\varepsilon_a \to \varepsilon_a + i\Gamma_a$), which will produce damping just near resonance. It is not presently known whether the intermediate resonances are eliminated due to damping, or due to spatial dispersion, or whether their non-observance is a deficiency of theory or experiment.

As a final stage of elaboration consider the "polariton" Hamiltonian:[6][10]

$$H_{POL} = \bar{H}_o + \bar{H}_{ER}, \qquad (3.10)$$

where $\bar{H}_{ER}$ as defined above is written in exciton variables, and $\bar{H}_o$ is

$$\bar{H}_o = H_{RAD} + H_{EX} \tag{3.11}$$

or, when written more explicitly:

$$\bar{H}_o = \sum_k \omega_k A_k^+ A_k + \sum_\lambda \varepsilon_\lambda a_\lambda^+ a_\lambda \quad . \tag{3.12}$$

Here λ denotes a full set of exciton quantum numbers including internal (n, or $\bar{K}$) and total (center of mass) k. Comparing (3.10), (3.6) and (3.12) it is immediately apparent that H_{POL} is a bilinear form in the operators A^+ and a^+. Consequently H_{POL} can be diagonalized by a linear transformation. Let us define new Bose operators by the transformation of form

$$B^+ = \sum_k \xi_k A_k^+ + \sum_\lambda \eta_\lambda a_\lambda^+ \tag{3.13}$$

where ξ_k and η_λ are "c" functions of the various quantum members Then the polariton Hamiltonian is brought to diagonal form by this transformation

$$H_{POL} = \sum WB^+B, \tag{3.14}$$

where W is the new eigenenergy. The eigenstates of H_{POL} are the linear combination kets of photon plus exciton: these are the quasiparticles called polaritons. Remark that at the present stage, the photons, excitons and ergo polaritons are all Bose objects. The inverse transformation to (3.13) permits us to write the interaction H_{EL} in the form

$$\bar{\bar{H}}_{EL} \sim \sum g'' (B_{k-q}^+ B_k b_q^+ + c.c.), \tag{3.15}$$

which represents a polariton scattering operator.

At this stage the conventional polariton picture of Raman scattering begins. [11] Several idealizations are invoked, of which the most important is that the physical Raman scattering intensity can be factorized into: a product of a transmission coefficient, representing all the boundary effects connected with transmission of the photon into and out of the finite crystal, times an "internal" cross-section for polariton scattering in an infinite crystal. If spatial dispersion is neglected, and attention is restricted to the coupling of a single dispersionless (1s) exciton to the photon then the polariton dispersion relation, which is the eigenvalue equation permitting the transformation (3.13) can be easily obtained. From Maxwell's equations, when propagating plane waves are sought in a (frequency dispersive) medium with classical dielectric susceptibility (neglecting damping)

$$\varepsilon(\omega) = \varepsilon_o + \frac{\varepsilon_1}{(\hat{\omega}_o^2 - \omega^2)} \tag{3.16}$$

where $\hat{\omega}_o$ is the resonance frequency of the classical oscillator. That is, the dispersion equation is given as

$$k^2 = (\omega/c)^2 \, \varepsilon(\omega) \tag{3.17}$$

The solutions of (3.17) are non-degenerate in the sense that for each frequency ω only at most one k can be a solution.

The theory of Raman scattering for non-degenerate polaritons has been simply given: the scattering is presumed to occur via $\bar{\bar{H}}_{EL}$, when a polariton on a single branch is scattered to a final state, producing a phonon ω_o. Owing to the interpretation of polaritons as quasiparticles (excitons dressed by photons in the crystal), one can use the simple "Golden Rule" of first order perturbation theory to calculate the cross-section or transition rate. Again, emphasizing that all states are restricted to lie on one polariton branch the total cross-section is given as

$$J^{POL}(\omega_1) \sim \frac{1}{v_g(i)v_g(f)} \, |\langle f|\bar{\bar{H}}_{EL}|\, i\rangle|^2 \, \delta(\omega_f - \omega_i). \tag{3.18}$$

In this expression $v_g(i,f)$ is the group velocity of the initial/final polariton, at energy $\omega_i = \omega_1$ or $\omega_f = \omega_2 = \omega_1 - \omega_o$ respectively. The measured external Raman cross-section is then presumed to be given by $J^{POL}(\omega_1)$ times a transmission factor,

$$I^{POL} = J^{POL}(\omega_1)\, T, \tag{3.19}$$

where T includes effects of geometry, boundaries, etc. This polariton picture was actually used to compute the Raman cross-section and it gave results which (for $\omega_1<\omega_{ex}$) were in general agreement with perturbation theory. [7][11] This one branch polariton theory must fail in the resonance region because v_g is not well defined there, and in addition, spatial dispersion effects should be included (*vide infra*), which complicates the theory by necessitating a multi-branch (degenerate polariton) picture.

To summarize the results of this theme: when intermediate state interactions are progressively included, the theory may change even qualitatively. This is dramatically illustrated in the progression from Bloch to exciton picture in which Coulomb interaction of electron and hole are included, owing to the appearance of bound states in the latter. This is reflected in a pole appearing in the scattering cross-section at resonance (ω_{ex}) rather than a numerical enhancement factor at ω_g, as in the non-interacting picture. On including the exciton-radiation interaction exactly, to produce polaritons, further quantitative changes may occur and, in the non-degenerate case, a quasi-particle description can be used.

4. Second Theme: Spatial Dispersion Effects

An active subject of research at present concerns the effect of exciton spatial dispersion on the optical properties of solids. Although there has been a considerable literature on this topic over the last decade and a half,[12] many important theoretical questions are only now being examined and resolved. Even more noteworthy is the absence of a convincing body of direct experimental confrontation of theoretical predictions and experimental measurement. This is particularly true in regard to absence of direct test of many optical properties of crystals which are predicted to be correlated with spatial dispersion.

From the viewpoint of our major theme of light scattering, interest attaches to spatial dispersion owing to the need to specify precisely the participating states: virtual states in Bloch and exciton pictures; real states for polaritons. When spatial dispersion is present (e.g. owing to center of mass motion of the excitons), the exciton and the polariton states resemble slightly the Bloch (uncorrelated) states insofar as a continuum density of states is available. Of more interest is the existence of additional channels for the scattering process, in particular in the case of multibranch (degenerate) polaritons, for frequencies near resonance. That is, the total scattering originates from a superposition of processes: the initial state and the final state can each be taken from one of the degenerate branches. Further, insofar as the polariton states, with spatial dispersion included, are considered as an exact set of eigenstates of the system, they are the correct states to use in the theory.

Recall that spatial dispersion may be introduced [13] into the theory for an infinite crystal by modifying the classical expression for the susceptibility (3.16). Interpret the $\hat{\omega}_o$ which appears in (3.16) as the characteristic frequency of an excitation; here an exciton. Then, to include spatial dispersion (center of mass motion of the exciton) write for $\hat{\omega}_o$:

$$\hat{\omega}_o \rightarrow \omega_{ex} + \hbar k^2/2M \equiv \omega_{ex} + Bk^2 , \tag{4.1}$$

where k is the total wave vector, M the total (electron plus hole) mass. The modified susceptibility for an infinite crystal (neglecting damping) is

$$\varepsilon(k,\omega) = \varepsilon_o + 4\pi A\,(\omega_{ex}^2 + Bk^2 - \omega^2)^{-1} . \tag{4.2}$$

Then the eigenvalue, or dispersion, equation for propagating plane electromagnetic waves in an infinite medium in which the susceptibility is (4.2) is

$$k^2 = (\omega/c)^2\,\varepsilon(k,\omega) . \tag{4.3}$$

Clearly this is an equation of 4th degree and for given ω it has two (in general independent) pairs of solutions $\pm k_1(\omega)$, $\pm k_2(\omega)$. For $\omega > (\omega_{ex}^2 + A/\varepsilon_o)^{1/2}$ the two solutions of the dispersion equation (4.3) are both real so that two plane waves may then propagate. It is convenient to work with the susceptibility $\chi(k,\omega) = [\varepsilon(k,\omega)-1]/4\pi$, and the spatial Fourier transform of it

$$\chi(\rho) = \chi_o\,\delta(\rho) + \chi_1 G_+(\rho) \tag{4.4}$$

where

$$\chi_o \equiv (\varepsilon_o - 1)/4\pi \;;\quad \chi_1 = \pi^{1/2}\,A/B\cdot\sqrt{2} \tag{4.5}$$

and

$$G_+(\rho) = [\exp(ik_+\rho)]\rho^{-1} . \tag{4.6}$$

Here k_+ is the wave number for which $\varepsilon(k,\omega)$ has a simple pole.

The case of interest is a semiinfinite crystal which we take bounded by the plane z = 0, and extending to infinity. Within the crystal the relation between polarization and electric field is non-local

$$P(r) = \int \chi(r,r')E(r')dr' \qquad (4.7)$$

As a non-local susceptibility in this case one can provisionally take [14]

$$\chi(r,r') = \chi(r-r')\,\Theta(z)\Theta(z') \qquad (4.8)$$

where $\chi(r-r') = \chi(\rho)$ is given in (4.4). This is the truncated (chopped off) translationally invariant Ansatz, frequently used. But in presence of the boundary plane the exciton will be reflected so that the proper eigenfunction in the crystal is a superposition of right and left running waves. To take this into account, an expression for the susceptibility was recently introduced[15]

$$\chi(r,r') = [\chi(r-r') + \chi^{+}(z + z')]\Theta(z)\Theta(z') \qquad (4.9)$$

where the first term is identical to (4.4) while the second which is proportional to the exciton reflectivity at the boundary is translationally invariant in the boundary (x,y) plane but depends on (z + z')[18].

It has only come to be realized recently that Maxwell's equations (in either integral or differential form) plus the specification of the non-local susceptibility in the medium are a determinate problem, in the presence of a boundary.[14][16] That is, the amplitude ratio of all waves can be found (without extra assumptions) including the reflected wave plus all propagating waves (including the case of degenerate polariton). The first case to be analyzed [14] used the susceptibility (4.4) and the integral formulation of electrodynamics. In this formulation, the basic equation of the theory is taken as

$$E(r,\omega) + 4\pi\, P(r,\omega) = E^{(i)}(r,\omega)$$

$$+ \text{curl curl} \int P(r'\omega)\; G_0(R)dV' \qquad (4.10)$$

where $E^{(i)}$ is the incident electric field, and

$$G_0(R) \equiv [\exp i(\omega R/c)]/R = [\exp i(k_0 R)]/R \tag{4.11}$$

with $R = |r-r'|$ and the validity of the Lorentz-Lorenz local field relation is assumed, in obtaining (4.11). The question of the meaning of the local field in a polariton theory is presently under examination.

When (4.4) - (4.8) is substituted into (4.10) and straightforward manipulations are performed three important results are obtained: (1) A generalized polariton extinction theorem: (2) The additional boundary condition: (3) The dispersion relation for polaritons in the semiinfinite crystal. Limitation of space only permits a brief remark to be made regarding these. The generalized extinction theorem prescribes the extinction of the incident electric field $E^{(i)}$ by the fields which originate from the polaritons. The a.b.c. gives just the additional constraint needed to determine all amplitude ratios. The dispersion relation for polaritons in the bounded medium with susceptibility (4.8) is exactly (4.3): identical to that in the infinite medium. Specifically, for the case of normal incidence, [14] the extinction theorem is

$$E^{(i)} - [(n_1+1)\, E_1 + (n_2+1)\, E_2]/2 = 0, \tag{4.12}$$

and the reflected field is given as

$$E^R + [(n_1-1)\, E_1 + (n_2-1)\, E_2]/2 = 0. \tag{4.13}$$

The Maxwell boundary conditions are

$$E^{(i)} + E^R = E_1 + E_2, \tag{4.14}$$

$$E^{(i)} - E^R = n_1 E_1 + n_2 E_2, \tag{4.15}$$

The a.b.c. (due to Sein)[14] is:

$$E_1\,(n_1 - \gamma)^{-1} + E_2\,(n_2 - \gamma)^{-1} = 0. \tag{4.16}$$

In these equations $k_o \equiv \omega/c$, $n_i = k_i/k_o$, and $\gamma \equiv k_+/k_o$. Notice that the extinction theorem is not independent of the Maxwell boundary conditions. Consequently, amongst the four independent amplitudes $E^{(i)}$, E^R, E_1, E_2 there are three equations which can be taken as the a.b.c. plus the two Maxwell boundary conditions: ergo the amplitude ratio is determined.

If the expression (4.9) is used for the susceptibility, one obtains again the same three types of results: (1), (2), (3) above. Noteworthy is that the dispersion relation obtained in this case is again identical to that above (i.e. (4.3)). Hence the dispersion of the propagating waves appears determined by the translationally invariant part of the susceptibility. Again the same extinction theorem result is found. But a major difference is in the a.b.c.. Taking the exciton reflectivity R = - 1 (phase shift of π) so there is an exciton node at z = 0, the a.b.c. becomes [15]

$$(P_1 - \chi_o E_1) + (P_2 - \chi_o E_1) = 0. \tag{4.17}$$

Crudely speaking, if the background polarization $\chi_o(E_1 + E_2)$ is small compared to the total polarization then the a.b.c.becomes $P_1 + P_2 = 0$, so that both polaritons are of equal magnitude in this case.

The significance of these results for resonant light scattering processes can now be explored: it is clear that by determining the amplitude ratios of all fields, the entire mode structure of the electromagnetic field is determined. A mode consists of the correct linear combination of external photon (incident and reflected) and internal polariton (photon plus exciton). Inelastic light scattering is considered as a scattering process of one such mode to another: under the influence of the interaction term $\underline{\underline{H}}_{EL}$ which couples the two polaritons via their exciton parts (in the crystal) and thus scatters one mode to another. What is observed is the photon part of initial and scattered mode. This unified picture of scattering has recently been applied to give a theory of Raman light scattering in insulators. The general expression which has been obtained for the scattering cross-section is fairly complex and owing to limitations of space we refer the reader to the original paper for details.[8] A result which seems to emerge from the general treatment is that for a bounded crystal, including spatial dispersion, the commonly assumed factorization of the cross-section into a transmission part (including reflection coefficient at the boundary) times an "internal" transition matrix element squared is not justified. Only when spatial dispersion is neglected does such a factorization arise. In the latter case (neglect of spatial dispersion) the general polariton result can be shown to go over to the perturbation theory result (3.8) for exciton intermediate states.

A very dramatic effect which is a result of taking spatial dispersion into account in the theory of Brillouin scattering has very recently been predicted. The essential feature of this predicted effect can be deduced from kinematic considerations. In the act of Brillouin scattering, an acoustic phonon of energy $\omega_A(q)$ is created where q is the quasimomentum or wave vector of the phonon. For acoustic phonons we have (in long wave limit)

$$\omega_A(q) = c_s\, q \quad , \tag{4.18}$$

where c_s is a sound velocity. On the other hand the polariton dispersion is specified by the solution of the equation (4.3) which we can write $\omega_p(k)$. Let the initial polariton state be of frequency ω_p, to which wave vectors k_1, k_2 correspond. For the final polariton state take ω'_p, and k'_1, k'_2 as the solution. Then in Stokes scattering to produce the acoustic phonon $\omega_A(q)$ we must have conservation of polariton psuedo-momentum in the crystal

$$k_i - k'_j = q \qquad (i,j) = 1,2 \tag{4.19}$$

and conservation of energy

$$\omega_p - \omega'_p = \omega_A = c_s q \quad . \tag{4.20}$$

(Note in case of backward scattering there will be the largest (absolute magnitude of) momentum transfer). The kinematics of an allowed scattering process requires a transition from the state $\omega_p(k_i)$ on one branch to $\omega'_p(k'_j)$ on a branch for which (4.19) and (4.20) hold. Hence the acoustic phonon dispersion "tunes" the permitted transition. A glance at the polariton acoustic photon dispersion characteristics (Fig. 1) shows us that very dramatic effects in Stokes Brillouin scattering can occur as the initial polariton energy, $\omega_p = \omega_i$ (the initial external photon energy) increases in the resonance region. Below ω_{ex}, for $\omega_i < \omega_{ex}$ only a single Brillouin shift δ_B will occur. When $\omega_i \to \omega_{ex}$, the magnitude of δ_B increases (perhaps sharply) depending on the change of group velocity on the dispersion branches. When $\omega_i > (\omega_{ex}^2 + A/\varepsilon_o)^{\frac{1}{2}}$, three additional Brillouin shifts become kinematically allowed. Consequently this theory predicts that for Brillouin scattering in an insulator with spatial dispersion, the Brillouin *doublet* (Stokes plus antiStokes) can become a Brillouin *octet*, for incident frequencies above a critical value. Calculation for CdS of the magnitudes of the expected ordinary and new shifts show them to be an observable range (5-15cm^{-1}) and of observable intensity, comparable to that of the usual shifts δ_B. At the time of writing these predicted effects connected to spatial dispersion have not yet been observed although active experimental work is under way.

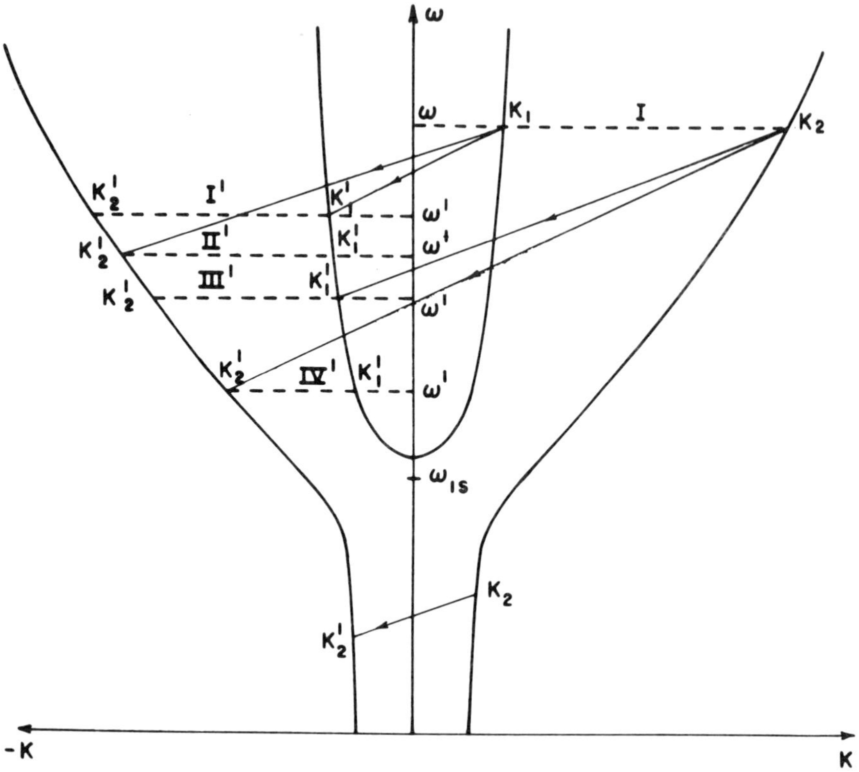

Fig. 1 A highly schematic illustration of the kinematics of backward Brillouin Stokes Scattering on the polariton picture including spatial dispersion, after ref. [15]. The initial state(s) is on the rightside of the Figure, the final state(s) on the left. For incident frequency less that ω_{1S} only one scattering process can occur: one Brillouin shift δ_B. As incident frequency increases, toward ω_{1S}, δ_B may increase sharply (when the knee of the lower polariton branch is raised). For incident frequency greater than ω_{1S}, two incident and two final states are allowed: thus 4 Brillouin shifts δ_B, δ'_B, δ''_B, δ'''_B may be observed. Note that the kinematics require that text equations (4.19) - (4.20) be satisfied so that the acoustic phonon will be on the acoustic branch "mass shell".

Observation of these effects would permit experimental verification of the a.b.c., as well as the extra waves, and determination of the exciton parameters especially the total mass. Actually the calculation was carried out for Sein a.b.c. (4.16) as well as the more recent ones (4.17) so a measurement may be able to distinguish amongst them.

A sketch of the semiclassical calculation of the Brillouin scattering cross section may be of interest to show how some of these factors appear. Let the incident photon be $\omega = c\,k_o$, the scattered photon $\omega' = c\,k_o'$ where c is the vacuum speed of light, and k_o, k_o' are photon momenta external to the crystal. The Maxwell equation for the electric field E' of the scattered wave is

$$-[\nabla^2 + k_o'^2(1 + 4\pi\chi(k_o'))]\,E' = 4\pi k_o'^2(\delta\chi\cdot E(k_o))\,(k_o') \tag{4.21}$$

(Note that (k_o') is the argument of E' and of $\delta\chi\cdot E$) .

Here we considered the modulation of the non-local susceptibility by a density fluctuation δn due to an acoustic photon

$$\delta\chi = (\delta\chi/\delta n)\,\delta n = -\chi(\delta\chi^{-1}/\delta n)\cdot\chi\delta n$$

$$= -\,\chi\,C\,\chi\,\delta n\,, \tag{4.22}$$

where C is taken as a constant. The right hand side of (4.21) is the source term for the scattered wave. For the differential scattering cross-section for back scattering (thermally averaged (<>)), take the ratio of flux of scattered/incident radiation which we write as

$$d^2\sigma/d\Omega d\omega' = (c/8\pi)\,(k_o')^4\,\langle|T_{k_o'\,k_o}|^2\rangle\,/\mathcal{J}_I \tag{4.23}$$

where $\mathcal{J}_I$ is the incident flux. To evaluate T (scattering) matrix element we take the initial and scattered polaritons as linear combinations of the two propagating waves

$$P = P_1\,e^{ik_1z} + P_2\,e^{ik_2z}\ . \tag{4.24}$$

The T matrix can be written as

$$T_{k'_o k_o} \sim C \cdot A \cdot \sum_{ij} P_i P'_j \quad \delta n\ (k\omega), \tag{4.25}$$

where A is an effective scattering volume and where $\delta n(k,\omega)$ is the Fourier Transform of δn. Taking $kT >> h\delta_B$, and neglecting interference we find the form of the result:

$$d^2\sigma/d\Omega d\omega' \sim C^2 \cdot (k'_o)^4 \cdot A^2 \sum_{ij} |P_i|^2 |P'_j|^2 \cdot kT \cdot \mathcal{L} \tag{4.26}$$

where $\mathcal{L}$ is a Lorenzian line shape factor

$$\mathcal{L} = [(k_i + k'_j - (\omega-\omega')/c_s)^2 + (\gamma_i + \gamma'_j)^2]^{-1} \quad ,$$

with γ, γ' representing damping of initial/scattered wave. The relevant point here is the dependence on $(P_i)^2(P'_j)^2$, for each channel. Recall that, according to an oversimplified version of the new a.b.c., (4.17) the magnitudes of the polariton components are equal. Consequently all 4 channels contribute equally, in this picture leading to the prediction of large intensity also for all the Brillouin components, including the new ones.

5. Conclusion

I hope that this discursive and rather personal survey of a field in process of evolution will serve as a useful introduction to the current literature. The few literature citations given (mostly to work of our group) should be taken as hopefully useful starting points for the reader who wishes to delve further.

References

1. *"Light Scattering in Solids"* ed. M. Balkanski (Flammarion, Paris 1971). Proceedings of the Second International Conference on Light Scattering held in Paris, July 1971. This contains papers dealing with most of the active topics. This volume can serve as a general reference for the present survey.

2. R. Loudon, Proc. Roy. Soc. *A275*, 218 (1963).

3. R.K. Chang, J.L. Lewis, R.L. Wadsack, Phys.Rev. Lett. *25*, 814 (1970); and ref. 1, pps. 41-46.
4. J.L. Birman and A.K. Ganguly, Phys. Rev. Lett. *17*,647 (1966); Phys. Rev. *162*, 806 (1967).
5. R.C.C. Leite, J.F. Scott, T.C. Damen, Phys. Rev. Lett. *22*, 780 (1969); Phys. Rev. *188*, 1285 (1969).
6. B. Bendow, and J.L. Birman in ref. 1, pp. 19-25.
7. B. Bendow, J.L. Birman, A.K. Ganguly, T.C. Damen, R.C. Leite, J.F. Scott, Optics Comm. *1*, 267 (1970).
8. R. Zeyher, C.S. Ting, J.L. Birman (in preparation).
9. Y. Oka, T. Kushida, Tech. Report A501, Inst. Solid State Physics, Tokyo, January 1972, to be published in J. Phys. Soc., Japan.
10. U. Fano, Phys. Rev. *103*,1202 (1956); J.J. Hopfield, Phys.Rev. *112*,1558 (1958); Phys. Rev. *182*,945 (1969); D. Mills, E. Burstein, Phys. Rev. *188*,1465 (1969).
11. B. Bendow and J.L. Birman, Phys. Rev. *B1*,1678 (1970); B. Bendow, Phys. Rev. *B2*,5051 (1970); ibid. *B4*,552 (1971).
12. Much of the literature, especially from Soviet authors is given in V.M. Agranovich,V.L. Ginzburg "*Spatial Dispersion in Crystal Optics and The Theory of Excitons*" J. Wiley,N.Y. 1966; and in *Progress in Optics*, Vol.IX, ed. E. Wolf, (North Holland, Amsterdam, 1971) Chap. VI, "Crystal Optics with Spatial Dispersion" by the same authors.
13. D.G. Thomas and J.J. Hopfield, Phys. Rev. *132*, 563 (1963); and also Ref. 12.
14. J.J. Sein, Ph.D. Thesis, New York University 1969, Phys. Letters *32A*, 141 (1970); J.L. Birman and J.J. Sein, Phys. Rev. *B6*, 2482 (1972).
15. R. Zeyher, J.L. Birman, W. Brenig, I and II, Phys. Rev. *B6*, 4613, 4617 (1972).
16. G. Agarwal, D. Pattanayak, E. Wolf, Phys. Rev. Letters *27*, 1022 (1971); Optics Comm. *4*, 255 (1971); ibid. *4*, 260 (1971).
17. A slowly varying factor of ω^4 is neglected in Eqs.(3.3), (3.4), (3.5), (3.8), (3.9) since this is not important near resonance.
18. The second term in Eq.(4.9) can be written: $\chi_1 R(o) G_+(\xi)$ where R is the reflectivity of the exciton, χ_1 and G_+ are defined in Eqs.(4.5), (4.6) and $\xi \equiv [(x-x')^2+(y-y')^2+(z+z')^2]^{\frac{1}{2}}$.

A GENERALIZED EXTINCTION THEOREM AND ITS ROLE IN SCATTERING THEORY*

Emil Wolf

University of Rochester, Rochester, N.Y.

When an electromagnetic wave is incident on a homogeneous medium with a sharp boundary, it is extinguished inside the medium in the process of interaction and is replaced by a wave propagated in the medium with a velocity different from that of the incident wave. A classic theorem of molecular optics due to P.P. Ewald (1912) and C.W. Oseen (1915) expresses the extinction of the incident wave in terms of an integral relation, that involves the induced field on the boundary of the medium. Various generalizations of this theorem have recently been proposed and it was also shown that the customary physical interpretation of the theorem is incorrect.

In this paper results of a recent investigation carried out in collaboration with D.N. Pattanayak are presented, which provide a generalization of the extinction theorem to any medium. Like the recent generalization due to J.J. Sein our derivation is based entirely on Maxwell's theory and not on molecular optics. A hypothesis is put forward as to the true physical significance of the extinction theorem and it is shown how the theorem may be used to solve scattering problems in a novel way. An analogous extinction theorem for non-relativistic quantum mechanics is also presented.

***

One of the most poorly understood theorems of classical electrodynamics is undoubtedly the so-called *extinction theorem* first formulated by P.P. Ewald [1] in 1912 in his basic investigations on the foundations of crystal optics and later by C.W. Oseen [2] in 1915 in his studies of dispersion of light in material media.

Let me first say a few words about the usual formulation of the theorem.

Suppose that a plane electromagnetic wave is incident from vacuo on a material medium with a sharp boundary. The medium will for the moment be assumed to be of the simplest kind - a linear, homogeneous, isotropic, non-magnetic dielectric.

We know that under the influence of the incident electromagnetic field another field will be generated inside the dielectric, which will have a different wave number and hence a different phase velocity. We may, therefore, say that inside the medium, the incident wave, propagated with the vacuum velocity of light c, is somehow *extinguished* by the interaction with the medium and is replaced by a new wave propagated with the velocity c/n, where n is the refractive index of the medium. The question then arises: how does the extinguishing of the incident wave come about? The Ewald-Oseen theorem provides an answer to this question. In mathematical terms the theorem may be expressed in the form: [3]

$$\underline{E}^{(i)}(\underline{r}_{<}) + \frac{1}{4\pi k^2} \nabla\times\nabla\times \int_S \Big\{\underline{E}(\underline{r}')\frac{\partial}{\partial n} G_o(\underline{r}_{<},\underline{r}') - G_o(\underline{r}_{<},\underline{r}')\frac{\partial}{\partial n}\underline{E}(\underline{r}')\Big\}\, dS = 0, \qquad (1)$$

valid at every point $\underline{r}_{<}$ inside the volume V bounded by a surface S (see Fig. 1) occupied by the medium. Here $\underline{E}^{(i)}$ and $\underline{E}$ represent the Fourier transforms (for frequency components $\omega = kc$, c being the vacuum velocity of light) of the incident electric field and of the total electric field generated inside the medium respectively (and taken in the integral in (1) in the limit as the surface is approached from inside V),

$$G_o(\underline{r},\underline{r}') = \exp\{ik|\underline{r}-\underline{r}'|\}/|\underline{r}-\underline{r}'| \qquad (2)$$

is the outgoing free-space Green's function of the Helmholtz equation and $\partial/\partial n$ denotes differentiation along the outward normal to the boundary surface.

The relation (1), which is essentially in the form as formulated by Oseen, was originally derived not from the macroscopic Maxwell theory, but rather from molecular optics, which is a microscopic theory. In this later theory the response of the medium to the incident field is expressed in terms of elementary dipole fields, generated by the interaction of the incident wave with the individual molecules of the medium. We note that in (1), the second term formally cancels the incident electric field at every point inside the medium. Since the second term involves the values of the total field $\underline{E}$ on the boundary surface S only it has been generally asserted that Eq. (1) implies that the incident field is extinguished entirely by those molecular dipoles that are situated on the boundary S of the dielectric. This is the original formulation of the Ewald-Oseen extinction theorem.

In the last few years the Ewald-Oseen extinction theorem has attracted a good deal of attention and various modifications and generalizations of it for more complicated media have been proposed and applied to numerous problems of current research interest. Here is a partial listing of the relevant publications, indicating the authors, year of publication and topics. The complete references are given in footnote 5.

(a)	A. Wierzbicki	(1961,1962)	Quadrupole radiation, reflection, refraction
(b)	N. Bloembergen and P.S. Pershan	(1962)	Non-linear optics
(c)	B.A. Sotskii	(1963)	Metals, optically active media
(d)	R.K. Bullough	(1968)	Many-body optics
(e)	J.J. Sein	(1969,1970)	Spatial dispersion, excitons
(f)	É. Lalor	(1969)	New formulation
(g)	É. Lalor and E.Wolf	(1971)	Interaction of charged particle with a dielectric
(h)	T. Suzuki (1971)		Diffraction
(i)	G.S. Agarwal, D.Pattanayak and E. Wolf	(1971	Spatial Dispersion
(j)	É. Lalor and E. Wolf	(1972)	Refraction and reflection
(k)	J.R. Birman and J.J. Sein	(1972)	Polaritons in bounded media

It is thus clear that the extinction theorem is playing an increasingly greater role in widely different areas. Of the numerous investigations those of J.J. Sein [4,5e] are of particular relevance to the subject matter of this talk. Sein showed that the extinction theorem which, as I already noted, was derived originally from molecular optics may also be derived from Maxwell's theory and he also showed that the traditional interpretation of the theorem is incorrect [4].

In this talk I will present results of a recent investigation that I carried out in collaboration with D. Pattanayak [6], (see also [7]), which we have attempted to answer the following two questions:

(1) Can the extinction theorem be generalized within the framework of Maxwell's theory to a medium of any kind, i.e. with arbitrary response?

(2) What is the true meaning of the theorem?

The first question has been partially answered already by Sein, but we will present quite a general and rigorous answer to it. Let me add that Sein's recognition that the extinction theorem follows also from Maxwell's theory represents an important contribution, since attempts to generalize it within the framework of molecular optics encounters formidable difficulties because of the local field corrections (associated with the Lorentz internal field).

As regards the second question - namely what is the true meaning of the theorem - we will put forward a hypothesis, supported by a few explicit solutions that we obtained with the help of the theorem.

We will also show that the extinction theorem has a strict analogue in potential scattering in non-relativistic quantum mechanics.

The full derivation of our main results is rather lengthy and I will only indicate the main steps.

Let us then consider the scattering of monochromatic electromagnetic wave incident from vacuo on a body with a sharp boundary. We assume the body to be of arbitrary kind; its response could be, for example, non-linear or non-local as in the case of spatial dispersion.

From Maxwell's equations for monochromatic fields, on eliminating, the electric displacement vector $\underline{D}$ and the magnetic induction vector $\underline{B}$ via the relations

$$\underline{D} = \underline{E} + 4\pi\,\underline{P}, \qquad \underline{B} = \underline{H} + 4\pi\,\underline{M}, \tag{3}$$

where $\underline{P}$ and $\underline{M}$ denote the polarization and magnetization vectors respectively, we obtain the four equations

$$\hat{L}\,\underline{E} = \underline{F}_e, \tag{4a}$$

$$\hat{L}\,\underline{H} = \underline{F}_h, \tag{4b}$$

$$\nabla . \underline{E} = 4\pi(\rho - \nabla . \underline{P}) \tag{4c}$$

$$\nabla . \underline{H} = -4\pi \nabla . \underline{M}, \tag{4d}$$

where $\hat{L}$ is the operator

$$\hat{L} = -k^2 + \nabla x \nabla x \tag{5}$$

and the source terms $\underline{F}_e$ and $\underline{F}_h$ are given by

$$\underline{F}_e = 4\pi[\frac{ik}{c}\underline{j} + k^2\underline{P} + ik\ \nabla x\ \underline{M}] \quad , \tag{6a}$$

$$\underline{F}_h = 4\pi[\frac{1}{c}\nabla x\underline{j} - ik\ \nabla x\ \underline{P} + k^2\underline{M}] \ . \tag{6b}$$

The vectors $\underline{E}$, $\underline{H}$, $\underline{j}$, $\underline{P}$ and $\underline{M}$ and the scalar ρ are, of course, functions of position ($\underline{r}$) and are taken at a fixed frequency ω.

The sources of the incident field are assumed to be in the domain $\tilde{V}$ exterior to V (see Fig. 1) and we take them to be at a finite distance [8] from V. It is clear then, since the exterior $\tilde{V}$ of V is vacuo, that

$$\underline{F}_e(\underline{r}) = 4\pi[\frac{ik}{c}\underline{j}_c + k^2\underline{P} + ik\ \nabla x\ \underline{M}] \qquad \text{if } \underline{r} \in V \tag{7a}$$

$$= \frac{4\pi ik}{c}\underline{j}_{ext} \qquad \text{if } \underline{r} \in \tilde{V}\ , \tag{7b}$$

$$\underline{F}_h(\underline{r}) = 4\pi[\frac{1}{c}\nabla x\ \underline{j}_c - ik\ \nabla x\ \underline{P} + k^2\ \underline{M}] \qquad \text{if } \underline{r} \in V \tag{8a}$$

$$= \frac{4\pi}{c}\ \nabla x\ \underline{j}_{ext} \qquad \text{if } \underline{r} \in \tilde{V}\ . \tag{8b}$$

In Eqs. (7) and (8), $\underline{j}_c$ denotes the conduction current density and $\underline{j}_{ext}$ denotes the external current density (representing the source). The corresponding charge densities are, of course, related to the current densities by the continuity equation. Further the polarization vector $\underline{P}$ and the magnetization vector $\underline{M}$ in (7a) and (8a) are assumed to be given functions of the electromagnetic field vectors $\underline{E}$ and $\underline{H}$, whose exact form depends on the nature of the medium. It is to be noted that because of this fact, the equations (4a) and (4b) are in general *coupled* to each other.

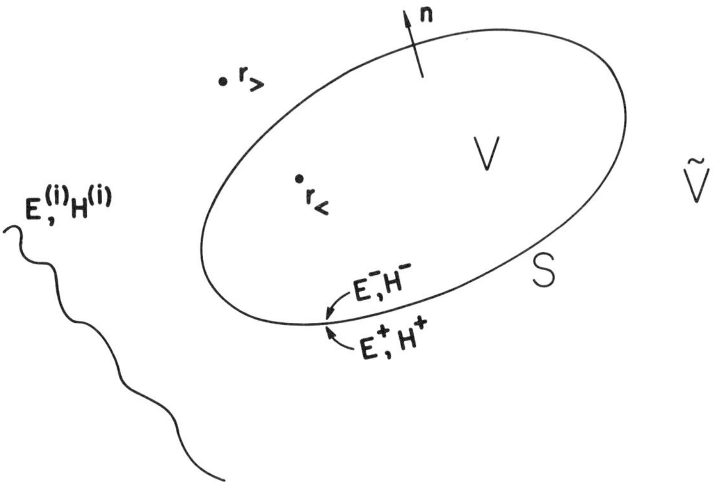

Fig. 1 Notation relating to scattering of an electromagnetic wave on a material medium.

To complete the formulation we must specify the behavior of the fields at the boundary. As is well known Maxwell's equations imply that across the boundary

$$\underline{n} \times (\underline{E}^+ - \underline{E}^-) = 0, \qquad \underline{n} \times (\underline{H}^+ - \underline{H}^-) = \frac{4\pi}{c} \underline{K} , \tag{9}$$

where the superscripts plus and minus denote limiting values as the boundary surface S is approached from outside and inside of the medium respectively (see Figure 1), $\underline{n}$ is the unit outward normal to the boundary surface S and $\underline{K}$ represents the surface current density which will be non-zero only for a perfect conductor. Although these conditions are generally referred to as *boundary conditions*, they are, actually jump conditions - or *saltus* conditions as they are called in the older literature. A clear appreciation of the difference between true boundary conditions and the saltus conditions is, as we shall see later, at the heart of the proper interpretation of the extinction theorem.

For later purpose we also note that the corresponding problem in non-relativistic quantum mechanics, namely the scattering from a finite step potential is mathematically much simpler since in place of several coupled equations involving the vector fields

$\underline{E}$ and $\underline{H}$, the quantum mechanical problem involves only a single equation for the Schrödinger scalar wave function $\psi(\underline{r})$. Moreover, since ψ and its normal $\partial\psi/\partial n$ must, according to basic postulates of quantum mechanics be *continuous* across a finite potential step we now have in place of the conditions (9) the conditions

$$\psi^+ - \psi^- = 0, \qquad \left(\frac{\partial\psi}{\partial n}\right)^+ - \left(\frac{\partial\psi}{\partial n}\right)^- = 0. \tag{10}$$

These are *continuity conditions* and not true boundary conditions either.

Returning to the electromagnetic problem we introduce a dyadic Green's function associated with the $\hat{L}$-operator,

$$\hat{L}\,\underline{\underline{G}} = 4\pi\delta(\underline{r}-\underline{r}')\underline{\underline{U}}, \tag{11}$$

($\underline{\underline{U}}$ = unit dyadic), which obeys the vectorial form of the Sommerfeld radiation condition at infinity:

$$\lim_{r\to\infty} r\,[\nabla\times\underline{\underline{G}} - ik\,\hat{\underline{r}}\times\underline{\underline{G}}] = 0, \tag{12}$$

where $\hat{\underline{r}}$ is the unit vector in the direction of $\underline{r}$ and $r = |\underline{r}|$. It is known that [9]

$$\underline{\underline{G}}(\underline{r},\underline{r}') = (\underline{\underline{U}} + \frac{1}{k^2}\nabla\nabla)\,G_o(\underline{r},\underline{r}') , \tag{13a}$$

where G_o is the outgoing free-space Green's function of the Helmholtz equation, viz.

$$G_o(\underline{r},\underline{r}') = \frac{e^{ik|\underline{r}-\underline{r}'|}}{|\underline{r}-\underline{r}'|} . \tag{13b}$$

Now our equation involving the electric field is of the form

$$\hat{L}\,\underline{E} = \underline{F}_e , \tag{14}$$

with a similar equation involving $\underline{H}$. From (11) and (14) one obtains, if one also uses the vectorial form of Green's theorem, the following identity valid for integration through any domain V' bounded by a closed surface S':

$$\int_{V'} \underline{E}(\underline{r}')\delta(\underline{r}-\underline{r}')d^3\underline{r}' = \frac{1}{4\pi}\int_{V'} \underline{F}_e(\underline{r}') \cdot \underline{\underline{G}}(\underline{r},\underline{r}')d^3\underline{r}' - \frac{1}{4\pi}\sum_e(\underline{r}), \qquad (15)$$

where

$$\sum_e(\underline{r}) = \int_{S'} \{[\underline{n} \times \nabla\times\underline{E}(\underline{r}')] \cdot \underline{\underline{G}}(\underline{r},\underline{r}') + [\underline{n} \times \underline{E}(\underline{r}')] \cdot \nabla\times\underline{\underline{G}}\}dS' \qquad (16)$$

and n is the unit normal to S' pointing outward from the volume V'.

Let us now take the volume V' to coincide either with the scattering volume V or with the exterior $\tilde{V}$ of it. Also we can take the field point $\underline{r}$ to be either in V or in $\tilde{V}$. Applying the theorem (15) separately to each of these four cases we obtain the following four relations:

(a) $\underline{r}\varepsilon V$, $\underline{r}'\varepsilon V$:

$$\underline{E}(\underline{r}_<) = \frac{1}{4\pi}\int_V \underline{F}_e \cdot \underline{\underline{G}}\, d^3\underline{r}' - \frac{1}{4\pi}\sum_e^{(-)}(\underline{r}_<), \qquad (17a)$$

(b) $\underline{r}\varepsilon V$, $\underline{r}'\varepsilon\tilde{V}$:

$$0 = \frac{ik}{c}\int_{\tilde{V}} \underline{j}_{ext} \cdot \underline{\underline{G}}\, d^3\underline{r}' + \frac{1}{4\pi} \cdot \sum_e^{(+)}(\underline{r}_<), \qquad (17b)$$

(c) $\underline{r}\varepsilon\tilde{V}$, $\underline{r}'\varepsilon\tilde{V}$:

$$\underline{E}(r_>) = \frac{ik}{c}\int_{\tilde{V}} \underline{j}_{ext} \cdot \underline{\underline{G}}\, d^3\underline{r}' + \frac{1}{4\pi}\sum_e^{(+)}(r_>), \qquad (17c)$$

(d) $\underline{r}\varepsilon\tilde{V}$, $\underline{r}'\varepsilon V$:

$$0 = \frac{1}{4\pi}\int_V \underline{F}_e \cdot \underline{\underline{G}}\, d^3\underline{r}' - \frac{1}{4\pi}\sum_e^{(-)}(\underline{r}_>). \qquad (17d)$$

Here

$$\sum_e^{(+)} = \int_{S^\pm} \{[n\times \nabla\times\underline{E}] \cdot \underline{\underline{G}} + [\underline{n}\times\underline{E}] \cdot \nabla\times \underline{\underline{G}}\}dS, \qquad (18)$$

and the upper or lower signs are taken on $\Sigma_e^{\pm}$ and on $S^{\pm}$ according as the limiting values are taken from outside (S^{+}) or inside (S^{-}) of the scattering volume. In deriving (17b) and (17c) we also used the radiation condition (12) which ensures that there is no contribution from a sphere of infinitely large radius (the outer boundary of the volume V).

Now the integral containing the external current, taken over the exterior $\tilde{V}$ of our scattering volume has a clear physical meaning. From the significance of $\underline{\underline{G}}$ as the outgoing free-space Green's function of the $\tilde{L}$-operator, this integral must evidently represent the unperturbed incident field, i.e.

$$\frac{ik}{c}\int_{\tilde{V}} \underline{j}_{ext}(\underline{r}').\underline{\underline{G}}(\underline{r},\underline{r}')d^3\underline{r}' = \underline{E}^{(i)}(\underline{r}), \tag{19}$$

irrespective whether the point $\underline{r}$ is situated inside or outside the scattering volume. Hence the equations (17b) and (17c) may be expressed in the compact form

$$\underline{E}^{(i)}(\underline{r}_<) + \frac{1}{4\pi}\sum\nolimits_e^{(+)}(\underline{r}_<) = 0, \tag{20a}$$

$$\underline{E}(\underline{r}_>) = \underline{E}^{(i)}(\underline{r}_>) + \frac{1}{4\pi}\sum\nolimits_e^{(+)}(\underline{r}_>) . \tag{20b}$$

We note that (20a) has some resemblance to the Ewald-Oseen extinction theorem, since it expressed the cancellation of the incident field at every point $\underline{r}$ inside the scattering medium V in terms of an integral involving the field on the boundary of the medium only. However because $\Sigma_e^{(+)}$ rather than $\Sigma_e^{(-)}$ appears in this surface integral, the integral involves the limiting values of the field taken from the outside, rather than from the inside of the scattering volume; but one can easily transform (20a) and also(20b) so as to involve the limiting values from the inside, since from the definition (18) of $\Sigma_e^{(-)}$ and $\Sigma_e^{(+)}$ and from the saltus conditions (9) one easily finds that

$$\frac{1}{4\pi}\left[\sum\nolimits_e^{(+)}(\underline{r}) - \sum\nolimits_e^{(-)}(\underline{r})\right] = -ik\int_{S^-}(\underline{n}\times\underline{M} - \frac{1}{c}\underline{K}) . \underline{\underline{G}}ds \tag{21}$$

Using this result in (20a) and (20b) and the expressions for $\Sigma_e^{(+)}$ one then obtains the following two relations:

$$\underline{E}^{(i)}(\underline{r}_<) + \frac{1}{4\pi}\,\underline{S}_e(\underline{r}_<) = 0, \qquad (22)$$

$$\underline{E}(\underline{r}_>) = \underline{E}^{(i)}(\underline{r}_>) + \frac{1}{4\pi}\,\underline{S}_e(\underline{r}_>), \qquad (23)$$

where

$$\underline{S}_e(\underline{r}) = \int_{S^-}\{[\underline{n}\times(\nabla\times\underline{E} - 4\pi ik\underline{M}) + \frac{4\pi ik}{c}\underline{K}].\underline{\underline{G}}(\underline{r}.\underline{r}')+[\underline{n}\times\underline{E}].\nabla\times\underline{\underline{G}}(\underline{r},\underline{r}')\}dS. \qquad (24)$$

The relation (22) must be satisfied at each point $\underline{r}_<$ inside the scattering volume V. It is one form of our *generalized* Ewald-Oseen theorem, valid for scattering by *any* medium, irrespective of the nature of the constitutive relations. I will indicate shortly how it reduces to the usual form of the Ewald-Oseen theorem when the medium is of the simplest kind. But first I want to say a little about what I believe is the true meaning of the theorem and also discuss briefly the significance of the complementary relation (23). For this purpose we also must note there is an analogous set of relations to (22) and (23), involving the *magnetic* rather than the electric field. They can be derived in a similar way and are

$$\underline{H}^{(i)}(\underline{r}_<) + \frac{1}{4\pi}\,\underline{S}_h(\underline{r}_<) = 0, \qquad (25)$$

$$\underline{H}(\underline{r}_>) = \underline{H}^{(i)}(\underline{r}_>) + \frac{1}{4\pi}\,\underline{S}_h(\underline{r}_>), \qquad (26)$$

where

$$\underline{S}_h(\underline{r}) = \int_{S^-}\{[\underline{n}\times(\nabla\times\underline{H} + 4\pi ik\underline{P}-(4\pi/c)\underline{j})].\underline{\underline{G}}+[\underline{n}\times\underline{H} +(4\pi/c)\underline{K}].\nabla\times\underline{\underline{G}}\}\,dS. \qquad (27)$$

We know that inside the medium, the $\underline{E}$ and $\underline{H}$ fields obey the equations (4a) and (4b),

$$\hat{L}\,\underline{E}(\underline{r}_<) = \underline{F}_e(\underline{r}_<)\ , \qquad \hat{L}\,\underline{H}(\underline{r}_<) = \underline{F}_h(\underline{r}_<)\ .$$

However these are *general field equations* valid inside the medium. They do not completely specify the scattered field since they involve neither the incident field, nor any boundary conditions. Our hypothesis is that the *two extinction theorems (22) and (25) represent boundary conditions subject to which the (generally coupled) field equations (4) provide unique solution for the fields* $\underline{E}$, $\underline{H}$ *inside the scattering medium (i.e. inside the volume* V), *when an electromagnetic field* $\underline{E}^{(i)}$, $\underline{H}^{(i)}$ *is incident on the medium.* Thus, according to this hypothesis, the two extinction theorems allow us to replace the original saltus problem - involving the solution both inside and outside the medium - by a *boundary value problem* for determining the field inside the scattering medium. The boundary conditions for this later problem are of a somewhat unusual kind, having the form of *non-local* relations. Once the solution inside the scattering medium has been obtained, the solution outside it may be determined from Eqs. (23) and (26) by substituting the boundary values into the surface integrals occuring in these formulae. Our new interpretation of the extinction theorems is supported by explicit solutions that were obtained for several special cases [10].

Up to this point we have considered only two of the four relations (17), namely (17b) and (17c). If the other two relations, viz. (17a) and (17d) are also used, as well as the relations (19) and (21) and the corresponding formulae involving the magnetic fields one obtains an alternative set of equations in the form of *integro-differential equations* valid both inside and outside the medium - for the unknown electromagnetic fields $\underline{E}$ and $\underline{H}$. Lack of time prevents a discussion of this point here but we will later consider briefly the analogous situation for the case of quantum mechanical potential scattering.

With the help of various vector identities and Maxwell's equations, our general extinction theorems may be expressed in many alternative but equivalent forms. The extinction theorem for the electric field may, for example, be transformed into the form

$$\underline{E}^{(i)}(\underline{r}_<) + \frac{1}{k^2}\nabla\times\nabla\times\left[\,\underline{I}^{(E)}(\underline{r}_<) + \underline{I}^{(P)}(\underline{r}_<) + \underline{I}^{(M)}(\underline{r}_<) + \underline{I}^{(J)}(\underline{r}_<)\right] = 0\ , \tag{28}$$

where

$$\underline{I}^{(E)}(\underline{r}_<) = \frac{1}{4\pi}\int_{S^-}\{\underline{E}(\underline{r}')\frac{\partial G_0(\underline{r}_<,\underline{r}')}{\partial n} - G_0(\underline{r}_<,\underline{r}')\frac{\partial \underline{E}(\underline{r}')}{\partial n}\}\,dS, \quad (29)$$

$$\underline{I}^{(P)}(\underline{r}_<) = -\int_{S^-}[\underline{n}\ \nabla.\ \underline{P}(\underline{r}')]\ G_0(\underline{r}_<,\underline{r}')\,dS, \quad (30)$$

$$\underline{I}^{(M)}(\underline{r}_<) = -\,ik\int_{S^-}[\underline{n}\times \underline{M}(\underline{r}')]\ G_0(\underline{r}_<,\underline{r}')\,dS, \quad (31)$$

$$\underline{I}^{(J)}(\underline{r}_<) = -\frac{i}{kc}\int_{S^-}[\underline{n}\ \nabla.\ \underline{j}(\underline{r}')-\underline{K}(\underline{r}')]\ G_0(\underline{r}_<,\underline{r}')\,dS. \quad (32)$$

Clearly if the medium is non-magnetic, $\underline{I}^{(M)}\equiv 0$, if it is a non-conductor $\underline{I}^{(J)}\equiv 0$. For a linear, homogeneous ,spatially non-dispersive non-magnetic dielectric, not only do these two terms vanish, but so does also the term $\underline{I}^{(P)}$, unless the frequency of the incident field coincides with a frequency at which the dielectric constant vanishes; for except in this case the polarization field is necessarily transverse [11] (i.e. $\nabla.\underline{P} = 0$). If we exclude this exceptional case, (28) reduces to

$$\underline{E}^{(i)}(\underline{r}_<) + \frac{1}{4\pi k^2}\ \nabla\times\nabla\times\int_{S^-}\{\underline{E}(\underline{r}')\frac{\partial G_0(\underline{r}_<,\underline{r}')}{\partial n} - G_0(\underline{r}_<,\underline{r}')\frac{\partial \underline{E}(\underline{r}')}{\partial n}\}dS=0, \quad (33)$$

which is seen to be identical with the Oseen formulation (1) of the extinction theorem. In this case ($\underline{M} = \nabla.\underline{P} = \underline{j} = \underline{K} = 0$, $\underline{P} = \chi\underline{E}$, χ being a constant), the equation of motion (4a) is not coupled to (4b) (since no magnetic term now occurs on the r.h.s. of (6a)) and reduces to

$$\nabla^2\underline{E}(\underline{r}_<) + n^2k^2\underline{E}(\underline{r}_<) = 0, \quad (34)$$

where

$$n^2 = 1 + 4\pi\chi\ . \quad (35)$$

According to our hypothesis the electric field $\underline{E}$ inside the medium is that solution of (34), which obeys the condition (33) at every point $\underline{r}$ inside the medium. Once this solution has been found, the field outside the medium is obtained from the formula

$$\underline{E}(\underline{r}_>) = \underline{E}^{(i)}(\underline{r}_>) + \frac{1}{4\pi k^2}\nabla\times\nabla\times\int_{S^-}\{\underline{E}(\underline{r}')\frac{\partial G_0(\underline{r}_>,\underline{r}')}{\partial n} - G_0(\underline{r}_>,\underline{r}')\frac{\partial \underline{E}(\underline{r}')}{\partial n}\}dS, \tag{36}$$

to which Eq. (23) may be shown to reduce in the present case.

One can also show that in some other special cases, our general extinction theorem (22) for the electric field reduces to various extinction theorems derived in recent years by other authors. Moreover, one can readily show that in the special case when no scattering medium is present at all our general extinction theorem for the electric field reduces to

$$\underline{E}^{(i)}(\underline{r}_<) + \frac{1}{4\pi}\int_{S^-}\{\underline{E}^{(i)}(\underline{r}')\frac{\partial G_0(\underline{r}_<,\underline{r}')}{\partial n} - G_0(\underline{r}_<,\underline{r}')\frac{\partial \underline{E}^{(i)}(\underline{r}')}{\partial n}\}dS = 0, \tag{37}$$

which will be recognized as the classic *integral theorem of Helmholtz and Kirchhoff* (cf. for example, ref. 3a, p.377).

Returning to the general forms (22) and (25) of the extinction theorems it seems quite remarkable that for a completely arbitrary medium - e.g. an inhomogeneous, anisotropic, non-linear or spatially dispersive medium - the cancelation of the incident field inside the medium is expressible entirely by the values that the field takes at the boundary of the medium

Finally I will show that the main results that we obtained for electromagnetic scattering have a strict analogue in non-relativistic quantum-mechanical potential scattering. Consider scattering of a free particle of momentum $\underline{p}$ on a three-dimensional potential barrier or potential well, characterized by a potential $\mathcal{V}(\underline{r})$ that vanishes outside a finite volume V, bounded by a surface S. (Fig. 2). For simplicity we assume that the potential $\mathcal{V}(\underline{r})$ has at most a finite discontinuity on S.

The Schrödinger equation for this problem may be written in the form

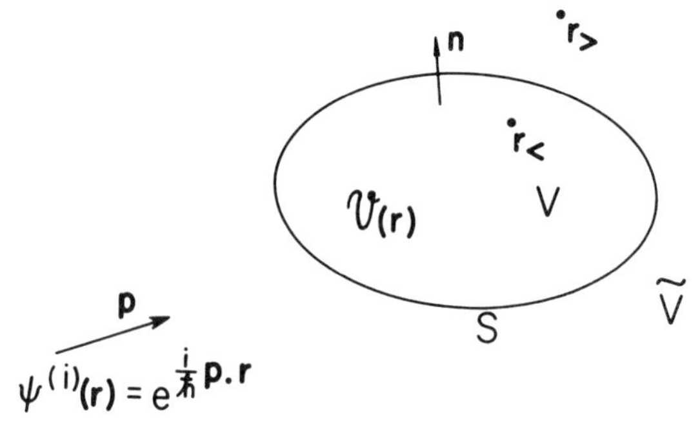

Fig. 2 Notation relating to quantum mechanical potential scattering.

$$(\nabla^2 + k^2)\ \psi(\underline{r}) = U(\underline{r})\psi(\underline{r}), \tag{38}$$

where

$$k^2 = \frac{2m}{\hbar^2} E, \qquad U(\underline{r}) = \frac{2m}{k^2}\mathcal{V}(\underline{r}) \quad \text{if } r \in V$$
$$= 0 \quad \text{if } r \in \tilde{V}. \tag{39}$$

Here m denotes the mass of the particle and $E = p^2/2m$ its energy, $\hbar$ is the Planck constant divided by 2π and $\tilde{V}$ denotes the (infinite) domain outside V.

The equation for the associated Green's function is

$$(\nabla^2 + k^2)\ G_o(\underline{r},\underline{r}') = -\ 4\pi\delta(\underline{r}-\underline{r}'), \tag{40}$$

the Green's function being, of course, the outgoing spherical wave [Eq. (13b) above].

From (38) and (39) we obtain, if we also use Green's theorem, the following identity valid for integration throughout any domain V' bounded by a closed surface S' :

$$\int_{V'} \psi(\underline{r}')\delta(\underline{r}-\underline{r}')d^3\underline{r}' = -\frac{1}{4\pi}\int_{V'} U(\underline{r}')\psi(\underline{r}')G_0(\underline{r},\underline{r}')d^3\underline{r}' - \frac{1}{4\pi}\sum(\underline{r}), \quad (41)$$

where

$$\sum(\underline{r}) = \int_{S'} \{\psi(r') \frac{\partial G_0(\underline{r},\underline{r}')}{\partial n} - G_0(\underline{r},\underline{r}') \frac{\partial \psi(\underline{r}')}{\partial n}\}dS \quad (43)$$

and $\partial/\partial n$ denotes differentiation along the outward normal to S'.

Let us now take the volume V' to coincide either with the scattering volume V or with the exterior $\tilde{V}$ of it. Again the field point $\underline{r}$ may be taken to be either in V or in $\tilde{V}$. Thus, in analogy with the electromagnetic case, we obtain four formulae:

(a) $\underline{r}\varepsilon V$, $\underline{r}'\varepsilon V$

$$\psi(\underline{r}_<) = -\frac{1}{4\pi}\int_V \psi(\underline{r}')U(\underline{r}')G_0(\underline{r}_<,\underline{r}')d^3\underline{r}' - \frac{1}{4\pi}\sum(\underline{r}_<), \quad (43a)$$

(b) $\underline{r}\varepsilon V$, $\underline{r}'\varepsilon\tilde{V}$

$$0 = -\frac{1}{4\pi}\int_{\tilde{V}} \psi(\underline{r}')U(\underline{r}')G_0(\underline{r}_<,\underline{r}')d^3r') + \frac{1}{4\pi}\sum(\underline{r}_<) - \frac{1}{4\pi}\sum^{(\infty)}(\underline{r}_<), \quad (43b)$$

(c) $\underline{r}\varepsilon\tilde{V}$, $\underline{r}'\varepsilon\tilde{V}$

$$\psi(\underline{r}_>) = -\frac{1}{4\pi}\int_{\tilde{V}} \psi(\underline{r}')U(\underline{r}')G_0(\underline{r}_>,\underline{r}')d^3\underline{r}' + \frac{1}{4\pi}\sum(\underline{r}_>) - \frac{1}{4\pi}\sum^{(\infty)}(\underline{r}_>), \quad (43c)$$

(d) $\underline{r}\varepsilon\tilde{V}$, $\underline{r}'\varepsilon V$

$$0 = -\frac{1}{4\pi}\int_V \psi(\underline{r}')U(\underline{r}')G_0(\underline{r}_>,\underline{r}')d^3\underline{r}' - \frac{1}{4\pi}\sum(\underline{r}_>), \quad (43d)$$

where

$$\sum(\underline{r}) = \int_S \{\psi(\underline{r}') \frac{\partial G_0(\underline{r},\underline{r}')}{\partial n} - G_0(\underline{r},\underline{r}') \frac{\partial \psi(\underline{r}')}{\partial n}\}dS, \quad (44a)$$

$$\sum^{(\infty)}(r) = \lim_{R\to\infty} \int_{S_R} \{\psi(\underline{r}') \frac{\partial G_o(\underline{r},\underline{r}')}{\partial n} - G_o(\underline{r},\underline{r}') \frac{\partial \psi(\underline{r}')}{\partial n}\} dS, \tag{44b}$$

$\Sigma^{(\infty)}$ being a contribution from a sphere of limitingly large radius $R\to\infty$ and $\partial/\partial n$ denotes differentiation along the outward normals to the respective volume regions. In the integral (44a) for $\Sigma(\underline{r})$ we need not distinguish between limits from inside and outside of V since ψ and $\partial\psi/\partial n$ must be continuous at the surface S [cf. Eq. (10) above].

The integrals over $\tilde{V}$ in (43b) and (43c) vanish since $U(\underline{r}) = 0$ in $\tilde{V}$. Also the contribution from the surface at infinity must evidently represent the incident wave, i.e.

$$-\frac{1}{4\pi}\sum^{(\infty)}(\underline{r}) = \psi^{(i)}(\underline{r}). \tag{45}$$

Hence (40b) and (40c) reduce to

$$\psi^{(i)}(\underline{r}_<) + \frac{1}{4\pi}\sum(\underline{r}_<) = 0, \tag{46}$$

$$\psi(\underline{r}_>) = \psi^{(i)}(\underline{r}_>) + \frac{1}{4\pi}\sum(\underline{r}_>) . \tag{47}$$

Further from (43a) we obtain if we also use (46)

$$\psi(\underline{r}_<) = \psi^{(i)}(\underline{r}_<) - \frac{1}{4\pi}\int_V \psi(\underline{r}')U(\underline{r}')G_o(\underline{r}_<,\underline{r}')d^3\underline{r}' , \tag{48}$$

and from (43d) if we use (47)

$$\psi(\underline{r}_>) = \psi^{(i)}(\underline{r}_>) - \frac{1}{4\pi}\int_V \psi(\underline{r}')U(\underline{r}')G_o(\underline{r}_>,\underline{r}')d^3\underline{r}' . \tag{49}$$

The relations (46) and (47) are strictly analogous to those that we found for the electromagnetic case. In particular (46) represents *an extinction theorem* expressing the cancellation of the incident wave function at every point inside the potential barrier or potential well in terms of the values of the total wave function ψ and its normal derivative $\partial\psi/\partial n$ at all points on the boundary of the potential barrier or potential well. We assert that this theorem has the same kind of significance as we postulated for the electromagnetic extinction theorem: It is a (non-local) boundary condition subject to which the Schrödinger equation (38) has to be solved inside the scattering volume V. Once this solution is known the wave function outside V can be determined from (47) by substitution. Note that (47) involves only a surface integral $\Sigma(\underline{r}_>)$, not a volume integral as it does in the usual formulation. The other two equations (47) and (48), which are seen to be of the same form irrespective whether the field point is inside or outside V will be recognized as the *usual integral equations for potential scattering*. Mathematically they are equivalent to the Schrödinger equation (38) together with the two equations (46) and (47).

We made use of the quantum mechanical extinction theorem (46) and the associated formula (47) to solve simple scattering problems, in order to verify the correctness of this new formulation of scattering. The results agree with those obtained by conventional methods based on the integral equations (48) and (49). I might add that the extinction theorem also leads correctly to *bound states* under appropriate conditions. One finds in these cases that the Schrödinger equation can then be solved subject to our non-local boundary condition (expressed by the extinction theorem (46))only when $\psi^{(i)}(\underline{r}) \equiv 0$.

Let me remark here that a quantum-mechanical extinction theorem, was also derived, from a different approach by Melvin Lax [12] in 1952 in his treatment of multiple scattering.

I will end by summarizing our main conclusions:

(1) We obtained on the basis of Maxwell's theory generalization of the classic Ewald-Oseen extinction theorem, valid rigorously for scattering from medium of any prescribed macroscopic response.

(2) We have put forward a hypothesis as to the true meaning of the theorem: it is a non-local boundary condition for the solution of the equation of motion for the *interior* scattering problem.

(3) We have shown that once the boundary values have been determined the *exterior* scattering problem can be solved in a novel way, involving only surface integrations.

(4) We have shown that these results have strict analogues in non-relativistic quantum-mechanical potential scattering and provide a new approach to solving such problems.

Finally let me say that scattering problems involving sharp boundaries, such as we have considered here are, in general, hard to solve even approximately, since the Born approximation cannot be used in such cases. It is possible that our new formulation might provide a basis for the development of useful approximate techniques for solving problems of this kind. For this new formulation takes very explicitly into account the sharp boundary, the very presence of which makes the usual perturbation methods inapplicable. We plan to discuss this and related problems in other publications.

References

* Research supported by the Air Force Office of Scientific Research and the Army Research Office (Durham).

1. P.P. Ewald, (a) Dissertation, Univ. of Munich, 1912;
   (b) Ann. Phys. *49*, 1 (1915).
2. C.W. Oseen, Ann. Phys. *48*, 1 (1915).
3. For a detailed account of the theorem see
   (a) M. Born and E. Wolf, *Principles of Optics* (Pergamon Press, Oxford and New York, 1970) 4th ed., §2.4,

   or

   (b) L. Rosenfeld, *Theory of Electrons* (North-Holland Publishing Co., Amsterdam, 1951) Chapt. VI, §4.
4. J.J. Sein, *An Integral-Equation Formulation of the Optics of Spatially-Dispersive Media*, Ph.D. Dissertation, New York University, 1969, Appendix III.
5. (a) A. Wierzbicki, Bul. Acad. Polonaise des Sciences *9*, 833 (1961); Acta Phys. Pol. *21*, 557 (1962), ibid *21*,575 (1962).
   (b) N. Bloembergen and P.S. Pershan, Phys. Rev. *127*,206 (1962).
   (c) B.A. Sotskii, Opt. Spectro. *14*, 57 (1963).
   (d) R.K. Bullough, J. Phys. A. (Proc. Phys. Soc.) Ser. 2, *1*, 409 (1968).
   (e) J.J. Sein, ref. 4 above and Opt. Comm. *2*, 170 (1970).
   (f) É. Lalor, Opt. Comm. *1*, 50 (1969).
   (g) É. Lalor and E. Wolf, Phys. Rev. Lett. *26*, 1274 (1971).
   (h) T. Suzuki, J. Opt. Soc. Amer. *61*, 1029 (1971).
   (i) G.S. Agarwal, D.N. Pattanayak and E. Wolf, Opt. Comm. *4*, 260 (1971).
   (j) É. Lalor and E. Wolf, J. Opt. Soc. Amer. *62*,1165 (1972).
   (k) J.L. Birman and J.J. Sein, Phys. Rev. *B6*, 2482 (1972).

6. Preliminary results of this investigation were presented in a lecture at the annual meeting of the Optical Society of America held in Ottawa in October 1971 (Abstr. WC16, J.Opt.Soc.Amer., *61*, 1560 (1971)) and in a note published in Optics Commun. *6*, 217 (1972).
7. While this manuscript was being prepared for publication a paper reporting some closely related results was published by J. de Goede and P. Mazur, Physica *58*, 568 (1972). This paper also contains some additional references to publications concerning the extinction theorem.
8. A slightly different argument to that given below is needed if the incident field is a plane wave (i.e. if the source is at infinity), but the final formulae remain the same. The case of plane wave incidence is discussed explicitly in connection with the quantum mechanical extinction theorem, in the last part of this paper.
9. See, for example, Chen-To Tai, *Dyadic Green's Functions in Electromagnetic Theory* (In-text Educational Publishers, Scranton and San Francisco, 1971).
10. One of them was presented in reference 5j .
11. That this is so follows at once from the Maxwell equation $\nabla . \underline{D} = 0$. For this implies that $0 = \nabla.(\varepsilon \underline{E}) = \varepsilon \nabla.\underline{E}$ (ε = dielectric constant), so that $\nabla.E = 0$ and hence also $\nabla.\underline{P} = 0$ (because of the linearity of the medium), unless $\varepsilon = 0$.
12. M. Lax, Phys. Rev. *85*, 646 (1952).

MODE ANALYSIS AND MODE COUPLING OF ELECTROMAGNETIC FIELDS IN SPATIALLY DISPERSIVE MEDIA*

D. N. Pattanayak

University of Rochester, Rochester, New York

The electrodynamics of a spatially dispersive medium filling the whole space has been the subject of many investigations. Pekar[1] in 1957 predicted many remarkable properties associated with such a medium. He showed that in a spatially dispersive medium the macroscopic polarization is connected to the electric field by a certain differential equation in spatial coordinates. Maxwell's equations along with this new differential constitutive relation showed clearly the possibility of several waves being propagated in the medium with the same polarization and frequency but with different velocities. In such media, longitudinal waves with non-zero group velocity may also be generated. These and other interesting properties of spatial dispersion attracted a good deal of attention from both theoretical and experimental physicists. The works of Agranovich, Ginzburg[2], Pekar[1], Hopfield and Thomas[3] and others, show the connection of the theory of excitons with the theory of spatial dispersion. However the electrodynamics for bounded spatially dispersive media has been only poorly understood. The need for additional boundary conditions (a.b.c's) to take account of the appearance of new waves in the medium resulted in a different set of a.b.c.'s, depending on various models employed.

In this paper we will present a self-consistent formulation of electrodynamics of spatially dispersive media. We will show that the a.b.c.'s are contained in the macroscopic Maxwell's equations together with the constitutive relations. Our analysis is a generalization of that recently developed by the author, in collaboration with G.S. Agarwal and E. Wolf[4].

*Research supported by the U.S. Army Research Office (Durham).

From the Maxwell's equations for non-magnetic medium, with no external charges and currents, one may obtain the following wave equation for the electric field in $(\underline{r}, \omega)$ space:

$$\nabla \times \nabla \times \hat{\underline{E}}(\underline{r},\omega) - k_o^2 \hat{\underline{D}}(\underline{r},\omega) = 0, \quad k_o = \omega/c \quad , \tag{1}$$

where

$$\hat{\underline{D}}(\underline{r},\omega) = \hat{\underline{E}}(\underline{r},\omega) + 4\pi \hat{\underline{P}}(\underline{r},\omega). \tag{2}$$

For a spatially dispersive medium, which is isotropic, stationary and spatially homogeneous the constitutive relation connecting the dielectric induction vector $\hat{\underline{D}}(\underline{r},\omega)$ and the electric field $\underline{E}(\underline{r},\omega)$ can be written as

$$\hat{\underline{D}}(\underline{r},\omega) = \varepsilon_o(\omega) \hat{\underline{E}}(\underline{r},\omega) + 4\pi \hat{\underline{P}}_{N.L.}(\underline{r},\omega), \tag{3}$$

where

$$\hat{\underline{P}}_{N.L.}(\underline{r},\omega) = \frac{1}{4\pi} \int_V \varepsilon^{(1)}(\underline{r}-\underline{r}',\omega) \hat{\underline{E}}(\underline{r}',\omega) d^3r' \ .$$

In Eq. (3) $\varepsilon_o(\omega)$ is the wave-vector independent background dielectric permittivity. Further $\varepsilon^{(1)}(\underline{r}-\underline{r}',\omega)$ is the non-local part of the dielectric permittivity, whose specific form for any medium can be obtained from a microscopic theory or from a phenomenological analysis and $\hat{P}_{N.L.}$ is the non-local part of the macroscopic polarization vector. In writing down Eq. (3) we have, of course, neglected inhomogeneities that may arise due to the boundary of the medium. For an analysis taking care of surface effects, one may follow Hopfield and Thomas's method[3] by introducing surface layers which are to be considered spatially non-dispersive.

The kernel $\varepsilon^{(1)}(\underline{r}-\underline{r}',\omega)$ can be assumed to satisfy the following propagator equation

$$L\left(\frac{\partial}{\partial x}, \frac{\partial}{\partial y}, \frac{\partial}{\partial z}\right) \varepsilon^{(1)}(\underline{r}-\underline{r}',\omega) = -4\pi\, \delta(\underline{r}-\underline{r}'), \tag{4}$$

where L is a linear operator of the general form

$$L = L_0 + L_1 + L_2 + \dots ,$$

$$L_0 = a_0 ,$$

$$L_1 = a_i \frac{\partial}{\partial x_i} , \tag{5}$$

$$L_2 = a_{ij} \frac{\partial^2}{\partial x_i \partial x_j} ,$$

$$\vdots$$

the a's being constants. In Fourier $(\underline{k},\omega)$ space Eq. (4) can be written as

$$L(\underline{K},\omega)\ \varepsilon^{(1)}(\underline{K},\omega) = -4\pi . \tag{6}$$

Eq. (6) is strictly true for infinite domain, but for finite domain may be assumed to be true at least as an approximation.

From Eq. (1) and Eq. (3) we may write

$$\nabla \times \nabla \times \hat{\underline{E}}(\underline{r},\omega) - k_0^2 \varepsilon_0 \hat{\underline{E}}(\underline{r},\omega) - k_0^2 \int_V \varepsilon^{(1)}(\underline{r}-\underline{r}',\omega)\ \hat{\underline{E}}(\underline{r}',\omega) d^3r' = 0. \tag{7}$$

Eq. (7) is the basic integro-differential equation of our theory. If we operate by L on Eq. (7) and use Eq. (4) we obtain the equation

$$L[\nabla\times\nabla\times\hat{\underline{E}}(\underline{r},\omega) - k_0^2\varepsilon_0\ \hat{\underline{E}}(\underline{r},\omega)] + 4\pi\ k_0^2\ \hat{\underline{E}}(\underline{r},\omega) = 0. \tag{8}$$

From Eq. (8) and (1) and (3) we can obtain the relation

$$L\ \hat{\underline{P}}_{N.L.}(\underline{r},\omega) = -\hat{\underline{E}}(\underline{r},\omega). \tag{9}$$

We shall solve Eq. (8) subject to the integro-differential equation (7). This is equivalent to solving equation (8) subject to the following generalized extinction theorem, recently obtained by Agarwal, Pattanayak and Wolf[5]

$$\int_S \underline{n} \cdot \underline{C} \left[\underline{\varepsilon}^{(1)}(\underline{r}-\underline{r}',\omega), \{ \underline{P}_{N.L.}(\underline{r}',\omega)\}_j \right] dS=0, \quad (j=1,2,3). \tag{10}$$

This theorem can readily be obtained from Eqs. (3) and (4) by applying Green's theorem, generalized to an arbitrary operator. In Eq. (10) $\underline{n}$ is unit vector pointing along the outward normal to the surface $\bar{S}$ bounding the volume V and $\underline{C}$ $[\phi, \psi]$ is defined by the relation

$$\nabla \cdot \underline{C} [\phi, \psi] = \phi L \psi - \psi L^{+} \phi , \tag{11}$$

L^{+} being the operator adjoint to L.

We now apply the above results to a spatially dispersive medium occupying a volume V. For the sake of simplicity we shall restrict ourselves to a special geometry, namely the case of a spatially dispersive medium occupying the domain $-\infty < x < \infty$, $-\infty < y < \infty$ and $0 < z < d$. The analysis that will follow can be applied to arbitrary domain but will involve theory of higher order partial differential equation. We can express the electric field $\hat{\underline{E}}(\underline{r},\omega)$ as a two-dimensional Fourier integral with respect to the x and y coordinates:

$$\hat{\underline{E}}(\underline{r}, \omega) = \iint_{-\infty}^{\infty} \hat{\underline{E}}(u,v; z,\omega)\, e^{i(ux+vy)} du dv. \tag{12}$$

Eq. (8) now becomes

$$L(\underline{S}) [\underline{S}\times\underline{S}\times\hat{\underline{E}}(u,v;z,\omega) - k_o^2\, \varepsilon_o \hat{\underline{E}}(u,v;z,\omega)]$$

$$+ 4\pi\, k_o^2\, \hat{\underline{E}}\, (u,v;z,\omega) = 0, \tag{13}$$

where

$$\underline{S} \equiv iu,\ iv,\ \frac{\partial}{\partial z}\ . \tag{14}$$

Eq. (13) can be decoupled in the components $\hat{E}_j(u,v,z;\omega)$ $(j = x,y,z)$ by standard mathematical methods to obtain the equation

$$\{L(\underline{S})(\underline{S}^2 + \varepsilon_o k_o^2) + k_o^2\}^2 [L(\underline{S}) + \frac{1}{\varepsilon_o}]\, \hat{\underline{E}}(r,\omega) = 0. \tag{15}$$

Not all the solutions of Eq. (15) are of interest to us. We shall keep only those solutions of Eq. (15) which satisfy Eq. (13). The most general solutions of Eq. (15) can be shown to be of the form

$$\underline{\hat{E}}(u,v;z,\omega) = \sum_j [\underline{\hat{E}}_j(u,v) + z\,\underline{\hat{E}}_j^{(R)}(u,v)]\, e^{iW_j z} , \tag{16}$$

where the W_j's, assumed to be all different and are obtained from the equations

$$\{L(\underline{\tilde{K}}_j)(\underline{\tilde{K}}_j^2 + k_o^2\varepsilon_o) + k_o^2\}\{L(\underline{\tilde{K}}_j) + \frac{1}{\varepsilon_o}\} = 0 , \tag{17}$$

where

$$\underline{\tilde{K}}_j \equiv iu,\ iv,\ iW_j$$

From Eqs. (15), (16) and (13) one can show that in order that (16) be a solution to Eq. (13) we must have

$$\underline{\hat{E}}_j^{(R)}(u,v) = 0 , \tag{18a}$$

and

$$\underline{\hat{E}}(u,v;z,\omega) = \sum_j \underline{\hat{E}}_j(u,v)\, e^{iW_j z} , \tag{18b}$$

which is the solution to equation (13). Hence from Eq. (12) one obtains the following angular spectrum representation of the electric field:

$$\underline{\hat{E}}(r,\omega) = \sum_j \iint_{-\infty}^{\infty} \underline{\hat{E}}_j(u,v)\, e^{i\underline{K}_j\cdot\underline{r}}\, du dv, \tag{19}$$

where

$$\underline{K}_j \equiv u,\ v,\ W_j .$$

Dispersion Relations

One can show from Eqs. (17), (19), (13) and (6) that for transverse waves the wave vectors $\underline{k}_j$ obey the dispersion relation

$$\underline{K}_j^2 - k_o^2\,[\varepsilon_o + \varepsilon^{(1)}(\underline{K}_j)] = 0,\ (\underline{K}_j\cdot\underline{\hat{E}}_j(u,v) = 0) . \tag{20}$$

For longitudinal waves we have the dispersion relation

$$[\varepsilon_o + \varepsilon^{(1)}(\underline{K}_j)] = 0 \ , \quad (\underline{K}_j \times \hat{\underline{E}}_j(u,v) = 0). \tag{21}$$

Our analysis so far is independent of any specific form of the dielectric permittivity. It can be shown from a theory of insulating isotropic dielectric that the general form of dielectric permittivity near a particular resonance may be expressed in the form[3]

$$\hat{\varepsilon}(\underline{K},\omega) = \hat{\varepsilon}_o(\omega) + \hat{\varepsilon}^{(1)}(K^2), \tag{22}$$

where

$$\hat{\varepsilon}^{(1)}(K^2) = \frac{\alpha_o(K^2)\omega_o^2(K^2)}{\omega_o^2(K^2)-\omega^2-i\omega\Gamma_o(K^2)}, \tag{23}$$

$\alpha_o(K^2)$ = oscillator strength associated with the oscillator.

$\Gamma_o(K^2)$ = phenomenological damping factor.

Let us expand both the numerator and denominator of Eq. (23) in powers of K^2

$$h\ \omega_o(K^2) = h\ \omega_o + \frac{h^2\ K^2}{2\ M_{ex}} + \dots \ ,$$

$$\alpha_o(K^2) = \alpha_o + \alpha_1\ K^2 + \alpha_2\ K^4 + \dots \ , \tag{24}$$

$$\Gamma_o(K^2) = \Gamma_o + \Gamma_1\ K^2 + \Gamma_2\ K^4 + \dots \ .$$

If we retain in each expansion a finite number of terms only and use the theory of partial fractions Eq. (22) may be expressed in the form

$$\varepsilon^{(1)}(K^2,\omega) = \sum_{n=1}^{N} \frac{C_n(\omega)}{K^2-K_n^2(\omega)}, \tag{25}$$

where $C_n(\omega)$ and $K_n(\omega)$ are expressible in terms of the coefficients in the expansions of $h\omega_0(K^2)$, $\alpha_0(K^2)$ and $\Gamma_0(K^2)$. In $(\underline{r},\omega)$ space, Eq. (25) can be written as

$$\hat{\varepsilon}^{(1)}(\underline{r}-\underline{r}',\omega) = \sum_{j=1}^{N} C_j(\omega)\varepsilon_j^{(1)}(\underline{r}-\underline{r}',\omega) , \tag{26a}$$

where

$$\varepsilon_j^{(1)}(\underline{r}-\underline{r}',\omega) = \frac{1}{4\pi} \frac{e^{i K_j |\underline{r}-\underline{r}'|}}{|\underline{r} - \underline{r}'|} . \tag{26b}$$

Now substituting Eq. (26a) in (9) we get the generalized extinction theorem

$$\int_S \underline{n}.\underline{C}\left[\sum_{\ell=1}^{N} \varepsilon_\ell^{(1)}(\underline{r}-\underline{r}',\omega),\ \{\underline{P}_{N.L.}(\underline{r}',\omega)\}_j\right] dS=0,\ (j-1,2,3). \tag{27}$$

Since each term in the surface integral in Eq. (27) obeys Helmholtz's equation with different parameters we must have

$$\int_S \underline{n}.\underline{C}\left[\varepsilon_\ell^{(1)}(\underline{r}-\underline{r}',\omega),\{\ \underline{P}_{N.L.}^{(\ell)}(\underline{r}',\omega)\}_j\right] dS = 0 ,$$

where

$$\underline{P}_{N.L.} = \sum_{\ell=1}^{N} \underline{P}_{N.L.}^{(\ell)};\ \underline{P}_{N.L.}^{(\ell)}(\underline{r},\omega)= \frac{1}{4\pi}\int_V \varepsilon_\ell^{(1)}(\underline{r}-\underline{r}',\omega)\underline{E}(\underline{r}',\omega)d^3r'. \tag{28}$$

Using the mode expansion for electric field Eq. (19) and Eqs. (3) and Eq. (26) we obtain from Eq. (28) after lengthy mathematical operations

$$\sum_{j=1}^{4N+2} \frac{\hat{\underline{E}}_j\ (u,v)}{W_j - \omega_\ell} = 0 ,$$

and

$$\sum_{j=1}^{4N+2} \frac{\hat{\underline{E}}_j(u,v)}{W_j + \omega_\ell} e^{iW_j d} = 0, \quad (\ell = 1,2,\ldots,N) \tag{29}$$

where

$$\omega_\ell = (K_n^2 - u^2 - v^2)^{\frac{1}{2}} \quad \text{for} \quad u^2 + v^2 < \text{Re } K_n^2 ,$$

$$= i(u^2 + v^2 - K_n^2)^{\frac{1}{2}} \quad \text{for} \quad u^2 + v^2 > \text{Re } K_n^2 .$$

Eqs. (29) are the mode coupling conditions. We could have obtained these conditions directly from the basic integro-differential equation (7) after explicit integrations using the mode expansion and dispersion relations for the electric field. The mode coupling conditions (29), which are equivalent to the generalized extinction theorem, play the same role as the a.b.c.'s. One can show that for a half space, i.e. in the limit as $d \to \infty$ Eq. (28) will be satisfied provided

$$\underline{P}^{(j)}_{N.L.}(0,\omega) + \frac{1}{i\omega_n} \frac{\partial}{\partial z} \underline{P}^{(j)}_{N.L.}(0,\omega) = 0 , \tag{30}$$

Eq. (30) may be considered as the boundary conditions on the nonlocal polarization. These a.b.c.'s differ from those of Pekar[1] and Hopfield and Thomas[3], but agree with those obtained by Sein[6] in the appropriate limit. One can readily verify that the mode coupling conditions or the a.b.c.'s along with the usual Maxwell boundary conditions lead to a unique solution of the problem of refraction and reflection[5] of an electromagnetic wave incident on a spatially dispersive half-space or a slab. The above analysis can be extended in principle to arbitrary domains.

To conclude, we have shown that Maxwell's theory is adequate to describe the electrodynamics of a spatially dispersive medium.

REFERENCES

1. S.I. Pekar, Zh. Eksp. Teor. Fiz. *33*, 1022 (1957), and *34*, 1176 (1958) [Sov. Phys. JETP *6*, 785 (1958), and *7*, 813 (1958)].
2. V.M. Agranovich and V.L. Ginzburg, USP. Fiz. Nauk *76*, 643 (1962), and *77*, 663 (1962) [Sov. Phys. USP *5*, 323, 675 (1962)], and *Spatial Dispersion in Crystal Optics and the Theory of Excitons* (Interscience, New York, 1966, Chapter X in particular).

3. J.J. Hopfield and D.J. Thomas, Phys. Rev. *132*, 563 (1963).
4. G.S. Agarwal, D.N. Pattanayak and E. Wolf, Phys. Rev. Lett. *27*, 1022 (1971). Optics Commun. *4*, 255 (1971, ibid. *4* 260 (1971).
5. G.S. Agarwal, D.N. Pattanayak and E. Wolf, Physics Letters, to be published.
6. J.J. Sein, Ph.D. thesis, New York University, 1969 (unpublished) and Phys. Lett. *32A*, 141 (1970).

AMPLITUDE STABILIZATION OF PULSES FROM A Q-SWITCHED RUBY LASER BY MEANS OF INTERACTION WITH A NON-LINEAR MEDIUM

M. Bertolotti, S. Martellucci, G. Vitali

Università di Roma, Roma, Italy

B. Crosignani, P. Di Porto,

Istituto Superiore P.T., Roma, Italy

1. Introduction

The possibility of improving the intensity stabilization of an electromagnetic wave, once it acts as fundamental field in the second harmonic generation process, has been recently examined. The two limiting cases of chaotic and "quasi-laser" sources have been examined in detail [1]. A common feature consists in an enhancement of intensity definition, up to a certain traveled length of nonlinear crystal where it reaches a maximum. While this distance is a function of both the non-linear susceptibility of the crystal and the features of the beam, the achieved maximum only depends on the latter ones. Here we report the extension of these calculations to the relevant situation of initial Gaussian intensity distribution centered around a given value. The theoretical predictions are compared with some preliminary experimental results.

2. Theory

The process of second harmonic generation is usually schematized by means of a simple model, in which two electromagnetic modes of frequency ω_1 and $\omega_2 = 2\omega_1$ are made to interact while propagating in a nonlinear crystal. A complete analytical treatment of this model has been given by Armstrong et al. [2]. In particular, whenever the second harmonic field is not present at the boundary z=0 of the crystal, the real amplitude $\rho_1(z)$ of the fundamental wave

$$E_1(z,t) = \rho_1(z)\exp[i(\kappa_1 z - \omega_1 t)]$$

evolves, in the case of perfect phase-matching and provided that $|\partial^2\rho_1/\partial z^2| << \kappa_1|\partial\rho_1/\partial z|$, as (using the Gaussian system of units)

$$\rho_1(z) = \rho_1(0)\ \mathrm{sech}[s\rho_1(0)z] \tag{1}$$

z being the length of the traveled path and

$$s = \frac{4\pi\ \omega_1^2}{\kappa_1 c^2}\ \chi_2 \tag{2}$$

where κ_1 labels the wave number of the fundamental field and χ_2 is the second order susceptibility of the medium.

Let us suppose the intensity distribution of the fundamental wave at z=0 to be

$$P(I_o) = \frac{1}{\sqrt{\pi}\ \sigma}\ \exp[-(I_o-\lambda)^2/\sigma^2] \tag{3}$$

where

$$\lambda = \langle I_o\rangle\ ;\quad \sigma^2 = 2[\langle I_o^2\rangle - \langle I_o\rangle^2]\ . \tag{4}$$

In order to give a correct physical meaning to Eq. (3) the relation $\exp[-\lambda^2/\sigma^2] << 1$ has to be verified due to the fact that $I_o \geqslant 0$. According to the relation $\rho_1^2(z)=I(z)$ (*) so that in particular $\rho_1^2(0) \equiv \rho_{10}^2 = I_o$ one has for the probability distribution of ρ_{10}

$$p(\rho_{10}) = \frac{2\rho_{10}}{\sqrt{\pi}\ \ \sigma}\ \exp[-(\rho_{10}^2 - \lambda)^2/\sigma^2]\ . \tag{5}$$

As a measure of intensity stabilization we choose the ratio $\langle I^2(z)\rangle/\langle I(z)\rangle^2$ a quantity which assumes a value equal to one for an ideal laser and two for a chaotic field. By means of Eqs.(1) and (5) one immediately obtains

* Actually ρ_1^2 is proportional to I, $\rho_1^2 = \alpha I$ with $\alpha = 2\pi/nc$(4). Obviously, one has to take into account this factor, which here is assumed to be equal to one in the numerical calculations for notational convenience.

$$\frac{\langle I^2(z)\rangle}{\langle I(z)\rangle^2} = \frac{\int_0^\infty p(\rho_{10})\rho_{10}^4 \operatorname{sech}^4(\rho_{10}\, sz)\, d\rho_{10}}{[\int_0^\infty p(\rho_{10})\rho_{10}^2 \operatorname{sech}^2(\rho_{10}\, sz)\, d\rho_{10}]^2} \tag{7}$$

or equivalently, in a dimensionless form,

$$\frac{\langle I^2(y)\rangle}{\langle I(y)\rangle^2} = \frac{1}{2}\sqrt{\pi}\, y^2 \frac{\int_0^\infty e^{-(x^2-\beta y^2)/y^4} x^5 \operatorname{sech}^4 x\, dx}{[\int_0^\infty e^{-(x^2-\beta y^2)/y^4} x^3 \operatorname{sech}^2 x\, dx\,]^2} \tag{8}$$

where $\beta = \lambda/\sigma$ and $y = \sigma^{\frac{1}{2}} sz$. We observe that from the definition of β and Eq. (5) it follows

$$\frac{\langle I_o^2\rangle}{\langle I_o\rangle^2} = \frac{1}{2\beta^2} + 1 \tag{9}$$

so that the ideal laser case is recovered for large β. The numerical evaluation of the quantity $H(y) = \langle I^2(y)\rangle/\langle I(y)\rangle^2 - 1$ furnishes the results reported in Table I for various values of β. These data are as well plotted in Fig. 1. It is obvious that a perfect stabilization would require $H(y) = 0$, so that the smallness of $H(y)$ furnishes a valid criterion of stabilization. For each value of β, $H(y)$ reaches a minimum for a given value y^*. It is worth noting that the ratio $H(0)/H(y^*)$ increases for increasing β, which shows that the relative stabilizing effect is more relevant for initially highly defined intensity.

3. Experimental set-up

We have used as the source to be stabilized a repetitively operated Q-switched ruby laser with pulses of varying maximum intensity.

The experimental set-up is whon in Fig. 2. The laser source was a commercial ruby laser (Barr & Stroud LU6), Q-switched by means of a rotating prism; single pulses lasting about 20 n sec

Table 1

y	H(y)				
	β = 1.5	β = 2	β = 2.5	β = 3	β = 5
0	0.222	0.125	0.080	0.055	0.020
0.1	0.209	0.120	0.076	0.052	0.0181
0.2	0.194	0.108	0.066	0.044	0.0133
0.3	0.172	0.109	0.053	0.033	0.0076
0.4	0.147	0.071	0.038	0.021	0.0029
0.5	0.120	0.052	0.024	0.011	0.0004
0.6	0.096	0.036	0.013	0.004	0.0005
0.7	0.075	0.023	0.006	0.001	0.0035
0.8	0.058	0.014	0.003	0.003	0.0091
0.9	0.047	0.011	0.005	0.008	0.0172
1.0	0.039	0.012	0.013	0.018	0.0280
1.1	0.038	0.020	0.025	0.033	0.0412
1.2	0.042	0.032	0.043	0.052	0.0571
1.3	0.053	0.051	0.066	0.076	0.0757
1.4	0.069	0.076	0.096	0.107	0.0972
1.5	0.091	0.109	0.133	0.143	0.1218
1.6	0.120	0.149	0.177	0.186	0.1497
1.7	0.156	0.198	0.231	0.237	0.1814
1.8	0.200	0.256	0.296	0.298	0.2171
1.9	0.253	0.327	0.372	0.369	0.2572
2.0	0.315	0.410	0.464	0.454	0.3022

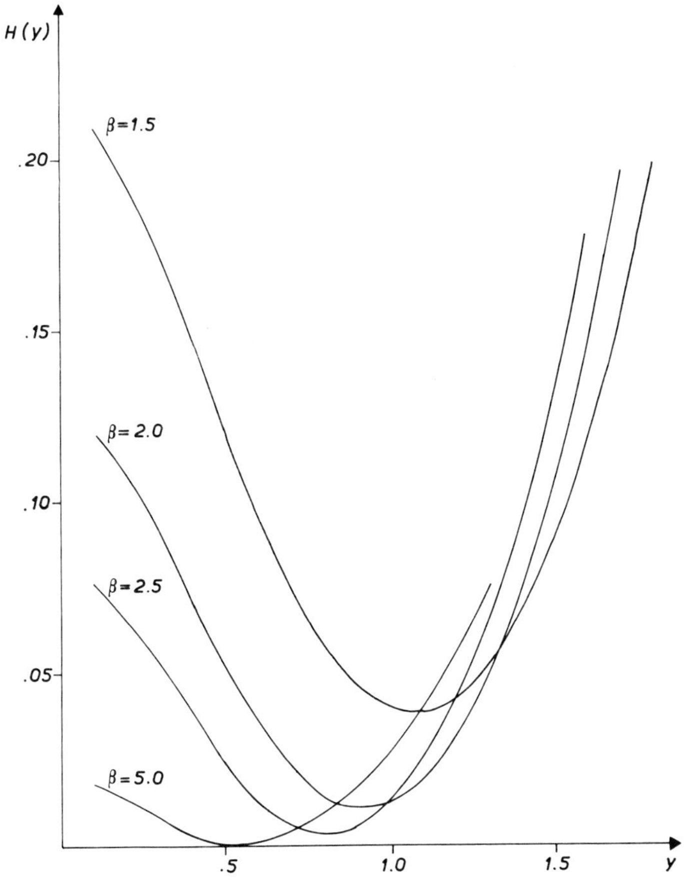

Fig. 1 Behaviour of H(y) at different numerical values of the β parameter.

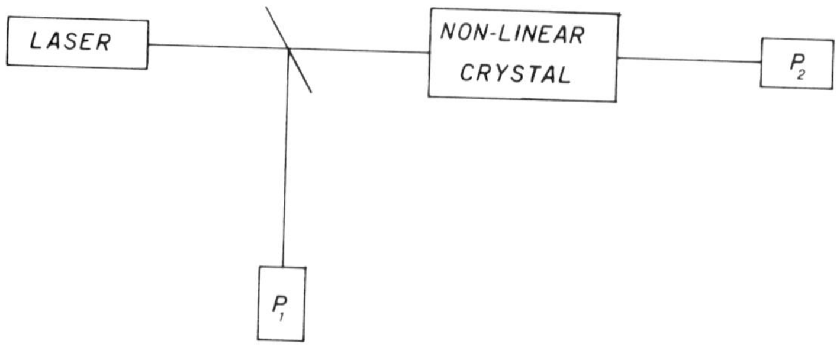

Fig. 2 Schematic drawing of the experimental set-up.

were obtained at each shot. By properly adjusting the repetition rate of laser firing (the maximum value being of the order of 10 shots/min), the amplitude of the light pulses was made to fluctuate by a chosen quantity. Partial focussing of the laser beam on the crystal surface has been used in order to get the desired value of the incident flux density. Unfortunately, the possibility of damaging the crystal (3) places an upper limit to the laser flux density obtainable at about 3.10^8 W/cm^2 for our experimental conditions.

Input and output monitor signals of the laser pulse were observed by two calibrated ITT photodyodes P_1 and P_2 on a two-beams cathode ray oscilloscope. The overall resolution time of the entire electronic circuit (photodiodes + connecting cables + preamplifiers) was of the order of 8 nsecs.

The non-linear material were KDP crystals of different thicknesses with high grade optical polished surfaces. We operated in phase-matching conditions for the production of second harmonics. A first crystal was 1.8 cm thick; however, the improper orientation of the entrance surfaces did not allow us to satisfy entirely the required phase-matching conditions, with a consequent loss in conversion efficiency. The second KDP crystal used was 0.7cm thick and was properly oriented for 6943 Å second harmonic generation.

4. Experimental results and discussion

Typical oscillograms of the input and output pulses are shown in Fig. 3. Note that the ouput signal is referred to the non-converted residual fraction of the 6943 Å light pulse transmitted through the KDP crystal. In Fig. 3 an incident pulse is

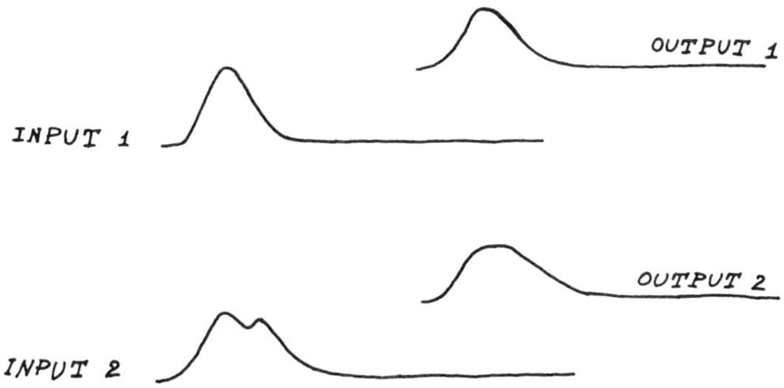

Fig. 3 Typical time dependence of laser pulse before (lower trace) and after (upper trace) travelling the non-linear crystal.

seen which has a very irregular shape; in the corresponding output pulse the irregularities are largely averaged as must be expected. The relative calibration of the two photodiodes used enable one to measure shot by shot the second harmonic efficiency conversion. Several series of a few hundreds shots have been performed with the different crystals, at laser firing repetition rates ranging from 1 shot/min up to 4 shots/min. The experimental results of each series have then been analyzed in order to put in evidence the change of intensity fluctuations between the input and output. In Fig. 4 are shown histograms of the input and output amplitude fluctuations for the case of the 0.7 cm thick KDP crystal. Note that both the histograms are normalized to the total input energy

$$E_{in} = \kappa \int N_i \, dV_i ,$$

where V_i is the photodiode maximum voltage pulse, $N_i dV_i$ the number of pulses of amplitude between V_i and $V_i + dV_i$ and κ is a constant depending on a) the quantum efficiency of the photodetector used, b) the geometrical factor of optical aperture of the same photodetector, c) the electrical impedance of the electronic circuit. As a consequence, the normalization factor in the output histogram is a function of the conversion efficiency. The total output energy may be expressed by

$$E_{out} = \kappa^1 \int N_o dV_o = \kappa \int N_o \, dV_i (1-\eta_i)$$

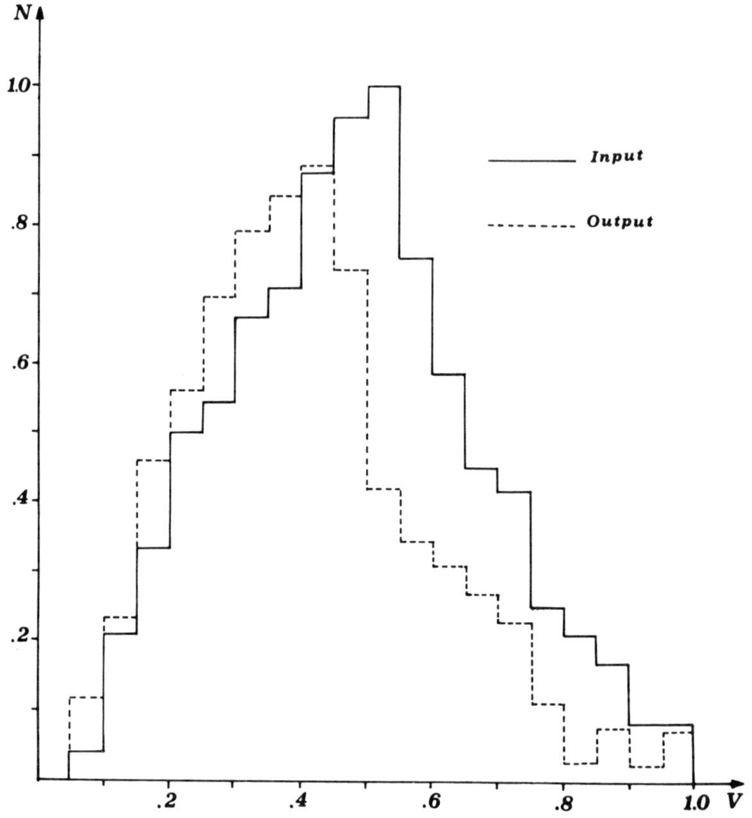

Fig. 4. Histograms of the input and output amplitude fluctuations. N is the pulse frequency and V is the photodiode voltage normalized at the maximum value ($3X10^8$ W/cm^{2}max. input flux).

where η_i is the value of the conversion efficiency. In our experiment η was fluctuating shot by shot between 5% and 40%.

The analysis of the histogram of Fig. 4 allows one to evaluate the mean intensity λ and the mean square deviation σ^2 for the input intensity distribution. We respectively find $\lambda = 1.4\times10^8$ Watt/cm^2 and $\sigma = 0.85\times10^8$ Watt/cm^2 and correspondingly $\beta\simeq1.7$. One can then estimate, for crystal length z of 0.7 cm, the dimensionless quantity y

$$y = z \left(\frac{8\pi^2\chi_2}{n\lambda_1}\right) \sqrt{\left(\frac{2\pi}{nc}\right)}\ \sigma^{\frac{1}{2}}$$

λ_1 being the wavelength of the laser light. By inserting the values $\lambda_1 = 6932$ Å and $\chi_2 = 6.10^{-9}$ e.s.u., we find $y \simeq 1$. Accordingly a rough interpolation of the data of Table I, gives a theoretical value of $\langle I^2(y)\rangle / \langle I(y)\rangle^2$ of about 1.04. On the other hand, the analysis of the experimental output distribution of Fig. 4 furnishes a value of $\langle I^2(y)\rangle / \langle I(y)\rangle^2$ of about 1.12. Since $\langle I_0^2\rangle / \langle I_0\rangle^2 = 1.18$, we see that there is an improvement of intensity stabilization less than the expected theoretical value. This discrepancy can be due to various factors, the main one being phase-mismatching. Anyway the purpose of this preliminary experiment is essentially to show in a qualitative way the presence of the stabilization effect.

References

1. M. Bertolotti, P. Di Porto, B. Crosignani - Phys. Rev. A *5*, 396 (1972).

2. J.A. Armstrong, N. Bloembergen, J. Ducuing and P.S. Pershan - Phys. Rev. *127*, 1928 (1962).

3. T.M. Christmas and J.M. Ley, Electronics Lett. *7*, 544 (1971).

4. N. Bloembergen, *Non-Linear Optics* (W.A. Benjamin Inc., New York, 1965) p. 89.

TWO-PHOTON TIME DISTRIBUTIONS IN MIXED LIGHT BEAMS

D. B. Scarl

Polytechnic Institute of Brooklyn, Brooklyn, N. Y.

Calculation of the second order correlation function for light that is a mixture of laser light and chaotic light of the same central frequency has been done by several groups. We have measured the two-photon counting rate in a mixed light beam and have found good agreement with these calculations. Half of the light in this beam originated in a stable single mode 633 nm He:Ne laser, and the other half in a low pressure He:Ne discharge tube. The d.c. discharge tube was about 20 cm long and was viewed end-on. Because the 633 nm Ne transition in a He:Ne discharge shows gain, the tube is a thick source, the line is slightly narrowed compared with a thin source, and the useful intensity is several times higher than any previously available narrow band chaotic source. For reasonable tube lengths, the amplitude distribution from this type of source is expected to be Gaussian.

The 1.3 nanosecond time resolution of the counting apparatus and the 1.9 GHz wide spectral density of the 633 nm line from the discharge tube were measured in separate experiments so that the two-photon time distribution measurements could be compared with the calculated distribution with no adjustments to the data.

In a mixed beam, the two-photon counting rate contains a term that is the product of the first order correlation function of the chaotic light with the (essentially constant) first order correlation function of the laser light. This "second order heterodyning" makes the counting rate linearly dependent on $f(\Delta t)$, the Fourier transform of the spectral density of the chaotic light, and allows direct measurement of $f(\Delta t)$ instead of $|f(\Delta t)|^2$, the quantity that is measured in two-photon experiments on chaotic light alone.

Recently developed photomultipliers with shorter time resolution and constant fraction timing discriminators may allow f(Δt) to be better resolved by reducing the spreading due to the detection system.

Two-Photon Experiment

In case of necessity, a photon can be defined as that particle that causes an electron to be emitted from the cathode of a photomultiplier. Although alternate descriptions of the emission process (and alternate definitions of the photon) are possible, this photon picture helps to unify our description of particles and may lead to a more concise understanding of the wave nature of other particles. One of the features of the photon as electron ejector is that the time at which an electron leaves a photocathode can be measured with high resolution. Time resolved photon detection experiments sometimes furnish useful examples for discussions of quantum mechanical measurement theory.

The average time between photons in a beam is inversely proportional to the intensity. The distribution of times between any two photons in a beam depends on the type of light being measured, its spatial and temporal coherence, its polarization, and on the time resolution, area, and noise of the detectors. Two-photon time measurements have been made on the 633 nm line from a He^3-Ne^{20} discharge for beams of chaotic light, laser light, and a mixture of 50% chaotic and 50% laser light. The two-photon time distribution in a beam of chaotic light shows the familiar photon bunching near zero time difference (the Hanbury Brown-Twiss effect); the distribution in mixed laser and chaotic light shows a similar bunching due mostly to the second order interference of laser and chaotic photons. In each case, the measured distributions agree well with predictions calculated using a delta function and Gaussian weight function for laser light and chaotic light respectively in a coherent state basis.

The d.c. discharge tube used as a source of narrow band chaotic light was a 22 cm long 2 mm bore laser tube without mirrors, viewed end on. The tube used in this way shows a gain of about 2% for 633 nm light entering one end, but has negligible spectral narrowing or departure from Gaussian statistics. This deep transparent source was able to produce a beam that, when collimated to 62% spatial coherence, had an intensity of $\sim 10^6$ photons per second. This high intensity is an improvement over previous thin chaotic sources and allows two-photon time distributions to be measured in shorter times. The full width at half maximum of the 633 nm line, measured with a scanning Fabry Perot interferometer was 1.9 GHz.

The laser was a conventional single mode He:Ne laser held near the center of the atomic 633 nm line by piezoelectric tuning of the cavity length. Its mode structure and position were monitored continuously during each run by a stable Fabry Perot inteferometer.

Figure 1 shows the optical arrangement. Details of the detectors, the electronics, and the calibration procedures have been published[1].

Results

For two noiseless, 100% efficient detectors with delta function resolution in space and time the two-photon counting rate at points separated by Δr and Δt is

$$W(\Delta r, \Delta t) = \mathrm{Tr}\rho\ E_1^+ E_1 E_2^+ E_2$$

where ρ is the field density matrix and E_1 and E_2 are the field operators evaluated at the points occupied by the detectors. If the coherent and Gaussian weight functions are convoluted to find ρ for a mixture of laser and chaotic light, and the resulting $W(\Delta r, \Delta t)$ is evaluated for $\Delta r = 0$, the two-photon counting rate can be written

$$W(0, \Delta t) = W_o\ [1 + a_1 f_1(\Delta t) + a_2 f_2(\Delta t)]$$

where W_o is the background rate at large Δt, f_1 is the squared Fourier transform of the line shape of the chaotic light normalized so that $f_1(0) = 1$, $f_2(\Delta t)$ is the similarly normalized Fourier transform unsquared, a_1 describes the (second order) interference between pairs of chaotic photons, and a_2 the interference between chaotic and laser photons.

The calculated values of a_1 and a_2 for a pair of efficient point (in space and time) detectors in laser, chaotic, and mixed beams are shown in Table 1.

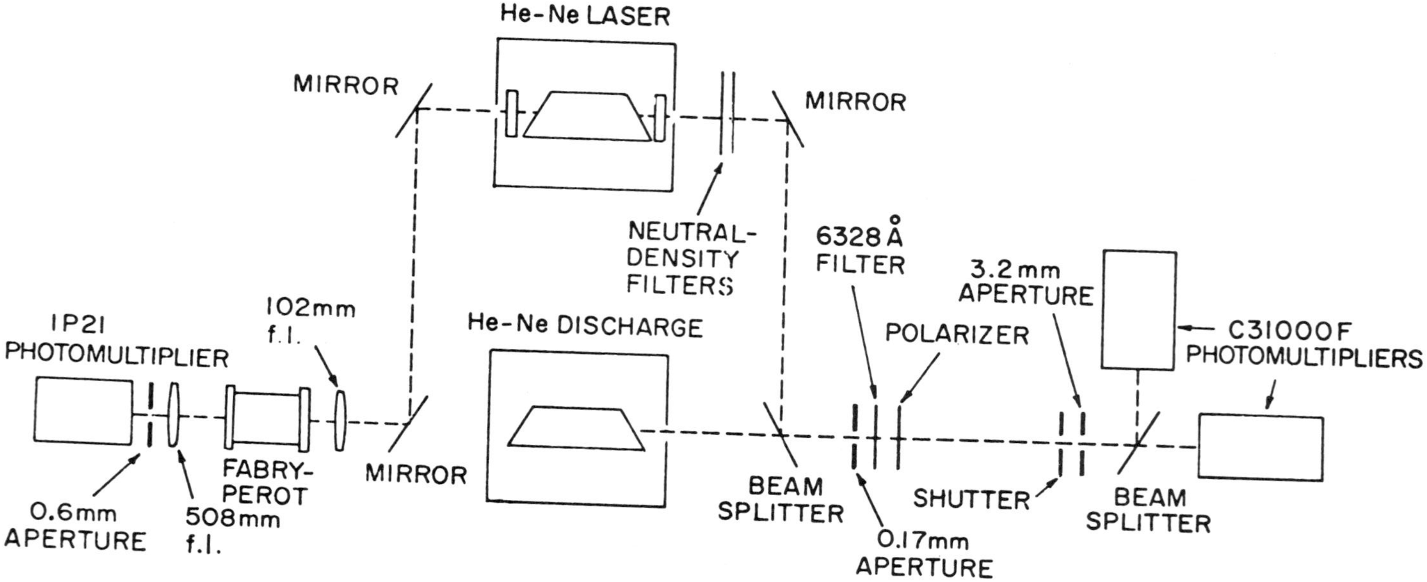

Fig. 1. The optical arrangement for two-photon time distribution measurements in a mixture of laser and chaotic light. Measurements in chaotic light alone and in laser light alone were made by removing the first beam splitter or replacing it by a mirror.

Point Detectors: $W(\Delta r, \Delta t) = W_0[1 + a_1 f_1(\Delta r, \Delta t) + a_1 f_2(\Delta r, \Delta t)]$

Real Detectors: $R(\Delta t) = R_0[1 + a_1' f_1'(\Delta t) + a_2' f_2'(\Delta t)]$

	Point Detectors a_1	Point Detectors a_2	Real Detectors $(a_1' + a_2')$	Experiment $(a_1' + a_2')$
Laser Light	0	0	0	0 ± .005
Chaotic Light	1	0	.15	.15 ± .01
50 % Laser Light 50 % Chaotic Light	.25	.50	.16	.17 ± .01

Table 1. Calculated and measured values of the coefficients describing second order interference between pairs of chaotic photons and chaotic and laser photons.

For real detectors, $W(\Delta r, \Delta t)$ must be integrated over the second spatial aperture, convoluted with the detector time resolution, and multiplied by the detector efficiencies. As a result, a_1 and a_2 are reduced to a_1' and a_2'; $f_1(\Delta t)$ and $f_2(\Delta t)$ are broadened to $f_1'(\Delta t)$ and $f_2'(\Delta t)$. The counting rate for real detectors is

$$R(\Delta t) = R_0[1 + a_1' f_1'(\Delta t) + a_2' f_2'(\Delta t)].$$

The value of $a_1' + a_2'$ for perfect detectors can be calculated from the counting rate at zero time difference and the counting rate at large (3 nanosecond) time difference. The experimentally measured values of $a_1' + a_2'$ are shown in Table 1. The errors quoted arise from the counting statistics (.005) and the uncertainty in the measured width of the chaotic light spectral line (.01).

Reference

1. G. Present and D. B. Scarl, Applied Optics *11*, 120 (1972).

PHOTO-COUNTING INVERSION IN PRESENCE OF DEAD TIME EFFECTS

C. L. Mehta

Indian Institute of Technology, New Delhi, India

The problem of obtaining information about the statistical properties of light from the photocounting statistics is of considerable interest and is referred to as the photocounting inversion problem. It is well known[1] that if one assumes the different photocounts to be statistically independent then the counting distribution p(n,T) is the Poisson transform of the probability density P(W) of the integrated intensity W:

$$p(n,T) = \int_0^\infty dW\, P(W)\, e^{-\alpha W} \frac{(\alpha W)^n}{n!}\ ; \quad W = \int_0^T I(t)dt. \tag{1}$$

Here α is the quantum efficiency and T is the counting interval of the detector. It has been shown earlier[2-4] that it is possible to invert the above relation and obtain the complete probability density of the fluctuating intensity of the light from the measurements of the statistical distribution of the photo-electrons. The modification of the counting formula in presence of the dead time effects, has been considered by Bedard[4,5] who obtained the following relationship between the counting distribution and the probability density of the fluctuating intensity of the light beam falling on the detector

$$P(n,T,\tau) = \int_0^\infty dW\, P(W)\, e^{-\alpha W} \frac{(\alpha W)^n}{n!} \left[1+\frac{\tau}{T} n(\alpha W-n+1)\right] + O\left(\frac{\tau^2}{T^2}\right) \tag{2}$$

Here τ is the characteristic dead time of the detector which is the recovery time of the detector after each registration of a count during which the detector does not respond to any external field. In the present investigation we obtain an explicit expression for P(W) in terms of $p(n,T,\tau)$ correct to the first order in τ/T. The integral relation (2) is first transformed into a differential equation satisfied by P(W) and this equation is then solved using Green's function techniques. Some specific cases are considered where $p(n,T,\tau)$ is known in the form of a closed analytic expression and P(W) is obtained for such cases. We also consider the approximation techniques used earlier[3,4] for obtaining P(W) for cases where only numerical values of $p(n,T,\tau)$ are known such as those obtained from experimental data. In such cases we assume that the experimental photocount distribution can be approximated as a sum of some smooth analytic functions, say exponentials of n,

$$p(n,T,\tau) = \sum_{j=1}^{m} a_j e^{-b_j n} . \qquad (3)$$

The constants a_j and b_j are determined numerically which best fit with the experimental data. Expression (3) is then used in the inversion formula and P(W) determined accordingly.

References

1. L. Mandel, Proc. Phys. Soc. (London) *72*, 1037 (1958); P. L. Kelly and W. H. Kleiner, Phys. Rev. *136*, A316 (1964); L. Mandel, E. C. G. Sudarshan and E. Wolf, Proc. Phys. Soc (London) *84*, 435 (1964).
2. E. Wolf and C. L. Mehta, Phys. Rev. Letters *13*, 705 (1964).
3. G. Bédard, Phys. Rev. *161*, 1304 (1967).
4. C. L. Mehta, *Progress in Optics* Vol. VIII, ed. E. Wolf, (North Holland Publishing Co., Amsterdam, 1970) p. 375.
5. G. Bédard, Proc. Phys. Soc. (London) *90*, 131 (1967).

IMAGE RECONSTRUCTION FROM THE MODULUS OF THE CORRELATION FUNCTION:

A PRACTICAL APPROACH TO THE PHASE PROBLEM OF COHERENCE THEORY*

D. Kohler and L. Mandel

University of Rochester, Rochester, N.Y.

1. Introduction

As is well known from the work of Michelson, van Cittert and Zernicke, the image of an incoherent luminous object can be reconstructed from measurements of the second order correlation function of the optical field far from the source. The correlation function is a complex function that contains information both about the amplitude and the phase of the electromagnetic field. However, in most practical situations the phase of the correlation function is difficult to measure, so that experiments are usually limited to a determination of the modulus alone. The lack of phase information has then to be overcome in other ways.

This phase problem, which has long been known in connection with the Michelson stellar interferometer, has its counterpart in other areas of physics, for example in x-ray crystallography, where the phase of the scattered field cannot be measured, and in particle scattering where the phase of the scattering amplitude remains unknown.

A number of suggestions have been made for tackling the phase problem in the optical domain. We have examined the practical feasibility of an interesting proposal due to Mehta[1], which is based on explicit use of the analytic properties of the correlation function $\Gamma(x)$, where x is a real parameter corresponding to the separation of two points in the optical field. If $|\Gamma(x)|$ could be determined experimentally for complex values of the argument x, then

*This work was supported by the Air Force Office of Scientific Research.

integration of the Cauchy-Riemann differential equations connecting the modulus and the phase of $\Gamma(x)$, would allow the phase of $\Gamma(x)$ to be derived from the modulus. Fortunately, such an analytic continuation of $\Gamma(x)$ into the complex x-domain is possible experimentally. It can be achieved by making measurements of the correlation function with exponential filters placed in front of the source.

We have made use of a folded wavefront interferometer in which the correlation function can be derived from a single photoelectric scan of the field. The information appears on computer tape, and is then processed by a computer, which reconstructs the image from the modulus of the correlation function alone.

2. Principle of the Method

For simplicity we shall confine ourselves to one-dimensional objects and images. According to a theorem of van Cittert and Zernicke[2], an incoherent, quasi-monochromatic light source of intensity distribution $I(y)$ in some object plane, will give rise to an optical field such that the second order correlation function $\Gamma(x)$ in the far field, at two points A and B separated by a distance x, is given by

$$\Gamma(x) = K \int_{-\infty}^{\infty} I(y)\, e^{ixyk_0/R}\, dy \quad . \tag{1}$$

Inversion of this relation yields the source distribution in terms of $\Gamma(x)$. Here $ck_0/2\pi$ is the mid-frequency of the light, R is the distance of the two (supposedly equidistant) points from the object plane and K is a real constant. The correlation function $\Gamma(x)$ is simply related to the interference fringes formed when light from the two points A and B is allowed to interfere[2]. In particular, the modulus of $\Gamma(x)$ is a measure of the modulation amplitude of the fringes. The phase of $\Gamma(x)$ is related to the positions of the fringe maxima and minima, and is much more difficult to determine in practice.

If the object is of compact support, i.e., if $I(y)$ vanishes outside some bounded region $a<y<b$, then $\Gamma(x)$ given by Eq.(1) is an analytic function of x over the entire complex x-plane. Hence, if we write

$$\left.\begin{aligned} x &= x_r + ix_i \quad , \\ \Gamma(x_r+ix_i) \equiv \Gamma(x_r,x_i) &= |\Gamma(x_r,x_i)| \exp i\phi(x_r,x_i) \quad , \end{aligned}\right\} \tag{2}$$

then $|\Gamma(x_r,x_i)|$ and $\phi(x_r,x_i)$ satisfy two differential equations of the Cauchy-Riemann type[1],

$$\frac{\partial|\Gamma(x_r,x_i)|}{\partial x_r} = -|\Gamma(x_r,x_i)| \frac{\partial\phi(x_r,x_i)}{\partial x_i} \tag{3}$$

$$\frac{\partial|\Gamma(x_r,x_i)|}{\partial x_i} = |\Gamma(x_r,x_i)| \frac{\partial\phi(x_r,x_i)}{\partial x_r} \quad . \tag{4}$$

It follows that, if we know $|\Gamma(x_r,x_i)|$ and its derivative $\partial|\Gamma(x_r,x_i)|/\partial x_i$, we can find $\phi(x_r,x_i)$ by integration of Eq.(4),

$$\phi(x_r,x_i) = \int_0^{x_r} |\Gamma(x_r',x_i)|^{-1} \frac{\partial}{\partial x_i} |\Gamma(x_r',x_i)| dx_r'$$

$$= \frac{\partial}{\partial x_i} \int_0^{x_r} \log|\Gamma(x_r',x_i)| dx_r' \quad . \tag{5}$$

The phase $\phi(x_r,0)$ of the correlation function on the real axis can therefore be found from a knowledge of the modulus $|\Gamma(x_r,x_i)|$ of the correlation function, provided this modulus is known also a little way off the real axis.

In order to determine $|\Gamma(x_r,x_i)|$ off the real axis, we observe that, if a filter of transmittance given by the exponential form $\exp(-x_i y k_0/R)$ is placed immediately in front of the source, then from Eq.(1), the correlation function is given by

$$K \int_{-\infty}^{\infty} I(y) \exp[iyk_0(x_r+ix_i)/R]dy = \Gamma(x_r+ix_i) \equiv \Gamma(x_r,x_i) \quad . \tag{6}$$

Hence we can obtain information about the correlation function off the real axis by making measurements with a number of different filters placed in front of the source. The exponent characterizing the filter need not be large, for we are generally interested in $|\Gamma(x_r,x_i)|$ only in the neighborhood of $x_i = 0$, where x_i is small. For this reason the exponential filter can generally be well approximated by a linear filter.

The modulus of $\Gamma(x_r, x_i)$ is found most conveniently from measurements of the visibility of interference fringes formed by light from the two points in question. Numerical integration of Eq.(5) then leads to the phase $\phi(x_r, x_i)$, from which the complex function $\Gamma(x_r, x_i)$, and hence the object $I(y)$, can be reconstructed. Some difficulties arise at points where $\Gamma(x_r, x_i)$ is zero, where the integrand has a logarithmic singularity and the phase undergoes a jump. The contributions from these points are best handled analytically rather than numerically.

3. Experimental Procedure

The apparatus for determining the correlation function of the optical field produced by the source is shown in Fig. 1. It

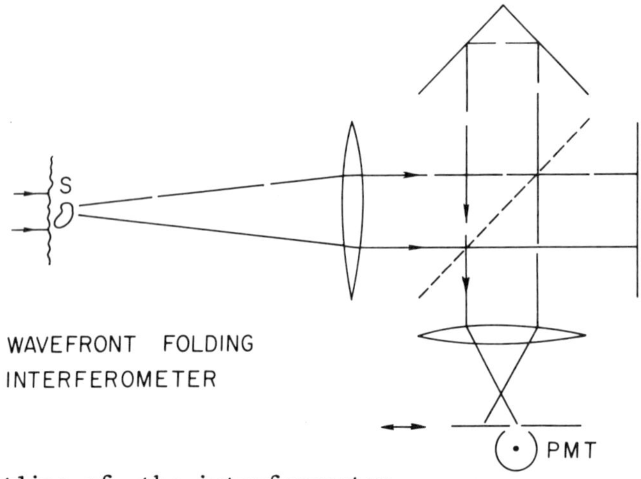

Fig. 1 Outline of the interferometer.

consisted of a folded wavefront interferometer[3], that permitted the visibility of the interference fringes, and therefore the modulus of the correlation function, to be determined from a single scan of a photodetector over the field of view. The source to be reconstructed was in the form of a transparency, that was illuminated from the rear by the laser light scattered from a rotating ground glass plate. This illumination simulated an incoherent source. The output signal from the photodetector was converted to digital form and registered on computer tape, and the data were then analyzed by computer.

The computer program itself was used to perform a certain amount of data smoothing. The Fourier transform of the squared envelope of the interferogram was truncated to eliminate high frequencies, and this function was then subjected to an inverse Fourier transformation, to generate a smoothed function $|\Gamma(x_r,x_i)|$. Figures 2a and 2b show the raw and smoothed data, respectively,

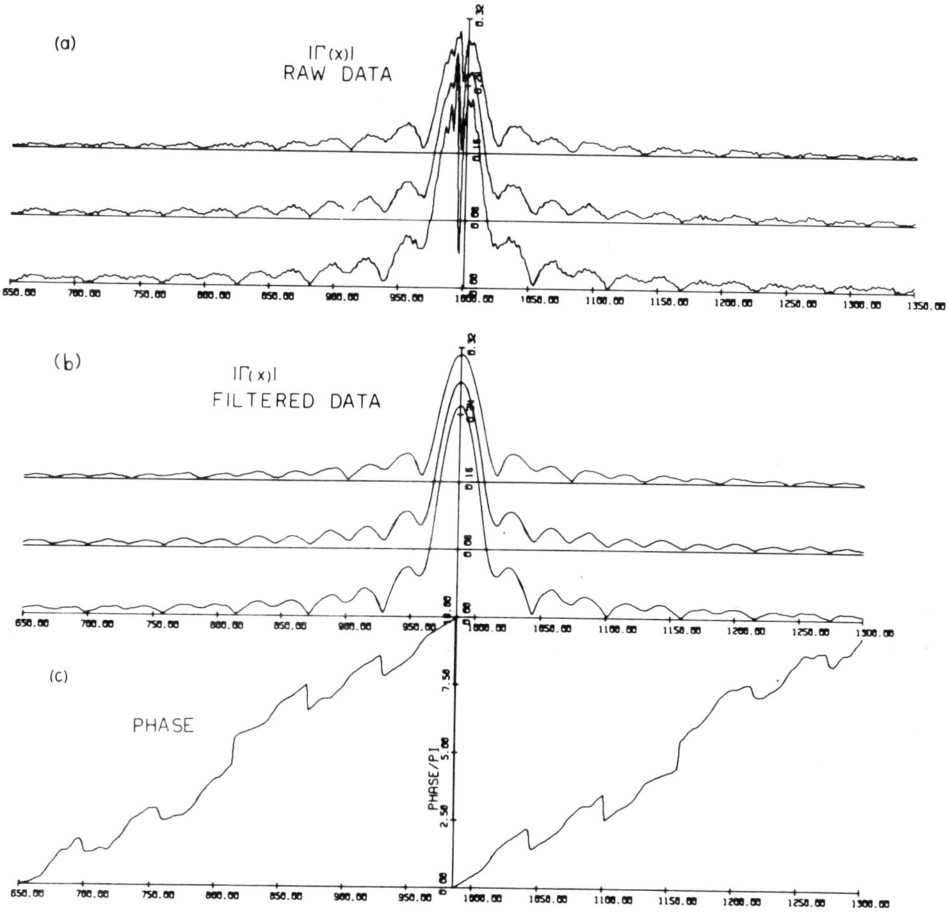

Fig. 2

(2a) Data for $|\Gamma(x_r,x_i)|$ derived from a scan of the interferometer field of view, once without filter, and twice with two different exponential filters. The source had a double step form as shown in Fig. 5.

(2b) Data for $|\Gamma(x_r,x_i)|$ after smoothing by computer.

(2c) The phase $\phi(x_r,0)$ derived from the smoothed $|\Gamma(x_r,x_i)|$.

corresponding to $|\Gamma(x_r,x_i)|$ for three different values of x_i, while Fig. 2c shows the phase function $\phi(x_r,0)$ that was derived from Eq.(5).

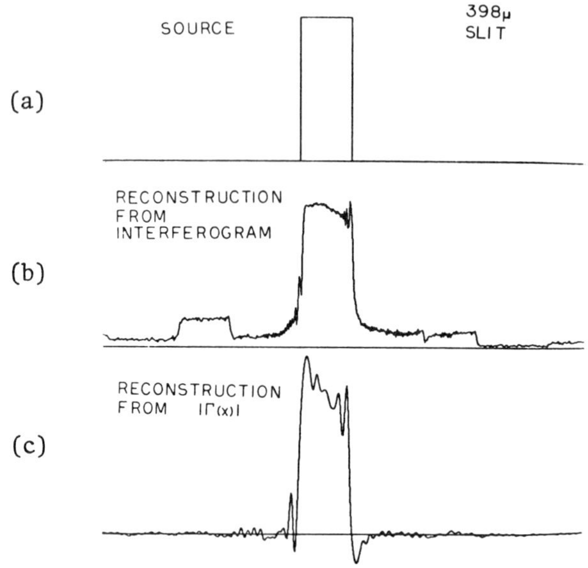

Fig. 3 An example of image reconstruction. (a) shows the object (idealized); (b) shows the image reconstructed from the measured complex correlation function $\Gamma(x)$; (c) shows the image reconstructed from the modulus of the correlation function by the exponential filter technique.

Figures 3, 4, and 5 show three examples of reconstructed images, together with the corresponding objects (under perfectly uniform illumination), and the images reconstructed from the complex correlation function as measured directly. The latter are prone to suffer from a number of ghost images, due to phase imperfections of the interferometer, which are noticeably absent from the images reconstructed from the modulus of the correlation function alone. Most of the remaining imperfections are almost certainly instrumental, and could be reduced by more careful attention to the design of the interferometer. It will be seen that the images reconstructed from $|\Gamma(x)|$ are superior in certain respects to those reconstructed from the complex correlation function $\Gamma(x)$. We feel that the principle of the method has been confirmed and its feasibility has been established.

(a)
(b)
(c)

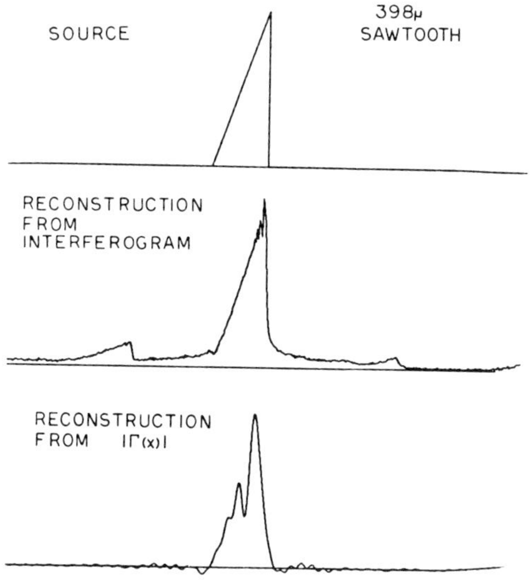

Fig. 4 An example of image reconstruction. (a) shows the object (idealized); (b) shows the image reconstructed from the measured complex correlation function Γ(x); (c) shows the image reconstructed from the modulus of the correlation function by the exponential filter technique.

(a)
(b)
(c)

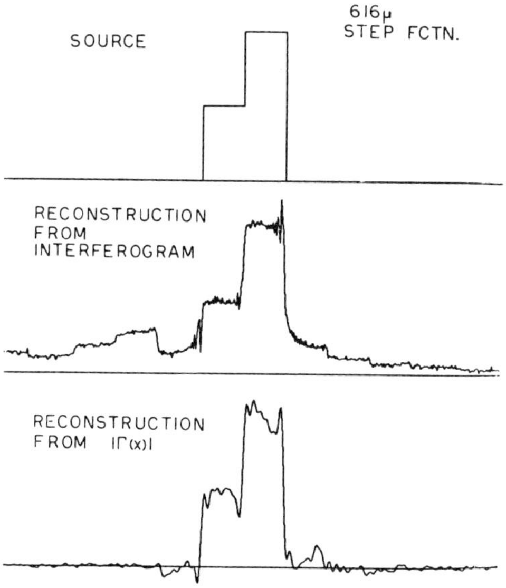

Fig. 5 An example of image reconstruction. (a) shows the object (idealized); (b) shows the image reconstructed from the measured complex correlation function Γ(x); (c) shows the image reconstructed from the modulus of the correlation function by the exponential filter technique.

A fuller account of this and other methods[4] of reconstructing images from the modulus of the correlation function will be published elsewhere.

References

1. C.L. Mehta, Nuovo Cimento *36*, 202 (1965).
2. See for example, M. Born and E. Wolf, *Principles of Optics, 4th ed.* (Pergamon Press, London, 1970) Chapter 10.
3. M.V.R.K. Murty, J. Opt. Soc. Am. *54*, 1187 (1964).
4. See also D. Kohler and L. Mandel, J. Opt. Soc. Am. *60*, 280 (1970).

PHOTON-CORRELATION SPECTROSCOPY

C. J. Oliver

Royal Radar Establishment, Malvern, Worcs, UK

1. Introduction

In order to put photon-correlation spectroscopy in its proper context, it is instructive to consider the various regimes of spectroscopy and the physical situations about which they provide information. Conventional spectroscopy of scattered light before the advent of the laser had two principal limitations. Firstly, the best resolution obtainable with interferometric methods was of the order of 0.1 to 1 cm^{-1}, in the case of a grating spectrometer and 10 MHz to 100 MHz in the case of a Fabry-Perot interferometer. Since the observed frequency shift of scattered light is essentially inversely proportional to the scale of the scattering process involved, it is apparent that these instruments were limited to studying atomic and molecular processes or, at best, such high frequency co-operative processes as scattering from acoustic phonons. Slower, long range processes, such as critical phenomena, processes involving large bodies, such as the diffusion of macromolecules, or processes involving bulk motion as in anemometry can only be studied by using the high resolution attainable with the application of the techniques of light beating, or its more recent derivative, photon correlation spectroscopy.

The second limitation of classical light scattering spectroscopy was that the sources used were of too large a bandwidth to achieve very high resolution and were of inadequate brightness to give a very strong scattered signal. A good single-isotope Mercury lamp, for example, would have a bandwidth of about 200 MHz[1] and would emit its power in all directions from a comparatively diffuse source area, making highly efficient collimation impossible. Thus the advent of the laser, providing an essentially single-frequency, well collimated, intense source, has enabled spectroscopists to study much weaker and much slower processes than were thus far possible.

Finally, the development of the electronic techniques of statistical spectroscopy, in which every photodetection is treated as an individual event, has allowed the greatest amount of spectral information possible to be derived from the scattered light.

2. Principles of Photon Correlation Spectroscopy

(i) Spectral Analysis Techniques

When optical radiation is incident on a dense medium, light can be considered to be scattered from dielectric constant fluctuations in the medium. This description is possible due to the long wavelength of optical radiation compared with the atomic spacing. For a small but finite volume, the wave equation for the scattered light from a general medium is given, using the notation of Nelson et al[2], in terms of a tensor operator $\underline{\underline{\alpha}}$ as

$$\underline{\underline{\alpha}}\,(-i\,\underline{\nabla},\omega)\,.\,\underline{E}\,(\underline{r},\,\omega)\;\equiv\;\left(\frac{c}{\omega}\right)^2\left[\underline{\nabla}\,\underline{\nabla}-\underline{\underline{1}}\,(\underline{\nabla}.\underline{\nabla})\right]\,.\,\underline{E}$$

$$-\;\underline{\underline{K}}(\omega)\,.\,\underline{E}=\underline{P}\,(\underline{r})\,\frac{e^{-i\omega t}}{\varepsilon_o} \qquad (1)$$

where $\underline{\underline{K}}$ is the dielectric tensor and P the polarization of the medium. Using Green's function techniques the solution of this equation for the field at a point $\underline{r}$ can be written as

$$\underline{E}(\underline{r},\omega_s)\;=\;\int_V\;\underline{\underline{G}}\,(\underline{r},\underline{r}')\,.\,\underline{P}(\underline{r}')\,d\underline{r}'\,\frac{e^{-i\omega_s t}}{\varepsilon_o} \qquad (2)$$

where ω_s is the frequency of the scattered light. The dyadic Green's function is given by

$$\underline{\underline{G}}(\underline{r},\underline{r}')\;=\;\int_{-\infty}^{+\infty}\frac{e^{i\underline{k}\,.\,(\underline{r}-\underline{r}')}\,d\underline{k}}{\underline{\underline{\alpha}}\,(\underline{k},\,\omega_s)\,(2\pi)^3}\;. \qquad (3)$$

The evaluation of this function for anisotropic media has been dealt with in detail by Lax. For the case of scattering from an isotropic medium the situation is simplified greatly so that the polarized component of the scattered field is given, in the far-field approximation, by the relation

$$E(\underline{r},t) \propto \delta\varepsilon\ (\underline{K},\ t)\ , \tag{4}$$

where $\underline{K}$ is the scattering vector defined by

$$\underline{K} = \underline{k}_s - \underline{k}_o\ , \tag{5}$$

$\underline{k}_o$ being the incident wave-vector and $\underline{k}_s$ the scattered wave vector. Thus, as is well known, a given scattering direction probes a given spatial Fourier component of the perturbation of the dielectric constant. Investigation of the correlation functions of the scattered field will therefore give information about the correlations present in the dielectric constant fluctuations.

Since photodetection can be described in terms of the quantum theory, the field $E(\underline{r}t)$ on the photodetector gives rise to a Poisson probability of obtaining a photodetection in a short time, Δt, proportional to the modulus squared of its positive-frequency component. In terms of an "instantaneous mean rate", $\bar{n}(t)$, of this Poisson distribution we have

$$\bar{n}(t) \propto \left|\ E^+(t)\ \right|^2 . \tag{6}$$

The output pulse train of photodetections thus consists of a Poisson process rate-modulated by temporal fluctuations in the scattered field.

A list of the various types of spectral analysis technique which could be applied to these photodetections is given in Table 1. These could operate in either the frequency or the time domain and could have either single- or parallel-channel data acquisition. The scanning electric filter, i.e. wave analyser, is a single-channel device, which has been widely used in light scattering spectroscopy since the early work of Cummins et al[3] in 1963. Delayed coincidence measurements, such as those of Morgan and Mandel[1], are single-channel time domain applications. Parallel channel wave analysers, i.e. banks of filters, are too inflexible to provide a satisfactory general purpose instrument over a wide

Type \ Domain	Time	Frequency
Single channel	Delayed coincidence	Wave Analyser
Parallel Channels	Autocorrelator	Bank of filters

TABLE 1. Different spectral analysis techniques which can be applied to the photodetector output signal.

range of frequencies. Temporal correlation was first applied in optical scattering spectroscopy by Cummins[4] in 1967, who used an analogue correlator in measurements of critical phenomena. Development of a parallel-channel correlator which utilized the digital nature of the photodetections obviously yields the greatest possible statistical accuracy since the information contained in each photodetection is used.

(ii) Photon-correlation

Let us consider the operation of such a photon-correlator in more detail. The discrete narrow pulses contained in the detector output preclude the use of sampling correlation techniques since these would be inordinately wasteful. The most convenient approach, included in Figure 1, is to count the number of detections occuring during contiguous sample times of duration T, chosen to be short compared with the correlation times under study. The normalized autocorrelation function of photon-counting fluctuations in this time, can be written as

$$g_T^{(2)}(t,t') = \frac{\langle n_T(t)\, n_T(t')\rangle}{\langle n_T(t)\rangle^2} \quad . \tag{7}$$

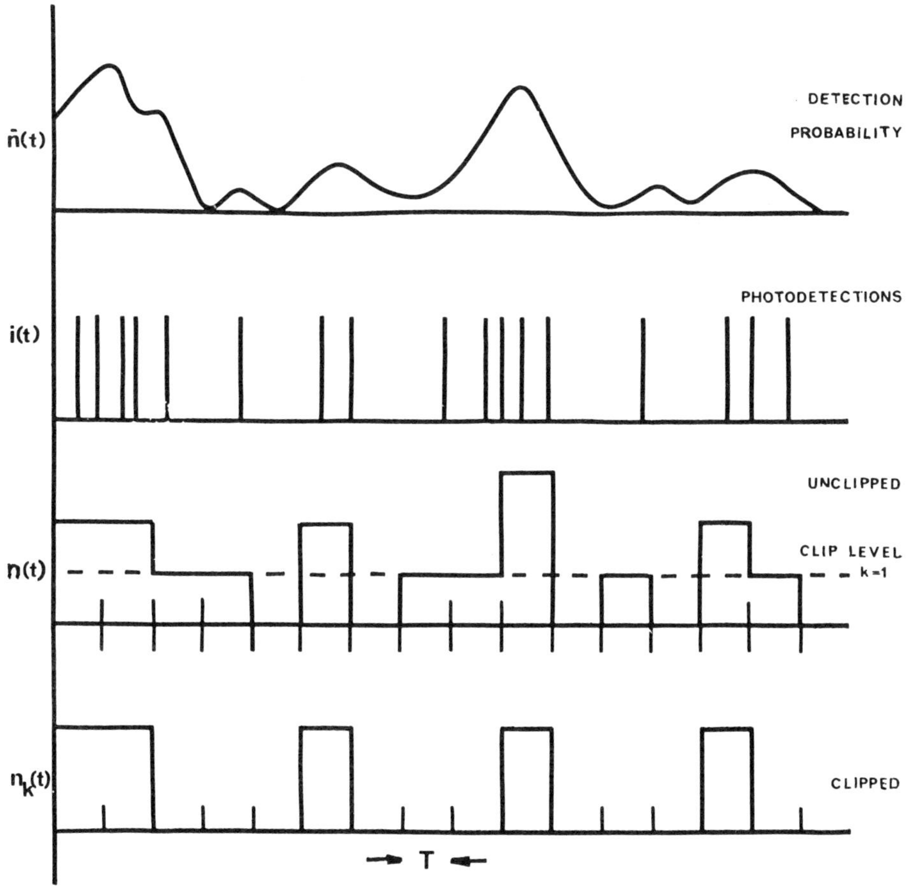

Fig. 1. Photodetection of scattered light together with the sampling and clipping scheme used in clipped autocorrelation.

For optical fields having Gaussian statistics and symmetrical spectra, as are generally encountered, this second-order correlation function is related to the first-order function by the equation

$$g^{(2)}(t,t') = 1 + \left| g^{(1)}(t,t') \right|^2 \tag{8}$$

due to Siegert[5]. Information about the first-order correlation function, which is the Fourier transform of the optical spectrum, can therefore be obtained from measurement of the second-order function in these cases. The limitations of Gaussian statistics and spectral symmetry can be overcome by using heterodyne detection, in which case the observed second-order autocorrelation function contains a term which is proportional to the first-order function and which can be made dominant. For stationary processes only the relative delay between the times t and t' is required to characterize the correlation function.

In principle therefore an auto-correlation function of photon counting fluctuations could be calculated by counting the photodetections occuring during each of a number of consecutive samples of duration T, and multiplying the signal channel by channel with the delayed information. Practically, one need use comparatively few delay channels (about 20) but it is advantageous to use many samples to increase the accuracy of the correlation coefficients obtained. The normalized correlation coefficient for the r'th delay channel, i.e. for a delay rT, would then be given by

$$g_T^{(2)}(rT) = \frac{N \sum_N n_T(rT)\, n_T(0)}{\left[\sum_N n_T(0)\right]^2} \tag{9}$$

where N is the number of samples. By recording simultaneously the coefficients for different delay times rT, the complete correlation function can be obtained.

(iii) Clipping

There is however a difficulty in taking the product $n_T(rT)n_T(0)$ associated with the fact that multiplication of numbers greater than one is time consuming and therefore limits the ultimate speed of such a correlator. In the micro-wave regime Van Vleck and Middleton[6] showed that the autocorrelation function for a signal with Gaussian statistics formed by considering only the zero crossing of the electric field was related to the true autocorrelation function by the equation

$$g^{(1)}_{\text{clipped}}(\tau) = \frac{2}{\pi} \arcsin\left[\, g^{(1)}(\tau)\right]. \tag{10}$$

Equivalent calculations in which clipping is applied to the analysis of photodetection in optical spectroscopy, were first performed by Jakeman and Pike[7]. As shown in Figure 1, a clip level k is selected so that the "clipped" signal $n_k(t)$ follows the conditions:

$n_k(t) = 1 \qquad \text{if } n(t) \geqslant k$

$$n_k(t) = 0 \qquad \text{if } n(t) < k,$$

where the suffix T denoting integration over the sample time has been dropped. By using the one-bit quantised clipped signal in the delay channel the complexity of the multiplication is reduced to a gating function, as shown in Figure 2, since multiplication

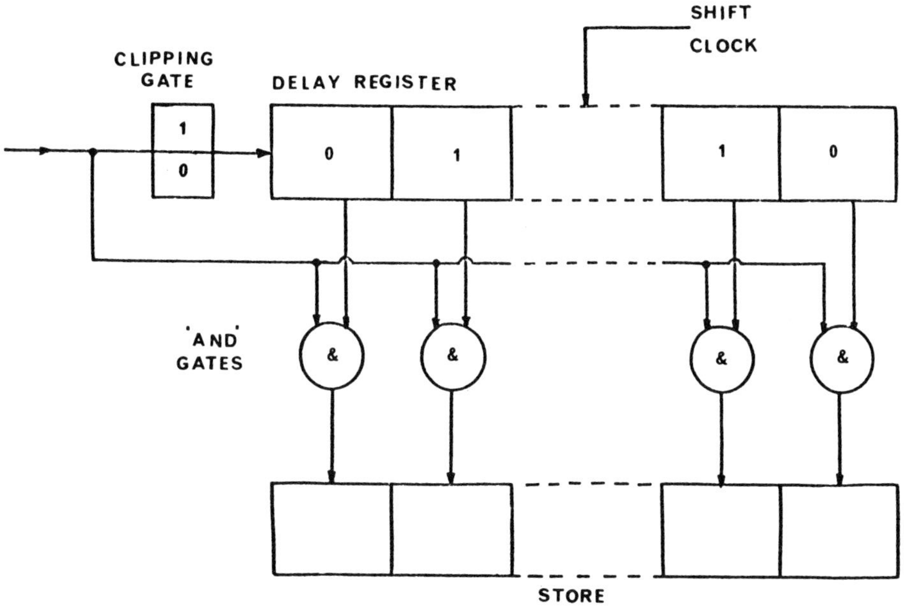

Fig. 2. A block diagram of the single-clipping correlator.

of n(0) by $n_k(rT) = 1$ or 0 is equivalent to gating n(0) by $n_k(rT)$. This reduction in complexity leads to a considerable increase in operating speed. The single-clipped autocorrelation coefficient for delay rT is given by

$$g_k^{(2)}\ (rT) = \frac{N \sum n_k\ (rT)\ n(0)}{\sum\limits_N n_k(0) \sum\limits_N n(0)} \;. \tag{11}$$

The single and double-clipped autocorrelation functions (the latter being formed by using the clipped signal in each channel) can be expressed, as derived by Jakeman and Pike[7], in terms of the optical field autocorrelation function $g^{(1)}(\tau)$ by the relations

$$g_k^{(2)}(\tau) = \frac{\langle n_k(\tau)\ n\ (0)\ \rangle}{\langle n_k(0)\rangle \langle n(0)\rangle} = 1 + f(A,\bar{n},k,T)\ \frac{1+k}{1+\bar{n}}\ \left|g^{(1)}(\tau)\right|^2 \qquad (12)$$

and

$$g_{oo}^{(2)}(\tau) = \frac{\langle n_o(\tau)\ n_o(0)\rangle}{[\ \langle n_o(0)\rangle\ ]^2} = \frac{1 + f'(A,\bar{n},k,T)\ \frac{1-\bar{n}}{1+\bar{n}}\ \left|g^{(1)}(\tau)\right|^2}{1 - \left(\frac{\bar{n}}{1+\bar{n}}\right)^2 \left|g^{(1)}(\tau)\right|^2} \qquad (13)$$

for single-clipping at k and double-clipping at zero respectively. The correction factors f,f'(A, $\bar{n}$, kT) are constants affecting the ratio of the spectral term to the background and are well understood from both theoretical and experimental investigations for the case of a single Lorentzian spectrum [7][8][9]. In addition, the accuracy with which linewidths can be obtained for single Lorentzian spectra have been studied under different conditions and good agreement has been obtained between theory and experiment [10][11][12]. A commercial photon-correlator using these principles is now available[13] and the technique is finding wider and wider application.

3. Applications

The wide range of sample times available with digital correlation techniques enables any phenomenon giving rise to a frequency shift between 1 Hz and 5 MHz to be studied. Some examples of such processes are given in Table 2. The use of laser scattering where possible to investigate physical processes offers three main advantages over other techniques. Firstly, the laser does not usually influence the process under study in the way normal probes would, unless too much power is used or the process absorbs the laser wavelength. Secondly, the laser transmitter and photodetector receiver can be remote from the region under study, so that places which are not readily accessible can be studied. Thirdly, very small regions can be investigated using a focussed laser beam and well collimated receiver optics.

One application of photon-correlation spectroscopy which has received much attention is the determination of diffusion coefficients of biological macromolecules by light scattering from the macromolecular Brownian motion. This approach was first adopted by Yeh and Cummins[14], who determined the translational diffusion coefficient of polystyrene latex spheres using heterodyne detection and a wave analyser. Intensity fluctuation measurements also using

REGIME	EXAMPLES
LARGE SCALE OR COLLECTIVE MOTION	Optical radar Vibration analysis Nerves and cells Critical phenomena Liquid crystals
ANEMOMETRY	Flow Turbulence
MOTILITY	Swimming algae, spermatozoa or bacteria
DIFFUSION	Macromolecules Polymers

TABLE 2. Examples of applications of photon-correlation spectroscopy.

wave analysis, on a range of samples were later reported[15][16][17], leading to the measurement of both translational and rotational diffusion coefficients for tobacco mosaic virus[18]. However the application of parallel-channel photon-correlation techniques to this type of experiment has enabled much lower macromolecule concentrations to be used, yielding results which are, generally speaking, more meaningful to the biochemist[19][20][21][22][23]. A typical result obtained recently from adenovirus[23] is shown in Figure 3. From this autocorrelation function, measured for 40 secs with a HeNe laser power of 25 mW and a solution concentration of 23 μgms/mℓ, an accuracy of 1% was obtained. Even with molecular weights of about 100, similar accuracy has been achieved in 20 minutes using a 500 mW argon ion laser with a solution concentration of 100 mgs/mℓ. Other available light scattering methods would have taken at least 100 times as long to achieve the same accuracy.

Another interesting field is the study of mobility of swimming bodies such as spermatozoa or algae by light scattering; such experiments were performed first by Berge et al[24]. The behaviour of these swimming bodies can be expressed in a way analogous to the kinetic theory of gases, since the bodies swim in straight lines with occasional random changes of direction. These swimming velocities also have approximately a Maxwellian distribution. Since

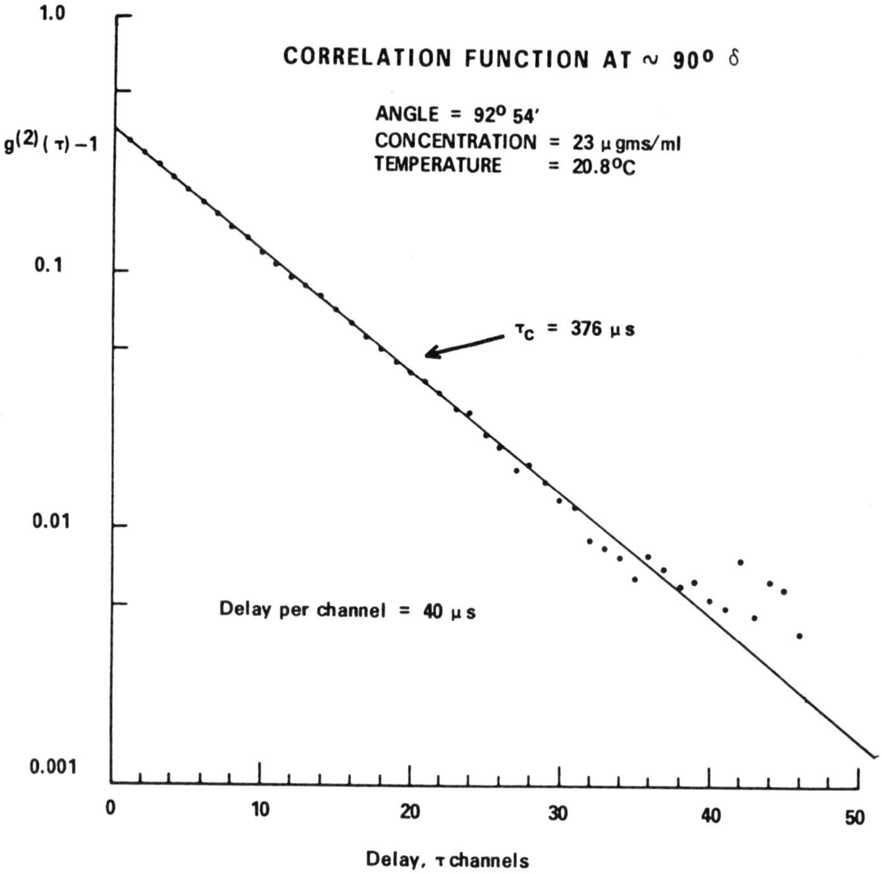

Fig. 3. The exponential part of the observed autocorrelation function $[g^{(2)}(\tau)-1]$ for the intensity fluctuations of laser light scattered from a adenovirus (conc 23 μgms/mℓ) at an angle of 92°54'.

the motion of individual bodies is with constant velocity over a time long compared with the Doppler shift frequency, the spectrum of the scattered light is related to the Doppler spectrum and hence the velocity distribution. For such a velocity distribution a Gaussian spectrum, and hence also a Gaussian autocorrelation function, would be observed, as shown in Figure 4. Here the autocorrelation function of the intensity fluctuations from scattering by marine algae (Duniella) is shown. Diminished mobility will be apparent as broadening of the correlation function. If the swimming motion is completely stopped by poisoning the algae with iodine, for example, a Brownian motion diffusion occurs from which the algae radius can be deduced. The increase in the time scale

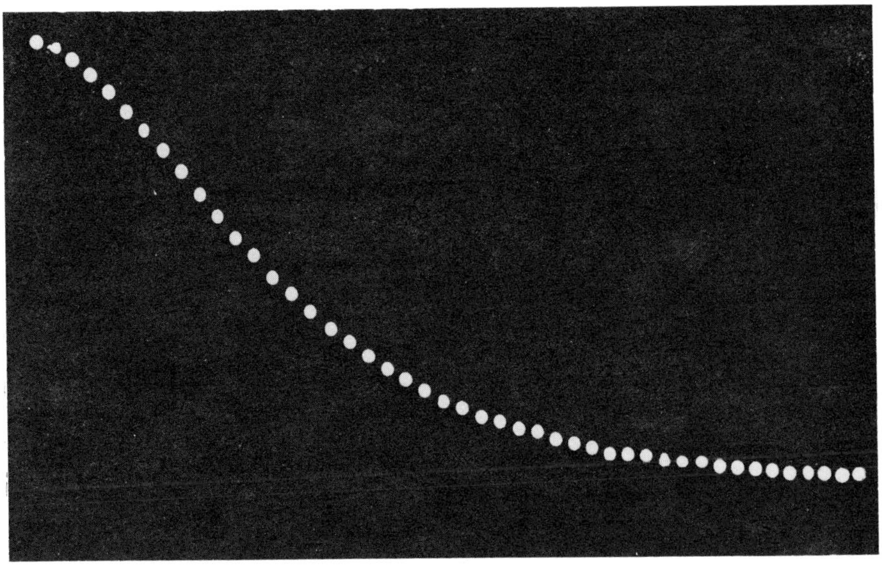

Fig. 4. The autocorrelation function of the intensity fluctuations of light scattered at 35° from swimming algae. The sample time was 100 μS.

is very marked as can be seen in Figure 5. In addition it can be seen that the form of the correlation function has changed from Gaussian to exponential.

Another interesting application is in the field of anenometry. Investigation of turbulent flow by both intensity-fluctuation and heterodyne (Doppler) spectroscopy[25] is now commonly used. More recently the intensity-fluctuations arising from unseeded air flowing through a fringe system formed by mixing two laser beams has been analysed with a photon-correlator[26]. The dust particles and natural aerosols present in the atmosphere provide adequate scattering intensity. The application of this technique to measure wind tunnel velocities is shown in Figure 6[27]. In the wind tunnel the air was actually filtered to be considerably cleaner than the atmosphere, so that the photodetection rate corresponded to an average of one count every one hundred fringes. The characteristic sinusoidal variation, shown in Figure 6, was obtained with a wind velocity of Mach 0.5. Sample times of 100 ns were used which illustrates the importance of fast correlators. The sinusoidal fluctuations are superimposed on a Gaussian type of background, which is dependent on the time taken for a particle to traverse the complete laser beam rather than the fringe spacing. With the existing geometry, and equipment having 50 nS resolution, velocities

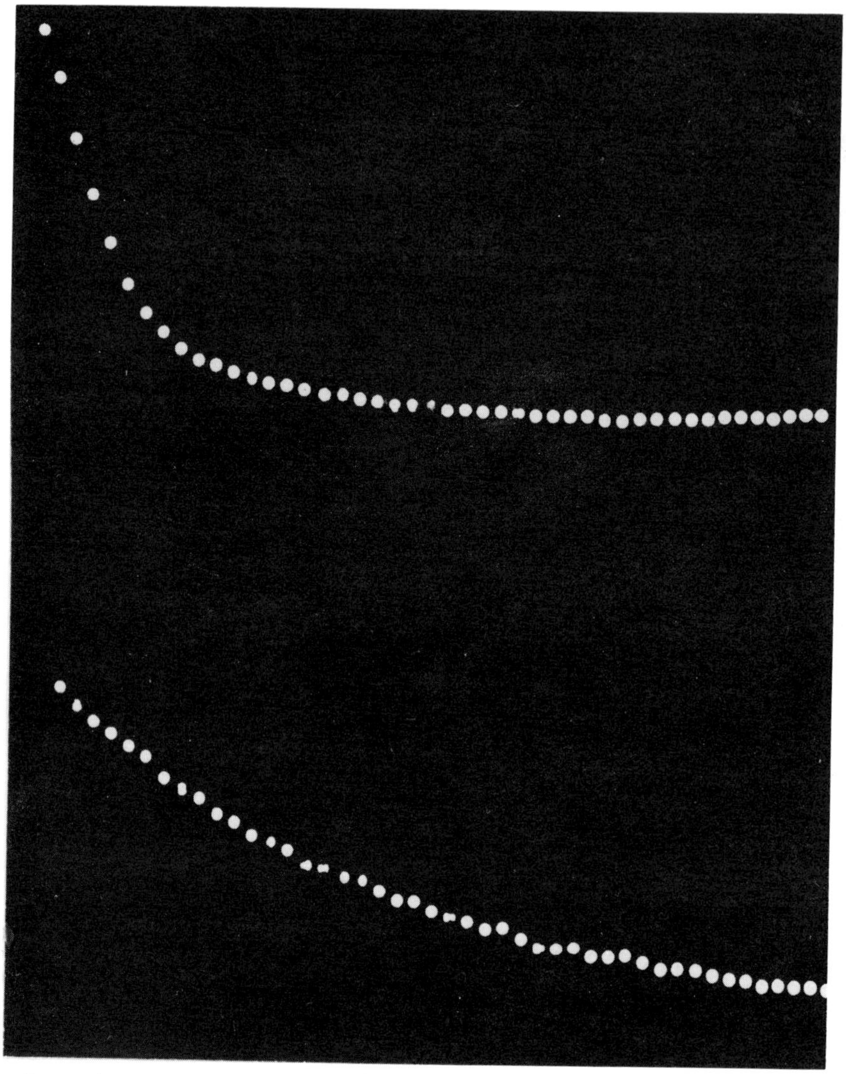

Fig. 5. The change of the observed autocorrelation function on adding iodine to kill the algae. The Gaussian shape corresponds to the swimming motion; the exponential to diffusion of dead algae after the iodine was added. The sample time was 100 μS and the scattering angle was 35°.

up to Mach 2 were measureable without seeding the air flow at all. The advantages of such a remote probe, which can study small regions (∿ 1 mm cube) in a wind tunnel without disturbing the flow in any way, are obvious.

Fig. 6. Measurement of velocity in a wind tunnel using the transit time through optical fringes. The wind speed was Mach 0.5 and the sample time was 100 nS.

4. Conclusions

A careful study of the factors relating to the measurement of intensity fluctuation spectra has shown that parallel-channel clipped photon correlation offers an optimum solution, in terms of speed and cost, to light scattering measurements in the fields of physics, chemistry, biology and engineering. The advantages of real-time processing of effectively all the incident signal, enabling measurements to be made in the minimum possible time, has opened still wider fields of application of light-scattering spectroscopy.

REFERENCES

1. B. L. Morgan and L. Mandel, Phys. Rev. Letters *16*, 1012 (1966).
2. D. F. Nelson, P. D. Lazay and M. Lax, *Proc. Int. Conf. Light Scattering in Solids, Paris*, ed. M. Balkanski (Flammarion Sciences, Paris) p. 477 (1971).
3. H. Z. Cummins, K. Knable, L. Gampel and Y. Yeh, Appl. Phys. Letters *2*, 62 (1963).
4. H. Z. Cummins, *International School of Physics' Enrico Fermi'*, Varenna, 1967, ed. R. Glauber (Academic Press, New York, 1968).
5. A.J.F. Siegert, MIT Rad. Lab Report No. 465 (1943).
6. J.H. VanVleck and D. Middleton, Proc. IEEE, *54*, 2 (1966).

7. E. Jakeman and E. R. Pike, J. Phys *A*, *2*, 115 (1969).
8. E. Jakeman, J. Phys *A*, *3*, 201 (1970).
9. E. Jakeman, C. J. Oliver and E. R. Pike, J.Phys *A*, *4*, 827 (1971).
10. E. Jakeman, E. R. Pike and S. Swain, J. Phys *A*, *3*, 155 (1970).
11. E. Jakeman, E. R. Pike and S. Swain, J. Phys *A*, *4*, 517 (1971).
12. A. J. Hughes, E. Jakeman, C. J. Oliver and E. R. Pike (1972), to be published.
13. Malvern Digital Correlator System, K7023, Manufactured by Precision Devices and Systems Ltd, Spring Lane, Malvern, Worcs, UK.
14. Y. Yeh and H. Z. Cummins, Appl. Phys. Letters *4*, 176 (1964).
15. N. C. Ford and G. B. Benedek, Phys. Rev. Letters *15*, 649 (1965).
16. F. T. Arecchi, Phys. Rev. *163*, 186 (1967).
17. S. B. Dubin, J. H. Lunacek and G. B. Benedek, Proc. N.A.S., *57*, 1164 (1967).
18. H. Z. Cummins, F. D. Carlson, T. J. Herbert and G. Woods, Biophys *J*, *9*, 518 (1969).
19. R. Foord, E. Jakeman, R. Jones, C. J. Oliver and E. R. Pike, *Southampton Conf. on Lasers and Optoelectronics*, IERE Conf. Proc. *14* (1969).
20. R. Foord, E. Jakeman, C. J. Oliver, E. R. Pike, R. J. Blagrove, E. Wood and A. R. Peacocke, Nature, 227, 242 (1970).
21. C. J. Oliver, E. R. Pike, A. J. Cleave, and A. R. Peacocke, Biopolymers, *10*, 1731 (1971).
22. E. Wood, W. H. Bannister, C. J. Oliver, R. Lontie and R. Witters, Comp. Biochem. Physiol. *40B*, 19 (1971).
23. C. J. Oliver, K. Shortridge and G. Belyavin (1972), to be published.
24. P. Berge, B. Volochine, R. Billard and A. Hamelin, C.R. Acad. Sci., Paris, 265, 889 (1967).
25. Bourke et al, Phys. Letters *28A*, 692 (1969).
26. E. R. Pike, J. Phys *D*, *5*, L23 (1972).
27. J. Abbis, T. Chubb, A.R.G. Mundell, P.R. Sharpe, C.J. Oliver and E.R. Pike, J. Phys *D*, to be published.

MULTI-TIME INTERVALS DISTRIBUTIONS OF PHOTOELECTRONS FROM STATIONARY PSEUDO-THERMAL LIGHT WITH GAUSSIAN SPECTRUM

Chérif Bendjaballah

Université de Paris, Orsay, France

1. Theory

We study the probability distribution $w_n(o/t)$ of intervals between photoelectrons, the first occuring at the origin of time and the (n+1) th at time t.

To solve this problem it is convenient to start with the recurrence relation between $w_n(o/t)$ and $p(n,t)$.

$p(n,t)$ is the probability of registering n photoelectrons in a time interval (o,t) and is given by the Mandel's formula

$$(1) \quad p(n,t) = \left\langle \left(\int_o^t I(\theta)d\theta \right)^n \frac{1}{n!} \exp\left(- \int_o^t I(\theta)d\theta\right) \right\rangle ,$$

where $I(\theta)$ is the instantaneous intensity.

It can be proven that $w_n(o/t)$ may be written as

$$(2) \quad w_n(o|t) = \frac{1}{\langle I \rangle} \left\langle I(o)I(t) \left(\int_o^t I(\theta)d\theta \right)^n \frac{1}{n!} \exp\left(-\int_o^t I(\theta)d\theta\right) \right\rangle ,$$

where $\langle I \rangle$ is the mean value of $I(\theta)$.

Then it can be easily shown for $n \geqslant 2$

$$(3) \quad \frac{d^2 p(n,t)}{dt^2} = \langle I \rangle \{ w_n(o|t) - 2w_{n-1}(o|t) + w_{n-2}(o|t) \} ,$$

with for n = 0 and 1,

$$\frac{d^2 p(o,t)}{dt^2} = \langle I \rangle \, w_o(o|t),$$

$$\frac{d^2 p(1,t)}{dt^2} = \langle I \rangle \{ w_1(o|t) - 2w_o(o|t) \} .$$

Here, we consider the time interval (o,t) to be arbitrary compared to the coherence time τ_c, of the light field. In this case, it is well known that the problem of finding p(n,t) consists in solving the Fredholm integral equation

$$(4) \quad \int_{-t/2}^{t/2} \gamma(\mu-\tau) \, \phi_k(\tau) \, d\tau = \lambda_k(t) \, \phi_k(\mu),$$

where λ_k is the eigenvalue and $\gamma(\tau)$ the normalized autocorrelation function of the field.

The probability distribution associated with each λ_k will be given by

$$(5) \quad p_k(n,t) = (\lambda_k \langle I \rangle t)^n \, [(1 + \lambda_k \langle I \rangle t)^{n+1}]^{-1} .$$

The procedure necessary to obtain the resultant p(n,t) will involve the discrete convolution of $p_k(n,t)$.

So we complete the calculations of $w_n(o|t)$ by summations which we can easily find from eq. (3):

$$(6) \quad w_n(o|t) = \langle I \rangle^{-1} \{ \sum_{m=0}^{n} \sum_{\ell=0}^{m} \frac{d^2}{dt^2} p(\ell,t) \} .$$

a) Exponential profile case

Here we have the exact solution of eq. (4). The expansion of the final result of p(n,t) was given by G. Bedard [1]. Then we can compute eq. (5) and we derive $w_n(o|t)$ by the use of eq. (6).

b) Gaussian profile case

In this case the integral equation (4) must be solved numerically. Using the recently described [2] Lachs' method that converts the integral equation to a matrix equation, we have obtained good approximate photocount statistics p(n,t). So we have computed the numerical values of $w_n(o|t)$ by the double summations indicated in eq. (6).

All the calculations in the case (a) and (b) have been performed with the aid of an IBM 360 Computer.

Of course, when we introduce an exponential profile for $\gamma(\tau)$ in the computer programm, we find again, with good accuracy, the results deduced from Bedard's calculations.

Fig. 1 shows the computed curves on semi-logarithmic scales for these two cases. The numerical values of the parameters are n=7; $\langle I\rangle = 5.10^5$pulses/s and τ_c = 5.65 μs.

2. Experimental verification

The experimental arrangement for the detection and recording system needed to measure $w_n(o|t)$ consists of a scaler of 12 fast bistables, where each output can be directed to a time-to-amplitude converter and the data are stored in a multichannel analyzer.

The laser used is an He-Ne Spectra Physics model 119 working above threshold.

The laser beam has been focused by a lens into a rotating ground glass disk with small irregularities of average size which are spaced approximately 1 μm for generating the pseudo-thermal light with Gaussian spectrum profile.

First of all, we consdier the particular case where the successive intervals are statistically independent.

As a matter of fact, when the rotation of a ground glass disk is stopped, we can still detect the laser light. One shows the experimental points of the $w_n(o/t)$ distribution with n = 63 and $\langle I\rangle = 5.10^6$ pulses/s are very close to the Gaussian law. Therefore

Fig. 1

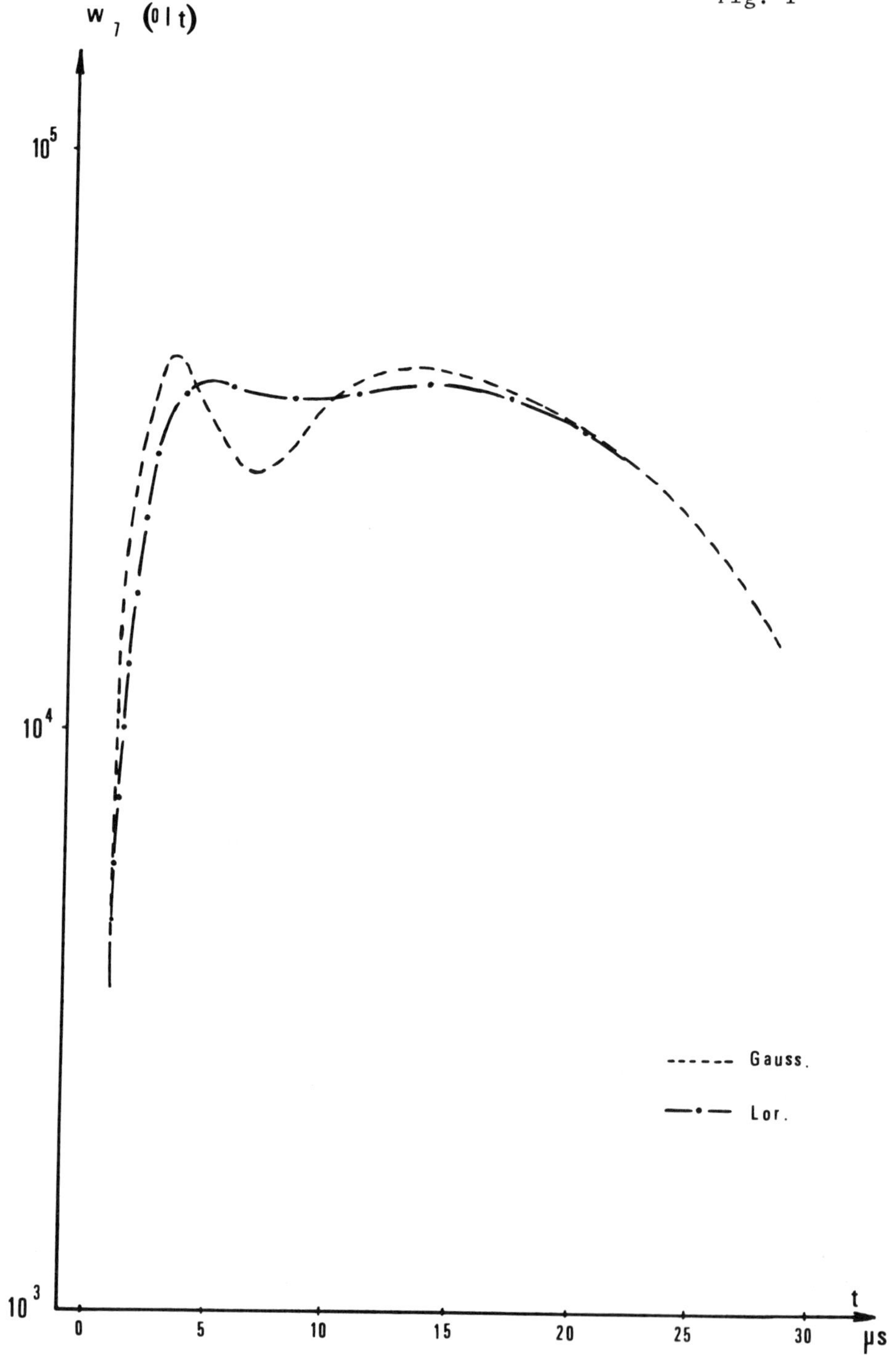

our results constitute a good agreement with the central limit theorem.

After that we can perform our measurements on pseudo-thermal light. The experimental results for n = 7, 15 and 31 and with $\langle I \rangle = 5.10^6$ pulses/s, $\tau_c = 2{,}5.10^{-6}$ s are recorded on semi-logarithmic scales in Fig. 2 and we compare it with the theoretical distributions represented with full lines.

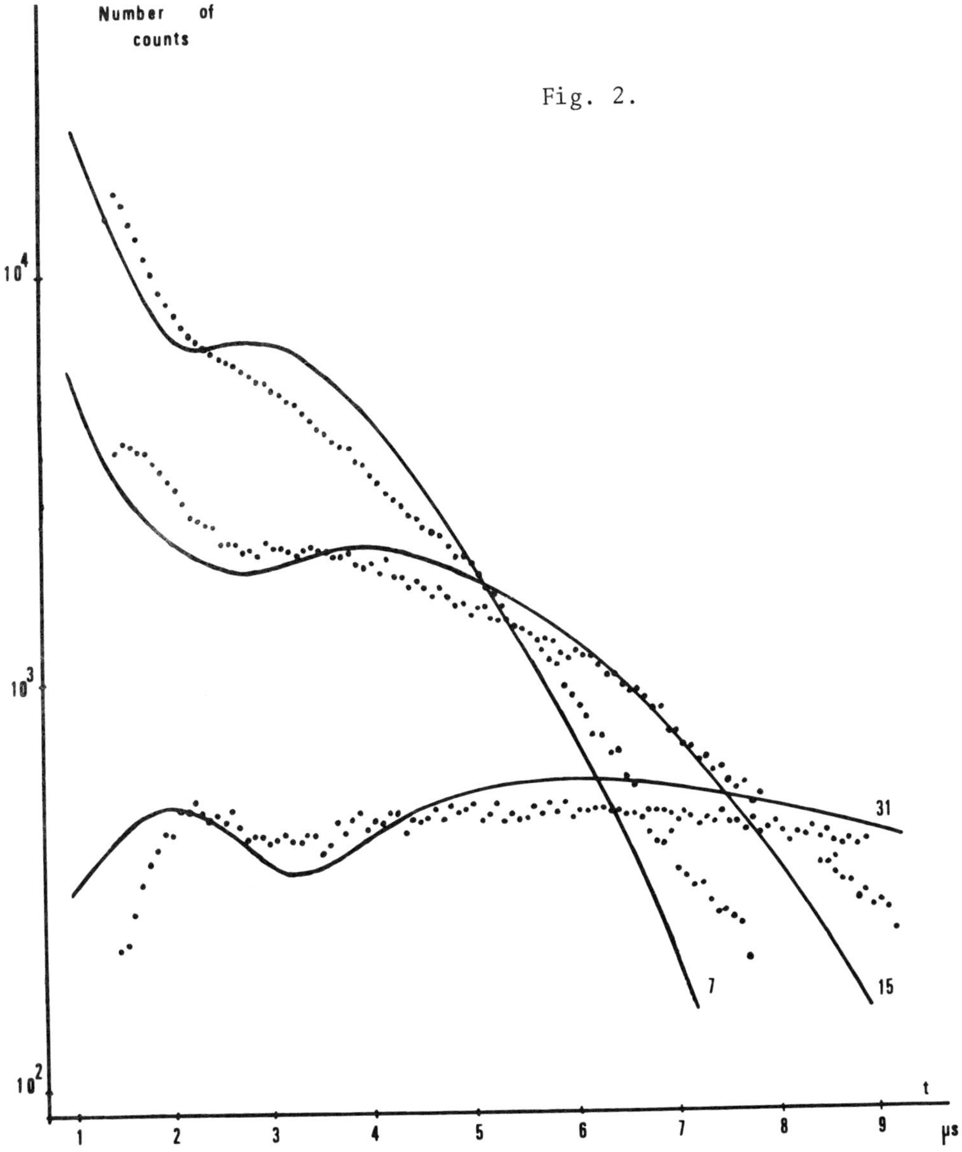

Fig. 2.

It may be concluded that the experimental points and the solutions of eq. (2) are sufficiently close.

Moreover it appears that the sum of many (31) dependent successive intervals is not Gaussian and it can be shown that the first and second peaks appearing in the $w_{31}(o|t)$ curve correspond approximately to the quasi-independent intervals (compound Poisson model) and to the independent intervals, respectively.

References

1. G. Bédard, Phys. Rev. *151*, p. 1038, (1966).

2. G. Lachs, Jour. Appli. Phys. *42*, p. 602, (1971).

ADVENTURES IN GREEN'S LAND: LIGHT SCATTERING IN ANISOTROPIC MEDIA

M. Lax

City College of New York, CUNY and
Bell Laboratories, Murray Hill, N.J.

and

D. F. Nelson

Bell Laboratories, Murray Hill, N.J.

1. Nature of the Problem

In all light scattering experiments, see Fig. 1, a source of radiation outside the sample enters the sample and is scattered in the source volume V_S by a fluctuation in the dielectric constant (induced by an acoustic phonon in Brillouin scattering, an optical phonon or polariton in Raman scattering, etc.). The generation of radiation at V_S and its propagation to the surface of a crystal must be calculated with due regard for crystalline anisotropy, in particular noncollinearity of the Poynting and propagation vectors.

At the surface, in addition to transmission loss, refraction produces an expansion in solid angle of the scattered beam and source volume demagnification. The detected power is proportional to the solid angle subtended at the detector. This is equivalent to a different solid angle inside the crystal. Since theoretical formulas usually apply inside the medium but experimental measurements are always made outside the medium, this solid angle ratio must be computed before a valid comparison can be made between theory and experiment.

To illustrate the difficulty associated with calculating solid angles, we consider the "trivial" case of an *isotropic medium*. One way to calculate the ratio of solid angles is to

SCATTERING GEOMETRY

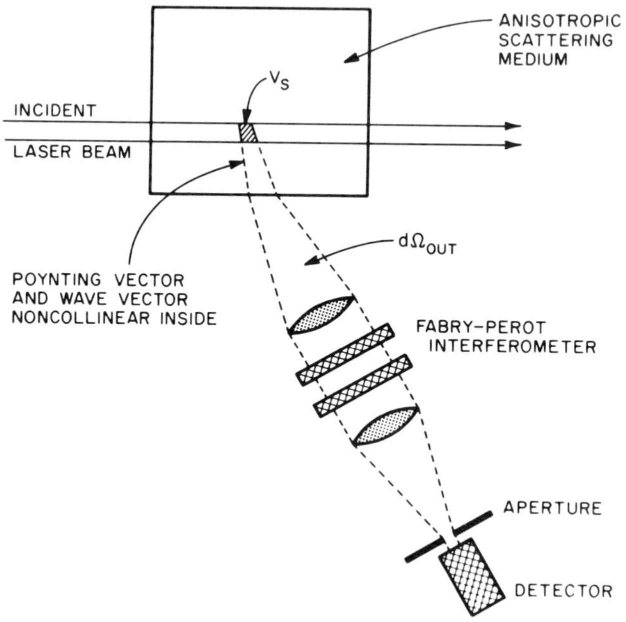

Fig. 1 A typical experimental set up for a light scattering experiment which displays the expansion of the solid angle of radiation on emerging from the crystal. The solid angle accepted by the detector is limited by the first lens. The aperture in front of the detector acts as a field stop and limits the length ℓ_S along the laser beam from which radiation is accepted. For a narrow laser beam, the cross-section area of the source volume V_S is limited only by the beam cross-section.

make use of the ratio

$$\frac{d\Omega_{in}}{d\Omega_{out}} = \frac{\sin\beta \; d\beta \; d\phi}{\sin\alpha \; d\alpha \; d\phi} = \frac{1}{n^2}\frac{\cos\alpha}{\cos\beta} \tag{1.1}$$

where the inside and outside azimuthal angles $d\phi$ are equal, and Snell's law

$$n \sin \beta = \sin \alpha \tag{1.2}$$

is used to evaluate $d\beta/d\alpha$.

It might be objected, however, that a solid angle is a projected area divided by the square of a distance. Since the apparent source or virtual image I_1 of the cone of refracted radiation outside the crystal is different from the origin S of the cone of radiation inside the crystal, different distances are involved. We can therefore calculate the ratio of solid angles by means of the formula (see Fig. 2)

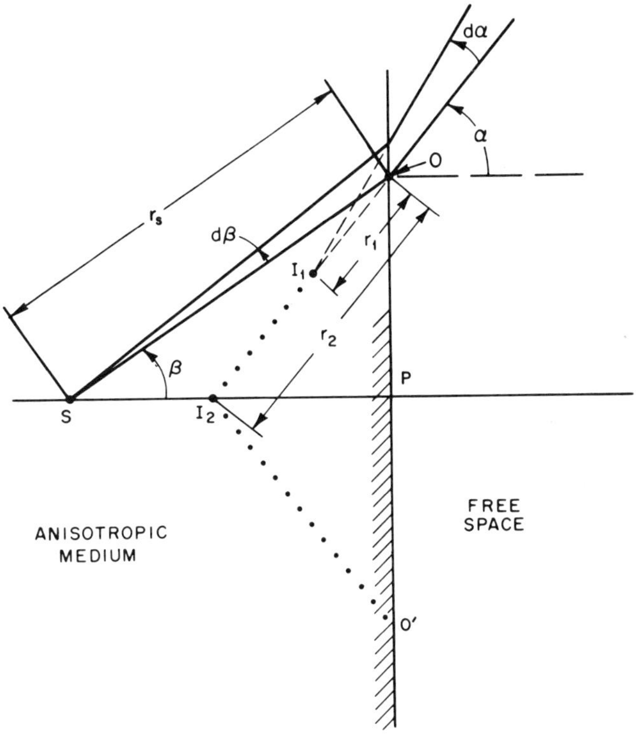

Fig. 2 Solid Angle Expansion. The source of illumination S has a virtual image at I_1 obtained by tracing rays an angle $d\alpha$ apart in the plane of incidence. If a pair of rays an angle $d\phi$ apart perpendicular to the plane are used, the virtual image is at I_2. In an isotropic medium I_2 occurs on a line from S normal to the surface at P. A circle drawn about P with radius Op yields a set of points equivalent to O. Rays from all these points differing only in ϕ converge at the same point I_2.

$$\frac{d\Omega_{in}}{d\Omega_{out}} = \frac{dA \cos \beta/(r_S)^2}{dA \cos \alpha/r_1^2} \tag{1.3}$$

where r_S is the distance from the source to the point on the surface, and r_1 the distance from the apparent source I_1 to O. The point O is the point at which the light ray hits the surface. A second light ray at angle $\alpha + d\alpha$ is traced back to determine the location of I_1 with the result

$$r_1 = \frac{r_s}{n} \left(\frac{\cos \alpha}{\cos \beta}\right)^2 , \tag{1.4}$$

from which we can conclude, using (1.3) that

$$\frac{d\Omega_{in}}{d\Omega_{out}} = \frac{1}{n^2} \left(\frac{\cos \alpha}{\cos \beta}\right)^p , \quad p = 3 . \tag{1.5}$$

If we compare Brillouin scattering in a liquid, with an exit angle α, rather than an exit angle 0°, the intensity should be reduced by a factor $(\cos \alpha/\cos \beta)^p$. Using $\alpha = 50^\circ$ in toluene, see Fig. 3, this reduction should be a factor 1.35 if $p = 1$ and 2.5 if $p = 3$. Within the experimental error of 5% we found no change in intensity.

Let us now take a vote on the correct value of p. You may choose p = 3, 2, 1, 0, -1 or other. The result of this vote is recorded in Table 1.

Which of the above calculations is correct? Both! What was tacitly overlooked in these calculations is that even the surface of an isotropic medium introduces astigmatism into the emerging pencil of rays as shown in Fig. 4. If we compute the apparent image using a pencil or rays spread in the $d\phi$ direction, i.e. perpendicular to the direction $d\alpha$, the rays yield an image at the point I_2 on the normal line (since all rays of the cone of angle α about the normal line intersect at the center). Thus we obtain the distance

$$r_2 = r_s \, (\sin\beta / \sin\alpha) = r_s/n \tag{1.6}$$

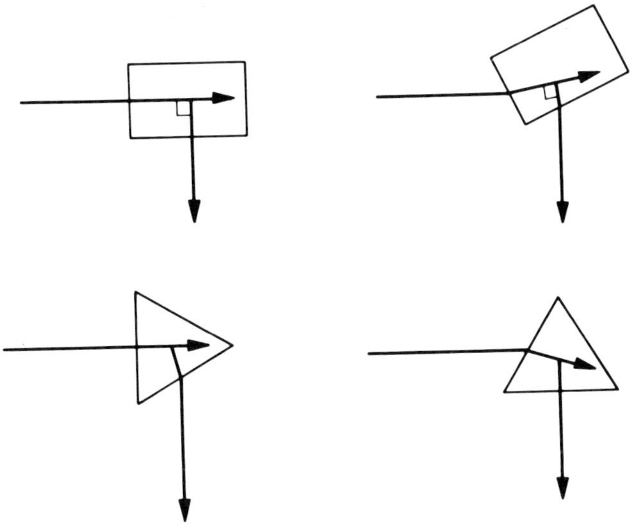

Fig. 3 A series of four Brillouin scattering experiments on a liquid (toluene) to test dependence of observed scattering intensity on incident laser direction and scattered light direction. All four experiments yielded the same ratio of scattered to input radiation intensity to within 5%.

and a solid angle ratio

$$\frac{d\Omega_{in}}{d\Omega_{out}} = \frac{dA \cos\beta/(r_s)^2}{dA \cos\alpha/(r_2)^2} = \frac{1}{n^2}\frac{\cos\beta}{\cos\alpha} \tag{1.7}$$

which is a p of -1.

One can now guess that the correct result is the geometric mean of Eqs. (1.5) and (1.7), yielding a p of +1 in agreement with the Jacobian calculation, Eq. (1.1), providing, the outside solid angle is now defined by

$$d\Omega_{out} = dA \cos\alpha/(r_1 r_2) \ . \tag{1.8}$$

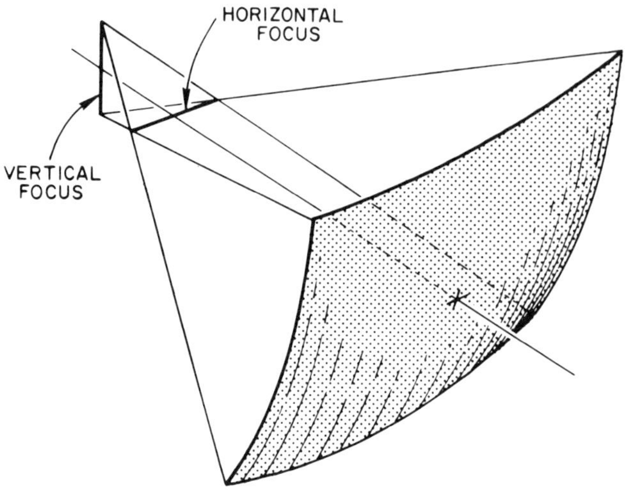

ASTIGMATIC PENCIL

Fig. 4 An astigmatic pencil of rays is displayed. The wave front is not spherical. The radius of curvature of a vertical arc is less than that of a horizontal arc. The set of vertical arcs converge onto a series of points along the horizontal focus, whereas the horizontal arcs converge at a set of points on the vertical focus.

The solid angle ratio calculation is even less obvious for an anisotropic medium. In using a pair of rays spread by $d\alpha$ to locate I_1, we must also take account of the change dn in the index of refraction. The symmetry argument used to simplify the calculation for a pencil of rays in $d\phi$ to obtain I_2 will no longer be generally applicable. Also, Snell's law applies to the propagation vector directions, but we shall show that, of course, one must use solid angles of Poynting vectors.

To be sure that we are using the correct definition of solid angle, as well as a check on the nontrivial algebra of ray tracing in the anisotropic case, we have made a calculation of Green's function *outside* an *anisotropic* medium for a point source inside. From a calculation of the Poynting vector and the detected power (which is what we are after anyway) we shall be able to deduce what is the appropriate way to modify an inside power calculation by solid angle corrections. We will also present in Sec. 4 a direct calculation of solid angle ratios for the general case.

To emphasize the nontrivial nature of the outside Green's function calculation, we note a related problem of radiation by a radio antenna above the earth (treated as *isotropic*). This problem was treated by Sommerfeld [1] in 1909, by Weyl [2] in 1919 and was subject to controversy until the work of van der Pol [3] in 1935 and Norton [4] in 1937.

If we assume now that for an isotropic medium, the detected power at angle α is reduced in the ratio

$$d\Omega_{in}/d\Omega_{out} = (1/n^2)\ (\cos\alpha/\cos\beta\ ), \tag{1.9}$$

how do we explain the equality observed between the two experiments in Fig. 3 using square cells? Could the reduction on exiting from the crystal be enhanced by a correction on entrance? We tested this idea by using the triangular cells shown in Fig. 3. In one case the incidence is normal, and the exit at an angle which should lead to a reduction in intensity. In the other case, the incident angle is not normal but the exit is normal which should lead to an increase in intensity if compensation is to explain the experiments with the rectangular cells. Result: All four experiments yielded the same intensity within the experimental accuracy.

Since we firmly believe in the solid angle correction (1.9), we must find a compensating factor elsewhere to overcome a 35% correction at 50^0 exit angle caused by the cosine ratio factor. We will show that demagnification of the source volume causes a change in the detected power which cancels the major part of the angular dependence of the solid angle correction. In order to elucidate these various factors, we shall obtain a more accurate Green's function inside an anisotropic crystal in Section 2, we shall evaluate the Green's function outside the crystal in Section 3, we shall examine the ratio of Poynting vector to $\underline{k}$ vector solid angles in Section 4 (as well as the inside to outside ratio of $\underline{k}$ vector solid angles) and we shall investigate the effect of demagnification factors on the source volume V_S in Section 5. Section 6 will provide an overall summary of what we believe to be the correct formulas for light scattering as examined outside an anisotropic crystal.

2. The Inside Green's Function

A. Formulation

The electric field produced by a given single frequency component of the nonlinear polarization $\underline{P}^{NL}(\underline{r})\exp(-i\omega t)$ can be expressed in Green's function form

$$\underline{E}(\underline{r},\omega) = \int \underline{\underline{G}}(\underline{r}-\underline{r}') \cdot \underline{P}^{NL}(\underline{r}')d\underline{r}' e^{-i\omega t} / \varepsilon_o \tag{2.1}$$

where the Green's function for an infinite medium of the wave equation obeyed by $\underline{E}$ has previously been shown [5] to be

$$\underline{\underline{G}}(\underline{R}) = \int \frac{\exp(i\underline{k}.\underline{R})}{\underline{\underline{\alpha}}(\underline{k},\omega)} \frac{d\underline{k}}{(2\pi)^3} \tag{2.2}$$

where

$$\frac{1}{\underline{\underline{\alpha}}(\underline{k},\omega)} = \frac{\omega}{c}^2 \sum_{\phi=1,2} \frac{\hat{\underline{E}}^{\phi}(\underline{s},\omega)\hat{\underline{E}}^{\phi}(\underline{s},\omega)}{(k/n^{\phi})^2 - (\omega/c)^2 - i0} - \frac{\underline{s}\underline{s}}{\underline{s}\cdot\underline{\underline{\kappa}}(\omega).\underline{s}} \tag{2.3}$$

Here, $\hat{\underline{E}}^{\phi}$ is an eigenvector of the wave equation for frequency ω and direction of propagation $\underline{s} = \underline{k}/k$,

$$(\underline{\underline{1}}-\underline{s}\underline{s}) \cdot \hat{\underline{E}}^{\phi} = (1/n^{\phi})^2 \underline{\underline{\kappa}}(\omega) \cdot \hat{\underline{E}}^{\phi} , \tag{2.4}$$

where $\underline{\underline{\kappa}}(\omega)$ is the dielectric tensor, and $n^{\phi} = n^{\phi}(\underline{s},\omega)$, the index of refraction at frequency ω for direction $\underline{s}$, is the associated eigenvalue.

In Eq. (2.3) the $\phi = 3$ or longitudinal eigenvector, $\underline{E}^3 || \underline{s}$, has been separated because $n^{(3)} = \infty$, and the last term in (2.3) will make no contribution in the asymptotic limit $|\underline{R}| = |\underline{r}-\underline{r}'| \to \infty$ (although we have previously shown that in nonlinear driven processes, the longitudinal contribution can be substantial) [6].

In Eq. (2.3) the eigenvectors are understood to be orthonormalized in accord with

$$\hat{\underline{E}}^{\phi} \cdot \hat{\underline{D}}^{\theta} = \delta^{\phi\theta} , \quad \hat{\underline{D}}^{\theta} \equiv \underline{\underline{\kappa}} \cdot \hat{\underline{E}}^{\theta} \tag{2.5}$$

as established in ref. 5. These conditions apply for $\phi,\theta = 1,2,3$. Thus $\hat{\underline{D}}^1$ and $\hat{\underline{D}}^2$ are necessarily perpendular to $\hat{\underline{E}}^3$ or the propagation direction $\underline{s}$ as shown in Fig. 5. Since the magnetic field $\underline{H}^1$ associated with solution 1 is in the direction $\nabla \times \hat{\underline{E}}^1$ or $\underline{s} \times \hat{\underline{E}}^1$ it is perpendicular to $\hat{\underline{E}}^3$ and $\hat{\underline{E}}^1$ and hence by (2.5) is in the direction $\hat{\underline{D}}^2$ as shown in Fig. 5. Moreover, $\underline{s} \times \hat{\underline{E}}^1$ or $\underline{H}^1$ is perpendicular to the plane of $\underline{s}$ and $\hat{\underline{E}}^1$ so that the Poynting vector $\hat{\underline{E}}^1 \times \hat{\underline{H}}^1$ is in this plane. Thus we see in Fig. 6 that the angle δ^ϕ between $\hat{\underline{E}}^\phi$ and $\hat{\underline{D}}^\phi$ is also the angle between the propagation and Poynting vectors $\underline{s}$ and $\underline{S}^\phi$.

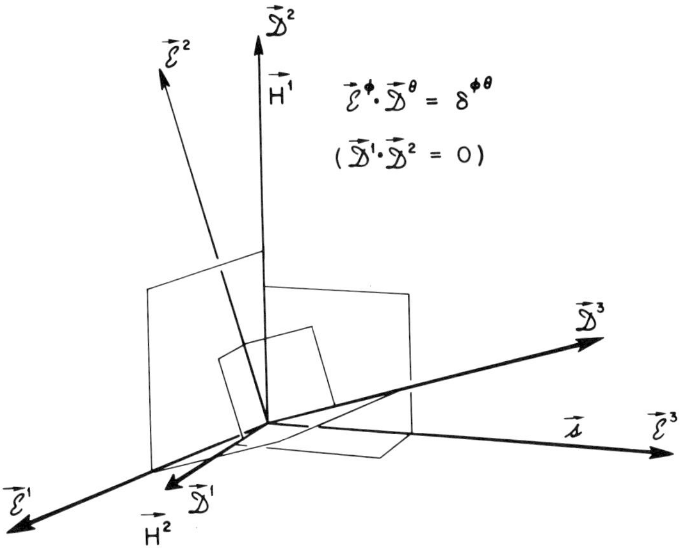

Fig. 5 The biorthogonality of the electric field eigenvectors $\hat{\underline{E}}^\phi$ and the associated electric displacement eigenvectors $\hat{\underline{D}}^1 = \underline{\underline{\kappa}} \cdot \hat{\underline{E}}^\phi$ is displayed, as well as the orthogonality of the vectors $\hat{\underline{D}}^1$, $\hat{\underline{D}}^2$ and the propagation vector $\underline{s} || \hat{\underline{E}}^3$. Since $\underline{H}^1 || \underline{s} \times \hat{\underline{E}}^1$, it is necessarily in the direction $\hat{\underline{D}}^2$ and vice versa.

B. Stationary Phase Method

Since we are interested in radiation at a large distance from the source ($kR >> 1$), only the asymptotic form of $\underline{\underline{G}}$ is needed. A stationary phase method is clearly appropriate. Our previous procedure (ref. 5, Sec. 14) applies a stationary phase integration over angles followed by a residue integration over the pole in the magnitude k. However, the work of Kogelnik [7] and Kogelnik and Motz [8] on magnetoionic media [8] made use of a stationary phase method of Lighthill [9] in which the residue integration is done first and the stationary phase method is applied second. In part C of this section we shall verify by doing one integral exactly that the Lighthill procedure is the correct order of operations.

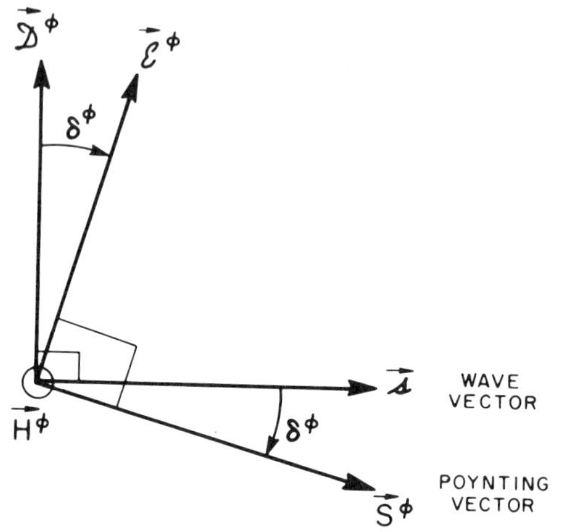

Fig. 6 The electric displacement, the electric field, the propagation and the Poynting vectors, $\hat{\underline{D}}^{\phi}$, $\hat{\underline{E}}^{\phi}$, $\underline{s}$ and $\underline{S}$ are all in one plane perpendicular to the magnetic field $\underline{H}^{\phi}$. See Fig. 5. Thus the angle δ^{ϕ} between $\hat{\underline{E}}^{\phi}$ and $\hat{\underline{D}}^{\phi}$ is the same as that between the propagation vector $\underline{s}$ and Poynting vector $\underline{S}^{\phi}$ appropriate to this mode ϕ.

We are concerned with evaluating an integral of the form

$$g(\underline{R}) = \int \frac{N(\underline{k})}{D(\underline{k})} e^{i\underline{k} \cdot \underline{R}} \, d\underline{k} \tag{2.6}$$

in which $N(\underline{k})$ has no singularities for finite $\underline{k}$ and $D(\underline{k})$ possesses a finite number of zeroes. Indeed our expansion of $\underline{\underline{\alpha}}^{-1}$ in terms of eigenvectors, (2.3), differs from the procedure of Kogelnik [7] and Kogelnik and Motz [8] by explicitly separating the complete Green's function integral (2.2) into terms in (2.3) each of which possesses only one pole in the upper half plane.

A residue integration of (2.6) over a component of $\underline{k}||\underline{R}$, yields

$$g(\underline{R}) = 2\pi i \int_{D=0} \frac{N(\underline{k}(u,v))}{|\underline{\nabla}D(\underline{k}(u,v))|} e^{i\underline{k}(u,v).\underline{R}} \, dudv \tag{2.7}$$

where a surface integral remains over the energy surface $D(\underline{k},\omega) = 0$. This surface which relates ω to $\underline{k}$ is parametrized by the surface parameters u and v, i.e. $\underline{k} = \underline{k}(u,v,\omega)$ is automatically obeyed on the surface. The stationary phase condition

$$\frac{\partial \underline{k}}{\partial u} . \underline{R} = \frac{\partial \underline{k}}{\partial v} . \underline{R} = 0 \tag{2.8}$$

selects a point $\underline{k}_o = \underline{k}(u_o,v_o)$ on the surface whose tangent vectors are perpendicular to $\underline{R}$, i.e. a point at which the surface normal is parallel to the direction of observation $\underline{R}$ as shown in Fig. 7. Since the energy surface is also the surface $\omega(\underline{k})$ = constant, the unit surface normal $\underline{t}$ is necessarily in the direction of the group velocity

$$v_g \underline{t} = \underline{\nabla}_k \omega(\underline{k}) \tag{2.9}$$

which can be shown to be in the direction of the Poynting vector [8,9,10]. Thus, the wave vectors which determine the intensity in the direction $\underline{R}$ come from the vicinity of that $\underline{k}_o$ on the energy surface whose Poynting vector is parallel to $\underline{R}$.

The conventional stationary phase evaluation which expands the exponent to second order about $\underline{k}_o$ reduces the integral (2.7) to

$$g(\underline{R}) = 4\pi^2 \frac{\exp(i\underline{k}_o.\underline{R})}{R} \frac{C_1}{K^{\frac{1}{2}}} \frac{N(\underline{k}_o)}{|\underline{\nabla}D(\underline{k}_o)|} \tag{2.10}$$

where K is the Gaussian curvature of the energy surface at $\underline{k}_o$ (the product of the two principal curvatures) and C = ±i or ±1 according to the signs of the curvatures and the direction of the group velocity. Since only $|C|^2$ enters the power, we shall ignore C.

In view of Eq. (2.3) we can rewrite the dyadic Green's function at large $\underline{R}$ in the form

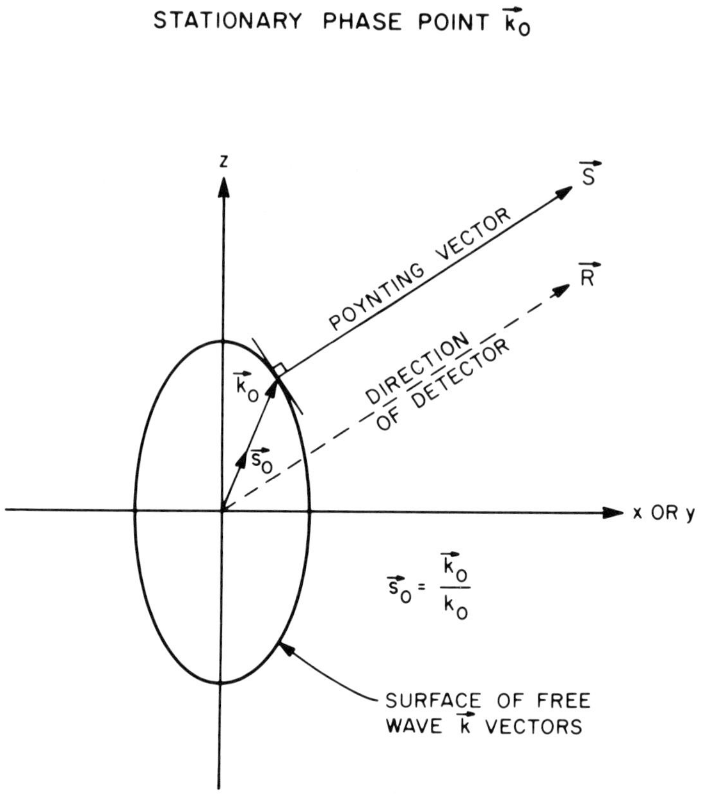

Fig. 7 The energy surface $\omega(\underline{k}) = \omega$ of a free wave is shown in $\underline{k}$ space. The stationary phase condition indicates that radiation in the direction $\underline{R}$ of the detector is produced primarily by a bundle of rays whose $\underline{k}$ vector, $\underline{k}_o$, is so chosen that the normal to the surface $\nabla_{\underline{k}}\omega(\underline{k})$ at $\underline{k}_o$ is parallel to $\underline{R}$, i.e., so that the Poynting vector $\underline{S}$ [which is necessarily parallel to the group velocity $\nabla_{\underline{k}}\omega(\underline{k})$] is in the direction of observation.

$$\underline{\underline{G}}(\underline{R}) = (\omega/c)^2 \sum_{\phi=1,2}{}' [n^{\phi}(\underline{s}_o)]^2 \, \hat{\underline{E}}^{\phi}(\underline{s}_o)\hat{\underline{E}}^{\phi}(\underline{s}_o) g^{\phi}(\underline{R}) \qquad (2.11)$$

where the dependence of n^{ϕ} and $\hat{\underline{E}}^{\phi}$ on ω is not explicitly indicated since ω remains constant in our discussion. What remains to be evaluated is the scalar Green's function

$$g^{\phi}(\underline{R}) = \frac{1}{n^{\phi}(\underline{s}_o)^2} \int \frac{\exp(i\underline{k}.\underline{R})\,d\underline{k}/(2\pi)^3}{[k/n^{\phi}(\underline{s})]^2 - (\omega/c)^2 - i\underline{0}} \tag{2.12}$$

which has been so defined that when $n^{\phi}(s)$ is a constant, as for the ordinary wave in a uniaxial crystal,

$$g^{ord}(\underline{R}) = \frac{\exp[iRn^{ord}\,\omega/c]}{4\pi R} \quad . \tag{2.13}$$

Whether a crystal is uniaxial or biaxial it is possible to show [11] that the gradient of the denominator D is given by

$$\underline{\nabla}_{\underline{k}} \left[\frac{k}{n(\underline{s})} \right]^2 = \frac{2}{n^2} \frac{k}{\cos\delta} \underline{t} \quad . \tag{2.14}$$

Equation (2.10) applied to (2.12) then yields

$$g^{\phi}(\underline{R}) = f^{\phi} \frac{\exp(i\underline{k}_o^{\phi} \cdot \underline{R})}{4\pi R} \tag{2.15}$$

where

$$f^{\phi} = \frac{\cos\delta^{\phi}}{k_o^{\phi}\sqrt{K}} \quad . \tag{2.16}$$

C. Exact Scalar Green's Function

A test of the accuracy of the stationary phase method can be made by comparing (2.15) with the following exact evaluation of the scalar Green's function $g^e(\underline{R})$ for an extraordinary wave in a uniaxial crystal. For this case, the denominator in (2.12) is particularly simple since

$$(1/n^e)^2 = (s_1^2 + s_2^2)/\kappa_{33} + s_3^2/\kappa_{11} \tag{2.17}$$

and

$$D = \frac{k^2}{(n^e)^2} - (\frac{\omega}{c})^2 = \frac{k_1^2 + k_2^2}{\kappa_{33}} + \frac{k_3^2}{\kappa_{11}} - (\frac{\omega}{c})^2 \qquad (2.18)$$

so that the group velocity is in a direction parallel to $\nabla_k D$ or

$$\underline{t} \propto [k_1^e/\kappa_{33}, k_2^e/\kappa_{33}, k_3^e/\kappa_{11}] \quad . \qquad (2.19)$$

If this is to be parallel to $\underline{R}$, we see that

$$\underline{k}_o^e \propto [\kappa_{33}R_1, \kappa_{33}R_2, \kappa_{11}R_3] \quad . \qquad (2.20]$$

Since $|k_o^e| = n^e\omega/c$, we see that

$$\underline{k}_o^e = \frac{\omega}{c} \frac{[\kappa_{33}R_1, \kappa_{33}R_2, \kappa_{11}R_3]}{[\kappa_{33}(R_1^2 + R_2^2) + \kappa_{11}R_3^2]^{\frac{1}{2}}} \qquad (2.21)$$

is the $\underline{k}$ vector that contributes to radiation in the direction $\underline{R}$.

The exact evaluation proceeds by noting that a rescaling of dimensions along the three coordinate axes can convert (2.18) to an isotropic quadratic form, in which case the integrations can be carried out explicitly to yield

$$g^e(\underline{R}) = \frac{\kappa_{33}\sqrt{\kappa_{11}}}{(n^e)^2} \frac{\exp\{i(\omega/c)[\kappa_{33}(R_1^2 + R_2^2) + \kappa_{11}R_3^2]^{\frac{1}{2}}\}}{4\pi[\kappa_{33}(R_1^2 + R_2^2) + \kappa_{11}R_3^2]^{\frac{1}{2}}} \quad . \qquad (2.22a)$$

The optical path in (2.22a) is *precisely* equal to

$$P \equiv (\omega/c)\,[\kappa_{33}(R_1^2 + R_2^2) + \kappa_{11}R_3^2]^{\frac{1}{2}} = \underline{k}_o^e \cdot \underline{R} \quad . \qquad (2.22b)$$

in addition to which

$$\underline{\nabla}_R P = \underline{k}_o^e \tag{2.23}$$

so that radiation in the direction $\underline{R}$ does use the propagation vector $\underline{k}_o^e$ as implied by the stationary phase method.

The exact result (2.22a) can be written in the form (2.15). The exact factor f, when reexpressed in terms of $\underline{s} = \underline{k}_o^e/k_o^e$ is given by

$$f^2 = (\underline{s} \cdot \underline{\underline{\kappa}} \cdot \underline{s})(\underline{s} \cdot \underline{\underline{\kappa}}^2 \cdot \underline{s})/\det \underline{\underline{\kappa}} . \tag{2.24}$$

The asymptotic result (2.16) can be reduced *exactly* to (2.24) by evaluating the Gaussian curvature K:

$$k^2 K = (\det \underline{\underline{\kappa}})(\underline{s} \cdot \underline{\underline{\kappa}} \cdot \underline{s})/(\underline{s} \cdot \underline{\underline{\kappa}}^2 \cdot \underline{s})^2 \tag{2.25}$$

and $\cos \delta = \underline{s} \cdot \underline{t}$

$$\cos \delta = (\underline{s} \cdot \underline{\underline{\kappa}} \cdot \underline{s})/(\underline{s} \cdot \underline{\underline{\kappa}}^2 \cdot \underline{s})^{\frac{1}{2}} \tag{2.26}$$

We may remark that $\cos \delta = 1$ in each principal direction.

In spite of the invariant appearance of the expressions (2.25), (2.26) they cannot be assumed to be valid in biaxial crystals. A truly general expression for δ is given, for example, by [11]

$$\tan^2 \delta^\phi = (\underline{\nabla}_s \ln n^\phi)^2 - [(\underline{s} \cdot \underline{\nabla}_s) \ln n^\phi]^2. \tag{2.27}$$

D. Scattered Power Inside

The electric field may now be calculated in terms of the nonlinear polarization by means of the Green's function formulas (2.1), (2.11) and (2.15). We shall assume as in ref. 5 that the nonlinear polarization can be approximated as a plane wave of amplitude $\underline{P}^{NL}$ over the small source volume V_S. Making the Fraunhofer approximation, we calculate the Poynting vector inside the crystal and the ratio of the scattered power P^{ϕ} to incident power P^{θ} following our earlier procedures [5,12]:

$$\frac{P^{\phi}}{P^{\theta}} = FB\ [(f^{\phi})^2 \ell_S d\Omega^r_{in}\ ] \tag{2.28}$$

where

$$B \equiv \frac{n^{\phi}\langle|\underline{e}^{\phi} \cdot \underline{P}^{NL}|^2\rangle\ \Omega}{\cos^3\delta^{\phi} n^{\theta} \cos\delta^{\theta} |E^{\theta}|^2}\ , \tag{2.29}$$

$$F \equiv \omega^4/(16\pi^2\varepsilon_o^2 c^4)\ ; \tag{2.30}$$

$\underline{e}^{\phi} = \hat{E}^{\phi}/|\hat{E}^{\phi}|$ is a unit electric field vector, $\underline{E}^{\theta}$ is the incident electric field (inside the crystal), ℓ_S is the linear dimension (along the laser beam) accepted by the detector and $d\Omega^r_{in}$ is the solid angle (in $\underline{r}$ space) inside the crystal subtended by the detector. The thermal average $\langle P^{NL} P^{NL}\rangle$ is proportional to $|\underline{E}^{\theta}|^2$ and inversely proportional to the volume Ω of the crystal since $\underline{P}^{NL}$ is the contribution to $\underline{P}^{NL}(r)$ of one normal mode of vibration.

The only deviation between (2.28) and previous results[5,12] is the presence of the extra factor $(f^{\phi})^2$.

3. The Outside Green's Function

A. Formulation

The Green's function (2.2) can be generalized to permit the detector point $\underline{r}$ to be outside the crystal by writing

$$\underline{\underline{G}}(\underline{r},\underline{r}') = \sum_{\phi=1,2} \left(\frac{\omega}{c}\right)^2 \int \frac{d\underline{k}}{(2\pi)^3} \frac{\underline{u}^{\phi}(\underline{k},\underline{r})\underline{u}^{\phi}(\underline{k},\underline{r}')^*}{[k/n^{\phi}(\underline{s},\omega)]^2 - (\omega/c)^2 - i0} \tag{3.1}$$

where

$$\underline{u}^{\phi}(\underline{k},\underline{r}')^{*} = \hat{\underline{E}}^{\phi}(\underline{s})\exp(-i\underline{k}\cdot\underline{r}')$$

and the dependence of $\hat{\underline{E}}^{\phi}$ on ω is suppressed. If the point $\underline{r}$ were inside the crystal, the *direct* (outgoing) wave with the appropriate singularity as $\underline{r} \to \underline{r}'$is obtained by setting

$$\underline{u}^{\phi}(\underline{k},\underline{r}) = \hat{\underline{E}}^{\phi}(\underline{s})\exp(i\underline{k}\cdot\underline{r}), \tag{3.2}$$

(which omits a wave reflected at the boundary). For $\underline{r}$ outside the crystal, we must replace (3.2) by the transmitted wave

$$\underline{u}\phi(\underline{k},\underline{r}) = t(\underline{k})|\hat{\underline{E}}^{\phi}(\underline{s})|\underline{e}^{\phi}(\underline{\ell}(\underline{k}))\exp[i\underline{\ell}(\underline{k})\cdot\underline{r}] \tag{3.3}$$

where $t(\underline{k})$ is the *amplitude* transmission coefficient, $\underline{\ell} = \underline{\ell}(\underline{k})$ is the outside propagation vector related by Snell's law to the inside propagation vector $\underline{k}$, and $\underline{e}^{\phi}(\underline{\ell})$ is the unit outside electric vector.

The procedures used by Weyl [2] and Sommerfeld [1] to evaluate similar integrals in the radio antenna problem make explicit use of isotropy and are no help to us when n^{ϕ} depends on orientation $\underline{s}$. The stationary phase procedure of Section 2B, however, can be applied by writing

$$\underline{\underline{G}}(\underline{r},\underline{r}') \approx \sum \underline{e}^{\phi}(\underline{\ell}_0)\hat{\underline{E}}^{\phi}(\underline{s}_0)|\hat{\underline{E}}^{\phi}(\underline{s}_0)|$$

$$\times\ t^{\phi}(\underline{k}_0)(\omega/c)^2[n^{\phi}(\underline{s}_0)]^2 g^{\phi}(\underline{r},\underline{r}') \tag{3.4}$$

where the scalar Green's function

$$g^{\phi}(\underline{r},\underline{r}') = \frac{1}{[n^{\phi}(\underline{s}_0)]^2}\int \frac{d\underline{k}}{(2\pi)^3}\ \frac{\exp[i\underline{\ell}(\underline{k})\cdot\underline{r} - i\underline{k}\cdot\underline{r}']}{[k/n^{\phi}(\underline{s})]^2 - (\omega/c)^2 - i0} \tag{3.5}$$

is no longer a function of $\underline{r}-\underline{r}'$.

B. Stationary Phase Evaluation

The exponent in (3.5) possesses a branch cut. Thus the integral has a stationary phase (or steepest descent) contribution plus a branch cut contribution. The latter has the significance of a guided surface wave [1-4] which is unimportant in the optical regime since the detector will always be many wavelengths from the surface.

The evaluation of the stationary phase contribution alone leads to considerable algebra, and we have only evaluated several special cases to be discussed below. All cases evaluated were found to be covered by the formula

$$g^{\phi}(\underline{r},\underline{r}')_{out} = \frac{\exp(iP)}{4\pi n^{\phi}(\underline{s}_o)} \frac{1}{[(r+r_1)(r+r_2)]^{\frac{1}{2}}} \frac{\cos\delta^{\phi}\cos\alpha}{\cos\beta} \tag{3.6}$$

where β is the angle between the Poynting vector and the normal to the surface, and $\alpha(>\beta)$ is the corresponding angle outside the crystal. For a given source position $\underline{r}'$ and detector position $\underline{r}$, the stationary phase condition determines the angles α and β (and the intersection point 0 on the surface) so as to minimize the optical path (Fermat's principle). See Fig. 2. In Eq.(3.6)

$$P = n(\underline{s}_o)r_S + r \tag{3.7}$$

is this minimum optical path, with r_S and r denoting the distance from the origin 0 on the surface to the source and detector respectively. The unit propagation vector $\underline{s}_o$ is that propagation vector direction consistent with the Poynting vector direction β.

The cases for which we have verified Eq. (3.6) are:

(1) any surface, isotropic medium, or ordinary ray, uniaxial crystal

$$r_1 = r_2(\cos\alpha/\cos\beta)^2, \quad r_2 = (r_S/n); \tag{3.8}$$

(2) surface z = 0, normal exit ($\alpha = \beta = 0$), extraordinary ray, uniaxial crystal

$$r_1 = r_2 = (z_S/n)\ (\kappa_{11}/\kappa_{33})\ ; \tag{3.9}$$

(3) surface x = 0, normal exit ($\alpha = \beta = 0$), extraordinary ray, uniaxial crystal

$$r_1 = x_S/n, \quad r_2 = r_1(\kappa_{33}/\kappa_{11})\ ; \tag{3.10}$$

(4) surface z = 0, incident Poynting angle β, exit angle α, extraordinary ray, uniaxial crystal

$$\begin{aligned} r_2 &= (r_S/n)\ (\kappa_{11}/\kappa_{33})(\cos\beta/\cos\theta), \\ r_1 &= r_2\cos\delta(\cos\alpha/\cos\beta)(\cos\alpha/\cos\theta), \end{aligned} \tag{3.11}$$

where θ is the angle between the propagation vector and the surface normal. The condition $r_1 \neq r_2$ indicates that the pencil of rays leaving a plane surface (even for normal exit in case 3) is an astigmatic pencil.

If we let $r \to 0$, our outside Green's function (3.6) should reduce to the inside Green's function (2.15) which requires that

$$\frac{1}{n^{\phi}}\ \frac{r_S}{(r_1 r_2)^{\frac{1}{2}}}\ \frac{\cos\delta^{\phi}\cos\alpha}{\cos\beta} = f^{\phi}\ . \tag{3.12}$$

In the above four cases, (3.12) is verified with the help of (2.24).

C. Huygens' Principle

For the first three cases, above, we have verified by integration to Fresnel accuracy that the outside Green's function can be calculated from the inside Green's function using a modified form of the Kirchoff-Huygens principle: [10]

$$g^{\phi out}(\underline{r},\underline{r}') = \frac{-2}{4\pi}\iint g^{\phi\ in}(\underline{r}_1-\underline{r}')\frac{\partial}{\partial n_1}\left[\frac{e^{i(\omega/c)|\underline{r}-\underline{r}_1|}}{|\underline{r}-\underline{r}_1|}\right] dS_1$$

$$= i(\omega/c)\iint g^{\phi\ in}(\underline{r}_1-\underline{r}')\frac{e^{i(\omega/c)|\underline{r}-\underline{r}_1|}}{4\pi\ |\underline{r}-\underline{r}_1|}(2\cos\alpha)\,dS_1 \tag{3.13}$$

with the obliquity factor 2 cos α rather than (cos α + cos β) or(cosα + n cos β).

The Huygens principle procedure has been successfully used by Kleinman, Ashkin and Boyd [13] to calculate the Green's function outside a crystal at normal incidence.

We have been able to show, by comparing the integral representation (3.5) for the external Green's function with (2.12) for the internal Green's function, that Huygens' principle in the form (3.13) is exactly obeyed for arbitrary (including biaxial) crystals with an arbitrarily oriented plane surface. Moreover, no stationary phase or other approximations are involved in the comparison.

D. Scattered Power Outside

The scattered power outside the crystal is easier to calculate from the Green's function than was the inside power since $\underline{E}$ is perpendicular to $\underline{s}$. Thus, inside

$$|\underline{E}\times(\underline{k}\times\underline{E})| = (n\omega/c)\ |\underline{E}\times(\underline{s}\times\underline{E})|$$

$$= (n\omega/c)\,|\underline{E}|^2\cos\delta \tag{3.14}$$

whereas outside

$$|\underline{E}\times(\underline{\ell}\times\underline{E})| = (\omega/c)\ |\underline{E}|^2. \tag{3.15}$$

Thus the outside form is smaller by a factor n cos δ as well as $|t|^2$. We find in fact

$$P^{\phi}_{out}/P^{\theta} = FB|4\pi g^{\phi}_{out}|^2 dA_D \cos\theta_D\ \ell_s |t^{\phi}|^2/(n^{\phi}\cos\delta^{\phi}) \tag{3.16}$$

where F and B are given in (2.29) and (2.30) where $dA_D \cos\theta_D$ is the area of the detector normal to the beam.

Before Eq. (3.16) can be compared with the corresponding expression (2.28) for the inside power, we need an expression for the transmission coefficient

$$T^{\phi} = \frac{P^{\phi}_{out}}{P^{\phi}} = \frac{|S_{out}|\,dA\cos\alpha}{|S_{in}|\,dA\cos\beta}$$

$$= \frac{\cos\alpha}{\cos\beta}\ \frac{\frac{1}{2}\varepsilon_0 c|E_{out}|^2}{\frac{1}{2}\varepsilon_0 n^{\phi} c\cos\delta^{\phi}|E_{in}|^2} = \frac{\cos\alpha}{\cos\beta}\ \frac{|t^{\phi}|^2}{n^{\phi}\cos\delta^{\phi}} \tag{3.17}$$

where dA is an element of surface area of the crystal and $dA\cos\alpha$ and $dA\cos\beta$ are projections of this area normal to the outside and inside Poynting vectors. The factor $n^{\phi}\cos\delta^{\phi}$ arises as shown in (3.14) and (3.15).

In view of the form (3.6), Eq. (3.16) yields a power that varies as

$$\frac{dA_D\cos\theta_D}{(r+r_1)(r+r_2)} = d\Omega_{out} \tag{3.18}$$

as the detector moves away from the source without changing direction. Our intuition tells us that the detected power will not change if the solid angle is maintained constant. Thus we conclude that (3.18) is an appropriate definition of solid angle which reduces when r = 0 to the choice (1.8) when the element of area is on the surface. This conclusion is strengthened by verifying that geometrical optics calculations of the virtual image locations I_1 and I_2 lead to the same distances r_1 and r_2 as found from the wave optics calculations (3.8)-(3.11). See Section 4.

If we insert (3.17) and (3.18) into (3.16) and use the form

(3.6) for the outside Green's function, we obtain for the power ratio (referred to the inside of the crystal)

$$P^{\phi}/P^{\theta} = FB \cos^2\delta^{\phi}(\ell_S \cos\alpha/\cos\beta)\,d\Omega_{out}/(n^{\phi})^2 \quad . \tag{3.19}$$

Comparison with the same ratio, (2.28), computed using the inside Green's function will only be consistent if

$$\frac{d\Omega^r_{in}}{d\Omega_{out}} = \frac{1}{(n^{\phi})^2}\frac{\cos^2\delta^{\phi}}{(f^{\phi})^2}\frac{\cos\alpha}{\cos\beta} \quad , \tag{3.20}$$

a conclusion we will verify in the next section. This will confirm the generality of the result (3.6) which we have only established in four special cases.

It is of interest to note that Eq. (3.19) contains the ratio $\cos\alpha/\cos\beta$ to the first power in agreement with our expectations based on (1.9). Since g^{ϕ}_{out} appears squared in the power, one might have expected the factor $(\cos\alpha/\cos\beta)$ in (3.6) to be squared. The extra factor of $\cos\alpha/\cos\beta$ is absorbed, however, into the power transmission coefficient (3.17).

4. Geometrical Optics Calculations

The preceding section has been a crutch. Wave optics has been used to calculate objects whose significance is geometrical because the geometrical calculations were nontrivial, and even the definitions such as (3.18) had to be determined. The cycle can be closed and our understanding verified by a number of geometrical optics calculations.

A. Virtual Image Distances

The distances r_1 and r_2 of the wave optics Green's function, Eqs. (3.6)-(3.11) are indeed the astigmatic virtual image distances of geometrical optics. They can be calculated by tracing a pair of rays back to their intersection. Agreement with the wave optics results will be obtained, however, only if one uses the Poynting vectors as the ray directions. Thus for case 4, Eq. (3.11), one must use Snell's law

$$n \sin\theta = \sin\alpha \tag{4.1}$$

with the angular dependence of the index (2.12),

$$(1/n)^2 = \sin^2\theta/\kappa_{33} + \cos^2\theta/\kappa_{11}, \qquad (4.2)$$

together with the relation [see (2.20)]

$$\tan\beta = (\kappa_{11}/\kappa_{33})\tan\theta \qquad (4.3)$$

between the propagation vector angle θ and the Poynting vector angle β. The reader is invited to verify that a pair of rays with $d\alpha$, $d\theta$ and dn related by (4.1)-(4.3) determine an r_1 in agreement with (3.11). The calculation of r_2 which involves only an azimuthal variation between the rays is much simpler.

B. Ratios of Inside and Outside Solid Angles: Ray Tracing

If dA is an element of surface area of the crystal, and $\cos\beta = \underline{n} \cdot \underline{t}$ is the cosine of the angle between the unit ray (Poynting) vector $\underline{t}$ and the unit surface normal $\underline{n}$, then the interior solid angle can be computed from

$$d\Omega^r_{in} = dA\cos\beta/(r_S)^2 \qquad (4.4)$$

where r_S is the source to surface distance along the ray (see Fig. 2). The outside solid angle, including the effects of astigmatism can be computed from

$$d\Omega_{out} = dA\cos\alpha/(r_1 r_2) \qquad (4.5)$$

where the distances r_1 and r_2 to the virtual images have been computed by ray tracing by the method discussed in part A of this section, in agreement with the values quoted in (3.8)-(3.11).

The solid angle ratio is then given by

$$d\Omega^r_{in}/d\Omega_{out} = (\cos\beta/\cos\alpha)(r_1 r_2/r_S^2) . \qquad (4.6)$$

The values of r_1 and r_2 may now be inserted from Eqs. (3.8)-(3.11). These values have been shown, however, to obey Eq. (3.12). The insertion of (3.12) in the form

$$r_1 r_2/r_S^2 = [\cos\delta^\phi \cos\alpha/(n^\phi f^\phi \cos\beta)]^2 \qquad (4.7)$$

into Eq. (4.6) leads to the ratio of solid angles (3.20) found by wave optics.

C. Solid Angle Ratios: Jacobian Procedures

The above derivation applies only to the four cases discussed. A strictly geometric proof is called for to verify the full generality of the wave optical results. But ray tracing is too complicated in the arbitrary case. We shall instead regard the solid angle ratio as a Jacobian problem which can be simplified by factoring it into two parts

$$\frac{d\Omega^r_{in}}{d\Omega_{out}} = \frac{d\Omega^r_{in}}{d\Omega^k_{in}} \frac{d\Omega^k_{in}}{d\Omega_{out}} \quad . \tag{4.9}$$

The first factor describes the ratio of solid angles in ray (Poynting) vector and $\underline{k}$ vector space and is independent of the existence of a boundary. The second factor is a pure boundary effect on $\underline{k}$ vector solid angles.

The first factor is easy to evaluate with the help of Gauss' *theorema egregium* [14] which can be written in our notation as

$$d\Omega^r = K dA^k \tag{4.10}$$

where dA^k is an element of area of the energy surface $\omega(\underline{k}) = \omega$ shown in Fig. 7, and K is the Gaussian curvature of that surface at the point $\underline{k}$. The theorem defines $d\Omega^r$ as the area on the surface of a unit sphere spanned by the unit normal vectors $\underline{t}$ to the energy surface whose origins (tails) have been shifted to a common origin. Since these vectors $\underline{t}$ are in the direction of observation $\underline{r}$ or of the Poynting vector $\underline{S}$, $d\Omega^r$ is simply the desired solid angle of the rays emanating from dA^k.

The solid angle of the associated bundle of $\underline{k}$ vectors is simply

$$d\Omega^k_{in} = dA^k \cos\delta/k^2 \tag{4.11}$$

since δ is the angle between $\underline{k}$ and the normal $\underline{t}$ to the surface element dA^k. Thus we arrive at our first solid angle ratio

$$d\Omega^r_{in}/d\Omega^k_{in} \;=\; Kk^2/\cos\,\delta\ . \tag{4.12}$$

We have previously obtained results in agreement with (4.12) for the special case of extraordinary rays in a uniaxial crystal using the relationship (2.19) to evaluate the Jacobian of the transformation from $\underline{k}$ to $\underline{r}$.

To evaluate the second solid angle ratio, we introduce a Cartesian coordinate system k_1, k_2, k_3 such that the 3 direction is the direction of the normal $\underline{n}$ to the surface of the crystal. The element of area (of the energy surface) can then be written

$$dA^k = dk_1 dk_2/\cos\,\beta \tag{4.13}$$

since β is the angle between the normal $\underline{t}$ to dA^k and the 3 or normal direction $\underline{n}$.

In a manner parallel to (4.11) and (4.13), the outside solid angle can be defined by

$$d\Omega_{out} \;=\; dA^{\ell}/\ell^2 \tag{4.14}$$

(since the outside propagation vector $\underline{\ell} = \ell(\underline{k})$ is parallel to the normal to the outside energy surface $\omega = c|\underline{\ell}|$ no factor cos δ appears) and

$$dA^{\ell} \;=\; d\ell_1 d\ell_2/\cos\ \alpha\ . \tag{4.15}$$

Since the tangential components are continuous,

$$dk_1 dk_2 = d\ell_1 d\ell_2\ . \tag{4.16}$$

The ratio of (4.11) to (4.14) can now be simplified to yield

$$d\Omega^k_{in}/d\Omega_{out} \;=\; \cos\ \delta\ \cos\,\alpha/(n^2\cos\,\beta), \tag{4.17}$$

since $k^2 = n^2\ell^2$.

If both media were anisotropic, Eq. (4.17) would be replaced by

$$\frac{d\Omega_1^k}{d\Omega_2^k} = \left(\frac{n_2}{n_1}\right)^2 \frac{\cos\delta_1}{\cos\delta_2}\,\frac{\cos\beta_2}{\cos\beta_1}\ , \tag{4.18}$$

where angle α is replaced by β_2 and β by β_1.

The product of (4.12) and (4.17) yields

$$\frac{d\Omega^r_{in}}{d\Omega_{out}} = \frac{k^2K}{n^2}\,\frac{\cos\alpha}{\cos\beta} = \frac{\cos^2\delta}{n^2f^2}\,\frac{\cos\alpha}{\cos\beta} = \left(\frac{\omega}{c}\right)^2 K\,\frac{\cos\alpha}{\cos\beta}\ , \tag{4.19}$$

where the second form makes use of (2.16). We see therefore that the correction f^2 to the absolute square of the inside Green's function produced by our present stationary phase procedure is cancelled by the solid angle correction. The angular factor $\cos\alpha/\cos\beta$ which also appears in (1.9) appears, however, to contradict the lack of angular dependence found in the experiments shown in Fig. 3. This factor will be cancelled by demagnification corrections calculated below.

D. Demagnification Corrections

The linear dimension ℓ_S along the laser beam accepted by the exit optics will depend on the angles α,β. In Fig. 8 we display ℓ_S and its virtual image ℓ_A produced by the crystal surface. The knife edges depict the image of the field stop at the virtual image of the scattering length ℓ_S. This stop is assumed to accept a fixed transverse dimension ℓ_D perpendicular to the *outside* beam direction. By means of the inset in Fig. 8 this is equivalent to admitting a transverse dimension ℓ perpendicular to the *inside* beam direction where

$$\ell/\cos\beta = \ell_D/\cos\alpha\ . \tag{4.20}$$

When ℓ is compared with the component of ℓ_S perpendicular to the beam direction, we find

$$\ell_S = \ell/\sin\theta_S = \ell_D \cos\beta/(\cos\alpha \sin\theta_S) \tag{4.21}$$

where θ_S is the scattering angle

The geometrical derivation given here is valid only when all the rays are in one plane - a customary but not universal condition.

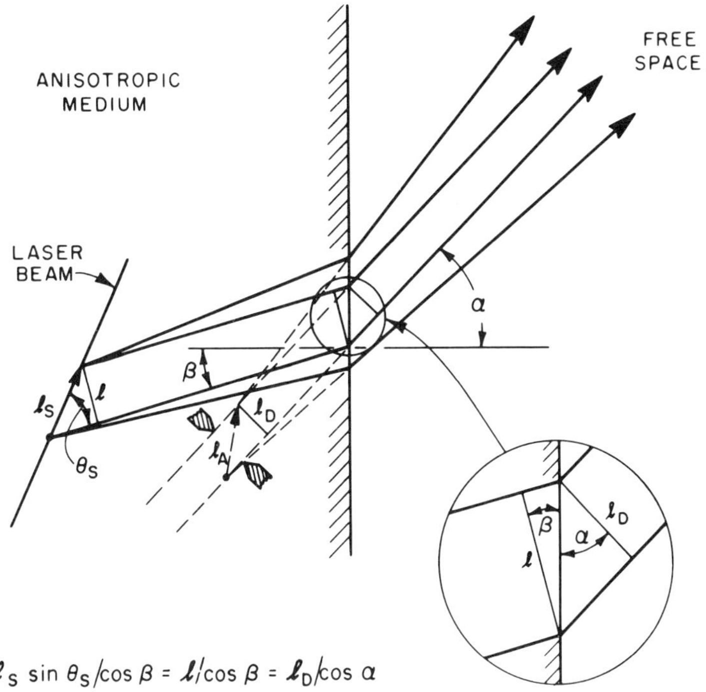

Fig. 8 Demagnification corrections when the incident ray, refracted ray and surface normal are all in one plane. The detector field stop (see Fig. 1) is represented in image space by the knife edges which accept a dimension ℓ_D perpendicular to the beam. The portion ℓ_S of the laser beam accepted is determined by the geometrical conditions shown, independent of the orientation of the virtual image ℓ_A of ℓ_S.

5. Formulas for Light Scattering Power

The experimentally measured incident and scattered powers outside the crystal can now be written

$$P_{out}^{\phi}/P_{out}^{\theta} = (P^{\phi}T^{\phi})/(P^{\theta}/T^{\theta})$$

$$= \frac{\omega^4}{16\pi^2\varepsilon_o^2c^4}\;\frac{\ell_D d\Omega_{out}}{(n^{\phi}\cos\delta^{\phi})(n^{\theta}\cos\delta^{\theta})}\;\frac{T^{\phi}T^{\theta}}{\sin\theta_S}\;\frac{\Omega<|\underline{e}^{\phi}\cdot\underline{P}^{NL}|^2>}{|E^{\theta}|^2} \tag{5.1}$$

where the theoretical result can be obtained equivalently from the outside Green's function (3.19) or from the inside Green's function (2.28) with the solid angle correction (4.18).

In a typical experiment [12,15] the power scattering ratio is measured for a sample and for a reference, and the two are compared. The first factor and the numerator of the second factor cancel in this comparison.

Equation (5.2) is still general in not having specified the nature of the nonlinear polarization. Thus it is applicable to Raman scattering, Brillouin scattering, second harmonic generation, etc. We shall illustrate the use of (5.1) by writing the result for the case of Brillouin scattering. In this case, the nonlinear polarization amplitude can be written in terms of the laser field amplitude $\underline{E}^{\theta}$ (inside the crystal) and the displacement gradient $u_{c,d}$:

$$P_i^{NL}(\omega_0+\omega_A) = \varepsilon_o\chi_{ijcd}E_j^{\theta}(\omega_0)u_{c,d}(\omega_A)e^{i(\underline{k}_0+\underline{k}_A)\cdot\underline{r}} \tag{5.2}$$

for the anti-Stokes component (and a complex conjugate contribution for the Stokes component) where ω_0 and ω_A are the optical (laser) and acoustic frequencies respectively and

$$u_{c,d} \equiv \partial u_c/\partial r_d = i(\underline{k}_A)_d\, u_c = ik_A a_d u_c \tag{5.3}$$

where $\underline{k}_A = k_A\underline{a}$ is the acoustic propagation vector and $\underline{a}$ is the associated unit vector with component a_d.

In Eq. (5.2), $P_i^{NL}(\omega_0+\omega_A)$, $E_j(\omega_0)$ and $u_c(\omega_A)$ are the positive frequency amplitudes denoted by $P_i(1,1)$, $E_j(1,0)$ and $u_c(0,1)$ in Eqs. (13.4), (13.12) or ref. 5. With this convention the displacement $u_c(\omega_A)$ associated with a single normal mode can be written in terms of the phonon amplitude $\mathcal{a}(k_A,\xi)$

$$u_c(\omega_A) = 2\left[\frac{\hbar}{2\rho\omega(\underline{k}_A,\xi)\Omega}\right]^{\frac{1}{2}} b_c(\underline{k}_A,\xi)\mathcal{a}(k_A,\xi) \tag{5.4}$$

where $\underline{b}(k_A,\xi)$ is the unit displacement vector for a phonon of wavevector $\underline{k}_A$ and type ξ, and ρ is the crystal density. The thermally averaged phonon amplitude is given by

$$<|\mathcal{a}(k_A,\xi)|^2> = [\exp(\hbar\omega_A/kT) - 1]^{-1} \approx kT/\hbar\omega_A \tag{5.5}$$

where $\omega_A = \omega(\underline{k}_A,\xi) = v_A k_A$ where v_A is the sound velocity $v_A(\underline{k}_A,\xi)$. Combining Eqs. (5.2)-(5.4), we obtain the thermal average

$$\Omega<|\underline{e}^\phi \cdot \underline{P}^{NL}|^2> \approx 2|M^{\phi\theta}|^2 \epsilon_o^2 kT|\underline{E}^\theta|^2/\rho v_A^2 \tag{5.6}$$

where the matrix M is given by

$$M^{\phi\theta} = e_i^\phi \chi_{ijcd} e_j^\theta b_c a_d \quad . \tag{5.7}$$

The final expression to be compared with experiment is

$$\frac{P_{out}^\phi}{P_{out}^\theta T^\phi T^\theta} = \left[\frac{\omega^4 kT\ell_D d\Omega_D}{8\pi^2 \rho v_A^2 c^4 \sin\theta_S}\right]\left[\frac{|M^{\phi\theta}|^2}{n^\phi \cos\delta^\phi n^\theta \cos\delta^\theta}\right] . \tag{5.8}$$

If one wishes to use the Pockels tensor p_{ijcd} rather than the photoelastic susceptibility χ_{ijcd}, Eq. (5.7) can be used with

$$M^{\phi\theta} = -\frac{1}{2}(n^\phi n^\theta)^2 d_i^\phi p_{ijgh} d_j^\theta b_g a_h \cos\delta^\phi \cos\delta^\theta \tag{5.9}$$

where $\underline{d}^\phi$ and $\underline{d}^\theta$ are unit electric displacement vectors.

The most thorough attempt to discuss Brillouin scattering in an anisotropic crystal by Fabelinskii [16] is based on the work of Motulevich [17] following a method of Ginzburg [18]. In this work , calculations are made only for light scattering inside the crystal. Moreover Motulevich does not distinguish between $d\Omega^r_{in}$ and $d\Omega^k_{in}$ nor between the source volume V_S and the crystal volume Ω. Furthermore, she ignored questions of solid angle expansion and source volume demagnification for the light emerging from the crystal. A careful consideration of these latter questions for anisotropic crystals has not been carried out previously.

Acknowledgment

The authors acknowledge with gratitude important suggestions made by G.A. Coquin, D.A. Kleinman and P.D. Lazay.

TABLE

Vote on the solid angle ratio $d\Omega_{in}/d\Omega_{out} = (1/n^2)(\cos\alpha/\cos\beta)^p$

p =	3	2	1	0	-1	Other	Abstain
Vote =	2	0	4	13	3	4	25

References

1. A. Sommerfeld, Ann. Physik *28*,665 (1909).
2. H. Weyl, Ann. Physik *60*, 481 (1919).
3. Balth van der Pol, Physica *2*,843 (1935).
4. K.A. Norton, Proc. Inst. Radio Engrs. *25*, 1192, 1203 (1937).
5. M. Lax and D.F. Nelson, Phys. Rev. *B4*, 3694 (1971).
6. D.F. Nelson and M. Lax, Phys. Rev. *B3*, 2795 (1971).
7. H. Kogelnik, J. of Research, Nat'l Bureau of Standards, Div. of Radio Propagation *64D*, 515 (1960).
8. H. Kogelnik and H. Motz, *Symposium on Electromagnetic Theory and Antennas*, Copenhagen, June 25-30, 1962, (Pergamon Press, New York, 1963) p. 477.
9. M.J. Lighthill, Phil. Trans. Roy. Soc. (London) Ser. A. *252*, 397 (1960).
10. M. Born and E. Wolf, *Principles of Optics*, (Pergamon Press, New York, 1970).
11. M. Lax and D.F. Nelson (to be published).
12. D. F. Nelson, P. D. Lazay and M. Lax, *Light Scattering in Solids*, M. Balkanski, ed., (Flammarion Sciences, Paris, 1971), p.477.
13. D.A. Kleinman, A. Ashkin and G. Boyd, Phys. Rev. *145*, 338 (1966).

14. J.J. Stoker, *Differential Geometry*, (Wiley-Interscience, New York, 1969).
15. D.F. Nelson, P.D. Lazay and M. Lax, Phys. Rev. *B6*, 3109 (1972).
16. G.P. Motulevich, Trudy Fiz. Inst. P.N. Lebedeva *5*, 9-62 (1950).
17. I.L. Fabelinskii, *Molecular Scattering in Light*, (Plenum Press, New York, 1968).
18. V.L. Ginzburg, ZH. Eks. i Teor. Fiz. *10*,601 (1940).

INTENSITY FLUCTUATION TECHNIQUES IN LIGHT SCATTERING

H.Z. Cummins

New York University, New York, N.Y.

During the last decade, the field of light scattering spectroscopy has experienced remarkable growth due to the availability of laser sources, and the introduction of "light beating" or "intensity fluctuation" spectroscopy.

In traditional (dispersive) spectroscopy, one measures the spectral power density of the optical field. In intensity fluctuation spectroscopy, the undispersed optical field illuminates a photodetector, and one measures the spectral power density or autocorrelation function of the photocurrent. Thus one studies $G^{(2)}$ rather than $G^{(1)}$. With this technique, spectral analysis of inelastic or quasi-elastic scattering in the range of several Hertz to several megaHertz from the incident light has become commonplace and has been applied to a wide range of problems in physics, chemistry, biology and engineering. Some of these experiments are reviewed with emphasis on recent developments.

Initially, all experiments relied on analog spectral power density measurements of the photocurrent. More recently, digital autocorrelation of the individual photoelectric pulses has come into general use, leading to improved precision and greater versatility. The analytic methods of statistical optics have been employed to analyze the signal to noise of the various experimental techniques. These calculations are reviewed with emphasis on the implications for new experiments.

COHERENCE PROPERTIES OF LIGHT THROUGH A LIQUID CRYSTAL

M. Bertolotti, S. Martellucci, F. Scudieri and

R. Bartolino

Università di Roma, Roma, Italy

1. Introduction

The aim of the present work is to study the spatial coherence of a laser beam travelling through a nematic liquid crystal cell under the application of an electric field. Depending on a) the applied d.c. electric field, b) the orientation of the liquid crystal, and c) the polarization of the incident light beam, the peculiar optical properties of the liquid crystal (induced via the interaction of the electric field with the dielectric anisotropy) allow one to modify almost at will the coherence of the light beam used.

Actually we have been able to get a variation of the spatial coherence, as measured in a near field pattern, from the nearly full coherence of the He-Ne laser used, when no d.c. electric field was applied to the liquid crystal, down to a totally incoherent field pattern, when the applied d.c. electric field was larger than a threshold value. Intermediate values of the degree of coherence have been obtained in a reproducible way as a function of the applied electric field. The possibility of obtaining a medium of desired dielectric anisotropy fluctuations by application of an electric field makes this new technique very useful when applied to the study of coherence properties of light beams.

It is well known[1] that the deformations of an oriented nematic liquid crystal induced by an external electric field are of different types: *normal* deformations governed only by the dielectric anisotropy (static deformations) and *anomalous* deformations (i.e. Williams domains, dynamical scattering, etc.) due to additional orientation processes that arise from the electrical conductivity. Deformations of one type may overcome the others depending on the actual nature of the liquid crystal used, the undisturbed optical axis orientation given to the nematic itself, and the value and direction of the applied electric field. In the present work we report the experimental results of the modifications suffered by a coherent light beam after travelling through a sample of nematic liquid crystal undergoing deformations of normal and anomalous type, at room temperature.

2. Experimental Set-Up

The coherence properties of the light beam have been studied by a Mach-Zehnder interferometer. The experimental set-up used is shown in Fig. 1. The light source is an unpolarized 1 mW He-Ne laser; E is a beam expander and P a polarizer; $M_{1,4}$ and $M_{2,3}$ are respectively the beam splitters and the mirrors of the Mach-Zehnder (mounted according to Kinder's arrangement); L and O the focussing lens (12.5 cm focal length) and the microscope 20x objective used to make an image of the liquid crystal cell S on the photographic film F (Ilford Pan F). Magnifications up to 200 times of the interference fringe patterns have been obtained. The Kinder's arrangement of the Mach-Zehnder allowed us to change the direction and the number of fringes in the field without changing their localization plane (where the cell S was placed). The degree of coherence of the near field pattern of the test beam after crossing the liquid crystal cell with respect to that of the reference beam is deduced from the visibility of the fringes as observed in the photograph. Typical exposure times in our

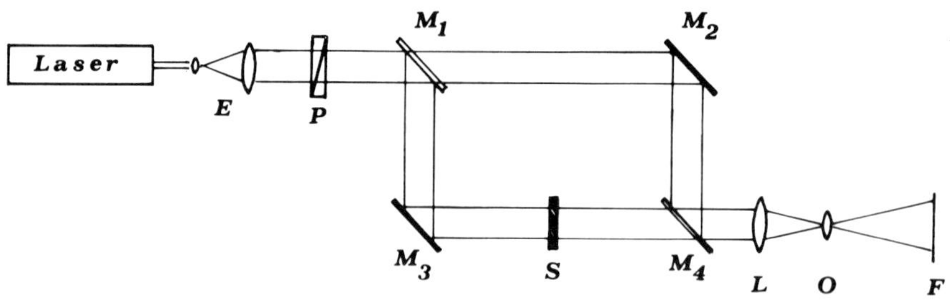

Fig. 1. Experimental set-up of the optical mounting.

measurements ranged between 0.25 and 20 secs. A schematic view of the liquid crystal cell is shown in Fig. 2a.

A thin film of MBBA nematic liquid crystal [n-p (Methoxybenziliden) p-n (butilanyline)] was placed between two optical flats taken apart by a 12.5 μm thick mylar spacer. The inner surfaces of the optical flats have been coated by a SnO layer, in order to work as a uniform, transparent, plane electrode. The electrodes were treated with a long chain surfactant [2] to make the undisturbed optical axis $\underline{n}_0$ of the liquid crystal parallel to the applied d.c. electric field $\underline{E}$ ($\underline{n}_0 \parallel \underline{E}$); alternatively, the experimental conditions characterized by $\underline{n}_0 \perp \underline{E}$ were accomplished by rubbing the SnO coated surfaces. The voltage across the electrodes was made to range from zero to values beyond the threshold voltages for normal deformations[1]

$$U_{o\varepsilon} = E_{o\varepsilon}\, x_0 = \pi\sqrt{\frac{K_{11}}{\varepsilon_0(\varepsilon_1-\varepsilon_2)}} \qquad \text{for } \phi_R = 0, \tag{1}$$

$$U_{o\varepsilon} = E_{o\varepsilon}\, x_0 = \pi\sqrt{\frac{K_{33}}{\varepsilon_0(\varepsilon_2-\varepsilon_1)}} \qquad \text{for } \phi_R = \frac{\pi}{2}, \tag{2}$$

where (see Fig. 3) x_0 is the thickness of the sample, ϕ_R the angle between $\underline{n}_0$ and the boundary surfaces (a plane geometry is assumed), K_{11} and K_{33} are the elastic constants of the material, ε_0 and $\varepsilon_1-\varepsilon_2 = \Delta\varepsilon$ (negative or positive) the dielectric constant and the electric field-induced dielectric anisotropy of the liquid crystal. By increasing the applied voltage, measurements of the loss of coherence have been made beyond the Carr-Helfrich[3] threshold $U_{o\sigma}$ for anomalous deformations. Note that in our experimental conditions (MBBA) it is $U_{o\varepsilon} < U_{o\sigma}$.

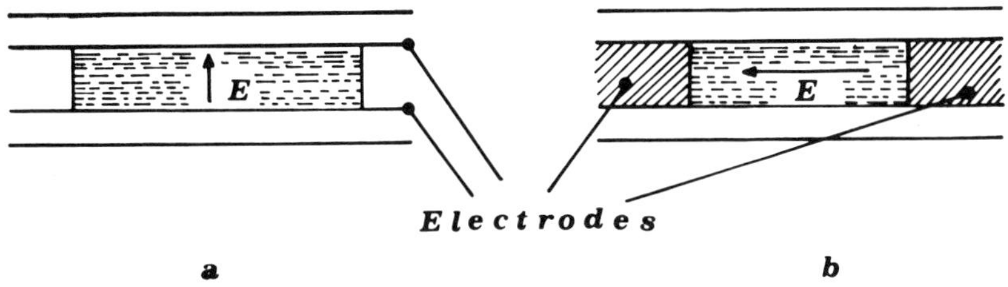

Fig. 2. Detailed view of the liquid crystal cell.

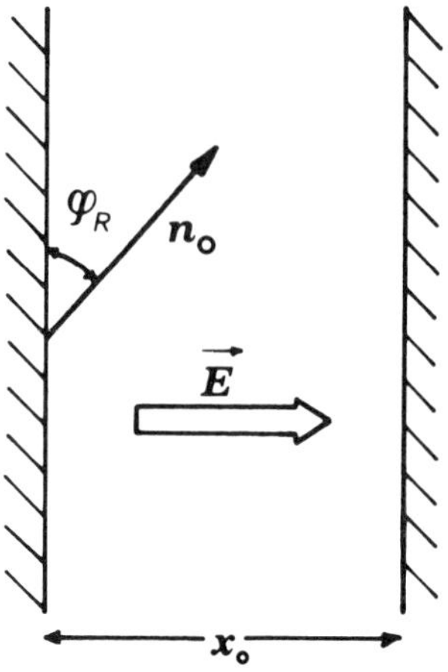

Fig. 3. Plane geometry of undisturbed orientation of the liquid crystal.

3. Experimental Results and Discussion

Let us discuss the experimental results obtained in the case $\underline{n}_0 \parallel \underline{E}$. When the electric field is applied in this geometry, the fringe position changes by increasing the field. In Fig. 4 the fringe shift , measured in number of fringes, is reported as a function of the applied voltage ($U < U_{0\sigma}$). The value of the applied voltage where a detectable fringe shift can be observed is in a very good agreement with the value obtained by substituting the numerical value of the constants in Eq. 2 for the MBBA case. The fringe shifts do not depend on the chosen polarization plane of the incident light beam. These results may be interpreted in a phenomenological way in terms of a collective bending of the molecular chains which tends to assume a configuration corresponding to a minimum of dielectrical potential energy. From an optical point of view the liquid crystal might be thought to work as a uniaxial crystal with the optical axis parallel to the direction of propagation of the light (of course [4], the optical anisotropy is a function of the applied electric field at $U_{0\varepsilon} < U < U_{0\sigma}$). The fringe shift saturates when the voltage value becomes of the order of $U_{0\sigma}$, and the process is governed by the anomalous deformations. The fringe pattern is then deforming more and more as the voltage further increases, again with no dependence from the light polarization plane. Small areas where spatial coherence is still preserved remains and at last, the visibility of the fringes goes to zero all over the field of view, when the voltage is of the order of 2 - 3 times $U_{0\sigma}$.

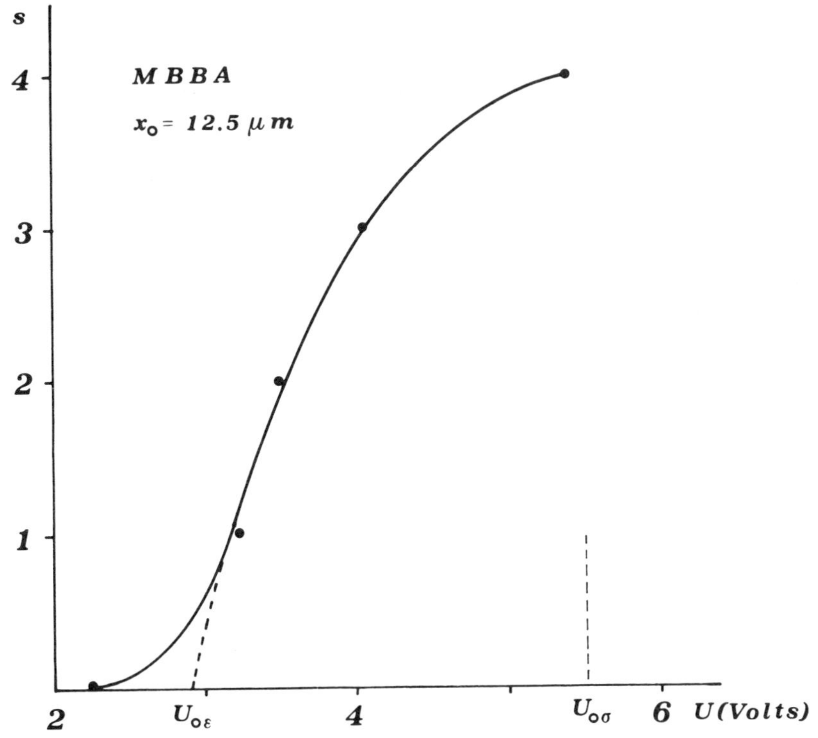

Fig. 4. Fringe shift vs. applied voltage for the $\underline{n}_0 \parallel \underline{E}$ case.

We observe that the fringe shift s = 4 fringes up to the start of the anomalous deformation, gives an index of refraction variation $\Delta n = 0.2$. This value must be compared with the maximum index change $\Delta n = n_\parallel - n_\perp$, which in MBBA is 0.3. This implies that the matter motion starts when the material is nearly all oriented. The same sample, when observed in conoscopic vision, shows the characteristic interference cross to move while the hues characteristics of positive uniaxial material appear in succession on the circular bands that contours the cross at increasing applied voltage. Correspondingly a minimum for the electrical conductivity ($\Delta\sigma > 0$) was observed[4]. The fringe shift we observe is in good agreement with the results reported also in a recent paper[1].

In the case $\underline{n}_0 \perp \underline{E}$ a completely different behavior is observed. Independently from the polarization plane of the incident light, no fringe shift is observed at $U > U_{0\varepsilon}$, that is the normal deformations do not play any role. When U approaches the $U_{0\sigma}$ value (see Fig. 5b) the anomalous deformations cause a local bending of the fringe pattern, provided that the polarization plane of the incident beam be parallel to the rubbing direction. At larger values of the applied voltage the visibility of the fringes deteriorates (even if it may be possible to localize several zones of partial coherence in the field), until the near field pattern becomes totally incoherent. A phenomenological interpretation of these results in terms of anomalous deformations[5] assumes that real space charge and fluid flows establish domains (Fig. 6) into the liquid crystal. The flow initiates when the $U_{0\sigma}$ threshold value is reached. The cellular motion is of vortex type, the dimensions of the vortices being of the order of the sample thickness[6]. When $U \gtrsim U_{0\sigma}$ the flow does not assume any more a collective character, so that it is possible to distinguish coherence areas in the field of view and over the quite long integration time (exposure time as long as 20 secs have been used). On the contrary, when $U >> U_{0\sigma}$ the entire material is flowing in a random way and the coherence is completely lost.

In the same experimental conditions ($\underline{n}_o \perp \underline{E}$), but with the polarization plane perpendicular to the rubbing direction, the phenomenon is much less evident even at very large values of the applied voltage. In other words the coherence property of the light is not affected in this case by the onset of vorticity into the material. The motion must therefore develop on the plane determined by the rubbing direction and that of the applied electric field; as

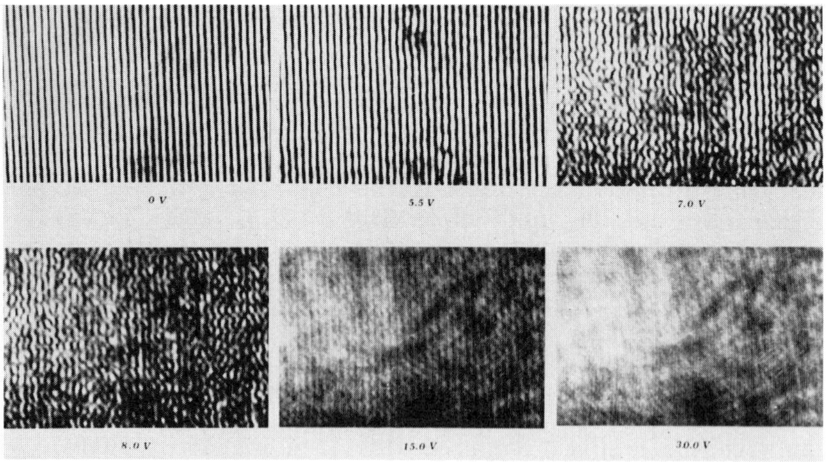

Fig. 5. Typical interferograms obtained in the $\underline{n}_o \perp \underline{E}$ case: a) undisturbed fringe pattern (U = 0); b) $U = U_{0\sigma} = 5.5$ volts; c) $U = 1.28\ U_{0\sigma}$; d) $U = 1.45\ U_{0\sigma}$; e) $U = 2.73\ U_{0\sigma}$ (threshold for dynamic scattering); f) $U = 5.46\ U_{0\sigma}$. Magnification 140×.

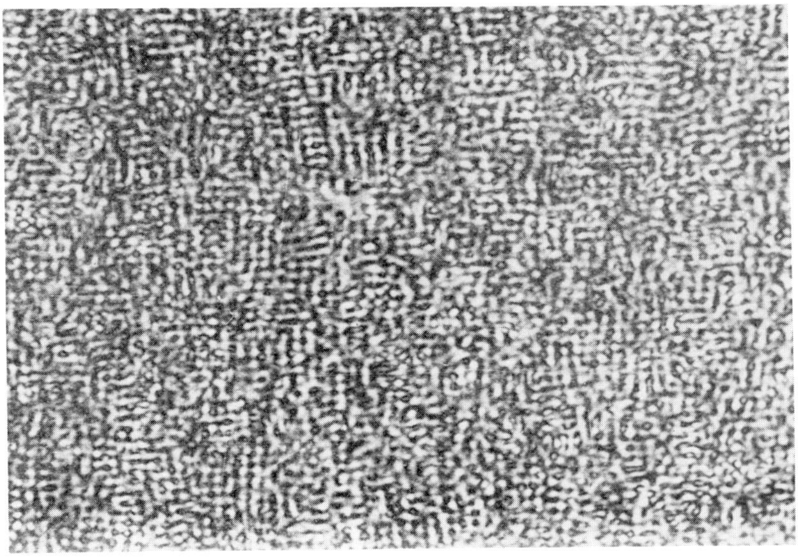

Fig. 6. Domains pattern for the $\underline{n}_o \perp \underline{E}$ case at U = 1.09 $U_{o\sigma}$ as obtained by recording on the photographic plate only the test beam. The magnification is the same as in Fig. 5a.

a consequence, the light wave sees always the same mean value of the refractive index and there are no fringe shifts. The small local deformations of the fringe patterns (see Fig. 7) represent the only visible effect of the vorticity, and they might be due to the gradient of the refractive index associated with the vortices themselves.

In order to evaluate the role of the applied electric field direction, measurements have been made by using a different set-up of the liquid cell (see Fig. 2b). The electrodes separation was of the order of 1 - 2mm and the sample thickness was 100 μm. For light propagation perpendicular to the direction of the applied field the phenomenon is now dependent on the chosen polarization plane of the light, for both the $\underline{n}_o \parallel \underline{E}$ and the $\underline{n}_o \perp \underline{E}$ case (note that for this geometry it is possible to give the initial orientation to the molecular chains by the rubbing technique). However the voltages corresponding to the $U_{o\varepsilon}$ and $U_{o\sigma}$ values are at least one order of magnitude larger than in the previous mountings.

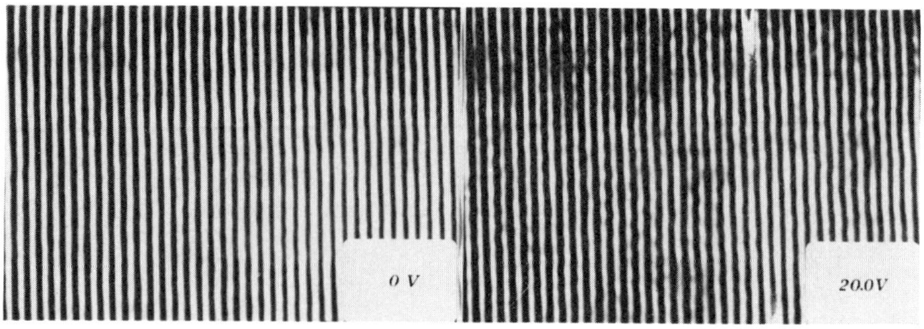

Fig. 7. Interferograms obtained for the $\underline{n}_o \perp \underline{E}$ case, but with incident light polarized at 90° with respect to $\underline{n}_o$ a) U = 0; b) U = 2.92 $U_{o\sigma}$.

4. Conclusions

As a concluding remark, we stress that liquid crystals may be efficiently used to affect appreciably the degree of coherence of an initially coherent light beam. Remarkable is the reproducibility of the phenomenon as a function of the applied electric field. In particular when the applied voltage is much larger than $U_{o\sigma}$, the liquid crystal may be considered as a turbulent medium with very large light scattering cross section[7]; therefore multiple scattering processes may occur, which strongly affect the coherence properties of the forward field of the light beam.

References

1. H. Gruler and G. Meier, Mol. Cryst. and Liq. Cryst. *16*, 299 (1972).
2. W. Haas, J. Adams and J. B. Flannery, Phys. Rev. Letters *25*, 1326 (1970).
3. W. Helfrich, J. Chem. Phys. *50*, 100 (1969).
4. M. Bertolotti, F. Scudieri, D. Sette and R. Bartolino; to appear in the September issue of J. Appl. Physics.
5. R. Williams, J. Chem. Phys. *39*, 384 (1963).
6. G. Durand, M. Veyssie, F. Rondelez and L. Leger, Comp. Rend. Acad. Scie. Paris *270B*, 97 (1970).
7. M. Bertolotti, B. Daino, P. Di Porto, F. Scudieri and D. Sette, J. Phys. A *4*, L97 (1971), and, the same authors, Journal de Physique, Supplement C1-1972, fasc. 2-3, Vol. 33 (1972), p. C1-63.

NON-CRITICAL RAYLEIGH SCATTERING FROM PURE LIQUIDS

C.J. Oliver, E.R. Pike and J.M. Vaughan

Royal Radar Establishment, Malvern, Worcs, UK

Introduction

The spectrum of light scattered by a non-critical pure liquid contains both Brillouin components, shifted from the centre frequency, and a non-shifted Rayleigh component. Acoustic phonon scattering gives rise to the Brillouin lines which are typically shifted by a few GHz from the laser line. Isobaric entropy fluctuations give rise to the non-shifted Rayleigh component whose width depends on the thermal diffusivity of the liquid. The half-width at half-height of the Rayleigh line is given by

$$\Gamma(\text{radians/sec}) = \frac{\Lambda}{C_p^*} \left(2n\, k_o \,\text{Sin}\, \frac{\theta}{2}\right)^2 , \qquad (1)$$

where k_o is the magnitude of the incident wave vector, θ the scattering angle, n the refractive index, C_p^* the specific heat per unit volume and Λ the thermal conductivity. The thermal diffusivity depends on the specific heat and the thermal conductivity and is given by Λ/C_p^*. Measurement of the infinite wavelength thermodynamic variables for pure liquids are subject to large discrepancies as can be seen from tables of constants. Measurement of the Rayleigh linewidth provides potentially a very accurate determination of the thermal diffusivity.

Replacing typical values of Λ/C_p^* ($\sim 10^{-3}$ cm^2 s^{-1}) in equation (1), the Rayleigh half-width at half-height can be seen to have a maximum value of about 20 MHz at an angle of 180° using a HeNe laser.

Even at this angle the optical resolution required ($\sim 10^8$) to measure this linewidth is at the limit of performance of conventional interferometers. Such measurements have, in fact, been made by Greytak [1], Oliver and Pike [2] and Searby [3] using Fabry-Perot interferometers. An alternative technique is to use heterodyne spectroscopy as was done by Lastovka and Benedek [4] and Berge and Dubois [5]. In addition, by using the newer parallel-channel digital techniques of photon correlation spectroscopy, optimum usage can be made of the heterodyne signal affording a considerable advance on the previous methods [4] [5]. This technique too is at the limit of its capabilities at backward angles. However by studying the scattering at more forward angles the situation can be improved. There are therefore attendant advantages and disadvantages of the two techniques which it is profitable to discuss.

Comparison of Interferometry with Heterodyne Spectroscopy

A spectrometer operating at its peak resolving power, as the Fabry-Perot would be here, introduces the problem of separating the instrument function from the actual linewidth. Where these are comparable the errors introduced in this process may be considerable. Using a correlator on the heterodyne spectrum on the other hand introduces no instrument function except for the effect of dead-time on the first channel. The only effect of aperture size, clip level, count rate, sample time and dead-time on the entire spectrum is to change the ratio of the spectral terms to the flat uncorrelated background. A second difficulty in the spectral analysis of the scattered light is caused by the inclusion of unshifted light scattered by the entrance and exit windows of the sample cell. Since pure liquids are weak scatterers this component can be considerable. With the Fabry-Perot interferometer this flare will contribute a component in the total spectrum having the shape of the instrument function. This will artifically narrow the observed width of the Rayleigh line leading to an incorrectly low value of the thermal diffusivity. In the case of heterodyne spectroscopy any flare only adds to the local oscillator signal and therefore does not effect the linewidth. A third problem in the spectral analysis arises from the scattering from dust contaminants in the liquid sample. This has a greater effect on the Fabry-Perot spectral measurements where again it gives rise to an artifically narrow line. In the case of heterodyne spectroscopy this scattering will largely be overcome by the local oscillator signal and so the apparent linewidth will hardly be affected.

However when we consider the light gathering power and statistical accuracy of the two techniques we find that the Fabry-Perot offers considerable advantages. The input aperture of the

Fabry-Perot is limited only by the spread of scattering vectors in the acceptance angle. However the accuracy of heterodyne spectroscopy is not improved significantly by using more than a few coherence areas; the coherence area is defined by the first minimum of the diffraction pattern of the source at the detector. Thus, since count-rate is proportional to area, the Fabry-Perot can accumulate the same number of counts in a shorter time. Now the accuracy with which the linewidth can be determined depends on the total counts and the total experimental time in a different way for the two techniques. For the Fabry-Perot, provided we ignore the effect of the instrument function, the fractional error in linewidth is given by

$$\frac{\Delta\Gamma}{\Gamma} = \frac{\pi}{\sqrt{2}} \frac{1}{\sqrt{N_c}} \tag{2}$$

for low count rates, where N_c is the total number of photodetections during the experiment. For the autocorrelator however under similar conditions the error can be written as

$$\frac{\Delta\Gamma}{\Gamma} = \frac{4.6}{\sqrt{N_c}} \frac{\sqrt{N_{coh}}}{\sqrt{N_c}} , \tag{3}$$

(based on Jakeman, Pike and Swain [6]) where N_{coh} is the total number of coherence times (reciprocal linewidths) taken for the experiment. Thus if the same number of photodections take longer to be collected the accuracy of the Fabry-Perot result is unchanged but that of the correlator is reduced. Since in this work the number of counts per coherence time, N_c/N_{coh}, is much less than one the accuracy achieved in heterodyne correlation spectroscopy is considerably less than that achieved with the Fabry-Perot. Some of this disadvantage will be overcome by the correlator system since it is a parallel-channel instrument whereas Fabry-Perot interferometers are normally single-channel devices.

To conclude the comparison one must weigh the advantages of obtaining a well-defined spectrum to which no corrections need to be made, as found in heterodyne spectroscopy, against the inherently greater statistical accuracy encountered with the Fabry-Perot interferometer. Since acceptable results can be obtained with heterodyne spectroscopy, particularly at forward angles where the count rate per coherence time is greater, we have concluded that the advantages in interpretation of this technique outweigh its disadvantages. The Rayleigh linewidth measurements described below were performed in this way. Optimum usage of the scattered light was achieved using a Malvern Digital Correlator Type K7023 [7], which gives real-time, parallel-channel data processing down to sample times of 50 ns.

Experimental Technique

The experimental arrangement is shown in figure 1.

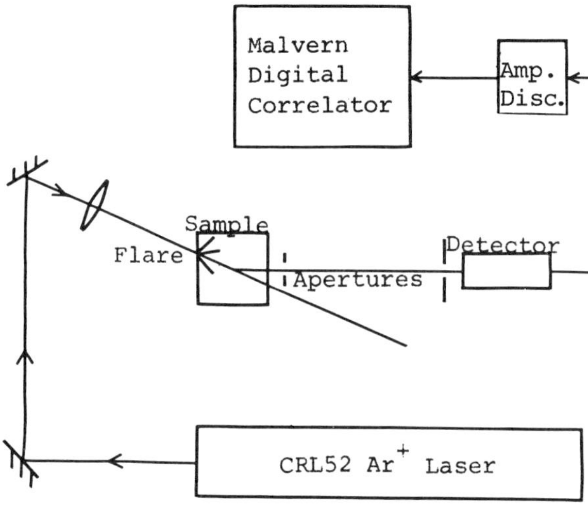

Fig. 1. Experimental arrangement for measuring the heterodyne photon-correlation spectrum for non-critical Rayleigh scattering of laser light by pure liquids.

Light from a CRL52 Argon ion laser, operated in single-frequency using an intercavity etalon, is scattered through an angle between $2\frac{1}{2}°$ and 5° to the detector, an ITT FW130 photomultiplier, where it is mixed with the local oscillator signal which is derived from the flare scattering off the walls of the glass cell. The photomultiplier output pulses were then standardised in an amplifier and discriminator and their autocorrelation function computed using the Malvern Digital Correlator. Samples of the liquid under study were transferred into clean spectrafluorimetry cells after purification by high-speed centrifugation. Any remaining dust introduced by handling was removed from the body of the liquid by centrifugation at 400g in the sample cell. Samples prepared in this fashion are sufficiently free from contaminants for scattering from dust not to be appreciable.

Since the sample is contained in a square cell the scattering angle θ is related to the measured angle in air, θ_{air} . For small scattering angles therefore equation (1) will be modified to

$$\Gamma(\text{radians/sec}) = \frac{\Lambda}{C_p^*} (k_o \, \theta_{air})^2 . \tag{4}$$

Several measurements of this linewidth for angles between 2½° and 5° were performed.

Theory

The observed normalised autocorrelation function, for heterodyne detection, has the form [8]

$$g^{(2)}(\tau) = 1 + \frac{\langle n_s \rangle}{\langle n_c \rangle} \{g^{(1)}(\tau) \exp(i\omega\tau) + g^{(1)*}(\tau)\exp(-i\omega\tau)\} + 0\left(\frac{\langle n_s \rangle}{\langle n_c \rangle}\right)^2 , \tag{5}$$

where $\langle n_s \rangle$ and $\langle n_c \rangle$ are the count rate from the scattered and coherent fields respectively. Thus by increasing $\langle n_c \rangle$ the terms in order of $(\langle n_s \rangle/\langle n_c \rangle)^2$ become negligible and, for a symmetrical spectrum,

$$g^{(2)} \tau \simeq 1 + \frac{2\langle n_s \rangle}{\langle n_c \rangle} g^{(1)}(\tau). \tag{6}$$

In this case

$$g^{(2)}(\tau) = 1 + C \exp\left[- \frac{\Lambda}{C_p^*} (k_o \, \theta_{air})^2 \tau\right] , \tag{7}$$

where the constant C contains correction factors for detector area, count rate, clip level, dead time and sample time. Thus a plot of the correlation time (time to 1/e) as a function of $(\theta_{air})^2$ can be used to calculate the thermal diffusivity Λ/C_p^*.

Results and Discussion

A typical heterodyne correlation function for Rayleigh scattering from Carbon-tetrachloride is shown in figure 2.

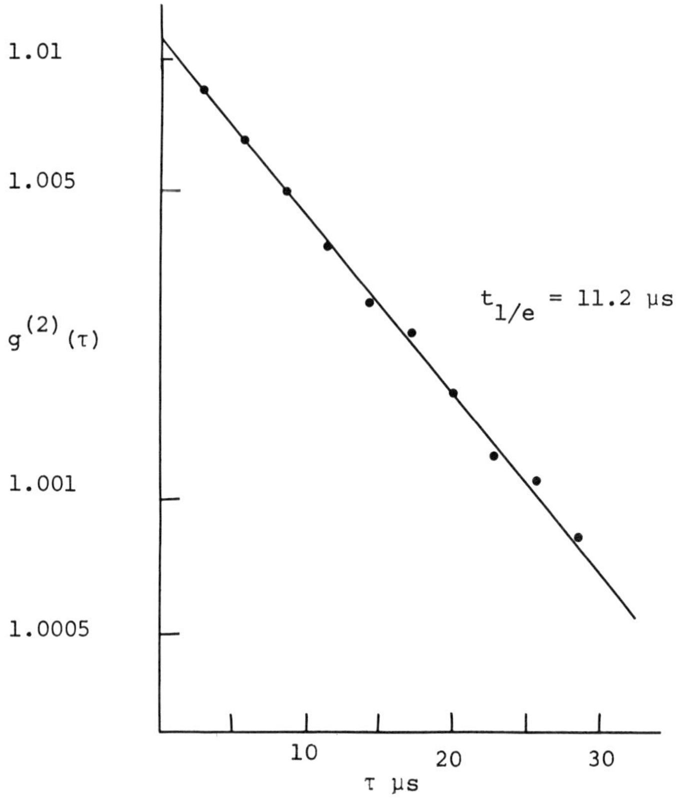

Fig. 2 A typical autocorrelation function for scattering off carbon-tetrachloride. The scattering angle is 3° 23' and the temperature 20°C.

The statistical accuracy can be seen from the scatter of the individual points about the straight line. The function is plotted on a logarithmic scale in the form $g^{(2)}(\tau)$. The intercept on the y axis is small because of the large fraction of the total light due to the local oscillator beam and also the large incoherent component due to Brillouin scattering. From the measurements at different angles the angular dependence of the Rayleigh linewidth of carbon-tetrachloride shown in figure 3, was obtained. The straight-line dependence lies well within the errors (shown as two standard deviations) of the individual points. From the gradient of this graph the thermal diffusivity of carbontetrachloride was calculated to be $7.6 \pm 0.3 \times 10^{-4} cm^2 S^{-1}$.

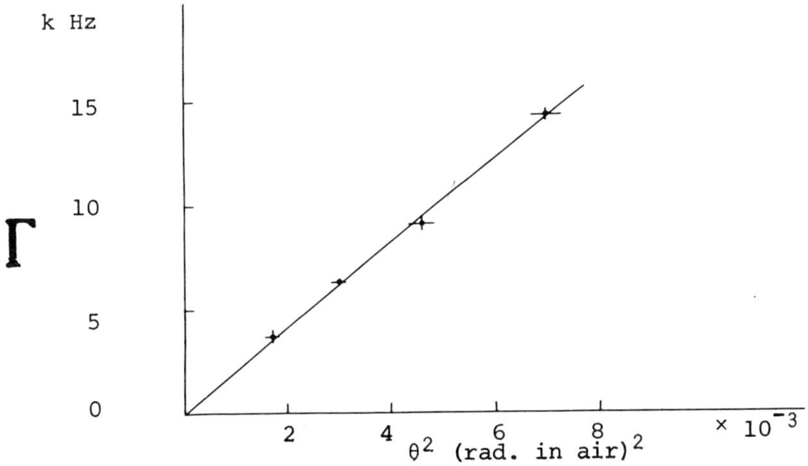

Fig. 3 The angular dependence of the Rayleigh linewidth for carbon-tetrachloride.

A comparison between the thermal diffusivity calculated from the infinite wavelength thermodynamic variables using data quoted by Oliver and Pike [2] and Searby [3] with the experimental measurements of the above authors, Greytak [1] and the present work is given in Table 1.

Calculated	Ref	Observed	Ref
10^{-4} cm^2 S^{-1}		10^{-4} cm^2 S^{-1}	
7.1 ± 0.2	(3)	6.98 ± 0.25	(3)
8.2 ± 0.3	(2)	8.0 ± 0.5*	(2)
		7.1 ± 1.5	(1)
		7.6 ± 0.3	present work

* This result has been corrected for K vector spread.

TABLE 1 Summary of the calculated and observed values of the thermal diffusivity of carbontetrachloride.

The spread in the calculated value of the thermal diffusivity from tables of thermodynamic variables is reflected in the 15% difference between the two values shown [2], [3]. It is therefore not possible to use this comparison to vindicate either technique compared with the other.

Conclusion

In conclusion it seems plausible that Fabry-Perot data could well include systematic errors due to flare in the scattering cell rendering thermal diffusivity results of dubious validity. We would suggest that the lack of problems in the interpretation of the data obtained by heterodyne photon-correlation spectroscopy outweighs the lower count rate and leads to results in which one can place greater reliance. That being so the technique offers a potentially very accurate method for determining thermal diffusivities of pure liquids.

References

1. See G.B. Benedek, *Polarisation Matiere et Rayonnement* (Presses Universitaires de France, Paris).
2. C.J. Oliver and E.R. Pike , Phys.Letts. *31A*,90 (1970).
3. G.M. Searby D. Phil. Thesis, Oxford University (1971).
4. J.B. Lastovka and G.B. Benedek,Phys. Rev.Letts. *17*, 1039 (1966).
5. P. Bergé and M. Dubois,C.R. Acad. Sci. (Paris) *269*, 842 (1969).
6. E. Jakeman, E.R. Pike and S. Swain,J. Phys. A *4*, 517 (1971).
7. Manufactured by Precision Devices and Systems Ltd, Spring Lane, Malvern, Worcs, UK.
8. E. Jakeman,J. Phys. A *3*, 201 (1970).

COOPERATIVE EFFECTS IN SPONTANEOUS EMISSION*

R. Bonifacio

Università di Milano, Milano, Italy

We discuss the cooperative superradiant decay of N two-level atoms contained in a pencil-shaped volume. The Dicke "coherence brightening" - i.e. the emission of a hyperbolic secant pulse whose intensity is proportional to N^2 and whose time duration is inversely proportional to N - occurs instead of the usual exponential decay under the following conditions:

i) the incoherent atomic decay, due to natural relaxation, is so slow that the individual atomic dipoles do not get out of phase with each other before the cooperative decay is over.

ii) the length of the pencil-shaped volume is smaller than a critical "cooperation length" inversely proportional to the square root of N. This condition guarantees that they cannot feed themselves back into atomic excitation to any appreciable amount, and that the envelope of the emitted pulse is essentially constant over the length of the sample, so that all the atoms radiate as a single macroscopic dipole rather than seeing and producing different values of the field at different points. The emitted light pulse is found to have different statistical properties for an incoherently and a coherently prepared "superradiant" atomic initial state. The former case is characterized by large quantum fluctuations and strong atom-atom and atom-field correlations. In the latter case quantum fluctuations are small and the system

---

*This report is based on recent work done in collaboration with F. Haake and P. Schwendimann and published in Phys. Rev. *A4*, 302 (1971) and Phys. Rev. *A4*, 854 (1971).

behaves essentially classically. By also solving for a class of coherently prepared intermediate initial states we show that large quantum fluctuations occur only if the initial total occupancy of the excited state differs from the total number of atoms at most by a number of order unity.

AMPLIFIED SPONTANEOUS EMISSION

L. Allen

University of Sussex, Brighton, Sussex, England

1. Introduction

In 1954 Dicke [1] discussed the possibility of a coherent co-operative process between an ensemble of two-level atoms. His paper has become a classic, and discusses the process now known as super-radiance. At the time the paper attracted relatively little attention except that of a few who worked in r.f. spectroscopy.

At the Quantum Electronics Conference in Paris in 1963, Dicke [2] adapted his theory to apply to a long 'pencil' of atoms and described how the radiation would be emitted into so-called "end-fire modes". It is fair to say that the paper was not well understood at the time, and that it was interpreted by some as a statement that the process Dicke had described in 1954 was the process now observed at work in the laser.

Following on from this work Carver [3] drew up an, at first sight, attractive scheme when he considered a column of two-level atoms of length L and square cross-section of dimension d. He considered that each mode would radiate into some diffraction limited solid angle $(\lambda/d)^2$ and that a particularly intense beam would result when the solid angle defined by the geometry of the column, $(d/L)^2$, was just filled by the zero-order diffraction, i.e.

$$\frac{d}{L} = \frac{\lambda}{d}$$

or

$$d^2 = L\lambda .$$

With this starting point and assuming that in some way, following Dicke, the intensity along the axis would be enhanced by a factor N, the number of atoms, he showed that the condition for emission into an end-mode was just the same as that of the laser threshold condition derived by Schawlow and Townes [4]. This seemed to be an attractive piece of work since it appeared to firmly link laser action with an enhancement factor N that seemed at that time to be a prediction of Dicke. Even if this analysis had been correct, and as we shall see it is not, one knows [5] with hindsight that the enhancement factor for such a geometry would not be N but μN where μ is a geometry dependent factor less than unity.

Meanwhile a number of workers had observed highly directional, intense radiation from certain "laser" systems even when one or both of the cavity mirrors had been removed. It is not certain who called such radiation "superradiance" although certainly White and Rigden [6] and Yariv and Leite [7] both used it in 1963 for just such radiation. It is significant that having used the word neither paper quotes the Dicke reference. Nor was any attempt made to connect their observations quantitatively with the Dicke theory. Nor did Carver connect his calculations with the experimental work that had already been published on the radiation from "mirror-less lasers", or "superradiance" as it was beginning to be called. This semantic confusion about superradiance continues to occur in the literature partly out of ignorance by those who know nothing of the collective process described by Dicke, and partly out of fear by those who do that 'real superradiance' is in reality only the same process in a disguised form.

The two regimes, however, seem incompatible. Dicke's superradiance would seem to depend explicitly on the creation of a coherently prepared atomic system, the wave-functions of whose atoms had some formal phase relationship with respect to one another. In the "high gain mirror-less laser system" on the other hand the mode of excitation is invariably random incoherent excitation by electron collisions and/or cascades from higher energy levels. Consequently while a number of workers [5], [8]-[10] were beginning to develop the theory appropriate to the Dicke situation for volumes of dimension large compared with λ, Allen and Peters were examining how the "high-gain mirror-less laser system"might be described and how, in particular, it was unnecessary to invoke any process other than stimulated emission [11]-[16]. Although it was true that mirror-less high gain systems had to some extent been investigated theoretically it was as laser amplifiers, and the signal generated within the medium was ignored. It transpires that to ignore this radiation is very much to throw away the baby with the proverbial bath water.

2. Threshold Condition for Amplified Spontaneous Emission

Allen and Peters [11] showed that Carver's happy agreement with the threshold condition for the laser predicted by Schawlow and Townes could be achieved without making assumptions that were either difficult to justify, such as $d^2 = L\lambda$, or by invoking superradiance enhancement by the factor N.

If a mode subtends a solid angle $\Delta\Omega = \lambda^2/d^2$ then a fraction $\lambda^2/4\pi d^2$ of the photons go into the end mode and if

$$\frac{\lambda^2}{4\pi d^2} N > 1$$

then an end mode is excited. If N is replaced by nLd^2, where n is the density of atoms, strictly of excited atoms, and allowance is made for the fact that only a fraction $\Delta\nu_N/\Delta\nu_D$ will go into the same mode, then the condition becomes

$$n_c > \frac{4\pi}{L\lambda^2} \frac{\Delta\nu_D}{\Delta\nu_N} .$$

This is the condition for one mode to be excited and, as the rate of stimulated emission into a mode is the same as the rate of spontaneous emission into the mode for one photon, this is the condition for stimulated emission to commence. It is not surprising that this result agrees with the Schawlow and Townes' formula, the only difference lies in the fact that here we consider the modes lying within the geometry of the atoms, so to speak, rather than of the modes of the cavity. It was obvious by inspection from numbers culled from published work that this result was, at the very least, compatible with what was actually observed.

In a subsequent paper [12] an alternative derivation of the threshold condition involving a simple rate equation for the photons and correctly attributing the threshold condition to inversion density rather than atom density led to the result

$$n_c = \frac{8\pi \Delta\nu_D \tau_2}{L\lambda^2\phi} ,$$

or

$$L_c = \frac{8\pi\ \Delta\nu_D\ \tau_2}{n\lambda^2\phi} \quad ,$$

for the critical length if n is the population inversion. This result is a factor of π different from the earlier derivation but has the advantage of introducing $\Delta\nu_D$, τ_2 and ϕ, as well as the inversion density, in a formal manner instead of phenomenologically as in the previous derivation.

The condition was experimentally verified in the c.w. 3.39 μm He-Ne and the pulsed output at 0.614 μm in Ne. The He-Ne was r.f. excited and plasma tubes of length 310 cm with bores varying between 2 and 4 mm in 0.5 mm stages, were used. The A.S.E. intensity was measured as a function of length by disconnecting the r.f. clips on the far end of the tube away from the detector so that the detector geometry remained constant. The measurements were carried out for various levels of excitation, care being taken to ensure that the inversion was constant in all parts of the tube by monitoring the spontaneous emission from the upper level as is seen at 0.633 μm through the side of the tube. The spontaneous emission could be controlled to an accuracy of 5%. Figure 1 shows $1/L_c$ plotted against inversion, the result in agreement with theory is a straight line through the origin.

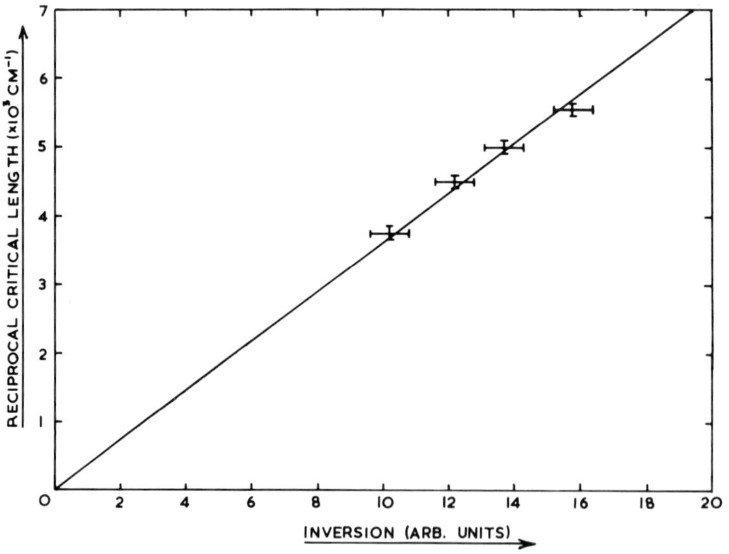

Fig. 1 Plot of $1/L_c$ against inversion density for the 3.39 μm transition in He-Ne.

The slope of the curve shows, assuming the appropriate value for $\phi, \lambda, \Delta\nu_N$, [17] that the inversions occurring were $\sim 10^8$ atoms/c.c. which is in agreement with values obtained by investigating the output intensity as a function of length for tube lengths just above L_c. The standard deviation in values of L_c for a given value of n for each of the different tube bores was less that 4%, verifying that the threshold condition is independent of bore and that Carver's condition $d^2 < \lambda L$ is not meaningful.

The pure neon system was pulse excited, and fourteen tubes with lengths between 44 and 110 cm and a bore of 2.5 mm were investigated. The neon was excited using a similar system to that of Geller et al. [18]. The gain of the system was maintained at a constant level for a chosen discharge voltage by arranging that the distance between the electrodes was kept constant even though the length of gas contributing to A.S.E. was varied. Figure 2 shows the results, L_c varied between 45 and 70 cm for the inversion densities used whereas the $d^2 < \lambda L$ criterion predicts a minimum length of $\sim$1000cm!

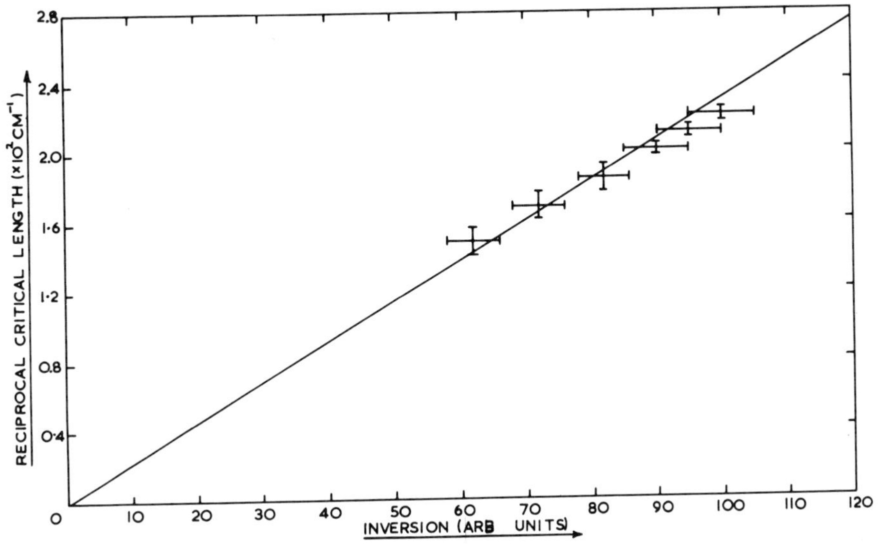

Fig. 2. Plot of $1/L_c$ against inversion density for the 0.614 μm transition in neon.

It is clear that the similarities between the derivation of the threshold condition for A.S.E. and that of the laser should lead to a simple relationship between the two [13]. Combining equations (12) and (98) of the paper by Stenholm and Lamb [19] and assuming a steady state such that $\dot{E} = 0$ yields

$$(1+I)^{\frac{1}{2}} = \frac{Q\pi^{\frac{1}{2}}\wp^2\overline{N}}{\hbar\ Ku\varepsilon_o}$$

for a laser near the centre of the gain curve. Re-writing this expression by replacing Q by $2\pi\ell_c/\delta_\ell\lambda$ where δ_ℓ is the fractional loss per pass in the laser resonator and replacing

$\overline{N}\ \ell_c$ by $\overline{n}\ L$,

where $\overline{n}$ is the inversion density in the excited part of the cavity and L is the length of excited region, we obtain,

$$(1+I)^{\frac{1}{2}} = \frac{1}{\delta_\ell}\ \frac{2\pi^{3/2}\wp^2}{\lambda\hbar\ Ku\varepsilon_o}\ L\overline{n}\ .$$

At threshold I = 0 and L = L_T, where L_T is the length of active discharge at which laser action begins, and so substituting Ku and $\wp^2$ in terms of the Doppler width and atomic lifetime of the transition

$$\overline{n}\ \ L_T = 3.2\pi^{3/2}\ \ \frac{\Delta\nu_D\ \tau_2\ \delta_\ell}{\lambda^2\ \phi}\ ,$$

we can easily show that

$$\frac{L_T}{L_c} = 0.71\ \delta_\ell .$$

In other words, for a given inversion density the length of discharge that has to be excited, L_c, to give A.S.E. is simply related to the length, L_T, that has to be excited to give laser action in a cavity categorized by the fractional loss per pass δ_ℓ.

The 3.39 μm transition in He-Ne was employed as it was known that it would operate easily both as a laser and as a source of A.S.E. A 125 cm long He-Ne discharge tube was placed between two highly reflecting concave mirrors R = 0.98 of radius of curvature 3m. The gas was r.f. excited and it was possible to excite a variable length of tube by removing or adding r.f. electrodes.

The output intensity was measured as a function of length of tube excited, for three inversion densities. For each of these inversions the mirror at the detector end was removed and the A.S.E. observed as a function of length of excited tube. (see Figure 3).

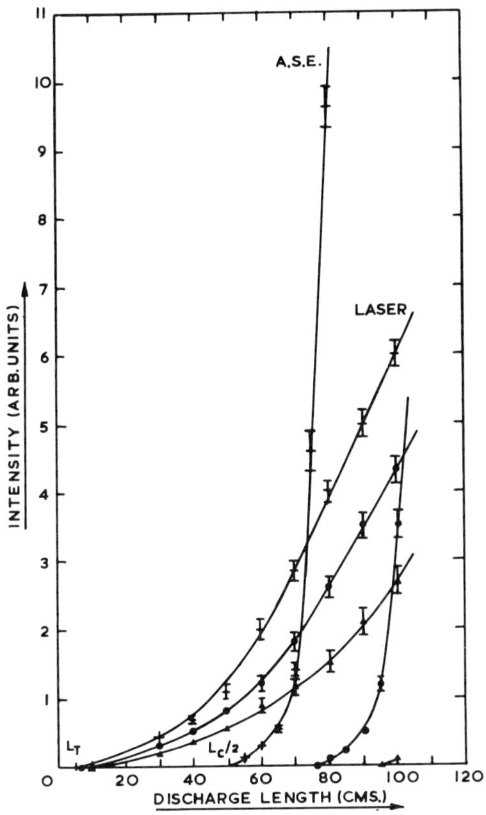

Fig. 3 Plots of laser and A.S.E. intensities against discharge length, for three inversion densities. The curve drawn with crosses has the highest inversion density and that with triangles the lowest.

It may be noticed that at a length of only $1.5L_c$ the A.S.E. output intensity exceeds that of the laser. This is simply due to the 2% transmission coefficient of the laser mirror; the nature of the outputs will, of course, be quite different. (See Section 5).

The above theory predicts that L_T should be linear with L_C. Figure 4 is in excellent agreement with this and shows a slope which seemed at the time to be slightly too large although only marginally so. Hindsight now allows us to realise that the use of one mirror could have played a larger role than had been anticipated (See Section 6). However the effect of the mirror near threshold is negligible and the theoretical prediction remains the same and is suitably verified. In previous verifications of laser theory [20], [21] considerable attention has been paid to the fact that near threshold it is important to take into account the quantum effects due to spontaneous emission. Consequently one should ordinarily

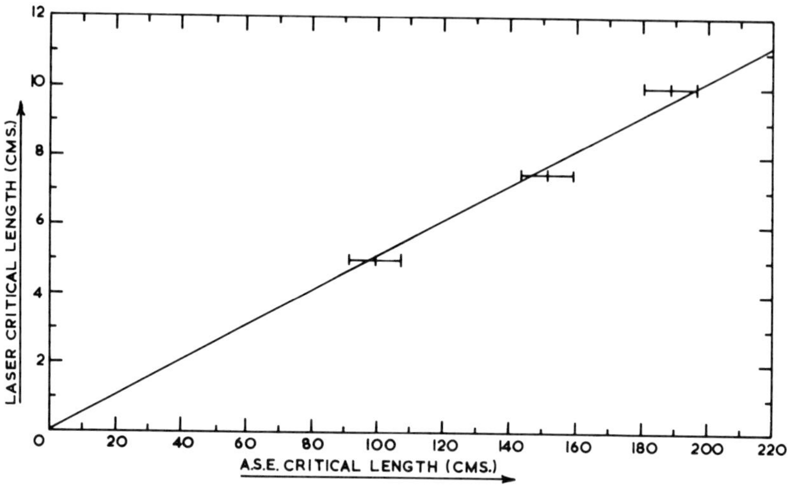

Fig. 4 Plct of L_T against L_c.

view the value of L_T predicted by Stenholm and Lamb theory with some caution, preferring perhaps to heed that of Scully and Lamb[22]. However, excellent results are achieved and it appears that the answer is that L_T is far below the lengths normally used in a laser, where in practice $L \sim L_c$, and the role of spontaneous emission introduces a very very small distortion of L_T unlike the situation previously investigated to verify Stenholm and Lamb's theory,(see Allen [21]).

3. Intensity and Saturation

The degree of success of the simple approach to A.S.E. threshold encouraged Allen and Peters to develop a full rate equation approach to the question of output intensity [14]. Yariv and Gordon [23] had previously considered amplification due to an inverted population of atoms but did not allow for saturation or for any of the effects of spontaneous emission. Kogelnik and Yariv [24] effectively considered the power output of A.S.E. by analysing the noise in a laser amplifier but again saturation was not taken into account.

The essential feature of the approach used [14] was to write complete rate equations for the populations of the two levels involved and for the photon transport equation. Careful account was taken of the fact that radiation spontaneously emitted at a particular point in the column of atoms only contributed to the observed output intensity, in either an amplified or unamplified form, provided it fell into a position dependent solid angle $\Delta\Omega(x)$. The photon flux at any position x in a column of atoms of length L is $[N(x) + N(L-x)]$. The appropriate equations are,

$$\frac{\partial n_2}{\partial t} = -(n_2-n_1)\frac{\sigma}{a}(N(x) + N(L-x)) - An_2 - A_2n_2 + R_2n_0 \qquad (1)$$

$$\frac{\partial n_1}{\partial t} = (n_2-n_1)\frac{\sigma}{a}(N(x) + N(L-x)) + An_2 + R_1n_0 - A_1n_1 \qquad (2)$$

$$\frac{\partial N}{\partial x} = \{(n_2-n_1) - \nu a\} N(x) + \frac{An_2\ a\Delta\Omega}{4\pi} \,. \qquad (3)$$

For a column with a length appreciably greater than the tube radius the variation over the tube cross-section may be ignored and,

$$\Delta\Omega = 2\pi\left[1 - \frac{(L - x)}{\{(L-x)^2 + d^2\}^{\frac{1}{2}}}\right] .$$

Solving equations (1) and (2) allow the saturation of the medium to be properly accounted for and allows, in turn, equation (3) to be integrated. The equations prove analytically insoluble unless an assumption is made. This is that $N(x) + N(L-x) = G$, a constant. Although this approximation was introduced with some trepidation it turned out to be completely justifiable. The integration of equation (3) is conducted by realising that for lengths in excess of $(L-L_c)$ the spontaneous emission will provide noise only and will not become a source of stimulated emission. Secondly, the photon number at L_c, $N(L_c)$ is effectively zero. The solution is,

$$N(L) = \frac{Y(L)d^2}{4} \int_{L_c}^{L} \frac{\exp(-x(L)y)}{y^2}\, dy \ ,$$

where

$$X(L) = \frac{K_1' G - K_2}{1 + K_3' G} \text{ and } Y(L) = \left(\frac{B + CG}{1 + EG}\right)\frac{Ad^2\pi}{4} \,.$$

In the range $L_c < x < (L-L_c)$ the total photon number is $N(L-L_c)$ and in the range $0 < x < L_c$ and $(L-L_c) < x < L$ the number varies between $N(L-L_c)$ and $N(L)$. The best value for G would thus seem to be the average value of the total photon number, $\overline{N}_L$.

The 3.39 μm He-Ne system was again investigated with a range of tube bores between 2 and 4 mm over a distance of 160 → 620 cms. For a particular inversion and bore the intensity .v. length results were fitted by computer to

$$I(L) = K \int_{L_c}^{L} \frac{\exp K_2 y/(1+K_3\overline{I}_L)}{y^2}\, dy \,,$$

where K has replaced the term in Y(L). This was certainly justified for He-Ne, since the spontaneous emission through the side-wall was independent of x and L to within the 5% that it could itself be maintained constant.

The tube was divided into three lengths each less than L_c. Each section was prepared separately to give a constant value of n_2. When all three sections were excited together, stimulated emission occurred along the tube and whether or not n_2 depended upon x or L in the range could be determined. As stated previously it did not within the 5% accuracy possible.

Figure 5 demonstrates the fit of theory and experiment. The value of the coefficient K holds little interest as it contains a scaling factor which translates a photon number into a photoelectric current, accounting also for the optics of the light collecting device. K_2 as fitted compares closely with the measured value of the gain coefficient as it should and K_3 relates to saturation. Figures 6 and 7 show the comparison of the predicted intensity as a function of discharge bore and inversion density using the same constants.

The 0.614 μm system in neon is unsuitable for the purposes of this part of the work. The initial exponential growth occurs over a very short range of tube lengths (∿ 1 cm) and the construction of discharge tubes to an accuracy of greater than ± ½ cm is formidable. Also the measurements are hindered by possible 'pulse chasing' effects in longitudinal pulse excited systems, where the exciting current pulse may be 'overtaken' by the A.S.E. light pulse along the tube before it reaches the earthed electrode. However, Leonard [25] has studied the build up of A.S.E. intensity with length for the N_2 second positive system at 0.337 μm. Figure 8 shows his data fitted to our theory and again K_2 agrees well with the small signal gain coefficient.

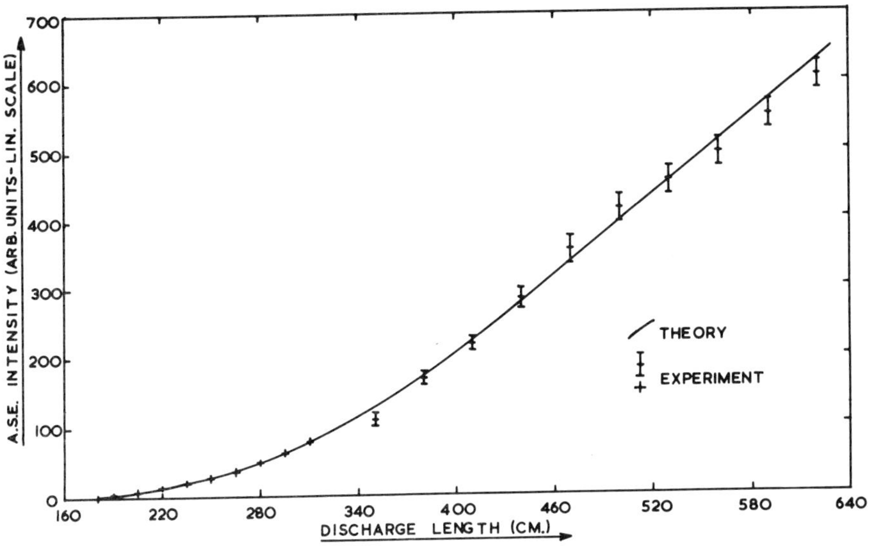

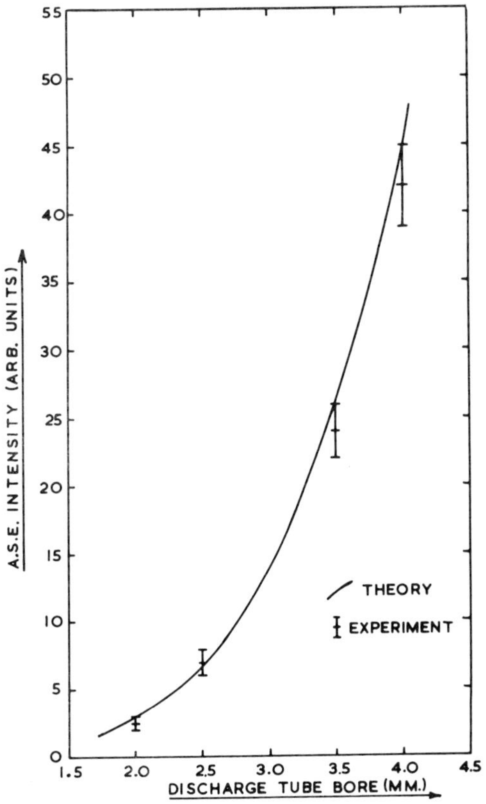

Fig. 5. Experimental points and theoretical curve (full line) for the variation of intensity with length for the 3.39 m, He-Ne A.S.E. transition. [Gain coefficient 1.58 x 10^{-2} cm^{-1}; bore 0.25 cm].

Fig. 6 Experimental points and theoretical curve (full line) for the variation of intensity as a function of tube bore for a gain coefficient of 1.58 x 10^{-2} cm^{-1} and tube length 205 cm, for the 3.39 μm He-Ne A.S.E. transition.

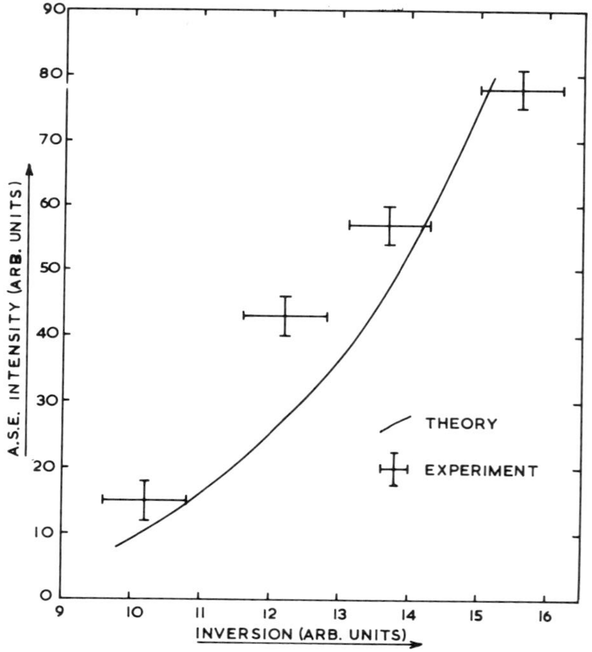

Fig. 7 Experimental points and theoretical curve (full line) for the variation of intensity as a function of inversion density for a tube bore of 0.25 cm and length 310 cm, for the 3.39 μm He-Ne A.S.E. Transition.

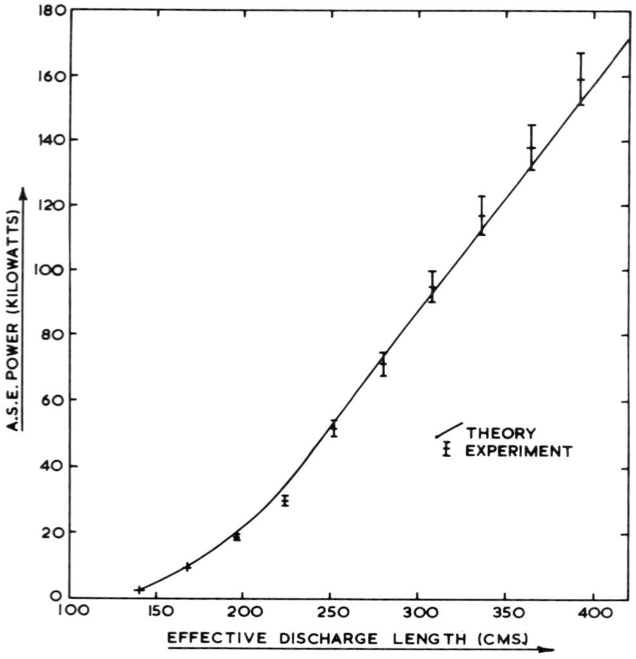

Fig. 8 Experimental points and theoretical curve (full line) for the variation of intensity with length for the 0.337 μm, A.S.E. transition in N_2.

4. Beam Divergence and Spatial Coherence

It seems likely that the beam divergence of the radiation from an A.S.E. system should arise because of one of the following considerations:

i. Diffraction at the aperture.
ii. The angle subtended by the geometry of the system
iii. The functional dependence between the angle into which radiation is emitted and the degree of amplification such radiation might expect to have.

One might expect (ii) and (iii) to be important provided they exceed the limiting value predicted by diffraction.

Peters and Allen [15] evaluated the role each of these components played by measuring the beam divergence as a function of length and inversion density in both the He-Ne 3.39 μm and the 0.614 μm neon system. The diffraction at the aperture is perhaps best described using Koppelmann's approach [26]. It may be shown that as $R \to 0$, $T \to 1$ and $\theta_0 = 0$ the complex amplitude of the A.S.E. due to diffraction would be

$$A_\theta = A_0 \sum_{m=0}^{\infty} \left(e^{-\gamma_m b}\right) \frac{\cos[(m+1)\pi-\phi] - \cos\phi}{[\phi^2 - \frac{(m+1)^2\pi^2}{4}]}$$

where $\gamma_m = \alpha_m + i\beta_m$, $\phi = \frac{\pi d}{\lambda}\sin\theta$

and

$$\alpha_m = \frac{0.512\ (b\lambda^3)^{\frac{1}{2}}(m+1)^2}{d^3}, \quad \beta_m = \frac{2\pi}{\lambda}\left[1 - \frac{(m+1)^2\lambda^2}{(2d)^2}\right]^{\frac{1}{2}}.$$

The purely geometric considerations lead to a divergence angle of d/L or possibly $d/(L-L_c)$. On the other hand it is possible to generate what we have called an "A.S.E. - geometric theory" by realising that spontaneous emission at $x = 0$ will give rise to an A.S.E. output with divergence d/L while emission along the axis at $x = (L-L_c)$ will give rise to an A.S.E. output with divergence d/L_c. But the contributions are not equal because the spontaneous emission originating at $x = 0$ has traversed the longer amplifying path. Thus it is necessary to sum the contributions due to all elements in the region $x = 0$ to $x = (L-L_c)$, for light emitted from any point in the tube cross-section, taking into account both the A.S.E. intensity of the emitted radiation due to spontaneous emission from these elements and also the angle into which the spontaneous radiation can be emitted and yet appear at the exit aperture of the tube. This has been done [27] using the previously determined constants K, K_2, and K_3 and applied to the experimental observations.

The beam divergence was measured in He-Ne by manually tracking an infrared detector, with a pinhole aperture, across the beam. In the neon system the A.S.E. output for a single pulse was photographed and the developed negative scanned with a microdensitometer to obtain the intensity distribution, after correction had been made for film non-linearities.

Figures 9 and 10 show the results for neon. In this system

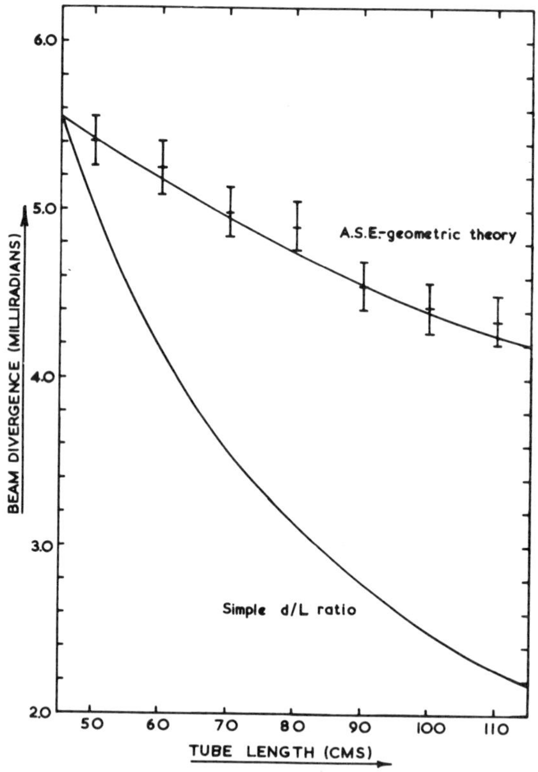

Fig. 9 Experimental points and theoretical curves for the A.S.E. geometric and simple d/L theories for the variation of beam divergence with tube length, for the 0.614 μm neon A.S.E. transition. The diffraction theory is not shown as it is 1.5 orders of magnitude smaller.

the Koppelmann theory is 1.5 orders of magnitude down on the geometric value and does not appear on the graph. It should be noted that the curves are predictions based on the previously determined constants, they are not fitted curves in any way. The correct

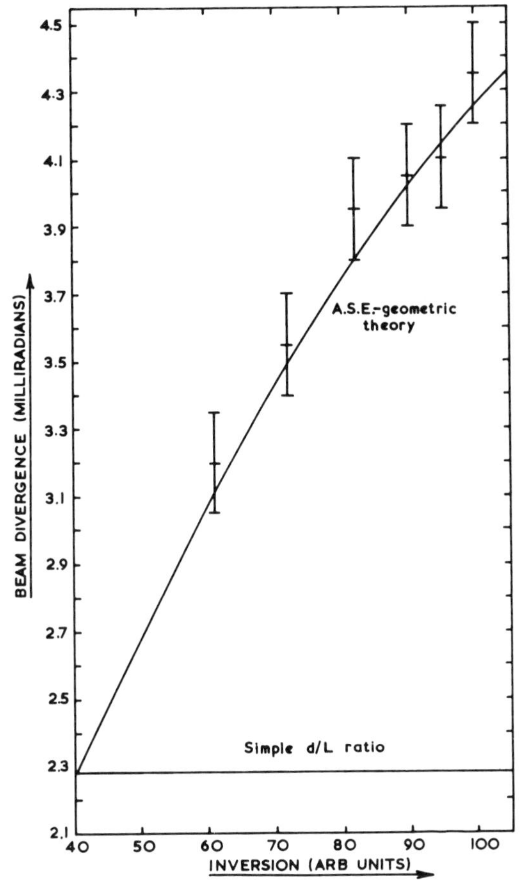

Fig. 10 Experimental points and theoretical curves for the A.S.E.-geometric and simple d/L theories for the variation of beam divergence with inversion density, for the 0.614 μm neon A.S.E. transition. The diffraction theory is not shown as it is 1.5 orders of magnitude smaller.

functional trend was also displayed at 3.39 μm both as a function of length and of inversion but the actual angle of divergence was out by a factor of ∿1.7. It is still not entirely clear why this should be the case.

The question of spatial coherence is one perhaps left for the moment in abeyance. As is shown in Figure 11, the spatial coherence develops with length for constant inversion density and we have also shown [15] that the degree of coherence increases with inversion density for a given length. However, Allen et al [28] have

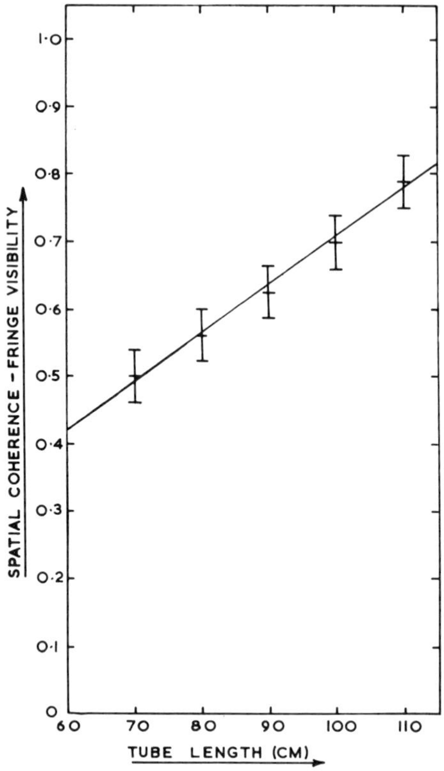

Fig. 11 Plot of the variation of spatial coherence with tube length using a pinhole separation of 0.5 mm and a tube of 2.5 mm bore for the 0.614 μm neon system. The pinholes are symmetrically displaced about the axis.

shown that the spatial coherence of light passed through similar tubes to those used in this work in the absence of an amplifying media, undergoes enhancement due to reflections at the walls. Thus it is necessary to carry out both parts of the experiment in one and the same set of tubes to determine what role the medium itself plays in determining the degree of spatial coherence. This work is now currently in hand at Sussex.

Glas [29] has shown that exposures for 50 and 2000 pulses in the 0.614 μm system produced fringes in the same position confirming that the fringe formation in the neon A.S.E. system is due to spatial rather than purely temporal coherence properties.

5. Spectral Distribution

In 1963 Yariv and Leite [7] showed that under certain approximations the relative linewidth of A.S.E., $\Delta\nu/\Delta\nu_D$, is given by

$$\Delta\nu/\Delta\nu_D = (\alpha L)^{-\frac{1}{2}}$$

for a column of atoms of length L and peak small signal gain coefficient α. However, they neglected i) critical length considerations ii) saturation; and iii) the geometry of the system which has the effect of varying the amount of spontaneous emission usefully emitted at any point in tube. Also it was necessary for them to assume that considerable narrowing occurred to obtain their final result.

The earlier theory for A.S.E. intensity [14] has been developed to take note of frequency [16]. When this is done it is found that

$$I_T(\nu, L) = I_{sp} d^2 g(\nu) \left\{ \int_{L_{c(\nu)}}^{L} \frac{\exp\left[\frac{-K_1 \bar{I}_L(\nu)\Delta\nu_D - K_2\Delta\nu_D g(\nu)}{1 + K_3 \bar{I}_L(\nu)\,\Delta\nu_D}\right] y}{y^2} \, dy + \frac{2}{L_c(\nu_o)} - \frac{1}{L_c(\nu)} \right\}.$$

Let the frequencies corresponding to the half intensity points of $I_T(\nu, L)$ be $\nu_o \pm \nu_{\frac{1}{2}}$, so that the width of the distribution, $\Delta\nu$, is $2\nu_{\frac{1}{2}}$. Then, by definition

$$I_T(\nu_o + \nu_{\frac{1}{2}}, L) = \tfrac{1}{2} I_T(\nu_o, L).$$

It is true that this definition has no real meaning when the emission profile consists of a Gaussian with a tiny component of A.S.E. superimposed on its centre. However as A.S.E. grows, although the resultant profile is complex such a definition becomes increasingly meaningful. Certainly when the A.S.E. emission is an order of magnitude larger than the spontaneous emission there can be no doubt as to the physical meaning of such a definition. It transpires that $\Delta\nu$ may be written in terms of integrals involving K_2, K_3 and L_c. In other words the results [14] discussed previously in Section 3 may be used to predict the linewidth of the output radiation.

For the 3.39 μm transition in the He-Ne, Figure 12a shows the distribution when $L = L_c(\nu_o) = 180$ cms. In 12b, $L=L_c(\nu_o) + 5$cms, A.S.E. occurs very near the line centre in a region less than 0.2 $\Delta\nu_D$ wide, but the width at half-height is still dominated by the spontaneous emission profile. In 12 c, $L = L_c(\nu_o) + 7$ cms but now the width at half-height is determined by A.S.E. rather than the spontaneous emission profile. So within 7 cms of $L_c(\nu_o)$ the predicted width at half-height has dropped to $\sim 0.2\ \Delta\nu_D$. Finally, 12d shows that when $L=2L_c(\nu_o)$, A.S.E. occurs over the whole range $\Delta\nu_D$.

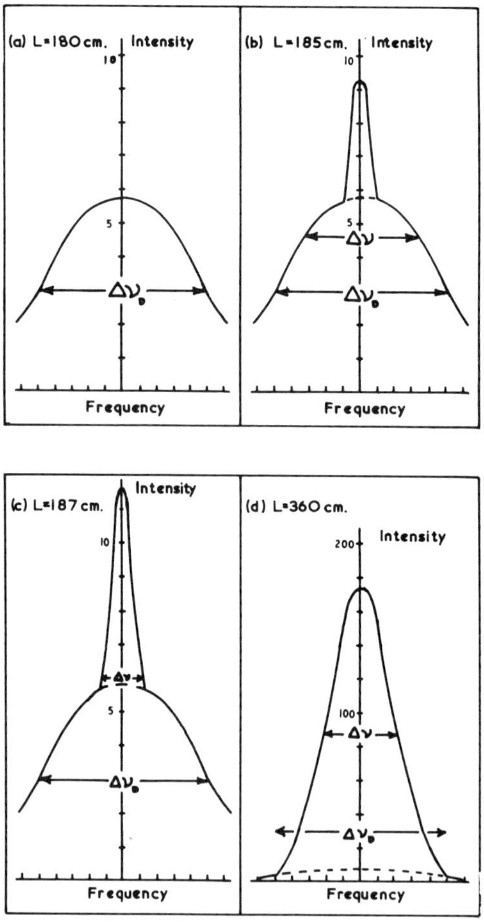

Fig. 12 Theoretical spectral profile of the 3.39 μm A.S.E. transition for $L_c(\nu_o)$ of 180 cms for tube lengths of a) 180 cms; b) 185 cms; c) 187 cns; and d) 360 cms.

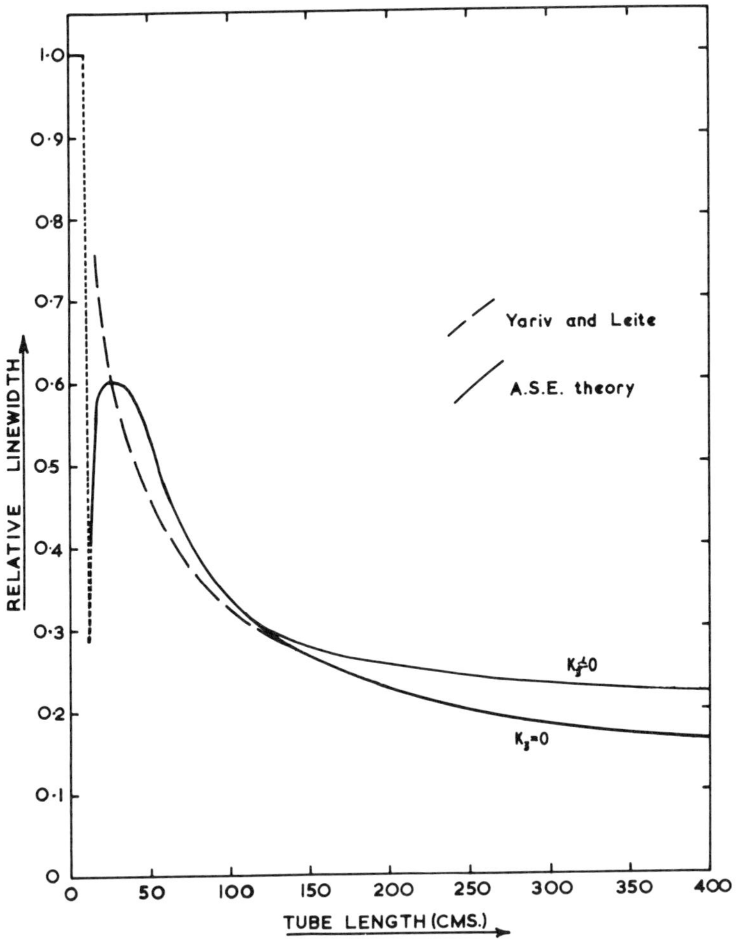

Fig. 13 Theoretical curves for the variation of relative radiation linewidth with tube length according to the A.S.E. and Yariv and Leite theories, for the 0.337 μm nitrogen transition.

Figure 13 shows the predicted relative linewidth as a function of length for the 0.337 μm line using the values of K_2, K_3 and $L_c(\nu_0)$ previously deduced. Just above threshold the linewidth narrows sharply and then begins to increase again. After $\sim 3L_c(\nu_0)$ the value reaches a maximum and begins to fall again. The region where the definition of linewidth is dubious is shown by a dotted line. At $\sim 10\ L_c$ the value approaches that given by Yariv and Leite but at greater lengths the two theories are again at variance because the A.S.E. theory has taken account of saturation. If we allow $K_3 \to 0$, corresponding to no saturation, our results for lengths >100 cms are consistent with Yariv and Leite to within 1%. However their approach does not predict the dip at lengths just in excess of L_c. Physically this dip comes about because of the inhomogeneous broadening of the spectral gain curve. For a given length of medium and inversion density, radiation at only certain frequencies is amplified. As the length increases a greater range of frequencies is involved but greater amplification now occurs at the line centre. Thus two competing processes are present, one tending to narrow and the other to broaden the range of frequencies involved. The result is the predicted dip. Work is at present in hand to experimentally verify these predictions.

6. A.S.E. and Laser Amplifiers

Recently Casperson and Yariv [29] have considered the problem of linewidth in laser amplifiers and conclude that, for an inhomogeneously broadened gain profile, increasing length would re-broaden the line back to its original Doppler width. They state that the same result should apply to A.S.E. under the belief that a self-generated signal would obey essentially the same equations. Litvak [30] has derived a similar result for amplification by inverted media and applied it to the problem of radiation from the interstellar medium. Both approaches, in addition, neglect any possiblity of a dip in the linewidth because critical length considerations are ignored.

While this type of broadening may occur in a laser amplifier it will not in general occur for A.S.E. which we characterize as radiation from a high gain medium in the absence of an external signal. In an amplifier, particularly when spontaneous emission is ignored, there is only a single travelling wave, whereas in an A.S.E. device there is in general two waves travelling in opposite directions [31]. Thus for a given length of medium L, the incremental gain coefficient at position z within the medium is $K_2/(1 + K_3I(z))$ whereas for an A.S.E. device it is independent of position within the medium, $K_2/(1 + K_3\bar{I}(L))$. This has the effect that radiation within an amplifier is at first exponential and then, as the saturation becomes important, linear. Whereas in an A.S.E. device of any length L the growth is always exponential throughout but with a saturation reduced gain coefficient. Casperson and Yariv explain that broadening occurs in the amplifier because growth at the line centre reaches the linear region (saturates) before the wings and consequently the continuing exponential growth in the wings, before they too saturate, causes the broadening. However within an A.S.E. system the growth of intensity is always exponential at all frequencies, although the gain coefficient is less in the wings, and so the broadening will not occur.

7. Astrophysics and A.S.E.

Papers written about the problem of intense emission from inverted media in the intersteller medium [32] leave doubts about whether a signal from another source is being amplified by the medium or whether one is looking at A.S.E. By looking at both intensity and spectral linewidth it may now be possible to distinguish between the two cases.

Allen and Peters [33] have applied their theory of A.S.E. explicitly to the question of radiation from OH molecules in the interstellar medium, making the assumption that the process is indeed, A.S.E. and not amplification of a more distant source. It turns out to be a pertinent thing to have done for the following reasons: a) no one in astrophysics had considered the possibility of a critical length being necessary for the process to start; b) Turner [32] believed that to account for the observed degree of polarization it was necessary for a high degree of saturation to occur, and that this would imply a broad linewidth at variance with the narrow linewidth observed.

When the best available value for inversion density is taken, it transpires that for the 1665 MHz transition in OH that the critical length for A.S.E. is $\sim 10^5$ cm, whereas geometrical considerations and interferometric measurements suggest that the maximum OH path length is in the range $\sim 10^{17}$- 10^{19} cm, which is only a factor of between 10^2 and 10^4 above the threshold length.

Although the gain α, L and L_c are vastly different for interstellar OH and for a laboratory N_2 system of the type analysed (See Section 3), the important parameters are αL and αL_c. and these are approximately the same in both systems, i.e. $(\alpha L)_{OH} \simeq (\alpha L)_{N2}$. Similarly, although dependent on several atomic factors the term which describes saturation also has the same order of magnitude in each case. Thus although the absolute values are not the same, the general shape of the N_2 intensity vs. length plot [14] should reproduce the interstellar OH case very well, particularly when evaluated for distances up to $\sim 10^3$ cm (i.e. 10^2 L_c). Even for distances up to 10^5 cm (i.e. 10^4 L_c) the output never "levels off" which is what many astrophysicists seem to expect when they write "fully saturated". However the output becomes linear indicating that the output is indeed "saturated". For the reasons discussed in Section 6, the linewidth, however, never broadens but continues to narrow, as shown in Table 1.

Table I

Length of Medium (in units of L_c)	$\Delta\nu/\Delta\nu_D$ (Relative spectral linewidth)
10	0.332
10^2	0.216
10^3	0.194
10^4	0.180
10^5	0.167

As the maximum degree of OH line-narrowing observed experimentally [34] is 0.22, this suggests that the 10^{17}cm value is more meaningful for the dimension of the active medium, a value which may be further justified by a combination of arguments about geometry and critical length [33].

The detailed behaviour of both intensity and linewidth, and of the concept of critical length should enable some very concrete analyses of certain current astrophysical theories, and further work on this problem is currently in hand.

8. Conclusions

It must be recognized that, as it stands at the moment, the theory of A.S.E. is a steady state theory. Although it has been applied to pulsed systems it has been done on the assumption that during the time of flight the population inversion in the tube remained sensibly constant. The problem of travelling wave excitation, or sequentially phased excitation in a pulsed cross-field device has yet to be solved although work is in hand on this topic.

The second limitation is also partly its strength. The original object of the work , as discussed in Section 1, was to see how far the radiation from high gain systems could be described by a rate equation approach which explicitly ignored coherent interactions. The approach is obviously very successful. Nonetheless the approach is also unarguably less than perfect. Even if the essential physics has been captured and superradiance effects correctly ignored it does not follow that it would not be useful to have for example the sort of insight into the behaviour of the induced polarisation in the medium that could be obtained from the off-diagonal terms in a density matrix using the sort of approach employed by Lamb for the laser [35]. Clearly much more formal and rigorous theoretical techniques could be applied than have been so far.

Although we have taken care to separate out the roles of coherent and incoherent interactions, it is clear that there is a regime where both may be important. After certain lengths of active media the field in a pulsed system could easily be large enough to act as a θ pulse [36]. In this case coherent pulse propagation effects could become important. It appears that this does not happen in C.W. systems because of the dephasing effects of continual electron-atom collisions. In pulse systems such as nitrogen the degree of inversion is high and with a pulse duration of $\sim$20 n sec and an upper level lifetime of $\sim$ 40 n sec, conditions seem at least partially suitable for the coherent pulse propagation [37,38]. Yet so far there seems to be no observations in this type of system that do not lend themselves to the rate equation approach. A proper treatment of the time development of A.S.E. in a pulsed system is, however, a prime necessity before one can really get to grips with this type of problem.

Although better theories may well soon appear [39] the essential ideas of considering the solid angle into which spontaneous emission is radiated, critical length, and the form of the saturation due to opposite going travelling ways would seem to be established and must be included. While helping to explain phenomena observed in the laboratory and contributing minimally to the debate about superradiance, the idea of A.S.E. seems to have real relevance to certain problems in astrophysics. Possibly, also, in the appropriate system the 'dip' in the spectral profile may allow useful well-defined, narrow band light sources in devices only marginally longer than L_c.

Acknowledgment

I am grateful to Gerry Peters not simply for running an eye over this manuscript, but for our enjoyable and fruitful collaboration on A.S.E.

References

1. Dicke, R.H., Phys. Rev. *93*, 99 (1954).
2. Dicke, R.H., *Quantum Electronics*, Vol. 1, ed. Grivet and Bloembergen (Columbia University Press, 1964).
3. Carver, T.R., App. Opt. *5*, 1090 (1966).
4. Schawlow, A. and Townes, C.H., Phys. Rev. *112*, 1940 (1958).
5. Rehler, N.E. and Eberly, J.H., Phys. Rev. *A3*, 1735 (1971).
6. White, A.D. and Rigden, J.D., Appl. Phys. Letters *2*, 211 (1963).
7. Yariv, A. and Leite, R.C.C., J. Appl. Phys. *34*, 3410 (1963).
8. Ernst, V. and Stehle, P., Phys. Rev. *176*, 1456 (1968).
9. Dialetis, D., Phys. Rev. *A2*, 599 (1970).
10. Lehmberg, R.H., Phys. Rev. *2*, 883 (1970).
11. Allen, L. and Peters, G.I., Phys. Letters *31A*, 95 (1970).
12. Peters, G.I. and Allen, L., J. Phys. *A4*, 238 (1971).
13. Allen, L. and Peters, G.I., J. Phys. *A4*, 377 (1971).
14. Allen, L. and Peters, G.I., J. Phys. *A4*, 564 (1971).
15. Peters, G.I. and Allen, L., J. Phys. *A5*, 546 (1972).
16. Allen, L. and Peters, G.I., J. Phys. *A5*, 695 (1972).
17. Faust, W.L. and McFarlane, R.A., J. Appl. Phys. *35*, 2010 (1964).
18. Geller, M., Altman, D.E. and De Temple, T.A., J. Appl. Phys. *37*, 3639 (1966).
19. Stenholm, S. and Lamb, W.E. Jr., Phys. Rev. *181*, 618 (1969).
20. Sayers, M.D., Allen, L. and Jones, D.G.C., J. Phys. *A2*, 102 (1969).
21. Allen, L., J. Phys. *A2*, 433 (1969).
22. Scully, M.O. and Lamb, W.E. Jr., Phys. Rev. *159*, 208 (1967).
23. Yariv, A. and Gordon, J.P., Proc. IEEE *51*, 4 (1963).
24. Kogelnik, H. and Yariv, A., Proc. IEEE *52*, 165 (1964).
25. Leonard, D.A., Appl. Phys. Letters *7*, 4 (1965).
26. Koppelmann, G., *Progress in Optics*, Vol. 7, ed. E. Wolf (North-Holland Publishing Co., Amsterdam, 1969) p. 2.
27. Peters, G.I., D. Phil. Thesis, University of Sussex, 1971 (unpublished).
28. Allen, L., Gatehouse, S., and Jones, D.G.C., Opt. Comm. *4*, 169 (1971).
29. Casperson, L.W. and Yariv, A., IEEE J. Quant. Electronics *8*, 80 (1972).
30. Litvak, M.M., Phys. Rev. *A2*, 2107 (1970).
31. Peters, G.I. and Allen, L., Phys. Letters *A39*, 259 (1972).
32. Turner, B.E., Pub. Asst. Soc. Pacific *82*, 996 (1970).
33. Allen, L. and Peters, G.I., Nature *235*, 143 (1972).
34. Barrett, A.H. and Rogers, A.E.E., Nature *210*, 188 (1966).
35. Lamb, W.E., Phys. Rev. *134*, 1429 (1964).
36. McCall, S.L. and Hahn, E.L., Phys. Rev. *183*, 457 (1969).
37. Icsevgi, A. and Lamb, W.E. Jr., Phys. Rev. *185*, 517 (1969).
38. Hopf, F.A. and Scully, M.O., Phys. Rev. *179*, 399 (1969).
39. Willis, C.R., Phys. Letters *A36*, 187 (1971).

NON-GAUSSIAN STATISTICS OF SUPERRADIANT RADIATION FROM SATURATED XENON 3.5μm LASER AMPLIFIER*

Hideya Gamo and Shih-Shung Chuang

University of California at Irvine, Irvine, California

Introduction

The intense stimulated emission, so-called superradiant radiation, can easily be observed in the case of a high gain laser amplifier such as the Xenon 3.5μm transition. The small signal gain of the 3.5μm transitions ($5d_{33}$ - $6p_{22}$) is of the order of 4 db/cm and is one of the highest gains among known gas lasers[1]. The intensity of amplified spontaneous emission can be comparable to that of ordinary cw gas laser. There are essential differences between the superradiant radiation and cw gas laser radiation; the spectral line-width of superradiant radiation is much broader than that of a cw laser. The center frequency of the superradiant radiation is more stable than in the case of usual cw laser, because no optical resonator is involved and the center frequency is determined by the energy levels of excited atoms[2,3].

The objective of the research described in this paper is to investigate the higher order coherence properties of the cw superradiant radiation beyond the spectral-line shape. The statistics of cw superradiant radiation lie between the Gaussian statistics of thermal radiation and the statistics of an ideal cw single mode laser. When the total gain of a laser amplifier is very small, the outgoing radiation consists of a random superposition of spontaneous

*This work was supported by the National Science Foundation (GK - 31856) and Air Force Office of Scientific Research (AFOSR 72-2155).

emissions by excited atoms. The statistics of spontaneous radiation are Gaussian based on the central limit theorem. According to the theory of stochastic processes, the statistics of a linearly amplified Gaussian random process are still Gaussian. When the intensity of amplified spontaneous radiation reaches a certain value, the gain saturation becomes significant. The laser amplifier in this stage can no longer be described as a linear transformation of the stochastic processes. A nonlinear transformation of a Gaussian process results in a non-Gaussian process. In the case of highly saturated superradiant radiation, the amplitude-statistics should be essentially different from Gaussian statistics. In order to determine the non-Gaussian statistics of radiation, we must measure the higher order correlations and evaluate the deviation of the observed quantity from the same quantity for the Gaussian process[4,5]. The variance and histogram of intensity fluctuations are higher order statistical quantities which will be treated in this paper.

There are several cases of a high gain medium in which superradiant radiations have been observed, such as Ne, 3.39μm, N_2, H_2 gas lasers, metal ion lasers, HF laser, GaAs semiconductor laser. We have chosen the Xenon 3.5μm transition, because several basic properties of the transition are known: the spontaneous emission rate was calculated by Faust and McFarlane[6] and Horrigan[7], life times of related atomic levels were measured by Schlossberg and Javan[8], and Allen, Jones and Schofield[9], the saturation parameter was determined by Klüver[1] and Chuang and Gamo[2,3] and the pressure effect was measured by Asami, Tako and Gamo[10] by means of the Lamb tuning dip measurement, by Chuang and Gamo[2] by applying the scanning Fabry-Perot interferometer to the superradiant radiation. A simple theoretical model of intensity fluctuations in the saturated laser amplifier was developed by Gamo[11].

We should also mention some difficulties of the Xenon laser, which required several years' experience to overcome. The clean-up of Xenon atoms by the wall and a cathode tends to make life-time of the Xenon plasma tube short. This problem has been solved by processing the plasma tube several days to saturate the wall and cathode by Xenon atoms. The Xenon superradiant radiation was usually regarded as too noisy for many practical applications. Slight reflection from a later stage will tend to make the high gain amplifier system unstable. This is especially serious when the radiation is analyzed by a scanning Fabry-Perot interferometer, or when the superradiant source is used for interferometric path-monitoring. This difficulty was eliminated by introducing the isolator[12] using the Faraday rotation in Yttrium Iron Garnet with Calcite dichroic polarizers[13]. The Xenon plasma has been known as unstable compared to the Helium-Neon plasma. The low frequency components of the intensity fluctuations of Xenon 3.5μm superradiant radiation may be made predominant by intensity modulation due to plasma instabilities. The significant part of plasma instabilities due to excitation waves or striations has been eliminated by using a large ballast resistor and minimizing the stray capacitance between anode and cathode of the discharge tube. Consequently, the low frequency components of the intensity fluctuations of superradiant radiation are no longer correlated with discharge current fluctuations[2]. The measurements of the average intensity, variance and histogram of intensity fluctuations of superradiant radiation may be seriously distorted by the saturation of the infrared detector. In order to eliminate the possible distortion due to saturation of detector we have performed a sequence of measurements with respect to various intensity levels for each case of output superradiant radiation.

After describing the experimental set-up and experimental results, we shall discuss the physical meanings of these experimental results. The experiments under preparation will also briefly be described.

Experimental Set-Up

The superradiant radiation source used for this experiment consists of three cascaded Xenon 3.5μm laser amplifiers. These plasma tubes are filled with pure Xenon 136 isotope of the pressure range from 80 to 160 m Torr. The characteristic features of these plasma tubes are: the first tube, i.d. 4mm, tube length 110cm, plasma length changeable from 10 to 100cm by 10cm step, the second tube, i.d. 4mm, tube length 100cm, plasma length 85cm, the third tube, i.d. 6mm, tube length 76cm, plasma length 60cm. All these tubes have Brewster windows made of infrasil. These plasma tubes are excited by voltage regulated DC power supply (Fluke 415A) with ballast resistors: the first tube, ballast resistor 300KΩ, power supply voltage 2,900V, excitation current 5mA, the second tube, 300Ω, 2,800V, 6mA, and the third tube, 300KΩ, 1,800 V, 5mA. These three tubes are separated by distances 75 and 22 cm, from the first to the second tube and from the second to the third tube respectively.

An InSb photovoltaic detector (Philco-Ford L 4541) cooled by liquid nitrogen was placed at the distance 180cm from the end of the third plasma tube after an interference filter and a CaF_2 lens with focal length 15cm and diameter 2.2cm. The photovoltaic detector was operated in the zero bias condition by compensating the average photo-voltage due to the incident radiation by an external bias voltage source. The detector noise is minimum in the zero bias condition. The linearity of the InSb detector was calibrated by using the superradiant radiation transmitted through the variable attenuator consisting of three Calcite dichroic polarizers where the transmittance can be varied in accordance with the $\cos^4\theta$ law by rotating the middle polarizer by the angle θ from the position of the maximum transmittance. The frequency bandwidth of the InSb photovoltaic detector under the zero bias condition was approximately 6 MHz. This was measured by using the radio frequency modulated He-Ne 6328 Å laser beam, and was also confirmed by the power spectrum of photocurrents generated by wide-band superradiant radiation analyzed by the scanning spectrum analyzer (Hewlett-Packard 141 T).

The average intensity of the Xenon 3.5μm superradiant radiation was measured by the experimental set-up illustrated in Fig. 1, which consists of the infrared detector mentioned above, a lock-in amplifier (Princeton Applied Research HR 8 with type C plug-in) and a light chopper (American Time, UR-L42C, 150 CPS). The disturbance due to the periodically reflected beam from the light chopper was eliminated by the YIG isolator mentioned above. The superradiant radiation propagating in the opposite direction was absorbed by a tilted polyethylene plate placed at the other end of the superradiant radiation source. The components of superradiant radiation subject to specular reflections at the plasma tube wall were effectively eliminated by the succeeding apertures formed by windows of tubes or opening of an isolator, because the angular distribution of these higher order propagating mode components are much larger than the lowest order propagating mode of interest.

In the case of an extremely high gain medium the component of superradiant radiation initiated at the further end of the first stage of the laser amplifier is predominant over the other components generated at the later stage. The number of predominant propagating spatial modes in the above superradiant radiation source consisting of three plasma tubes is approximately one with respect to the predominant component mentioned above. The number of the degrees of freedom of the radiation produced by the incoherent light source with area dA and subtending the solid angle $d\Omega$ is given by $dA d\Omega/\lambda^2$, where λ is the wavelength of radiation. The area of the incoherent source dA in the above calculation is assumed to be equal to the cross-section of the plasma tube, $\pi(0.2)^2$ cm^2, and $d\Omega = 0.44 \times 10^{-6}$ limited by the aperture of the YIG isolator with 3mm diameter at the distance 4 meter.

The superradiant radiation source consisting of three cascaded plasma tubes, is spatially coherent, because the measured

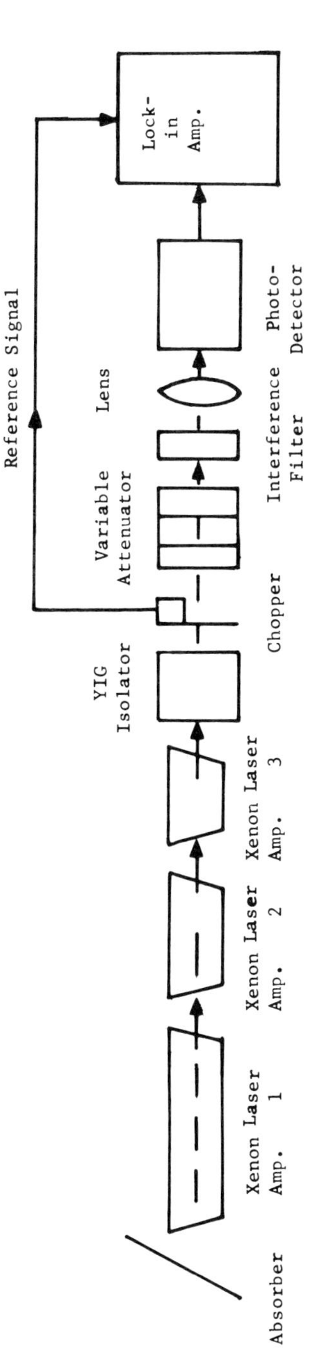

Fig. 1. Experimental set-up for measuring the average intensity of the Xenon 3.5μm superradiant radiation.

diffraction angle of the radiation through the YIG isolator 1.43×10^{-3} radian is slightly larger than the calculated value 1.25×10^{-3} radian. The latter value corresponds to the diffraction limit from a single propagation mode. The difference may be due to specular reflection within the YIG crystal (6mm long).

Superradiant radiation with various intensity levels was obtained by changing the plasma length of the first tube mentioned above. The maximum intensity of the superradiant radiation from three cascaded laser amplifiers under the above excitation condition was 2.1mW/cm^2, according to the measurement by the Eppley thermopile, and the output voltage of the InSb photovoltaic detector mentioned above was 600 mV for the same radiation, through the reading of the lock-in amplifier and corrected by using the calibration curve of the detector.

The spectral line-width of the superradiant radiation was measured by using a scanning Fabry-Perot interferometer with instrumental line-width 7 MHz and mirror separation 51cm (Modified Lansing Scanning Interferometer 130-215). The experimental set-up for line-width measurement is illustrated in Fig. 2. It should be noted that the isolator between the superradiant radiation source and the scanning Fabry-Perot interferometer is absolutely necessary for the experiment, since the reflectance of the interferometer becomes almost 100% at the anti-resonance condition. In the case of a superradiant source consisting of three cascaded plasma tubes, two cascaded isolators with total isolation of 64db and a variable attenuator were used for eliminating the interfering feedback from the scanning interferometer. The intensity of output radiation was kept within the linear region of the InSb photovoltaic detector by adjusting the variable attenuator.

The measurement of variance, histogram and power spectrum of intensity fluctuations produced by the superradiant radiation was performed by the system illustrated in Fig. 3. The electronic amplifier for amplifying the output voltage of the infrared detector consists of wide-band amplifiers (Keythley 110, 180 MHz and Keythly 106, 180MHz), and the Tektronix 1A1 plugin unit as an amplifier (50 MHz). The variance of fluctuating photocurrents was measured by a true RMS voltmeter (Hewlett-Packard 3400A, 10MHz bandwidth). The waveform of the fluctuating photocurrent was monitored by the cathode ray oscilloscope (Tektronix 555). The power spectrum of the fluctuating photocurrents was measured by a wave analyzer (Hewlett-Packard 310A). The histogram of the intensity fluctuations is obtained by means of the data acquisition system illustrated in Fig. 4. The fluctuating output voltage of the amplifier mentioned above is converted into a binary signal using an analog-to-digital converter (Preston scientific 8500 VHS,

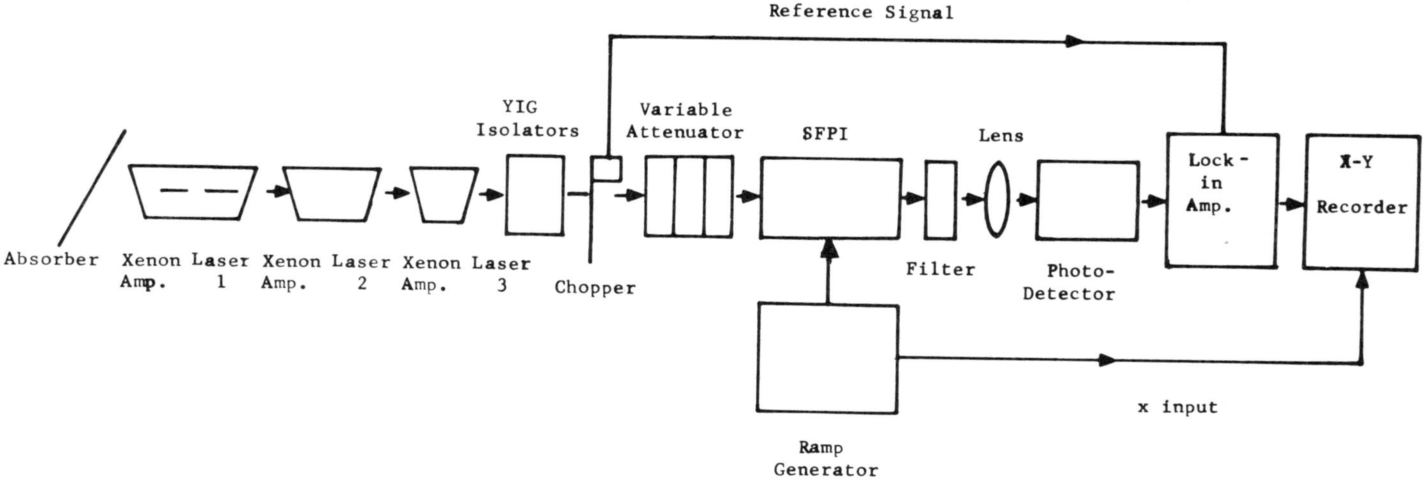

Fig. 2. Experimental set-up for measuring the spectral line-width of the Xenon 3.5μm superradiant radiation by the scanning Fabry-Perot Interferometer (SFPI).

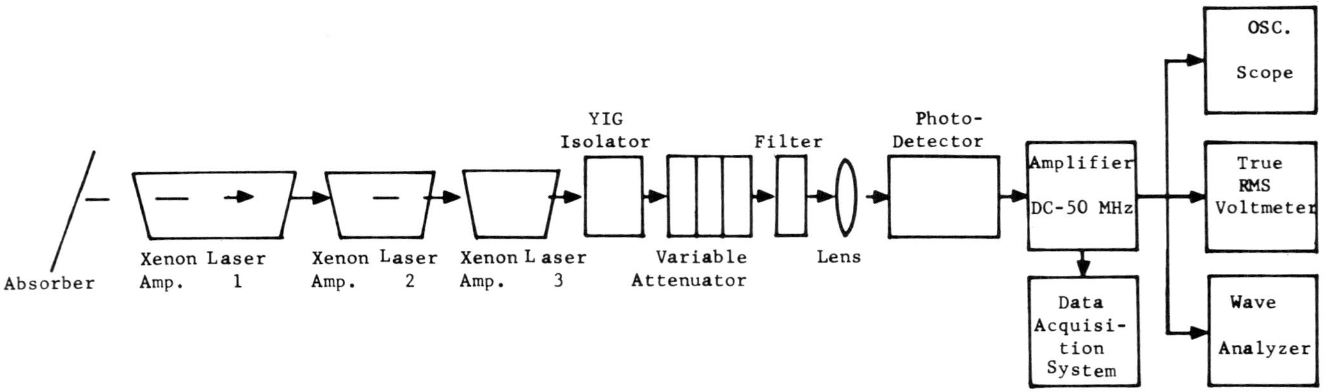

Fig. 3. Experimental set-up for measuring the variance, histogram and power spectrum of fluctuating photocurrents.

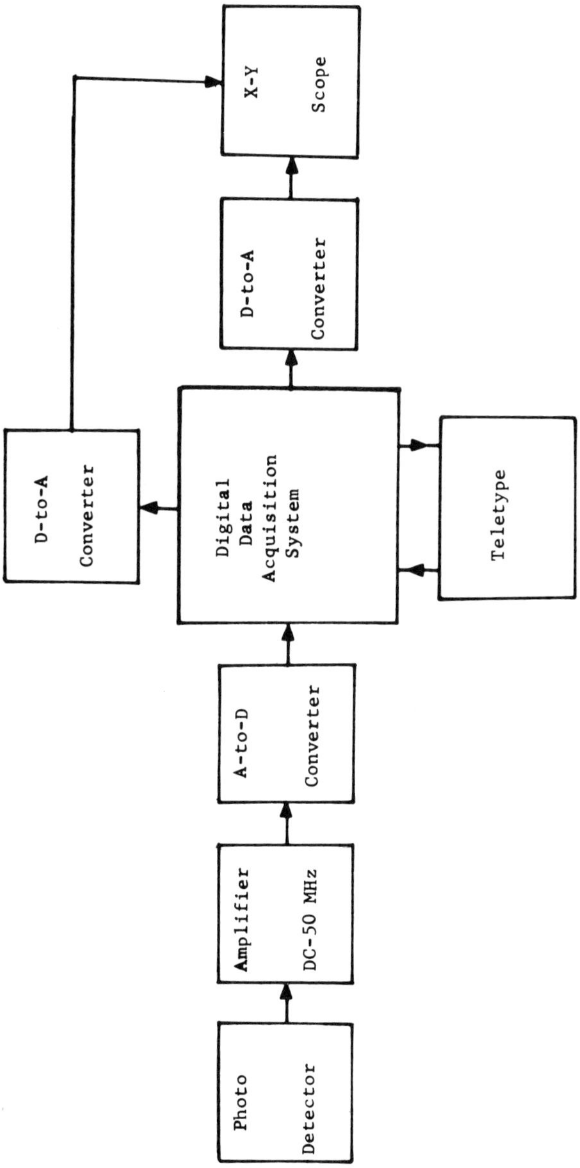

Fig. 4. Statistical data acquisition system.

14 bits plus sign, 100kc sample rates). The main digital data acquisition system is a PDP-15/20 (18 bits word, 16k core-memory and 0.8μ sec memory cycle time). The accumulated digital information of the histogram was converted into analog signals by a digital-to-analog converter (Phoenix Data DAC 1170-R, 10 bits plus sign) and was displayed on the memory scope (Hewlett-Packard 141A).

Experimental Results

The spectral line-width and variance of intensity fluctuations of the Xenon 3.5μm superradiant radiation for various levels of the average intensity measured by the above experimental set-up are illustrated in Fig. 5. The data point for the lowest average intensity was obtained by using the superradiant source consisting of the second and third plasma tubes whose characteristic features are described in the preceeding section. All other data points were obtained by using the superradiant radiation source consisting of three plasma tubes, where the plasma length of the first tube was varied by a 10cm step.

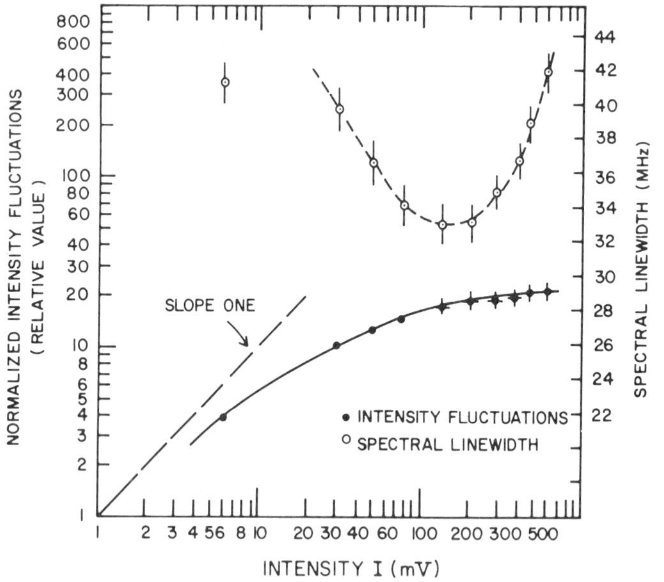

Fig. 5. The spectral line-width and variance of intensity fluctuations of Xenon 3.5 μm superradiant radiation. The variance of fluctuating photocurrents divided by the average photocurrent, $\langle(\Delta I)^2\rangle/I$, is shown in arbitrary scale. The intensity 2.1 mW/cm^2 is shown as 600mV of InSb photovoltaic detector.

The spectral linewidth Δf of the superradiant radiation was derived from the full width Δf_0 at half-intensity of the interferogram of the scanning interferometer with the instrumentation line width Δf_i = 7 MHz by using the relation: $(\Delta f)^2 = (\Delta f_0)^2 - (\Delta f_i)^2$. The minimum line-width 33 MHz was observed for the above bilateral superradiant radiation, where two superradiant radiations are propagating mutually in opposite directions. The coherence length defined by $\Delta l = c/\Delta f$ of the radiation with spectral line-width 33MHz is 9 meters. The re-broadened line-width 42MHz was observed in case of the longest plasma length of the first tube. This clearly indicates a rather heavy saturation in bilateral superradiant radiation. The spectral line-width of the lowest intensity for two cascaded plasma tubes deviates systematically from a sequence of spectral line-widths measured for the radiation from three cascaded plasma tubes. This may be qualitatively explained by considering the rather heavy saturation due to oppositely propagating superradiant radiation in the case of three plasma tubes.

The variances of intensity fluctuations with line-widths 40 and 42 MHz were measured as a function of the average intensities by using the variable filter consisting of three dichroic polarizers. The observed variances closely follow the lines of slope two. The variances at higher intensities tend to deviate from these slope-two lines due to the saturation of the InSb photovoltaic detector. The variances extrapolated by the slope-two line at the intensity level without attenuation are used in Fig. 5.

For convenience in considering the physical meaning of experimental results, the variance of the fluctuating photocurrent divided by the average photocurrent is illustrated in Fig. 5 on an arbitrary scale as a function of the average photocurrent. The RMS intensity fluctuations are 20% and 2% of the average intensity for unsaturated and saturated superradiant radiations respectively.

In order to clarify the statistical properties of highly saturated superradiant radiation we shall compare the waveforms, the variance, histogram, and power spectrum of the intensity fluctuations observed for unsaturated and saturated superradiant radiations with approximately equal spectral line-widths and equal average intensities. The waveforms of the fluctuating photocurrents produced by unsaturated and saturated superradiant radiations with spectral line-widths 40 and 42 MHz, respectively, are illustrated in Fig. 6A and B. The lower traces are waveforms of the detector-amplifier noise. These cathode ray oscilloscope displays indicate the rather small intensity fluctuations in saturated superradiant radiation. This can be investigated further quantitatively by measuring the histogram of fluctuating photocurrents as well as the variances illustrated in Fig. 5. The observed histograms for these

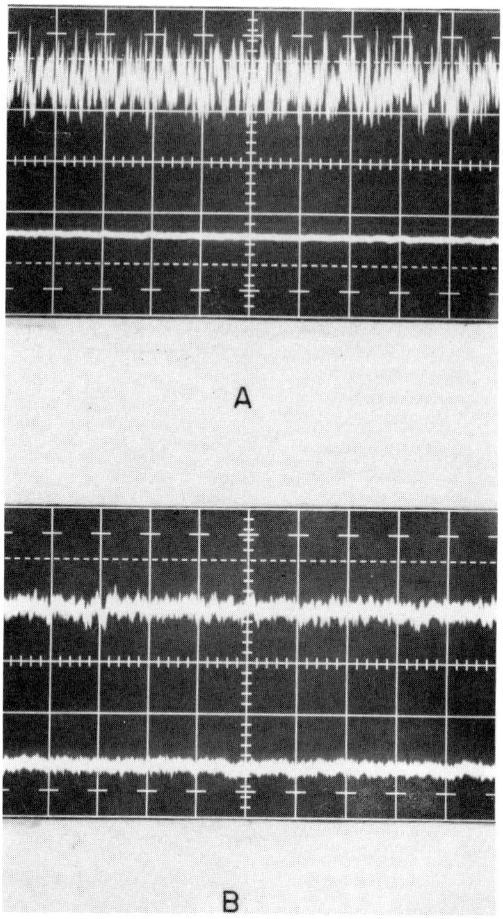

Fig. 6. Waveforms of fluctuating photocurrents and detector-amplifier noise. A: Unsaturated superradiant radiation with spectral line-width 40 MHz, B: Saturated superradiant radiation with spectral line-width 42 MHz.

two typical superradiant radiations are illustrated in Fig. 7A and B. The histograms were obtained by using 10,000 samples taken at 50 kHz rate. In the measurement of histograms we have adjusted the variable attenuator in order to obtain the same RMS photovoltage fluctuations reading for both the saturated and the unsaturated cases. The above histograms are the convolutions of photocurrent fluctuations and detector-amplifier noise, since detector-amplifier noise ratio is rather small. The signal-to-noise ratio was approximately 10.

The power spectrum of the intensity fluctuations of the unsaturated and saturated superradiant radiation are illustrated in Fig. 8. The integral of these curves should be equal to the variance of the fluctuating photocurrents. The intensity fluctuations of saturated superradiant radiation are definitely smaller than those of unsaturated radiation.

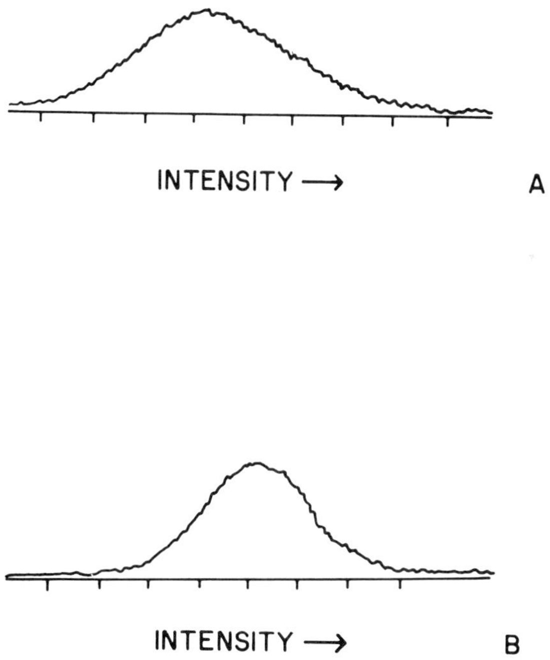

Fig. 7. Histograms of fluctuating photocurrents. A: Unsaturated superradiant radiation with spectral line-width 40MHz. B: Saturated superradiant radiation with spectral line-width 42MHz.

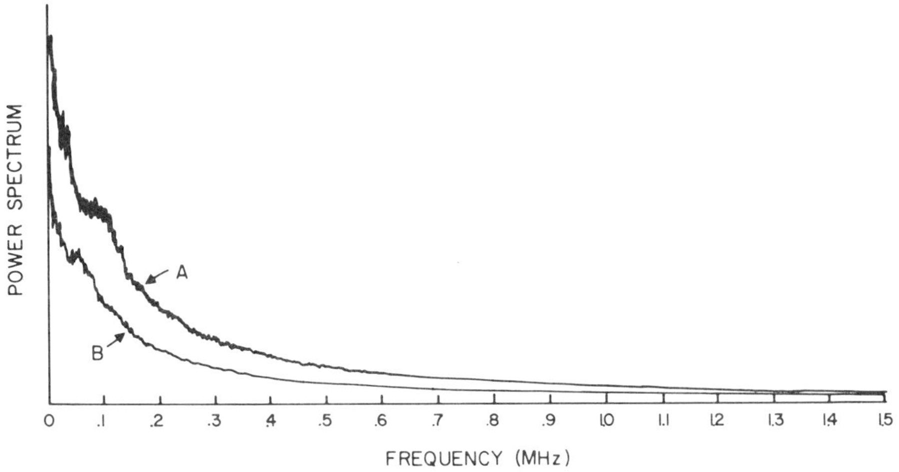

Fig. 8. Square-root of power spectrum of fluctuating photocurrents. A: Unsaturated superradiant radiation with spectral line-width 40MHz. B: Saturated superradiant radiation with spectral line-width 42MHz.

Discussion

From the experimental results described above we can conclude that the statistics of saturated superradiant radiation are essentially different from the Gaussian amplitude-statistics satisfied by thermal radiation or ordinary spontaneous emission. This is clearly demonstrated in the histogram of intensity fluctuations in Fig. 7B, which is closer to the Gaussian distribution. In the case of Gaussian amplitude-statistics we obtain exponential intensity statistics. It should be noted that below the oscillation threshold of a Xenon 3.5μm laser with feedback, we observed a histogram close to the exponential distribution. The histogram of unsaturated superradiant radiation (Fig. 7A) is somewhat similar to the Γ-distribution, which is obtained by superposing several independent intensity fluctuations with exponential distributions. This can be justified by considering that the spectral line-width 40MHz is approximately 7 times the detector band-width.

The characteristic features of the cw saturated superradiant radiation may be most effectively explained by comparing those of a cw single mode laser. This is demonstrated by the dependence of the normalized variance of the intensity fluctuations upon the intensity of the outgoing superradiant radiation in Fig. 5. The variance of the intensity fluctuations divided by the output intensity approaches a constant value when the laser amplifier is saturated by superradiant radiation. In the case of a cw single mode laser the same quantity becomes maximum at the oscillation threshold, and then decreases above the threshold[14,15]. This makes a significant distinction between superradiant radiation and a cw single mode laser. In order to confirm this fact by using the more heavily saturated superradiant radiation, we are preparing additional plasma tubes. The constant value of the normalized variance of the intensity fluctuations should not be confused with the shot noise generated by the photocurrent, which is approximately 40 db below the variance mentioned above.

The situation that the spectral line-width of the superradiant radiation is broader than the frequency band-width of the infrared detector makes the analysis of variance, the histogram and the power spectrum of the intensity fluctuations somewhat complicated. For instance, the observed variance of the intensity fluctuation for Gaussian amplitude-statistics is dependent upon the ratio of the spectral-line width W and the detector band-width W_0 as follows:

$$< (\Delta I)^2 > = I^2/\sqrt{1 + 2W^2/W_0^{\,2}} \, ,$$

where Gaussian spectral line-shape and frequency response function of the detector-amplifier are assumed. When $W_0 \ll W$, this is reduced to

$$< (\Delta I)^2 > = I^2 \, W_0/(\sqrt{2} \, W).$$

The latter corresponds to the formula $< (\Delta I)^2 > = I^2 \, W_0/W$ for the case of a rectangular spectral line-shape and frequency response function of the detector-amplifier[16]. The correction to the variance for variable spectral line-width from 30 MHz to 40 MHz is rather small when it is displayed on the logarithmic scale in Fig. 5.

We are investigating the statistics of superradiant radiation filtered by a Fabry-Perot interferometer with the line-width comparable to the detector-amplifier band-width in order to clarify the essential difference between the coherence properties of superradiant radiation and a cw single mode laser radiation. We have observed that the intensity fluctuations of the filtered saturated superradiant radiation satisfy approximately Gaussian intensity statistics. A more quantitative analysis of these experiments will be performed in the near future.

Compared to the bilateral superradiant radiation described in this paper, the unilateral superradiant radiation from laser amplifiers cascaded by isolators between the laser amplifiers has interesting features, such as significantly enhanced line-narrowing and ease of comparison with theoretical calculation, although the saturation effect is much more pronounced in the bilateral superradiant radiation. We wish to report on the statistics of unilateral superradiant radiation in the future.

The low frequency components of the power spectrum of intensity fluctuation of superradiant radiation are predominant, as is illustrated in Fig. 8. These components are most likely to be originated in radiation processes, such as the nonlinear interaction between different propagating modes or cross relaxation. The contribution of the 1/f noise of the InSb photovoltaic detector to the power spectrum of fluctuating photocurrents was very small. This was confirmed by observing the scattered diagram displayed in X - Y scope with output photocurrents from two infrared detectors illuminated by the same superradiant radiation through a beam splitter. The characteristic features of low frequency components of intensity fluctuation will be investigated further by restricting the number of propagating modes.

We hope that our results of Xenon 3.5μm superradiant radiation will be helpful in understanding the nature of other types of superradiant radiation from high-gain amplifying media in a wide range of wavelengths, and may also be useful in understanding the physical mechanism of astronomical OH and H_2O maser radiation[17,18]. We are convinced that the superradiant source described above will be conveniently used for interferometric measurements[2,19], because of its stable center frequency and its low intensity fluctuations.

Acknowledgment

The authors wish to thank Dr. Royal E. Rostenbach, Dr. Marshall C. Harrington and Dr. George A. Vanasse for their encouragement and support for this research. Thanks are also due to Donald G. Lubnau for his help in the histogram measurements. We wish to dedicate this paper to the memory of the late Dr. Thomas J. Walter, who processed some of Xenon plasma tubes used for this experiment, developed the basic statistical data processing system, and performed preliminary experiments on the Xenon superradiant radiation when we were at the Electrical Engineering Department, the University of Rochester. Dr. Walter, who was employed by the Air Force

Cambridge Research Laboratories, passed away in an airplane crash during the Air Force Expedition in the South Pacific, June, 1971.

References

1. J. W. Klüver, J. Appl. Phys. *37*, 2987-2999 (1966).
2. H. Gamo and S. S. Chuang, *Study of Superradiant Source for Path-Monitoring References in Fourier Spectroscopy*, Final Report AFCRL-71-0612, August 15, 1971.
3. S. S. Chuang, Infrared Isolator using Yttrium Iron Garnet and Xenon Unilateral Superradiant Radiation, Ph.D. Dissertation, Electrical Engineering Department, University of Rochester, September 1970.
4. H. Gamo and T. J. Walter, *Investigation of Operational Possibility of Laser Radiation in Partial Coherence Region*, AFCRL-68-0354, June 28, 1968; AD678554.
5. Thomas J. Walter, Instrumentation for Higher Order Statistics of Radiation and Experimental Study of Superradiant Radiation, Ph.D. Dissertation, Electrical Engineering Department, University of Rochester, August 1968.
6. W. L. Faust, and R. A. McFarlane, J. Appl. Phys. *35*, 2010-2015 (1964).
7. F. Horrigan, *Xenon Laser Research*, AFAL-TR-65-221, February 1966.
8. H. R. Schlossberg and A. Javan, Appl. Phys. Letters *11*, 1242-1244 (1966).
9. L. Allen, D. G. C. Jones and D. G. Schofield, J. Opt. Soc. Am. *59*, 842 (1969).
10. S. Asami, T. Tako and H. Gamo, *Pressure Effect in a Xenon 3.5μm Gas Laser*, Paper I-2, 1970 International Quantum Electronic Conference, Kyoto, Japan, September 9, 1970.
11. H. Gamo, IEEE J. Quant. Electronics *2*, 7 (April, 1966).
12. H. Gamo and S. S. Chuang, *Infrared Isolator Using Yttrium Iron Garnet*, Scientific Report AFOSR 70-1956TR, June 30, 1970.
13. T. J. Bridges and J. W. Klüver, Appl. Opt. *4*, 1121 (1965).
14. H. Risken, Zeit. f. Phys. *186*, 85 (1965).
15. R. D. Hempstead and M. Lax, Phys. Rev. *161*, 350 (1967).
16. H. Gamo, J. Opt. Soc. Am. *56*, 441 (1966).
17. N. H. Dieter and H. Weaver, Sky and Telescope *31*, No. 3, 2-6 (1966).
18. A. H. Barrett, Science *157*, 881-889 (1967).
19. Guy Guelachvili and Jean-Pierre Maillard, *Proceedings of Aspen International Conference on Fourier Spectroscopy*, 151-160 Special Reports, No. 114, AFCRL-71-0019, January 1971.

ON THE THEORY OF RADIATING ELECTRONS

Robin Asby

Royal Holloway College, University of London, England

In 1938 Dirac[1] showed that the self force on a point electron could be calculated in a consistent manner by assuming that it was due to just part of the field of the electron. Thus by splitting the field into two parts, a radiation field, which gave rise to the self force, and a bound field containing the diverging part of the field, he derived the Lorentz-Dirac equation governing the motion of the electron in an electromagnetic field. We discuss how this theory and the Lorentz-Dirac equation may be derived for each particle of a system of particles using the principle of least action from the action integral,

$$I = \int d^4x \; \{ -\frac{1}{8\pi} \frac{\partial A_\mu}{\partial x_\nu} \frac{\partial A_\mu}{\partial x_\nu} + \frac{1}{c} j_\mu A_\mu \}$$

$$\sum_{\substack{i \text{ over all} \\ \text{particles}}} m_i c^2 \int \sqrt{dz_\mu^{(i)} \, dz_\mu^{(i)}} \qquad (1\text{-}1)$$

The charged particles of this system are considered to be point charges, charge e. The world line of the ith particle is given by the equation,

$$z_\mu = z_\mu^{(i)} \, (\lambda_i) \quad , \qquad (1\text{-}2)$$

where the position coordinates z_μ are functions of the monotonically increasing parameter λ_i, the proper time. Each particle gives rise to a current

$$j_\mu^{(i)}(x) = ec \int d\lambda_i \frac{dz_\mu^{(i)}}{d\lambda_i} \delta^4(x - z^{(i)}(\lambda_i)) \, , \tag{1-3}$$

where

$$\delta^4(x) = \delta(x_0)\, \delta(x_1)\, \delta(x_2)\, \delta(x_3) \, ,$$

the four dimensional Dirac delta function, and the convention $x_0 = ct$, $x_4 = ict$ is used.

The total current $j_\mu(x)$ due to the particles of the system may therefore be written

$$j_\mu(x) = ec \sum_{\substack{i \text{ over all} \\ \text{particles}}} \int d\lambda_i \frac{dz_\mu^{(i)}}{d\lambda_i} \delta^4(x - z^{(i)}(\lambda_i)) \, . \tag{1-4}$$

The action integral, Eq. (1), for the coupled system is constructed from the two action integrals which describe the motion of the field due to prescribed charges and the motion of the charges due to prescribed field. Thus starting from Eq. (1.1) holding the particle coordinates fixed and varying the field we may derive in the conventional way the following equations which govern the motion of the electromagnetic field,

$$\Box^2 A_\mu = - \frac{4\pi}{c} j_\mu \, . \tag{1-5}$$

Here the differential operator $\Box^2$ is given by

$$\Box^2 = \frac{\partial}{\partial x_\mu} \frac{\partial}{\partial x_\mu} = \frac{\partial^2}{\partial x_1^2} + \frac{\partial^2}{\partial x_2^2} \quad \frac{\partial^2}{\partial x_3^2} - \frac{1}{c^2} \frac{\partial^2}{\partial t^2}$$

These equations (1-5) together with the Lorentz condition are equivalent to Maxwell's equations which are the equations of motion of the electromagnetic field which are solved most easily for a given charge-current density. In the derivation of equation (1-5) only the first two terms of the action integral contribute.

Similarly holding the field fixed and varying the coordinates describing a particular particle gives the equation of motion of that particular particle in a given field,

$$m_i \frac{d^2 z_\mu^{(i)}}{d\lambda_i^2} = \frac{e}{c} F_{\mu\nu}(z^{(i)}) \frac{dz_\nu^{(i)}}{d\lambda_i} \tag{1-6}$$

where

$$F_{\mu\nu} = \frac{\partial A_\nu}{\partial x_\mu} - \frac{\partial A_\mu}{\partial x_\nu} \tag{1-7}$$

In this derivation only the last two terms of the action integral contribute.

These two equations (1-5) and (1-6), when taken together are not sufficient to describe the coupled system. It might be expected that carrying out the variational procedure with the appropriate coordinates for the coupled system (which have to be determined) that all three terms of the action integral will contribute.

In order to derive the equation of motion of a particle in the coupled system we use the solution of the field equations, eqs. (1-5), to substitute for the field in each of the first two terms of the action integral. We find on splitting the field into two parts, following Dirac, that the action integral may be simplified and that under two assumptions, carrying through the variational procedure with the particle coordinates, leads to the Lorentz-Dirac equation for each of the particles.

This procedure thus suggests that the appropriate coordinates for the coupled system are the particle coordinates and the free field (radiation field) potentials[2]. The general solution of the electromagnetic field equations, eq. (1-5), may be written,

$$A_\mu = A_\mu^{(B)} + A_\mu^{(F)} , \tag{1-8}$$

where $A_\mu^{(B)}$ is a particular integral of equation (1-5) and $A_\mu^{(F)}$ is the general solution of the associated homogeneous equation. The particular integral is chosen to be

$$A_\mu^{(B)} = \frac{1}{c^2} \int d^4x^1 \; j_\mu(x^1) \; G^{(b)}(x - x^1) , \tag{1-9}$$

where $G^{(b)}(x)$ is the time symmetric Green's function of the electromagnetic field given by

$$G^{(b)}(x) = \frac{1}{2}\left\{\frac{\delta(t - \frac{r}{c})}{r} + \frac{\delta(t + \frac{r}{c})}{r}\right\} \tag{1-10}$$

half the sum of the retarded and advanced Green's functions. The field defined by Eq. (1-9) will be referred to as the 'bound' field.

The general solution of the homogenous equation, the free field, may be written in the form

$$A_\mu^{(F)}(x) = A_\mu^{(i)}(x) + \frac{1}{c^2}\int d^4x^1 j_\mu(x^1)\, G^{(f)}(x-x^1) \tag{1-11}$$

where the function $G^{(f)}(x)$ is given by

$$G^{(f)}(x) = \frac{1}{2}\left\{\frac{\delta(t - \frac{r}{c})}{r} - \frac{\delta(t + \frac{r}{c})}{r}\right\} \tag{1-12}$$

half the difference of the retarded and advanced Green's functions. Hence the *total* field may be considered as an externally imposed field plus the retarded field as is conventional. However in general the precise form of the free field term, eq. (1-11), will be determined by boundary and other conditions.

In order to derive the equation of motion of a particular particle labeled by coordinates $Z_\mu^{(i)}$, the electromagnetic potential $A_\mu(x)$ is written in the form

$$A_\mu(x) = A_\mu^{(i)}(x) + A_\mu^{(b)}(x) + A_\mu^{(F)}(x)\ . \tag{1-13}$$

In the equation $A_\mu^{(i)}$ is the bound field due to the ith particle, and $A_\mu^{(b)}$ is the bound field due to the rest of the particles; $A_\mu^{(F)}$ is the free field (eq. (1-11)) as before. The total current is written as

$$j_\mu(x) = j_\mu^{(i)}(x) + j_\mu^{(b)}(x)\ , \tag{1-14}$$

where $j_\mu^{(i)}$ is the current due to the ith particle and $j_\mu^{(b)}$ is the current due to the rest of the particles.

On substituting these equations (1-13) and (1-14) into equation (1-1) the terms which concern the ith particle may be separated out. They divide into four groups:

(a) the matter term

$$I_m = - m_i c^2 \int d\lambda_i \left[\frac{dz_\mu^{(i)}}{d\lambda_i} \frac{dz_\mu^{(i)}}{d\lambda_i}\right]^{1/2} \tag{1-15}$$

(b) the self interaction terms

$$I_s = \int d^4x \left\{- \frac{1}{8\pi} \frac{\partial A_\mu^{(i)}}{\partial x_\nu} \frac{\partial A_\mu^{(i)}}{\partial x_\nu} + \frac{1}{c} j_\mu^{(i)} A_\mu^{(i)}\right\} \tag{1-16}$$

(c) the interaction with the free field

$$I_f = \int d^4x \left\{- \frac{1}{4\pi} \frac{\partial A_\mu^{(i)}}{\partial x_\nu} \frac{\partial A_\mu^{(F)}}{\partial x_\nu} + \frac{1}{c} j_\mu^{(i)} A_\mu^{(F)}\right\} \tag{1-17}$$

(d) the interaction with the bound field of the other particles

$$I_b = \int d^4x' \left\{- \frac{1}{4\pi} \frac{\partial A_\mu^{(i)}}{\partial x_\nu} \frac{\partial A_\mu^{b}}{\partial x_\nu} + \frac{1}{c} (j_\mu^{(i)} A_\mu^{(b)} + j_\mu^{(b)} A_\mu^{(i)})\right\} . \tag{1-18}$$

On carrying out the variational procedure with the coordinates of the ith particle only these terms will contribute.

2. Interaction with the Bound Field

We consider first the contribution to the action integral from the group of terms (d) eq. (1-18). These terms of the action integral will give rise to the force on the particle due to the bound field of the rest of the particles. Using eqs. (1-9) and (1-14), equation (1-18) may be written in the form

$$I_b = -\frac{1}{4\pi c^4} \int d^4x \int d^4x' \int d^4x'' \, j_\mu^{(i)}(x') \, j_\mu^{(b)}(x'') \frac{\partial G^b}{\partial x_\nu}(x-x') \frac{\partial G^b(x-x'')}{\partial x_\nu}$$

$$+ \frac{1}{c^3} \int d^4x \int d^4x' (j_\mu^{(i)}(x) \, j_\mu^{(b)}(x') + j_\mu^{(i)}(x') \, j_\mu^{(b)}(x)) G^b(x-x'). \tag{2-1}$$

The second integral of this equation may be simplified by interchanging the primed and the unprimed variables of integration in the second term and making use of the symmetric nature of the Green's function,

$$G^{(b)}(x-x') = G^{(b)}(x'-x) . \tag{2-2}$$

Carrying this out and also interchanging the variables of integration in the first term of eq. (2-1), the whole expression may be rewritten in the form

$$I_b = \frac{1}{c^3} \int d^4x \int d^4x' j_\mu^{(i)}(x) \, j_\mu^{(b)}(x') \; \{2G^{(b)}(x-x')$$

$$- \frac{1}{4\pi c} \int d^4x'' \frac{\partial G^{(b)}}{\partial x''_\nu}(x''-x) \frac{\partial G^{(b)}}{\partial x''_\nu}(x''-x')\} . \tag{2-3}$$

The integration over the Green's functions in the second term of this equation may be carried out by parts. The surface terms resulting from the partial integration do not contribute and we find using,

$$\Box^2 \, G^{(b)}(x-x') = -4\pi \;\; \delta^4(x-x'), \tag{2-4}$$

that

$$- \frac{1}{4\pi c} \int d^4x'' \frac{\partial G^{(b)}}{\partial x''_\nu}(x''-x) \frac{\partial G^{(b)}}{\partial x''_\nu}(x''-x') = G^{(b)}(x-x') . \tag{2-5}$$

Using this result eq. (2-3) reduces simply to

$$I_b = \frac{1}{c^3} \int d^4x \int d^4x' \; j_\mu^{(i)}(x) \, j_\mu^{(b)}(x') \; G^{(b)}(x-x'), \tag{2-6}$$

which is just

$$I_b = \frac{1}{c} \int d^4x \; j_\mu^{(i)}(x) \; A_\mu^{(b)}(x) \quad . \tag{2-7}$$

Thus although both the first two terms of equation (1) have contributed the resultant expression is in exactly the form of the second term, the 'interaction' term.

Substituting for the currents from equations (1-3) and (1-4), eq. (2-6) may be written in the form

$$I_b = \sum_{j \neq i} \frac{e^2}{c} \int d\lambda_i \int d\lambda_j \frac{dz_\mu^{(i)}}{d\lambda_i} \frac{dz_\mu^{(j)}}{d\lambda_j} G^{(b)}(z^{(i)} - z^{(j)}) \quad . \tag{2-8}$$

Using the standard result,

$$G^{(b)}(x-x') = c \; \delta(c^2t^2-r^2) \quad , \tag{2-9}$$

We can reduce this term of the action integral to

$$I_b = \sum_{j \neq i} e^2 \int\int dz_\mu^{(i)} \; dz_\mu^{(j)} \delta\{(z_\nu^{(i)} - z_\nu^{(j)})(z_\nu^{(i)} - z_\nu^{(j)})\} \; , \tag{2-10}$$

which is exactly the form of the interaction term of the Fokker-Wheeler-Feynman[3] action-at-a-distance theory.

On carrying out the variational procedure we find that this term leads to the force on the particle due to the bound field of the other particles. We may therefore regard this force as being just due to the direct action of the other particles in the same sense as in the action-at-a-distance theory. The bound field has been eliminated from the action integral and thus it has the same secondary nature as the field in the action-at-a-distance theory.

3. Interaction with the Free Field

The interaction of the ith particle with the free field is

governed by the part (c) eq. (1-17) of the action integral (1-1).

$$I_f = \int d^4x \left\{ - \frac{1}{4\pi} \frac{\partial A_\mu^{(i)}}{\partial x_\nu} \frac{\partial A_\mu^{(F)}}{\partial x_\nu} + \frac{1}{c} j_\mu^{(i)} A_\mu^{(F)} \right\} \tag{3-1}$$

Substituting for the bound field due to the ith particle from eq.(1-9) this equation becomes

$$I_f = - \frac{1}{4\pi c^2} \int d^4x \int d^4x' \; j_\mu^{(i)} (x') \frac{\partial G^{(b)}}{\partial x_\nu} (x-x') \frac{\partial A_\mu^{(F)}}{\partial x_\nu} (x)$$

$$+ \frac{1}{c} \int d^4x \; j_\mu^{(i)} (x) \; A_\mu^{(F)} (x) \quad . \tag{3-2}$$

The integral

$$J = \frac{1}{2\pi c} \int d^4x \; \frac{\partial G^{(b)}}{\partial x_\nu} (x-x') \frac{\partial A_\mu^{(F)}}{\partial x_\nu} (x)$$

has been evaluated explicitly[4] writing the Green's function as an angular spectrum of plane waves[5] and expanding the free field as an arbitrary integral sum of homogeneous plane waves. After a straightforward calculation we find that

$$\frac{1}{2\pi c} \int d^4x \; \frac{\partial G^{(b)}}{\partial x_\nu} (x-x') \frac{\partial A_\mu^{(F)}}{\partial x_\nu} (x) = A_\mu^{(F)} (x') \quad . \tag{3-3}$$

Using this result equation (3-2) becomes

$$I_f = - \frac{1}{2c} \int d^4x \; j_\mu^{(i)} (x) \; A_\mu^{(F)} (x) + \frac{1}{c} \int d^4x \; j_\mu^{(i)} (x) \; A_\mu^{(F)} (x) \quad . \tag{3-4}$$

In order to derive the Lorentz-Dirac equation for the ith particle on carrying out the variational procedure the first term of these two terms must not contribute.

It may be easily shown that this condition will be satisfied if, and only if, the condition

$$I' = \int d^4x\, j_\mu(x)\, A_\mu^{(F)}(x) = 0 \qquad (3\text{-}5)$$

is satisfied; where $j_\mu(x)$ is the *total* current of the system. If this is satisfied the first term of equation (3-5) may be discarded.

Substituting for eq. (1-11) the integral I' may be written in the form

$$I' = \int d^4x\, j_\mu(x)\, A_\mu^{in}(x) + \frac{1}{c^3} \int d^4x \int d^4x'\, j_\mu(x)\, j_\mu(x')\, G^{(f)}(x-x'). \qquad (3\text{-}6)$$

Using the antisymmetric nature of the function $G^f(x)$,

$$G^f(x-x') = -G^f(x'-x) \qquad (3\text{-}7)$$

the second term can be seen to be identically zero and the condition eq. (3-5) reduces to

$$I' = \int d^4x\, j_\mu(x)\, A_\mu^{in}(x) = 0. \qquad (3\text{-}8)$$

$A_\mu^{in}(x)$ is an arbitrary prescribed field satisfying the homogeneous wave-equation, therefore in order to satisfy this requirement the condition

$$A_\mu^{in}(x) = 0 \qquad (3\text{-}9)$$

is imposed.

Under these conditions the part of the action integral describing the interaction of the ith particle with the free field may be reduced to

$$I_F = \int d^4x\, j_\mu^{(i)}(x)\, A_\mu^{(F)}(x) \qquad (3\text{-}10)$$

The need for the condition (3-9) and its implications has yet to be investigated fully, however it appears to be connected with the problem of eliminating the unphysical solutions of the Lorentz-Dirac equation[6].

If eq. (3-10) is written out in the form

$$I_F = \frac{1}{c^3} \int d^4x \int d^4x' \; j_\mu^{(i)}(x) \; j_\mu^{(b)}(x') \; G^f(x-x')$$

$$+ \frac{1}{c^3} \int d^4x \int d^4x' \; j_\mu^{(i)}(x) \; j_\mu^{(i)}(x') \; G^f(x-x') \; , \qquad (3\text{-}11)$$

and the variational procedure carried out with respect to the coordinates of the ith particle the second term gives no contribution to the equation of motion. This confirms that the free field $A_\mu^{(F)}$ is in some sense independent of the particle motion. This is consistent with the finding that only the interaction term of the *total* action integral contributes in describing the interaction of the particle with the free field.

Hence we are led to the conclusion that the interaction of the ith particle with the field consists of two parts. The first part is the interaction through the bound field with the other particles, the bound field being a secondary concept without degrees of freedom of its own. The second part is just the interaction with the free field, this part of the total field being independent of the particle motion and containing the degrees of freedom associated with the electromagnetic field.

Apart from the self-interaction terms the part of the action integral which concerns the ith particle may therefore be reduced to

$$I_i = \frac{1}{c} \int d^4x \; j_\mu^{(i)}(x) \left(A_\mu^{(b)}(x) + A_\mu^{(F)}(x)\right) - m_i c^2 \int d\lambda_i \left[\frac{dz_\mu^{(i)}}{d\lambda_i} \frac{dz_\mu^{(i)}}{d\lambda_i}\right]^{\frac{1}{2}} . \qquad (3\text{-}12)$$

On substituting for the current from equation (1-3) and carrying out the variational procedure with respect to $z_\mu^{(i)}$ we obtain the Lorentz-Dirac equation of motion for the ith particle with of course no externally imposed field.

$$m_i \frac{d^2 z_\nu^{(i)}}{d\lambda_i^2} = \frac{e}{c} F^r_{\mu\nu}(z^i) \frac{dz_\nu^{(i)}}{d\lambda_i}$$

$$+ \frac{2}{3}\frac{e^2}{c^3}\left[\frac{d^3 z_\nu^{(i)}}{d\lambda_i^3} - \frac{1}{c^3}\frac{dz_\mu^{(i)2}}{d\lambda_i^2}\frac{dz_\mu^{(i)2}}{d\lambda_i^2}\frac{dz_\nu}{d\lambda_i}\right] , \tag{3-13}$$

where

$$F^r_{\mu\nu} = \frac{\partial A^r_\nu}{\partial x_\mu} - \frac{\partial A^r_\mu}{\partial x_\nu} , \tag{3-14}$$

and

$$A^r_\nu(x) = \frac{1}{c^2} \int d^4x' \; j_\mu^{(b)}(x') \{G^b (x-x') + G^f (x-x')\} .$$

A^r_ν and $F^r_{\mu\nu}$ are the conventional retarded potential and field respectively due to all the other particles of the systems.

4. The Self Interaction

The self interaction terms group (b), eq. (1-16), are the terms which contain the divergence problems,

$$I_s = \int d^4x \left\{ -\frac{1}{8\pi} \frac{\partial A_\mu^{(i)}}{\partial x_\nu} \frac{\partial A_\mu^{(i)}}{\partial x_\nu} + \frac{1}{c} j_\mu^{(i)} A_\mu^{(i)} \right\} . \tag{4-1}$$

Substituting for the field of the ith particle using equation (1-9) and carrying out the integral over the Greens functions, Eq. (2-5), these terms of the action integral reduce to

$$I_s = - \frac{1}{2c} \int d^4x \int d^4x' \; j_\mu^{(i)}(x) \; j_\mu^{(i)}(x') \; G^b(x-x')$$

$$+ \frac{1}{c} \int d^4x \int d^4x' \; j_\mu^{(i)}(x) \; j_\mu^{(i)}(x') \; G^b(x-x') \quad . \tag{4-2}$$

Both terms are thus of the same form but the integrand of each diverges. This may easily be seen since on substituting for the current, each term is of the form

$$I = \frac{e^2}{c} \int d\lambda_i \int d\lambda_i' \frac{dz_\mu^{(i)}}{d\lambda_i} \frac{dz_\mu^{(i)}}{d\lambda_i'} G^{(b)}\{z^{(i)}(\lambda_i) - z^{(i)}(\lambda_i')\} \tag{4-3}$$

and the Greens function diverges when $\lambda_i = \lambda_i'$. Hence although this approach contains the traditional divergence difficulties it can be seen that by introduction of an *ad hoc* factor of two in the interaction term the two terms could be regarded as 'cancelling' leaving the required finite action integral.

It is well known from quantum mechanical considerations that the interaction between two particles through the electromagnetic field changes in form as the particles approach each other so that it must be expected that the interaction term of equation (1-1) has only the correct form for 'large' distances. The 'non-cancellation' of the two divergent terms could be interpreted as being indicative of this breakdown.

The action integral which corresponds to equation (1-1) in quantum electrodynamics differs in the interaction term from the classical case if the various quantities concerned are regarded as operators[7].

$$I = \int d^4x\{-\frac{1}{2}\frac{\partial A_\nu}{\partial x_\mu}\frac{\partial A_\nu}{\partial x_\mu} + \frac{ie}{2}\left(\bar{\psi}\gamma_\mu A_\mu \psi - \psi \gamma_\mu^T A_\mu \bar{\psi}\right)$$

$$- \frac{\hbar c}{2}\bar{\psi}(\gamma_\mu \frac{\partial}{\partial x_\mu} + \kappa_0)\psi - \frac{\hbar c}{2}\psi(\gamma_\mu^T \frac{\partial}{\partial x_\mu} - \kappa_0)\bar{\psi}\} . \tag{4-4}$$

Here the field A_μ is defined exactly as in the classical case and the current j_μ is given by

$$j_\mu(x) = \frac{iec}{2}\{\bar{\psi}(x)\gamma_\mu \psi(x) - \psi(x)\gamma_\mu^T \bar{\psi}(x)\} . \tag{4-5}$$

On carrying through the previous procedure with this action integral we find the following equation of motion for the electron-position field

$$\gamma_\mu \frac{\partial\psi(x)}{\partial x_\mu} + \kappa_0\psi(x) - \frac{ie}{\hbar c}\gamma_\mu A^r_\mu(x)\,\psi(x) =$$

$$\frac{e^2}{2c}\int d^4x'\, G^b(x-x')\,\{\gamma_\mu S(x-x')\,\gamma_\mu\psi(x') - \gamma_\mu S_1(x-x')\,\gamma^T_\mu\,\overline{\psi}(x')\}. \tag{4-6}$$

where

$$S(x-x') = \psi(x)\,\overline{\psi}(x') + \overline{\psi}(x')\,\psi(x) \tag{4-7}$$

and

$$S_1(x-x') = \psi(x)\,\psi(x') + \psi(x')\,\psi(x)\ . \tag{4-8}$$

A detailed derivation of this equation will be published elsewhere. It may be seen that using this approach two extra terms appear in the field equation, over the usual form, both of which depend on anticommutators of the fields at two different space time points. These terms make the field equation both nonlinear and nonlocal. It is of interest that they are of the form of the usual renormalization terms in quantum electrodynamics.

References

1. P.A.M. Dirac, Proc. Roy. Soc. *A167*, 148 (1938).
2. F. Rohrlich, Phys. Rev. Letters *12*, 375 (1964).
3. A.D. Fokker, Z. f. Physik *58*, 386 (1929); J.A. Wheeler and R.P. Feynman, Rev. Mod. Phys. *17*, 157 (1945) and *21*, 425 (1949).
4. If an integration by parts analagous to that used in equation (2-5) is attempted it may be shown that the surface terms *do contribute*.
5. R. Asby and E. Wolf, J. Opt. Soc. Am. *61*, 52 (1971).
6. F. Rohrlich, *Classical Charged Particles* (Addison Wesley Publishing Co., Reading, Mass., 1965) p. 136, "The Asymptotic Conditions".
7. All quantities are defined as J. Schwinger, Phys. Rev. *74*, 1439 (1948).

ATOMIC & MOLECULAR SPECTROSCOPY BY SATURATION WITH COHERENT LIGHT

A.L. Schawlow

Stanford University, Stanford, California

For many studies in the interaction of light with matter, we need sources of powerful, coherent, monochromatic light tunable to the wavelengths absorbed by simple atoms and molecules. Gas lasers, although limited to a few discrete lines with a limited tunability across each, have been used in the Hänsch-Borde modulation method of saturation spectroscopy to study hyperfine structures of a number of iodine lines (Levenson, Sorem and Hänsch). Recently Sorem has extended this work to investigate the Zeeman effect. Also, a new method of intermodulated fluorescence permits us to use saturation narrowing at pressures so low that collision effects are negligible. Lines as narrow as 3 Mhz (one part in 2×10^8) have been observed.

Hänsch has also been able to make a nitrogen-pumped dye laser quite monochromatic, with a width of about 300 Mhz. Further narrowing, when needed, is obtained by a passive confocal resonator scanned in synchronism with the laser. With the filter, there is considerable amplitude noise, but the output is as narrow as 7 Mhz and the pulse length is stretched from 5 ns to 30 ns. With this laser, hyperfine structures in sodium and fine structure of the hydrogen H line have resolved. When excited atoms are produced in a gas discharge, Stark broadening can be minimized by suddenly interrupting the discharge current and delaying the laser pulse until some of the ions and electrons have recombined.

A number of interesting problems can be investigated with the new techniques, since it is now possible to tune a laser to whatever atomic or molecular system is expected to exhibit a particular phenomenon in its simplest form.

RESPONSE FUNCTIONS FOR STRONGLY DRIVEN SYSTEMS

B. R. Mollow

The University of Massachusetts, Boston, Mass.

The coupling of an atom to the quantized electromagnetic field, in the electric dipole approximation, may be described in terms of the correlation or "response" function

$$\langle \mu(t)\mu(t')\rangle \tag{1}$$

where μ is the atomic electric dipole moment operator. The emission of photons, for example, is described in lowest order by an electromagnetic field-correlation function [1] which is simply proportional to the function [2] $\langle \mu^{(-)}(t)\mu^{(+)}(t')\rangle$, the superscripts denoting frequency-signatures according to the usual convention. The response of the atom to a weak perturbing field

$$E'(t) = E_0' e^{-i\nu t} + E_0'^{*} e^{i\nu t} , \tag{2}$$

on the other hand, is governed by the complex linear susceptibility [3]

$$\chi(\nu) = \frac{i}{\hbar} \int_{-\infty}^{t} dt' \langle [\mu(t),\mu(t')]\rangle \, e^{i\nu(t-t')} , \tag{3}$$

in terms of which the oscillating dipole moment induced by the perturbing field may be expressed as

$$\langle \mu(t)\rangle' = \chi(\nu)\, E_0' \, e^{-i\nu t} + \text{c.c.} , \tag{4}$$

and in terms of which the rate at which energy is absorbed from the perturbing field may be expressed as

$$W'(\nu) = -i\nu E_0'^{*}(\chi(\nu) - \chi^{\dagger}(\nu))E_0' \quad , \tag{5}$$

the symbol † representing complex conjugation and transposition of the (suppressed) vector indices.

If the unperturbed atom is governed by a stationary process, the function $<\mu(t)\mu(t')>$ will depend only upon the time difference $t - t'$. In the case we wish to consider, however, the atom is driven throughout the process under consideration by a (possibly) strong time-dependent "pump" field. [2,4-6] The correlation function in (1) therefore has, in addition to a stationary component, components with a more complicated time-dependence, which give rise to time-dependent components in the complex susceptibility in addition to the time-independent component. We shall begin by discussing the stationary components, and then briefly discuss the features which arise from the explicit time-dependence of the dynamics of the pumped atom.

The stationary part of the function which describes the emission of photons during atomic transitions from the state $|j>$ to the state $|k>$ (where $E_j > E_k$) is proportional to the function

$$g_{e(jk)}(t-t') = <a_{jk}^{\dagger}(t')\ a_{jk}(t)> , \tag{6}$$

where $a_{jk} \equiv |k><j|$ and $a_{jk}^{\dagger} \equiv |j><k|$ are the atomic lowering and raising operators, with expectation values

$$<a_{jk}> = \rho_{jk} \qquad\qquad <a_{jk}^{\dagger}> = \rho_{kj} \tag{7a}$$

and moments

$$<a_{jk}^{\dagger}\ a_{jk}> = \rho_{jj} \equiv \bar{n}_j \tag{7b}$$

$$<a_{jk}\ a_{jk}^{\dagger}> = \rho_{kk} \equiv \bar{n}_k \quad . \tag{7c}$$

The Fourier transform

$$\tilde{g}_{e(jk)}(\nu) = \int_{-\infty}^{\infty} d\tau e^{i\nu\tau} g_{e(jk)}(\tau) \tag{8}$$

of $g_{e(jk)}(\tau)$ is proportional to the spectral density of the radiation emitted by the atom during transition from $|j>$ to $|k>$, and is thus appreciable for values of ν near the resonance frequency $\omega_{jk} \equiv (E_j - E_k)/\hbar$. The total intensity of the radiation emitted during such transitions is consequently, by virtue of Eq. (7b), proportional to the population of the upper state,

$$\int \tilde{g}_{e(jk)}(\nu)\, d\nu/2\pi = \bar{n}_j \ . \tag{9}$$

The stationary part of the response of the atom to a weak perturbing or "signal" field [5,6] (Eq. (2)) during atomic transitions between the states $|k\rangle$ and $|j\rangle$ (where $\nu \approx \omega_{jk}$) is described by the correlation function

$$g_{a(jk)}(t-t') = \langle [a_{jk}(t),\ a^{\dagger}_{jk}(t')] \rangle \ . \tag{10}$$

The complex susceptibility $\chi(\nu)$ is proportional to the quantity

$$i \int_{0}^{\infty} d\tau e^{i\nu\tau}\ g_{a(jk)}(\tau) \ , \tag{11}$$

while the rate at which energy is absorbed from the signal field is proportional to the Fourier transform

$$\tilde{g}_{a(jk)}(\nu) = \int_{-\infty}^{\infty} d\tau e^{i\nu\tau}\ g_{a(jk)}(\tau) \ . \tag{12}$$

The absorption rate from a broadband signal field, by virtue of Eqs. (12), (10), and (7), is proportional to the population difference between the levels in question,

$$\int \tilde{g}_{a(jk)}(\nu)\, d\nu/2\pi = \bar{n}_k - \bar{n}_j \ . \tag{13}$$

The atomic correlation functions we have described may be evaluated in a straightforward manner by means of the *quantum regression theorem*.[7] This theorem enables us simply to express, by means of the Markoff approximation, multitime expectation values in terms of the time-dependent solutions for the elements of the atomic matrix and the corresponding equilibrium values.

These quantities are easily evaluated in the case in which the pump field

$$E(t) = E_0 e^{-i\omega t} + E_0^* e^{i\omega t} \tag{14}$$

induces resonant transitions between a single pair of states $|0>$ and $|1>$, of energy separation $E_1 - E_0 \equiv \hbar\omega_{10} \approx \hbar\omega$, if relaxation mechanisms connecting either of these states to the other states of the atom can be ignored. The equilibrium solutions $\rho_{11}(t) = \bar{n}_1$, $\rho_{00}(t) = \bar{n}_0$, $\rho_{10}(t) = \bar{\alpha}_{10} e^{-i\omega t}$ and $\rho_{01}(t) = \bar{\alpha}_{10}^* e^{i\omega t}$ (all other matrix elements are unaffected in equilibrium by the driving field) may then be obtained, in the resonant approximation, from the relations

$$\bar{\alpha}_{10} = iE_0 \cdot \mu_{10}(\bar{n}_0 - \bar{n}_1)/z^*$$

$$\bar{n}_0 - \bar{n}_1 = (\bar{n}_0^{(0)} - \bar{n}_1^{(0)})\kappa|z|^2/(\kappa'\Omega^2 + \kappa|z|^2)$$

$$\bar{n}_0 + \bar{n}_1 = \bar{n}_0^{(0)} + \bar{n}_1^{(0)} , \tag{15}$$

where κ and κ' are the diagonal and off-diagonal damping constants, [6] respectively, $\bar{n}_1^{(0)}$ and $\bar{n}_0^{(0)}$ are the equilibrium occupation numbers in the absence of the pump field, $z = \kappa' + i\Delta\omega$, $\Delta\omega = \omega - \omega_{10}$, and $\Omega = 2|E_0 \cdot \mu_{10}|/\hbar$.

For atomic transitions between the same pair of states $|1>$ and $|0>$ as are coupled by the pump field, the emission spectrum has the general form [2]

$$\tilde{g}_{e(10)}(\nu) = 2\pi|\bar{\alpha}_{10}|^2 \delta(\nu-\omega) + \tilde{g}_{e(10)}^{inc.}(\nu) , \tag{16}$$

the first term on the right-hand side representing coherent scattering of light from the pump field and the second, incoherent, term representing what may be described as resonance fluorescence.

In the limit of low pump field intensity ($\Omega \ll \kappa, \kappa', |\Delta\omega|$), the spectral density is well approximated, for $\bar{n}_0^{(0)} = 1$, $\bar{n}_1^{(0)} = 0$, by the relation

$$\tilde{g}_{e(10)}(\nu) = \frac{\frac{1}{4}\Omega^2}{|z|^2} \{2\pi\delta(\nu-\omega) + (\frac{2\kappa'-\kappa}{\kappa}) [\frac{2\kappa'}{(\nu-\omega_{10})^2 + \kappa'^2}]\} . \tag{17}$$

An incoherent contribution is thus present even for weak pump fields, the integrated incoherent intensity being equal to $(2\kappa'-\kappa)/2\kappa'$ times the total (coherent plus incoherent) intensity. The incoherent component is absent in this limit only if the off-diagonal decay rate κ' is exactly equal to one half of the diagonal decay rate κ, as it is when the damping mechanism is purely radiative, or, more generally, arises from "soft" coupling of the atom to a bath of harmonic oscillators. [4] In the strong-collisional model of atomic relaxation [8] (where $\kappa' = \kappa$), by contrast, the coherent and incoherent components are of equal intensity, originating from the steady state and the transient parts, respectively, of the solution for the time-dependent atomic dipole moment.

In the limit of very high pump intensity ($\Omega >> \kappa, \kappa', |\Delta\omega|$), the emission spectrum for the $|1\rangle \to |0\rangle$ transition is almost completely incoherent. It consists of the sum of three terms, one centered at the pump frequency ω, and one centered at each of the displaced frequencies $\omega + \Omega$ and $\omega - \Omega$, each of the latter two terms having half of the integrated intensity of the central term.[9-11] The spectral density in this limit is well approximated by the relation

$$\tilde{g}_{e(10)}(\nu) = \frac{\frac{1}{2}\kappa'}{(\nu-\omega)^2+\kappa'^2} + \frac{\frac{1}{4}\bar{\kappa}}{(\nu-\omega-\Omega)^2+\bar{\kappa}^2} + \frac{\frac{1}{4}\bar{\kappa}}{(\nu-\omega+\Omega)^2+\bar{\kappa}^2} \qquad (18)$$

where $\bar{\kappa} = \frac{1}{2}(\kappa + \kappa')$.

The absorption of energy from a weak signal field which induces resonant transitions between the same pair of states as are coupled by the pump field (i.e., one with frequency $\nu \approx \omega \approx \omega_{10}$) is described by a function whose stationary part is rather complicated in form. [6,12] It is simply related to the emission spectrum only in the limit of low pump intensity, where it is proportional to the incoherent part of the emission spectral density as given by Eq. (17). (The absorption spectrum would have a coherent part proportional to $\delta(\nu-\omega)$ only if the signal and pump fields were correlated in phase.) No simple proportionality exists, on the other hand, between the emission and absorption spectra in general. The difference between the two is most pronounced at high pump intensity, where the absorption line-shape function for certain values of the relevant parameters takes on negative values[6], representing stimulated emission rather than absorption. (The consequent amplification of the signal field is the result of photons which are transferred to it from the pump field.)

In the case of resonant transitions between one of the two states coupled by the pump field and one other, uncoupled state of the atom, the stationary parts of the emission and absorption spectra are again represented by quite different functions. [5] In this case, however, the two functions are essentially identical both in the limit of very weak and of very intense pump fields, differing from one another only for pump field intensities of intermediate magnitude. In the limit of very intense pump fields, for example, in the case of resonant transitions between an atomic state $|j\rangle$ and the upper driven state $|1\rangle$, the emission and absorption spectra, for $\omega_{j1}>0$, are both proportional to the function [5]

$$\frac{\kappa_j'}{(\nu-\omega_{j1}-\frac{1}{2}\Omega)^2+\kappa_j'^2}+\frac{\kappa_j'}{(\nu-\omega_{j1}+\frac{1}{2}\Omega)^2+\kappa_j'^2}, \tag{19}$$

where κ_j' here is the mean of the off-diagonal decay rates for the density matrix elements $\rho_{j1}(t)$ and $\rho_{j0}(t)$.

So far we have discussed explicitly only the stationary part of dynamics of the system. As we have mentioned above, however, the oscillations of the pump field (14) induce nonstationary processes as well. [13] These lead to an atomic correlation function of the form

$$\langle\mu(t)\mu(t')\rangle = G_0(t-t')+e^{-i\omega t}G_+(t-t')+e^{i\omega t}G_-(t-t') , \tag{20}$$

and consequently to a time-dependent complex susceptibility, with oscillating as well as constant components,

$$\chi(\nu;t) = \chi_0(\nu) + e^{-i\omega t}\chi_+(\nu) + e^{i\omega t}\chi_-(\nu) . \tag{21}$$

When the atom is stimulated by the signal field (2) these terms give rise, according to Eq. (4), to components in the atomic dipole moment oscillating at the frequencies $\pm\nu\pm\omega$, and hence to parametric up- and down-conversion, and to stimulated subharmonic generation, processes which may also be described in terms of nonlinear dielectric susceptibilities. [14]

For signal field frequencies ν much greater than the pump field frequency ω, the rate of absorption of energy from the signal field is a well-defined time-dependent quantity, having the general form [12]

$$W'(\nu;t)= W_0'(\nu) + e^{-i\omega t}W_+'(\nu) + e^{i\omega t}W_+'^*(\nu) . \tag{22}$$

It should be noted, however, that no very simple physical meaning can be attached to the oscillating absorption components in Eq.(22) if ω is not small compared to ν, since the absorption of energy from the signal field is a well-defined concept only in the sense of a time-average over many periods of its oscillation.

The oscillating components in Eq. (22) are appreciable, and

in fact are comparable in magnitude to the d.c. component, for frequencies ν near the resonance frequencies ω_{j1} and ω_{j0}, each of which, in accordance with the preceding remark, must be taken as large compared to ω_{10}. This is the situation in optical pumping and double resonance experiments, [15] where the states $|0\rangle$ and $|1\rangle$ are typically members of a Zeeman or hyperfine atomic multiplicity driven by an r.f. pump, and where an optical signal field induces transitions to another atomic state $|j\rangle$. The signal field is normally broadband in such experiments, however, with a spectrum relatively constant over the resonance peaks in the absorption line-shape-function. The evaluation of the oscillating components in the absorption rate (and of similar components in the emission rate when the states $|0\rangle$ and $|1\rangle$ are members of an excited atomic level) can therefore be carried out by the elementary methods of perturbation theory, which lead to a simple expression for $W_+^!$ proportional to the product of the signal-field spectral density at the resonance frequency times the off-diagonal matrix element $\bar{\alpha}_{10}$. The operation of the pump field plays no very detailed role in such calculations, its effect being taken into account only in the determination of $\bar{\alpha}_{10}$. In the case of a coherent signal field, on the other hand, the correct treatment of the problem is rather more complicated, since the oscillating components, like the d.c. components, depend upon ν in a complicated way near the resonance frequencies. The evaluation of the time-dependent terms in the coherent case can however be carried out in a straightforward way by means of the correlation function approach we have outlined. [13]

As an example of a nonstationary process somewhat different from those we have mentioned, let us consider, finally, one which occurs in the emission of light during transitions from an excited state $|j\rangle$ to both of the pump-field-coupled states $|0\rangle$ and $|1\rangle$, which may be taken as members of a ground-state multiplicity. In addition to the terms which lead to the stationary emission spectrum in Eq. (19) (plus a similar expression with ω_{j1} replaced by ω_{j0}), we find a nonstationary term in the correlation function (1) which is equal to

$$\frac{2i\bar{n}_j}{\Omega'} e^{-\kappa_j^!|t-t'|} \sin\tfrac{1}{2}\Omega'(t-t')\{(E_0\cdot\mu_{10})\mu_{j1}\mu_{0j}e^{i\bar{\omega}_{j1}t-i\bar{\omega}_{j0}t'} + (E_0^*\cdot\mu_{01})\mu_{j0}\mu_{1j}e^{i\bar{\omega}_{j0}t-i\bar{\omega}_{j1}t'}\}, \tag{23}$$

where $\bar{\omega}_{j1} = \omega_{j1} - \tfrac{1}{2}\Delta\omega$, $\bar{\omega}_{j0} = \omega_{j0} + \tfrac{1}{2}\Delta\omega$, and $\Omega' = [\Omega^2+(\Delta\omega)^2]^{\frac{1}{2}}$. This expression vanishes at $t = t'$, and hence gives no contribution (not even an oscillating one) to the overall emission rate. It represents a time-dependent, nonstationary correlation between two fields of different frequencies, one emitted by the atom during transitions from $|j\rangle$ to $|1\rangle$, the other during transitions from $|j\rangle$ to $|0\rangle$. The correlation exists because of the coupling between the states $|1\rangle$ and $|0\rangle$ which is produced by the pump field.

References

1. R.J. Glauber, Phys. Rev. *130*, 2529 (1963); L. Mandel and E. Wolf, Rev. Mod. Phys. *37*, 231 (1965).
2. B. R. Mollow, Phys. Rev. *188*, 1969 (1969); Phys. Rev.*A2*, 76 (1970).
3. D. W. Ross, Ann. Phys. (N.Y.) *36*, 458 (1966).
4. B. R. Mollow and M. M. Miller, Ann. Phys. (N.Y.) *52*, 464 (1969).
5. B. R. Mollow, Phys. Rev. *A5*, 1522 (1972).
6. B. R. Mollow, Phys. Rev. *A5*, 2217 (1972).
7. See M. Lax, Phys. Rev. *172*, 350 (1968) and related references.
8. R. Karplus and J. Schwinger, Phys. Rev. *73*, 1020 (1948).
9. The emission spectrum has been discussed for the case of strong collisional relaxation by M. Newstein, Phys. Rev. *167*, 89 (1968), who finds results in agreement with ours in the limit of strong applied fields.
10. C. R. Stroud, Jr., Phys. Rev. *A3*, 1044 (1971) has obtained spectra which roughly resemble the ones we have found in Ref. (2a) in the limit of strong applied fields, though differing even in that limit from our results in certain important respects.
11. An interesting extension of the methods of Ref. 2 which allows for the possibility of "unimolecular decay" has been made by M. F. Goodman and E. Thiele, Phys. Rev. *A5*, 1355 (1972).
12. For the case $\nu \approx \omega$, slowly varying time-dependent components oscillating at the frequency $2(\nu-\omega)$ are present in the absorption spectrum in addition to the d.c. components found in Ref. 6, and in addition to the oscillating components in Eq. (22).
13. The nonstationary processes will be treated in greater detail by the author in a subsequent publication.
14. See for example N. Bloembergen, *Nonlinear Optics*, (W. A. Benjamin, Inc., N. Y., 1965).
15. An excellent review of this subject is given by G. W. Series in *Quantum Optics, Proceedings of the Tenth Session of the Scottish Universities Summer School in Physics, 1969*, edited by S.M. Kay and A. Maitland (Academic Press, London and New York, 1970).

MECHANISM OF INTERACTION OF PICOSECOND LIGHT PULSES WITH MATTER YIELDING SELF-TRAPPING

Arkadiusz H. Piekara

Warsaw University, Warszawa, Poland

Recent achievements in the field of light trapping within spherical molecules using picosecond pulses [1,2] compelled physicists to search for another mechanism of self-trapping than optical Kerr effect. Thus, electronic hyperpolarizability of molecules was discussed.[3] However, this process seems to be not adequate to explain self-trapping, giving no saturation of the Kerr constant. [4,5] Instead, another kind of interaction between the intense electric field of a light wave and lattice vibrations in liquids and solids was proposed in a series of papers.[6] This process consists in a dipole-dipole interaction, both dipoles being induced by the electric field of a light wave. Even in a molecular lattice with molecules of spherical symmetry this interaction, called "pull effect", provides attractive forces between neighboring molecules. This results in vibration enhancement followed by an increase of molecular polarizability, increasing thereby the electric permittivity for the optical frequency of the medium:

$$\varepsilon = \varepsilon_o + \varepsilon_2 E^2 + \varepsilon_4 E^4 + \ldots, \qquad (1)$$

where the term $\varepsilon_2 E^2$ is positive and predominant.

More insight to this problem has been achieved by studying the optical behavior of a one-dimensional crystal lattice in solids and liquids.

Simplifying the problem consider a row of equidistant spherical molecules, of mutual distance a, perpendicular to the light beam direction. Denoting the polarizability of an isolated molecule by α and taking into account the interaction of point dipoles induced by electric field E of the light wave in nearest neighboring molecules, one obtains for the changed value α' of the polarizability an expression of the form

$$\alpha' = \alpha \{1 - 2\alpha [(a - 2x)^{-3} + (a + 2x)^{-3}]\}^{-1} , \tag{2}$$

x being the elongation of molecular vibrations of cut-off frequency. These vibrations will be most substantially enhanced owing to the force F acting on each molecule which, in linear approximation, will be

$$F = K \overline{E^2} x, \tag{3}$$

with

$$K = 192 \left(\frac{\varepsilon+2}{3}\right)^2 \left(1 - \frac{4\alpha}{a^3}\right)^{-3} \alpha^2 a^{-5} . \tag{4}$$

Thus, the equation of motion takes the form

$$\ddot{x} + \left(\omega_o^2 - \frac{K\overline{E^2}}{m}\right) x = 0 , \tag{5}$$

ω_o being the frequency of molecules without field and m - the molecular mass. Hence, owing to the elctric field the frequency decreases and at the same time the amplitude increases, increasing the polarizability of molecules. Calculations of $\langle\Delta\alpha\rangle=\langle\alpha'(E)\rangle - \langle\alpha'(0)\rangle$ for different liquids lead for ε_2 to a value of an order of magnitude of 10^{-13}, which is of a right order required by experiment.[4] Admitting a weak anharmonicity of molecular vibrations one obtains a wrong sign for ε_4, viz. $\varepsilon_4 > 0$. On the other hand, strong anharmonicity due to the assumed interaction of a Lennard-Jones potential type yields $\varepsilon_4 < 0$, ensuring self-focusing of a light beam [7].

The above discussed pull effect seems to be the fundamental mechanism causing the polarizability increase and thus starting the light trapping. It may be accompanied by other processes, e.g. by the molecular hyperpolarizability and particularly, for non-spherical molecules, by libration of molecules, and - if some ten nanosecond light pulses are used - also by optical Kerr effect and steric effects. [8]

References

1. R. R. Alfano and S. L. Shapiro, Phys. Rev. Letters *24*, 592 (1970).
2. J. P. McTague, C. H. Lin, T. K. Gustafson, and R. Y. Chiao, Phys. Letters *32A*, 82 (1970).
3. R. G. Brewer and A. D. McLean, Phys. Rev. Letters *21*, 271 (1968).
4. A. H. Piekara, Appl. Phys. Letters *13*, 225 (1968).
5. S. Kielich and S. Wozniak, Acta Phys. Polon. *A39*, 233 (1971).
6. A. H. Piekara, Phys. Status Solidi *42*, 43 (1970); Japan Jour. Appl. Phys. *10*, 266 (1971); Appl. Optics *10*, 2563 (1971).
7. This last result has been obtained together with J. Goldhar during a stay of the present author at M.I.T. with Professor Ali Javan, to be published.
8. T. K. Gustafson and C. H. Townes, "Influence of Steric Effects and Compressibility of Nonlinear Response to Laser Pulses and the Diameters of Self-trapped Filaments", Phys. Rev. A6, 1659 (1972).

RESONANCE FLUORESCENCE LINESHAPES WITH INTENSE APPLIED FIELDS*

C. R. Stroud, Jr.

University of Rochester, Rochester, New York

1. Introduction

Resonance fluorescence is an old subject, which received a great deal of attention through the twenties and even into the 30's and 40's. R. W. Wood's book presents beautiful pictures of it taken in the twenties. The theory was treated in great detail by Wigner and Weisskopf, and by Heitler [1,2]. It is perhaps a little startling to see the subject getting so much attention at a meeting today. There seem to be two reasons for the renewed interest. First, the development of tunable laser sources allows one to do experiments with intense, accurately resonant sources which may be coherent over many spontaneous decay lifetimes. Previous theoretical treatments of the effect have for the most part used perturbation theory which assumes that the population remains almost entirely in the ground state. This assumption is clearly not valid for the intense resonant fields now available. Further, the perturbation theory is not appropriate for exciting fields which are coherent over long periods of time. The second reason for looking at this area is that this is just the area in which it has been suggested that fruitful experiments might be carried out to test quantum electrodynamics.[3] The neoclassical theory has been worked out in some detail for these problems and predictions have been made which appear to be at variance with those of QED. In truth however, a careful, rigorously quantum, treatment of these problems does not appear to have been worked out. Those recent treatments which have been carried out have been rate equation approximations, contained phenomenological decay constants, or used Markov approximations,

*Research supported in part by the National Science Foundation.

or other assumptions whose range of validity is difficult to determine. [4] In addition they have all neglected the effect of the radiative frequency shifts which, as I will show, are important in determining the lineshape as well as the line center in general. If one is to test QED then one should have a proper QED treatment of the phenomena to be studied.

The purpose of the present paper is to carry through as carefully as possible this dynamic problem including radiative frequency shifts. While the present treatment does not represent the complete final treatment of the problem, it does represent progress toward that end, and points out the limited range of validity of some of the commonly used approximations.

We will assume that, initially the system is prepared in a state which is a product of the atomic ground state and a Fock state for the field, with $n \gg 1$ photons in one mode which is resonant or very nearly resonant with the transition from the atomic ground state to the first excited state. All other field modes are assumed initially unoccupied. The system is then assumed to absorb and re-emit photons from the resonant field mode an arbitrary number of times by stimulated absorption and emission before spontaneously emitting a photon into some other field mode. In addition to these real transitions, the atom is allowed to make virtual transitions from the ground state and the excited state into the various other states of the atom, both bound and ionized. It is important to include such virtual processes in order to properly describe the radiative frequency shifts.

2. Method of Calculation

The multiplicity of possible processes available to the system makes the set of possible states just enumerated not the most desirable set for basis states to describe the evolution of the system. It is more convenient to determine the effect of the applied field on the atom first and then to determine the nature of the fluorescence from the atom in the presence of the field. Thus, we will use as basis states eigenstates of the atom plus resonant field mode system. This is not a perturbation theory approximation, it is simply a judicious choice of basis states so that as much as possible of the physics is already contained in them. The eigenstates of the system of an atom plus a highly excited field mode exactly resonant with a transition from the ground state to the first excited state of the atom are illustrated in Fig. 1. Neglecting the interaction Hamiltonian, the energy eigenvalues of the atom plus resonant field mode system would be a series of atomic spectra displaced one above the other by the energy of a single photon of the resonant field mode, ie. by the difference in

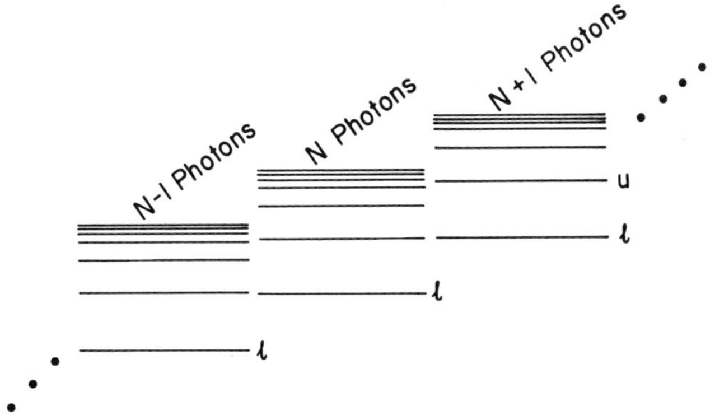

Fig. 1. The energy levels of the atom-plus-resonant-field-mode-system neglecting the interaction energy.

energy between the ground state and the first excited state. In this system the active levels are double degenerate. The lower state with n photons has exactly the same energy as the upper state with n-1 photons. This spectrum neglects the effect of the interaction Hamiltonian. The inclusion of the interaction Hamiltonian just mixes the degenerate pairs and slightly splits them into doublets whose separation is directly proportional to the amplitude of the electric field in the applied field mode, ie. to the square root of the number of photons in the resonant field mode. This is the so-called ac Stark effect. [5] The state vectors for the doublet which is a mixture of the state with n photons and the atom in its ground state, and the state with n-1 photons and the atom in its first excited level are

$$|n\pm\rangle = \frac{1}{\sqrt{2}} (|n-1\rangle|u\rangle \pm |n\rangle|\ell\rangle) \quad , \tag{1}$$

and their eigenvalues are

$$E_{n\pm} = n\hbar\Omega \pm \hbar\varepsilon_n \quad , \tag{2}$$

where $\hbar\varepsilon_n$ is the interaction matrix element between the two degenerate states.[6] We will use these states as basis vectors for describing the problem of resonance fluorescence. The transitions which will occur between these states are illustrated in Fig. 2.

The solid line indicates the real transition in which one photon is scattered from the excited field mode into some other

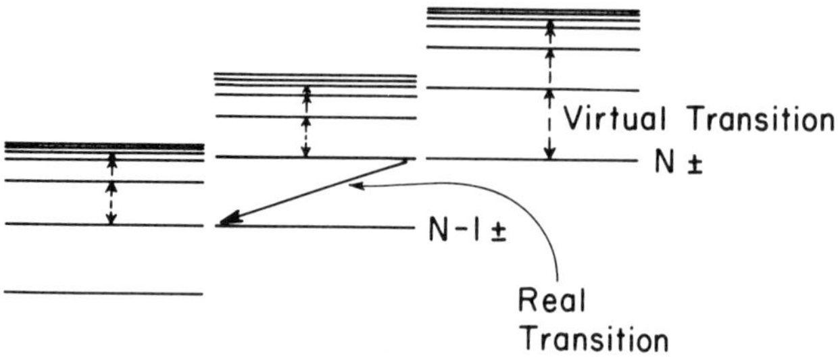

Fig. 2. The transitions allowed in a single fluorescent decay.

mode. The dotted lines indicate the non-energy conserving virtual transitions which are allowed according to the theory and which must be included to properly treat the radiative frequency shifts. We assume that the state vector for the system can be written as a linear combination of the initial states $|n\pm\rangle$ with no photons in any of the other modes, and the other states which may be reached from that initial state with their accompanying photons in other modes. (See Table 1) If we substitute a linear combination of

BASIS STATES	
INITIAL STATES	$\|N,+\rangle$ $\|N,-\rangle$
FINAL STATES	$\|N-1,+,1_{k,\lambda}\rangle$ $\|N-1,-,1_{k,\lambda}\rangle$
STATES REACHED BY VIRTUAL TRANSITIONS FROM INITIAL STATES	$\|N,J,1_{k,\lambda}\rangle$ $\|N-1,J,1_{k,\lambda}\rangle$
STATES REACHED BY VIRTUAL TRANSITIONS FROM FINAL STATES	$\|N-1,J,1_{k,\lambda},1_{k',\lambda'}\rangle$ $\|N-2,J,1_{k,\lambda},1_{k',\lambda'}\rangle$

Table 1

these basis states with unknown coefficients into Schrödinger's time dependent equation

$$
\begin{aligned}
|\psi(t)\rangle = {} & a_{n,+}(t)\, e^{-i\varepsilon t}\, |n,+\rangle + a_{n,-}(t)\, e^{i\varepsilon t}\, |n,-\rangle \\
& + \sum_{k,\lambda} \sum_{j} b_{n,j}(k,t) e^{-i(kc + \Omega_j)t} |n,j,1_{k\lambda}\rangle \\
& + \sum_{k,\lambda} \sum_{j} b_{n-1,j}(k,t) e^{-i(kc-\Omega+\Omega_j)t} |n-1,j,1_{k\lambda}\rangle \\
& + \sum_{k,\lambda} \{b_{n-1,+}(k,t) e^{-i\varepsilon t} |n-1,+,1_{k\lambda}\rangle \\
& \qquad + b_{n-1,-}(k,t) e^{i\varepsilon t} |n-1,-,1_{k\lambda}\rangle\} e^{-i(kc-\Omega)t} \\
& + \sum_{k,\lambda} \sum_{k',\lambda'} \sum_{j} c_{n-1,j}(k,k',t) e^{-i(k+k')ct} \\
& \qquad e^{-i(\Omega_j-\Omega)t} |n-1,j,1_{k\lambda}, 1_{k',\lambda'}\rangle \\
& + \sum_{k,\lambda} \sum_{k',\lambda'} \sum_{j} c_{n-2,j}(k,k',t) e^{-i(k+k')ct} \\
& \qquad e^{-i(\Omega_j-2\Omega)t} |n-2,j,1_{k\lambda}, 1_{k',\lambda'}\rangle
\end{aligned}
\tag{3}
$$

we can obtain the equations of motion for the time dependent coefficients. These equations of motion can be solved by use of Laplace transforms in a straightforward but rather tedious calculation. The only approximations which must be made in carrying out this solution are to neglect the small (one part in 10^8) correction to exponential decay which enters into all spontaneous emission calculations, and to neglect some corrections which are also demonstrably one part in 10^8 of the line shift and linewidth.

3. Results

The result of this calculation is an expression for the amplitudes of the final states,

$$b_{n-1,\pm}(k,t) = \mp \frac{\varepsilon V_{\ell u}(k)}{2\sqrt{2}\,\hbar} e^{i(kc-\Omega\pm\varepsilon)t} \left\{ \frac{(S_2^+ + ikc - i\Omega + i\Delta_u \mp i\varepsilon)}{i\gamma(S_2^+ - S_1^+)(S_2^+ - S_1^-)} e^{S_2^+ t} \right.$$

$$- \frac{(S_2^- + ikc - i\Omega + i\Delta_u \mp i\varepsilon)}{i\gamma(S_2^- - S_1^+)(S_2^- - S_1^-)} e^{S_2^- t} + \frac{(S_1^+ + ikc - i\Omega + i\Delta_u \mp i\varepsilon)}{\delta(S_1^+ - S_2^+)(S_1^+ - S_2^-)} e^{S_1^+ t}$$

$$\left. - \frac{(S_1^- + ikc - i\Omega + i\Delta_u \mp i\varepsilon)}{\delta(S_1^- - S_2^+)(S_1^- - S_2^-)} e^{S_1^- t} \right\} \tag{4}$$

where

$$V_{\ell u}(k) \equiv \langle \ell | \langle 1_{k\lambda} | H_{int} | vac \rangle | u \rangle$$

$$\Delta_j \equiv \text{Lamb Shift of } j^{th} \text{ level}$$

$$\gamma \equiv \frac{1}{2} [(\Delta_u - \Delta_\ell)^2 + 4\varepsilon^2]^{1/2}$$

$$\delta \equiv \frac{1}{2} [(i\Delta_u - i\Delta_\ell + A/2)^2 - 4\varepsilon^2]^{1/2}$$

$$S_1^\pm \equiv -(i/2)(\Delta_u + \Delta_\ell) - A/4 \pm \delta$$

$$S_2^\pm \equiv -ikc + i\Omega - \frac{i}{2}(\Delta_u + \Delta_\ell) \pm i\gamma \quad . \tag{5}$$

This is rather complicated, but some general features can be discerned. First it is made up of four components which, in general, decay at different rates as $t \to \infty$. Thus in general no single decay rate can be used to describe the system, and one certainly cannot describe the process by ordinary rate equation approximations. Secondly, we can see that there is a wealth of detail in the solutions of this relatively simple problem. The spectrum is in general a very complicated thing. A relatively simple special case is obtained by going to the asymptotic limit of a very intense applied field, in which case the Stark splitting is large compared with the linewidth or the Lamb shift. [7] If we go to that limit and look at the probability that a photon is scattered into a mode with frequency kc we find

Probability of scattered photon with frequency kc =

$$= \lim_{t\to\infty} \{|b_{n-1,+}(k,t)|^2 + |b_{n-1,-}(k,t)|^2\}$$

$$= \frac{|V_{\ell u}(k)|^2}{8\hbar^2}\{\frac{1}{(kc-\Omega+2\varepsilon)^2+A^2/16} + \frac{2}{(kc-\Omega)^2+A^2/16} + \frac{1}{(kc-\Omega-2\varepsilon)^2+A^2/16}\}.$$

The lineshape is made up of three separate Lorentzians, the center one located at the atomic resonance frequency, and the other two located symmetrically on either side at a distance equal to the Stark shift. (See Fig. 3). Each of the sidebands is one half as

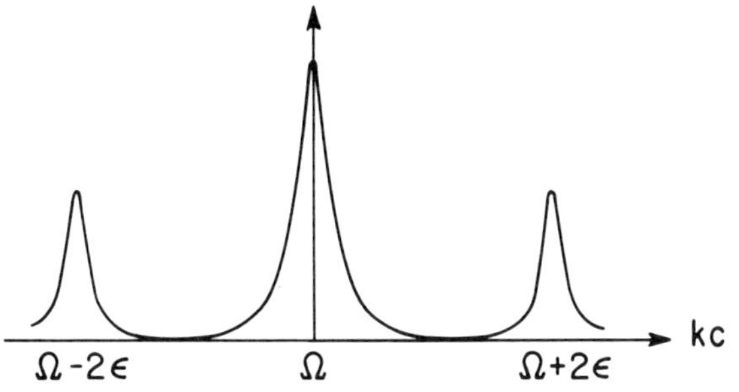

Fig. 3. The spectrum of the scattered photons in the strong field limit. The splitting is due to the AC stark effect. Each component is only half as broad as in ordinary spontaneous emission.

intense as the central peak. There are two surprising things about this lineshape. First, each component is only one half as wide as that obtained in ordinary spontaneous emission without the strong applied field. The reason for this is the rapid oscillation which induced absorption and emission cause in the population of the excited state. This means that the atom is effectively in the upper level only half of the time and thus the spontaneous emission rate which is proportional to the population of the upper level is halved. The other interesting feature of this lineshape is that it is centered at the unshifted natural frequency of the atom. This result is not so significant as it might first seem however, since the applied field is also at this frequency. If we repeat the whole calculation for an applied field at some other frequency we find that in the strong field limit we again get this same lineshape but centered about the applied field frequency. For the applied field frequency more than a few natural linewidths away from resonance the necessary field strength to reach the asymptotic region becomes quite large.

For weaker fields the lineshapes go over into those predicted by Heitler, [2] while in the intermediate region the lineshapes are extremely complex. For lack of time we will not present a catalog of the lineshapes for various parameters, those may be obtained from eq. (4), and will be published elsewhere. Instead, let me make a few comments on the methods used, their range of validity and their applicability to other problems of interest, and finally compare the results with those of the Neoclassical theory.

4. Discussion

It should be pointed out that the use of Fock states in this calculation for the applied field is not essential. The calculation can be carried through as well for a monochromatic coherent state. The results are the same for fields strong enough to cause appreciable Stark splitting. The calculation has actually been carried through in detail for that case.

It should also be pointed out that the technique of using Laplace transforms is not new for such problems.[8] What is unusual in these calculations is the inclusion of these nonenergy conserving processes in a dynamic problem. The inclusion of these virtual transitions is necessary not only to obtain the quantitative results we have gotten, but even to get the same qualitative results. If the same calculation had been carried out making the usual two-level approximation and neglecting "counter-rotating" terms then the results would have been different.[9] In fact, in that case in the strong field limit the lineshape would have been the same but it would have been centered at the atomic resonance frequency plus one half the Lamb shift. This result is in error and points up one of the points which this paper set out to show: one cannot safely go to the two level atom approximation and neglect non-energy conserving transitions if one wishes to calculate radiative level shifts. This result is certainly true in quantum electrodynamics and would appear to also be true in the Neoclassical theory. Certainly before the values of the "dynamic Lamb shift" calculated by Crisp and Jaynes can be compared with experiment their neglect of the "counter-rotating" terms should be closely investigated.[3]

The techniques of calculation which we have used can be applied to far more complex systems than the simple two level resonance fluorescence calculation which we have treated here. One can just as well treat three or four level problems, or even double resonance problems in which two separate resonant fields are applied to separate atomic transitions sharing a common level. Calculations of all of these types have been carried out and the details will be published elsewhere in the near future. In spite of the wide range of problems which can be treated efficiently by these techniques, the method has a deficiency whose seriousness is yet to be determined. While we have allowed any number of stimulated absorptions and emissions we have described only a single fluorescence. Since the interaction with the applied field was turned on at t = 0 and remains on thereafter, continually pumping the system back to the excited state, one would expect that there might be some transient period after which the atom settled down into a steady state behavior. The steady state might have the atom oscillating back and forth between the ground state and the exicted state and simply repeating the transient behavior over and over, or perhaps even settling down into some definite linear combination of the ground state and excited state absorbing and emitting radiation at the same rate. The long term solutions to such problems do not appear to have been obtained in QED without resort to phenomenological damping or Markovization. The present calculations are correct if each successive fluorescent decay occurs before the next fluorescence begins so that there is no interference from one to the next, and in any case are correct for the initial transient period after the field is suddenly turned on. It should be noted that this objection does not apply to the more complicated three or four level problems in which the state reached by the fluorescence is not the same as the initial state. In that case there will be only one fluorescence.

The comparison of these results with those of the neoclassical theory is not as straightforward as one might hope because of this lack of knowledge of the long term solutions in QED. According to the semiclassical theory there is an initial transient period after which the atom settles down to a steady state in which the dipole moment oscillates periodically giving rise to a sharp line spectrum. [10] The transient portion has some similarity to the QED results given here, but depends critically on the way the exciting field is turned on. A field which is slowly increased to a steady state value will cause a quite different transient response of the dipole moment from that produced by a field which is suddenly turned on. Any comparison which can be made between the two theories will not be too meaningful until we obtain long term solutions to the quantum equations.

Footnotes

1. V. Weisskopf, Z. fur Physik *85*, 451 (1933). V. Weisskopf and E. Wigner, Z. fur Phyzik *63*, 54 (1930).
2. W. Heitler, *The Quantum Theory of Radiation*, 3rd ed. (Oxford University Press, London, 1954) p.20.
3. M. D. Crisp and E. T. Jaynes, Phys. Rev. *179*, 1253 (1969) C. R. Stroud, Jr. and E. T. Jaynes, Phys. Rev. *A1*, 106 (1970) E. T. Jaynes, Phys. Rev. *A2*, 260 (1970).
4. See for example the paper of B. R. Mollow in this volume,p.529, or M. Newstein, Phys. Rev. *167*, 89 (1968).
5. The Stark effect in all of its forms is discussed in the review article by A. M. Bouch - Bruevich and V. A. Khodovoi, Usp. Fiz. Nauk 93, *71* (1967) [Soviet Phys. Usp. *10*, 637 (1968)].
6. The explicit form for ε is given in C. R. Stroud, Jr., Phys. Rev. *A3*, 1044 (1971), eq. (9).
7. This limit is easily reached with a dye laser in the optical region. The requirement for alkali metals is only of the order of Watts/cm^2.
8. The earliest use of this technique seems to be by G. Kallen in *Handbuch der Physik*, ed. S. Flugge (Springer Verlag, Berlin, 1958), Vol. V.
9. We have published such a calculation (reference (6)), which contains results which are correct in the two-level approximation but not in agreement with the present more complete analysis.
10. See the papers of reference (3) for detailed analytic solutions to the neoclassical equations for resonance fluorescence.

ELECTRON BEAMS AS CARRIERS OF OPTICAL COHERENCE

L.D. Favro, D.M. Fradkin, P.K. Kuo and W.B. Rolnick

Wayne State University, Detroit, Michigan

Recently there was reported an unusual experiment in which an electron beam appears to be able to transport laser light from one place to another.[1] In this experiment an electron beam first interacts with a laser field in the presence of a thin dielectric film and then subsequently emits light at the laser frequency when it impinges on a second (non-flourescent) dielectric film. While this result is still unconfirmed, it has stimulated much theoretical interest in the possibility of using electrons as carriers of optical coherence. The fact that electron beams can carry coherence in the microwave region has been known since the advent of the kylstron. In this frequency region the coherence time of the electrons is short compared to the period of oscillation of the electromagnetic waves, and the classical theory which treats the electrons as point charges is completely satisfactory. In the optical region, however, the electron coherence time can be long compared to the period of the oscillation and the electron wave packets are modified by the presence of the field. Thus the classical point charge approximation is no longer valid.

In a common semiclassical approach to this problem, the initial electron-radiation interaction is described by treating the electrons quantum mechanically and the radiation field classically. As a result of this interaction, the individual electron wavefunctions are amplitude modulated at the frequency of the classical field. (It should be noted that the same modulation occurs if the field is treated as a *coherent* quantum state of the radiation). In the second stage of the semiclassical approach the modulated charge-current densities resulting from the modulated wavefunctions are used as a classical source in Maxwell's equations. Radiation is obtained preferentially at the modulation frequency during any subsequent acceleration of the electrons. While this approach gives believable results in some cases, the practice of using a *quantum* charge-current density as a *classical* source conflicts with usual textbook prescription which uses *matrix elements* rather than expectation values as the source. The textbook prescription is a reasonable approximation to a fully quantum mechanical theory, but the one which uses expectation values is qualitatively in disagreement with the principle of conventional quantum mechanics. Since classical radiation intensities are quadratic in the source currents this use of expectation values gives intensities which are proportional to the *fourth* power of an electron's wavefunction. This disagrees with the quantum mechanical superposition principle which requires amplitudes to be linear and intensities to be quadratic in the initial wavefunction. This error results in the prediction of incorrect results in two cases which have been checked with quantum mechanical calculations. First, it predicts that a single electron can emit radiation preferentially at the modulation frequency. There exist rigorous proofs that this is impossible within the framework of quantum mechanics. Second, if one uses wavefunctions which result from the interaction with an incoherent quantum field, there is no density modulation and hence no information is carried in the charge current densities. Again, this is in contradiction to quantum mechanical calculations.

A correct quantum mechanical understanding of the problem involves the recognition that electron-electron phase correlations are induced by the first interaction with the radiation field. [2] This amounts to a transfer of coherence information from the field to the electron beam. These correlations then cause the electrons to act *collectively* in subsequent emission processes. While single electron radiation is also present, it can be shown that only the collective radiation is emitted preferentially at the modulation frequency. The coherence properties of the original field are transferred intact to the collective part of the secondary radiation. This can be demonstrated with a model consisting of two (single mode) optical cavities traversed by an electron beam. The Hamiltonian of the system in second quantized form is then

$$H = Ho + H'$$

$$Ho = \hbar\omega(a^{\dagger}a + b^{\dagger}b) + \sum_{\underline{k}} \hbar\omega_{\underline{k}}\, \eta^{\dagger}_{\underline{k}}\, \eta_{\underline{k}}$$

$$H' = \sum_{\underline{q}\,\underline{q}'} \frac{\hbar}{V} A(\underline{q},\underline{q}')a^{\dagger}\eta^{\dagger}_{\underline{q}}\,\eta_{\underline{q}'} + e^{-i(\underline{q}-\underline{q}')\cdot\underline{R}}B(\underline{q},\underline{q}')b^{\dagger}\eta^{\dagger}_{\underline{q}}\,\eta_{\underline{q}'}$$

+ hermitian conjugate .

Here $a^\dagger$ and $b^\dagger$ are the photon creation operators of the first and second cavities and $\eta_{\underline{q}}^\dagger$ is the electron creation operator in the momentum state $\underline{q}$ quantized in a volume V. The quantities A and B are essentially the Fourier transforms of the mode functions of the cavities. The distance between the cavities has been taken to be $\underline{R}$. The phase factor resulting from the propagation of the electron waves between the cavities has been written in explicitly. The initial state of the electron beam is taken to be a collection of wave packets with a total energy spread much less than the photon energy $\hbar\omega$ and the initial state of the second cavity is taken to be the vacuum. The state of the first cavity is left arbitrary. With these assumptions it can be shown that the expectation values of products of photon operators for the second cavity at time t are given by

$$\langle b^{\dagger n} b^m \rangle_t = (ft)^{n+m} (C_k^*)^n (C_k)^m \langle a^{\dagger n} a^m \rangle_{t=0} \left[1 + 0\left(\frac{1}{N}\right) + 0\left(\frac{e^2}{\hbar c}\right)\right]$$

where f is the flux of the electron beam, N is the number of electrons which have traversed both cavities and where C_k is given by

$$C_k = \frac{im^2}{2\pi^2\hbar^2 k^2} \int d^2 q_\perp A(\underline{k}_+ + \underline{q}_\perp, \underline{k}) B^*(\underline{k}_+ + \underline{q}_\perp, \underline{k}) \; e^{i(k_+ - k)R} \sin(k_+ + k_- - 2k)R$$

in Fresnel (near region) limit

$$C_k = \frac{im^2}{\pi\hbar^2 kR} A(\underline{k}_+, \underline{k}) B^*(\underline{k}_+, k) e^{i(k_+ - k)R} \cos(k_+ + k_- - 2k)R$$

in Fraunhofer (far region) limit

where $\underline{k}$ is the wave vector of the beam, $k_\mp^2 = k^2 \pm 2m\omega/\hbar$, all three vectors $\underline{k}$, $\underline{k}_\pm$ are in the direction of $\underline{R}$ and $\underline{q}_\perp$ is perpendicular to $\underline{R}$. This amounts to saying that for the collective radiation (leading term in N), any homogeneous identity satisfied by a is also satisfied by b. Thus the radiation in the second cavity has all the properties of the radiation in the first one except that the amplitude is changed by the factor ftC_k. In particular, when the field in the first cavity is classical, the field in the second one is also classical with a phase shift equal to the phase of C_k. It should be pointed out, however, that the source of the energy supplied to the second cavity is not the first cavity but the kinetic energy of the electron beam, since the electron beam does not take energy out of the first cavity. As long as the power supplied by the beam is more than the rate of energy loss by the cavities an optical amplifier is realized (an optical analogue of a klystron). However, it is interesting to contrast the two different ways a beam of electrons can carry a signal. In the microwave region, it is the particle nature that serves as the carrier, somewhat analogous to the propagation of sound waves through air by means of a fluctuating particle density. In the optical region it is the wave nature of the beam that manifests itself as the carrier, in analogy with amplitude-modulated radiowaves.

References

(1) H. Schwarz and H. Hora, Appl. Phys. Letters *15*, 349 (1969); H. Schwarz, Trans. N.Y. Acad. Sci. *33*, 150 (1971); H. Schwarz, Appl. Phys. Letters *19*, 148 (1971).

(2) L.D. Favro, D.M. Fradkin, and P.K. Kuo, Lett. Nuovo Cimento *4*, 1147 (1970).
C. Becchi and G. Morpurgo, Phys. Rev. *D4*, 288 (1971).
Jun Kondo, J. Appl. Phys. *42*, 4458 (1971).
L.D. Favro, D.M. Fradkin, and P.K. Kuo, Phys. Rev. *D3*, 2934 (1971).

DETUNED SINGLE MODE LASER AND DETAILED BALANCE

H. Risken

Universität Ulm, Ulm, Germany

1. Introduction

The statistical properties of laser light have been investigated both theoretically (e.g. [1-20]) and experimentally (e.g. [21-30]) in great detail. For a single mode laser not too far from threshold the theory is rather simple (it is essentially governed by only one parameter) and it agrees with experiments very well. In the experiments [21-30] only the statistical properties of the intensity are compared with the theory. Measurements giving information on the statistical properties of the phase of the light field, for instance the usual linewidth, have been made in the threshold region only quite recently [31], see fig. 1.

In the statistical properties of the intensity, the detuning (difference between atomic and resonator frequency) enters only in a trivial way via a change of the scaling paramters and, depending on definition, via a shift of the pump paramter. Therefore, the results without detuning may immediately be used. In the statistical properties of the phase of laser light, e.g. the usual linewidth, however, the detuning enters in a more complicated way. Since now one is able to measure the linewidth in the threshold region without detuning [31], a measurement of the linewidth with an appreciable detuning seems to be possible. Therefore a theory of the linewidth in the threshold region including detuning is needed for further possible comparison. Using a linearization procedure, a calculation of this effect was made already some time ago outside the threshold region [5,8]. Results in the threshold region have been obtained recently [32].

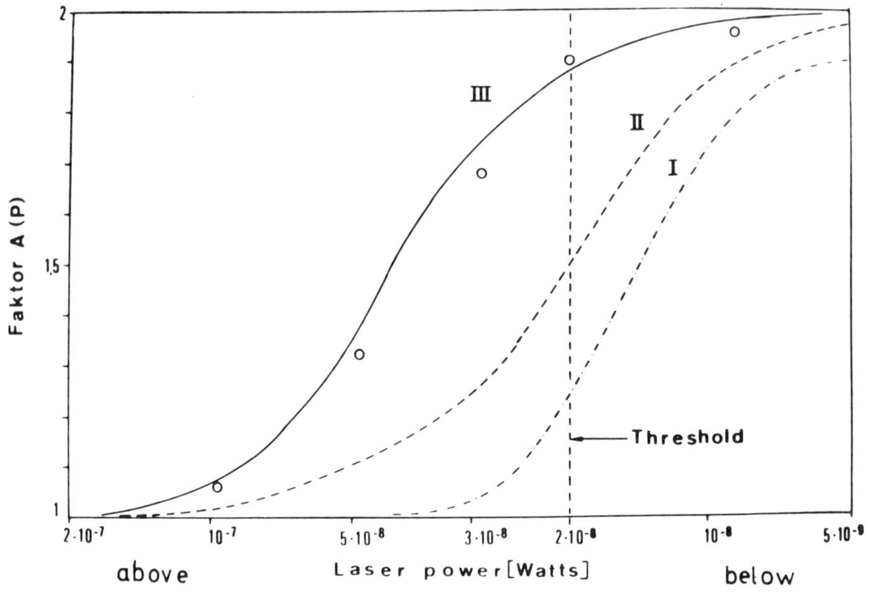

Fig. 1 Experimentally determined values of the linewidth factor $A(P)=\alpha_L$ (circled dots). Curve III represents a theoretical result given by Hempstead and Lax [12] and Risken [10]. Curves I and II were predicted by Grossmann and Richter [20]. This figure is reproduced from the paper of Gerhardt, Welling and Guttner [31]. I wish to thank Prof. Welling for sending me their results before final publication.

Summary of the results

The linewidth $\Delta\nu$ may be written as

$$\Delta\nu = \frac{C}{P} \quad \alpha_L\ (a,\delta)\ . \tag{1.1}$$

Here, C is a laser constant containing the detuning parameter δ, which is defined in (3.4), (3.5) respectively, P is the power output and α_L is a linewidth factor. Without detuning ($\delta=0$), $\alpha_L(a,0)$ varies from 2 to 1 by passing through the threshold region from below to above threshold, i.e. by varying the pump parameter a from negative to positive values. With detuning this factor is increased, see fig. 2. If the detuning is not too large, $\alpha_L(a,\delta)$

may be approximated by the value without detuning in the following way

$$\alpha_L(a,\delta) = \alpha_L(a,0) + [2 - \alpha_L(a,0)]\delta^2 . \tag{1.2}$$

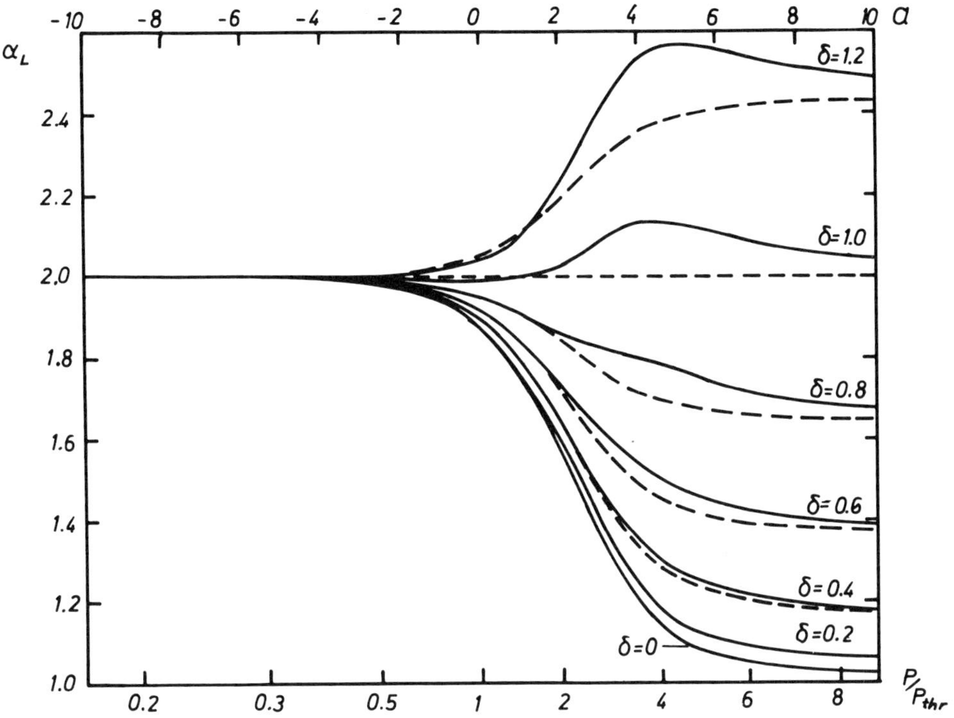

Fig. 2 The linewidth factor $\alpha_L(a,\delta)$ as a function of the pump parameter resp. as a function of the normalized power output P/P_{thr} for various detuning parameters δ (solid curve). The approximation (4.11)

$$\alpha_L(a,\delta) = \alpha_L(a,0) + [2 - \alpha_L(a,0)]\ \delta^2$$

is also shown (dashed line). Note that in this figure the pump power is increased to the right, whereas in fig. 1 it is increasing when going from right to left.

This enhancement of the linewidth due to detuning may be explained qualitatively in the following way (see fig. 3): without detuning the ratio of 2 to 1 occurs because above threshold the laser amplitude

$$E(t) = u(t) e^{-i\Omega t} + u^*(t) e^{i\Omega t}$$

δ = 0   phase diffusion produced
         only by Γ_φ

δ ≠ 0   phase diffusion produced
         by Γ_φ and Γ_τ

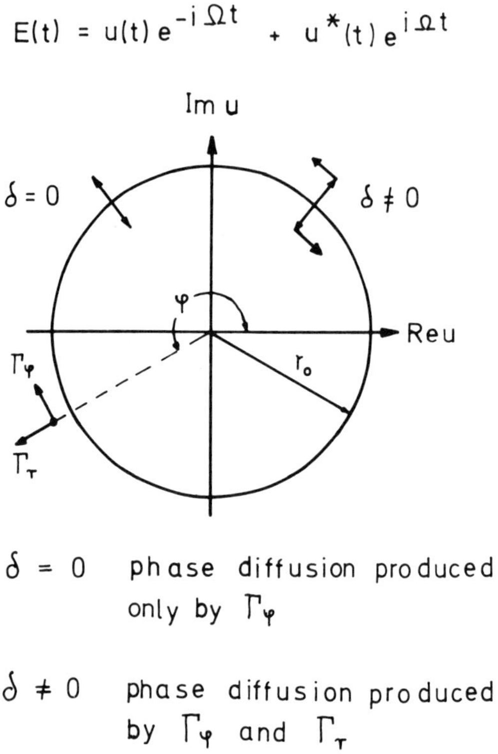

Fig. 3   Explanation of the enhancement of the linewidth factor due
         to detuning.

is stabilized and therefore only half of the noise power (only that
in φ-direction) contributes to the linewidth. Including detuning,
however, the small fluctuations in the r-direction around r_o lead
to a $(r-r_o)$-dependent additional motion of the phase ρ, thus leading
to an additional phase diffusion above threshold. Below threshold,
no   such additional diffusion occurs (if the trivial change of the
scaling paramters and the shift of the pumping parameter is not
taken into account).

     Without detuning the linewidth may be expressed by an eigen-
value of a certain hermitian operator. Taking detuning into account
one has to solve a non-hermitian eigenvalue problem. The problem is
simplified to some extent, because detailed balance is still valid

2 Detailed Balance

The statistical properties of laser light can be obtained by solving a proper Fokker-Planck equation, see § 3. The validity of detailed balance simplifies the solution of the Fokker-Planck equation. In this paragraph, we derive an eigenvalue expansion of the Fokker-Planck equation, if detailed balance is valid (following closely ref. [34], the solution of the stationary distribution in detailed balance was derived by Graham and Haken [33]).

The Fokker-Planck equation for the conditional probability $P(\{x\},\{x^0\},\tau)$ of the variables $(\{x\}) = (x_1,x_2,\ldots,x_N)$ is given by

$$\frac{\partial P}{\partial \tau} = L_F(\{x\})\ P \tag{2.1}$$

where L_F is the Liouville Fokker-Planck operator

$$L_F(\{x\}) = - \sum_i \frac{\partial}{\partial x_i} D_i(\{x\}) + \sum_{i,j} \frac{\partial^2}{\partial x_i \partial x_j} D_{ij}(\{x\}) \ . \tag{2.2}$$

The diffusion coefficient D_{ij} is assumed to be symmetric.

The variables $\{x\}$ are classified into even and odd variables according to their transformation with respect to time reversal $\{x\} \to \{\tilde{x}\}$

$$\left.\begin{array}{l} \tilde{x}_i = x_i \\ \tilde{x}_i = - x_i \end{array}\right\} \tilde{x}_i = \varepsilon_i x_i \qquad \begin{array}{ll} \varepsilon_i = 1 & \text{even variables} \\ \varepsilon_i = - 1 & \text{odd variables} \end{array} \ .$$

The drift coefficient $D_i(\{x\})$ may be split into an irreversible and a reversible drift coefficient, $D_i = D_i^{ir} + D_i^{rev}$, defined by

$$D_i^{ir}(\{x\}) = \frac{1}{2}[D_i(\{x\}) + \varepsilon_i\ D_i(\{\varepsilon x\})] \tag{2.4}$$

$$D_i^{rev}(\{x\}) = \frac{1}{2}[D_i(\{x\}) - \varepsilon_i\ D_i(\{\varepsilon x\})] \ .$$

For later purpose we split the stationary probability current

$$S_i = D_i W_{st} - \sum_j \frac{\partial}{\partial x_j} (D_{ij} W_{st})$$

$$= W_{st} [D_i - \sum_j \frac{\partial D_{ij}}{\partial x_j} - \sum_j D_{ij} \frac{\partial \ln W_{st}}{\partial x_j}] . \quad (2.5)$$

(W_{st} is the stationary solution $P(\{x\},\{x^o\},\infty)$) into a reversible and and irreversible part

$$S_i^{rev} = W_{st} D_i^{rev} \quad (2.6)$$

$$S_i^{ir} = W_{st} [D_i^{ir} - \sum_j \frac{\partial D_{ij}}{\partial x_j} - \sum_j D_{ij} \frac{\partial \ln W_{st}}{\partial x_j} . \quad (2.7)$$

Detailed balance is expressed by

$$W_2(\{x\},\{x^o\},\tau) = W_2(\{\varepsilon x^o\},\{\varepsilon x\},\tau) \quad (2.8)$$

where W_2 is the stationary joint distribution

$$W_2(\{x\},\{x^o\},\tau) = P(\{x\},\{x^o\},\tau) \cdot W_{st}(\{x^o\}) . \quad (2.9)$$

It may be shown that the necessary and sufficient conditions for detailed balance to be fulfilled are [34]

$$W_{st}(\{x\}) = W_{st}(\{\varepsilon x\}) \quad (2.10)$$

and the operator equation

$$L_F(\{x\}) \quad W_{st}(\{x\}) = W_{st}(\{x\}) \, L_F^+(\{\varepsilon x\}) . \quad (2.11)$$

Eq. (2.11) is equivalent to

$$D_{ij}(\{x\}) = \varepsilon_i \varepsilon_j D_{ij}(\{\varepsilon x\}) \tag{2.12}$$

$$S_i^{ir} = 0 \tag{2.13}$$

$$\sum_i \frac{\partial S_i^{rev}}{\partial x_i} = 0 . \tag{2.14}$$

The stationary solution may be obtained from (2.13) and (2.7). This requires that the integrability relation

$$\frac{\partial}{\partial x_j} \sum_k (D^{-1})_{ik} [\sum_L \frac{\partial D_{kL}}{\partial x_L} - D_k^{ir}] = \frac{\partial}{\partial x_i} \sum_k (D^{-1})_{jk} [\sum_L \frac{\partial D_{kL}}{\partial x_L} - D_k^{ir}] \tag{2.15}$$

is satisfied. Using (2.13) and (2.7), eq. (2.14) becomes

$$\sum_i \frac{\partial D_i^{rev}}{\partial x_i} = \sum_{ij} D_i^{rev} (D^{-1})_{ij} [\sum_k \frac{\partial D_{jk}}{\partial x_k} - D_j^{ir}] . \tag{2.16}$$

If the coefficients of the Fokker-Planck equation are given, eq. (2.12), (2.15) and (2.16) may be used to see whether detailed balance is valid or not.

Introducing the operator

$$L = (W_{st})^{-\frac{1}{2}} L_F (W_{st})^{\frac{1}{2}} , \quad L^+ = (W_{st})^{\frac{1}{2}} L_F^+ (W_{st})^{-\frac{1}{2}} \tag{2.17}$$

the key relation (2.6) simplifies to

$$L(\{x\}) = L^+(\{\varepsilon x\}) . \tag{2.18}$$

If even and odd variables are present, the operator L will generally not be hermitian. The operator L is explicitly given by

$$L = L_H + L_A \tag{2.19}$$

$$L_H = \sum_{i,j} \frac{\partial}{\partial x_i} D_{ij} \frac{\partial}{\partial x_j} - V \tag{2.20}$$

$$V = (W_{st})^{-\frac{1}{2}} \sum_{i,j} \frac{\partial}{\partial x_i} D_{ij} \frac{\partial (W_{st})^{\frac{1}{2}}}{\partial x_j} \tag{2.21}$$

$$L_A = - \sum_i D_i^{rev} \frac{\partial}{\partial x_i} + \sum_i D_i^{rev} (W_{st})^{-\frac{1}{2}} \frac{\partial (W_{st})^{\frac{1}{2}}}{\partial x_i} . \tag{2.22}$$

L_H is the hermitian part, L_A is the antihermitian part of L. The operator L is only hermitian, if the reversible drift-coefficient vanishes.

In order to obtain an eigenfunction expansion of the conditional probability P of the Fokker-Planck equation (2.1) usually eigenfunctions of L and L^+ (respectively of L_F and L_F^+) must be used, which form a biorthogonal set. Because of the key relation (2.11) or equivalently (2.18), however, the eigenfunctions of L^+ are simply $\psi_\mu(\{\varepsilon x\})$ where $\psi_\mu(\{x\})$ are the eigenfunctions of L

$$\left.\begin{aligned} L(\{x\})\, \psi_\mu(\{x\}) &= - \lambda_\mu\, \psi_\mu(\{x\}) \\ L^+(\{x\})\, \psi_\mu(\{\varepsilon x\}) &= - \lambda_\mu\, \psi_\mu(\{\varepsilon x\}) \end{aligned}\right\} \tag{2.23}$$

The stationary eigenfunction ($\lambda_0=0$) is given by $\psi_0=(W_{st})^{\frac{1}{2}}$. We assume that the eigenfunctions ψ_μ exist and form a complete biorthogonal set. Then the expansion of the conditional probability P of the Fokker-Planck equation is

$$P(\{x\},\{x^0\},\tau) = \frac{\psi_0(\{x\})}{\psi_0(\{x^0\})} \sum_\mu \psi_\mu(\{x\})\, \psi_\mu(\{\varepsilon x^0\})\, e^{-\lambda_\mu \tau}, \tag{2.24}$$

if the eigenfunctions are normalized according to

$$\int \ldots \int \psi_\mu(\{x\})\, \psi_\mu(\{\varepsilon x\})\, dx_1 \ldots dx_N = \delta_{\mu\nu} . \tag{2.25}$$

Thus the problem is reduced to that of calculating the (generally complex) eigenfunctions of the operator L.

3. Detuned Single Mode Laser

The statistical properties of the light amplitude E(t) of a single mode laser are described by its slowly varying complex amplitude u(t), i.e.

$$E(t) = u(t)\, e^{-i\Omega t} + u^*(t)\, e^{i\Omega t} . \tag{3.1}$$

Not too far from threshold, the equation of motion for the slowly varying amplitude u(t) is the laser Langevin equation [1,2,5,6,8, 11-17]

$$\dot{u} - \beta(1 + i\delta)[a\sqrt{g/\beta} - |u|^2]\, u = \Gamma(t) \tag{3.2}$$

$$\langle\Gamma(t)\,\Gamma^*(t')\rangle = 4g\,\delta(t-t'); \quad \langle\Gamma(t)\Gamma(t')\rangle = 0 . \tag{3.3}$$

In (3.2) δ is a detuning parameter which is defined for a homogeneously broadened line by

$$\delta = \frac{\omega_c - \omega_A}{\gamma} \tag{3.4}$$

where ω_c is the cavity frequency, ω_A is the atomic frequency, and γ is the atomic linewidth (cavity linewidth << γ). For an inhomogeneous Gaussian distribution of the frequencies of the atoms (halfwidth = $\sqrt{\ln 2}\,\alpha$) δ is given by

$$\delta = \frac{2\gamma}{\alpha} \quad \frac{\omega_c - \omega_A}{\alpha} , \tag{3.5}$$

see for instance [6] (again cavity linewidth << δ). The laser constants β and g depend on the detuning parameter. For a homogeneously broadened line we have

$$\beta = \beta(\delta) = \frac{\beta(0)}{1 + \delta^2} \quad ; \quad g = g(\delta) = \frac{g(0)}{1 + \delta^2} . \tag{3.6}$$

Because the $\Gamma(t)$ are δ-correlated and because of the large number of photons, we have a continuous Markov process. Therefore the conditional probability of u may be determined by a Fokker-Planck equation. If the real part of u is denoted by x_1, the imaginary part of u by x_2

$$u = x_1 + i\, x_2 \tag{3.7}$$

x_1 will be an even variable, x_2 an odd variable

$$\varepsilon_1 = 1 \quad , \quad \varepsilon_2 = -1 \,. \tag{3.8}$$

The irreversible and reversible drift coefficients and the diffusion coefficients are then given by [2,5,6,8,11-18]

$$\left.\begin{aligned} D_1^{ir} &= \beta[a\sqrt{g/\beta} - (x_1^2 + x_2^2)]\, x_1 \\ D_2^{ir} &= \beta[a\sqrt{g/\beta} - (x_1^2 + x_2^2)]\, x_2 \end{aligned}\right\} \tag{3.9}$$

$$\left.\begin{aligned} D_1^{rev} &= -\beta\, x_2[a\sqrt{g/\beta} - (x_1^2 + x_2^2)]\, \delta \\ D_2^{rev} &= \beta\, x_1[a\sqrt{g/\beta} - (x_1^2 + x_2^2)]\, \delta \end{aligned}\right\} \tag{3.10}$$

$$D_{11} = D_{22} = g \quad ; \quad D_{12} = D_{21} = 0 \,. \tag{3.11}$$

The Fokker-Planck equation with these coefficients is the leading part of a general Fokker-Planck equation (with an infinite number of derivatives). This general Fokker-Planck equation follows from the fully quantum mechanical master equation, see refs. [16,17,35].

It is convenient to use scaled variables $\bar{x}_i$ and scaled times $\bar{t}$ defined by

$$\bar{x}_i = \sqrt[4]{\beta/g}\; x_i \quad ; \quad \bar{t} = \sqrt{g\beta}\; t \,. \tag{3.12}$$

Then the laser constants β and g are normalized to 1.

It may easily be checked that the conditions of detailed balance i.e. eqs. (2.12) (2.15) and (2.16), are fulfilled.* Therefore,

* M. Lax has treated both x_1 and x_2 as even variables. He found that detailed balance is violated in the presence of detuning. [36]

the results of §2 are applicable. In polar coordinates $\overline{x}_1 = r\cos\phi$, $\overline{x}_2 = r\sin\phi$, we obtain

$$L_H = \frac{\partial^2}{\partial r^2} + \frac{1}{r}\frac{\partial}{\partial r} + \frac{1}{r^2}\frac{\partial^2}{\partial\phi^2} - V \tag{3.13}$$

$$V = a + [\frac{a^2}{4} - 2]r^2 - \frac{a}{2}r^4 + \frac{1}{4}r^6 \tag{3.14}$$

$$L_A = -(a - r^2)\delta \cdot \frac{\partial}{\partial\phi} . \tag{3.15}$$

The probability current $S_i = S_i^{rev}$ in the stationary state has only a ϕ- component:

$$S_\phi = r(a - r^2)\ \delta . \tag{3.16}$$

Because $e^{in\phi}$ is an eigenfunction of L with respect to ϕ, we may write (double index notation)

$$\psi_\mu(\overline{x}_1,\overline{x}_2) = \frac{\psi_{nm}(r)}{\sqrt{r}}\ \frac{e^{in\phi}}{\sqrt{2\pi}} \tag{3.17}$$

$$\psi_\mu(\varepsilon_1\overline{x}_1, \varepsilon_2\overline{x}_2) = \psi_\mu\ (\overline{x}_1,-\overline{x}_2) = \frac{\psi_{nm}(r)}{\sqrt{r}}\ \frac{e^{-in\phi}}{\sqrt{2\pi}} \tag{3.18}$$

(the factor $1/\sqrt{r}$ was split off for convenience).

The generally complex eigenfunctions ψ_{nm} and eigenvalues λ_{nm} are then obtained by the non-hermitian eigenvalue equation

$$\psi''_{nm} + [\lambda_{nm} - V_n - in(a-r^2)\delta]\ \psi_{nm} = \sigma \tag{3.19}$$

$$V_n = (n^2 - \frac{1}{4})\frac{1}{r^2} + a + (\frac{a^2}{4} - 2)r^2 - \frac{a}{2}r^4 + \frac{r^6}{4} . \tag{3.20}$$

The stationary solution ($\lambda_{oo} = 0$) follows from (2.13) and (3.17) by

$$\psi_{oo}(\tau) = \sqrt{C_n r}\ \exp\ [\frac{ar^2}{4} - \frac{r^4}{8}] . \tag{3.21}$$

The conditional probability distribution (2.21) specializes to ($\tau \geq 0$)

$$P(r,\phi;r^o,\phi^o,\tau) = \frac{1}{2\pi r}\frac{\psi_{oo}(r)}{\psi_{oo}(r^o)}\sum_{m=0}^{\infty}\sum_{n=-\infty}^{\infty}\psi_{nm}(r)\psi_{nm}(r^o)e^{in(\phi-\phi^o)}e^{-\lambda_{nm}\tau} . \tag{3.22}$$

The stationary joint distribution is given by

$$W_2(r,\phi;r^o,\phi^o;\tau) =$$

$$\frac{\psi_{oo}(r)}{2\pi r}\frac{\psi_{oo}(r^o)}{2\pi r^o}\sum_{m=0}^{\infty}\sum_{n=-\infty}^{\infty}\psi_{nm}(r)\psi_{nm}(r^o)\, e^{in(\phi-\phi^o)}e^{-\lambda_{nm}\tau} \tag{3.23}$$

for ($\tau \geq 0$) and by

$$W_2(r,\phi;r^o,\phi^o;\tau) =$$

$$\frac{\psi_{oo}(r)}{2\pi r}\frac{\psi_{oo}(r^o)}{2\pi r^o}\sum_{m=0}^{\infty}\sum_{n=-\infty}^{\infty}\psi_{nm}(r)\psi_{nm}(r^o)\, e^{in(\phi^o-\phi)}e^{-\lambda_{mn}|\tau|} . \tag{3.24}$$

for ($\tau \leq 0$).

4. Linewidth

The correlation function of the amplitude u(t) is defined by

$$g(a,\tau) = \langle u^*(t+\tau)\, u(t)\rangle$$

$$= \iint u^* u^o\, W_2(u,u^o,\tau)\, d^2u\, d^2u^o . \tag{4.1}$$

Inserting the explicit expressions (3.23) and (3.24), we obtain in scaled units for $\tau \geq 0$

$$\bar{g}(a,\bar{\tau}) = \sum_{m=0}^{\infty}\left[\int_o^{\infty} r\,\psi_{oo}(r)\psi_{1m}(r)\,dr\right]^2 e^{-\lambda_{1m}\tau} \tag{4.2}$$

and for $\tau \leq 0$

$$\bar{g}(a,\bar{\tau}) = \bar{g}^* (a,|\bar{\tau}|) \quad . \tag{4.3}$$

A calculation of the matrix elements in (4.2) shows that to an accuracy of a few percent

$$[\int_0^\infty r\,\psi_{oo}(r)\ \psi_{1m}(r)\,dr]^2 \sim$$

$$\int_0^\infty r^2\psi_{oo}^2\ (r)\ dr\delta_{mo} = <|\bar{u}|^2>\ \delta_{mo} \quad . \tag{4.4}$$

Therefore the correlation function is approximately [32]

$$g(a,\tau) = <|u|^2>\exp\{-\sqrt{g\beta}\,[\lambda_{10}^{(r)}|\tau| + i\lambda_{10}^{(i)}\tau]\} \ . \tag{4.5}$$

The imaginary part $\lambda_{10}^{(i)}$ of λ_{10} determines the frequency shift relative to Ω, whereas the real part $\lambda_{10}^{(r)}$ gives the linewidth. This linewidth may be written as

$$\Delta\nu = \lambda_{10}^{(r)}\ \sqrt{g\beta} = \alpha_L\ g/<|u|^2> \ , \tag{4.6}$$

where

$$<|u|^2> = \sqrt{g/\beta}<|\bar{u}|^2> \tag{4.7}$$

$$<|\bar{u}|^2> = \int_0^\infty r^2\psi_{oo}^2\ dr = a + \frac{2}{\sqrt{\pi}}\ \frac{e^{-\frac{1}{4}a^2}}{1+\mathrm{erf}(\frac{a}{2})} \quad . \tag{4.8}$$

The power output P divided by the power output at threshold, i.e, P_{thr} is

$$\frac{P}{P_{thr}} = \frac{<|\ \bar{u}|^2>}{<|\ \bar{u}|^2>_{thr}} = \frac{\sqrt{\pi}}{2}\ <|\bar{u}|^2> \ . \tag{4.9}$$

The linewidth factor α_L is defined by

$$\alpha_L = \alpha_L(a,\delta) = \lambda_{10}^{(r)} \cdot <|\bar{u}|^2> \ . \tag{4.10}$$

Fig. 2 shows the results of the numerical calculation of this linewidth factor α_L [32]. Without detuning, α_L varies from 2 to 1 by passing through the threshold region. With increasing detuning this ratio is diminished. For δ larger than 1, the linewidth factor above threshold is even larger than that below threshold.

If the detuning δ is not too large, the linewidth factor $\alpha_L(a,\delta)$ may approximately be expressed by the factor $\alpha_L(a,0)$ without detuning

$$\alpha_L(a,\delta) = \alpha_L(a,0) + [2 - \alpha_L(a,0)]\,\delta^2 . \tag{4.11}$$

For instance for $\delta = 1$, we have according to (4.11) $\alpha_L(a,1) = 2$, which is correct to an error of less than 7%, see fig. 2. Using a proper linearization of eq. (3.2) well below and well above threshold, it may be shown that outside the threshold region eq. (4.11) is an exact result, in agreement with refs. [5.8]. In the linearized region below and above threshold, the linewidth for a detuned laser was also calculated by Richter and Grossmann [37]. Below threshold their result agrees with ours, above it does not.

References

1. H. Haken, Z. Physik *181*, 96 (1964).
2. H. Risken, Z. Physik *186*, 85 (1965).
3. H. Sauermann, Z. Physik *188*, 480 (1965), *189*, 312 (1966).
4. M. Lax and R.D. Hempstead, Bull. Am. Phys. Soc. *11*, 111 (1966).
5. H. Risken, C. Schmid and W. Weidlich, Z. Physik *193*, 37 (1966).
6. V. Arzt, H. Haken, H. Risken, H. Sauermann, C. Schmid and W. Weidlich, Z. Physik *197*, 207 (1966).
7. J.A. Fleck, Phys. Rev. *149*, 309 (1966), *149*, 322(1966).
8. M. Lax (QV) in: *Physics of Quantum Electronics*, eds. P.L. Kelley, B. Lax and P.E. Tannenwald (McGraw-Hill Book Co., Inc., New York 1966) p. 735.
9. M. Scully, and W.E. Lamb, Phys. Rev. Letters *16*, 853 (1966), Phys. Rev. *159*, 208 (1967).
10. H. Risken, Z. Physik *191*, 302 (1966).
11. H. Risken and H.D. Vollmer, Z. Physik *201*, 322 (1967), *204*, 240 (1967).
12. R.D. Hempstead, and M. Lax, Phys. Rev. *161*, 350 (1967).
13. M. Lax, and W.H. Louisell, (QIX) IEEE J. Quantum Electron. QE-3, 47 (1967).
14. H. Risken, Fortschr. Physik *16*, 261 (1968).
15. M. Lax, in *Statistical Physics Vol 2*, Brandeis University Summer Institute in Theoretical Physics, page 269, (Gordon and Breach, New York-London-Paris, 1968) .

16. H. Haken, *Laser Theory*, in Encyclopedia of Physics Vol. XXV/2c, eds. S. Flugge (Springer Verlag, Berlin-Heidelberg-New York, 1970)
17. H. Risken, in *Progress in Optics*, Vol. VIII, ed. E. Wolf (North Holland, Amsterdam, 1970) p. 239.
18. J.P. Gordon, and E.W. Aslaken, IEEE J. Quantum Electron. *QE-6* 428 (1970).
19. M. Lax, and M. Zwanziger, Phys. Rev. Let. *24*, 937 (1970).
20. S. Grossmann, and P.H. Richter, Z. Physik *242*, 458 (1971).
21. A.W. Smith, and J.A. Armstrong, Phys. Rev. Letters *16* , 1169 (1966).
22. J.A. Armstrong, and A.W. Smith, in *Progress in Optics*, Vol. 6, ed. E. Wolf (North Holland Publishing Co., Amsterdam; John Wiley and Sons, New York 1967), p. 211.
23. F.T. Arecchi, G.S. Rodari, and A. Sona, Phys. Letters *25A*, 59 (1967).
24. R.F. Chang, R.W. Detenbeck, V. Korenmann, C.O. Alley, and U. Hochuli, Phys. Letters *25A*, 272 (1967).
25. F.T. Arecchi, M. Giglio and A. Sona, Phys. Letters *25A*, 341(1967).
26. F. Davidson, and L. Mandel, Phys. Letters *25A*, 700 (1967).
27. E.R. Pike, in: *Quantum Optics*, eds. S.M. Kay and A. Maitland, (Academic Press, London-New York 1970), p. 127.
28. D. Meltzer, W. Davis and L. Mandel, Appl. Phys.Letters *17*, 242 (1970).
29. D. Meltzer, and L. Mandel, Phys. Rev. Letters *25* , 1151 (1970); Phys. Rev. *3A*, 1763 (1971).
30. F.T. Arrecchi, and V. Degiorgio, Phys. Rev. *3A* , 1108 (1971).
31. H. Gerhardt, H. Welling and A. Güttner, Z. Physik *253*, 113 (1972).
32. H. Risken and K. Seybold, Phys. Lett. *38A*, 63 (1972).
33. R. Graham, and H. Haken, Z. Physik *243*, 289 (1971).
34. H. Risken, Z. Physik 251, *231*(1972).
35. H. Haken, H. Risken, and W. Weidlich, Z. Physik *206*, 355 (1967)
36. Ref. 15, p. 386.
37. P.H. Richter, and S. Grossman, Z. Physik *248*, 244 (1972).

TUNABLE RAMAN LASERS

C. K. N. Patel

Bell Telephone Laboratories, Holmdel, New Jersey

1. Introduction

Various techniques exist for producing tunable laser radiation in the infrared portion of the spectrum. These include the spin-flip Raman (SFR) lasers[1], parametric oscillators[2] and the tunable diode lasers [3]. In the present paper, we will concentrate only on the SFR lasers without attempting to compare the three different techniques of generating tunable laser radiation.

The SFR laser dates back to the early 1970 when the first SFR laser action in InSb pumped with a 10.6μ CO_2 laser was reported [1]. In the last two years, the progress in the field of SFR lasers has been extremely rapid and a large number of innovations and new ideas have appeared.[4-11] In the present paper, we will not attempt to discuss all the aspects of SFR lasers but concentrate mainly on the recent developments. The early work on the SFR lasers has been summarized in the review articles by Patel and Shaw, [8] and Patel.[12] These articles describe the earlier work on spontaneous Raman scattering as well as the details of the SFR laser.

In Section 2, we will very briefly summarize some of the early findings. We will describe the operation of Q-switched CO_2 laser operated SFR laser and the cw operating CO pumped SFR laser. A brief discussion of the power output characteristics for both the cw as well as the Q-switched SFR lasers will be given. The SFR laser operated with a Q-switched CO_2 laser at 10.6μ has the tunability extending from 9μ to 14.6μ. The power output is approximately 1 kW. For the cw mode of operation of the SFR laser at 5.3μ, the tunability extends from around 5μ to about 6.5μ with a cw tunable power output of the order of 1 W. The conversion efficiency for the cw SFR laser has been as high as 70%, while that for the Q-switched SFR lasers is about 10%. In Section 3 we will describe the low field operation [13] of the SFR lasers. We have now extended the tuning range of the SFR lasers to magnetic fields as low as 200G. This operation requires the use of very low carrier concentration InSb sample and allows us to do away with the superconducting magnet for tuning the SFR laser. In Section 4, we will discuss the measurement of the SFR laser linewidth [14] carried out by heterodyning the SFR output with a cw gas laser. Experimentally the measured linewidth was 1 kHz while the instrument resolution was also 1 kHz indicating that the linewidth of the SFR laser is considerably narrower than 1 kHz. This linewidth is in agreement with the calculated linewidth. In this section we will also discuss the problem of stability of the SFR laser which has to be considered in order to completely use the very narrow linewidth that is available from the present tunable source. In Section 5 we will discuss some of the recent applications of the SFR lasers. It can be seen that the SFR laser has a large number of applications at the present time as well as a very large future. These applications include high resolution spectroscopy, pollution detection etc. We will not discuss the earlier results but give only the recent developments in this area. Finally we will conclude the paper by pointing out some techniques for improving the SFR laser operation and extending the tuning range.

2. Summary of Earlier Work

In the earlier studies of spontaneous light scattering from electrons in semiconductors it was found that three distinct processes gave rise to tunable Raman scattering from electrons in InSb.[15-17] These were (i) the $\Delta\ell = 2$ process where an electron in the lowest Landau level is excited to the second Landau level and the energy of the input photon is changed by an amount approximately equal to $\sim 2\omega_c$ (ii) the $\Delta\ell = 1$ process where the Landau level quantum number is changed by one, and (iii) the SFR process where no change in the Landau level quantum number was involved but only the spin state of the electron was changed. The output frequency for these three Raman scattering processes is given by:

$$\omega_s = \omega_o - g\mu_B B \quad \text{(SFR)} \tag{1}$$

$$\omega_s = \omega_o - 2\omega_c \quad (\Delta\ell = 2) \tag{2}$$

$$\omega_s = \omega_o - \omega_c \quad (\Delta\ell = 1) \tag{3}$$

where ω_s is the frequency of the Stokes Raman scattered photon, ω_o is the frequency of the input photon, g is the g-value of the electrons, μ_B is the Bohr magneton, B is the magnetic field, ω_c is the cyclotron frequency given by $\omega_c = eB/m^*c$, m^* is effective mass of the electrons in InSb, and c is the velocity of light.

Of these three processes for obtaining tunable Raman scattering, the SFR process was the most intense, and the Raman scattering cross-section, σ, was about ten times larger than that for either the $\Delta\ell = 1$ or the $\Delta\ell = 2$ processes. In addition the spontaneous Raman scattering linewidth was at least a factor of 10 narrower for the SFR process as opposed to that for the two other processes. Thus it was felt that the SFR process held the greatest promise for obtaining stimulated Raman scattering and a tunable laser.

The gain arising from a Raman scattering process is given by

$$g_s = \frac{16\pi^2 c^2 (S/\ell d\Omega)}{\hbar\omega_s^3 n_p^2 (\bar{n} + 1)\Gamma} \; I \tag{5}$$

Where $S/\ell d\Omega$ is the Raman scattering efficiency given by $n_e \times \sigma$, n_e is the electron concentration, n_p is the refractive index at pump frequency, $(\bar{n} + 1)$ is the Boltzmann factor, Γ is the spontaneous Raman scattering linewidth and I is the intensity of the pump radiation. It is clear that we need to have as large a Raman scattering efficiency and as small a linewidth as possible to produce a large Raman gain. The SFR Raman scattering process has a Raman scattering cross-section of approximately 10^{-23} cm^2 sr^{-1}. Typical linewidths are ~ 1 cm^{-1}. The expected Raman gain[1] at 10.6μ is $\sim 10^{-5} cm^{-1}/Wcm^{-2}$. Thus a Q-switched CO_2 laser producing a power output of $\sim$ 10kW can easily give us SFR gains exceeding 100 cm^{-1}, and spin-flip Raman laser action should be possible. Patel and Shaw reported such a laser action in Reference 1. It was seen that beyond the threshold input power of $\sim$100W the output scattered light increased very rapidly for a very small increase in the input power. This was recognized as the threshold for stimulated SFR scattering or the threshold for the SFR laser as seen in Figure 1. Corresponding to the tremendous increase in the scattered light output beyond threshold there was a significant narrowing of the spectrum as seen in Figure 2 where we show the spectra below threshold and above threshold for the SFR laser. This clearly demonstrates SFR laser action in InSb. The linewidth that is seen for the SFR laser in Figure 2 is determined by spectrometer resolution. The calculated linewidth for this situation is expected to be extremely narrow. Later in the paper we will describe a measurement of the linewidth of the SFR laser using heterodyne techniques.

Subsequent work on the 10.6μ pumped SFR laser showed that it is possible to obtain the second Stokes [10] SFR laser as well as the anti-Stokes SFR [6] laser. These three processes result in the tuning range for the SFR laser which extends from around 9μ to 14.6μ. The tuning characteristic of the anti-Stokes SFR laser is given by

$$\omega_s = \omega_o + g\mu_B B \qquad (6)$$

and that for the second Stokes SFR is given by

$$\omega_s = \omega_o - 2g\mu_B B \qquad (7)$$

The power output in the Q-switched operation of the SFR laser on the first Stokes line is approximately 1 kW, for the second Stokes it is approximately 100 W and for the anti-Stokes SFR laser the power output is 30 W. The total tunability of the SFR pumped at 10.6μ is shown on Figure 3 for a magnetic field up to 100 kG.

In order to obtain cw laser action is is necessary that the threshold be reduced considerably from the 100 W described for the

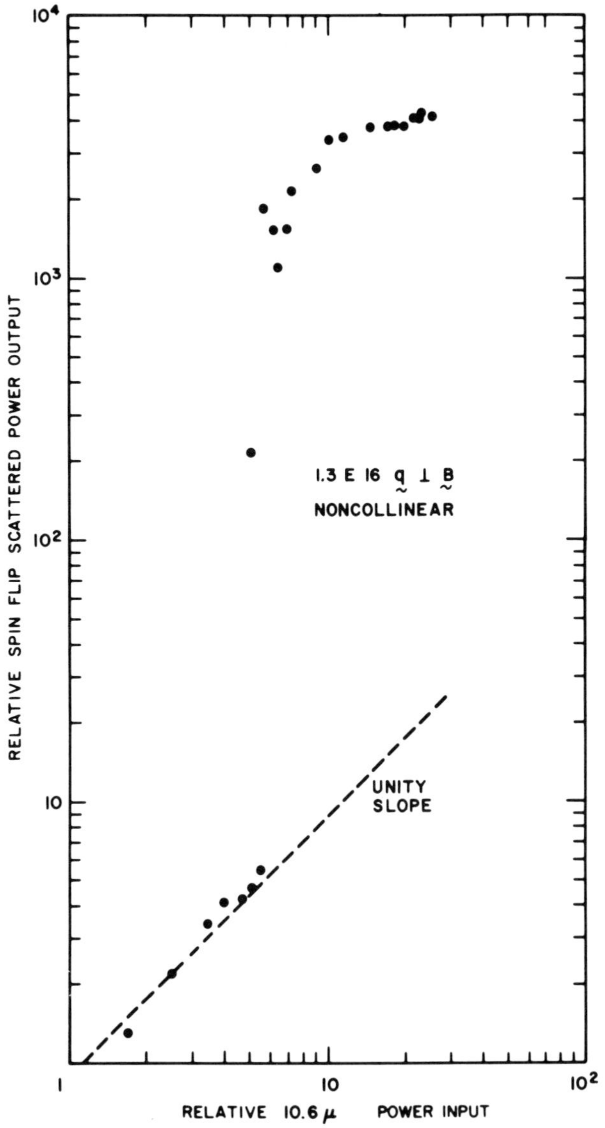

Figure 1. Spin-Flip Raman scattered light output at a wavelength 11.5μ as a function of 10.6μ Q-switched pump power from a CO_2 laser (in arbitrary units) for $n_e \sim 1.3 \times 10^{16}$ cm^{-3} InSb sample at a temperature of 18K and a magnetic field of 40 kG. Maximum pump power is 1 kW inside the sample and the maximum spin-flip Raman laser output power is 100 W (peak).

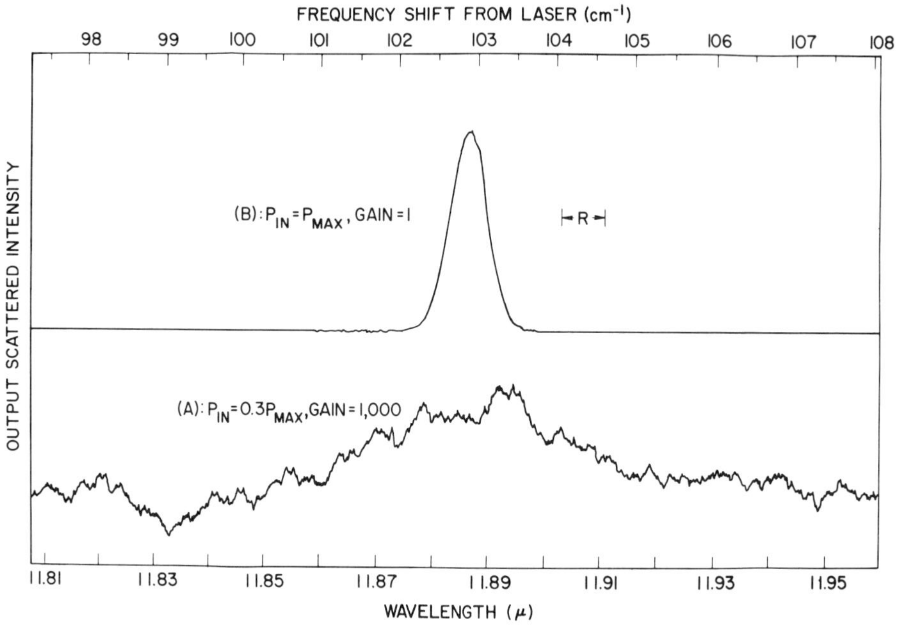

Figure 2. Spectral analysis of the spin-flip Raman scattered light output below and above spin-flip Raman laser threshold at a magnetic field of ∿52 kG. R is the spectrometer resolution.

CO_2 laser pumped SFR laser. This has been achieved by using resonance enchancement of the cross-section [17] in InSb. By choosing a pump laser frequency ω_0 such that it is very close to the bandgap, Eg, of InSb it is possible to obtain a sizeable increase in the Raman scattering cross-section and thereby a sizeable increase in the Raman gain. This results in a reduction in the threshold for the SFR laser. The bandgap in InSb is at 5.3μ thus the CO laser at 5.3μ is an ideal choice. Mooradian, Brueck and Blum [5] demonstrated the cw SFR laser in InSb using the CO laser. The lowest observed threshold for the cw SFR laser at 5.3μ is ∿ 5–10 mW. Figure 4 shows a partial tuning curve for the cw SFR laser pumped with the CO laser.

It should be pointed out that since the Raman laser frequency is determined by the pump frequency ω_0 in Eq. (1), a change of ω_0 also gives rise to a change in the Raman laser frequency. Thus by choosing different pump laser frequencies we can increase the tuning range of the SFR laser. Both the CO and the CO_2 lasers generate a large number of transitions[18] in the 5μ and 10μ range,

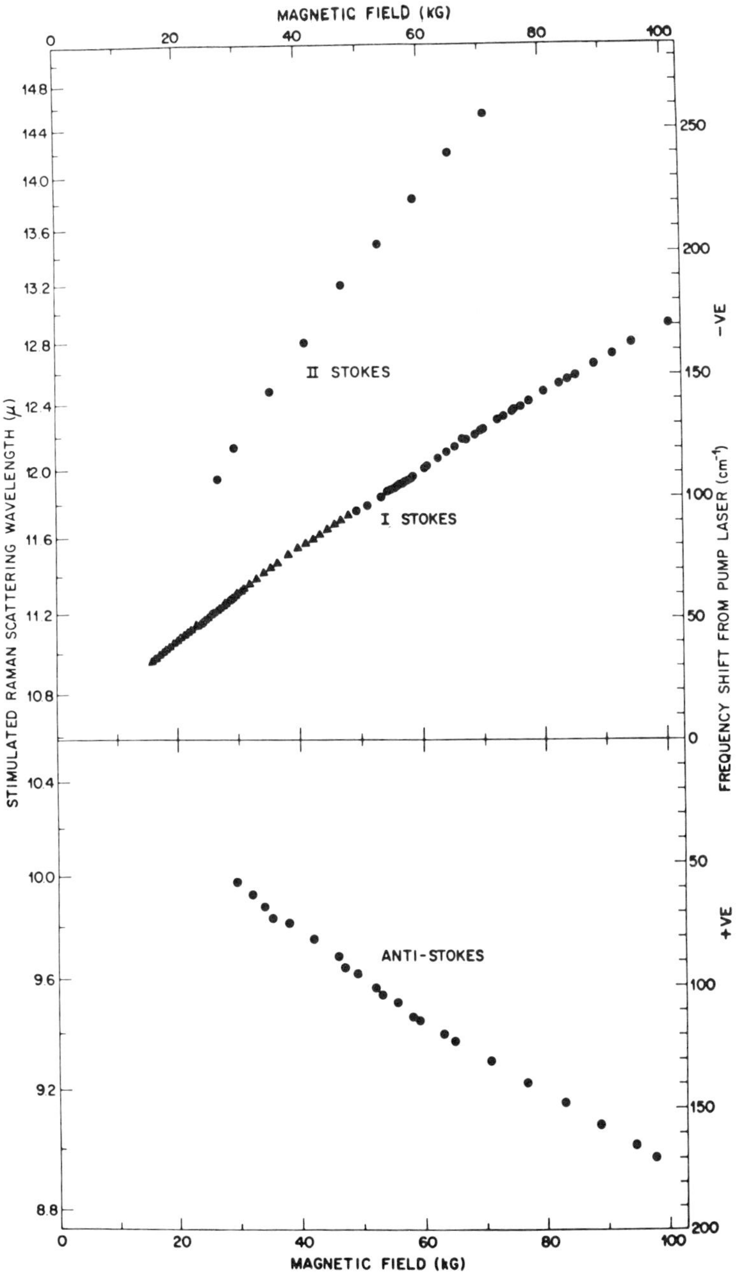

Figure 3. Overall tuning curve for the SFR laser pumped with the Q-switched CO_2 laser at 10.5915μ showing the tunability of the I-Stokes, II-Stokes and Anti-Stokes SFR laser.

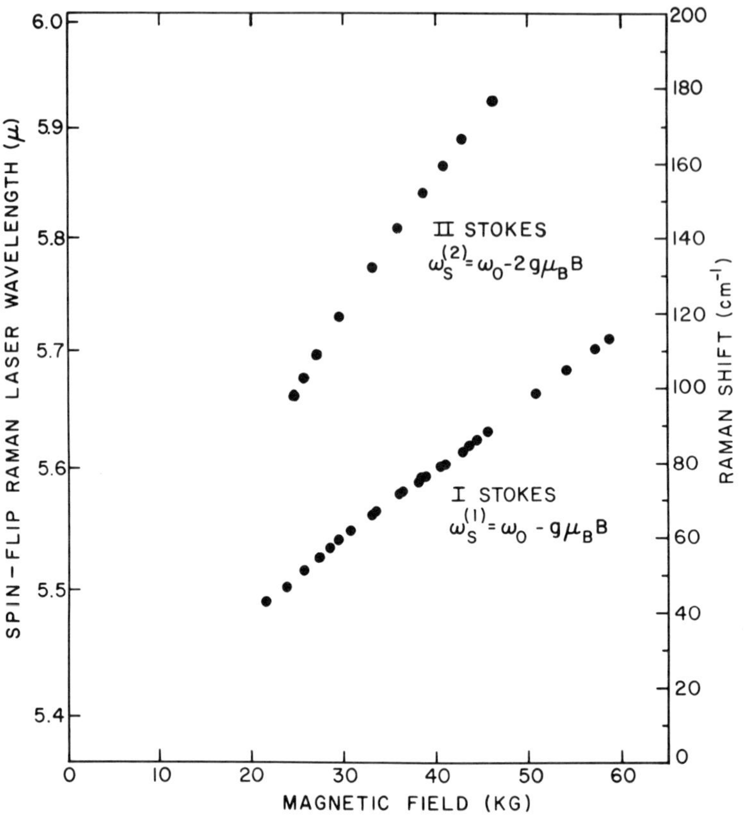

Figure 4. Overall tuning curve for the cw SFR laser pumped with CO laser at 5.3648μ.

respectively. Using the various transitions as pump frequencies, the tuning range of the SFR laser now covers from 5μ to 6.5μ and and from 9μ to 14.5μ.

The output power from the SFR laser is limited from two considerations. The first is the pump depletion i.e. output power stops increasing when it depletes the pump power. The second consideration for the SFR laser is that under some conditions the SFR laser power output saturates even before any pump depletion has been observed. This is because of the fact that for every SFR scattered photon one electron is excited from the lower spin sublevel to the upper spin sublevel.[1,8] This electron has to return to the lower spin-state before it can scatter another pump photon into a SFR photon. The spin-relaxation time, τ_{SR}, in InSb is long and it is of the order 10^{-9} sec. Thus it is possible to create a bottleneck in which the electrons cannot relax to the lower spin flip state fast enough to maintain Raman gain for producing SFR laser action. At 10.6μ such a saturation has been observed. To obtain high powers at 10.6μ it is necessary to use larger volumes of InSb samples.

The earlier work has been carried out with superconducting magnets for providing the magnetic field. It was found that in all of these experiments the lowest magnetic field at which the SFR laser action was obtained was limited ∿ 15 kG. This was obtained with a $n_e \approx 4.5 \times 10^{15}$ cm^{-3}. This low field limit is set by the quantum limit considerations[8] which indicate that in order to obtain maximum Raman gain all the electrons must be in the lowest spin state. Since both the Fermi energy as well as the separation between the spin-levels is the function of magnetic field, a certain minimum magnetic field is required before the quantum limit is reached. This is the minimum magnetic field at which SFR laser action can be obtained. We will see in the next section that it is possible to lower the minimum magnetic field requirement by using InSb sample having lower carrier concentrations. This will considerably increase the versatility and the usefulness of the SFR laser because we can do away with the superconducting magnet and use electromagnets if the low magnetic field limit is brought down to fields of the order of a few kG.

In addition to the gross tuning characteristics of the SFR laser seen in Figs. 3 and 4, the fine tuning characteristics of the SFR lasers have also been investigated in detail. It has been found that because of the existence of the cavity modes of the SFR laser, the tuning of the laser frequency is not continuous. [7] In another experiment at 10.6μ where operation at pump powers ∿ 20 times above SFR laser threshold has been carried out, we find that a continuous tuning is possible. The details of these experiments can be found in Reference 12. It is clear, however, that even at threshold the mode pulling effects result in a tuning over a sizeable frequency range.

The optimum electron concentration [8] for maximum SFR laser power output has been seen to be approximately 1.3×10^{16} cm^{-3}. With this carrier concentration the quantum limit indicates that the minimum magnetic field for the SFR laser operation is ∿ 23 kG. Lowering the carrier concentration to go to lower magnetic fields results in a reduction in the maximum power output available under Q-switched operation, and under some conditions under cw operation as well. This will be seen in greater detail in the next section.

3. Low Field Operation of Spin-Flip Raman Lasers

Since the SFR laser can be gross tuned by changing the pump wavelength [See Eq.(1)] in discrete jumps, it is clear that to obtain a continuously tunable radiation, the SFR laser has to be able to tune over a range that covers only the spacing between the discrete pump wavelengths. As can be seen from Reference 18, the CO as well as the CO_2 lasers generate discrete lines separated by approximately 4 cm^{-1} and 2 cm^{-1} covering a range from approximately 5μ to about 7.5μ and 9μ to about 11μ respectively. The inclusion of isotopically substituted CO_2 as well as the N_2O laser extends this discrete tuning range from about 5 microns to 12μ in the infrared. Since the SFR laser tuning rate is approximately 2 cm^{-1}/kG (as determined from the g-value of -50 for the electrons in InSb), the maximum magnetic field needed to cover the entire range from 5μ to 12μ is approximately 2kG which can be easily obtained using a small electromagnet. This removes the necessity of having a superconducting solenoid which has been used for SFR laser experiments so far. Thus it is important that we investigate the possibility of operation of the SFR laser at low magnetic fields.

It can be seen from the discussion in Section 2 that the earlier SFR laser experiments had been limited to a minimum magnetic field of about 15 kG. This limit was explained in terms of the requirement of reaching quantum limit [8] for the electrons in InSb. The quantum limit considerations indicate that the lowest magnetic field of 15 kG (observed for SFR laser operation in an InSb sample having $n_e \approx 4.5 \times 10^{15}$ cm^{-3}) can be lowered if we operate the SFR laser with lower electron concentration InSb samples. Lower electron concentration at first sight would indicate that the SFR gain will be reduced proportionately since the SFR scattering efficiency, $S/\ell d\Omega$, is proportional to the number of electrons [see Eq.(5)]. Operation of the SFR laser at magnetic fields of 5 kG or less would require an electron concentration of $\lesssim 1 \times 10^{15}$ cm^{-3} implying that the Raman gain would be reduced by a factor of 5 compared to that in the 4.5×10^{15} cm^{-3} sample. This simple consideration neglects the change in other properties of InSb as carrier concentration is changed. One of the most important considerations with lowering carrier concentration is that the linewidth of the spontaneous SFR scattering is expected to go down. Studies of spin resonance in InSb having different electron concentrations[19] have shown that $n_e \approx 1 \times 10^{16}$ cm^{-3} gives a spin resonance linewidth of approximately 0.5 cm^{-1}. This linewidth reduces to approximately .15 cm^{-1}

with $n_e \sim 2 \times 10^{16}$ cm^{-3}, and to about .05 cm^{-1} with $n_e \sim 4 \times 10^{14}$cm^{-3}. As seen from the Raman gain expression in Eq.(5), the reduction in the linewidth will partly compensate the loss in the Raman gain attributable to the reduction in carrier concentration. Thus for operation at 10.6μ the expected threshold for the SFR laser will go up by approximately a factor of two when the electron concentration is reduced from $\sim 1 \times 10^{16}$ cm^{-3} (optimum concentration for maximum power output from the SFR laser) to about 4×10^{14} cm^{-3} which will allow us to extend the lowest magnetic field operation to fields in the region of 2 kG. Moreover, the operation at 5.3μ using the CO laser has additional contribution to the gain which arises from the resonance enhancement which will be larger as one reduces the carrier concentration. This is because the reduced concentration gives rise to an increase in the electron collision time, τ_e, resulting in an increase in the maximum possible resonance enhancement as discussed in Reference 12. The reduction in electron concentration also results in a sharper and well defined band edge (as a result of the increase in the τ_e). Thus the operation of the 5.3μ SFR laser in lower n_e samples can be carried out with the pump frequency considerably closed to the bandgap and thus improve the contribution from the resonance enhancement. [The sharpening of the absorption edge with InSb with reducing carrier concentration has also allowed us to obtain the first cw operation of the anti-Stokes spin-flip Raman laser since there is a low loss region between the pump wavelength and the absorption edge. This has been discussed in greater detail in Ref.[13] where the first operation of the low field spin-flip Raman laser has been reported.] These considerations indicate that for cw SFR laser operated with a pump wavelength near 5.3μ the reduction in carrier concentration should not give rise to any appreciable change in the threshold for SFR laser action.

Experimentally the low field operation with lower electron concentration InSb samples has been demonstrated. It has been shown[13] that an InSb sample having an $n_e \sim 1\times10^{15}$cm^{-3} allows us to tune the SFR laser at magnetic fields of the order of 400 gauss. The operation at 5.3μ with a $n_e \sim 1\times10^{15}$cm^{-3} sample yielded a SFR laser threshold which was not different from that for the 1×10^{16}cm^{-3} sample. At 10.6μ the SFR threshold for the $n_e \sim 1\times10^{15}$cm^{-3} sample was approximately a factor of three higher[12] than that in the $n_e \sim 1\times10^{16}$cm^{-3} sample, in agreement with the above discussions. Recent work[20] has allowed us to further reduce this lowest magnetic field to about 200G. At the present time, it is possible to operate the SFR laser with a small permanent magnet or with a small electromagnet.

In the cw operation mode at 5.3μ the power output on the first Stokes SFR laser is approximately 1W for an input power of approximately 6 watts. (See Fig. 5 for a typical spectrum). It should be pointed out that a Stokes power leaves the InSb sample from both the ends so the total power output in the tunable range is approximately 2W. It should also be pointed out that the low field limit of 200G as observed for $n_e \approx 1\times10^{15}cm^{-3}$ sample is considerably lower than what one would predict for the quantum limit considerations[8]. The reason for this behavior is not understood at the present time and further work is clearly necessary to explain completely the low field operation of the SFR laser.

Figure 5 shows the spectral analysis of the power output from one such low field SFR laser. The conversion efficiency from pump to the various SFR lasers is estimated to be $\sim$70%. Figure 6 shows the low field extension of the tuning range of the 5.3μ SFR laser using a $n_e \approx 1\times10^{15}cm^{-3}$ sample. We have carried out these experiments at a sample temperature of approximately 1.6K. However, we have observed that the low field SFR laser action can be obtained at temperatures as high as 20-25K indicating that liquid helium immersion is not a necessary requirement. A closed cycle cooler which delivers a temperature in the range of 10-20K should be more than adequate. Thus the low field operation has now converted the spin-flip Raman laser from a bulky and heavy device into a small device which can be operated with a small electromagnet and a closed cycle cooler, considerably increasing its usefulness.

There are, however, penalties paid for the reduced carrier concentration of InSb. The penalty in terms of the loss gain is not serious. However, the serious problem comes from the reduction in the level at which the output power no longer increases with the input power. This saturation discussed in Refs. 1 and 8 arises from the fact that the electron spins are not able to relax back from the upper spin level to the lower spin level after scattering a photon at pump frequency into a photon at Stokes SFR frequency. The spin-relaxation time, τ_{SR}, is $\sim10^{-9}$ sec for $n_e \approx 10^{16}cm^{-3}$. The τ_{SR} is expected to become longer as the carrier concentration is reduced. Thus the reduced carrier concentration hurts doubly in the maximum saturation level at which the SFR laser can be operated. However, for the cw SFR laser we do not yet see the effect of reduced carrier concentration and increased spin relaxation time on the conversion efficiency. These effects are very clearly observed when we operate the low carrier concentration SFR laser with a Q-switched CO_2 laser where the output power is seen to decrease markedly as the carrier concentration is reduced. This, however, can be compensated by using a larger volume of the InSb for SFR lasers. Another problem associated with lower carrier concentration samples, is that at the low temperatures necessary for the SFR laser action the high field operation is limited because of the freeze out of electrons. We have seen that the 1×10^{15} cm^{-3} sample operates reasonably well only up to a magnetic field of around 40 kG. Beyond 40 kG the operation is very much weaker and the difficulties arising from freeze out of

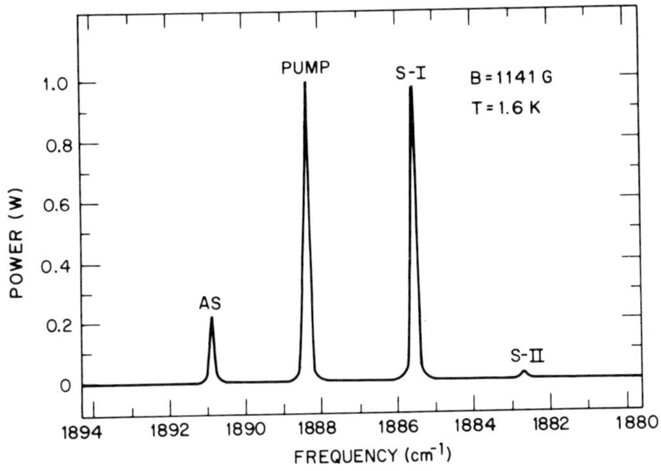

Figure 5. Spectrum of the low field spin-flip Raman laser at a magnetic field of 1141 G. (Input pump power is 5.5 W, carrier concentration is 1×10^{15} cm^{-3}).

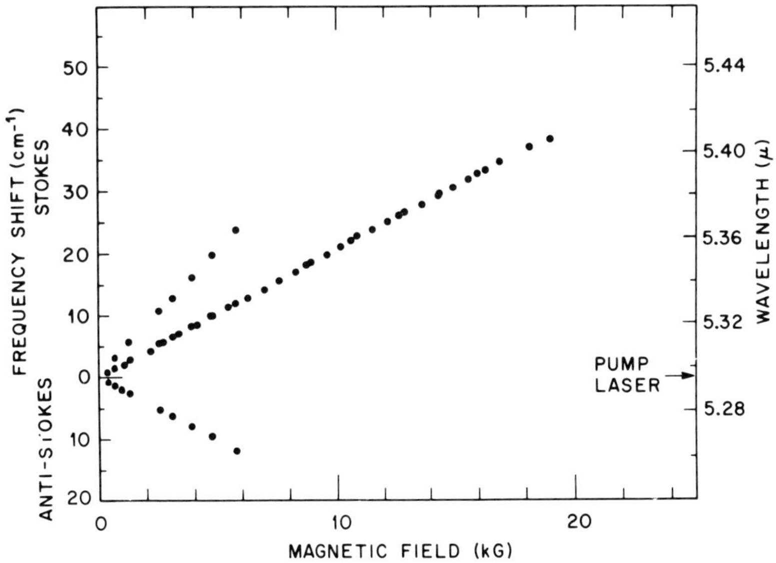

Figure 6. Low field SFR laser tuning curve for $n_e \approx 1 \times 10^{15}$ cm^{-3} InSb sample using P_{9-8} (12) transition of the cw CO laser as the pump.

carriers becomes evident. Operation with lower carrier concentration samples in the range $n_e \approx 1 \times 10^{14}$ to 5×10^{14} cm^{-3} will have even greater difficulties arising from freeze out, since the freeze out occurs at lower magnetic fields as n_e is reduced. There are a number of possible solutions by which the freeze out can be avoided. One solution is to use the bias field which impact ionizes the donors giving rise to the carriers in the conduction band even at higher magnetic fields. However, the primary utility of the lower carrier concentration samples is the low field operation rather than the high field operation. Thus the disadvantage arising from freeze out may not be very serious.

The low field operation with a small electromagnet allows us a very fine control over the magnetic field and thus over the wavelength of the SFR laser. This increases the utility of the SFR laser as a spectroscopic tool. In addition, the stability of the electromagnets and the ease of resetting the magnetic field to a given value indicates that the heterodyne experiments which require precise positioning of the local oscillator frequency can be easily carried out with the low field SFR laser using electromagnets rather than the high field SFR laser using superconducting magnets. This particular advantage has allowed us to measure the linewidth of the SFR laser by heterodyning its output with a fixed frequency CO laser line as discussed in the next section. Early experiments to accomplish this using a superconducting magnet had failed for reasons which are not clearly understood at the present time.

Using the electromagnet as the source of variable magnetic field of ∿3kG, we have been able to tune the SFR laser from about 5.1μ to excess of 6μ on a continuous basis by using different transitions from a CO laser, and from about 10.5μ to around approximately 11μ using the different transtions from the CO_2 laser in the Q-switched operation.

4. Linewidth of the Spin-Flip Raman Laser

The linewdith of a laser is ultimately determined by the amplitude and the phase fluctuations in the radiation phenomenon which leads to stimulated emission. The Schawlow-Townes expression for laser linewidth, [21] $\Delta\nu_L$, (far above threshold) is given below.

$$\Delta\nu_L \approx 8\pi h \nu_L \left[\Delta\nu_c \Delta\nu_{sp} / (\Delta\nu_c + \Delta\nu_{sp})\right]^2 P^{-1} \qquad (8)$$

Here the ν_c is the cavity linewidth, [22] ν_{sp} is the spontaneous linewidth of the gain process responsible for the laser action, ν_L is the frequency of the laser oscillator, P is the power output, and h is Planck's constant. It can be seen that for typical lasers such as gas lasers, this linewidth is of the order of 1 Hz. For the SFR lasers, the spontaneous linewidth depends upon carrier concentration and the sample temperature. This linewidth is approximately 0.5 cm^{-1} for $n_e \approx 1.3 \times 10^{16}$ cm^{-3}, 0.15 cm^{-1} for $n_e \approx 1 \times 10^{15}$ cm^{-3} InSb sample and 0.05 cm^{-1} for $n_e \approx 2 \times 10^{14}$ cm^{-3} InSb sample. In addition the temperature dependence of this linewidth has also been measured by EDR-ESR study [19] which shows that at temperatures below approximately 15-20 K the linewidth is narrow and is given by the above numbers for an InSb sample. At higher temperatures the linewidth increases very rapidly as seen in Fig. 7 where we have replotted McCome and Wagner's data [19] to show the temperature dependence of the EDR-ESR resonance linewidth (also the SFR spontaneous scattering linewidth [23] where the incident and scattered photons are both normal to the magnetic field). For the $n_e \approx 1 \times 10^{15}$ cm^{-3} InSb sample, we see that the low field operation as described in previous section we will dictate that the sample temperature be kept below 15K for low threshold power for stimulated Raman scattering. [See Eq.(5) for gain expression.] Thus to obtain the calculated SFR laser linewidth we will need to know the carrier concentration as well

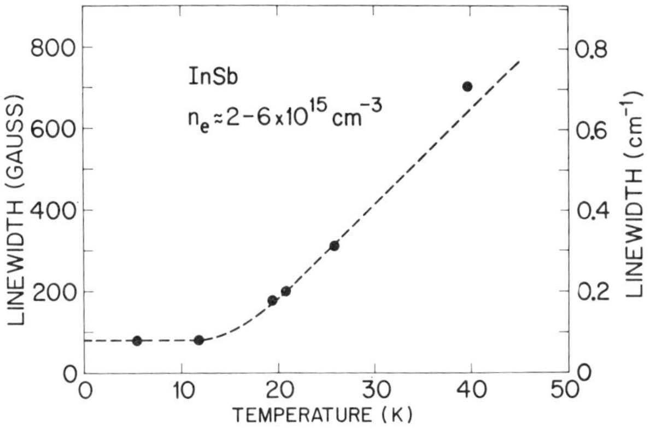

Figure 7. Spin resonance linewidth (also the spontaneous SFR scattering linewidth) for 2×10^{15} cm^{-3} InSb sample as a function of sample temperature (data obtained from Ref. 19).

as temperature of the particular SFR laser sample. The cavity linewidth [22] ν_c is determined by the absorption losses in the sample as well as the reflection losses. For the typical SFR lasers which have been considered so far, the laser cavity is formed by uncoated InSb surfaces which have a reflectivity of 36%. The absorption and reflection losses amount to approximately 1 neper for a typical SFR laser but could be considerably less. Using typical values of the spontaneous SFR linewidth of 0.15 cm^{-1}, the SFR laser linewidth is evaluated to be 1 Hz for cw operation at 5.3μ with an output power of ∿ 1W. The SFR laser linewidth will be approximately 10^{-3} Hz at 10.6μ in the Q-switched operation when the power output is approximately 1 kW. In the Q-switched operation, however, the linewidth will be determined by the Fourier components of the pulse which under typical conditions [1] is approximately 150 nsec. Thus the SFR laser linewidth for the Q-switched operation at 10.6μ will not be limited by amplitude and phase fluctuations in the gain process but will be limited by the Fourier components of the pulse and will be approximately 1 MHz.

In order to measure narrow linewidths of the order of a few Hz for 5.3μ cw operation or of the order of 1 MHz for Fourier spectrum limited linewidth of 10.6 μ , conventional spectrometers such as grating spectrometers or interferometers are clearly not indicated. As seen from similar measurements with gas lasers, we see that heterodyne spectroscopy i.e., heterodyning the power output of the SFR laser with a gas laser (whose linewidth is independently known) appears to be the only meaningful way of determining the linewidth. The low field operation of the SFR laser has allowed us to carry out the heterodyne measurements of the linewidth of the SFR laser [14] at 5.3μ. Figure 8 shows the experimental setup. Instead of using two separate lasers, one for pumping the SFR laser and the second as the local oscillator (for heterodyning), we adjusted the grating of the CO laser such that the CO laser oscillated on two separate transitions of CO separated by 4 cm^{-1}. This scheme has a possible advantage of removing the problems associated with pump frequency fluctuations (see later). The magnetic field for the SFR laser, provided by an electromagnet, was adjusted to give a magnetic field of 1713 gauss which gives a SFR laser shift of ∿4 cm^{-1}. The $P_{9-8}(12)$ transition of CO which is the pump transition was adjusted to have about 10 times the power output of the $P_{9-8}(13)$ transition of CO. The latter was used as the local oscillator. Thus the power output from the InSb SFR laser sample consists of the SFR laser due to $P_{9-8}(12)$, the $P_{9-8}(13)$ transition and the unused $P_{9-8}(12)$power output. At the magnetic field of 1713 G the $P_{9-8}(13)$ and the SFR laser due to $P_{9-8}(12)$ are nearly coincident. Focussing this power output through a spectrometer to remove all but the SFR laser due to $P_{9-8}(12)$ and the $P_{9-8}(13)$ line on a fast Cu:Ge detector allows us to measure the rf beat between the SFR laser and a local oscillator

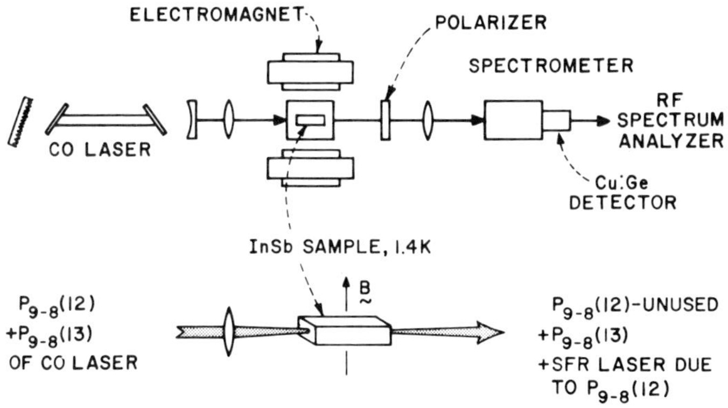

Figure 8. Experimental setup for measuring the SFR linewidth by heterodyne technique.

to determine the width of the SFR laser.

The experimental results are shown in Fig.9 which show the rf beat described above, centered at a frequency of ∿ 80MHz. It can be seen that the measured linewidth of the SFR laser is ∿1 kHz. This is limited by the rf spectrum analyzer resolution which was 1 kHz. Thus, it can be concluded that the actual linewidth of the spin-flip Raman laser is considerably smaller than 1 kHz in agreement with what we would expect for such an InSb SFR laser. (The sample was 1.1×10^{15} cm^{-3} InSb, 9 mm long, immersed in liquid helium at 1.6 K.) Thus it can be seen that the SFR laser linewidth is extremely narrow, and appears to be in agreement with the calculated linewidth of ∿ 1 Hz.

The narrow linewidth should not be confused with the stability of the SFR laser. The stability is determined by several considerations described below. The first one is the stability of the pump laser. The random fluctuations in the length of the pump laser cavity will result in the pump frequency which fluctuates with time and since the SFR laser frequency is determined by the pump frequency, a similar variation in the output frequency of the SFR laser will occur. Because there is significant frequency pulling of the SFR laser frequency due to existence of the SFR cavity modes, the fluctuations in the SFR laser frequency will not be as large as that for the pump laser frequency. However, in our situations where we are using the same CO laser to give the pump as well as the local oscillator wavelength, the frequency

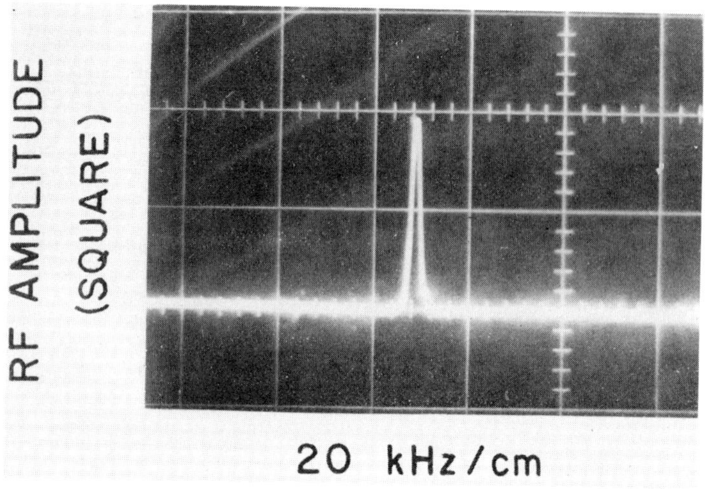

Figure 9. Spectrum analyzer display of the rf beat between the $P_{9-8}(13)$ transition of the CO laser and the SFR laser pumped with the $P_{9-8}(12)$ transition of the CO laser at a magnetic field of 1713 G. Beat frequency is the 82 MHz, sweep width is 20 kHz, and a spectrum analyzer resolution is 1.0 kHz.

pulling will be detrimental because it removes some of the advantage of using the same laser for providing both the pump and the local oscillator. These fluctuations in our experimental situation were judged to be small and estimated to be of the order of a few kHz. The second source of instability of the SFR laser frequency comes from the fluctuation in the magnetic field. When we use the electromagnet and a special stabilized power supply, we estimate that the SFR laser frequency fluctuations due to variation in the magnetic field amount to $\sim$ few kHz over a period of several seconds. Yet another factor contributing to SFR laser frequency instability is the thermal fluctuations of the SFR laser cavity which changes the position of the SFR laser cavity modes and gives rise to laser frequency fluctuations due to mode pulling. In the situation where the SFR laser is operating on a cw basis (as in the above measurements) and where the sample is immersed in a liquid helium bath at a constant temperature this source of frequency fluctuations can be neglected. Experimantally we found that the output frequency as seen in the rf beat fluctuated by approximately 20 kHz over a period of several minutes. As mentioned above this comes essentially from variations in magnetic field and variations in the pump laser cavity mode which affects the beat frequency through the mode pulling effects.

In order to utilize the extremely narrow linewidth of the order of a few Hz one has to take extreme care stabilizing pump laser frequency, the magnetic field and the temperature of the InSb sample. It can be seen that this is within the reach of present day technology and stable sources having linewidths of the order 1 Hz, a comparable stability, and a tunability over wide frequency range are possible. We will see in the next section that in many of the practical experiments of high resolution spectroscopy and other applications of the SFR laser, this extreme stabilization may not be required. It is possible to obtain a SFR laser frequency stability of the order of a few kHz on a routine basis.

5. Applications of the Spin-Flip Raman Lasers

With the high pulsed as well as cw powers available from the tunable SFR lasers, it is clear that a number of hither to impossible experiments can be carried out. The linewidth measurements described in the previous section clearly indicate that we have an extremely monochromatic source which is tunable. Such sources have tremendous implications in high resolution spectroscopy, pollution detection, local oscillators in communications systems and radar, nonlinear optics etc. Some of the applications have been previously described in the review article.[12] High resolution studies have been carried out of ammonia absorption spectrum [24] and water vapor absorption spectrum.[14] Pollution detection [25] of NO in air has also been carried out. In the present paper we will describe some of the recent developments in both the high resolution spectroscopy and pollution detection which have become largely possible because of the low field operation of the SFR laser.

(a) High Resolution Spectroscopy

Since the output frequency of the SFR laser is determined by the magnetic field, an application of a small time dependent magnetic field will give rise to an output frequency of the SFR laser which is modulated. This scheme can be used in carrying out modulation spectroscopy which allows one to measure small absorptions without sacrificing resolution. Since the g-value for the electrons InSb is $\sim$50 the tuning rate for the SFR laser is $\sim$60-70 MHz/Gauss.

Thus the ac magnetic field and the drive current to produce such a time dependent magnetic field is quite small. Figure 10 shows

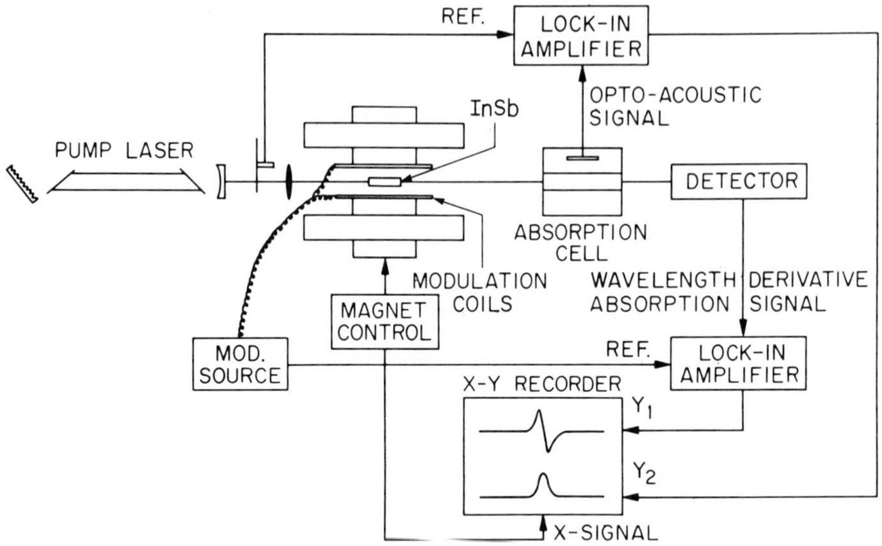

Figure 10. Experimental setup for absorption and modulation spectroscopy.

an experimental setup for modulation spectroscopy. In addition to the SFR laser tuning magnetic field provided by the electromagnet there are two small air cored coils which provide a small ac magnetic field for modulating output frequency of the SFR laser. The transmission through the cell is then measured either through an acousto-optic method[26] or by measuring the transmitted signal and extracting the time dependent variation in synchronism with the modulating magnetic field. Figure 11 shows a spectrum of water at a frequency of 1885.24 cm^{-1}. It can be seen that the modulation spectroscopy is indeed a useful way of exploiting the characteristic the SFR laser and allows us to measure very small absorptions. In this experiment the linear absorption through the cell was approximately 10^{-5}. The linewidth that we see on Figure 11 is approximately 140 MHz which is the Doppler width of water vapor at room temperature. The gas pressure is < 1 Torr in order to keep the pressure broadening as low as possible. Using this technique we can study the pressure broadening of the H_2O absorption

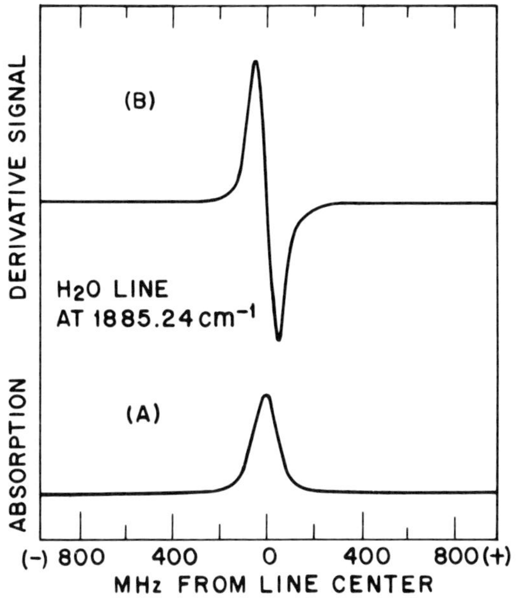

Figure 11. Absorption and modulation derivative signal for H_2O vapor line at 1885.24 cm^{-1}. Air pressure $\lesssim$1.0 Torr.

line with additional buffer gases. This has been carried out and it is possible to compare experiment with theory of Bennedict and Calfee.[27] These results will be published elsewhere.

A further study of the high resolution capability of the SFR laser is shown in Figure 12 where we see the absorption modulation spectroscopy of NO at low pressure. Two transitions are seen. One belongs to the Ω = 1/2 vibrational level of NO and the other one to the Ω = 3/2 vibrational level. For the Ω = 1/2 vibrational level we see the expected Λ-doubling, which has not been previously observed in infrared spectroscopy [28] because of the expected small splitting. Splitting observed is approximately 350 MHz which is in reasonable agreement with theory.

The high power capabilities of the SFR laser have allowed us to carry out saturation spectroscopy experiments. The experimental setup is shown in Fig.13. The detector signal shows the familiar Lamb-dip at the center of the Doppler broadened transition causing absorption. The Lamb-dip spectroscopy has allowed us to measure linewidths considerably smaller than the Doppler widths of the gaseous absorbers. Lamb-dip widths which are determined by the homogeneous broadening of the absorption process are much narrower than the Doppler widths at low pressures. We have measured Lamb-dip

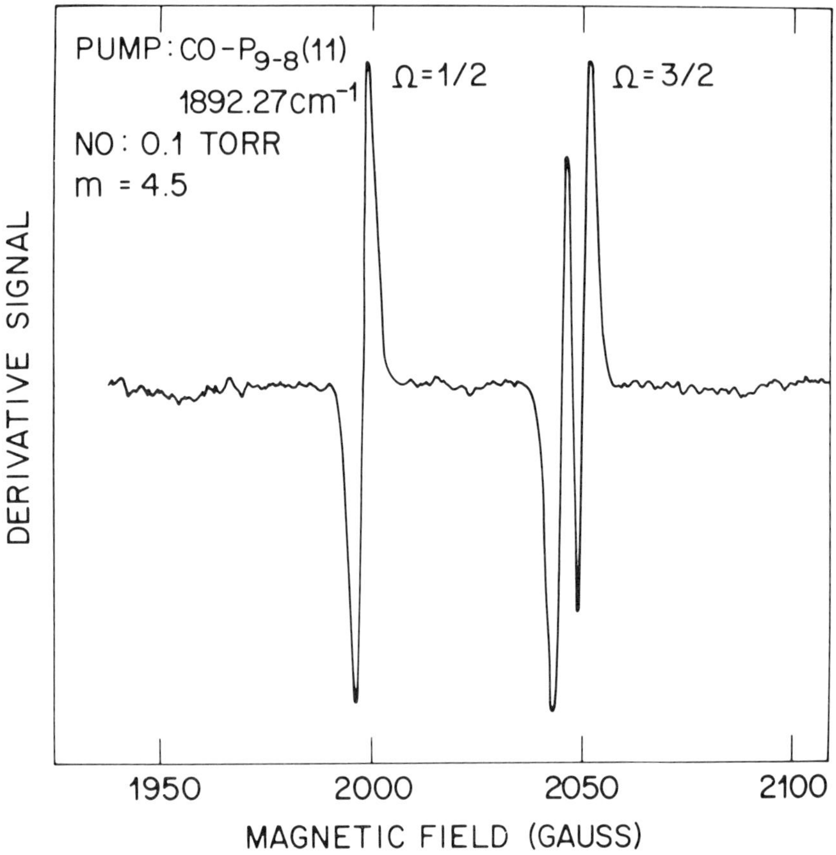

Figure 12. Modulation derivative signal for absorption due to NO showing the Λ-doubling of the Ω = 1/2 vibrational transition.

widths of ∿50-100 kHz.

(b) Pollution Detection

Earlier we have used the SFR laser for measuring NO concentration in air [25] by measuring the absorption through optoacoustic spectroscopy. We have improved the experimental techniques considerably by using the low field SFR laser operation.

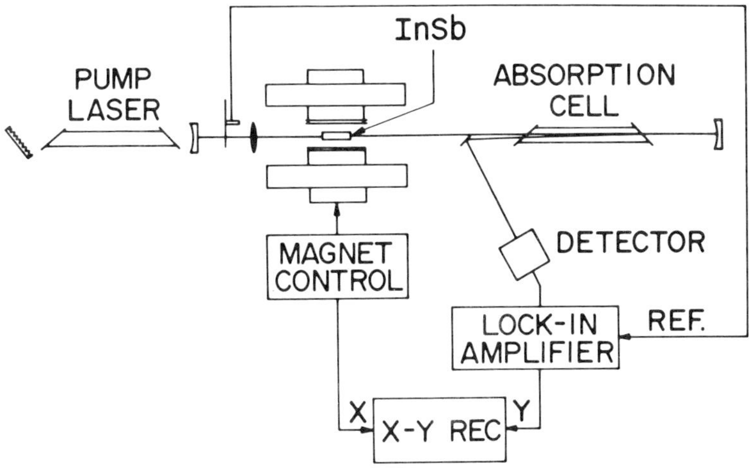

Figure 13. Experimental setup for using SFR laser to carry out Lamb-dip spectroscopy.

Figure 14 shows the spectrum from the opto-acoustic cell as the magnetic field is tuned from 0 to 6 kG using the electromagnet. The opto-acoustic cell contained 20 ppm NO in a background pressure of nitrogen at a total pressure of 76 Torr. (The experimental setup is the same as Figure 10). Here we see clearly that the SFR laser is an exceptional tool for pollution studies as well. In Figure 15 we show the two particular lines of NO which we use for measuring NO concentration. The upper trace shows 20 ppm NO calibration sample at a pressure of 90 Torr, and the lower trace shows signal from the optoacoustic cell over the same range of the tuning of the SFR laser with dry nitrogen gas at a pressure of 90 Torr. The electronic sensitivity of the detection apparatus has been increased by a factor of 100. From the amount of noise seen on the lower trace we deduce that we have the capability of detecting less than 1 ppB of NO in air using an integration of 1 second which was used for the obtaining of the data in Figure 15. Thus we have improved detection capabilities by a factor of 10 over the previous results. It can be seen that either increasing the time constant or by integration of the area under NO absorption curve will allow us to improve the sensitivity by another factor of 10 making the detection capabilities to better than 0.1 ppB.

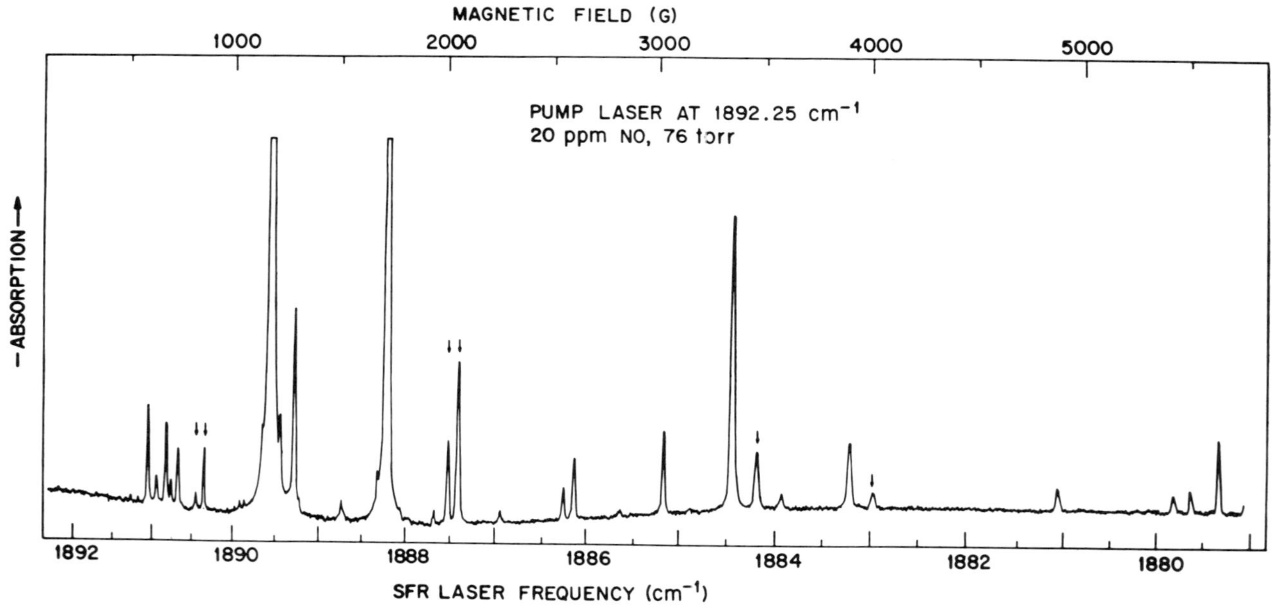

Figure 14. Absorption spectrum (obtained with opto-acoustic setup) for a 20 ppm NO sample in 76 Torr of dry nitrogen. The absorption lines indicated by arrows belong to NO, and the other absorption lines arise from residual water vapor in the calibation sample.

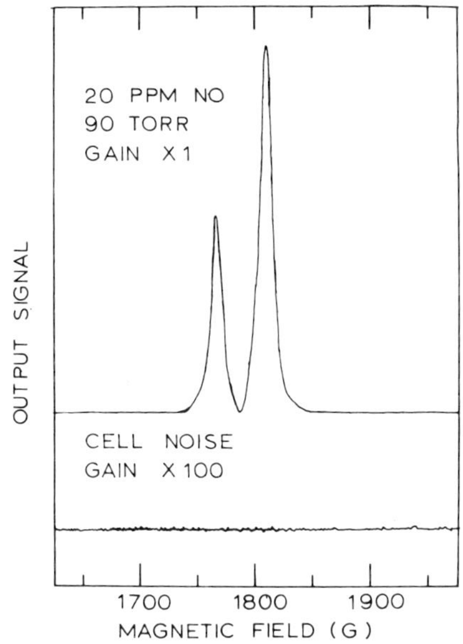

Figure 15. Details of the absorption spectrum covering the NO absorption lines (a) 20 ppm calibration sample of NO in 90 Torr in dry nitrogen (b) The absorption signal from the cell containing 90 Torr of dry nitrogen with the electronic sensitivity increased by a factor of 100 over that in (a) showing the noise on the trace being equal to 0.1 ppB of NO signal. Integration time for trace (b) is 1 sec.

(c) Communications and Radar

Because of the narrow linewidth and the capabilities of tuning the SFR laser by very small magnetic fields, it is felt that the SFR laser could play an important role as a local oscillator in a communications and radar systems. The magnetic fields required for sizeable shifts in the output frequency are quite small because of the large tuning rate. To obtain a frequency modulation of 60 MHz one needs only a magnetic field of approximately 1G.

6. Conclusion

In the above we have detailed some of the operating characteristics and the most recent advances in the field of spin-flip Raman lasers. While it is true that the tuning range of the spin-flip Raman lasers is rather limited viz. from 5μ to about 6.5μ and from 9μ to about 14.6μ, it is felt that further work both on materials for spin-flip Raman lasers, as well as a new pump transitions will allow us to extend the SFR laser tunability to the entire infrared spectrum. In conjunction with nonlinear optics the tuning range can be considerably increased as indicated in the review article in Reference 12. It is, thus, reasonable to conclude that the spin-flip Raman laser appears to be an important source of tunable coherent narrow frequency radiation in the infrared which has and will have very large number of potential applications.

References

1. C. K. N. Patel and E. D. Shaw, Phys. Rev. Letters *24*, 451 (1970).
2. S. E. Harris, Proc. IEEE *57*, 2096 (1969).
3. See for example E. D. Hinkley, Appl. Phys. Letters *16*, 351 (1970); K. W. Nill, F. A. Blum, A. P. Calawa and T. C. Harman, Appl. Phys. Letters *19*, 79 (1971).
4. C. K. N. Patel, Appl. Phys. Letters *18*, 274 (1971).
5. A. Mooradian, S. R. J. Brucck and F. A. Blum, Appl. Phys. Letters *17*, 481 (1970).
6. E. D. Shaw and C. K. N. Patel, Appl. Phys. Letters *18*, 215 (1970).
7. S. R. J. Brueck and A. Mooradian, Appl. Phys. Letters *18*, 229 (1971).
8. C. K. N. Patel and E. D. Shaw, Phys. Rev. *B3*, 1279 (1971).
9. R. L. Allwood, S. D. Devine, R. G. Mellish, S. D. Smith and R. A. Wood, J. Phys. *C3*, L186 (1970).
10. R. B. Aggarwal, B. Lax, C. E. Chase, C. R. Pidgeon, D. Limpert and F. Brown, Appl. Phys. Letters *18*, 383 (1971).
11. R. L. Allwood, R. B. Dennis, S. D. Smith, B. S. Wherrett and R. A. Wood, J. Phys. *C4*, L63 (1971).
12. C. K. N. Patel, *Proceedings of the ESfahan Symposium on Lasers Physics and Applications*, (to be published).
13. C. K. N. Patel, Appl. Phys. Letters *19*, 400 (1971).
14. C. K. N. Patel, Phys. Rev. Letters *28*, 649 (1972).
15. R. E. Slusher, C. K. N. Patel and P. A. Fleury, Phys. Rev. Letters *18*, 77 (1967).
16. P. A. Wolff, Phys. Rev. Letters *16*, 225 (1966).
17 Y. Yafet, Phys. Rev. *152*, 858 (1966).
18. See for example C. K. N. Patel, in *Advances in Lasers*, Ed.E. K. Levine (M. Dekker, N.Y. 1969) pp. 1-183.
19. B. D. McCombe and R. J. Wagner, Phys. Rev. *B4*, 1285 (1971); B. D. McCombe (Private Communication).
20. C. K. N. Patel, (to be published).

21. A. L. Schawlow and C. H. Townes, Phys. Rev. *112*, 1940 (1958); C. H. Townes in *Advances in Quantum Electronics* ed. J. R. Singer (Columbia University Press, N. Y. 1961) p.3; See also W. R. Bennett, Jr., Appl. Optics Supplement *1*, 24 (1962).
22. H. Kogelnik and T. Li, Proc. IEEE *54*, 1312 (1966).
23. S. R. J. Brueck and F. A. Blum, Phys. Rev. Letters *28*, 1458 (1972).
24. C. K. N. Patel, E. D. Shaw and R. J. Kerl, Phys. Rev. Letters *25*, 8 (1971).
25. L. B. Kreuzer and C. K. N. Patel, Science *173*, 45 (1971).
26. L. G. Kreuzer, J. Appl. Phys. *42*, 2934 (1971).
27. W. S. Benedict and R. F. Calfee, *Line Parameters for 1.9 and 6.3 micron Water Vapor Bands*, Environmental Science Services Administration Professional Paper 2 (U. S. Dept. of Commerce, Washington, D. C. 1967).
28. See for example, T. C. James, J. Chem. Phys. *40*, 762 (1964).

DYE LASERS: RECENT ADVANCES IN TUNABLE SYSTEMS

M. Hercher

University of Rochester, Rochester, New York

B. Snavely

Eastman Kodak Co., Rochester, New York

The laser-pumped dye laser comes close to being an ideal laser. It is efficient (better than 40% of the optical pump power can be obtained from the dye laser), tunable over a broad range, can be operated in a single mode without undue difficulty, and it can be quite accurately described in terms of a simple model. In this paper we review the operating characteristics of current continuous dye lasers, and we discuss techniques for tuning and controlling the bandwidths of these lasers.

Figure 1 shows a simplified energy level diagram for a typical organic dye, and Table 1 lists general parameters used to describe

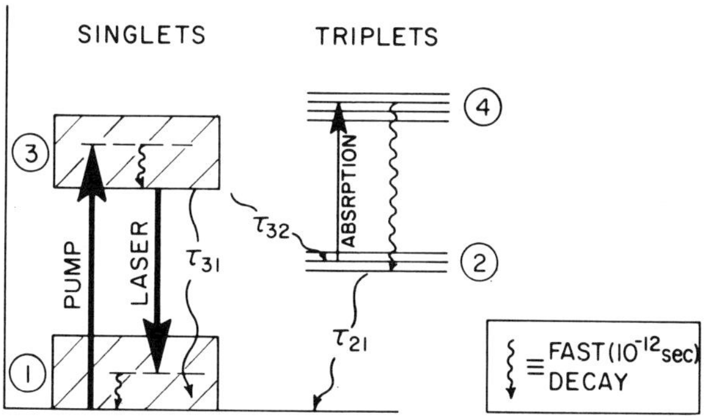

Fig. 1. Energy Levels in a Dye Laser.

Table 1

Dye Laser Parameters

N_1	Population density of electronic level number 1. (etc.)
W	Optical pumping rate
τ_{31}	Spontaneous decay time for $3 \rightarrow 1$ transition (etc.)
τ_3	Net lifetime of excited singlet ($\sim$5 nsec)
σ_{31}	Stimulated emission cross-section for $3 \rightarrow 1$ transition at dye laser wavelength ($\sim 2.5 \times 10^{-16} cm^2$)
σ_{13}	Absorption cross-section for $1 \rightarrow 3$ transition at dye laser wavelength ($\sim 2.5 \times 10^{-18} cm^2$)
I_L	Intracavity dye laser irradiance (photons/cm^2 sec).
ℓ	Length of active medium (dye cell thickness) (1 mm)
γ	Passive resonator loss per transit (nepers) (0.1) (includes reflection loss, excited state absorption, etc.)
t_1	Cavity transit time $\equiv$ L/C, where L is optical length of cavity

dye lasers and gives their approximate values in the case of the dye rhodamine 6G. The rate equations for the populations of the relevant electronic levels, and for the photon irradiance inside the laser cavity may be written in the following form:

$$\dot{N}_1 = -WN_1 + N_2/\tau_{21} + N_3/\tau_{31} + I_L(\sigma_{31}N_3 - \sigma_{13}N_1) \quad (1)$$

$$\dot{N}_2 = N_3/\tau_{32} - N_2/\tau_{21} \quad (2)$$

$$\dot{N}_3 = WN_1 - N_3/\tau_3 - I_L\ (\sigma_{31}N_3 - \sigma_{13}N_1) \quad (3)$$

$$\dot{I}_L = I_L\ [(N_3\sigma_{31} - N_1\sigma_{13})\ \ell - \gamma]/t_1 \quad (4)$$

As shown in Fig. 1, a typical laser dye exhibits absorption at laser wavelength corresponding to a transition between excited triplet states (e.g. the states labelled 2 and 4 in Fig. 1). Moreover, the first excited triplet (labelled 2) generally has a long lifetime, typically on the order of milliseconds, unless steps are taken to quench the triplet. This long lifetime leads to the accumulation of excited molecules in state 2 (sometimes referred to as "triplet hang-up"), thereby reducing the number of molecules which can participate in laser emission. Both to eliminate this latter effect, and to minimize excited triplet absorption, it is highly desirable to prevent the build-up of a significant population in the triplet manifold. Fortunately, there are a variety of effective quenching agents -- such as oxygen and cyclooctatetraene -- which serve to reduce the triple life τ_{21} and possibly to decrease the intersystem crossing rate, τ_{32}^{-1}.

If we consider a dye laser which is optically pumped by the focused beam from an argon ion laser, and we assume (1) that ground state absorption at the laser wavelength is negligible, (2) about 2/3 of the pump light is absorbed by the dye, (3) excited state triplet absorption is negligible, and (4) most of the dye molecules are in the ground state at any given instant, then the single-pass gain of the dye laser can be simply expressed as:

$$G(\text{nepers}) \simeq \frac{P_p \sigma_{31} \tau_3}{A\, h\nu_p} \tag{5}$$

where P_p is the pump power (watts), $h\nu_p$ is the energy of a pump photon (joules) and A is the cross-sectional area of the focused pump beam. The assumptions made in arriving at this result are all reasonable for a typical continuous dye laser. We can use this expression to find out to what cross-sectional area the pump beam should be focused in order to produce a specified small signal gain for the dye laser: e.g. if P_p is 250 mwatt @ 5145Å, σ_{31} is $2.5 \times 10^{-16}\text{cm}^2$, τ_3 is 5 nsec, and the desired gain is 10%, then we find that the pump beam must be focused to a spot which is about 25 microns in diameter. This small beam waist is a major factor in the design of dye laser resonators: the pump beam must be tightly focused, and the beam waist of the dye laser beam must lie within the focused pump beam. This in turn means that the alignment of the dye laser resonator is likely to be critical, and that one must contend with the possibility of optical damage in the vicinity of the focused pump beam (e.g. damage to the dye cell windows).

Figure 2 shows two different dye laser resonator configurations, each of which provides for a focused pump beam and an intra-cavity dispersing element for coarse spectral tuning. Although gratings have been used as tuning elements in flashlamp-pumped dye lasers, we prefer the use of prisms in continuous dye lasers because of their very small insertion loss (less than 1%). Both of the resonators shown in Fig. 2 have regions in which the beam is collimated: these regions can be used for the insertion of fine tuning elements and/or miscellaneous components such as sample cells or mode-lockers.

The spectral tuning range of a given dye depends, of course, on the dye. At this time the only dyes which have been operated

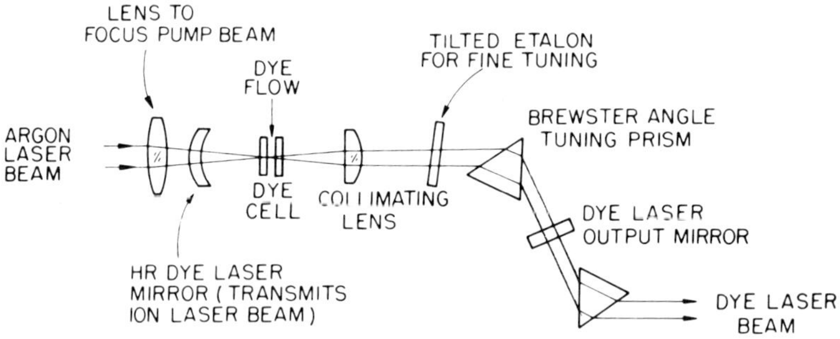

Fig. 2a Tunable CW dye laser configuration, system of Hercher and Pike (Ref. 1).

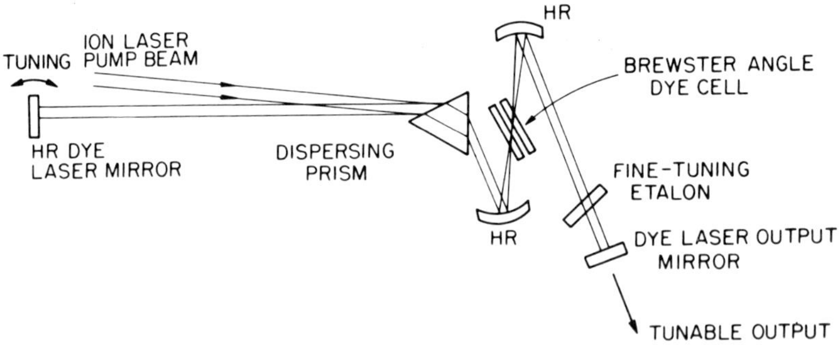

Fig. 2b Tunable CW dye laser configuration, modified system of Kogelnik et al (Ref. 2).

continuously are rhodamine dyes, similar to rhodamine 6G (R6G), and sodium fluorescein. Rhodamine 6G in water plus a deaggregating agent can be tuned from approximately 5600Å to 6600Å, and sodium fluorescein can be tuned from 5200Å to 5800Å. The fluorescence (and hence laser emission) of R6G can be shifted a few hundred angstroms to the green by using hexafluoroisopropanol as a solvent for the dye. The tuning range of all these dyes is sharply limited at the short wavelength end by ground state singlet absorption, and is less sharply limited at the long wavelength end by a decreasing stimulated emission cross-section and by excited triplet absorption.

We have been particularly concerned with the problem of obtaining stable single axial mode operation of a continuous dye laser. By itself, an intracavity tuning prism produces a dye laser spectral bandwidth which is approximately given by:

$$\Delta \lambda \sim \frac{\lambda}{20D\ (dn/d\lambda)}\ , \tag{6}$$

where D is the diameter of the collimated beam at the prism, $dn/d\lambda$ is the dispersion of the prism, and where we assume that the laser emission can be quenched when the dispersion of the prism introduces an additional 5% loss per transit. Using typical values (D=1.5 mm, $dn/d\lambda = 10^3 cm^{-1}$, λ = 6000Å) we obtain a spectral bandwidth of about an angstrom, which is what we find experimentally. (Using either a number of prisms in series or a more exotic prism made of lead molybdate, rather than SF18 glass, this bandwidth can be reduced to below 0.25Å.) In order to further reduce this spectral bandwidth we have used a number of uncoated intracavity tilted etalons. Each of these acts as an off-axis Fabry-Perot filter with a free spectral range of $(1/2nd)cm^{-1}$, where n and d are the refractive index and thickness of the solid etalon. The shift in the wavelength of maximum transmission is related to the angle of tilt by:

$$\Delta \lambda = \frac{\lambda \theta^2}{2n^2}\ . \tag{7}$$

Because this relationship between tuning and tilt angle does not depend on the thickness of the etalon, a number of different etalons will 'track' as long as they are tilted together -- and they can be tuned relative to one another by tilting about an axis orthogonal to the common axis of tilt. It should, in principle, be possible to design a single coated etalon to obtain dye laser emission in a single axial mode.

Figure 3 shows a typical spectrum of a single mode dye laser in the form of 60 superposed scans obtained during a one second interval. During this interval the dye laser changed its frequency, more or less randomly, within a range of approximately 100 to 150 MHz. We are interested in using the dye laser to explore the structure of resonance lines with a precision of at least 30 MHz, so that the performance illustrated in Fig. 4 was unacceptable. In seeking to reduce the bandwidth of a continuous single mode dye laser we have considered a number of separate factors contributing to its bandwidth. These included spontaneous emission, thermal fluctuations in the resonator length, density fluctuations in the liquid dye, and miscellaneous perturbations to the resonator length (such as acoustic pick-up, vibration etc.). These various contributions are summarized in Table 2 together with estimates of their values in the case of a dye laser. It is clear from these data that the observed bandwidth of the dye laser is due to "miscellaneous" rather than fundamental considerations. This is hardly surprising. The frequency shift due to a change ΔL in the optical length L, of a resonator is given by:

$$\Delta \lambda = \frac{\lambda \Delta L}{L} , \tag{8}$$

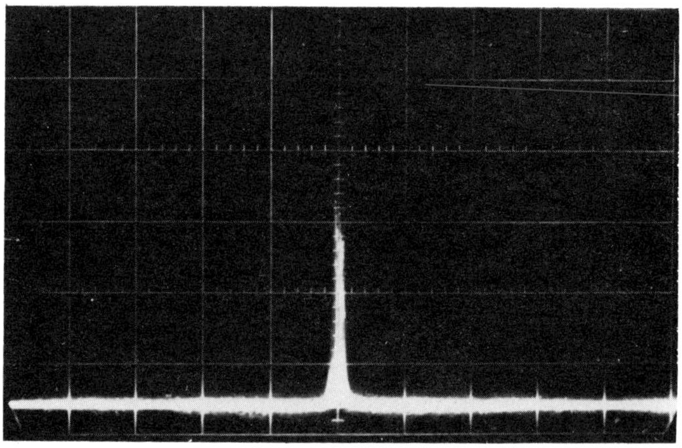

Fig. 3. Single mode dye laser spectrum: 800 MHz/cm; 1 sec. exposure (60 superposed scans); dye laser power approx. 80 mW.

Table 2

Contributions to Dye Laser Bandwidth

I. Spontaneous Emission

$$\Delta\nu_1 \sim \frac{\Delta\nu_c}{N} \sim \frac{2\pi h\nu(\Delta\nu_c)^2}{P_{out}}$$

$\Delta\nu_c \sim 50$ MHz

$P_{out} \sim 50$ mwatt $\qquad \Delta\nu_1 \sim 0.1$ Hz

$h\nu \sim 3 \times 10^{-19}$ j

II. Thermal Fluctuations in Resonator Length

$$\Delta\nu_2 \sim \nu\ (2kT/YV)^{1/2}$$

$kT \sim 4 \times 10^{-21}$ j

$Y \sim 10^5$ j/cm^3 (bulk modulus) $\qquad \Delta\nu_2 \sim 5$ Hz

$V \sim 10^3$ cm^3

$\nu \sim 5 \times 10^{14}$ Hz

III. Density Fluctuations in Dye

$$\Delta\nu_3 \sim \nu\mu(kT\beta d/a)^{1/2}/L$$

$\beta \sim 5 \times 10^{-4}$ cm^3/j (Vol. compressibility)

$kT \sim 4 \times 10^{-21}$ j $\qquad \Delta\nu_3 \sim 10$ KHz

$d \sim 0.1$ cm (dye cell thickness)

$a \sim 2 \times 10^{-6}$ cm^2 (mode cross-section)

IV. Miscellaneous Changes in Resonator Length

$$\Delta\nu_4 = \nu\Delta L/L = \nu(d\Delta\mu + (\mu-1)\Delta d + \Delta L_o)/L$$

where resonator length is L_o, dye cell thickness is d and dye index is μ.

for $\Delta\nu_4 < 100$ MHz
with $L = 30$ cm
and $\nu = 5 \times 10^{14}$ Hz

$\Delta\mu < 3 \times 10^{-5}$
$\Delta d < 6 \times 10^{-6}$ cm $= \lambda/10$
$\Delta L_o < \lambda/20$

so that with a 20 cm resonator length, a change in the optical length of the resonator of only 4×10^{-6}cm ($\lambda/15$) produces a 100 MHz frequency shift. Perturbations of this size can arise from vibration of resonator components, temperature changes in the dye solution, or even instability in the location of the laser beam within the cavity. Clearly one must reduce the magnitude of these perturbations in order to reduce the laser frequency fluctuations. But in addition one can reduce the effect of given perturbations simply by increasing the resonator length, as shown in Eq. (8) above. When this increase was combined with steps to reduce the perturbations in resonator length, we were able to reduce the short term (i.e. less than 30 sec) fluctuations in the dye laser frequency to less than 20 MHz.

Figure 4 shows the optical layout of a continuous dye laser with a 20 MHz bandwidth. The resonator components are rigidly attached to a massive steel plate which sits on a vibration-isolated table. The motor which drives the pump for the dye flow is in a sound-proof box, and a ballast tank is used between the pump and the dye cell. A water-cooled heat exchanger is used to maintain the dye solution at a constant temperature, and the laser is enclosed in a plastic box to isolate it from air currents. The resonator length is 90 cm. Three dispersing prisms were used both to reduce the bandwidth prior to etalon insertion, and to fold the resonator into a convenient configuration.

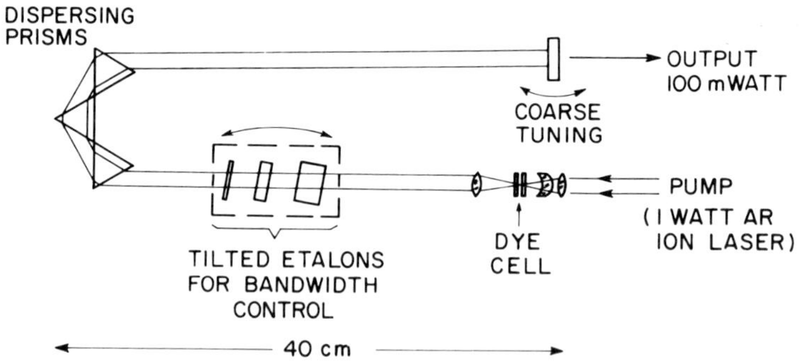

Fig. 4. Stable single mode dye laser

References

1. M. Hercher and H.A. Pike, Optics Commun. *3*, 346 (1971).

2. H. Kogelnik, E.P. Ippen, A. Dienes and C.V. Shank, I.E.E.E. J. Quant. Electronics *8*, 373 (1972).

A THEORY OF LIGHT WITHOUT PHOTONS: IMPLICATIONS FROM FERMIS TO LIGHT YEARS

Mendel Sachs

State University of New York, Buffalo, N.Y.

A new theory of elementary matter [1], that rejects the photon concept of light, has led to generalizations that have physical implications and have made numerical predictions in the domains of elementary particle physics, atomic physics and astrophysics.

The theory is based on three essential axioms: 1) the principle of general relativity, 2) a generalized Mach principle, and 3) the correspondence principle. The first axiom asserts that the laws of nature described in frames of reference that are in arbitrary types of relative motion, must be in one-to-one correspondence. This axiom implies a space-time symmetry that entails an underlying symmetry group whose most primitive representation implies that, in their most general form, the laws of nature must be expressed with the continuous field concept and a mathematical structure in terms of a two-component spinor field formalism - *not quantized.*

The second axiom, a generalized Mach principle, essentially postulates that any physically realistic system is, fundamentally, a *closed system.* The implication follows that not only the inertial manifestation of matter, but all of the physical manifestations of a material system, are features of the dynamical coupling of the components of the closed system - a physical system that is fundamentally one, without actually separable parts. The mathematical consequences of this axiom are a) the underlying field equations are intrinsically *nonlinear* and b) "free fields" are not included in the class of solutions of the underlying field equations.

A physical consequence of the generalized Mach principle is that not only the mass m of an elementary particle, but also its electric charge, e, magnetic moment, etc. are not to be taken as intrinsic properties of a "thing." They are rather related to the coupling of the components of the closed system. Thus, if the environment of an electron, say, should be continuously depleted, the mass, electromagnetic coupling, etc., should correspondingly tend to vanish. An interesting implication is that while the electric charge e is not an invariant, fundamental property of an elementary particle, the fine structure constant, $\alpha = e^2/\hbar c$, which is a measure of the strength of the electromagnetic *coupling*, is a fundamental constant. Thus, α may factor in any way, so long as it remains invariant (e.g. one can relate this constant to the coupling of an elementary particle with charge e/3 to one with charge 3e, etc. without changing the prediction of the theory).

The third axiom of the theory, the correspondence principle, requires for empirical reasons that the field equations must asymptotically approach the forms of the theories that are superseded, in the domains where the latter were numerically successful. In the case of the theory of light, then, it must be structured to asymptotically approach the formal structures of nonrelativistic quantum mechanics and the Maxwell description of electromagnetism. Thus, while the "free field solutions" do not occur in this theory, the field equations and their solutions do *approach* those of the conventional quantum mechanical and electromagnetic theories, even though the actual limit cannot be reached exactly. The asymptotic case corresponds to the linear limit of the theory, which occurs when the quantity of energy-momentum transfer between the interacting matter is sufficiently small.

Thus, this field theory contains all of the successful predictions of nonrelativistic quantum mechanics, as well as those of the conventional electromagnetic theory applied in the macroscopic domain. But in its unapproximated form, the theory does not correspond with relativistic quantum field theory. Quantization, nondeterminism, the uncertainty relations, linear superposition, etc. are rejected as conceptual basis elements of the theory. They only appear, in their mathematical form, in a linear approximation for this theory. In addition to yielding an entirely finite description of elementary matter (e.g. without any need to "renormalize") this theory also has the advantage of bridging the gap to the *formal structure* of nonrelativistic quantum mechanics in a smooth logical and mathematical fashion that is demonstrably consistent.

According to the axiomatic basis of this theory, the phenomenon of light does not refer to a "thing-in-itself." Light is rather a manifestation of the coupling of matter. There is no quantization and the vector feature of the Maxwell field is reducible to a factorized (spinor) version. Thus, the photon, *per se*, plays no role in the theory.

The interpretation of light as a manifestation of coupled matter is not new. In the nineteenth century it was implied by Faraday in his writings on the interpretation of his electromagnetic field as a field of force of one bit of charged matter in acting on other charged matter (the "test bodies"). One of the first to discuss the notion in the twentieth century was G.N. Lewis, (in 1926)[2]. He said that such an interpretation of light

> "requires us to consider the process as a perfectly symmetric one, so that we can no longer regard one atom as an active agent and the other as the accidental and passive recipient, but both atoms must play coordinate and symmetrical parts in the process of exchange."

Similar views of light were investigated later by several others, including Wheeler and Feynman [3], who refer to this as an "action-at-a-distance" approach, since it requires that light would not be emitted by some quantity of matter unless there would be some other quantity of matter, somewhere else to absorb this light, at the later time when it gets there. But the "light," here, is not a thing that moves on its own, toward the absorber. It is rather the name for a mutual relation between matter and matter, described in space and time.

My view of the phenomenon of light agrees with much of the conceptual content of the ideas of Faraday, Lewis and with Wheeler and Feynman. It differs however, with Faraday, in its relation to a fundamentally closed system, rather than Faraday's open system of additive fields to represent the superposition of the potential effects of charged matter on some test body in space and time. A resulting mathematical difference is that mine is necessarily a nonlinear field theory while Faraday's approach implies a linear field theory.

My approach differs from Lewis' and from the Wheeler-Feynman approach, in my fundamental requirement of the field concept. One starts here with a closed system, described by n coupled fields in a single space-time. On the other hand, the Wheeler-Feynman "action-at-a-distance" theory of electrodynamics is based, strictly, on the particle concept. They describe a composite system as n separable particles, with their trajectories in a 4n-dimensional space-time.

An important difference with the other theories also follows from my assertion of the principle of general relativity as a fundamental axiom of a theory of light. The mutual interaction of matter necessarily entails *non-uniform motion* - therefore, a curved space-time. That is to say, in its exact formulation, light is the propagation of an electromagnetic interaction between charged components of a physical system in a curved space-time. The use of the special relativistic limit of the Maxwell theory is only meant here as a mathematical approximation, where one replaces a section of a curved space-time with a flat space-time that is tangent to the actual space-time at the point where the "observer" is located. The validity of using this "special relativity" approximation is indeed indicated by the empirical facts. Nevertheless, one should not consider it as exact, nor logically acceptable, since the flat space geometry (special relativity) corresponds to a matterless universe, while "light" is *defined* here as a mutual interaction of matter.

In the initial stages of my investigation (as well as in the earlier comments of Wheeler and Feynman, and the more recent comments of Scully and Sargent [4]), it has been pointed out that the photon concept is not actually required to explain most of the experimental data that one normally claims to require it. But one physical phenomenon that Scully and Sargent claim necessarily requires the photon concept is that which would require the role of vacuum fluctuations. One example is the Lamb shift in hydrogenic atoms. In my investigation, however, I have derived the Lamb splitting of the hydrogenic energy levels, in quite good numerical agreement with the experimental data - but without the use of the photon concept, or even classical electromagnetic radiation[5]. It follows here from a natural generalization of the electromagnetic potential that comes from a factorization of Maxwell's equations into a pair of two-component spinor field equations[6]. It was shown that the resulting spinor equations reproduce all of the predictions of the conventional Maxwell formalism. But because this is a generalization, it predicts more physical consequences (i.e. some of the conservation equations of this formalism are in one-to-one correspondence with all of those of the usual formalism, while some of the conservation equations have no counterpart).

One of the extra predictions that follows from this generalization appears (in a natural way) as an addition to the ordinary Coulomb potential term in the Dirac Hamiltonian for hydrogen. The symmetry of this extra term is such that the accidental degeneracy is no longer present in the eigenstates of this Hamiltonian. Thus, the Lamb splitting is predicted as due to a generalization of the Maxwell formalism. The factorization from the vector to the spinor form of electromagnetism also leads to a new fundamental constant in the theory, which we have called g_M, that has the dimension of length. But the first numerical calculation that was carried out on the hydrogen spectrum, according to this theory, was the ratio of Lamb splittings, $(3S_{\frac{1}{2}} - 3P_{\frac{1}{2}})/(2S_{\frac{1}{2}} - 2P_{\frac{1}{2}})$. To the accuracy required to compare the theoretical predictions with the data, *this ratio is independent of* g_M. It does not depend on any adjustable parameters. Once it was seen that the theoretical ratio was in good agreement with the experimental data, the $(2S_{\frac{1}{2}} - 2P_{\frac{1}{2}})$ theoretical splitting was fit to the experimental value, giving the value, $g_M = 2.087 \times 10^{-14}$cm. This constant was then fixed for all future applications of this generalization to electrodynamic problems[7].

Thus, the field theory I have been investigating predicts the entire hydrogen spectrum from the features of the electron-proton bound system alone. There are no electromagnetic radiation, vacuum fluctuations or photons involved. The spectrum is strictly a property of the mutual electromagnetic interaction of this two-particle system - just as the Bohr spectrum or the Schrödinger spectrum were so. The better agreement with the observations, than the latter two-body models of hydrogen predict, is a consequence of a natural generalization of the Maxwell field of force that is dictated by the axioms of this theory, rather than being inserted *ad hoc*.

In addition to the fine structure of hydrogen, some of the other successful predictions of this theory have been the following:

1. An exact bound state solution for the closed system of nonlinear field equations for the particle-antiparticle pair predicts all of the data that are usually attributed to an actual annihilation (or creation) of matter along with the creation of photons[6b].

2. This solution leads to the prediction that ordinary observed matter is actually embedded in a "physical vacuum" that is an ideal gas of such pairs. This is a *countable set* of particle-antiparticle pairs in a particular bound state (which is its actual ground state - with intrinsic energy $2mc^2$ below that of the state in which the particle and antiparticle are (asymptotically) free of each other). The "physical vacuum," according to this theory, is then in terms of an ideal gas of atoms, in the same sense that an ordinary gas of atoms (of a normal gas) is a *countable set* of things, except, of course, for the "field" versus "thing" description.

3. The coupling of this gas of pairs (the "physical vacuum") in a finite cavity, to other matter (say the detecting apparatus that "looks at" this gas through a small window in the cavity) leads to the prediction of the Planck spectral distribution for blackbody radiation - without the need of photons[8].

4. The coupling of this physical vacuum to ordinary radiating atoms leads to the correct predictions for the lifetimes of the excited states of the latter[1d].

5. With the assumption of a particular specific volume for this physical vacuum of pairs, the curvature of space-time in a microscopic region (fermis) predicts the correct magnitudes for the inertial masses of the electron and the muon. These appear in the theory as members of a mass doublet. The excited mass state - the muon - follows here as a heavy electron which results from an alteration of the space-time in the region of an electron, that is caused by a quadrupolar excitation of the physical vacuum. It is found that the *ratio* of mass values of this doublet (which is independent of the specific volume of the physical vacuum) is the order of α (the fine structure constant). A plausibility argument, based on dynamical considerations of the bound states of particle-antiparticle pairs, for this problem, further leads to a ratio that is the order of $2\alpha/3 \sim 1/205$, which is close to the observed electron:muon mass ratio[9].

It is also shown how the lifetime of the muon state - which decays by *electromagnetic coupling* with the physical vacuum of pairs, is suppressed to the order of 200 times longer than a corresponding (quadrupolar) de-excitation of positronium, giving $\tau \sim 2 \times 10^{-6}$ sec - in agreement with the data. The theory also provides an explanation for the interpretation of the decay products to include a neutrino-antineutrino pair, from a decay that is purely electromagnetic[10].

6. The model of the physical vacuum of this theory implies that in regions of the universe where ordinary matter would have sufficiently high temperature ($T \sim 10^9$ °K), a large number of these pairs would dissociate, thereby yielding effects of jets of high energy accelerated electrons and high frequency radiation, as observed in the astrophysical observations of quasars.

References

1. The theory with its physical predictions, thus far, has recently been summarized in a review by the author in the series of papers called "A New Theory of Elementary Matter",(a) Part I: Philosophical Approach and General Implications, Int. J. Theor. Phys. *4*, 433 (1971);(b) Part II: Electromagnetic and Inertial Manifestations, Int. J. Theor. Phys. *4*, 453 (1971);(c) Part III: A Self-Consistent Field Theory of Electrodynamics and Correspondence with Quantum Mechanics, Int. J. Theor. Phys. *5*, 35 (1972);(d) Part IV: Two-Particle Systems: The Particle-Antiparticle Pair and Hydrogen, Int. J. Theor. Phys. *5*, 161 (1972).

2. G.N. Lewis, Proc. Nat. Acad. Sci. (USA) *12*, 22 (1926).
3. J.A. Wheeler and R.P. Feynman, Rev. Mod. Phys. *17*, 157 (1945).
4. M.O. Scully and M. Sargent III, Physics Today *25*, No. 3, 38 (1972).
5. M. Sachs and S.L. Schwebel, Suppl. Nuov. Cim. *21*, 197 (1961); also see Ref. 1(d).
6. M. Sachs, (a) Nuov. Cim. *31*, 98 (1964); (b) Int. J. Theor. Phys. *1*, 387 (1968); Int. J. Theor. Phys. *4*, 145 (1971).
7. For example, e-p scattering is analyzed in M. Sachs and S.L. Schwebel, Nucl. Phys. *43*, 204 (1963) and e-He^4 scattering in M. Sachs, Nuov. Cim. *53A*, 56 (1968).
8. M. Sachs, Nuov. Cim. *37*, 977 (1965) and Ref. 1(d).
9. M. Sachs, Nuov. Cim. *7B*, 247 (1972).
10. M. Sachs, Nuov. Cim. (in press).

BLACK BODY HEAT CAPACITY AND ACTION-AT-A-DISTANCE ELECTRODYNAMICS

A. M. Gleeson†

University of Texas, Austin, Texas

In this paper we are attempting the review of a dead language. We are presenting it for two reasons. The primary reason is purely aesthetic and the satisfaction derives directly from our ability to describe at least some system completely and consistently. The other reason is of course pragmatic and related to extracting from this dead language a lesson which may apply in the realm of our modern experience.

The paper reports on work that was carried on by us but also in a sense, reporting on the activities of our several colleagues at the Center for Particle Theory at The University of Texas. As you may have become aware from Sudarshan's earlier paper in these Proceedings [1], we are very much interested in developing a consistent theory of matter, especially the electron. In his paper he discussed some aspects of our present approach to the development of a consistent quantum field theory. The present paper will discuss and review a classical theory of matter, and like the previous paper, it will emphasize the theory of electrodynamics. The methods and the restrictions as in that case can be applied to other systems.

The fundamental entities of any physical theory are the ultimate depositories of dynamical information. The present method of description for the transfer of dynamical information from one set of fundamental entities to another involves local field variables. The earlier Newtonian method of direct action at a distance interaction between particles was replaced by the action by contact

†Work supported by the U.S. Atomic Energy Commission.

methods of Faraday and Maxwell. With Maxwell the types of fundamental entities increased and these new field variables now had an independent dynamical significance. The fantastic success of the Maxwell approach was proof positive of its natural superiority over the Newtonian method.

But at the same time that it supplied a successful basis for the description of the transfer of dynamical information from particle to particle. This new method brought about grave difficulties for the development of consistent theory of the fundamental entities. It is important to note that the classical action at a distance mechanics of point particles is the only consistent theory of matter available. The standard theory of matter using local fields does not consistently admit fundamental entities in its present formulation. It is interesting to note that in all his/her pursuit of rigor and clarity, the modern physicist has moved away from a mode of description that was at least apparently consistent to a newer mode, which without essential modification is unable to handle a discussion of fundamental objects. Yet the field quantities furnish a very successful phenomenology of interactions. The probable explanation for this escapist approach is that modern physics has advanced by redefining the problem to be solved. When atoms were the fundamental entities they were explained by assuming that they were composed of smaller entities. There are now experimental indications that this route for advancement may have come to a deadend. The present theory of the electron is a theory of point electrons, and is well known for its associated difficulties.

We review the program of action at a distance electric forces. This review will provide both an aesthetic structure and some insight into the nature of fields. This leads us to the pragmatic aspects of the paper to attempt to extract the lessons which may apply in the development of new languages. [2]

The formulation of the theory follows a rather conventional path. The basic approach to dynamics being by means of a variational principle of an action. There are minor technical complications associated with the inclusion of delayed effects. Let us study a simple nonrelativistic model which displays these complications.

Consider two equal masses, each with one degree of freedom, which are interacting by means of delayed contact interaction. The Lagrangian for this system is:

$$L = \frac{m}{2}\{\dot{q}_1(t)\dot{q}_1(t) + \dot{q}_2(t)\dot{q}_2(t)\}$$
$$+ \alpha\{q_1(t)q_2(t+\Delta) + q_2(t)q_1(t+\Delta)\}$$
$$+ \alpha\{q_1(t)q_2(t-\Delta) + q_2(t)q_1(t-\Delta)\}$$

and

$$I = \int_{t_1}^{t_2} L dt$$

where

$$T \equiv \frac{m}{2}\{\dot{q}_1(t)\dot{q}_1(t) + \dot{q}_2(t)\dot{q}_2(t)\} \quad .$$

Under the usual variations, allowing end point time translations, one obtains the Hamiltonian and equations of motion. That is

$$\delta I = \int_{t_1}^{t_2} \{- \frac{d}{dt}(\frac{\partial T}{\partial \dot{q}_1})\delta q_1 - \frac{d}{dt}(\frac{\partial T}{\partial \dot{q}_2})\delta q_2$$
$$+ \alpha\{q_2(t+\Delta)\delta q_1(t) + q_2(t)\delta q_1(t+\Delta)$$
$$+ q_1(t)\delta q_2(t+\Delta) + q_1(t+\Delta)\delta q_2(t)$$
$$+ (\Delta \to -\Delta)\}dt$$
$$+ \{\frac{\partial T}{\partial \dot{q}_1}\Delta q_1 + \frac{\partial T}{\partial \dot{q}_2}\Delta q_2 - \{\frac{\partial T}{\partial \dot{q}_1}\dot{q}_1(t) + \frac{\partial T}{\partial \dot{q}_2}\dot{q}_2(t)$$
$$- T(\dot{q}_1\dot{q}_2) - \alpha\{q_1(t)q_2(t+\Delta) + q_2(t)q_1(t+\Delta)$$
$$q_1(t)q_2(t-\Delta) + q_2(t)q_1(t-\Delta)]\}\Delta t \Big|_{t_1}^{t_2} \quad ,$$

where Δq_1 and Δq_2 are the total end point variations and Δt is the time translation.

Since the arbitrary variations δq_i are at different times, it is necessary to shift the limits of the appropriate integrals to isolate the equations of motion. Setting the end point variations to zero, this is now

$$\delta I = \int_{t_1}^{t_2} [\{- \frac{d}{dt} (\frac{T}{\partial \dot{q}_1}) + 2\partial(q_2(t+\Delta) + q_2(t-\Delta))\} \delta q_1$$

$$+ \{ - \frac{d}{dt} (\frac{T}{\partial \dot{q}_2}) + 2\alpha(q_1(t+\Delta) + q_1(t-\Delta))\} \delta q_2 dt$$

$$+ \alpha \int_{t_2}^{t_2-\Delta} (q_1(t-\Delta)\delta q_2(t) + q_2(t-\Delta)\delta q_1(t))dt$$

$$+ \alpha \int_{t_1-\Delta}^{t_1} (q_1(t-\Delta)\delta q_2(t) + q_1(t-\Delta)\delta q_1(t))dt + (\Delta \to -\Delta)$$

$$- \{\frac{\partial T}{\partial \dot{q}_1} \dot{q}_1(t) + \frac{\partial T}{\partial \dot{q}_2} \dot{q}_2(t) - T(\dot{q}_1,\dot{q}_2) - \partial\{q_1(t)q_2(t+\Delta)$$

$$+ q_2(t)q_1(t+\Delta) + q_1(t)q_2(t-\Delta) + q_2(t)q_1(t-\Delta)\} \Delta t \Big|_{t_1}^{t_2} .$$

These equations now yield the expected time independent Hamiltonian and equations of motion,

$$\frac{\partial}{\partial t} (\frac{\partial T}{\partial \dot{q}_i}) = 2\alpha(q_j(t+\Delta) + q_j(t-\Delta)) \quad i = 1, 2 \quad j \neq i.$$

An interesting feature of the force equation is the natural time symmetry. This feature is manifest by the 2α in the force law. Both the time advanced and time retarded interactions in the Lagrangian contribute a time symmetric force. The physical source of this symmetry is of the symmetry between the particles and the natural action-reaction force of the Lagrangian approach.

The finite end point integrations are in a sense leakage terms due to the delay of the interaction and a proper identification of terms requires them to vanish. These terms are dropped by taking t_1 to $-\infty$ and t_2 to $+\infty$.

It is important to note that this theory is finite and well defined. A self interaction could be added which would also still be finite and well defined.

This simple model contains all the essential features of modern action at a distance formalism and obviously contains no fields. The elements of the theory are only the coordinate degrees of freedom.

The field, however, can be easily introduced. This is done in a simple two-step process. First instead of this simple approach, use the multiple time formalism for action at a distance. The action is now written as

$$I = \frac{m}{2} \int_{t_1}^{t_2} (\dot{q}_1(t)\dot{q}_1(t) + \dot{q}_2(t)\dot{q}_2(t))dt + \sum_{1,2} \int_{t_1}^{t_2} dt \int_{t_1'}^{t_2'} dt'\alpha q_i(t)\, K_{ij}(t,t')q_j(t')$$

$$\text{where } K_{ij} = \begin{pmatrix} 0 & \delta(t-t'-\Delta) + \delta(t-t'+\Delta) \\ \delta(t-t'+\Delta) + \delta(t-t'-\Delta) & 0 \end{pmatrix} .$$

This action is identical to the previous one. Its variation yields exactly the same results but now expressed as

$$\frac{d}{dt}\left(\frac{\partial T}{\partial \dot{q}_i}\right) = 2\alpha \int_{-\infty}^{\infty} K_{ij}(t,t')q_j(t')dt' .$$

It is now obvious how to define a field at point q_i at time t. Define

$$F_i(t) = \int_{-\infty}^{\infty} K_{ij}(t-t')q_j(t')dt' .$$

q_j is the source of this field.

In this formalism, it is obvious that the field is operating only as a device which serves to keep track of the activities of its source. In other words, the degrees of freedom on the field are the past or future behavior of the other particles in the problem.[2] A self-interaction can be added by including diagonal terms to K_{ij}. These terms do not really modify the interpretation. With this interpretation of the field as the bookkeeper of the system of particles past and future, the infinities of normal field theory take on a new light. If you have a company with a bookkeeping scheme which is bankrupting the company then you should change the bookkeeping scheme: not get rid of the company!

This formalism can be applied to classical electricity and magnetism. In this case the action is

$$I = \frac{1}{2}\sum_a \int m_a \dot{x}_a^2 ds_a + \frac{1}{2}\sum_{ab}{}' \iint K_{ab} ds_a ds_b$$

$$K_{ab} = e^2 \dot{x}_a^{\mu} \dot{x}_{b\mu} \delta((x_a - x_b)^2) \quad .$$

A variation of this action yields the following

$$\ddot{x}_{a\mu} = e\, \dot{x}_a^{\nu} F^{+}_{a\nu\mu} \quad ,$$

where

$$F^{+}_{a\nu\mu} = \frac{\partial A^{+}_{a\nu}}{\partial x_a^{\mu}} - \frac{\partial A^{+}_{a\mu}}{\partial x_b^{\nu}}$$

and

$$A^{+}_{a\mu} = \sum_b{}' e \int_{-\infty}^{\infty} \delta((x_a - x_b)^2)\dot{x}_{b\mu} ds_b \quad .$$

We have gone directly to the pseudo field formalism.

As is always the case, the variation yields the expected conserved quantities. The form of the conservation laws is not local in time as it is in field theory but contains integrals over the pasts and futures of the particle trajectories.[3] The + on $F^{+}_{a\mu\nu}$ and $A^{+}_{a\mu}$ indicates the time symmetry of the law of force. This is of course characteristic of our action at a distance approach. Although this theory looks like conventional classical electricity and magnetism except for the fact that it is finite and time symmetric.

This equation as it stands of course does not yield the desired equations of motion. The equations of motion for a single electron contain the well known radiation reaction term. This effect is not time symmetric and therefore takes some arranging. In fact a successful theory based on an action-at-a-distance approach is due to the separation of the problem into parts. It will be seen that the radiation reaction is the result of the rest of the system on the charge under analysis. This is the general lesson of this method, if you are interested in the behavior of parts of the system then a field theory phenomenology is simplest to formulate, but, if you are concerned about the fundamental form of the interaction then the action at a distance approach is more easily handled. These equations can be manipulated into the correct form provided the absorber assumption holds.[4] Define the "field"

$$F^{+}_{\nu\mu} = \frac{\partial A^{+}_{\nu}}{\partial x_{\mu}} - \frac{\partial A^{+}_{\mu}}{\partial x_{\nu}}$$

$$A^{+}_{\mu} = \sum_{b} e \int_{-\infty}^{\infty} \delta((x - x_b)^2)\dot{x}_{b\mu} ds_b \ .$$

$F^{+}_{\nu\mu}$ and A^{+}_{μ} differ only slightly from $F^{+}_{a\nu\mu}$ and $A^{+}_{a\mu}$. Similarly $F^{ret}_{\nu\mu}$, $F^{adv}_{\nu\mu}$ and $F^{ret}_{a\nu\mu}$, $F^{adv}_{a\nu\mu}$ can be defined. The absorber condition is then that

$$F^{ret}_{\nu\mu} - F^{adv}_{\nu\mu} = 0 \text{ everywhere.}$$

This condition can be motivated by assuming an absorbing universe such that

$$\frac{1}{2}\{F^{ret}_{\nu\mu} + F^{adv}_{\nu\mu}\} = 0 \ ,$$

outside some region. In other words, there is a region beyond which no force is felt. Since $F^{ret}_{\nu\mu}$ and $F^{adv}_{\nu\mu}$ have very different flux behavior they cannot sum to zero in some region unless both are zero in that region. Therefore

$$\Rightarrow F^{-}_{\nu\mu} \equiv F^{ret}_{\nu\mu} - F^{adv}_{\nu\mu} = 0$$

outside some region. The $F^{-}_{\nu\mu}$ satisfies the source free Maxwell's equations and therefore vanishes everywhere, which is the required absorber condition.

Manipulation of the time symmetric equation of motion using the absorber condition yields the desired equation of motion for a single charged particle assuming the rest of the universe is inert but absorbing. This equation is no longer time symmetric although its basis was and the reason is based on a desire to develop a phenomenology where you treat the motion of only one electron and do not take into account the reaction of the rest of the universe. It is in this sense that the radiation reaction of the equation of motion is a useful phenomenology when discussing the behavior of part of a large system.

The black body problem is often cited as an example of a phenomena which proves the existence of a field. The radiation from a heated cavity has been a problem whose resolution has often been the turning point of many approaches to physical phenomena. The modern approach to the black body problem is based on the derivation of Debye.[5] In this case the field is the entire dynamical source of the heat capacity, the modes of the field being excited thermally. The success of this derivation is the source of the belief that this phenomena verifies the physical reality of the field. As a matter of fact, a simple scaling argument reinforces this belief.

The number of degrees of freedom within a fixed frequency interval found in black body radiation scales as the volume of the cavity. This is consistent with the field interpretation since the field fills the volume. In an action at a distance theory all dynamical information is contained in the particle degrees of freedom. The particles are in the walls of the cavity and these scale as the $(\text{volume})^{2/3}$.

This argument is specious and is really a "straw man" whose knocking down provides the necessary insight to the absorber hypothesis. We first decide what is an "ideal" black body.

In the black body situation an infinity of charges are contained in the walls of a cavity. The arrangement is such that charges outside the cavity are not affected by the charges inside. This is shown in Fig. 1.

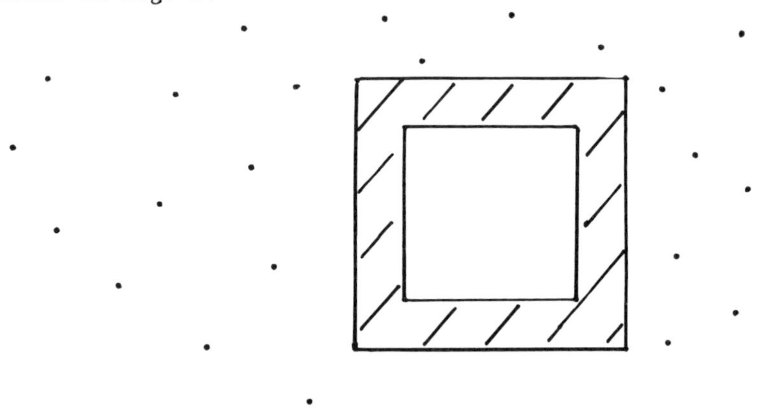

Outside Cavity

Fig. 1. Configuration of Black Body Heat Capacity.

In other words, $F^{(+)}_{(\text{outside})\mu\nu}$ has two sources, the inside charges and the other outside charges but $F^{(+)}_{(\text{outside})}$ due to the inside charges must vanish by the condition of what we mean as a Black Body experiment. The Black Body situation is thus one in which a sub-system is in an absorber configuration. It is an arrangement of infinity of charges such that the absorber condition holds on its "field" at the surface. This condition then yields the correct set of harmonic modes to yield the Rayleigh-Jeans formula. Actually we know this to be but an idealized description of the actual situation. The perfectly reflecting surface is actually a material that provides charges which drive the field in the cavity. As in calculations of absorption the charges in a half wave length are principly responsible for reflecting the wave. Thus for each mode the number of charges involved scales as the surface area times half wavelength which then goes as the volume. Thus for the idealized cavity there are just the correct infinity of charges to account for the requisite degrees of freedom.

This result not only implies that there is no inconsistency between black body radiation and action at a distance but as often seems to be the case, the Black Body problem contains the germ of the solution to the larger problem as to what a field actually is.

We now actually reverse the usual argument combining both the previous lines of thought. The equation of motion of any charge depends on the situation in the past and in the future of all other charges in the universe (action at a distance). If the problem requires a description of parts of the system this can be handled phenomenologically by the introduction of a field. The field will have infinite degrees of freedom if there is an infinity of charges in the universe. In all probability it only has a large number of degrees of freedom anyway. The arrangement of the field boundary conditions correspond to conditions on the infinity of charges being ignored. If you use the action at a distance approach you must absorb if you want to discuss radiation. If you use the field you are actually bookkeeping the particle degrees of freedom you have ignored. This is summarized in Fig. 2.

The field has its advantages when viewed now as a phenomenology. Radiation is simple and the description is local in time.

The pragmatic lesson alluded to at the beginning of this paper refers to the use of general action at a distance forces to generate fields. This will generate fields with a broad range of properties. The inclusion of finite self interactions is possible and their existence would imply local fields that differ from the conventional types. The shadow states are examples of states obtained from a naturally time symmetric field generated in this fashion.

Acknowledgment

The author wishes to thank E. C. G. Sudarshan with whom all these questions have been discussed over the past few years.

Large Number of Degrees of Freedom

Action at a distance ↔ Allows a Field

Action at a distance	Allows a Field
↓	↓
Time Symmetric	Local in Time
↓	↓
radiation only possible if absorbed (you are really only interested in what happens in part of system) fixed boundary conditions	radiation simple and boundary conditions independently determined by situation.

Fig. 2. Conditions for Models of Interactions.

References

1. E. C. G. Sudarshan, "Finite Quantum and Classical Electrodynamics", this volume, p. 289.
2. E. C. G. Sudarshan, Fields and Quanta *2*, 175 (1972).
3. J. W. Dettman, A. Schild, Phys. Rev. *95*, 1057 (1954).
4. J. A. Wheeler, R. P. Feynman, Rev. Mod. Phys. *17*, 157 (1945); *21*, 425 (1949).
5. P. Debye, Ann. Physik *33*, 1427 (1910).

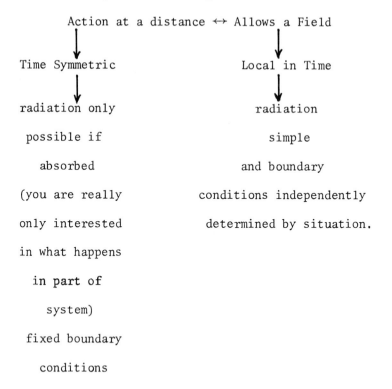

THE IMPLICATIONS OF RADIATIVE EQUILIBRIUM IN JAYNES' EXTENSION OF SEMICLASSICAL RADIATION THEORY

J. P. Gordon and F. R. Nash

Bell Laboratories, Holmdel, New Jersey

The development of the quantum theory of matter and radiation hinged rather importantly upon two experimental facts, namely, the stability of atoms in their ground states, and the blackbody radiation law. The present quantum electrodynamics (QED) contains explanations of these facts along with virtually all others that atomic spectra can offer. However, many awkward divergences (such as the infinite contribution the transverse electromagnetic field makes to the electron mass) exist in the theory which must be explained away before theory and experiment agree. For this and other more philosophical reasons, further development of the theory is certainly desirable.

In recent years E. T. Jaynes and his coworkers [1] have attempted to find possible new paths for the theory by retreating toward semiclassical theory [2], using Schroedinger's equation as a description of atomic systems, but keeping a classical description for electromagnetic fields. They have labelled their theory neoclassical and, by inclusion of the radiation reaction field, have analyzed the spontaneous emission phenomenon in its context. While such a theory may indeed suggest new insights it contains what seems to us to be a fatal flaw, namely it is inconsistent with Planck's black-body radiation law and/or Boltzmann's law of statistical mechanics, both of which have been quite well verified experimentally. This point is the theme of the present paper, and of a more detailed companion paper [3] to be published elsewhere.

One can gain some insight into the problem by considering how the different theories regard the radiation field of a small bound system of charged particles such as an atom. The basic retarded solution of Maxwell's equations, given below in the nonrelativistic dipole approximation, is a valid expression in all theories. The total radiation field can be expressed as [4]

$$\underline{E}(\underline{r},t) = \underline{E}_o(\underline{r},t) + \{r^{-3}c^{-2}[\underline{r}\times(\underline{r}\times\ddot{\underline{p}})]\}_{t'}, \tag{1}$$

where

$$\underline{p} = \sum e_i\underline{r}_i \tag{1a}$$

is the dipole moment of the atom, t' = t - r/c is the retarded time, $\underline{r}$ is the position vector of an observation point, and finally $\underline{E}_o(\underline{r},t)$ is an arbitrary external field, namely the field that would have been at $\underline{r}$ in the absence of the atom.

The different theories may be distinguished by their treatments of the dipole moment vector. In quantal theory (QED) the position vectors $\underline{r}_i$ become operators, and $\underline{p}_i$ may be expressed, by means of a completeness relation for the atomic energy states, in the form

$$\underline{p} = \sum_{m,n} |m\rangle\langle m| \sum_i e_i\underline{r}_i|n\rangle\langle n| \; ; \text{ (QED)} \tag{2}$$

Equation (2) is an identity, showing how the dipole moment operator is transformed into a set of raising and lowering operators among the energy states. In the neoclassical theory (NCT) only the expectation value of the quantal moment is assumed to have import. Hence an atom in some general state $|\Psi\rangle$ is assumed to have a dipole moment given by

$$\underline{p} = \langle\Psi|\underline{p}|\Psi\rangle = \sum_{m,n} a_m^* a_n \underline{p}_{mn} ; \text{ (NCT)} \tag{3}$$

where

$$a_n \equiv \langle n|\Psi\rangle \; ; \; \underline{p}_{mn} \equiv \langle m| \sum_i e_i \underline{r}_i |n\rangle. \tag{3a}$$

Let us now consider the energy exchange between the atom and the field. If W_a is the energy of the atom, then the rate of work done on the atom by the external field is given by

$$\left(\frac{dW_a}{dt}\right)_{abs.} = (\underline{E}_o \cdot \dot{\underline{p}})_{av} \tag{4}$$

where the subscript av on the right-hand side indicates that an average should be taken over a few cycles of the field frequency. Equation (4) gives the absorption rate; if it is negative then its absolute value is the induced emission rate. This equation poses a difficult problem for the neoclassical theory. If its right-hand side is a classical expression, in accord with NCT's treatment of fields and dipole moments, then W_a must also be a classical (as opposed to an operator) quantity. But W_a must be an operator if one is to achieve Schroedinger's equation and a quantized picture of the atom. In other words, the statistical energy uncertainty quantal theory attributes to atoms, exemplified by the equation

$$\langle\Psi|W_a^2|\Psi\rangle \geqslant \langle\Psi|W_a|\Psi\rangle^2$$

with the equality holding only if $|\Psi\rangle$ is a pure energy state or a mixture of degenerate energy states, is not maintained by (4) in the neoclassical theory. This problem is further illustrated in a thought experiment below.

There is another method of obtaining the absorption rate, using (1). If we evaluate at time t the energy flow outward through a sphere of radius r with the atom at its center, and extract the terms linearly proportional to the dipole moment, the result by conservation of energy should equal the induced emission rate of the atom at the earlier time t'. The appropriate relation is

$$\left(\frac{dW_a}{dt}\right)_{\text{abs., at } t'} = -\frac{c}{4\pi}\int_{\text{sphere, at } t} [\underline{E}_p\times\underline{H}_o+\underline{E}_o\times\underline{H}_p]_{av}\cdot\hat{n}\, dS \qquad (5)$$

where E_p is the dipolar field of (1), namely

$$\underline{E}_p(\underline{r},t) = [r^{-3}c^{-2}\underline{r}\times(\underline{r}\times\ddot{\underline{p}})]_{t'}, \qquad (5a)$$

H_o and H_p are the magnetic fields corresponding to E_o and E_p, and $\hat{n}^o$ is the outward unit normal vector on the sphere. Within a quasi-harmonic, or slowly varying envelope, approximation the right-hand sides of (5) and (4) are equal, as can be readily shown by taking the external field in the form of a Gaussian beam. For our present discussion we may consider the external field to be a classical field; this is acceptable in the quantal theory since the external field is divorced from its source, provided one exercises proper care in the ordering of operators. Again we see from (5) that in NCT the energy W_a of the atom cannot develop its proper quantal uncertainty since the right-hand side of (5) in NCT is a completely classical quantity.

An example of this difficulty is contained in the following thought experiment. An atom, initially in its ground state, is acted upon by a "$\pi/2$" pulse of incident radiation which raises it half way to a resonant excited state. The neoclassical theory predicts that an amount of energy equal to exactly $h\nu/2$ has been transferred from the field to the atom. On the other hand the atom must be regarded as having the energy either of the ground state or of the excited state, each with probability one-half, in order to have a theory conceptually compatible with the photoelectric effect or the results of Stern-Gerlach type experiments. For example, if the excited state of the atom had sufficient energy to decay by emitting a photoelectron, and did this with probability p, then after the aforementioned $\pi/2$ pulse, the atom would emit a photoelectron with probability p/2, indicating that it had acquired the energy of the excited state with probability 1/2. In NCT's picture energy is conserved on the average, but in any particular experiment, such as a $\pi/2$ pulse experiment mentioned above with a resultant photoelectron, energy is *not* conserved. Thus one may reach the conclusion that either the theory is incomplete, or that conservation of energy must be given up on a microscopic basis.

Let us go on now to consider spontaneous emission. Using (1) again, we see that the outward energy flow contains a term

$$\frac{c}{4\pi} \int_{\text{sphere}} (\underline{E}_p \times \underline{H}_p)_{av} \cdot \hat{n}\, dS$$

which in independent of the external field and in the quasiharmonic approximation is proportional to the square of the atomic dipole moment. This is the spontaneous emission rate [6]. Squaring the dipole moment is a straightforward matter in the classical and neoclassical theories in accordance with their separate definitions of the dipole moments. In particular, in the neoclassical theory the power spontaneously radiated in a nondegenerate transition is given from (3) as proportional to

$$2|\underline{p}_{nm}|^2 |a_n|^2 |a_m|^2 \quad ; \text{ (NCT)} \tag{6}$$

In the quantal theory the ordering of operators is important when they are multiplied together, and classical correspondence cannot tell us how to do it. In taking the square of (2), however, the necessary ordering is dictated by the fact that all spontaneous transitions must be downward. This ensures the stability of the ground state. In the correctly ordered form of $|\underline{p}|^2$, all lowering operators must appear to the right of all raising operators. Thus the squared dipole moment for the downward transition $n \rightarrow m$ rather directly achieves the form

$$2|\underline{p}_{nm}|^2 (|n\rangle\langle n|) \tag{7}$$

This ordering of the dipole operators is consistent with the normal ordering of field operators, the ordering which makes the energy of the vacuum state equal to zero and is consistent with treating $\underline{E}_0$ as a classical field. Equation (7) is the usual quantal result. For an atom in state $|\Psi\rangle$ the expectation value of (7) is

$$2|\underline{p}_{nm}|^2 |a_n|^2 \tag{8}$$

The fact that (8) is bigger than (6) can be interpreted heuristically to mean that in the quantal theory the radiating dipole moment has an intrinsic quantal uncertainty, so that its mean square, (8), is larger than the square of its mean, (6). This must be so if the quantal uncertainty of the energy of the atom is to show up in the radiation field subsequent to their interaction. Consider a simple "atom" having only two nondegenerate energy levels of importance. In accord with (6) and (8), the equations of motion for spontaneous emission in neoclassical and quantal theories, are respectively

$$d|a_2|^2/dt = -A|a_1|^2|a_2|^2; \text{ NCT} \tag{9a}$$

$$d|a_2|^2/dt = -A|a_2|^2 \;; \quad \text{QED} \tag{9b}$$

where A is the usual Einstein coefficient. Equations (9) are graphed in Fig. 1. Applied to a dilute collection of such two-level atoms, we can sum (9a) and (9b) over the collection, and so achieve the relations

$$-(dN_2/dt)_{\text{spont}} \leqslant AN_1N_2/(N_1+N_2); \text{ NCT} \tag{10a}$$

$$-(dN_2/dt)_{\text{spont}} = AN_2 \quad ; \quad \text{QED} \tag{10b}$$

where

$$N_j \equiv \sum_{\text{atoms}} |a_j|^2 \qquad j = 1,2$$

defines the mean level populations. The inequality in (10a) stems from the nonlinearity of (9a) and reflects the convex curvature of the neoclassical curve in Fig. 1. For example, the right-hand side of (10a) could go to zero in case the atoms were distributed some in each of the two pure energy states. The equality holds only if

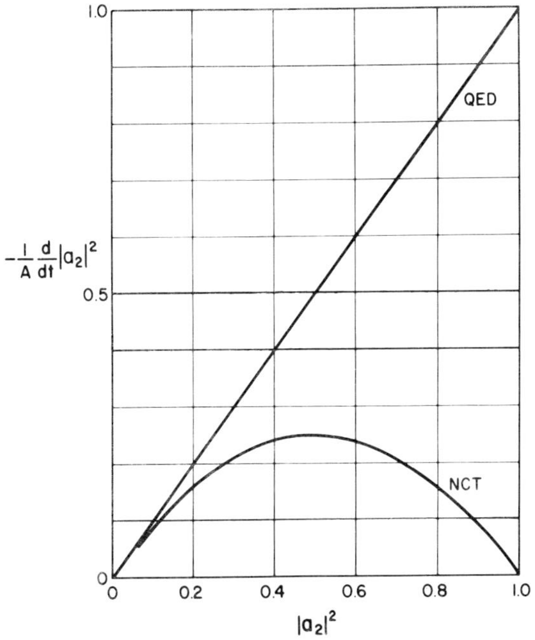

Figure 1.

all atoms are in the *same* mixed state.

We consider now the implications of the differing spontaneous emission rates in two situations, namely noise generation in a linear amplifier and thermal equilibrium with a black-body radiation field.

The equation of motion for field intensity in a linear maser type amplifier composed of nondegenerate two-level systems is

$$dI(\nu)/dz = \alpha(\nu)\ [I(\nu) + N(\nu)] \tag{11}$$

where $I(\nu)$ is the spectral intensity of a transverse field mode traversing the amplifier, $\alpha(\nu)$ is the gain coefficient, and $N(\nu)$ is the noise spectral intensity generated by the amplifier which enters that mode. Induced emission is responsible for the gain, while spontaneous emission is responsible for the noise. From quantal theory one finds the result

$$N(\nu) = h\nu \frac{N_2}{N_2 - N_1} \;;\; \text{QED} \tag{12}$$

while from the neoclassical theory because of the reduced spontaneous emission rate one finds correspondingly

$$N(\nu) \leq h\nu \frac{N_2}{N_2 - N_1} \cdot \frac{N_1}{N_1 + N_2} \;;\; \text{NCT.} \tag{13}$$

One can also show that $N(\nu)$ is the minimum energy that a high gain amplifier of this kind can be used to detect. The quantal theory result shows a mimimum detectable energy of $h\nu$ (for $N_1=0$, or complete inversion), in agreement with the requirements of the uncertainty principle. The neoclassical result (13) has no such limitation , which is basically in concert with its premise that the fields are classical; hence the field need not be subject to an uncertainty principle. However, it was shown by Bohr and Rosenfeld [5] that the uncertainty principle applied to a test particle used to measure a radiation field strength prevents an exact measurement of two fields and hence reveals their nonclassical character.

Finally we turn to the black-body radiation law. Under conditions where we may expect rate equations to hold; namely a dilute system of atoms with rapid phase relaxation (one may envisage some nearly elastic collision mechanism to ensure the latter), we may safely write an Einstein type equation for the rate of change of of the upper state population N_2, namely

$$\frac{dN_2}{dt} = -\Gamma N_2 + \rho[\beta_{21}N_1 - \beta_{12}N_2] \tag{14}$$

where $-\Gamma N_2$ is the spontaneous emission rate, ρ is the spectral radiation density at the transition frequency, and the β's represent induced absorption and emission coefficients. Derivations of (14) based on the quantal theory agree with that of Einstein, and give

$$\Gamma = A,\ \beta_{21} = B_{21},\ \beta_{12} = B_{12};\ \text{QED} \tag{15}$$

where

$$B_{21} = B_{12} = [c^3/8\pi h\nu^3]A \tag{15a}$$

are the usual Einstein A and B coefficients which are constants independent of the level populations. If one then inserts in (14) the radiation density according to Planck's law, namely

$$\rho = 8\pi h\nu^3/c^3[\exp(h\nu/kT)-1], \tag{16}$$

along with the level population ratio according to Boltzmann's law, namely

$$N_1/N_2 = \exp(h\nu/kT), \tag{17}$$

one finds equilibrium, as identified by $dN_2/dt = 0$.

Let us see now how the change to the spontaneous emission rate of the neoclassical theory affects these matters. The descriptions of induced processes differ in neoclassical and quantal theories *only* when one considers quantal uncertainty phenomena. The B coefficients which pertain to average transition rates are identical in both theories, and in fact the population dependences of the induced processes have been nicely verified in some laser amplifier experiments [7]. The second term on the right of (14) is therefore the same in all theories. If we insert the relations (15) and (15a) for the β's into (14) we obtain for the equilibrium case the relation

$$\frac{\Gamma}{A} = \rho\left(\frac{N_1-N_2}{N_2}\right) \frac{c^3}{8\pi h\nu^3} \tag{18}$$

The neoclassical expression for Γ for a gas of simple non-degenerate two-level atoms may be found by comparing (14) with (10a). We thus obtain

$$\Gamma/A \leq N_1/(N_1+N_2); \quad \text{NCT} \tag{19}$$

in contrast with the QED result $\Gamma = A$.

Upon inquiry into the possible consistency of (19) with the thermal equilibrium relation (18), we see that it is impossible without modification of Planck's law (16) and/or Boltzmann's law (17). The difference between (19) and the QED result is most acute in the high temperature limit, where the right-hand side of (19) approaches 1/2. However, the consequent necessary reduction of the right-hand side of (18) by a factor of at least two in that limit through modification of (16) and/or (17) is altogether precluded by the accuracy (a few percent) with which these laws have been verified experimentally[7].

Thus while the motivation for the neoclassical theory has been the desire to eliminate some conceptual difficulties found in quantum electrodynamics, one finds that as presently conceived the neoclassical theory introduces its own conceptual difficulties which are illustrated by the nonconservation of energy on a microscopic scale. More importantly, the neoclassical theory badly fails to account quantitatively for the thermal equilibrium state of radiation and matter.

These results, along with some more direct experimental evidence as presented at this Conference, lead inexorably to the conclusion that the spontaneous emission rate found in the neoclassical theory is incorrect, and hence that the phenomenon of spontaneous emission remains outside the realm of validity of semiclassical theory.

References

1. E. T. Jaynes and F. Cummings, Proc. IEEE *51*,89 (1963); M. D. Crisp and E. T. Jaynes, Phys. Rev. *179*, 1253 (1969); C. R. Stroud and E. T. Jaynes, Phys. Rev. *A1*, 106 (1970); E. T. Jaynes, Phys. Rev. *A2*, 260 (1970).
2. L. I. Schiff, *Quantum Mechanics*, (McGraw-Hill Book Company, Inc., New York 1955), 2nd ed., Chap. 10.
3. F. R. Nash and J. P. Gordon, to be submitted to the Physical Review.
4. L. Landau and E. Lifshitz, *The Classical Theory of Fields*, (Addison-Wesley Company, Inc., 1951), Chap. 9.
5. W. Heitler, *The Quantum Theory of Radiation*, (Oxford University Press, London, 1944), 2nd ed., Chap. 2.
6. To find the effect of spontaneous emission directly on the atom in the spirit of (4) requires the evaluation of the radiation reaction field, an exercise which has been carried out in the NCT, but which is beyond the scope of the present discussion.
7. These points are discussed at some length in reference 3.

A QUANTUM ELECTRODYNAMIC INVESTIGATION OF THE JAYNES-CRISP-STROUD APPROACH TO SPONTANEOUS EMISSION*

Jay R. Ackerhalt†

University of Rochester, Rochester, New York

Joseph H. Eberly‡ and Peter L. Knight

Stanford University, Stanford, California

1. Introduction

Motivated by the work of Jaynes and co-workers, [1] the old problem of atomic level shifts and widths has recently begun to be re-examined. While QED is a highly successful theory (indeed the only workable field theory we have), it is still beset with self-energy infinities ultimately associated with the point-like nature of the electron. These are removed from the public view by the process of renormalization, in the hope that eventually some high frequency cut-off will be found to make the renormalization constants finite: such modifications proposed usually imply a radius to the electron. In the usual approach to QED, perturbation theory is used. Jaynes, prompted by the great progress made in theory and experiments on atoms interacting with electromagnetic fields, has re-investigated the problem of spontaneous emission by solving the relevant semi-classical equations of motion for the interacting field-atom system directly. An interesting product of such an approach (other than

*Research partially supported by the U.S. Army Research Office (Durham) and the U.S. Atomic Energy Commission

†National Science Foundation Predoctoral Trainee

‡Permanent Address: Department of Physics and Astronomy, University of Rochester, Rochester, New York 14627

the novel time-dependence) is the non-appearance of divergences: the finite size of the atomic charge distribution removes the point like singularity when retardation is taken into account.

We have adopted the approach advocated by Jaynes, but have applied it to the integration of the relevant quantum electrodynamic equations of motion instead of the Jaynes-Crisp-Stroud semiclassical equations. The field operators are then derived in terms of the source operators. We find the following results of this analysis: (1) our predicted level width agrees with the semiclassical width in the way Jaynes has emphasized; (2) the level shift calculated by Crisp and Jaynes (called a Lamb shift by Stroud and Jaynes) is reproduced exactly when their same approximations are used; (3) the time evolution of the system agrees with the more usual methods of QED. In particular, there is no frequency chirping of the emitted light, and the level shift is a static, not a dynamic function of time. Finally, the time evolution characteristics of semiclassical theory are obtained from the QED theory by decorrelating consistently the atom from the field operators when taking expectation values. It is interesting that the approach to semiclassical radiation theory, so long advocated by Jaynes, the direct integration of the equations of motion over long times, is relatively easily taken over to QED.

2. The Equations of Motion

We take the Hamiltonian for a single bound two-level atom interacting with a radiation field in pseudospin notation as

$$H = \frac{\hbar\omega_0}{2}\,\sigma_3 - \frac{\omega_0\mu}{c}\,\sigma_2 \int g(r)A_Z(r,t)\,d^3r + H_F \tag{1}$$

where $g(r)$ represents the retardation effect [2] over the atomic charge distribution. In particular, for a hydrogenic atom transition from the $2P_Z$ to the 1S state (which represents the two levels, say, of our atom), $g(r)$ is

$$g(r) = \frac{1}{(2\pi)^3}\int d^3k\, e^{i\underline{k}\cdot\underline{r}}\,\tilde{g}(k) = \frac{e^{-3r/2a}}{8\pi(2a/3)^3} \tag{2}$$

and $\tilde{g}(k) = (1 + \frac{4}{9}k^2a^2)^{-2}$.

The Heisenberg equations of motion for the atomic operators are

$$\dot{\sigma}_3 = \sigma_1 \tilde{A}(t) \tag{3}$$

$$\dot{\sigma}_2 = \omega_0 \sigma_1 \tag{4}$$

$$\dot{\sigma}_1 = -\omega_0 \sigma_2 - \sigma_3 \tilde{A}(t) \tag{5}$$

where we have defined $\tilde{A}(t) = (2\mu\omega_0/\hbar c) \int g(r) A_Z(r,t) d^3r$ for convenience. Defining the positive and negative frequency operators by

$$\sigma_1 \equiv \sigma^+ + \sigma^- \qquad \sigma_2 \equiv i(\sigma^- - \sigma^+) \tag{6}$$

$$A^{\pm}(r,t) \equiv \sum_\lambda \sqrt{\hbar}\, \frac{c}{2\pi} \int \frac{d^3k}{\sqrt{\omega_{k\lambda}}}\, \hat{\varepsilon}_{k\lambda} a_{k\lambda}^{\mp}(t)\, e^{\pm i\underline{k}\cdot\underline{r}} \tag{7}$$

(where $[a^-_{k\lambda}(t), a^+_{k'\lambda'}(t)] = \delta^3(k-k')\delta_{\lambda\lambda'}$, etc.), then (3) to (5) become, in the RWA (rotating wave approximation):

$$\dot{\sigma}_3 = \sigma^+ \tilde{A}^+(t) + \tilde{A}^-(t)\sigma^- \tag{8}$$

$$\dot{\sigma}^+ = i\omega_0 \sigma^+ - \tfrac{1}{2}\tilde{A}^-(t)\sigma_3 \tag{9}$$

$$\dot{\sigma}^- = -i\omega_0 \sigma^- - \frac{1}{2}\sigma_3 \tilde{A}^+(t) \quad . \tag{10}$$

We derive the field, as do Jaynes and Crisp, from Maxwell's wave equation

$$\nabla^2 A^{\pm} - \frac{1}{c^2} \ddot{A}^{\pm} = -\frac{4\pi}{c} J_{\perp}^{\pm} \tag{11}$$

where the transverse current is defined by

$$J_{\perp}^{\pm} = \pm \frac{i\omega_0 \mu}{(2\pi)^3} \sum_\lambda \sigma^{\mp} \int \hat{\varepsilon}_{k\lambda}\, \hat{\varepsilon}_{k\lambda}\cdot \hat{z}\, \tilde{g}(k)\, e^{i\underline{k}\cdot\underline{r}}\, d^3k \quad . \tag{12}$$

The formal solution

$$A_Z^{\pm}(r,t) = \pm \frac{i4\pi c\mu\omega_0}{(2\pi)^3} \int_0^t dt' \sigma^{\mp}(t') \sum_\lambda \int d^3k\, \frac{e^{i\underline{k}\cdot\underline{r}} (\hat{\varepsilon}_{k\lambda}\cdot \hat{z})^2 \tilde{g}(k) \sin\omega(t-t')}{\omega} \tag{13}$$

gives for the integral we need in the equations of motion the approximate result [3]

$$\tilde{A}^{\pm} \simeq A_0^{\pm} \pm 2i\Delta_c \sigma^{\mp}(t) - A\sigma^{\mp}(t) \qquad (14)$$

where

$$A \equiv 4\omega_0^3 \mu^2/3\hbar c^3 \quad \text{and} \quad \Delta_c \equiv \frac{5\omega_0^2 \mu^2}{16\hbar a c^2}$$

where $A_0^{\pm}$ stand for the free-field homogeneous solutions.

3. Solutions

We may now use (14) to write the atomic equations in terms of source operators entirely, except for the necessity of keeping the homogeneous solutions. The σ^- equation becomes:

$$\dot{\sigma}^- = -i\omega_0 \sigma^- - \frac{\sigma_3}{2}[A_0^+ + (2i\Delta_c - A)\sigma^-] ,$$

using $\sigma_3\sigma^- = -\sigma^-$, valid for equal time operators, we obtain

$$\dot{\sigma}^- = [-i(\omega_0 - \Delta_c) - \frac{A}{2}]\sigma^- - \frac{\sigma_3}{2} A_0^+ .$$

We take vacuum expectation values to eliminate the free fields, and get

$$\langle\dot{\sigma}^-\rangle_{vac} = \{-i(\omega_0 - \Delta_c) - \frac{A}{2}\}\langle\sigma^-\rangle_{vac} \qquad (15)$$

where the significance of both the Einstein A coefficient and the Crisp-Jaynes frequency shift is obvious. The solution of (15) is trivial:

$$\langle\sigma^-(t)\rangle_{vac} = \langle\sigma^-(0)\rangle_{vac}\, e^{-At/2}\, e^{-i(\omega_0 - \Delta_c)t} , \qquad (16)$$

and shows that Δ_c apparently *plays the role of a transition frequency shift in QED* as well as in the neoclassical theory. It is, however, not time-dependent in QED. The solution for $\langle\sigma_3(t)\rangle$ follows in the same way:

$$\langle\sigma_3(t)\rangle_{vac} = \{\langle\sigma_3(o)\rangle_{vac} + 1\}e^{-At} - 1 \tag{17}$$

showing the typical QED exponential decay of upper state occupation.

In semiclassical theory, on the other hand, the operator *expectation* values are used *throughout* and atom-field correlations are neglected: one takes

$$\langle\sigma^+ a_{k\lambda}\rangle_{vac} = \langle\sigma^+\rangle_{vac}\langle a_{k\lambda}\rangle_{vac}$$

and the equations of motion become, in the absence of external fields,

$$\langle\dot{\sigma}^-\rangle = -i\omega_o\langle\sigma^-\rangle - (i\Delta_c - \frac{A}{2})\langle\sigma_3\rangle\langle\sigma^-\rangle \tag{18}$$

$$\langle\dot{\sigma}_3\rangle = -2A\langle\sigma^+\rangle\langle\sigma^-\rangle . \tag{19}$$

It is immediately obvious that the coefficients A and Δ_c are the same in both treatments. The semiclassical solutions [1] for $\langle\sigma_3\rangle$ and $\langle\sigma^-\rangle$are

$$\langle\sigma_3(t)\rangle = - \tanh \frac{A}{2}(t-t_o) \tag{20}$$

$$\langle\sigma^-(t)\rangle = \frac{1}{2} \operatorname{sech} \frac{A}{2}(t-t_o)\ e^{-i(\omega_o t+\theta(t))} \tag{21}$$

where $\theta(t)$ represents the effects of a time-dependent frequency shift

$$\delta\omega(t) = \frac{d\theta}{dt} = - \Delta_c \tanh \frac{A}{2}(t-t_o) . \tag{22}$$

Thus QED and semiclassical theory agree that the level width is A/2; and furthermore QED reproduces the Jaynes-Crisp shift Δ_c by making their assumptions (RWA, etc.). Because of the correlated dynamics in QED there is no frequency chirping of the emitted light. The level shift is a static, *not* a dynamic function of time. At long times, of course, when the effects of the atom-field correlations are no longer important, the theories agree in their time-evolution. However, these correlations play the role at short times of removing the metastable singular point of the semiclassical system. It is interesting that the shift Δ_c, common to both theories, does not agree with the usual Bethe formula, and more strikingly does not need renormalization or cut-offs. To see why this should be so, we analyze the relationship between Δ_c and the result of a more conventional QED calculation of level shifts.

4. QED Frequency Shifts via Perturbation Theory

The usual QED treatment of level shifts uses the $\underline{p}\cdot \underline{A}$ interaction in second order perturbation theory and omits the A^2 terms as shifting all levels equally. [4] For our two-level atom, the QED shift of the upper (+) and lower (−) states would be

$$\hbar\Delta^{\pm} = P\sum_{\lambda}\int d^3k \frac{<0_{k\lambda};\pm|\frac{e}{mc}\,\underline{p}\cdot\underline{A}|\mp;1_{k\lambda}><1_{k\lambda};\mp|\frac{e}{mc}\,\underline{p}\cdot\underline{A}|\pm;0_{k\lambda}>}{(E_{\pm} - E_{\mp} - \hbar ck)} . \tag{23}$$

In the RWA $\Delta^- = 0$; and Δ^+ may be evaluated easily using a plane wave expansion of the field:

$$\hbar\Delta^{+} = + \left(\frac{e}{mc}\right)^2\left(\frac{\hbar c}{4\pi^2}\right) P\sum_{\lambda}\int \frac{d^3k}{k} \frac{<+|\underline{p}\cdot\hat{\varepsilon}_{k\lambda}e^{i\underline{k}\cdot\underline{r}}|-><-|\underline{p}\cdot\hat{\varepsilon}_{k\lambda}e^{-i\underline{k}\cdot\underline{r}}|+>}{\hbar ck_o - \hbar ck} \tag{24}$$

where $\hbar ck_o = E_+ - E_-$.

In the conventional approach, the dipole approximation is used. Then (24) may be written (after doing the polarization sum and the angular integral)

$$\hbar\Delta^{+} = -\frac{2}{3\pi}\frac{\alpha}{m^2c^2}\,|<-|p|+>|^2\int \left(1 - \frac{1}{1-k/k_o}\right)\hbar c\, dk \tag{25}$$

$$\equiv \delta m + \Delta_L$$

where we have followed common notation conventions. These terms are divergent: the linearly divergent δm term represents that part of the shift independent of binding and will be present for a free electron; it is concealed by renormalization of the electron's mass. The binding-dependent logarithmically divergent integral Δ_L is the Bethe part of the Lamb shift; it is usually cut-off on physical grounds [5] at $mc/\hbar$.

It is not immediately obvious how these divergent usual results relate to the finite shift Δ_c. Let us follow the Jaynes-Crisp prescription and include retardation, then write Δ^+ in our previous notation as

$$\hbar\Delta^+ = -\frac{\omega_o^2\mu^2}{(2\pi)^2c^3}\cdot\frac{8\pi}{3}\,P\int_0^\infty\left(1-\frac{\omega_o}{\omega_o-\omega}\right)\left(1+\frac{4}{9}\frac{\omega^2a^2}{c^2}\right)^{-4}d\omega\,. \tag{26}$$

(The extra factor $[1+(4/9)\omega^2a^2/c^2]^{-4}$ comes from the inclusion of retardation in the matrix elements, and we have converted matrix elements of $\underline{p}$ to those of the dipole moment $\underline{\mu} \equiv e\underline{r}$). So

$$\hbar\Delta^+ = -\frac{2}{3}\frac{\omega_o^2\mu^2}{\pi c^3}\;\frac{3c}{2a}\;P\int_0^\infty\left(1-\frac{x_o}{x_o-x}\right)(1+x^2)^{-4}\,dx \tag{27}$$

where $x_o = 2a\omega_o/3c$. Using the fact that $x_o \sim 10^{-3} \ll 1$, we find

$$\hbar\Delta^+ = -\frac{\omega_o^2\mu^2}{\pi c^2 a}\left[\frac{5\pi}{32}-\frac{11}{12}x_o + x_o\ln\left(\frac{1}{x_o}\right) + O(x_o^2)\right]. \tag{28}$$

Thus due to the strong convergence supplied by retardation, [6] Δ^+ is now finite.

At this point two comments are required: (1) We know that the RWA has eliminated from consideration any integrals like (26), but with *anti*-resonant denominators $(\omega+\omega_o)^{-1}$. We have seen that the resonant integrals have been cut off by retardation at $\omega\sim c/a\sim 10^3\omega_o$. Clearly if frequencies a *thousand times* greater than ω_o are important, we can't continue to ignore $2\omega_o$ contributions. The RWA is without justification in a level shift calculation and must be removed. We will do so promptly. (2) Note that the dominant term in the convergent Δ^+, the analog of δm in (25), is just

$$\Delta^+ \Rightarrow -\frac{5\omega_o^2\mu^2}{32ac^2\hbar} = -\frac{1}{2}\Delta_c, \tag{29}$$

exactly half the Jaynes-Crisp shift. In other words Δ_c seems to be most closely related to the part of the perturbative level shift which *does not* conventionally show up in the physical end result, due to renormalization. That this is so in our case too is easily verified by abandoning the RWA, calculating Δ^-, and finally taking the difference $\Delta^+ - \Delta^-$. The Δ_c parts cancel, and we are left with:

$$\delta\omega_{\text{pert.}} = \Delta^+ - \Delta^- = -\frac{A}{\pi}\left[\ln\frac{3c}{2a\omega_o} - \frac{11}{12}\right] + O(x_o^{\,2}). \tag{30}$$

The failure of our integration of the Heisenberg equations in Sec. (3) to give a frequency shift in agreement with (30) is interesting to consider. It is not due to a basic flaw in the method, but to our adoption of a conventional and superficial and incorrect indentification of the positive and negative frequency parts of the field. The equation (14) should be replaced by

$$\tilde{A}^{\pm} \simeq A_o^{\pm} \pm 2i\Delta_c\sigma^{\mp}(t) - A\sigma^{\mp}(t) \mp i\{\Delta_c + \delta\omega_{\text{pert.}}\}\sigma_1 \tag{31}$$

where retardation has been retained, the RWA abandoned, and terms up to $O(x_o^{\,2})$ kept in the evaluation of integrals. This expression, which one may obtain directly from the Hamiltonian given in (1) by use of the canonical commutation rélations for the electromagnetic field, then leads to a solution like that in (16), but with Δ_c replaced by $-\delta\omega_{\text{pert}}$. Both our approach, and the usual perturbative analysis, give the same quantum electrodynamic frequency shift.

5. Discussion and Conclusions

We have shown that Jaynes' program to obtain atomic level shifts and widths by integrating directly the semiclassical equations of motion, without resorting to perturbation theory, may be taken over to quantum electrodynamics fairly simply. A casual analysis of those QED equations (based on common assumptions, including the RWA) which correspond to the Jaynes-Crisp-Stroud equations, gives a frequency shift for the two-level transition identically equal to the Jaynes-Crisp shift. The shift is not time dependent however.

A comparison with the usual perturbative QED result for the level shift (whose difference is the frequency shift in question) then is seen to pose a number of questions. First of all, the perturbative shift is *not* equal to the Jaynes-Crisp shift. Second, one sees clearly that the effective cut-off provided by retardation is so high as to invalidate the RWA. Third, the closest correspondence to the Jaynes-Crisp shift formula is provided by the mass renormalization term in QED, just that term usually considered to have *nothing to do with* the transition frequency shift.

The conflict between the perturbative and non-perturbative QED results is resolved by giving up the Jaynes-Crisp approximations (RWA, etc.). A correct identification of positive and negative frequency parts of the field operators shows that the shift obtained from the QED equation-of-motion approach agrees with the perturbative approach, and is essentially Bethe's result, albeit in the context of a fictitious two-level atom.

After all this is said, however, it must be recognized that Jaynes and co-workers are consciously constructing a new theory, not merely trying to do quantum electrodynamics differently. Thus the usual calculational procedures, and the usual interpretation of the various terms calculated need have no relevance for them. Apart from comparison with experiment, the question of internal consistency of the theory becomes the only criterion for judgement. It seems to us that the use of the RWA is an important inconsistency. Apart from that objection, renormalization appears to raise the most urgent unanswered questions bearing on both the internal consistency of neoclassical theory, and on the magnitude of calculated level shifts.

It will be essential in any full appraisal of neoclassical theory to know at least whether neoclassical theory subscribes to Kramers' dictum that *all* theories are subject to renormalization of their "bare" component parts. It is well to keep in mind that the force of this dictum is *independent* of whether the "bare" quantities are finite or infinite. To be specific, for example, does the finite neoclassical level shift still contain a neoclassical free-electron part, presumably different from the δm part of (25), to be subtracted out in future refinements of the calculation? Perhaps it is still too early to expect detailed answers to questions of this kind.

References and Footnotes

1. M.D. Crisp and E.T. Jaynes, Phys. Rev. *179*, 1253 (1969); C.R. Stroud, Jr., and E.T. Jaynes, Phys. Rev. *A1*, 106 (1970).

2. We have directly verified by explicit calculation that all of the usual results of QED (including the divergent behavior of the level shift) are obtained using the present method but neglecting retardation. However, retardation is the crucial factor in deriving the QED analogue of the Jaynes-Crisp level shift.
3. The integrals occurring in the evaluation of $A^{\pm}$ are of the same form as that in equation (26) and are performed using the same approximations. Such a derivation closely follows those of reference 1.

4. E.A. Power, *Introductory Quantum Electrodynamics* (American Elsevier Publishing Company, New York, 1965).

5. This is justified by the reduction in the degree of divergence found using a relativistic treatment, since at $\hbar ck \gtrsim mc^2$, pair states remove the high energy contribution to Δ_L which then becomes convergent; for the same reason δm becomes logarithmically divergent.

6. A result apparently well known in the older literature due to Waller shows that if one takes the non-relativistic formalism seriously and includes retardation and recoil energies, the free electron self-energy diverges only logarithmically. It is known that the corresponding Δ_L for a real hydrogen atom converges; Lamb has pointed out that this effectively cuts the integrals off at twice Bethe's cut-off and disagrees with experiment.
I. Waller, Zeits. fur Phys. *62*, 673 (1930).
N.M Kroll and W.E. Lamb, Jr., Phys. Rev. *75*,388 (1949).

7. It is amusing, but without physical significance, that if the RWA is imposed consistently on (31) then the (incorrect) result found by using the RWA in perturbation theory, $\delta\omega_{RWA}=\Delta^+$, is duplicated by our Heisenberg equation result.

CRITICAL COMMENTS ON RADIATION THEORY

W.E. Lamb, Jr.

Yale University, New Haven, Conn.

A number of theoretical alternatives to quantum electrodynamics have been proposed in recent years, and are discussed at this Conference. They include theories in which the radiation field is treated as a classical c-number field, and others in which the field is an operator function of its sources. These alternatives are critically examined.

COOPERATIVE PHENOMENA IN RESONANT PROPAGATION

E. Courtens

IBM Research Laboratory, Zürich, Switzerland

The cooperative emission of two atoms constrained to radiate along the line joining them can be calculated exactly. [1] One finds that for an atomic separation smaller than $\sim c/2\gamma$ the atoms radiate cooperatively, whereas for a greater separation they essentially radiate as if they were alone. This rather academic example can serve to introduce the notion of cooperation in resonant emission. The solution also emphasizes that a *multi-mode* quantum mechanical treatment is required to arrive at the proper answer. The notions of *maximum cooperation time* and *maximum cooperation number*, valid for a large assembly of radiating atoms, can however be arrived at on the basis of perturbation theory, provided a self-consistent argument is used. [1] The cooperation time fixes an *upper* limit to the duration of superradiant emission. It is the emission time when the superradiant system is essentially emitting in the absence of an applied field, as is the case for a "long" 2π-pulse, i.e., one for which the velocity of propagation v is much smaller than c/n. The maximum cooperation number is the maximum number of atoms that can possibly contribute to a *coherent* superradiant pulse.

These definitions can be extended to systems of finite extent, to inhomogeneous lines, and to systems with a broadband distributed loss. The quantum mechanical results are then found to be in entire agreement with those of semiclassical calculations. [1] For instance, one finds that the duration of π and 2π-pulses is never longer than an appropriately calculated maximum cooperation time. Also the spatial extent of these pulses is always shorter or equal to the maximum cooperation length. These considerations allow some interesting predictions for the time evolution of a system of very large extent initially pumped to the upper state. One expects that the system will exhibit a spiking emission, and will divide into random superradiant regions whose average length will be the cooperation length.

As another case of interest, a semiclassical calculation of coherent Raman propagation in the presence of both laser and Stokes radiation has been developed. [2] If one assumes that the frequency separation between the laser and the transition from the ground state to the intermediate state of the Raman process is sufficiently large (so that the intermediate state population can be neglected), but also sufficiently small (so that higher-order Stokes processes are much weaker than the first-order Stokes process), such coherent propagation can occur. The conditions for the theory to apply can be satisfied by many systems, in particular by the well-known K-vapor-ruby laser- and ruby laser Stokes from nitrobenzene system, in which propagation effects have recently been observed. [3] The equations describing the coupled system can be cast into quite an elegant form which is very similar to the Bloch-Maxwell equations. One of the results of the theory is that no steady-state solution exists for which both the laser and the Stokes fields are in the form of pulses.

References

1. F.T. Arecchi and E. Courtens, Phys. Rev. *2A*, 1730 (1970).

2. E. Courtens, in *The Laser Handbook* F.T. Arecchi and E.O. Shulz-Dubois, eds.,(North Holland Publ. Company),to appear in the fourth quarter of 1972, ch. E. 5, Section 5.4.

3. N. Tan-no, K. Yokoto and H. Inaba, reported at the 7th International Quantum Electronics Conference, Montreal, Canada, May 1972, paper P. 7.

THEORY OF SUB-COOPERATION-LIMIT OPTICAL PULSES IN RESONANT ABSORBERS*

Joseph H. Eberly†

Stanford University, Stanford, Calif.

It has been known for some time that the interaction of spins or two-level atoms with a classical radiation field can depend in an interesting way on the existence of a certain coherence time τ_a, defined by [1,2]

$$\frac{1}{\tau_a^2} = \frac{\pi}{2} N \hbar \omega \kappa^2 . \tag{1}$$

Here N and ω are the density and transition frequency of the atoms or spins, and $\hbar\kappa/2$ is the dipole matrix element of the transition. Following Arecchi and Courtens,[1] I will call τ_a the "cooperation time" for the system of atoms.

For my purposes here, the significance of τ_a is that it establishes a three-part time scale for the interaction of atoms and pulses of radiation. Let's see how that happens. Clearly a pulse might be either longer or shorter than τ_a itself. But the mere coexistence of a lot of atomic dipoles, necessary for τ_a to be defined, necessarily also implies dipole-dipole forces which will act to interrupt dipole-pulse coherence. The associated incoherent interruption time T is easily shown to be related to τ_a:

---

*Research supported by the National Science Foundation and U.S. Atomic Energy Commission.

†Permanent Address: Department of Physics and Astronomy, University of Rochester, Rochester, New York

$$T = \omega\, \tau_a^2 . \tag{2}$$

Thus there are three pulse length regimes to consider: Pulses longer than both τ_a and T, pulses shorter than both, and pulses with intermediate lengths.

It is easy to show[1] that pulses of the sort encountered in self-induced transparency[3] must be of the intermediate variety. That is,

$$\omega\, \tau_a^2 > \tau_{sit} > \tau_a \tag{3}$$

must be satisfied by s.i.t. pulse lengths τ_{sit}.

Now we all understand that unless some Maxwell demon can be recruited to organize the chaotic dipole-dipole interactions, *all* coherent pulses, not just those of the s.i.t. type, will have to be shorter than $\omega\tau_a^2$. It is the *right-hand* inequality in (3) that interests me here. Let me begin by writing a more extended chain of times:

$$\omega\tau_a^2 > \tau_{sit} > \tau_a > \tau_{scl} > \omega^{-1} , \tag{4}$$

where τ_{scl} is the length of a hypothetical "subcooperation-limit" pulse. The last two inequalities make it clear that I imagine such subcooperation-limit pulses to be shorter than the co-operation time τ_a, but also long enough to contain a number of cycles of the carrier.

In fact, in order to sustain my two-level atom assumption, I must require the s.c.l. pulses to satisfy

$$\tau_a > \tau_{scl} >> \omega^{-1}. \tag{5}$$

In other words, I imagine subcooperation-limit pulses which are nevertheless slowly-varying pulses.

The first point to settle is the size of the domain left open for s.c.l. pulses. In order for an s.c.l. pulse to be slowly varying, containing at least hundreds of cycles of the carrier, it is necessary that $\tau_{scl} \gtrsim 0.1$ psec. The remaining question is the size of τ_a. As it turns out,[2,4] for many common resonant optical absorbers, (ruby and alkali metal vapors, for example), $\tau_a \sim 0.1$ nsec.

Thus one finds a three-order-of-magnitude range of pulse lengths available for slowly-varying subcooperation-limit pulses:

$$0.1 \text{ nsec} > \tau_{scl} > 0.1 \text{ psec} \quad , \tag{6}$$

a range comparable to that available for optical self-induced transparency pulses:

$$0.1 \ \mu\text{sec} > \tau_{sit} > 0.1 \text{ nsec} \quad . \tag{7}$$

Marth and I have taken the point of view that it should be interesting, and perhaps important, to understand these subcooperation-limit pulses. They are obviously within the ballpark of current experimental capability. We have approached a study of s.c.l. pulses with the idea of seeing in what ways they may be the same as, or different from, the familiar longer s.i.t. pulses which have been studied in great detail lately by Gibbs and Slusher[5].

Let me borrow some results from Marth's thesis research[4] to show some of the answers he has found to the questions we have raised. Let me mention first two temporary but important assumptions made. Since $\tau_{scl} < \tau_a$ by definition, and since $T_2^* \sim \tau_a$ in many systems of interest, he assumes all interesting pulse lengths τ are enough shorter than T_2^* to make inhomogeneous broadening superfluous for steady-state propagation. Furthermore, for simplicity here, we assume that the pulse carrier frequency and the common atomic resonance frequency are identical. (This latter assumption by itself is removed in Ref.4. As I shall mention below, the first assumption can also be removed.)

The important experimental quantities associated with steady-state pulses can be identified as follows. The electric field strength is written:

$$\underline{E}(t,z) = \mathcal{E}(t-z/V) \text{ Re } \{(\hat{x}+i\hat{y})e^{i\Phi(t,z)}\} \quad , \tag{8}$$

where $\mathcal{E}$ is the real steady-state amplitude, and the phase Φ is made up of carrier and pulse contributions:

$$\Phi(t,z) = \omega t - Kz + \phi(t-z/V) \quad . \tag{9}$$

Because of the strong field-atom interaction, the pulse velocity V may be quite different from c, and the wave vector K may differ from ω/c. Note that the instantaneous frequency $d\Phi/dt$ is allowed to differ from ω; that is, $\dot{\phi}$ may be a function of time, leading to the possibility of chirping.

In Figs. 1 and 2 (taken from Ref.4) are shown V and $\dot{\phi}$ as a function of pulse length τ for different optically resonant absorbers. The velocity curve is relatively unexciting, since when $\tau < \tau_a \sim 10^{-10}$ sec one sees that $V \sim c$. This is an important

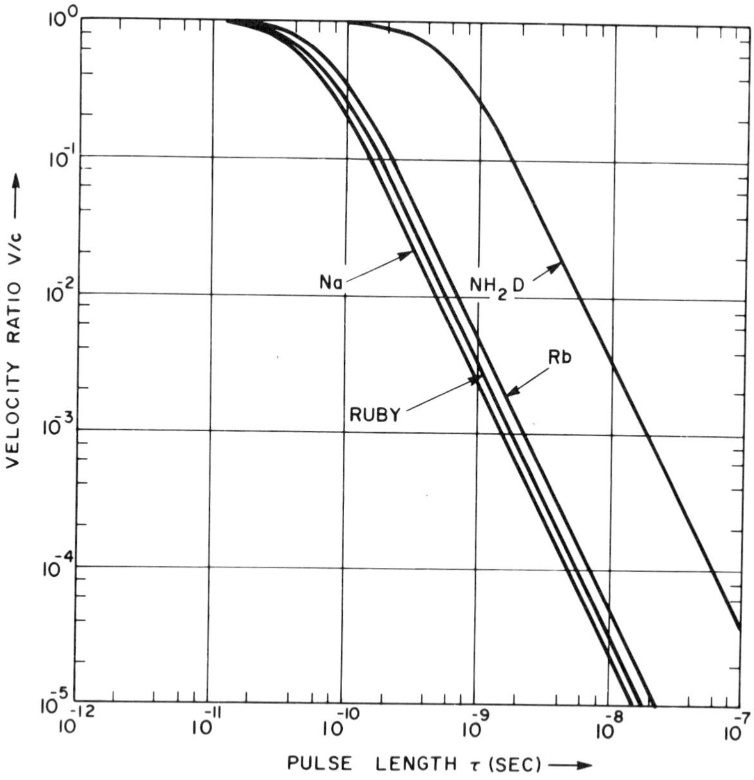

Fig.1. A plot of the velocity ratio , V/c, versus the pulse length, τ, for different media. The curves correspond, left to right respectively, to Na vapor, ruby, Rb vapor, and NH_2D.

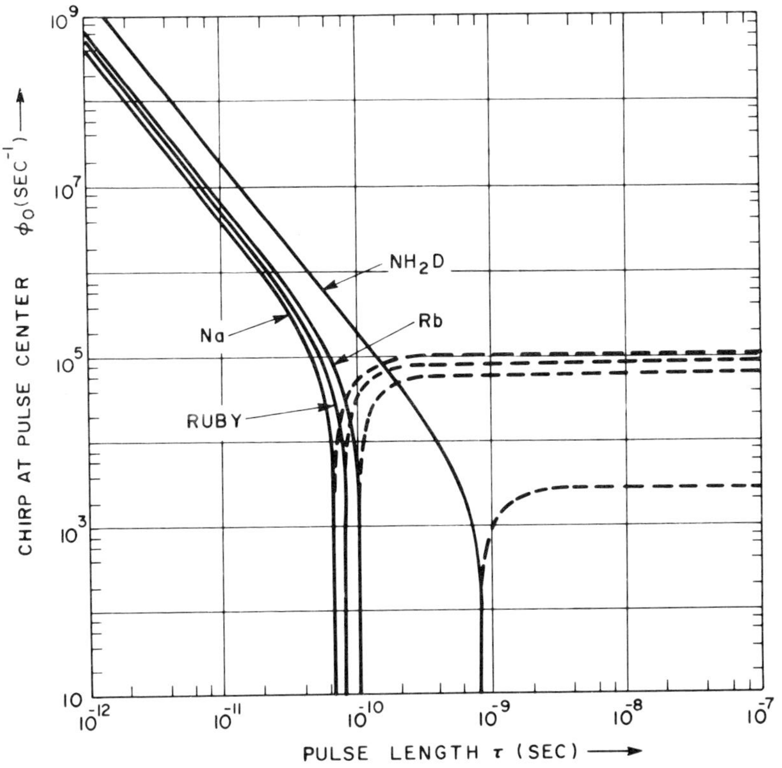

Fig. 2 A plot of the chirp at pulse center $|\dot{\phi}_0|$, versus the pulse length, τ, for different media. The curves correspond, left to right respectively, to Na vapor, ruby, Rb vapor, and NH_2D. The solid portions of the curves indicate positive values of $\dot{\phi}_0^2$ while the dashed portions correspond to negative values.

result, however, in two respects. First, it is obviously quite different from the very slow velocities associated with s.i.t. Second, it ensures that our neglect of back-scattering is an excellent approximation[6]. It also means that the pulse is not sharing large fractions of its energy with the atoms.

The chirping curve is obviously something new. We see that *all* pulses, not just s.c.l. pulses, are chirped. In fact, one can easily find from Marth's results[4] a very simple formula for the chirp at pulse center:

$$\dot{\phi}_0 \sim \begin{cases} 3/2\omega\tau^2, & \tau < \tau_a \\ -3/2\omega\tau_a^2, & \tau_a < \tau < \omega\tau_a^2 \\ -3/4\tau, & \tau > \omega\tau_a^2 \end{cases} \tag{10}$$

There are no previous quantitative results on chirping in absorbers. As Slusher and Gibbs point out[5], even estimates of the importance of chirping (especially for s.c.l. pulses) have been difficult to make.

Of course the results for $\tau > \tau_a$ (i.e., for non-s.c.l. pulses) given in (10) and Figs. 1,2 are to be ignored because inhomogeneous broadening is quite important for such pulses. However, the s.c.l. regime's expression can be taken seriously, and it imparts new information. In the first place, it says that $\dot{\phi}_0\tau \ll 1$ since $\omega\tau \gg 1$. That is, the chirp definitely can never be large enough to affect the bandwidth of the pulse. On the other hand, the chirp is certainly measureable interferometrically, at least in principle. The accumulated phase shift over a one-foot propagation length is about $\pi/2$ for a 1 psec pulse.

Of course, when chirping is present there is no area theorem available to guide the area of the steady-state pulses. In Fig.3, also taken from Ref. 4, one can see how little the relation between $1/\tau$ and the maximum of the envelope is changed by chirping.

Now, having shown some of the results, I must sketch the theory. Because the s.c.l. pulses are very slowly varying, the two-level model is a good one for our atoms. Thus the optical Bloch equations suffice:

$$du/dt = \dot{\phi}v \tag{11}$$

$$dv/dt = -\dot{\phi}u + \kappa\mathcal{E}w \tag{12}$$

$$dw/dt = -\kappa\mathcal{E}v \tag{13}$$

Notice that chirping has been allowed for explicitly, but not detuning, and that incoherent relaxation processes have been ignored as unimportant (for pulses sufficiently short compared to $\omega\tau_a^2 = T$). In the familiar way, Bloch's equations must be made compatible with Maxwell's equations, but we'll ignore that temporarily.

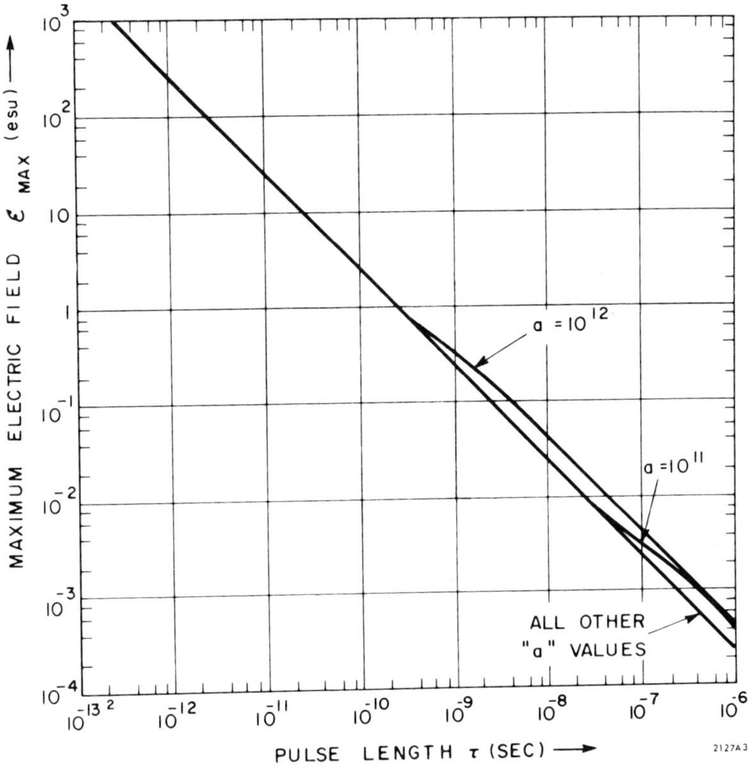

Fig. 3 A plot of the maximum electric field envelope, $\mathcal{E}_{max}$, versus the pulse length, τ, for several values of $a \equiv \tau_a^{-1}$. The curves use $\omega = 2.35 \times 10^{15}$ sec^{-1}, $\rho = 4.35 \times 10^{-18}$ esu-cm and $\gamma = 0$.

The theory begins by letting the *Bloch* equations say what the *field* must be, as follows: First, we recognize that $\kappa\,\mathcal{E}/\omega$ must be very small for slowly varying pulses. Let us denote this ratio of Rabi frequency to carrier frequency by ρ. Then we assert that

$$\dot{\phi} = \sum_{m=0}^{\infty} \chi_m \rho^{2m} \tag{14}$$

$$w = \sum_{m=0}^{\infty} w_m \rho^{2m} \quad . \tag{15}$$

There is some support for each assertion, but we won't belabor that point here.

The point is that Eqs.(11)-(15) can be combined to yield (to lowest interesting order):

$$\rho(t,z) = \rho_0 \text{ sech } \left(\frac{t - t_0}{\tau}\right) \tag{16}$$

as well as

$$\begin{aligned} w_0 &= -1 \\ w_1 &= \frac{1}{2}(\omega\tau)^2 \\ w_2 &= \frac{1}{6}(\omega\tau)^2 \left[\rho_0^{-2} - (\omega\tau/2)^2\right] . \end{aligned} \tag{17}$$

It is now the task of the *Maxwell* equations to pin down the various *pulse parameters*, such as maximum amplitude ρ_0, velocity V, chirp coefficients χ_0, χ_1, ..., etc. As I implied at the beginning, Maxwell's equations are up to the task, and among the results are those shown graphically in Figs. 1-3.

At this point let me recall my earlier remark that it *is* possible to accommodate inhomogeneous broadening into the power series scheme of Eqs.(14) and (15). It is amusing that this is the case, because I believe it is the first time that inhomogeneous broadening and chirping have coexisted self-consistently within a steady-state pulse theory for either amplifiers or absorbers.

Although the inhomogeneous broadening results will form the core of another paper[7], let me at least give here the results corresponding to Eqs.(17) to show how naturally a detuning frequency γ enters into the theory:

$$\begin{aligned} w_0(\gamma) &= -1 \\ w_1(\gamma) &= \frac{1}{2}(\omega\tau)^2[1 + (\gamma\tau)^2]^{-1} \\ w_2(\gamma) &= \frac{3}{2}(\omega\tau)^2 \frac{\rho_0^{-2} - (\omega\tau/2)^2 + \frac{2}{3}(\gamma\tau)(\chi_1\tau)}{[1 + (\gamma\tau)^2][9 + (\gamma\tau)^2]} . \end{aligned} \tag{18}$$

The $w_0(\gamma)$ and $w_1(\gamma)$ expressions are identical to those derived in the theory of self-induced transparency[3]. Furthermore a 2π unchirped pulse has $\rho_0 = 2/\omega\tau$ and $\chi_1 = 0$, leading to $w_2(\gamma) = 0$, again in agreement with well known s.i.t. results[3].

A few concluding remarks are in order. I have sketched a theory of pulses shorter than τ_a. Several results can be emphasized: (1) Slowly-varying steady-state pulses shorter than the cooperation limit are predicted to exist; (2) s.c.l. pulses travel at velocities on the order of c; (3) all s.c.l. pulses are chirped; (4) the chirp can never be large enough to contribute to the pulse bandwidth, but may be detectable interferometrically; and (5) inhomogeneous broadening can be added to the theory relatively easily.

Finally, a remark about the theory itself: Because the presence of chirping demolishes the s.i.t. area theorem[3], the usual identity of pulse area with dipole turning angle is invalid. The usual theories of steady-state pulses[3], based upon exploitations of this identity, are not very helpful guides to a chirped-pulse theory. We have used a new approach, embodied in the expansions (14) and (15), which is closer to perturbation theory in spirit.

ACKNOWLEDGEMENTS

I am pleased to acknowledge an enormous number of discussions and arguments with R. A. Marth on the subject of this paper. Occasional conversations with H. M. Gibbs, E. L. Hahn, S. L. McCall, and R. E. Slusher have also tended to keep me honest. In addition, I must express my appreciation for the hospitality of Profs. S. D. Drell and A. L. Schawlow at Stanford during 1971-72.

REFERENCES

1. There is a very nice semi-quantitative discussion of the significance of τ_a due to F. T. Arecchi and E. Courtens, Phys. Rev. *A2*, 1730 (1970). See also A. I. Alekseev *et al.*, JETP *19*, 220 (1964).
2. L. Davidovich in "Electromagnetic Interactions of Two-Level Atoms," J. H. Eberly, Ed.,Proc. 1970 Rochester Symposium (University of Rochester, January 1970), p.65 (unpublished); and L. Davidovich and J. H. Eberly, Optics Comm. *3*, 32 (1971).
3. S. L. McCall and E. L. Hahn, Phys. Rev. *183*, 457 (1969). See the review by G. L. Lamb, Jr., Rev. Mod. Phys. *43*, 99 (1971)

for many other references.
4. R. A. Marth, Ph.D. Thesis, Dept. of Electrical Engineering, Polytechnic Institute of Brooklyn (unpublished) (1972).
5. R. E. Slusher and H. M. Gibbs, Phys. Rev. *A5*, 1634 (1972). See also the paper by Slusher to this conference, *Proc. of the Third Rochester Conference on Coherence and Quantum Optics*, (Plenum Press, New York, 1973) p. 3, this volume.
6. This can be verified directly by adapting suitably a result in M. D. Crisp, Optics Comm. *1*, 59 (1969). Arguments leading to qualitatively similar conclusions are given by E. Courtens and A. Szöke, Phys. Letters *28A*, 296 (1968).
7. J. H. Eberly and R. A. Marth (unpublished).

MULTIPLE PULSE CHIRPING IN SELF-INDUCED TRANSPARENCY*

Ljubomir Matulic

St. John Fisher College, Rochester, New York

Joseph H. Eberly†

Stanford University, Stanford, California

The interest in phase modulation of short optical pulses has continually increased in the last few years[1]. In this note we will restrict ourselves to the first order theory of steady-state pulses. That is to say, we will work in the common "slowly varying" approximation according to which the envelope and the phase of the pulse vary very little within an optical cycle[1a,c,d]. This approach leads to the prediction that the single optical pulses (solitons) propagating in linear absorbers must show no phase modulation[2]. It predicts also the possible existence of a variety of non-chirped as well as chirped *pulse trains*[3]. Here we will present a unified view of this problem.

Let us remark that the theory which makes no use of the slowly varying approximation shows that even the single pulses are phase modulated[4]. However, the mathematical complexities inherent in the dynamics of phase modulated pulses makes the treatment of multiple pulses very difficult without the benefit of the slowly varying approximation.

We are interested in steady state pulses traveling through nonlinear absorbers without distortion and without loss of energy

*Research partially supported by the National Science Foundation.

†Permanent Address: Department of Physics and Astronomy, University of Rochester, Rochester, New York.

in the spirit of McCall and Hahn self-induced transparency[1a]. The nonlinear absorbers consist of a passive background material (which may be just the vacuum) in which active atoms, resonant with the light, are suspended.

We will represent the electric field of the pulse as a product of an envelope and a carrier in the following way:

$$\underline{E}(z,t) = \varepsilon(z,t)[\hat{x}\cos\Phi(z,t) + \hat{y}\sin\Phi(z,t)], \qquad (1)$$

$$\Phi(z,t) = \omega t - (k + \Delta k)z + \phi(z,t)\,. \qquad (2)$$

The real envelope $\varepsilon(z,t)$ and the real phase function $\phi(z,t)$ are slowly varying functions of z and t, changing little over many wavelengths $2\pi/k$ and periods $2\pi/\omega$. We have set $k = \eta_0\omega/c$, where η_0 is the index of refraction of the host medium, ω the carrier frequency and c the velocity of light in free space. We will neglect the dispersion of the host medium[3,5]. The dispersive effects arising in the interaction between the light and the resonant atoms are accounted for by Δk[3].

The interaction of the pulse in Eq. (1) with the resonant two level atoms is described by the hamiltonian

$$H = H_0 - \underline{P}\cdot\underline{E}\,, \qquad (3)$$

where H_0 is the atom's unperturbed hamiltonian and $\underline{P}$ its electric dipole moment. We will assume that the transition frequencies of these atoms are inhomogeneously broadened, exhibiting a distribution about the field carrier frequency ω described by $g(\gamma)$, with $\int_{-\infty}^{\infty} g(\gamma)d\gamma = 1$. The parameter $\gamma = \omega_0 - \omega$ represents the detuning of one atom's frequency ω_0 relative to the carrier frequency ω.

We will limit ourselves to the discussion of the propagation of undistorted, shape-preserving pulses, for which the envelop and phase functions ε and ϕ depend on their arguments z and t only through the local time variable $\zeta = t - z/V$. V is the velocity of these shape-preserving pulses. The components of the Bloch vector representing the two level resonant atoms, u, v and w also depend only on ζ.

The dynamical equations of motion for the undistorted pulses of Eq. (1) interacting with the atoms described by the hamiltonian of Eq. (3) can be written in the following relatively simple way[1a]:

$$\dot{u} = -(\gamma - \phi)v \qquad (4)$$

$$\dot{v} = (\gamma - \dot{\phi})u + \kappa\,\varepsilon\, w \qquad (5)$$

$$\dot{w} = -\kappa\ \varepsilon\ v \tag{6}$$

$$\dot{\varepsilon} = -(1/m^2) \int_{-\infty}^{\infty} v(\zeta,\gamma) g(\gamma)\ d\gamma \tag{7}$$

$$(\dot{\phi}+\Delta k/\delta)\varepsilon = (1/m^2) \int_{-\infty}^{\infty} u(\zeta,\gamma) g(\gamma)\ d\gamma\ . \tag{8}$$

The notation is the same as in Ref. 3. In particular $1/m^2 = \pi N\hbar\omega/c\eta_0\delta$, $\delta = 1/V - 1/c_0$ and $\kappa = 2p/\hbar$.

The existence of solutions of this system of nonlinear coupled differential equations is obviously of great interest from a physical as well as a mathematical point of view. We will exhibit the general solution (general in the sense that we will make precise presently) of this quantum system, and we will see that it represents a variety of multipulse trains as well as the single hyperbolic secant pulse of McCall and Hahn. We will be especially interested, however, in the study of multiple pulse trains.

The difficulty in solving the system Eqs. (4)-(8) lies in the fact that the pulse $\underline{E}$ is supported by an inhomogeneous atomic line, $g(\gamma)$. To arrive at analytic solutions we will make the so called "factorization assumption", according to which one assumes that the absorptive component of the atomic polarization may be written as

$$v(\zeta,\gamma) = F(\gamma)\ v(\zeta,0), \tag{9}$$

with an implied normalization $F(0) = 1$. To the best of our knowledge, only under this special assumption[1a,c], or its equivalent[1d], have analytic solutions been found for arbitrary $g(\gamma)$. [See, however, J. H. Eberly, Ref. 4.] Having in mind this restriction, the solution of the system (4)-(8) which we shall derive here is its general solution.

The basic tactic in solving a system of coupled nonlinear differential equations such as ours is to obtain a sufficient number of first integrals, or conservation laws, which enable us to reduce it to a quadrature. Thus, from the Eqs. (4), (5) and (6) we obtain immediately the following first integral

$$u^2 + v^2 + w^2 = 1, \tag{10}$$

which states that the Bloch vector representing the atom is of unit magnitude.

Using Eqs. (6), (7) and (9) a second first integral is derived:

$$\varepsilon^2 = (2/\mu^2 F)(w - w_0). \tag{11}$$

Here $1/\mu^2 = (1/m^2) \int_{-\infty}^{\infty} F(\gamma)g(\gamma)d\gamma$ and $w_0 \leq 1$ is an integration constant.

Differentiating (8) and using (4), (7) and (9) we obtain the following important relation between the envelope of the pulse and the modulation of its phase[3,6]:

$$2\dot{\phi}\dot{\varepsilon} + \ddot{\phi}\varepsilon = (\bar{\gamma} - \Delta k/\delta)\dot{\varepsilon}, \tag{12}$$

where $\bar{\gamma} = \int_{-\infty}^{\infty} \gamma F(\gamma)g(\gamma)d\gamma / \int_{-\infty}^{\infty} F(\gamma)g(\gamma)d\gamma$.

Equation (12) is a first order differential equation in $\dot{\phi}$ which can be integrated at once to yield

$$\dot{\phi} = \frac{1}{2}(\bar{\gamma} - \frac{\Delta k}{\delta}) + \frac{C_1}{\varepsilon^2}, \tag{13}$$

C_1 being a new integration constant.

We now determine Δk in such a way as to make the constant part of $\dot{\phi}$ vanish so that ω in (2) is the part of the carrier frequency independent of the pulse envelope. Therefore, we have to have

$$\Delta k = \bar{\gamma}\delta \tag{14}$$

and

$$\dot{\phi} = C_1/\varepsilon^2. \tag{15}$$

Thus, the chirp in first order self-induced transparency, when it is present, must be inversely proportional to the light intensity[1d,2,7a,b]. Since only when $C_1 \neq 0$ do we have non-trivial phase modulation, we will refer to C_1 as the chirping constant.

Substitution of Eqs. (7) and (15) into Eq. (4) leads to another first integral

$$u = \frac{\mu^2 F\gamma}{\kappa}\,\varepsilon + \frac{C_1\mu^2 F}{\kappa}\,\frac{1}{\varepsilon} + C_3 \tag{16}$$

with an integration constant C_3.

Differentiating (7) and using (5), (11), (15) and (16) we find a nonlinear differential equation for ε

$$\ddot{\varepsilon} = -\frac{\kappa}{2}\,\varepsilon^3 - \left(\frac{\kappa^2 w_o}{\mu^2 F} + \gamma^2\right)\varepsilon + C_1^2\,\varepsilon^{-3} + \frac{C_1 C_3 \kappa}{\mu^2 F}\,\varepsilon^{-2} - \frac{C_3\kappa\gamma}{\mu^2 F}\,. \tag{17}$$

The left side of this equation cannot depend on γ, therefore its right side must also be γ-independent. Now if the quantity $C_3/\mu^2 F$ in the coefficient of ε^{-2}in Eq. (17) does not depend on γ, then there is no way of making the last term in (17) independent of γ. Moreover, it can be shown that there are no factorable solutions for which C_3 is different from zero. Therefore, without loss of generality we will set hereafter $C_3 = 0$. We can also show a posteriori that the γ dependence of the remaining coefficients in Eq. (17) properly cancels out. Thus, the solution of the system (4)-(8) obtained by us is the most general solution satisfying the factorization assumption, Eq. (9).

An integration of Eq. (17) followed by a multiplication by ε^2 yields

$$\varepsilon^2\dot{\varepsilon}^2 = \frac{\kappa^2}{4}\left(-\varepsilon^6 + M\varepsilon^4 + N\varepsilon^2 + Q\right), \tag{18}$$

where M and Q represent the coefficients in (17) in an obvious way, and N is a constant arising in the integration of (17). It is convenient to write $N = 4C_2/\kappa^2$.

It is now obvious that we have reduced our problem to a quadrature, since (18) can be written as

$$\int_{\varepsilon_0}^{\varepsilon} \frac{\varepsilon\, d\varepsilon}{\left(-\varepsilon^6 + M\varepsilon^4 + N\varepsilon^2 + Q\right)^{1/2}} = \frac{\kappa}{2}\,\zeta\,, \tag{19}$$

with the initial condition: $\varepsilon = \varepsilon_o$ when $\zeta = 0$.

When (19) is solved for ε, then (15) can be used to obtain ϕ while (7), (11) and (16) give v, w and u respectively. Finally, we employ (10) to determine the spectral response function $F(\gamma)$ in a self-consistent way[3].

It is obvious from Eq. (19) that the nature of the solutions for ε will depend on the values of the coefficients of the sixth degree polynomial $P_6(\varepsilon) = -\varepsilon^6 + M\varepsilon^4 + N\varepsilon^2 + Q$, which in turn depends on the integration constants C_1 and C_2. In this case the sixth degree polynomial reduces to a cubic in ε^2. The solutions will be elliptic functions, and their form will depend on the relative positions of the three zeros of the cubic polynomial with respect to 0. Although these solutions are superficially rather different[1d], we have shown[3] that all of them can be *described by a single analytic expression.*

Let us denote $\varepsilon^2 = S$, and let S_1, S_2 and S_3 be the roots of the cubic polynomial mentioned above. Then the steady state form $\varepsilon^2 = S$ is given by:

$$S(\zeta) = S_3[1 - \ell^2 \, \mathrm{sn}^2(\zeta/\tau;k)] \tag{20}$$

where

$$\frac{1}{\tau} = \frac{\kappa}{2}(S_3 - S_1)^{\frac{1}{2}} \tag{21}$$

$$k^2 = \frac{S_3 - S_2}{S_3 - S_1} \leq 1 \tag{22}$$

$$\ell^2 = \frac{S_3 - S_2}{S_3} \leq 1, \tag{23}$$

and we have assumed $S_1 < 0 < S_2 < S_3$.

Clearly, the character of these solutions will depend as much on the parameter ℓ, which we may call the chirping parameter, as on the modulus k. We will give the physical interpretation of both of them later on. We know that k satisfies the relation $k^2 \leq 1$. On the other hand, it is easy to show that the chirping parameter ℓ satisfies the following relation

$$k^2 \leq \ell^2 \leq 1 \quad . \tag{24}$$

Using (21) and (23) we obtain $S_3 = 4k^2/\ell^2\kappa^2\tau^2$, with the help of which (20) can be written as:

$$\varepsilon(\zeta) = \varepsilon_0 \left(\frac{k}{\ell}\right) [1 - \ell^2 \mathrm{sn}^2(\zeta/\tau;k)]^{\frac{1}{2}} \tag{25}$$

where $\varepsilon_0 = 2/\kappa\tau$. Equation (25) contains as a special case the envelopes of all undistorted pulses previously found[8], including both of Dialetis' instances of chirped pulse trains[1d]. In particular, when $\ell = k = 1$ the hyperbolic secant single pulse of McCall and Hahn results[1a].

All functions given by Eq. (25) are positive with period 2K[K=K(k) is the complete elliptic integral of the first kind]. Because of possible phase variation we still have the possibility of finding "zero-π" pulses even though ε is always positive according to our convention. We will have more to say about this question later on.

Corresponding to our solution (25) for the real field envelope, we obtain very complicated expressions for the components of the atomic polarization and for the energy of the atom. These will be published elsewhere[3].

Substituting (25) into (15) we obtain the most important relation for our present purpose, i.e., an expression for the chirp of the self-induced transparency pulses:

$$\dot{\phi}(\zeta) = \pm \frac{1}{\tau} \frac{[(1 - \ell^2)(\ell^2 - k^2)]^{\frac{1}{2}}}{\ell(1 - \ell^2 \mathrm{sn}^2)} \quad . \tag{26}$$

An integration of (26) yields the phase function itself:

$$\phi(\zeta) = \pm \frac{1}{\ell} [(1 - \ell^2)(\ell^2 - k^2)]^{\frac{1}{2}} \, \Pi(\zeta/\tau;\ell^2;k^2), \tag{27}$$

where $\Pi(u;\ell^2;k^2)$ is the incomplete elliptic integral of the third kind. We have set without loss of generality $\phi(0) = 0$.

We shall now discuss in some detail the functions (26) and (27). In the first place we see that $\dot{\phi} \equiv 0$ only when $\ell = k$ or $\ell = 1$, and these are the only unchirped multipulse solutions.

They have been discovered by Arecchi-DeGiorgio-Someda[8a], by Crisp[8b] and by Eberly[8c]. The shape of the envelope for these pulses can be obtained from our general expressions, Eq. (25). When $\ell = k$, we get $\varepsilon(\zeta) = \varepsilon_0\, \mathrm{dn}(\zeta/\tau;k)$, while for $\ell = 1$ we obtain $\varepsilon(\zeta) = \varepsilon_0\, k\, \mathrm{cn}(\zeta/\tau;k)$. All other multipulse trains exhibit chirping in greater or lesser degree.

In order to discuss further the phase modulation, we define the following two quantities:

$$M(\ell) = \frac{1}{\ell}\left(\frac{\ell^2 - k^2}{1 - \ell^2}\right)^{\frac{1}{2}} \tag{28}$$

$$m(\ell) = \frac{1}{\ell}\left[(1 - \ell^2)(\ell^2 - k^2)\right]^{\frac{1}{2}} . \tag{29}$$

Equation (28) gives the maximum and Eq. (29) the minimum of $|\tau\, \dot{\phi}|$ as function of the chirping parameter ℓ. Figure 1 is a plot of these functions for k = 1/4.

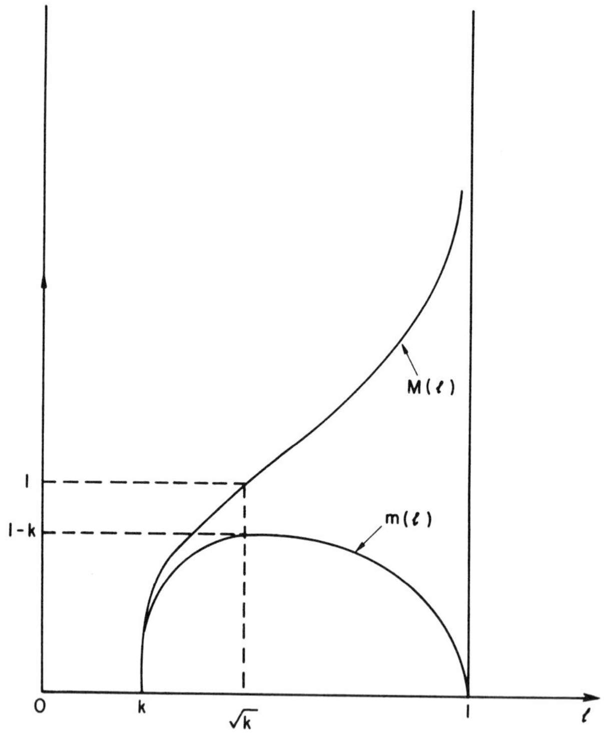

Fig. 1. Plot of $M(\ell)$ and $m(\ell)$ as a function of ℓ, for k = 1/4.

The function $M(\ell)$ is defined in the interval $[k,1)$ and increases monotonically from zero to infinity. For $\ell = 1$, $\dot{\phi}(\zeta) = 0$ for all ζ except for $\zeta = (2n + 1)K\tau$, at which points it has singularities. This is precisely the situation for which $\varepsilon(\zeta)$ vanishes. The $\ell = 1$ case is the only case in which the field envelope is not smooth, approaching zero with sharp cusps as $\zeta \to (2n + 1)K\tau$. See Fig. 2.

In order to elucidate further this behavior we calculate the change in ϕ, given by (27), during one period of oscillation from 0 to 2K. This change is given by

$$\Delta\phi = \pm \frac{2}{\ell} [(1 - \ell^2)(\ell^2 - k^2)]^{\frac{1}{2}} \Pi(K;\ell^2;k^2),$$

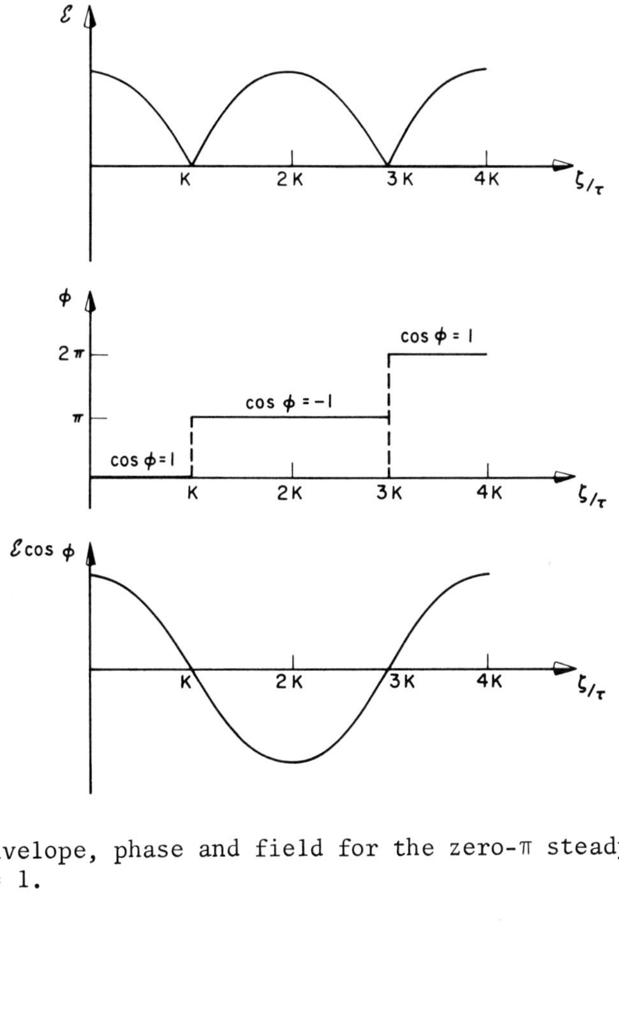

Fig. 2. Envelope, phase and field for the zero-π steady state pulses, $\ell = 1$.

and for $\ell = 1$ it reduces to $\Delta\phi = \pm\,\pi$. Then, when the field envelope has zeros ($\ell = 1$) there is no chirping strictly speaking (Eq. (26)), the phase staying constant during the whole period. The phase, however, changes abruptly by 180° at the points at which the envelope has its cusps, a possibility that has been suggested previously by computer studies of optical pulse propagation[9].

Thus we find ourselves able to investigate analytically one of the interesting phenomena of computer studies, namely "zero π" pulses, pulses in which the envelope becomes negative periodically, allowing thereby zero net area under the envelope. In our approach we can avoid the problems of interpretation associated with negative envelopes because we find none[10]. However, the real part, $\varepsilon \cos \phi$, of the *complex* envelope does become negative periodically, in just such a way that $\varepsilon \cos \phi$ is a smooth, physically realistic, function. Neither ε nor ϕ is always smooth, but no physical significance requires smoothness of either of them separately.

To see how this comes about, let us consider the x-component of the field Eq. (1)

$$E_x(T) = \varepsilon(T) \cos[\omega t + \phi(T)]. \qquad (30)$$

(For simplicity we have set $z = 0$ and $T = t/\tau$.) For $\ell = 1$, Eq. (30) reduces to

$$E_x(T) = \varepsilon(T) \cos\phi(T) \cos\omega t,$$

since $\phi(T)$ is a step function that increases in steps of π and we choose the origin of $\phi(T)$ at zero, i.e., $\phi(0) = 0$. Now, with reference to Fig. 2, we see that $E_x(T)$ and its derivatives are indeed continuous for all T, because the discontinuities in ε and ϕ counteract each other. Thus the imposition of the physical requirement of smoothness on the physical field strength $\varepsilon \cos \phi$ eliminates the need for the introduction of "negative envelope" solutions.

Finally, we show in Fig. 3 the graphs of the envelope ε, the "real field" $\varepsilon \cos\phi$ and the phase function ϕ for a near-critical case for which $\ell = 0.991$. The envelope is obviously positive, while the field goes negative at the appropriate values of ϕ ($\pi/2$, $3\pi/2$,..). It is interesting to notice how in this case the "slowly" varying envelope is modulated by an even more slowly varying function $\cos \phi$. Of course, inside the "envelope" $\varepsilon \cos\phi$ we must still imagine the fast oscillations of the factor $\cos\omega t$.

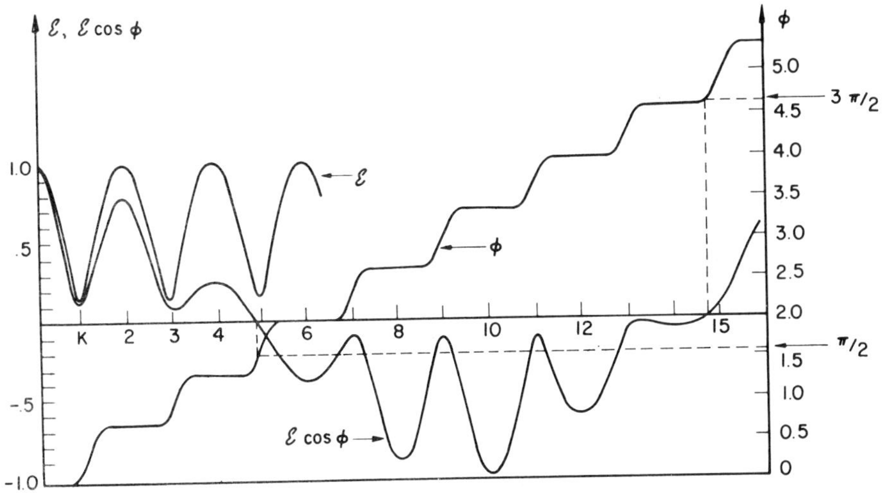

Fig. 3. Envelope, phase and "field" for k = 0.99 and ℓ = 0.991.

Fig. 4 shows the function $\Delta w(\gamma)$, which represents the variation of the energy of the atom as a function of its detuning

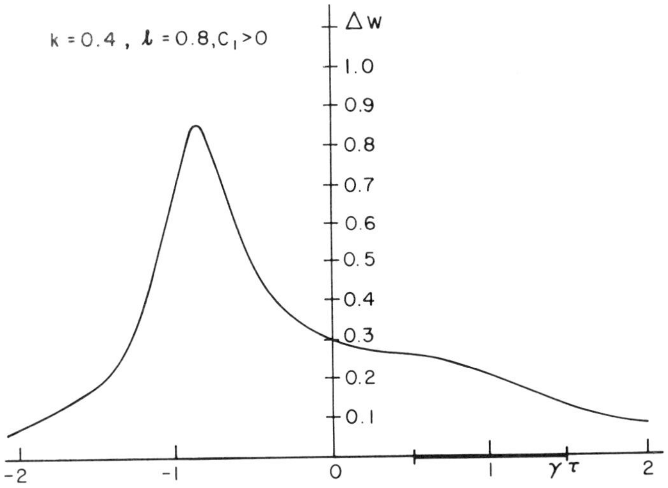

Fig. 4. The variation of the energy of the atoms, $\Delta w = w_{max} - w_{min}$, as a function of $\gamma\tau$, for k = 0.4 and ℓ = 0.8, and C_1>0. The heavy line on the $\gamma\tau$-axis represents the excursion of $\dot{\phi}\tau$.[1]

frequency γ, for a case when the chirping constant C_1 [Eq. (15)] is positive. We see that the atoms which are most strongly interacting with the field have their transition frequencies slightly below the carrier frequency. When the chirp has opposite sign ($C_1<0$), the atoms with the largest excursion in their energies are those with transition frequencies slightly above the field's carrier frequency. The pulse velocity V is affected by the sign of the chirping constant C_1, unless the atomic line shape $g(\gamma)$ is symmetric in γ.

We shall finally give a physical interpretation of the amplitude modulation parameter k and the phase modulation parameter ℓ. We shall also indicate a possible way of producing experimentally the amplitude and phase modulated trains in self-induced transparency.

Figure 5 shows a period, from $\zeta = -K\tau$ to $\zeta = K\tau$, of the pulse intensity $\varepsilon^2(\zeta)$, where $\varepsilon(\zeta)$ is given by (25). As shown in this

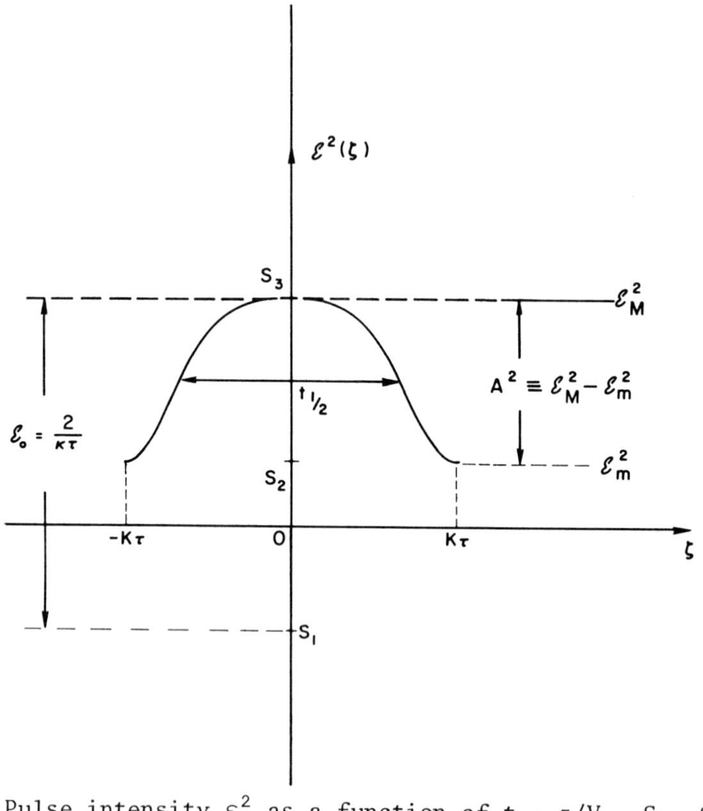

Fig. 5. Pulse intensity ε^2 as a function of t - z/V. S_1, S_2 and S_3 are the zeros of the cubic polynomial $P_3(S)$.

Figure we define

$$A^2 \equiv \varepsilon_M^2 - \varepsilon_m^2 \tag{31}$$

to be the amplitude of the modulation of the intensity function $\varepsilon^2(\zeta)$. Then, using Eqs. (21) to (23) we obtain

$$\begin{aligned} k &= A/\varepsilon_0 \\ \ell &= A/\varepsilon_M \\ \varepsilon_o &= 2/\kappa\tau \quad . \end{aligned} \tag{32}$$

In terms of these new physical parameters, A, ε_M and ε_0 (or τ), the envelope of our oscillatory pulses and their chirp (Eqs. (25) and (26)) can be written as follows:

$$\varepsilon(\zeta) = \varepsilon_M \; [1 - (A/\varepsilon_M)^2 \mathrm{sn}^2(\zeta/\tau; A/\varepsilon_0)]^{\frac{1}{2}} \tag{33}$$

$$\dot{\phi}(\zeta) = \frac{\kappa}{2\varepsilon_M} \; \frac{[(\varepsilon_M^2 - A^2)(\varepsilon_0^2 - \varepsilon_M^2)]^{\frac{1}{2}}}{1-(A/\varepsilon_M)^2 \mathrm{sn}^2(\zeta/\tau; A/\varepsilon_0)} \quad . \tag{34}$$

It is seen, therefore, that the nature of these oscillatory solutions depend on three physical parameters which, in principle, can be experimentally controlled: the amplitude of the intensity modulation A^2, the peak value ε_M of the pulse, and the pulse width $\tau = 2\kappa/\varepsilon_0$. These parameters are arbitrary except for the limitations implied by Eq. (24), namely:

$$A \leq \varepsilon_M \leq \varepsilon_0 .$$

Thus, if we want to excite only one of these waves, we must impose on a resonant medium the dynamical conditions characteristic of this special wave. Note that it is sufficient to control only the pulse envelope, since all the atomic variables as well as the phase are determined by k and ℓ, and thus by the envelope parameters A, ε_M and ε_o (or τ). In practice this might be done by using as a source a tunable cw dye laser whose output could be modulated, for instance with the help of a Pockels cell, the required frequency of this modulation being well within the realm of experimental capabilities (of the order of 10^9Hz if sodium vapor is used for the resonant medium). It also seems possible, once such steady state pulses are achieved, to detect directly their phase modulation by interferometric experiments.

Let us notice that when the peak value of the pulse is in the relation $\varepsilon_M = 2/\kappa\tau$ to its width we obtain the dn unchirped pulses, while when this peak value is equal to the amplitude modulation A, one gets the cn unchirped pulses. The solitary sech pulse results when $\varepsilon_M = A = 2/\kappa\tau$, being therefore characterized by a single free parameter.

The exact physical meaning of the parameter τ can be given by the transcendental expression

$$t_{\frac{1}{2}}/2\tau = sn^{-1}[(\tfrac{1}{2})^{\frac{1}{2}};A\kappa\tau/2] \ ,$$

where $t_{\frac{1}{2}}$ is the full width at half *amplitude* of one oscillation of the intensity pattern (Fig. 5).

References

1. (a) S. L. McCall and E. L. Hahn, Phys. Rev. Letters *18*, 908 (1967); Phys. Rev. *183*, 457 (1969); (b) Recent experimental work has been reported by H. M. Gibbs and R. E. Slusher, Phys. Rev. Letters *24*, 638 (1970); Appl. Phys. Letters *18*, 505 (1971) and Phys. Rev. *A5*, 1634 (1972); (c) A review of ultrashort pulse phenomena, and many other references, are given by G. L. Lamb, Jr., in Rev. Mod. Phys. *43*, 99 (1971); (d) D. Dialetis, Phys. Rev. *A2*, 249 (1970).
2. L. Matulic, Optics Comm. *2*, 249 (1970).
3. L. Matulic and J. H. Eberly, Phys. Rev. *A6*, 822 (1972).
4. R. A. Marth and J. H. Eberly (unpublished); R. A. Marth, Ph.D. thesis (Polytechnic Institute of Brooklyn, 1972)(unpublished); J. H. Eberly, this volume, p. 649.
5. J.H. Eberly and L. Matulic, Optics Comm. *1*, 241 (1969).

6. L. Matulic, P. W. Milonni and J. H. Eberly, Optics. Comm. *4* 181 (1971).
7. (a) J. H. Eberly, Ph.D. thesis, Stanford University (unpublished); (b) S. R. Barone, Phys. Letters *32A*, 8 (1970).
8. (a) F. T. Arecchi, V. DeGiorgio and C. G. Someda, Phys. Letters *27A*, 588 (1968); (b) M. D. Crisp, Phys. Rev. Letters *22*, 820 (1969); (c) J. H. Eberly, Phys. Rev. Letters *22*, 760 (1969).
9. F. A. Hopf and M. O. Scully, Phys. Rev. *179*, 399 (1969). F. A. Hopf, G. L. Lamb, Jr., C. K. Rhodes and M. O. Scully, Phys. Rev. *A3*, 758 (1971).
10. Note that it is not as simple as just a matter of choice. If one prefers to talk about periodically negative envelopes and thereby avoid having to introduce explicit $\pm\,\pi$ phase jumps, then one is left without insight into the wide variety of cases in which the phase jumps are *not* equal to $\pm\,\pi$. An interesting example of such a situation is shown in Fig. 3.

COUPLED SUPERRADIANCE MASTER EQUATIONS: APPLICATION TO FLUCTUATIONS IN COHERENT PULSE PROPAGATION IN RESONANT MEDIA *

Charles R. Willis

Boston University, Boston, Mass.

R.H. Picard

Air Force Cambridge Research Laboratories, Bedford, Mass.

Cooperative phenomena in radiation-matter interactions range from coherent pulse propagation on the one hand to superradiant emission on the other. The question of fluctuations in cooperative phenomena has received little consideration despite its importance both for stability studies of pulse propagation and for elucidating the connection between superradiance and pulse propagation. The usual procedure is to treat coherent pulse propagation and superradiance as completely different problems in spite of the fact that the interaction Hamiltonian is the same and the effects proportional to N^2 dominate in both cases. One of the main purposes of this paper is to present and discuss the results of a master-equation formalism that provides a unified treatment of coherent pulse propagation and fluctuation phenomena by splitting off and treating exactly the self-consistent part of the matter-field interaction. A second main purpose of the paper is to show how easily one may generalize existing formulations of the superradiance problem by treating dynamically the radiation field, as well as the matter,thus leading to a system of two coupled master equations for the reduced density operators of the matter and of the field.

However, before discussing the new equations we would like to briefly review some aspects of the usual treatment of coherent pulse

* A preliminary report on the results of this paper was presented at the 1971 winter meeting of the American Physical Society, Cambridge, Mass., Bull. Am. Phys. Soc. *16*, 1402 (1971).

propagation and superradiance to provide background for our derivation. One usually describes coherent pulse propagation [1] by the coupled Bloch-Maxwell equations which are obtained from the quantum-Liouville equation assuming that, at all times, there are no correlations between radiation and matter variables. Consequently the Bloch-Maxwell equations are just the self-consistent field approximation (SCFA) to the quantum-Liouville equation [2]. The resultant equations contain no fluctuations. We would like to emphasize that the SCFA is a completely nonperturbative theory. No dimensionless parameters are required to be small nor are there any limits on the magnitudes of the field and atomic polarization.

On the other hand, the problem of superradiance is approached by using first-order perturbation theory in most analyses-- for example, those due to Dicke [3], Fain [4], Ernst and Stehle [5], Rehler and Eberly [6], and Agarwal [7]. The existence of a time range which is "long enough, but not too long", required to justify the use of Fermi's Golden Rule of perturbation theory is equivalent to demanding that the dimensionless parameter $\varepsilon \equiv \tau_{int}/\tau_{rel}$ be small, where τ_{int} is the interaction time (or effective duration of an interaction) and τ_{rel} is the relaxation time (or effective time between interactions). In the superradiance problem $\tau_{int} \simeq \omega_0^{-1}$ (ω_0 is the atomic resonance frequency) when the atoms radiate into the continuum of vacuum modes, and $\tau_{rel} \simeq T_s$, the superradiant decay time [3,6]. Typical values for ε are then 10^{-5}- 10^{-6}, so that perturbation theory is expected to be valid. A different dimensionless parameter ε results from the perturbative approach of Bonifacio, Schwendimann, and Haake [8] and of Haake and Glauber [9], where the atomic system radiates, not into the free-space vacuum but into a single longitudinal mode of the "cavity" delimited by the sample's end faces. [10].

A further important distinction between the usual approaches to coherent pulse propagation and superradiant emission is that the former necessarily consider stimulated absorption and emission processes whereas the latter universally ignore them.

We will show how one may achieve significant improvements in the description of coherent optical phenomena by combining these two approaches, doing perturbation theory about the self-consistent field (SCF) rather than about the uncoupled matter and field. Such an approach permits us both to make contact with semiclassical treatments of superradiance [3,6] and coherent pulse propagation, and to determine the modifications of these treatments due to fluctuations. It also leads one to consider the effect in superradiant problems of a radiation field which is not trivially determined by the matter but is an independent dynamical entity coupled to the matter through stimulated emission and absorption terms in the theory.

One of us [11] has used such an approach to attack the problem of the single-mode steady-state laser, employing Bogoliubov's theory of kinetic equations [12] to derive coupled kinetic equations for the reduced matter and radiation density operators including both SCF and fluctuation terms. In the present paper we generalize this model to allow for particle-particle correlations, multimode radiation fields, and nonstationary behavior.

Our model consists of a continuum of plane-wave radiation modes $e^{i(\underline{k}\cdot\underline{r}-\Omega_k t)}$ interacting with N two-level atoms, the αth atom at position $\underline{X}_\alpha$ having upper and lower states $|\pm>_\alpha$ separated in energy by $\hbar\omega_0$. The Hamiltonian of the model is a sum of matter, field and interaction Hamiltonians,

$$H = H_m + H_f + H_{int},$$

where

$$H_m = \frac{\hbar\omega_0}{2} S_0 = \frac{\hbar\omega_0}{2} \sum_{\alpha=1}^{N} s_\alpha , \tag{1}$$

$$H_f = \hbar \sum_k \Omega_k (a_k^+ a_k^- + \frac{1}{2}) ,$$

and

$$H_{int} = - \int d\underline{r} P(\underline{r}) \cdot \underline{E}(\underline{r}) = \hbar \sum_k (\mu_k P_k^+ a_k^- + \mu_k^* P_k^- a_k^+) .$$

In Eqs. (1) we have used the following notation: The population inversion operator for the αth atom is

$$s_\alpha \equiv |+>_{\alpha\alpha}< +| - |->_{\alpha\alpha}<-| = [\sigma_\alpha^+ , \sigma_\alpha^-] ,$$

where $\sigma_\alpha^\pm \equiv |\pm>_{\alpha\alpha}<\mp|$ are the atomic raising and lowering operators. The electric field operator is

$$\underline{E}(\underline{r}) = \frac{1}{2}[\underline{E}^+(\underline{r}) + \underline{E}^-(\underline{r})] ,$$

which has the modal decomposition

$$\underline{E}^\pm(\underline{r}) = - \sum_k \left(\frac{2\pi\hbar\Omega_k}{V}\right)^{1/2} \left\{ \begin{matrix} \underline{\varepsilon}_k^* \\ \underline{\varepsilon}_k \end{matrix} \right\} e^{\mp i\underline{k}\cdot\underline{r}} a_k^\pm ,$$

where $a_k^{\pm}$ are the creation and annihilation operators for the kth mode, satisfying $[a_k^-, a_\ell^+] = \delta_{k\ell}$, and $\underline{\varepsilon}_k$ is its polarization vector. The collective atomic variables P_k and S_κ are defined by [13]

$$P_K^{\pm} = \sum_\alpha e^{\pm i\underline{k}\cdot\underline{X}_\alpha} \sigma_\alpha^{\pm} \tag{2}$$

and

$$S_\kappa = \sum_\alpha e^{i\underline{\kappa}\cdot\underline{X}_\alpha} s_\alpha$$

and satisfy

$$[P_k^+, P_\ell^-] = S_{k-\ell},\ [P_k^-, S_\kappa] = 2P_{k-\kappa}^-,\ [S_\kappa, P_k^+] = 2P_{k+\kappa}^+ \tag{3}$$

In the second form of H_{int} we have used the rotating-wave approximation and μ_k is the coupling constant defined by

$$\hbar\mu_k = \underline{\wp}^*\cdot\underline{\varepsilon}_k (2\pi\hbar\Omega_k/V)^{1/2} \tag{4}$$

where $\underline{\wp} \equiv \langle +|e\underline{r}| - \rangle$ is the transition dipole moment. We will also need the collective variables in the position representation, whose form and algebra follow from Eqs. (2) and (3) by Fourier transformation,

$$P^{\pm}(\underline{r}) = \sum_\alpha \delta(\underline{r} - \underline{X}_\alpha)\, \sigma_\alpha^{\pm},$$

$$S(\underline{r}) = \sum_\alpha \delta(\underline{r} - \underline{X}_\alpha)\, s_\alpha, \tag{5}$$

$$[P^+(\underline{r}), P^-(\underline{r}')] = \delta(\underline{r}-\underline{r}')S(\underline{r}),\ \text{and so forth.}$$

In terms of the collective variables in the position representation the atomic polarization is

$$\underline{P}(\underline{r}) = \underline{\wp}^* P^+(\underline{r}) + \underline{\wp}\, P^-(\underline{r}).$$

We split the Hamiltonian as discussed above into an unperturbed part H^o, containing the self-consistent part of the interaction, and a perturbation μV in the following way,

$$\begin{aligned} H &= H^o + \mu V, \\ H^o &= H^o_m + H^o_f, \\ H^o_m &= H_m + \langle H_{int} \rangle_f, \\ H^o_f &= H_f + \langle H_{int} \rangle_m , \end{aligned} \tag{6}$$

and

$$\mu V = H_{int} - \langle H_{int} \rangle_f - \langle H_{int} \rangle_m .$$

In terms of the full field-plus-matter density operator $F_N(f,m)$ and the reduced density operators of the field and of the matter,

$$R(f) \equiv Tr_m F_N(f,m)$$

and

$$\rho_N(m) \equiv Tr_f F_N(f,m),$$

respectively, the definitions of $\langle H_{int} \rangle_f$ and $\langle H_{int} \rangle_m$ are

$$\langle H_{int} \rangle_f = Tr_f [H_{int}(f,m) R(f)]$$

and

$$\langle H_{int} \rangle_m = Tr_m [H_{int}(f,m) \rho_N(m)],$$

Thus H^o_m describes the interaction of the atomic polarization with the mean field, H^o_f describes the interaction of the field with the mean polarization, and μV describes the interaction of the atoms with field fluctuations and of the field with atomic polarization fluctuations.

From the quantum-Liouville equation for F_N, using the method of Bogoliubov [12] or Zwanzig's projection techniques [14], one derives coupled master equations for R and ρ_N. The resultant equations in the first Born approximation are

$$\frac{dR}{dt} + \frac{i}{\hbar}[H_f^o(t),R] = -\frac{\mu^2}{\hbar^2}\int_0^\infty d\tau Tr_m[V(t), [V(t,t-\tau),R\rho_N]]$$

and

$$\frac{d\rho_N}{dt} + \frac{i}{\hbar}[H_m^o(t),\rho_N] = -\frac{\mu^2}{\hbar^2}\int_0^\infty d\tau Tr_f[V(t),[V(t,t-\tau),R\rho_N]] \ , \qquad (7)$$

where

$$V(t,t-\tau) \equiv U^{o\dagger}(t-\tau,t)V(t-\tau)U^o(t-\tau,t).$$

The time-dependent operator $U^o(t-\tau,t)$ is a solution of the time-dependent Schrodinger equation for the Hamiltonian $H^o(t)$, subject to the initial condition $U^o(t,t) = 1$; that is

$$U^o(t-\tau,t) = \exp[-\frac{i}{\hbar}\int_t^{t-\tau} dt' \, H^o(t')].$$

The appearance is complicated slightly by the fact that H^o(and hence V also) have explicit time dependences due to their dependence on the density operators R(t) and $\rho_N(t)$. In Bogoliubov's method two conditions are needed to justify the existence of the master equation: (1) the functional Ansatz, or synchronization condition, whereby there exists an intermediate stage during the evolution of a dynamical system (the kinetic stage) during which the details of an interaction are forgotten and the time dependence of F_N comes entirely from the reduced density operators R and ρ_N,

$$F_N(f,m,t) = F_N[R(f,t), \rho_N(m,t)],$$

and (2) the asymptotic condition, whereby asymptotically in the infinite past the system evolving under the influence of H^o loses memory of atom-field correlations,

$$F_N(f,m,t-\tau) \xrightarrow[\tau \to \infty]{U^o} R(f,t-\tau)\rho_N(m,t-\tau).$$

In general, $V(t,t-\tau)$ depends on the SCF

$$\langle \underline{E}(\underline{r})\rangle_t \equiv \mathrm{Tr}_f[\underline{E}(\underline{r})R(t)]$$

through the occurrence of the SCF in H_m^o. Nevertheless, we choose to ignore the dependence of $V(t,t-\tau)$ on the SCF at this time, even though it introduces qualitatively new and interesting terms into the master equation, because its effect is often small.

When Eqs. (6) and (1) are substituted into Eqs. (7) and the result is simplified, one obtains the master equations in modal language,

$$\dot{R} + i\sum_\ell \Omega_\ell[a_\ell^+a_\ell^-,R] + i\sum_\ell(\mu_\ell\langle P_\ell^+\rangle[a_\ell^-,R]-\mathrm{h.c.})$$

$$-\left.\frac{\delta R}{\delta t}\right|_{res} = -\sum_{\ell,m}\mu_\ell \quad C_\ell\{\mu_m^*\langle\Delta P_\ell^+\Delta P_m^-\rangle\ [a_\ell^-,[a_m^+,R]]$$

$$+\ \mu_m\langle\Delta P_\ell^+\Delta P_m^+\rangle\ [a_\ell^-,[a_m^-,R]]$$

$$-\ \mu_m^*\ \langle S_{\ell-m}\rangle[a_m^+,\Delta a_\ell^- R]\}$$

$$+\ \mathrm{h.c.} \tag{8}$$

and

$$\dot{\rho}_N + i\omega_o[S_o,\rho_N] + i\sum_\ell(\mu_\ell\langle a_\ell^-\rangle[P_\ell^+,\rho_N] - \mathrm{h.c.}) - \left.\frac{\delta\rho_N}{\delta t}\right|_{res}$$

$$= \sum_\ell\ |\mu_\ell|^2\ C_\ell[P_\ell^-,\rho_N\Delta P_\ell^+]$$

$$-\sum_{\ell,m}\mu_m C_m\ \{\mu_\ell^*\langle\Delta a_\ell^+\Delta a_m^-\rangle[P_\ell^-\ ,[P_m^+,\rho_N]]$$

$$+\mu_\ell\langle\Delta a_\ell^-\Delta a_m^-\ \rangle[P_\ell^+,[P_m^+,\rho_N]]\}$$

$$+\ \mathrm{h.c.}\quad , \tag{9}$$

where $C_\ell = C_\ell(\omega_0-\Omega_\ell)$ is the lineshape resulting from the τ integration and the fluctuation in an operator O is defined by $\Delta O \equiv O-\langle O\rangle = O - Tr_{m,f}(OF_N)$. We have added radiation and matter reservoir terms to the equations to allow for radiation losses and atomic relaxation, which are important in certain cases. The lineshape C_ℓ will also be a function of the relaxation parameters when relaxation is important.

There are five types of terms appearing in Eqs. (8) and (9). In order of appearance there are (a) free field (matter) time development terms, (b) SCF interaction terms, (c) reservoir interaction terms, (d) fluctuation terms describing spontaneous emission (both coherent and incoherent), and (e) fluctuating terms describing induced emission and absorption. When we neglect right members of Eqs. (8) and (9), we obtain the modal decomposition [15] of the coupled Maxwell-Bloch equations of coherent propagation in density-operator form. On the other hand, existing formulations of super-radiance master equations [7-9] are obtained by demanding that the field always remains in a vacuum state. As a consequence, one eliminates the R equation entirely and eliminates from the ρ_N equation both the SCF term proportional to μ and the induced terms proportional to μ^2.

We now show how it is possible to obtain a formal solution for the radiation density operator R. This is easier to do if we first expand R in terms of coherent photon states $|\{\alpha_k\}\rangle$, that is

$$R(t) = \int \Pi_k\, d^2\alpha_k\, \mathcal{R}(\{\alpha_k\},t)\,|\{\alpha_k\}\rangle\langle\{\alpha_k\}|\ ,$$

where $\mathcal{R}$ is a c-number weight function[16]. Upon substitution of this expression into Eq. (8) we obtain an equation of Fokker-Planck type for the weight function,

$$\frac{\partial\mathcal{R}}{\partial t} = \sum_{\ell,m} \{- \frac{\partial}{\partial\alpha_\ell^*} [b_{\ell m}(\alpha_m^*,t)\mathcal{R}]$$

$$+ \frac{1}{2}D_{\ell m}(t) \frac{\partial^2\mathcal{R}}{\partial\alpha_\ell^*\partial\alpha_m} + \frac{1}{2} D'_{\ell m}(t) \frac{\partial^2\mathcal{R}}{\partial\alpha_\ell\partial\alpha_m}\}$$

$$+ \text{c.c.}, \tag{10}$$

where

$$D_{\ell m}(t) \equiv 2\mu_\ell\mu_m^* C_\ell \langle\Delta P_\ell^+\Delta P_m^-\rangle_t,$$

$$D'_{\ell m}(t) \equiv -2\mu_\ell^* \mu_m^* C_\ell \langle \Delta P_\ell^- \Delta P_m^- \rangle_t,$$

$$b_{\ell m}(\alpha_m^*, t) \equiv f_{\ell m}(t)(\alpha_m^* - \langle a_m^+ \rangle_t) + g_\ell(t)\delta_{\ell m},$$

$$f_{\ell m}(t) \equiv \mu_\ell \mu_m^* C_m^* \langle S_{\ell - m} \rangle_t + i\,\Omega_\ell\,\delta_{\ell m},$$

$$g_\ell(t) \equiv i(\Omega_\ell \langle a_\ell^+ \rangle_t + \mu_\ell \langle P_\ell^+ \rangle_t).$$

We have omitted the reservoir relaxation term from this equation. Its solution is then a displaced multivariate Gaussian distribution in $\{\alpha_k\}$,

$$\mathcal{R} = \exp\{\lambda(t) - \frac{1}{2}\sum_{\ell,m} [A_{\ell m}(t)(\alpha_\ell^* - \langle a_\ell^+ \rangle_t)(\alpha_m - \langle a_m^- \rangle_t)$$

$$+ B_{\ell m}(t)(\alpha_\ell^* - \langle a_\ell^+ \rangle_t)(\alpha_m^* - \langle a_m^+ \rangle_t)$$

$$+ \text{c.c.}]\},$$

where $A_{\ell m}$ and $B_{\ell m}$ are inverses of covariance matrices and λ is a normalization factor related to $D_{\ell m}$ and $D'_{\ell m}$, $f_{\ell m}$ and g_ℓ by ordinary differential equations in time. Hence the solution of Eq. (10) is completely determined if we know the time dependence of the first and second moments of the matter density operator ρ_N.

On the other hand, when we examine the equations of motion for these matter moments we see that they are coupled to field moments. For lack of space, we omit the complete listing of all the important matter and field moment equations, and limit ourselves to illustrating their form by showing one first-moment and one second-moment equation. Although we have worked until now in modal language, we will write these moments in the conjugate position language. Despite the complete equivalence of the two descriptions, for some problems one language seems more natural or convenient than the other. For example, coherent propagation problems are customarily written in terms of space variables, while spontaneous emission problems seem more natural in terms of modes. When both types of effects are present in the same problem, neither representation has a clear preference.

Among the first-moment equations we select the one for the

mean population difference $\langle S(\underline{r})\rangle$ which reads

$$\left(\frac{\partial}{\partial t} - \frac{\delta}{\delta t}\Big|_{res}\right) \langle S(\underline{r})\rangle$$

$$= - \frac{2i\wp}{\hbar} \hat{\underline{\wp}} \cdot \langle \underline{E}^{+}(\underline{r})\rangle \langle P^{-}(\underline{r})\rangle$$

$$- \frac{2\wp^{2}}{\hbar^{2}} \int d\underline{y} \; [g_{s}(\underline{y}-\underline{r}) \langle \Delta P^{+}(\underline{y}) \Delta P^{-}(\underline{r})\rangle$$

$$+ g_{I}(\underline{r}-\underline{y}) \langle S(\underline{r})\rangle \hat{\underline{\wp}} \cdot \langle \Delta \underline{E}^{+}(\underline{r}) \Delta \underline{E}^{-}(\underline{y})\rangle \cdot \hat{\underline{\wp}}^{*}]$$

$$+ \text{c.c.} \tag{11}$$

We have written $\underline{\wp} = \wp\hat{\underline{\wp}}$, $\hat{\underline{\wp}}$ being a unit vector satisfying $\hat{\underline{\wp}} \cdot \hat{\underline{\wp}}^{*} = 1$, and we define the kernels in the $\underline{y}$ integration by

$$g_{I}(\underline{r}) = \frac{\pi}{V} \sum_{k} e^{i\underline{k}\cdot\underline{r}} \delta_{-}(\omega_{o} - \Omega_{k}) \tag{12a}$$

and

$$g_{s}(\underline{r}) = \frac{2\pi^{2}}{V} \sum_{k} e^{i\underline{k}\cdot\underline{r}} \hbar\Omega_{k} w(\theta_{k}) \delta(\omega_{o} - \Omega_{k}) , \tag{12b}$$

where

$$w(\theta_{k}) = \sum_{s=1}^{2} \left| \hat{\underline{\wp}} \cdot \underline{\varepsilon}_{ks} \right|^{2} . \tag{13}$$

The sum in Eq. (13) is a sum over the two possible polarization states of the field and the k summation in Eqs. (12) are sums over only the possible wave vectors $\underline{k}$ for the field, whereas in prior equations k has stood for the composite index $k \equiv (\underline{k},s)$. The lineshapes C_{k} appearing in Eq. (12) have been evaluated for the case where relaxation can be neglected; that is $C_{k} = \pi\delta_{+}(\omega_{o} - \Omega_{k})$.

To illustrate the form of the second moment-equations we write the dynamical equation for the second matter moment appearing on the right side of Eq. (11),

$$\left(\frac{\partial}{\partial t} - \frac{\delta}{\delta t}\bigg|_{res}\right)\langle\Delta P^{+}(\underline{r})\Delta P^{-}(\underline{r}')\rangle = \frac{i\wp}{\hbar}\hat{\underline{\wp}}\cdot\langle\underline{E}^{+}(\underline{r})\rangle\langle\Delta S(\underline{r})\Delta P^{-}(\underline{r}')\rangle$$

$$-\frac{\wp^2}{\hbar^2}g_s(0)\langle\Delta P^{+}(\underline{r})\Delta P^{-}(\underline{r}')\rangle$$

$$+\frac{\wp^2}{\hbar^2}\int d\underline{y}[g_s(\underline{y}-\underline{r})\langle S(\underline{r})\Delta P^{+}(\underline{y})\Delta P^{-}(\underline{r}')\rangle$$

$$-2g_I(\underline{r}-\underline{y})\hat{\underline{\wp}}\cdot\langle\Delta\underline{E}^{+}(\underline{r})\Delta\underline{E}^{-}(\underline{y})\rangle\cdot\hat{\underline{\wp}}^{*}\langle\Delta P^{+}(\underline{r})\Delta P^{-}(\underline{r}')\rangle$$

$$+2g_I(\underline{r}'-\underline{y})\hat{\underline{\wp}}^{*}\cdot\langle\Delta\underline{E}^{-}(\underline{y})\Delta E^{-}(\underline{r}')\rangle\cdot\hat{\underline{\wp}}^{*}\langle\Delta P^{+}(\underline{r})\Delta P^{+}(\underline{r}')\rangle$$

$$+g_I(\underline{r}'-\underline{y})\hat{\underline{\wp}}\cdot\langle\Delta\underline{E}^{+}(\underline{r})\Delta\underline{E}^{-}(\underline{y})\rangle\cdot\hat{\underline{\wp}}^{*}\langle S(\underline{r})S(\underline{r}')\rangle]$$

$$+\text{ c.c. with } \underline{r}\leftrightarrow\underline{r}'. \tag{14}$$

Although Eq. (14) appears to be much more complicated than Eq. (11), it has the same basic structure; namely, reservoir and SCF interaction terms, followed by spontaneous and induced terms involving fluctuations from SCF. One feature of Eqs. (11) and (14) is worth noting; in contrast to the field where moments of different orders are not interdependent, first moments of the matter depend on second moments, second moments of the matter depend on third moments and so on. If we truncate the equations by breaking up the third moments in a plausible, although *ad hoc* way - for example,

$$\langle S(\underline{r})\Delta P^{+}(\underline{r}')\Delta P^{-}(\underline{r}'')\rangle = \langle S(\underline{r})\rangle\langle\Delta P^{+}(\underline{r}')\Delta P^{-}(\underline{r}'')\rangle \quad ,$$

and so forth, we obtain a closed set of equations for the first and second moments of the matter and of the radiation. The coupled moments then consist of the first moments $\langle\underline{E}^{+}(\underline{r})\rangle$, $\langle P^{+}(\underline{r})\rangle$, $\langle S(\underline{r})\rangle$ and the second moments, $\langle\Delta\underline{E}^{+}(\underline{r})\Delta E^{\pm}(\underline{r}')\rangle$, $\langle\Delta P^{+}(\underline{r})\Delta P^{\pm}(\underline{r}')\rangle$, $\langle\Delta P^{+}(\underline{r})\Delta S(\underline{r}')\rangle$, and $\langle\Delta S(\underline{r})\Delta S(\underline{r}')\rangle$.

In order to apply the above formalism to the problem of fluctuations in coherent propagation, we introduce the customary slowly varying envelope approximation (SVEA). Moreover, we assume the field is exactly resonant. In the SCF case where there are no fluctuations, this assumption allows one to eliminate the phases of the field and polarization as dynamical variables [1,17], reducing the usual five coupled (real) variables to three. In our notation these assumptions mean that

$$\langle \underline{E}^{-}(\underline{r})\rangle_t \cdot \hat{\boldsymbol{\wp}}^{*} = \frac{1}{2}\langle \mathcal{E}(\underline{r})\rangle_t \, e^{-i\Phi(\underline{r},t)} \tag{15a}$$

and

$$\langle P^{-}(\underline{r})\rangle_t = -\frac{i}{2}\langle \Pi(\underline{r})\rangle_t \, e^{-i\Phi(\underline{r},t)} , \tag{15b}$$

where

$$\Phi(\underline{r},t) \equiv \phi + \omega_o t - \underline{k}_o \cdot \underline{r},$$

ϕ is constant, and $\langle \mathcal{E}\rangle$ and $\langle \Pi\rangle$ are real and slowly varying. We will assume further that this behavior carries over to the fluctuating case, where the phase may also be eliminated as a dynamical variable with a similar reduction in the number of second-moment equations. This procedure has the additional feature of segregating effects proportional to N^2 from those proportional to N. A typical second-moment reduction is

$$\langle \Delta P^{+}(\underline{r})\Delta P^{-}(\underline{r}')\rangle_t = \frac{1}{4} e^{-i\underline{k}_o \cdot (\underline{r}-\underline{r}')} \langle \Delta\Pi(\underline{r})\Delta\Pi(\underline{r}')\rangle_t$$

$$+ \delta(\underline{r}-\underline{r}') \cdot \frac{1}{2}[n(\underline{r}) + \langle S(\underline{r})\rangle_t] , \tag{16}$$

where $n(\underline{r}) \equiv \Sigma_\alpha \delta(\underline{r}-\underline{X}_\alpha)$ is the atomic number density. The new moment defined in Eq. (16), $\langle \Delta\Pi \, \Delta\Pi\rangle$ is real, slowly varying, and proportional to N^2 (Strictly, N(N-1)), whereas the second term in the equation is proportional to N. Similarly, we replace other second moments by real slowly varying quantities.

After making the SVEA and the constant-phase assumption, Eq. (11) assumes the somewhat simpler form

$$\left(\frac{\partial}{\partial t} - \frac{\delta}{\delta t}\bigg|_{res}\right)\langle S(\underline{r})\rangle = -\frac{\wp}{\hbar}\langle \mathcal{E}(r)\rangle\langle \Pi(r)\rangle \quad +$$

$$- \frac{\wp^2}{\hbar^2} G_s{}^R(0) \cdot \frac{1}{2} [n(\underline{r}) + \langle S(\underline{r}) \rangle]$$

$$- \frac{\wp^2}{\hbar^2} \int d\underline{y} [G_s^R(\underline{y}-\underline{r}) \langle \Delta\Pi(\underline{y}) \Delta\Pi(\underline{r}) \rangle$$

$$+ G_I(\underline{y}-\underline{r}) \langle S(\underline{r}) \rangle \langle \Delta\mathcal{E}(\underline{r}) \Delta\mathcal{E}(\underline{y}) \rangle \quad , \tag{17}$$

where

$$G_s(\underline{r}) = g_s(\underline{r}) \, e^{-i\underline{k}_o \cdot \underline{r}} \quad ,$$

with a similar definition for G_I, and the superscript R designates the real part of the function. The kernel G_s in the SVEA has a simple relation to both the single-atom decay time T_1 and the superradiant decay time T_s, namely

$$\frac{1}{T_1} = 2 \frac{\wp^2}{\hbar^2} G_s{}^R(0)$$

and

$$\frac{1}{T_s} = \frac{N}{2} \frac{\wp^2}{\hbar^2} \int \frac{d\underline{r}}{V} \int \frac{d\underline{y}}{V} \quad G_s{}^R(\underline{r}-\underline{y}) . \tag{18}$$

The various terms in Eq. (17) are the same as in Eq. (11) except that the spontaneous emission term involving fluctuations has been split into an incoherent emission and a coherent emission term.

As a simple illustration we show how the set of moment equations including Eq. (17) may be used as the basis of a perturbative scheme for obtaining corrections due to fluctuations in the problem of steady-state 2π pulse propagation in a lossless attenuator [1]. For our zeroth-order theory we assume that there are no fluctuations about the SCF, resulting in the equations

$$\eta_o \langle \dot{\mathcal{E}} \rangle_o = - 2\pi\omega_o \wp \langle \Pi \rangle_o ,$$

$$\langle \dot{\Pi} \rangle_o = (\wp/\hbar) \langle \mathcal{E} \rangle_o \langle S \rangle_o \quad , \tag{19}$$

and

$$\langle\dot{S}\rangle_o = -(\mathcal{P}/\hbar)\langle\mathcal{E}\rangle_o\langle\Pi\rangle_o ,$$

where it is assumed that the solution is static in a frame moving in the +z direction with speed v and $\eta \equiv (c/v)-1$. The pulse solution of Eq. (19) is the well-known hyperbolic secant with area 2π,

$$\langle\mathcal{E}\rangle_o = \frac{2\hbar}{\mathcal{P}\tau_p} \operatorname{sech}\left(\frac{t - v^{-1}z}{\tau_p}\right) , \tag{20}$$

and the velocity is determined by the pulsewidth τ_p through the condition

$$\eta_o = 2\pi\hbar\omega_o(N/V)(\mathcal{P}\tau_p/\hbar)^2 . \tag{21}$$

In the next order we add terms $O(\mathcal{P}^2/\hbar^2)$ onto the SCF equations but use the zero-order values for the variables appearing in these terms. Thus the first-order equations are

$$\eta_1\langle\dot{\mathcal{E}}\rangle_1 + 2\pi\omega_o\mathcal{P}\langle\Pi\rangle_1 = 0 ,$$

$$\langle\dot{\Pi}\rangle_1 - (\mathcal{P}/\hbar)\langle\mathcal{E}\rangle_1\langle S\rangle_1 = -(\mathcal{P}^2/\hbar^2)\, G_s^R(0)\langle\Pi\rangle_o , \tag{22}$$

and

$$\langle\dot{S}\rangle_1 + (\mathcal{P}/\hbar)\langle\mathcal{E}\rangle_1\langle\Pi\rangle_1 = -(\mathcal{P}^2/\hbar^2)G_s^R(0)\,\frac{1}{2}\,[n+\langle S\rangle_o] .$$

We note that the right members of Eq. (22) contains only incoherent emission terms of O(N) and are the same type of terms one would obtain by introducing phenomenological relaxation times T_2' and T_1. Using the first of Eqs. (18) the solution of Eqs. (22) is

$$\langle\mathcal{E}\rangle_1 = [1 - \frac{\tau_p}{T_1}\, f_1\left(\frac{t-v^{-1}z}{\tau_p}\right)]\, \langle\mathcal{E}\rangle_o , \tag{23}$$

with the plausible interpretation the corrections will be small unless τ_p is an appreciable fraction of T_1.

A similar perturbation solution can be carried out for the case of steady-state π pulse propagation in a lossy amplifier [18]. If the atomic system is not allowed to radiate except when the pulse is present, one obtains corrections of order N to the SCF equations with a result analogous to Eq. (23),

$$\langle \mathcal{E} \rangle_1 = [1 - \frac{2\nu_R}{3\pi\omega_o} (\frac{N}{V} \lambda^3)^{-1} f_2(\frac{t-v^{-1}z}{\tau_p})] \langle \mathcal{E} \rangle_o ,$$

where ν_R is the radiation dissipation rate. This hypothesis is rather artificial, however, and in a more realistic analysis of a long amplifier, particle-particle correlations giving rise to superradiant correction terms of $O(N^2)$ develop. These terms become more important as the pulse propagates further down the amplifier. One is then faced with an interesting competition between the coherent evolution through Bloch states described by the Maxwell-Bloch equations and the increasingly coherent spontaneous emission process.

In summary, we have derived a formalism that combines the usually separate problems of coherent pulse propagation and superradiance. Our resultant coupled master equations for R and ρ_N treat both radiation and matter as distinct dynamical and statistical entities. Consequently, in addition to the usual coherent and incoherent spontaneous emission terms the equations for both R and ρ_N contain stimulated absorption and emission processes. A further consequence of the coupled master equations is that we can characterize the general form of the radiation statistics without a complete solution of the problem. By splitting off and treating exactly the self-consistent part of the interaction, we are able to make close contact with the SCF equations used in coherent pulse propagation. We derive correction terms linear in N in these equations due to incoherent damping. The more interesting problem of determining the effect of coherent damping terms proportional to N^2 on the SCF solutions is presently under study.

We have also derived an analogue for the inhomogeneously broadened case of the theory presented above. The inclusion of inhomogeneous broadening has a minor effect on the formal structure but complicates details considerably, e.g. second moments such as $\langle P_k^+(\omega) P_{k'}^-(\omega') \rangle$ depend on two different frequencies.

In conclusion, we would like to point out an unsolved problem that pertains to all superradiance master equations (including ours); that is they suffer from a divergence problem for large N. If any of the theories is carried to the next higher order in the interaction (second Born approximation), the ratio of the second Born and first Born emission terms is $N(\omega_o T_S)^{-1}$; each succeeding order brings in another power of N. This divergence is due to the essentially "long-range" nature of the interaction. What is needed to avoid the divergence is a selective summation of divergent diagrams. We are working on this program but feel the correct answer will look very similar to our present model of a SCF with fluctuations. The details of the results in the present paper and further results will be published elsewhere.

Footnotes and References

1. S.L. McCall and E.L. Hahn, Phys. Rev. *183*, 457(1969). G.L. Lamb Jr.,Rev. Mod. Phys. *43*, 99 (1971).
2. C.R. Willis, J. Math. Phys. *5*, 1241 (1964).
3. R.H. Dicke, Phys. Rev. *93*, 99 (1954).
4. V.M. Fain and Ya. I. Khanin, *Quantum Electronics* (MIT Press, Cambridge, Mass., 1969).
5. V. Ernst and P. Stehle, Phys. Rev. *176*, 1456 (1968).
6. N.E. Rehler and J.H. Eberly, Phys. Rev. *A3*, 1735 (1971).
7. G.S. Agarwal, Phys. Rev. *A2*, 2038 (1970); *A4*, 1783 (1971); *A4*, 1791 (1971).
8. R. Bonifacio, P. Schwendimann and F. Haake, Phys. Rev. *A4*, 302 (1971) and Phys. Rev. *A4*, 854 (1971).
9. F. Haake and R.J. Glauber, Phys. Rev. *A5*, 1457 (1971).
10. Instead they obtain a small parameter ε by giving the photons a finite lifetime in the cavity equal to the one-way transit time, $\tau_{int} = L/c$ and choosing $\tau_{rel} \equiv (\mu^2 NL/c)^{-1}$ where μ is the atom-field coupling constant defined in Eq. (4).
11. R.H. Picard, Ph.D. Thesis, Boston University (1968).
12. N.N. Bogoliubov in *Studies in Statistical Mechanics*, edited by J. deBoer and G.E. Uhlenbeck (North Holland Publishing Co., Amsterdam, 1962) pp. 5-118.
13. Our operators $P_K^{\pm}$ are identical with Dicke's operators $R_k^{\pm}$ in ref. 3.
14. R.H. Zwanzig, Physica *30*, 1109 (1964).
15. R.H. Picard and C.R. Willis, Phys. Letters *37A*, 301 (1971).
16. R.J. Glauber, Phys. Rev. *131*, 2766 (1963). The weight function is denoted by $P(\{\alpha_k\})$ in this reference.
17. F.A. Hopf and M.O. Scully, Phys. Rev. *179*, 399 (1969).
18. F.T. Arecchi and R. Bonifacio, IEEE J. Quantum Electron, *1*, 169 (1965).

THE LASER-PHASE TRANSITION ANALOGY - RECENT DEVELOPMENTS*

Marlan O. Scully ≠

University of Arizona, Tucson, Arizona

I. The Analogy

Considerations involving the analogies between phase transitions in ferromagnets, superfluids and superconductors have emphasized the similarities between these systems near their critical temperatures [1]. For example, the order parameter varies near the critical temperature as $\eta \sim \varepsilon^{\beta}$, where β is a simple number (1/2 for a mean field theory) while η and ε are defined in table 1.

Recently, it has been demonstrated that a useful comparison can be made between thermodynamic phase transition phenomena and laser threshold behavior [2],[3],[4]. The physics behind this analogy [5],[6] is contained in the observation that the laser analysis is a self-consistent field theory in which the atomic dipoles are induced to emit in sympathy with a self-consistent field produced by the entire atomic ensemble. As mentioned in an earlier work [7] and developed in Ref.5, this is analogous to the ferromagnetic situation in which each spin interacts with all of the other spins through the mean self-consistent magnetization. The similarity between these two systems is summarized in table 2. The free energy of a thermodynamic phase transition in the Landau approximation to second order phase transitions leads to a free energy F(m) of the form shown in Fig. 1. The probability density P(m) then goes like exp(-F/kT). In the corresponding laser

---

* Work supported by the United States Air Force Office of Scientific Research.

≠ Alfred P. Sloan Fellow.

System	Order parameter η	Symmetry breaking mechanism μ	Reservoir variable ϵ
Magnet	M	H	$\frac{T - T_c}{T_c}$
Superconductor	$\lvert\Delta\rvert e^{i\theta}$	Related to Josephson current $j = j_0 \sin(\theta_1 - \theta_2)$	
Superfluid	$\sqrt{\rho_s}\ e^{i\theta}$		
Laser	E	S	$\frac{\sigma - \sigma_t}{\sigma_t}$

Table 1. This table outlines the similarities between the different systems near the critical temperature T_c. M and H are the Magnetization and external field. $|\Delta|$ and $\sqrt{\rho_S}$ are amplitude of the superfluid components having phase Θ. The current j $\sin\phi$ is set up when a tunnel junction experiment is carried out. The laser system is further discussed in the text and Table 2.

analysis the probability density for the electromagnetic field, P(E), (as obtained from the quantum theory of the laser) has the form $\exp(-G/K\sigma)$, where K is one-fourth the spontaneous emission rate, see table 1. The G factor has exactly the same form as that obtained from the Landau theory of second order phase transitions, as shown in Fig.1.

It should be emphasized that in the thermodynamic treatment of the ferromagnetic order disorder transition, there are three variables required: M, H and T. In order to have a complete analogy, it is important to realize that in addition to the electric field-magnetization, population inversion-temperature correspondences, there must exist a further correspondence between the external magnetic field and a corresponding symmetry-breaking mechanism in the laser analysis. As shown in Ref. 5 and illustrated in Fig. 2, this symmetry breaking mechanism in the laser problem corresponds to an injective classical signal S. This leads to a skewed effective free energy.

An example of how the analogy can provide us with deeper insight is contained in the fact that we are able to *guess* correctly

Summary of comparison between the laser and a ferromagnetic system treated in a mean field approximation.

	Order parameter	Reservoir variable	Coexistence curve	Symmetry breaking mechanism	Critical isotherm (value of order parameter at critical point)	
Ferromagnet	M	T (Temperature)	$M = \begin{cases} 0, & T > T_c \\ [\frac{c}{d}(\frac{T-T_c}{T})]^{1/2}, & T < T_c \end{cases}$	H External field	$M = (\frac{H}{dT_c})^{1/3}$	(Cont.)
Laser	E	σ (Population inversion) σ_t (Threshold inversion)	$E = \begin{cases} 0, & \sigma < \sigma_t \\ [\frac{a}{b}(\frac{\sigma-\sigma_t}{\sigma})]^{1/2}, & \sigma > \sigma_t \end{cases}$	S Injected classical signal	$E = (\frac{2S}{b\sigma_t})^{1/3}$	(Cont.)

	Zero field susceptibility	Thermodynamic potential	Statistical distribution
Ferromagnet (Cont.)	$X \equiv \frac{\partial M}{\partial H}\Big\vert_{H=0} = \begin{cases} [c(T-T_c)]^{-1}, & T > T_c \\ [2c(T_c-T)]^{-1}, & T < T_c \end{cases}$	$F(M) = \frac{1}{2} c(T-T_c)M^2 + \frac{1}{4} dTM^4 - HM + F_0$	$P(M) = N'' \exp(-\frac{F(M)}{kT})$
Laser (Cont.)	$\xi \equiv \frac{\partial E}{\partial S}\Big\vert_{S=0} = \begin{cases} [\frac{1}{2} a(\sigma_t - \sigma)]^{-1}, & \sigma < \sigma_t \\ [a(\sigma - \sigma_t)]^{-1}, & \sigma > \sigma_t \end{cases}$	$G(xy) = -\frac{1}{4} a(\sigma - \sigma_t)(x^2+y^2) + \frac{1}{8} b\sigma(x^2+y^2)^2 - Sx + G_0$	$P(E) = N' \exp(-\frac{G(E)}{K\sigma})$

Table 2. Laser-Ferromagnet Comparison

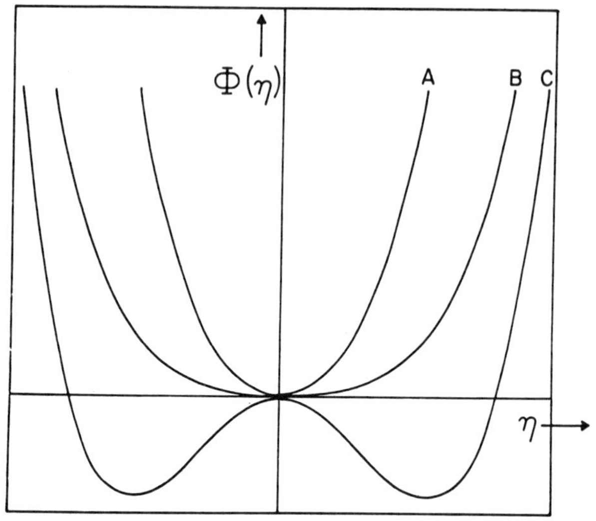

Figure 1. In this figure we plot the function $\Phi(\eta)$ for the laser, $\phi \equiv G$, below (A) at (B) and above (C) threshold and for the magnet $\phi \equiv F$ above (A) at (B) and below T_c.

Figure 2. Figure depicting the broken symmetry mode of operation for both a ferromagnet and a superconductor.

the P(E) for a laser influenced by an injected signal, by analogy with the corresponding magnetic problem in the broken symmetry mode of operation.

2. Statistical Dynamics of a Second Order Phase Transition Near T_c

Having seen then that the analogy provides deeper insight into laser operation, one is tempted to ask: "Can the time dependent statistical dynamics of a system near its phase transition be profitably studied in light of the laser analysis"? This has been investigated [8], [9] with Goldstein for a ferromagnet coupled to a thermal reservoir (as illustrated in Fig. 3)and the question seems to be answered in the affirmative. An equation of motion for the magnetization probability density where $\rho(t)$ is the total density operator for the entire system is found [8] to be

$$P(m,t) = Tr[\rho(t)\ \delta(m-\sum_i m_i)] \tag{1}$$

$$\dot{P}(m,t) = \left\{\frac{N}{a}\left[(e^{-\partial_m}-1)\ \Gamma_2(m) + (e^{\partial_m}-1)\ \Gamma_1(m)\right] + \left[(e^{\partial_m}-1)\ \Gamma_1(m) - (e^{-\partial_m}-1)\ \Gamma_2(m)\right]m\right\} P(m,t). \tag{2}$$

Here $\partial_m = \partial/\partial_m$ and the exponential of this differential operator implies an infinite order differential equation. The two damping functions $\Gamma_1(m)$ and $\Gamma_2(m)$ are defined by (see Fig. 3)

$$\Gamma_1(m) = 2g^2 Re \int_{-\infty}^{t} dt' \left\langle \sum O_\alpha^\dagger(t) \sum_\beta O_\beta(t') \right\rangle_R \exp\{i\omega_{21}(t-t')\}\ , \tag{3a}$$

and $\Gamma_2(m)$ is given by a similar expression found by replacing all 0's ($0^\dagger$'s) by $0^\dagger$'s (0's) and ω_{21} by $-\omega_{21}$.

Near T_c the equation of motion (2) is well approximated by a

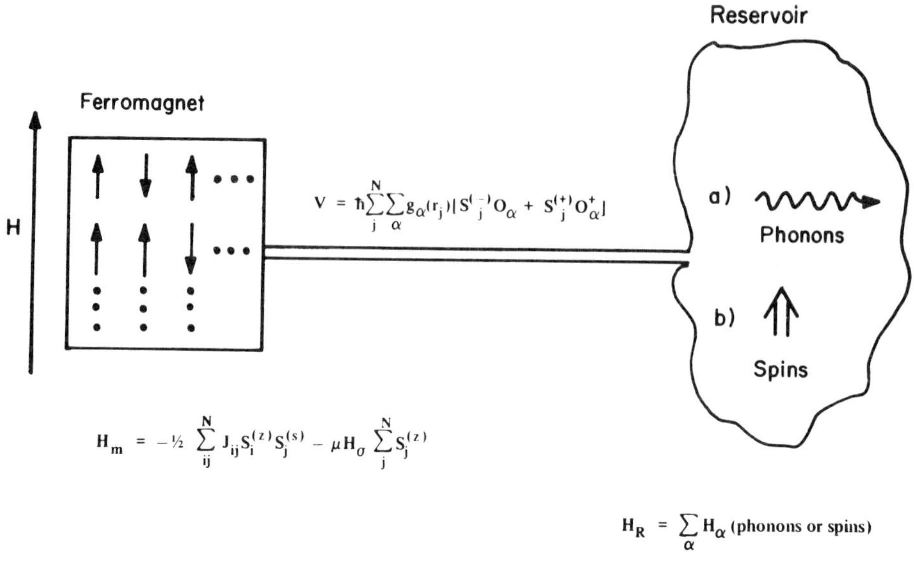

Figure 3. Interaction of ferromagnet with reservoir consisting of phonons or spins in thermal equilibrium.

Fokker-Planck equation of motion obtained by expanding the exponential factors up to second derivatives only:

$$\dot{P}(m,t) = \left\{ \frac{\partial}{\partial m} \left[\frac{N}{2} (\Gamma_1(m) - \Gamma_2(m)) + (\Gamma_1(m) + \Gamma_2(m))\, m \right] \right.$$

$$\left. + \frac{1}{2} \frac{\partial^2}{\partial m^2} \left[\frac{N}{2} (\Gamma_1(m) + \Gamma_2(m)) + (\Gamma_1(m) - \Gamma_2(m))\, m \right] \right\} P(m,t) \; . \tag{3b}$$

These equations of motion for the mean magnetization when the system is coupled to reservoirs of type (a) or (b), as in Fig.3, are then given by

$$(1/\gamma^{(a)}) \frac{d}{dt} \langle m(t) \rangle = 1 - \langle m(t) \rangle \coth[\tau(h + \langle m(t) \rangle)] \tag{4}$$

$$(1/\gamma^{(b)})\frac{d}{dt}\langle m(t)\rangle = \tanh[\tau(h+\langle m(t)\rangle)] - \langle m(t)\rangle \quad , \tag{5}$$

where $\gamma^{(a)}[\gamma^{(b)}]$ is a decay rate characteristic of the phonon [spin] reservoir and $h = H_o/(\lambda N\mu/\Omega)$; $\tau = T_C/T$; $k_B T_C = \lambda N\mu^2/\Omega$, with λ the Weiss internal field constant and Ω is the volume of the magnet. It is of interest to note that there are comments in the literature to the effect that irreversible effects may be nearly insensitive to the precise model of the reservoir used. In fact the mean equations of motion are very different for two reservoirs considered. It is amusing to note that, since for the laser $\dot{E}\propto\partial G/\partial E$, one might have thought that $\dot{M}\propto\partial F/\partial M$ for the ferromagnet (at least near T_c). This is not the case as is evidenced by Eqs. (4) and (5).

Let us next consider the steady state expression for P(m) as implied by Eq. (3). Calling the drift and diffusion coefficents A and B respectively, as in (3.03), we see that the steady-state solution $P^{(a)}(x)$ satisfies

$$0 = \frac{\partial}{\partial x}\left[-A\,P^{(a)}(x) + \frac{\partial}{\partial x}B\,P^{(a)}(x)\right] \tag{6}$$

and we obtain the steady-state solution:

$$P^{(a)}(x) = \eta\left\{\frac{1}{B(x)}\exp\left[\int_0^x dx'\,\frac{A(x')}{B(x')}\right]\right\} , \tag{7}$$

where η is a normalization constant. Noting that N is an extremely large number ($N \sim 10^{23}$), we find the steady-state solution to be

$$P^{(a)}(x) = N\exp -N\int_0^x dx'\,\frac{x'\coth[\tau(h+x')] - 1}{\coth[\tau(h+x')] - x'} . \tag{8}$$

For temperatures near T_c ($\tau \sim 1$), x will be small and we may expand the solution (8) (since the peak of the distribution occurs at the molecular field value of x, which will be small near T_c) to obtain (for zero external field)

$$P^{(a)}(x) \simeq C \exp -N \int_0^x dx'([(1-\tau)]\, x' + [\tau(1-\tau) + \frac{1}{B}\tau^3]\, x'^3)$$

$$= C \exp\left\{ - \frac{1}{K_B T}\left[C(T-T_c)\frac{m^2}{2} + d'T\frac{m^4}{4}\right]\right\} , \quad (9)$$

where

$$C = \frac{K_B}{N\mu^2} ; \; d' = [\tau^3 + 3\,(1-\tau)]d; \; d = \frac{K_B}{3N^3\mu^4} . \quad (10)$$

The thermodynamic theory of fluctuations predicts that the probability density for the fluctuations of a thermodynamic variable η is given by

$$P(\eta) = C\, e^{-G(\eta)/k_B T} , \quad (11)$$

where $G(\eta)$ is the appropriate thermodynamic potential predicted by Landau theory, namely,

$$G(\eta) = G_0 + \alpha(T - T_c)\;\eta^2 + \Delta\eta^4 , \quad (12)$$

where η is the order parameter of the transition. That is, η is a thermodynamic variable which is assumed to vanish above the transition temperature T_c (corresponding to the disordered state) and be non-zero below T_c (in the ordered state). In the case of a ferromagnet the order parameter is the magnetic moment m. The precise values of the constants α and Δ given by Eqs. (9) and (10) are not predicted by the Landau theory.

3. A "New" Approach to Laser Theory

Having established and investigated the laser-ferromagnet analogies from a macroscopic vantage point, we next turn to the microscopic question.

> "What in the laser problem is analogous to the individual microscopic spins in the sense that the macroscopic quantities E and M are analogous?"

The answer to this question requires taking a new look at the laser problem. This work has been carried out in collaboration with Lamb and Lang and comprises the substance of three papers [10], [11], [12] to be published in the Physical Review.

We consider the theory of an optical maser based on a model of a laser cavity which is coupled to the outside world. In such a model there are an infinite number of modes (of the universe) corresponding to each of the Fox-Li modes. The coupling to the outside world provides an effective quality factor Q of the cavity. In this way we investigate the conditions under which the ("microscopic") multi-mode approach reduces to the usual single ("macroscopic") quasi-mode treatment and obtain a deeper understanding of the mechanism leading to the very narrow line width of laser radiation.

Details of the calculation are given in Ref. 9. For those normal modes having frequency Ω_k (=ck) close to a Fox-Li "resonant" frequency $\Omega(=ck_o)$, the cavity eigenfunctions are found to be

$$u_k(z) = M_k \sin[k(z-\ell)] \qquad (z > 0) \tag{13a}$$

$$= \xi_k \sin[k(z+L)] \qquad (z < 0) \,, \tag{13b}$$

where ξ_k is a phase factor which alternates between 1 and -1 as k increases from one value to the next.

The coefficients M_k in (13a) are defined as

$$M_k = \Gamma \Lambda \{ (\Omega_k - \Omega)^2 + \Gamma^2 \}^{-\frac{1}{2}} \,, \tag{14}$$

where Γ is the band width associated with the mirror transparency. The density ρ of states of Ω_k space is found to be

$$\rho = L/c\pi \;. \tag{15}$$

An arbitrary undriven field in the entire cavity can be expressed as the real part of the complex field

$$E(z,t) = \sum_k E_k(0)\, u_k(z)\, e^{-i\Omega_k t} = e^{-i\Omega t} \sum_k E_k(t)\, u_k(z), \tag{16}$$

which is to be understood as a sum over modes of the large cavity; i.e., "the universe." The semitransparency of the mirror leads to a damping of free oscillations in the laser cavity. If at t=0, the laser cavity contains a field (in the complex notation) of the form

$$E(z,0) = |E_o| \, e^{-i\phi} \sin k_o(z-\rho) \quad , \tag{17}$$

while no field exists outside the cavity, at later times we find

$$E(z,t) = |E_o| \sin k_o(z-\ell) \exp\left[-i(\Omega t + \phi) - \Gamma t\right] \quad . \tag{18}$$

The equation (18) indicates that the field localized in the maser cavity decays exponentially due to leakage through the mirror at a rate Γ .

When the excited atoms are injected into the laser cavity and allowed to interact with the electromagnetic field, the atoms acquire a time dependent dipole associated with the off diagonal elements of the atomic density matrix. The collective effect of many atoms produces a macroscopic polarization, which acts as the driving force for the radiation field.

The projection of the macroscopic polarization onto the k-th mode has the following form

$$P_k(t) = \sum_{\mu} a_{k\mu} E_{\mu}(t) + \sum_{\mu\rho\sigma} b_{k\mu\rho\sigma} E_{\mu} E_{\rho}^{*} E_{\sigma} + \ldots . \tag{19}$$

The coefficients $a_{k\mu}$ and $b_{k\mu\rho\sigma}$ depend on properties of the lasing atoms and (in the case of no atomic motion) the field can now be written as follows:

$$\dot{E}_k(t) + i(\Omega_k - \Omega) E_k(t) = \alpha M_k \sum_{\mu} M_{\mu} E_{\mu}(t)$$

$$- \beta M_k \sum_{\mu\rho\sigma} M_{\mu} M_{\rho} M_{\sigma} E_{\mu}(t) \, E_{\rho}^{*}(t) \, E_{\sigma}(t) \quad , \tag{20}$$

where

$$\alpha = \nu a/\varepsilon_o L, \; \beta = \nu b/\varepsilon_o L. \quad (21)$$

It is to be noted that the range of k as well as μ, ρ and σ has been restricted to the neighborhood of one Fox-Li mode corresponding to the single quasi-mode laser operation. Equation (20) constitutes our working equation. In ref.[9] it is shown that the coupled multimode equations (20) can be transformed into a single equation which corresponds to the conventional quasimode theory of laser behavior. In terms of the quantity $A(t) = \sum M_k E_k(t)$ which can be identified as the complex amplitude of a Fox-Li it is "seen" (after considerable analysis) that A(t) obeys the equation

$$\dot{A}(t) + \Gamma A(t) = \alpha M^2 A(t) - \beta M^2 |A(t)|^2 A(t) \quad . \quad (22)$$

Separating real and imaginary parts we obtain the usual equations for phase and amplitude of single quasi-mode operation. Hence we see all the modes of the universe to "lock" at a single frequency with a definite phase relationship between them. The close analogy between the "mode-locking" here and the cooperative interaction between the spins in the ferromagnet or the electron pairs (in time reversed Block states) in the superconductor is evident. This and other considerations will be discussed in more detail elsewhere.

REFERENCES

1. H. E. Stanley, *Introduction to Phase Transition and Critical Phenomena*, (Oxford University Press, Oxford, 1971).
2. The quantum laser theory as developed by Scully and Lamb is reviewed in *Quantum-Optics, The Proceedings of the International School of Physics "Enrico Fermi,"* Course XLI, ed. R.J. Glauber (Academic Press, New York, 1969) p. 586.
3. M. Lax's work is reviewed in his *Brandeis Lectures in Theoretical Physics*, Chretien, Gross, and Deser (Gordon and Breach Science Publishers, New York 1968) Vol.2.
4. The work of the Stuttgart group is reviewed by W. Weidlich and H. Haken in *Quantum Optics, The Proceedings of the International School of Physics "Enrico Fermi,"* Course XLI, ed. R.J. Glauber (Academic Press, New York, 1969) p. 630.

5. V. DeGiorgio and M. O. Scully, Phys. Rev. *A2*, (1970) 1170.
6. R. Graham and H. Haken, Z. Physik *237*, (1970) 31.
7. M. Scully and W. E. Lamb, Jr., Phys. Rev. *159*, 203 (1967).
8. J. Goldstein, M. Scully, and P. Lee, Phys. Lett. *35A*, 317 (1971).
9. J. Goldstein and M. Scully, Phys. Rev. (to be published); see also, F. Haake (to be published).
10. R. Lang, M. Scully and W. Lamb, Phys. Rev. (to be published).
11. R. Lang and M. Scully, Phys. Rev. (to be published).
12. R. Lang and M. Scully, Phys. Rev. (to be published).

MULTIPLE-TIME-SCALE ANALYSIS OF COHERENT RADIATION

Paul S. Lee and Y. C. Lee

State University of New York, Buffalo, New York

The Multiple-Time-Scale Perturbation Theory (MTSPT) was first developed by Krylov and Bogoliubov[1] for solving non-linear mechanical problems. In 1963, Frieman[2] and Sandri[3] extended this method to study the kinetic theory of gases, making use of the existence of several distinct time scales in that theory.

In the ordinary time-dependent perturbation theory, it often happens that the n-th order term diverges like t^{α} at large t, where α is generally a function of n. For example, this secular behavior at large t occurs in the calculation of transition rates in quantum mechanics. When t is small, the first few terms in the expansion may suffice for the description of the behavior of the system; but at large t, an appropriate sum over an infinite number of terms must be carried out in order to obtain a finite result. In problems concerned with radiative processes, this defect can be corrected by realizing the existence of two very distinct time scales, one corresponding to the inverse of the frequency of the transition, ω_0^{-1}, the other corresponding to the inverse of the radiative linewidth, γ^{-1}. Based on these different time scales, a multiple-time-scale perturbation expansion can be constructed. Mathematically, this expansion is actually a rearrangement of the terms in the conventional perturbation series to eliminate the secular behavior in each order by exploiting the extra degrees of freedom afforded by the introduction of the multiple time variables. Physically, the detailed evolution of the system in the fine time scale of ω_0^{-1}, which is not of much interest, is essentially averaged over in the rougher time scale of γ^{-1} in this expansion.

We consider a system of two-level atoms interacting with a transverse radiation field. The interaction Hamiltonian is given by

$$H' = \sum_{i}\sum_{\underline{k}} \{ \varepsilon A^{*}_{\underline{k}} e^{i\underline{k}\cdot\underline{x}_i} C_{\underline{k}} R_{+}(i) + \varepsilon A_{\underline{k}} e^{-i\underline{k}\cdot\underline{x}_i} C^{+}_{\underline{k}} R_{-}(i) \} ,$$

where $C_{\underline{k}}$ and $C^{+}_{\underline{k}}$ denote, respectively, the photon annihilation and creation operators; $R_{+}(i)$ and $R_{-}(i)$ represent, respectively, Dicke's raising and lowering operators[4] for the i-th atom; $A_{\underline{k}}$ is essentially the matrix element of the atomic dipole moment. In accordance with the resonance approximation[4], terms in H' which do not lead to energy-conserving first order transitions have been dropped. The coupling strength parameter ε is to be set to unity at the end of the calculation.

To illustrate the method, let us first consider just one excited atom in the $|\uparrow\rangle$ state, separated from the ground state $|\downarrow\rangle$ by ω_0. The Schrödinger equation for

$$|\psi(t)\rangle = a(t)|\uparrow; 0_{\underline{k}}\rangle + \sum_{\underline{k}} b_{\underline{k}}(t)|\downarrow; 1_{\underline{k}}\rangle ,$$

can be written as

$$i\frac{d}{dt} a(t) = \sum_{\underline{k}} \varepsilon A^{*}_{\underline{k}} e^{i\omega_{0k}t} b_{\underline{k}}(t) ,$$

$$i\frac{d}{dt} b_{\underline{k}}(t) = \varepsilon A_{\underline{k}} e^{-i\omega_{0k}t} a(t) .$$

As is well known, the ratio of the radiation linewidth γ_0 to the radiation frequency ω_0 is $\sim \varepsilon^2 e^2v^2/\hbar c^3$, where v is roughly the electron velocity within the atom. Accordingly we now introduce the multiple-time-scale and replace the original time variable t by τ, which represents collectively the variables $\tau_0, \tau_1, \tau_2, \ldots$, defined by $\partial\tau_n/\partial t = \varepsilon^n$, n = 0,1,2,... . The original time derivative d/dt now becomes

$$\frac{d}{dt} \rightarrow \frac{\partial}{\partial\tau_0} + \varepsilon\frac{\partial}{\partial\tau_1} + \varepsilon^2\frac{\partial}{\partial\tau_2} + \ldots .$$

Thus the rate of change of the amplitudes a and b_k will be analyzed according to the different time scales τ_n. By expanding $a(\tau)$ and $b_k(\tau)$ into power series of ε and by exploiting the extra degrees of freedom due to the introduction of the additional variables τ_1, τ_2,.... we can impose, for example, conditions on $\partial a^{(m)}(\tau)/\partial\tau_n$ and on $\partial b_k^{(m)}(\tau)/\partial\tau_n$ for $n \geq 1$ to suit our convenience. Thus from the considerations of the 0-th and the first order equations we can easily establish that both amplitudes $a(\tau)$ and $b_k(\tau)$ must be independent of the variables τ_0 and τ_1 in order that the secular behavior be eliminated. Similar considerations in higher orders yield

$$\frac{\partial}{\partial\tau_2} a^{(o)}(\tau_2,\ldots) = i\,(\omega_s + i\gamma_s)\, a^{(o)}(\tau_2,\ldots) \quad ,$$

$$a^{(1)}(\tau) = 0,$$

$$\frac{\partial}{\partial\tau_0} a^{(2)}(\tau_0,\tau_2) = -\sum_{\underline{k}}|A_{\underline{k}}|^2\Big[\int_0^{\tau_0} du\, e^{i\omega_{ok}u} - i\zeta(\omega_{ok})\Big]\, a^{(o)}(\tau_2,\ldots),$$

where

$$\zeta(x) \equiv \lim_{\delta\to o} \frac{1}{x+i\delta} = P\,\frac{1}{x} - i\pi\delta(x),$$

and

$$\omega_s + i\gamma_s \equiv -\sum_{\underline{k}}|A_{\underline{k}}|^2\,\zeta(\omega_{ok}) \; .$$

The solution at large τ_0 is

$$a^{(o)}(\tau) = e^{i\omega_s\tau_2-\gamma_s\tau_2} \quad ,$$

$$a^{(1)}(\tau) = 0 \; ,$$

$$a^{(2)}(\tau) \to 0,$$

$$a^{(3)}(\tau) = 0,$$

$$a^{(4)}(\tau) \to 0,$$

.....

This means that in the limit of large τ_0, the amplitude $a(\tau)$ is

$$a = a^{(0)}(\tau_2) = e^{i\omega_s \tau_2 - \gamma_s \tau_2} .$$

Here we see that, indeed, only the two physically meaningful time scales τ_0 and τ_2 enter into the final result in accordance with our previous discussion and that, in the long time limit, the fine details in the scale of τ_0 has been smoothed out so that only the averaged behavior is manifested in the rougher time scale of τ_2. Similarly, the other amplitude $b_k(\tau_0, \tau_2)$ can also be found exactly in the limit of large τ_0. After setting ε back to unity, the result is

$$b_{\underline{k}}(t) = A_{\underline{k}} \frac{e^{-i\omega_{ok}t} \, e^{i(\omega_s + i\gamma_s)t} - 1}{\omega_{ok} - \omega_s - i\gamma_s} ,$$

which leads to the usual Lorentzian line shape.

This same method of MTSPT has now been applied to a system of N atoms, with and without random motion. Again, exact solutions can be found and the effect of many-body (including 2-, 3-, 4-, ... etc.) correlations can be examined in detail in many cases, in which the interparticle distance is not necessarily much smaller or larger than the spontaneous radiation wavelength. From these solutions the following qualitative results can be extracted:

i) There are in general, in the radiation spectrum, of the order of N peaks which are grouped around ω_0.

ii) When the interparticle distance r is small compared to the spontaneous radiation wavelength λ, the N-body system can be looked upon as forming many overlapping groups. These groups all have radii $R \sim \lambda/2$. Particles inside each group are strongly correlated while the effect from particles outside the group is negligibly small. This indicates that a coherence range $l_{coh} \sim \lambda/2$ can be associated with such a system.

iii) When the interparticle distance r is not small but of the order of $\lambda/2$, which is the regime of most experimental works[5], the coherence range l_{coh} loses its meaning and some striking results are found. For example, let us consider a one-dimensional model where all atoms are lying along a straight line. Under the condition that only the two-body correlation is included and initially there is only one atom in the excited state, the many-body effect on the frequency shift and the linewidth broadening of the coherent peak is plotted in Fig. 1.

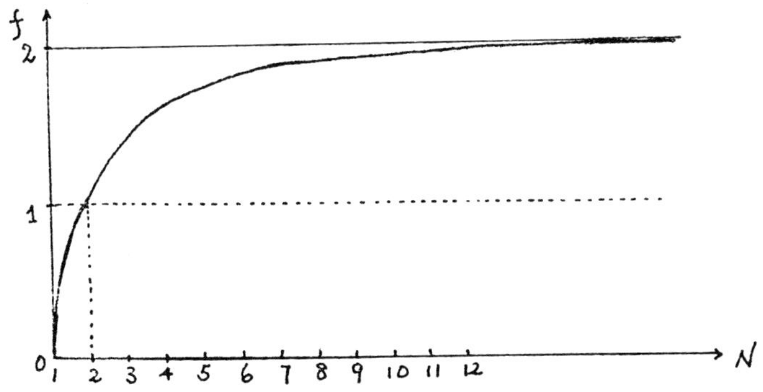

Fig. 1.

where N is the number of particles in the system and f is a quantity which is defined by

$$f \equiv \frac{\omega(N) + \omega_s}{\omega(2) + \omega_s} \approx \frac{\gamma(N) - \gamma_s}{\gamma(2) - \gamma_s} ,$$

where $\omega(N)$ is the frequency shift and $\gamma(N)$ is the linewidth of the coherent spectral line of the N atom system. When the three-, four-, ... body correlations are included, it is found that the quantity f fluctuates around 2 and behaves similarly as that plotted in Fig. 1.

iv) Dicke's result[4] is a special case of the general solution under the condition that $kr \ll 1$.

References

1. N.M. Krylov and N.N. Bogoliubov, *Introduction to Nonlinear Mechanics*, (trans. by Kraus Reprint Co., Millwood, N.Y., 1943).

2. E.A. Frieman, J. Math. Phys. *4*, 410 (1963).

3. G. Sandri, Phys. Rev. Letters *11*, 178 (1963).

4. R.H. Dicke, Phys. Rev. *93*, 99 (1954), Y.C. Lee and D.L. Lin, Phys. Rev. *183*, 147 (1969), ibid. *183*, 150 (1969) and references contained therein.

5. H.G. Kuhn and J.M. Vaughan, Proc. Roy. Soc. (London) *A277*, 297 (1964).

ORTHOGONAL OPERATORS AND PHASE SPACE DISTRIBUTIONS IN QUANTUM OPTICS*

S. Billis and E. A. Mishkin

Polytechnic Institute of Brooklyn, Brooklyn, N. Y.

1. Introduction

Electromagnetic fields at optical frequencies are excited by indeterministic sources. Their statistical description is usually given by means of the density operator $\hat{\rho}$ or a phase-space distribution which is, in general, a quasi-probability distribution. Expansions of the density matrix $\hat{\rho}$ in terms of a given complete set of orthogonal operators establishes a one-to-one correspondence between the expanded operator $\hat{\rho}$ and the quasi-probability phase-space distribution that appears in said expansion. The time evolution of the field's statistics are obtained then as a solution of the equation of motion of the density operator $\hat{\rho}$ or the differential equation of the corresponding phase-space distribution. The latter method has the advantage of being unburdened by problems of commutativity associated with the $\hat{\rho}$ operator. With the statistical information vested in the phase-space distribution, in lieu of the density matrix $\hat{\rho}$, the expected value of an observable $\hat{F}$ is given by an integral of the phase-space distribution multiplied by a weighting function which is representative of the observable $\hat{F}$. This is essentially the method first introduced by Wigner[1] and Moyal[2].

In this presentation the dynamical observables of the quantum system are statistical variates. The phase-space distribution which provides a complete dynamical specification of the system involves the two complementary complete sets of observables. Each

* Work supported by the Office of Naval Research

complete set of operators, associated with these observables, does not commute with the other.

For simplicity, we limit the following computations to systems with a single degree of freedom. In quantum optics this corresponds to the single mode electromagnetic field. It usually involves then the photon creation and annihilation operators $\hat{a}^\dagger$, $\hat{a}$ in lieu of $\hat{q}, \hat{p}$.

The unitary Weyl operators[3] $\hat{D}(\alpha) = \exp(\alpha\hat{a}^\dagger - \overset{*}{\alpha}\hat{a})$, which induces the coherent field $|\alpha\rangle = \hat{D}(\alpha)|0\rangle$ maps the complex plane onto a vector space of operators, orthogonal in the sense, $(\hat{D}(\alpha), \hat{D}(\alpha')) = \pi\, \delta^2(\alpha-\alpha')$, where the inner product of operators implies the trace of the product of the first operator with the adjoint of the other. This allows for expansions of a wide class of arbitrary operators in a manner analogous to the well known expansions in terms of orthogonal functions. The operator expansion establishes a one-to-one correspondence between the expanded operator and its weighting function which is, in general, a quasi-probability distribution.

We introduce in this paper several methods for finding other complete orthonormal operator sets which afford alternate integral, or series, expansions for a weakly constrained set of operators. We define in Section 2 the orthogonality concept and its use in operator integral expansions in terms of known complete sets of orthogonal operators and extend the well known Parseval theorem to these operator expansions. In Section 3 we obtain through an isomorphic mapping on sets of orthogonal functions complete one and two dimensional sets of orthonormal operators of discrete and continuous parameters. In Section 4 we consider operator sets derived by means of orthogonality conserving transformations and consider the completeness. The properties of two particular orthogonal operators which depend upon a discrete and continuous parameter are presented then in Section 5 and 6, respectively. In Section 7, equivalent Heisenberg equation of motion in the phase-space is derived for an arbitrary operator which is implicitly time independent in the Schrodinger picture. In the last section, by means of an orthogonal operator expansion, we reduce the operator equation of motion for the reduced system density matrix, in the interaction picture, to an equivalent integro-differential equation of the corresponding phase-space distribution.

2. Inner Product and Orthogonality

We define the trace of the product of the two operators $\hat{A}$ and

$\hat{B}$ as their inner product

$$(\hat{A},\hat{B}) = \mathrm{Tr}\,[\hat{A}\ \hat{B}^{+}] \quad . \tag{2.1}$$

The usual conjugation relation

$$(\hat{A},\hat{B}) = (\hat{B},\hat{A})^{*} \tag{2.2}$$

follows from this definition. We term the two operators $\hat{A}$ and $\hat{B}$ orthogonal to each other when the trace of their product vanishes

$$\mathrm{Tr}(\hat{A}\ \hat{B}^{+}) = (\hat{A},\ \hat{B}) = 0 \quad \hat{A} \neq \hat{B} \tag{2.3}$$

Let $\hat{A}(\alpha)$, α a complex number, be an infinite continuous set of operators in an inner product space V. Its orthogonal complement $\hat{B}(\alpha')$ is that set of operators in V for which[4]

$$(\hat{A}(\alpha),\ \hat{B}(\alpha')) = \delta^{2}(\alpha-\alpha'), \tag{2.4}$$

$\delta^{2}(\alpha)$ is a two-dimensional Dirac function, $\delta^{2}(\alpha-\alpha') = \delta(\alpha_r-\alpha'_r)\,\delta(\alpha_i-\alpha'_i)$. $\hat{B}(\alpha')$ is a subspace of V whether or not $\hat{A}(\alpha)$ is one. The operator set $\hat{A}(\alpha)$ is complete when the subspace spanned by $\hat{A}(\alpha)$ is the whole space V. Equivalently, if $\hat{C}$ is any vector operator in V the set $A(\alpha)$ is complete if, and only if,

$$(\hat{C},\ \hat{B}(\alpha')) = 0 \tag{2.5}$$

implies that C = 0.

The orthogonality and completeness properties, Eq. (2.4) allow integral expansions for a wide class of weakly constrained operators in terms of the operator sets $A(\alpha)$, $B(\alpha')$. Consider the weighted integral $\hat{F}$, taken over the whole complex plane ξ, of the operators $\hat{A}(\xi)$ with the weighting function $f(\xi)$

$$\hat{F} = \int f(\xi)\ \hat{A}\ (\xi)\,d^{2}\xi \tag{2.6}$$

Parseval's identity for the inner product of 2 elements of a finite inner product space may be extended now to integral operator expansions of the form Eq. (2.6)

$$(\hat{F},\hat{G}) = \int f(\xi)\ (\hat{A}(\xi),\ \hat{G})\ d^{2}\xi \quad . \tag{2.7}$$

When $\hat{G}$ is the orthogonal complement $\hat{B}(\alpha')$, of the operator set $\hat{A}(\alpha)$ of the integral expansion (2.6)

$$f(\alpha) = (\hat{F},\hat{B}(\alpha)) \tag{2.8}$$

The Hilbert Schmidt norm of the operator $\hat{F}$[5] is then a special case of eq. (2.7) and it is finite when the functions $f(\alpha)$, $f'(\alpha)$

$$\begin{aligned} f(\alpha) &= (\hat{F},\hat{B}(\alpha)) \\ f'(\alpha) &= (\hat{F},\hat{A}(\alpha)) \end{aligned} \tag{2.9}$$

are continuous, square integrable, but not necessarily bounded functions

$$(\hat{F},\hat{F}) = (f,f') \tag{2.10}$$

3. Orthogonal Operators Derived from Complete Sets of Orthogonal Functions

We consider the observable $\hat{L}$ with the real continuous set of characteristic values ℓ and complete orthonormal set of characteristics kets $|\ell\rangle$

$$\hat{L}|\ell\rangle = \ell|\ell\rangle; \quad \langle\ell|\ell'\rangle = \delta(\ell-\ell'). \tag{3.1}$$

$\psi(q,\ell)$ and its Fourier transform $\phi(p,\ell)$ are the position and momentum representations of the characteristic kets $|\ell\rangle$, respectively

$$\psi(q,\ell) = \langle q|\ell\rangle \; ; \quad \phi(p,\ell) = \langle p|\ell\rangle \tag{3.2}$$

The one-to-one mapping τ of the complete set of orthogonal functions $\psi(q,\ell)$ onto the operator set $\hat{\psi}(\hat{q},\ell)$

$$\tau\colon\ \psi(q,\ell) \rightarrow \hat{\psi}(\hat{q},\ell) \tag{3.3}$$

results in a complete orthogonal set of operators

$$(\hat{\psi}(\hat{q},\ell),\ \hat{\psi}(\hat{q},\ell)) = \mathrm{Tr}\ [\hat{\psi}(q,\ell)\hat{\psi}^{\dagger}\ (\hat{q},\ell')]$$

$$= \int \langle q|\hat{\psi}^{\dagger}\ (\hat{q},\ell')\hat{\psi}(\hat{q},\ell)|q\rangle\ dq \qquad (3.4)$$

$$= \delta(\ell-\ell')$$

The mapping τ is not an isomorphism[6] since the sets of functions $\psi(q,\ell)$ are not closed under the algebraic operations of multiplication and addition and do not form a group.

Let C_g denote the set of functions $g(q)$ with valid integral representations in terms of the set of orthogonal functions $\psi(q,\ell)$ and $\hat{C}_F$ the set of operators which result from the mapping τ on the set C_g

$$\tau :\quad Cg \to \hat{C}_{\hat{F}}$$
$$g(q) \to \hat{g}(\hat{q}) = \hat{F} \qquad (3.5)$$

The operators $\hat{C}_{\hat{F}}$ may now be represented as weighted integrals in terms of the set of orthogonal operators $\hat{\psi}(q,\ell)$ ensuring the completeness of the set $\hat{\psi}(\hat{q},\ell)$ of Eq. (3.3). The set C_g forms a group under the binary operation of addition and the mapping (3.7) preserves the algebraic structure of the group

$$\tau\ [g_1(q) + g_2(q)] = \tau[g_3(q)]$$
$$= \hat{g}_3(\hat{q}) \qquad (3.6)$$
$$= \hat{g}_1(\hat{q}) + \hat{g}_2(\hat{q})$$
$$= \tau[g_1(q)] + \tau[g_2(q)]$$

for all $g_1(q)$ and $g_2(q)$ which are elements of C_g and can be viewed as an isomorphism.

When the spectrum of the observable $\hat{L}$ is discrete, the Dirac delta functions of Eq. (3.1) and (3.4) reduce to Kroenecker deltas and the expansions of the operators $\hat{F}$ in terms of the complete discrete sets $\psi_{\ell}(\hat{q})$ and $\phi_{\ell}(\hat{p})$ read

$$\hat{F} = \sum_{\ell} A_{\ell} \hat{\phi}_{\ell} (\hat{p})$$
$$\hat{F} = \sum_{\ell} B_{\ell} \hat{\psi}_{\ell} (\hat{q}) \tag{3.7}$$

with the coefficients A_{ℓ} and B_{ℓ} given by the inner products

$$A_{\ell} = (\hat{F}, \hat{\phi}_{\ell}(\hat{p}))$$
$$B_{\ell} = (\hat{F}, \hat{\psi}_{\ell}(\hat{q})) \tag{3.8}$$

Operators $\hat{U}(\ell,k)$, which are elements of the set $\hat{U}_{\hat{p}\hat{q}}$ defined as the ordered product of the two sets $\hat{\phi}_{\hat{p}}$ and $\hat{\psi}_{\hat{q}}$

$$\hat{U}_{\hat{p}\hat{q}} = \sqrt{2\pi\hbar}\ \hat{\phi}_{\hat{p}} \times \hat{\psi}_{\hat{q}} \tag{3.9}$$

have the form

$$\hat{U}\ (\ell,k) = \sqrt{2\pi\hbar}\ \hat{\phi}\ (\hat{p},\ell)\ \hat{\psi}\ (\hat{q},k) \tag{3.10}$$

The operators $\hat{U}(\ell,k)$ are easily shown to form a complete orthonormal set

$$(\hat{U}(\ell,k),\ \hat{U}(\ell',k')) = 2\pi\hbar\ \mathrm{Tr}[\hat{U}(\ell,k)\ \hat{U}^{\dagger}\ (\ell',k')] =$$
$$\delta(\ell-\ell')\ \delta(k-k') \tag{3.11}$$

see Eq. (3.4), they allow, therefore, integral expansion for a set of weakly constrained ordered operators in $\hat{p}$ and $\hat{q}$. A similar set of conclusions could be drawn for the set of operators $\hat{U}_{\hat{q}\hat{p}}$.

4. Orthogonality Conserving Transformations

The coherent state representations $\phi(\alpha,\ell)$ of the characteristic kets $|\ell\rangle$ of the observable $\hat{L}$

$$\phi(\alpha,\ell) = \langle \ell|\alpha\rangle \tag{4.1}$$

form a complete orthonomal set

$$(\phi(\alpha,\ell),\ \phi\ (\alpha,\ell')) = \pi\delta(\ell-\ell') \tag{4.2}$$

We used here the completeness property of the coherent states[7,8]

$$\pi^{-1}\int|\alpha\rangle\langle\alpha|d^2\alpha = 1 \tag{4.3}$$

With the unitary Weyl displacement operations $\hat{D}(\alpha)$[3]

$$\hat{D}(\alpha) = e^{\alpha\hat{a}^\dagger-\alpha^*\hat{a}}\ ;\ \hat{D}^\dagger(\alpha) = \hat{D}^{-1}(\alpha) = \hat{D}(-\alpha) \tag{4.4}$$

forming a complete orthonormal set, as defined by Eq. (2.4),

$$(\hat{D}(\alpha),\ \hat{D}(\beta)) = \pi\ \delta^2(\alpha-\beta), \tag{4.5}$$

but for the constant π, we define their transformation

$$\hat{F}(\ell) = \pi^{-1}\int d^2\xi\ \phi(\xi,\ell)\hat{D}(\xi) \tag{4.6}$$

which leads to the new set of operators $\hat{F}(\ell)$ dependent on the real parameter ℓ. Clearly, the transformation (4.6) conserves the orthogonality property of the $\hat{D}(\alpha)$ operations since

$$\begin{aligned}(\hat{F}(\ell),\ \hat{F}(\ell')) &= \pi^{-2}\int d^2\xi d^2\zeta\ \phi(\xi,\ell)\ \phi^*(\zeta,\ell)(\hat{D}(\xi),\ \hat{D}(\zeta)) \\ &= \delta(\ell-\ell')\end{aligned} \tag{4.7}$$

see equations (4.2) and (4.5).

The uniqueness of the functions $\phi(\alpha,\ell)$ ensures the completeness property of the operator set $\hat{F}(\ell)$. The operator set $\hat{F}(\ell)$

is "one-dimensional". It affords weighted integral representations for operators $\hat{Q}$ over the least largest interval of the real axis which includes all the characteristic values of $\hat{L}$

$$\hat{Q} = \int q(\ell)\ \hat{F}(\ell)\ d\ell\ :\quad q(\ell) = (\hat{Q},\hat{F}(\ell))\quad . \tag{4.8}$$

The two-dimensional delta function

$$\delta^2(\alpha) = \delta(\alpha_r)\ \delta(\alpha_i)\ ;\quad \alpha = \alpha_r + i\alpha_i \tag{4.9}$$

is alternately expressable in integral form in terms of the orthogonal functions ψ,ϕ

$$\int d^2\xi\ \psi(\xi_r,\alpha_r)\ \psi^*(\xi_r\alpha'_r)\ \phi(\xi_i,\alpha_i)\phi^*(\xi_i,\alpha'_i)$$
$$= \delta^2(\alpha-\alpha') \tag{4.10}$$

We consider now the operator $\hat{H}(\alpha)$ of the complex variable α defined by the integral

$$\hat{H}\ (\alpha) = \int d^2\xi\psi(\xi_r,\alpha_r)\ \phi\ (\xi_i,\alpha_i)\ \hat{D}\ (\xi) \tag{4.11}$$

The inner product of the operator $\hat{H}(\alpha)$ with $\hat{H}(\alpha')$ is given by

$$(\hat{H}(\alpha),\ \hat{H}(\alpha')) = \int d^2\xi d^2\zeta\quad \psi(\xi_r,\alpha_r)\ \phi(\xi_i,\alpha_i)\psi^*(\zeta_r,\alpha'_r)\phi^*(\zeta_i,\alpha'_i)$$
$$(\hat{D}(\xi),\ \hat{D}(\zeta)) = \pi\delta^2(\alpha-\alpha') \tag{4.12}$$

which may be regarded as a species of orthogonality rule for the "two-dimensional" operator set $\hat{H}(\alpha)$.

5. Discrete Sets of Orthogonal Operators

We presently identify the observable $\hat{L}$ with the single mode photon number operator $\hat{N} = \hat{a}^\dagger\hat{a}$ and generate thereby, with the aid of the orthogonal transformation (4.6), the discrete set of orthogonal operators $\hat{F}_n$

$$\begin{aligned}\hat{F}_n &= \pi^{-1} \int d^2\xi \langle\xi|n\rangle \hat{D}(\xi) \\ &= (\pi\sqrt{n!})^{-1} \int d^2\xi \; e^{-1/2|\xi|^2} \; \xi^{*n} \hat{D}(\xi)\end{aligned} \tag{5.1}$$

where $|n\rangle$ denotes the constant photon number state.

The operators $\hat{F}_n$ form a complete orthogonal set

$$(\hat{F}_n, \hat{F}_m) = \pi^{-1} \int d^2\xi \langle m|\xi\rangle\langle\xi|n\rangle = \delta_{nm} \tag{5.2}$$

with the following properties

$$\begin{aligned}\hat{F}_n|\beta\rangle &= \pi^{-1} \int d^2\xi \langle\xi|n\rangle e^{-1/2|\xi|^2} \; e^{\xi\hat{a}^\dagger} \; e^{-\xi^*\beta}|\beta\rangle \\ &= \pi^{-2} (\sqrt{n!})^{-1} \int d^2\xi d^2\gamma \; e^{-|\xi|^2 - \xi^*\beta} \xi^{*n} \; e^{\xi\gamma^*} |\gamma\rangle\langle\gamma|\beta\rangle\end{aligned} \tag{5.3}$$

We integrate first with respect to the variable ξ, with the aid of the identity[5]

$$\pi^{-1} \int d^2\xi \; f(\xi) \; e^{\xi^* y - z|\xi|^2} = \frac{1}{z} f\,(y/z) \quad \text{Re } z > 0 \tag{5.4}$$

and obtain

$$\begin{aligned}&\pi^{-1} \int d^2 \xi^{*n} \; e^{-|\xi|^2} \; e^{-\xi^*\beta} \; e^{\xi\gamma^*} \\ &= (-1)^n d^n_\beta \;\; \pi^{-1} \int d^2 \xi \; e^{-|\xi|^2 - \xi^*\beta + \xi\gamma^*} \\ &= (-1)^n d^n_\beta \; \{e^{-\beta\gamma^*}\} = \gamma^{*n} e^{-\beta\gamma^*}\end{aligned}$$

Eq. (5.3) now takes the form

$$\hat{F}_n|\beta\rangle = (\pi\sqrt{n!})^{-1} \int d^2\gamma \; e^{-\beta\gamma^*} \; \gamma^{*n} \; |\gamma\rangle\langle\gamma|\beta\rangle$$

$$= (\pi \sqrt{n!})^{-1} \int d^2\gamma \; \gamma^{*n} \; e^{-\frac{1}{2}|\beta|^2 \; -\frac{1}{2}|\gamma|^2} |\gamma>$$

$$= e^{-\frac{1}{2}|\beta|^2} \; \pi^{-1} \int d^2\gamma \; |\gamma><\gamma|n> \tag{5.5}$$

$$= e^{-\frac{1}{2}|\beta|^2} \; |n> = |n><0|\beta>$$

A similar computation shows the effect of the adjoint operator $\hat{F}_n^{\dagger}$ on the coherent state $|\beta>$

$$\hat{F}^{\dagger}_n |\beta> = |0><n|\beta> \tag{5.6}$$

Suggesting the off diagonal matrix element form for the operator $\hat{F}_n$

$$\hat{F}_n = |n><0| = \frac{\hat{a}^{\dagger n}}{\sqrt{n!}} \; \hat{F}_o \tag{5.7}$$

With the aid of the coherent state representation of the operators $\hat{F}_n$

$$<\alpha|\hat{F}_n|\beta> = \frac{e^{-\frac{1}{2}|\beta|^2 \; -\frac{1}{2}|\alpha|^2} \; \alpha^{*n}}{\sqrt{n!}} \tag{5.8}$$

we arrive at another formal expression of the $\hat{F}_n$ operator

$$\hat{F}_n = (n!)^{-\frac{1}{2}} \; \hat{a}^{\dagger\, n} 0^{\hat{a}^{\dagger}\hat{a}} \tag{5.9}$$

by comparing the series

$$G = \frac{(1-\hat{a}^{\dagger}\hat{a} + \hat{a}^{\dagger 2}\hat{a}^{2} \; \ldots)}{2!} \tag{5.10}$$

and its matrix elements

$$<\alpha|\hat{G}|\beta> = \left(1 - \alpha^*\beta + \frac{\alpha^{*2}\,\beta^2}{2!} \cdots \right) <\alpha|\beta>$$

$$= e^{-\frac{1}{2}|\alpha|^2}\, e^{-\frac{1}{2}|\beta|^2} \tag{5.11}$$

with the matrix elements (5.8). The operator $\hat{F}_n$ therefore can be written in the ordered form

$$\hat{F}_n = \frac{\hat{a}^{\dagger n}}{\sqrt{n!}} \quad 1 - \hat{a}^\dagger\hat{a} + \frac{\hat{a}^{\dagger 2}\hat{a}^2}{2!} \cdots \tag{5.12}$$

With the normally ordered form of the operator $e^{x\hat{a}^\dagger\hat{a}}$[9]

$$:e^{x\hat{a}^\dagger\hat{a}}: = \sum_{r=0}^{\infty} \frac{(e^x-1)^r\, \hat{a}^{\dagger r}\hat{a}^r}{r!} \tag{5.13}$$

suggesting the formal expression (5.9).

$$\hat{F}_o = 0^{\hat{a}^\dagger\hat{a}} = |0><0| \tag{5.14}$$

in agreement with Eq. (5.9).

We may define now another set of discrete orthogonal operators $\hat{G}_{n,m}$ which depend on two discrete parameters n,m

$$\hat{G}_{n,m} = \hat{F}_n\hat{F}_m^\dagger \ ; \ (\hat{G}_{n,m},\ \hat{G}_{n',m'}) = \delta nn'\ \ \delta mm' \tag{5.15}$$

The photon annihilation and creation operators, expanded in terms of the set $\hat{G}_{n,m}$ read

$$\hat{a} = \sum_n \sqrt{n+1}\quad \hat{G}_{n,n+1} = \sum_n \sqrt{n+1}\ |n><n+1| \ ; \ \hat{a}^\dagger = \sum_n \sqrt{n+1}|n+1><n| \tag{5.16}$$

Additional complete sets of discrete orthogonal operators can be constructed by identifying L with other known observables with discrete spectra.

6. The $\hat{T}(\varepsilon,\xi)$ Orthogonal Operators[10]

As an example of an orthogonal operator which is a function of a continuous parameter we consider the Hermetian operator $\hat{T}(\varepsilon,\alpha)$ of the complex variable α, and real parameter $-1<\varepsilon<1$

$$\hat{T}(\varepsilon,\alpha) = \pi^{-1} \int d^2\xi \quad e^{\xi^*(\alpha - \frac{\hat{\gamma}}{1-\varepsilon^2}) - \xi(\alpha^* - \frac{\hat{\gamma}^+}{1-\varepsilon^2})} \; ; \; \hat{T}^\dagger(\varepsilon,\alpha)=\hat{T}(\varepsilon,\alpha) \tag{6.1}$$

For, $\varepsilon=0$, it coincides with the parity operator $\hat{T}(\alpha)$ developed by Cahill and Glauber[5]. This operator arises in conjunction with the minimum uncertainty states[11,12] $|\gamma\rangle$ which are the characteristic states of the operator

$$\begin{aligned} \hat{\gamma} &= \hat{a} + \varepsilon\, \hat{a}^\dagger ; \quad -1<\varepsilon<1 \\ \hat{\gamma}|\gamma\rangle &= \gamma|\gamma\rangle ; \\ [\hat{\gamma},\hat{\gamma}^\dagger] &= (1-\varepsilon^2) \end{aligned} \tag{6.2}$$

and display anti-correlation effects for $\varepsilon>0$[14,15] Gaussian functions.

The operators $\hat{T}(\varepsilon,\alpha)$ form a complete orthonormal set

$$(\hat{T}(\varepsilon,\alpha),\ \hat{T}(\varepsilon,\beta)) = \frac{\pi\ \delta^2(\alpha-\beta)}{1-\varepsilon^2} \tag{6.3}$$

which therefore, can serve as another basis for an integral representation for arbitrary weakly constrained operators[13,16].

By recalling the familiar formula for the Fourier integral representation of the two-dimensional δ-function

$$\delta^2(\alpha) = \pi^{-2} \int d^2\xi\ e^{\alpha\xi^* - \alpha^*\xi}$$

we may express the operator $\hat{T}(\varepsilon,\alpha)$ in form of a δ-function involving the operator γ in its argument

$$\hat{T}(\varepsilon,\alpha) = \pi\ \delta^2\left(\alpha - \frac{\hat{\gamma}}{1-\varepsilon^2}\right) \tag{6.4}$$

In terms of this notation we may write the representation (2.8) in the form

$$\hat{F}(\gamma) = \int d^2 \xi \; f(\varepsilon,\xi) \; \delta^2 \left(\xi - \frac{\hat{\gamma}}{1-\varepsilon^2}\right) \tag{6.5}$$

The weight function associated with this representation bears a particularly direct relationship to the operator being expanded.

The operator $\hat{T}_A(\varepsilon,\xi)$ formed by an antinormal ordering of the integrand in Eq. (4.1) is the projection operator upon the minimum uncertainty state[10]. The expansion in terms of the $\hat{T}_A(\varepsilon,\xi)$ operators leads to a representation similar to the diagonal or P, representation of the density matrix operator $\hat{\rho}$ which has been widely discussed in the literature on quantum optics[17-23].

The minimum uncertainty states $|0\rangle_\gamma$ corresponding to the characteristic value $\gamma=0$ is not the ground state of the harmonic oscillator. The energy associated with this state is

$$\langle E\rangle = \frac{1+\varepsilon^2}{1-\varepsilon^2}\frac{\hbar\omega}{2} \tag{6.6}$$

The states $|n\rangle_\gamma$ follow from the $|0\rangle_\gamma$ state

$$|n\rangle_\gamma = \frac{\hat{\gamma}^{\dagger n}}{(1-\varepsilon^2)^{n/2}(n!)^{\frac{1}{2}}} |0\rangle_\gamma \tag{6.7}$$

They form a complete orthonormal discrete basis

$$\begin{aligned} {}_\gamma\langle m|n\rangle_\gamma &= \delta_{nm} \\ \sum_n |n\rangle_\gamma\, {}_\gamma\langle n| &= 1 \end{aligned} \tag{6.8}$$

and satisfy the characteristic equation

$$\hat{\gamma}^\dagger \hat{\gamma}|n\rangle_\gamma = (1-\varepsilon^2)\, n|n\rangle_\gamma \tag{6.9}$$

The constant photon states $|n\rangle$ and the coherent states $|\alpha\rangle$ are subsets of the broader sets $|n\rangle_\gamma$ and $|\gamma\rangle$ respectively.

The unitary operator $\hat{D}(\varepsilon,\alpha)$

$$\hat{D}(\varepsilon,\alpha) = e^{\frac{\alpha\hat{\gamma}^\dagger - \alpha^*\hat{\gamma}}{1-\varepsilon^2}}$$

$$\hat{D}^\dagger(\varepsilon,\alpha) = \hat{D}^{-1}(\varepsilon,\alpha) = \hat{D}(\varepsilon,-\alpha) \tag{6.10}$$

which appears in the integrand of Eq. 4.1 forms an orthogonal set

$$(\hat{D}(\varepsilon,\alpha),\ \hat{D}(\varepsilon,\beta)) = \frac{\pi}{1-\varepsilon^2}\,\delta^2\,(\alpha-\beta) \tag{6.11}$$

It is a generalization of the corresponding displacement operator $\hat{D}(\alpha)$ and obeys the multiplication rule

$$\hat{D}(\varepsilon,\beta)\,\hat{D}(\varepsilon,\alpha)\,\hat{D}^\dagger(\varepsilon,\beta) = e^{\frac{\beta\alpha^* - \alpha\beta^*}{1-\varepsilon^2}}\,\hat{D}(\varepsilon,\alpha) \tag{6.12}$$

which when substituted into the definition (6.1) yields the unitary transformation of the $\hat{T}(\varepsilon,\alpha)$ operations

$$\hat{T}(\varepsilon,\alpha) = \pi^{-1}\int d^2\xi\,\hat{D}(\varepsilon,\alpha(1-\varepsilon^2))\,\hat{D}(\varepsilon,\xi)\,\hat{D}^\dagger(\varepsilon,\alpha(1-\varepsilon^2)) = \hat{D}^\dagger(\varepsilon,\alpha(1-\varepsilon^2))$$

$$\hat{T}(\varepsilon,0)\,\hat{D}^\dagger(\varepsilon,\alpha(1-\varepsilon^2)) \tag{6.13}$$

The α dependence of the operator $\hat{T}(\varepsilon,\alpha)$ is induced by the transformation (6.13) involving the operator $\hat{D}(\varepsilon,\alpha(1-\varepsilon^2))$.

To find the eigenvalues of the operator $\hat{T}(\varepsilon,\alpha)$ we examine the ε dependence of the operator $\hat{T}(\varepsilon,\alpha)$ and note that the operator $\hat{T}(\varepsilon,0)$ is defined by Eq. (6.15) as the integral

$$\hat{T}(\varepsilon,0) = \pi^{-1}\int d^2\xi\,\hat{D}(\varepsilon,\xi) \tag{6.14}$$

Since the operators $\hat{D}(\varepsilon,\xi)$ form a complete orthogonal set, Eq. (6.14) is a displacement operator expansion with the weighting function given by

$$(\hat{T}(\varepsilon,0),\ \hat{D}(\varepsilon,\beta)) = \frac{1}{\pi}\int d^2\xi\ (\hat{D}(\varepsilon,\xi),\ \hat{D}(\varepsilon,\beta)) = \frac{1}{1-\varepsilon^2} \tag{6.15}$$

see Eq. (6.11).

The completeness relation of the minimum uncertainty states

$$\frac{1}{\pi(1-\varepsilon^2)}\int |\gamma><\gamma|\ d^2\gamma = 1 \tag{6.16}$$

allows us to express the weighting function (6.15) alternately as

$$(\hat{T}(\varepsilon,0),\ \hat{D}(\varepsilon,\beta)) = \frac{1}{\pi(1-\varepsilon^2)}\int d^2\gamma <\gamma|\ \hat{T}(\varepsilon,0)\ \hat{D}(\varepsilon,-\beta)\ |\gamma>$$

$$= \frac{1}{\pi(1-\varepsilon^2)}\int d^2\gamma <\gamma|\ e^{\frac{|\beta|^2}{2(1-\varepsilon^2)}}\ e^{\frac{-\beta\hat{\gamma}^\dagger}{1-\varepsilon^2}}\ \hat{T}(\varepsilon,0)\ e^{\frac{\beta^*\hat{\gamma}}{1-\bar{\varepsilon}^2}}|\gamma> \tag{6.17}$$

or

$$\frac{1}{1-\varepsilon^2} = \frac{e^{|\beta|^2/2(1-\varepsilon^2)}}{\pi(1-\varepsilon^2)}\int d^2\gamma\ \frac{e^{-\beta\gamma^* + \beta^*\gamma}}{1-\varepsilon^2} <\gamma\ |\ \hat{T}(\varepsilon,0)|\gamma> \tag{6.18}$$

The identity (5.4) implies now the following expression for $<\gamma|\hat{T}(\varepsilon,0)|\gamma>$

$$<\gamma|\hat{T}(\varepsilon,0)|\gamma> = \frac{2}{1-\varepsilon^2}\ e^{\frac{-2|\gamma|^2}{1-\varepsilon^2}} \tag{6.19}$$

The operator $\hat{T}(\varepsilon,\ 0)$ equals then

$$\hat{T}(\varepsilon,0) = \frac{2}{1-\varepsilon^2} : e^{-2/1-\varepsilon^2 \hat{\gamma}^\dagger\hat{\gamma}} :$$

$$\hat{T} = \frac{2}{1-\varepsilon^2} \{1 - \frac{2}{1-\varepsilon^2}\hat{\gamma}^\dagger\hat{\gamma} + \frac{4}{(1-\varepsilon^2)^2}\frac{\hat{\gamma}^{\dagger 2}\hat{\gamma}^2}{2!} - \ldots\} \tag{6.20}$$

where : : denotes normal ordering.

We show in Appendix A that if f is any function of $\hat{\gamma}^\dagger\hat{\gamma}'$, its normal form : $f(\hat{\gamma}^\dagger\hat{\gamma})$: is given by the double series

$$: f(\hat{\gamma}^\dagger\hat{\gamma}): = \sum_{r=0}^{\infty}\sum_{s=0}^{\infty} \frac{(-1)^s \, f[(r-s)(1-\varepsilon^2)]\hat{\gamma}^{\dagger r}\hat{\gamma}^r}{(r-s)! \quad s! \; (1-\varepsilon^2)^{r-s}} \tag{6.21}$$

In particular, the normal form of the operator

$$\hat{f}(\hat{\gamma}^\dagger\hat{\gamma}) = e^{y\hat{\gamma}^\dagger\hat{\gamma}}$$

is given by the series expansion

$$: e^{y\hat{\gamma}^\dagger\hat{\gamma}} : = \sum_{r=0}^{\infty} \frac{(e^{y(1-\varepsilon^2)} - (1-\varepsilon^2))^r \hat{\gamma}^{\dagger r}\hat{\gamma}^r}{(1-\varepsilon^2)^r \; r!} \tag{6.22}$$

see Appendix A. Eq. (A.8)

Since y is an arbitrary parameter we may select y so that

$$e^y = (-2 + (1-\varepsilon^2))^{\frac{1}{1-\varepsilon^2}}$$

or

$$e^{y(1-\varepsilon^2)} - (1-\varepsilon^2) = -2 \tag{6.23}$$

which upon substitution into Eq. (6.22) yields the normal form of the operator

$$\{-(1+\varepsilon^2)\}^{\frac{\hat{\gamma}^\dagger\hat{\gamma}}{1-\varepsilon^2}} \quad \text{as}$$

$$\{-(1+\varepsilon^2)\}^{\frac{\hat{\gamma}^\dagger\hat{\gamma}}{1-\varepsilon^2}} = \sum_{r=0}^{\infty} \frac{(-2)^r \hat{\gamma}^{\dagger r}\hat{\gamma}^r}{(1-\varepsilon^2)^r\, r!} \tag{6.24}$$

Since the expansion (6.22) is unique, a comparison of Eq. (6.24) with the expansion coefficients for the operator $\hat{T}(\varepsilon,0)$ as given by Eq. (6.20) suggests that we may rewrite the operator $\hat{T}(\varepsilon,0)$ formally as

$$\hat{T}(\varepsilon,0) = \frac{2}{1-\varepsilon^2}\{-(1+\varepsilon^2)\}^{\frac{\hat{\gamma}^\dagger\hat{\gamma}}{1-\varepsilon^2}} \tag{6.25}$$

By using Eq. (6.13) and the displacement property of the unitary operators $\hat{D}(\varepsilon,\alpha(1-\varepsilon^2))$

$$\hat{D}^\dagger(\varepsilon,\alpha(1-\varepsilon^2))\,\hat{\gamma}\,\hat{D}(\varepsilon,\alpha(1-\varepsilon^2) = \hat{\gamma} + \alpha(1-\varepsilon^2)$$

$$\hat{D}^\dagger(\varepsilon,\alpha(1-\varepsilon^2))\,\hat{\gamma}^\dagger\,\hat{D}(\varepsilon,\alpha(1-\varepsilon^2)) = \gamma^\dagger + \alpha^*(1-\varepsilon^2) \tag{6.26}$$

we obtain the following expressions for the operator $\hat{T}(\varepsilon,\alpha)$

$$\hat{T}(\varepsilon,\alpha) = \frac{2}{1-\varepsilon^2}\hat{D}(\varepsilon,\alpha(1-\varepsilon^2))\,\{-(1+\varepsilon^2)\}^{\frac{\hat{\gamma}^\dagger\hat{\gamma}}{1-\varepsilon^2}}\,\hat{D}^\dagger(\varepsilon,\alpha(1-\varepsilon^2)) \tag{6.27}$$

$$= \frac{2}{1-\varepsilon^2}\{-(1+\varepsilon^2)\}^{\frac{[\gamma^\dagger-\alpha^*(1-\varepsilon^2)][\gamma-\alpha(1-\varepsilon^2)]}{1-\varepsilon^2}}$$

The expansion of $\hat{T}(\varepsilon,\alpha)$ in terms of the eigenstates $|n\rangle_\gamma$ of the operator $\hat{\gamma}^\dagger\hat{\gamma}$ follows then from Eqs. (6.8) and (6.9)

$$\hat{T}(\varepsilon,\alpha) = \frac{2}{1-\varepsilon^2}\sum_{n=0}^{\infty}\hat{D}(\varepsilon,\alpha(1-\varepsilon^2))|n\rangle_\gamma\{-(1+\varepsilon^2)\}^n{}_\gamma\langle n|\hat{D}^\dagger(\varepsilon,\alpha(1-\varepsilon^2)) \tag{6.28}$$

The states $\hat{D}(\varepsilon,\alpha(1-\varepsilon^2))\ |n>_\gamma$ thus form a complete orthonormal set of characteristic states of the operator $\hat{T}(\varepsilon,\alpha)$

$$\hat{T}(\varepsilon,\alpha)\ \hat{D}(\varepsilon,\alpha(1-\varepsilon^2))|n>_\gamma = \frac{2}{1-\varepsilon^2}\ \{-(1+\varepsilon^2)\}^n\ \hat{D}(\varepsilon,\alpha(1-\varepsilon^2))|n>_\gamma \tag{6.29}$$

with eigenvalues

$$\frac{2}{1-\varepsilon^2}\ \{-\ (1+\varepsilon^2)\ \}^n \tag{6.30}$$

which are independent of α.

7. Equivalent Heisenberg Equations of Motion of the Phase-Space Quasi-Probability Distributions

The time evolution of the statistics of the fields generated by optical sources is vested in the density matrix $\hat{\rho}(t)$ which satisfies an operator equation of motion. The solution of this equation of motion, and of Heisenberg equations of motion of operators in general is characterized by an additional degree of difficulty associated with the non-commutative character of the equation's unknown. We derive, therefore, equations of motion of the phase-space distributions that correspond to the Heisenberg equations of motion of the representative operators. We start with the equation of motion of the observable $\hat{A}$, in the Heisenberg picture,

$$d_t\hat{A}(t) = -\frac{i}{\hbar}\ [\hat{A},\hat{H}];\ d_t \equiv \frac{d}{dt} \tag{7.1}$$

and expand the time dependent operator $\hat{A}$ in terms of the unitary displacement operators $\hat{D}(\alpha)$ with the time dependence vested in the weighting function $a(\alpha,t)$

$$\begin{aligned} \hat{A}(t) &= \pi^{-1} \int d^2\xi\ a(\xi,t)\ \hat{D}(\xi) \\ a(\alpha,t) &= (\hat{A}(t),\ \hat{D}(\alpha)) \end{aligned} \tag{7.2}$$

We obtain hence, with the aid of Eq. (7.1),

$$i \hbar \dot{a}(\alpha,t) = ([\hat{A}(t), \hat{H}], \hat{D}(\alpha)) \tag{7.3}$$

Expanding now the Hamiltonian $\hat{H}$ in terms of the $\hat{D}(\alpha)$ operators we obtain

$$\begin{aligned} \hat{H} &= \pi^{-1} \int d^2\xi \; h(\xi) \; \hat{D}(\xi) \\ h(\alpha) &= (\hat{H}, \hat{D}(\alpha)) \end{aligned} \tag{7.4}$$

The weighting function $h(\alpha)$ is time independent since the system is assumed conservative. Eq. (7.3) reads now

$$i \, h \, \dot{a} \, (\alpha,t) = \pi^{-2} \int d^2\xi \; d^2\tau \; a(\tau,t) h(\xi) \; ([\hat{D}(\tau),\hat{D}(\xi)], \hat{D}(\alpha)) \tag{7.5}$$

In light of the Abelian group character, but for a phase factor, of the displacement operators

$$\hat{D}(\tau) \; \hat{D}(\xi) = \hat{D}(\tau+\xi) \exp \frac{1}{2}(\tau\xi^* - \text{c.c.}) \tag{7.6}$$

the commutator in Eq. (7.5) takes the form

$$[\hat{D}(\tau), \hat{D}(\xi)] = \hat{D}(\tau+\xi) \; 2i \sin \, (\underline{\xi} \times \underline{\tau})_z \tag{7.7}$$

where the vectors $\underline{\xi},\underline{\tau}$ have x, y components ξ_r, ξ_i and τ_r, τ_i respectively. The equation of motion (7.5) has now the structure of a convolution integral but for the sine term,

$$\alpha(\alpha,t) = \frac{2}{\pi\hbar} \int d^2\tau \; a(\tau,t) \; h(\alpha-\tau) \sin \, (\underline{\alpha} \times \underline{\tau})_z \tag{7.8}$$

For many operators it is possible to find a considerably simpler representation as an expansion in terms of the coherent state projection operators $|\alpha\rangle\langle\alpha|$. This is the familiar P-representation and the Heisenberg equation of motion for an observable $\hat{A}$ in the Heisenberg picture may be written in terms of this real-valued weight-function. Expanding the time dependent operator $\hat{A}(t)$ then in terms of the operator.

$$\hat{T}_A(\alpha) = |\alpha\rangle\langle\alpha| \tag{7.9}$$

which is the Fourier transform of the antinormally ordered operator $\hat{D}_A(\alpha) = e^{-\alpha^*\hat{a}} \, e^{\alpha a^\dagger}$, we obtain

$$\hat{A}(t) = \pi^{-1} \int d^2\xi \; a_A(\xi,t) \; | \; \xi\rangle\langle\xi \; |$$

$$a_A(\alpha,t) = (\hat{A}(t), \; \hat{T}_N(\alpha)) \tag{7.10}$$

where $\hat{T}_N(\xi)$ is the Fourier transform of the normally ordered operator

$$\hat{D}_N(\alpha) = e^{\alpha\hat{a}^\dagger} \; e^{-\alpha^*\hat{a}} \quad \text{and}$$

$$(\hat{T}_A(\alpha), \; \hat{T}_N(\beta)) = \pi \; \delta^2(\alpha-\beta) \tag{7.11}$$

The time derivative of Eq. (7.10), using Eq. (7.1), leads to the integro-differential equation for the distribution $a_A(\alpha,t)$

$$\dot{a}_A(\alpha,t) = -\frac{i}{h} \; ([\hat{A}(t), \; \hat{H}], \; \hat{T}_N(\alpha))$$

$$= + \frac{\pi^{-2}}{ih} \int d^2\tau \; d^2\xi \; a_A(\tau,t) h_A(\xi) [(\hat{T}_A(\tau), \; \hat{T}_A(\xi)], \hat{T}_N(\alpha)). \tag{7.12}$$

where $h_A(\xi)$ is

$$h_A(\xi) = (\hat{H}, \; \hat{T}_N(\xi)) \tag{7.13}$$

Expanding the commutator in Eq. (7.12) and using the relation (7.9) we obtain

$$= + \frac{\pi^{-2}}{ih} \int d^2\xi \; d^2\tau \; a_A(\tau,t) \; h_A(\xi) \; \{\langle\tau|\xi\rangle\langle\xi| \; \hat{T}_N(\alpha)|\tau\rangle \; - \; c.c. \; \} \tag{7.14}$$

The coherent state matrix elements of the operator $\hat{T}_N(\alpha)$ are

$$\langle\xi|\hat{T}_N(\alpha)|\tau\rangle = 2\langle\xi|\tau\rangle \lim_{s\to 1} \frac{e^{\frac{2}{s-1}(\xi-\alpha)^* \; (\tau-\alpha)}}{1-s} \tag{7.15}$$

where the limit must be taken as s approaches unity from smaller real values. Substituting this expression into Eq. (7.14) we obtain the equation of motion

$$\dot{a}_A(\alpha,t) = \frac{2\pi^{-2}}{ih} \lim_{s\to 1} \int a_A(\tau,t) \; h_A(\xi) \; e^{-|\xi-\tau|^2} \{ \frac{e^{\frac{2}{s-1}(\xi-\alpha)^*(\tau-\alpha)}}{1-s}$$

$$- \; c.c.\} \; d^2\xi \; d^2\tau \tag{7.16}$$

Eq. (7.16) suggests the equation of motion for the diagonal representation of the density operator $\hat{\rho}(t)$ in the Schroedinger picture

$$\frac{\partial P(\alpha,t)}{\partial t} = \frac{2i}{\pi^2\hbar} \lim_{s\to 1} \int P(\tau,t)\, h_A(\xi)\, e^{-|\xi-\tau|^2} \left\{ \frac{e^{\frac{2}{s-1}(\xi-\alpha)^*(\tau-\alpha)}}{1-s} - c.c. \right\} d^2\xi\, d^2\tau \tag{7.17}$$

$$P(\alpha,t) = (\hat{\rho}(t), \hat{T}_N(\alpha)) \tag{7.18}$$

in lieu of the Heisenberg equation of motion of the density matrix.

$$\frac{\partial\hat{\rho}}{\partial t} = -\frac{1}{ih}[\hat{\rho}(t), \hat{H}] \tag{7.19}$$

8. The Master Equation of the Reduced System Density Operator in the Phase-Space

We consider a system of two interacting parts with the Hamiltonian $\hat{H}$ in the form

$$\hat{H} = \hat{H}_A + \hat{H}_B + \hat{V} \tag{8.1}$$

$\hat{H}_A$ and $\hat{H}_B$ are the Hamiltonian operators of the two free constituent systems and $\hat{V}$ is a weak interaction term.

To simplify the following computations we assume system A with a single degree of freedom. Being concerned in the main with quantum optics, this would correspond to a single mode electromagnetic field interacting with atomic system B.

The density operator for the overall system $\hat{\rho}(t)$, in the Schroedinger picture, satisfies the equation

$$i\hbar\frac{\partial\hat{\rho}}{\partial t} = [\hat{H}_A + \hat{H}_B + \hat{V}, \hat{\rho}] \tag{8.2}$$

We are interested in mean values of the operator $\hat{M}$ which is a function of the system A, or B, operators only

$$<\hat{M}(t)> = Tr_{B,A}[\hat{M}\hat{\rho}(t)] = Tr_A[\hat{M}Tr_B\hat{\rho}(t)] \quad (8.3)$$

where we have traced over the variables of both system parts A and B. In its final form it involves only information contained in the reduced density matrix of system A

$$\hat{\rho}_A(t) = Tr_B\hat{\rho}(t) \quad (8.4)$$

Where we traced $\hat{\rho}(t)$ over the system B variables. $\hat{\rho}_A(t)$ is a function of the system A operators only and following Eq. (8.2) we obtain below an equation of motion for $\hat{\rho}_A(t)$.

To proceed, it is most convenient to work in the interaction picture. To this end we let

$$\hat{\rho}_i(t) = e^{\frac{i}{\hbar}[\hat{H}_A+\hat{H}_B]t}\,\hat{\rho}(t)\,e^{-\frac{i}{\hbar}[\hat{H}_A+\hat{H}_B]t} \quad (8.5)$$

To simplify the following computations we assume system A with a single degree of freedom. Being concerned in the main with quantum optics this would correspond to a single mode electromagnetic field interacting with atomic system B. Systems A and B are independent of each other before coupling.

$$[\hat{H}_A, \hat{H}_B] = 0 \quad (8.6)$$

In fact, all systems A operators in the Schrodinger picture commute with all system B operators in the same picture.

The reduced density matrix of system A, in the interaction picture, reads then

$$\hat{S}(t) = Tr_B\,\hat{\rho}_i(t) \quad (8.7)$$

We assume the interaction to be turned on at, t=o; prior to this time systems A and B are independent of each other. The density operator factorizes then into the direct product

$$\hat{\rho}(0) = \hat{\rho}_A(0)\,\hat{\rho}_B(0) = \hat{S}(0)\,\hat{\rho}_B(0) = \hat{\rho}_i(0) \quad (8.8)$$

We also assume that system B is so large that its statistical properties are unaffected by its weak coupling to system A.

The reduced density operator $\hat{S}(t)$ which determines the statistical properties of the system A satisfies what is known as the master equation of motion[25]

$$\frac{\partial \hat{S}}{\partial t} = - \frac{1}{\hbar^2} \int_0^t \mathrm{Tr}_B[\hat{V}(t), [\hat{V}(t'), \hat{S}(t)\, \hat{\rho}_B(0)]]\, dt' \quad (8.9)$$

derived through an iterative technique. $\hat{V}(t)$ is the coupling operator $\hat{V}$ in the interaction picture.

The usual difficulties associated with integro-differential equations are compounded in the case of the master equation (8.9) for the reduced system density operator by the non-commutative nature of the operator variables. The one-to-one correspondence between the operator $\hat{S}(t)$ and its weighting function $s(\xi,t)$, in the coherent state representation, permits the derivation of an equivalent master equation of motion in terms of the phase-space quasi-probability distribution $s(\xi,t)$ unburdened by the problem of commutativity.

To this end, we expand the time dependent reduced density matrix $\hat{S}(t)$ in terms of the coherent states with the time dependence vested in a weighting function $s(\xi,t)$ given by

$$s(\xi,t) = \mathrm{Tr}_A \{ \hat{S}(t)\, \hat{T}_N(\varepsilon) \}$$
$$\hat{S}(t) = \pi^{-1} \int s(\xi,t) |\xi\rangle\langle\xi| d^2\xi \quad (8.10)$$

Differentiating both sides of Eq.(8.10) with respect to the variable t, by virtue of Eq.(8.9) we obtain

$$\frac{\partial s(\xi,t)}{\partial t} = \mathrm{Tr}_A\{ \frac{\partial \hat{S}}{\partial t}\, \hat{T}_N(\xi) \} \quad (8.11)$$

$$= \mathrm{Tr}_A \{ - \frac{1}{\hbar^2} \int_0^t \mathrm{Tr}_B\, [\hat{V}(t), [\hat{V}(t'), \hat{S}(t)\, \hat{\rho}_B(0)]]\, \hat{T}_N(\xi)\, dt' \}$$

We investigate one term in the double commutator of the past equation

$$I = \mathrm{Tr}_A \{ - \frac{1}{\hbar^2} \int_0^t \mathrm{Tr}_B \{ \hat{V}(t)\, \hat{V}(t')\, \hat{S}(t)\, \hat{\rho}_B(0)\}\, \hat{T}_N(\xi)\, dt' \} \quad (8.12)$$

The coupling operator in the interaction picture $\hat{V}(t)$ which appears in Eq. (8.12) may be written in the form of a direct product

$$\hat{V}(t) = \hat{V}_A(t)\ \hat{V}_B(t) \tag{8.13}$$

where $V_A(t)$ and $V_B(t)$ are operator functions of the system A and B operators respectively since the system A and B operators usually commute with one another.

The reduced density operator $\hat{S}(t)$ is a function of the system A operators only so that we may rewrite Eq. (8.12) as

$$I = Tr_A \{ -\frac{1}{\hbar^2} \int_0^t \hat{R}_A(t,t')\ \hat{S}(t)\ \hat{T}_N(\xi)\ dt' \} \tag{8.14}$$

where $R_A(t,t')$ is an operator correlation function of the system A operators only

$$\begin{aligned}
\hat{R}_A(t,t') &= Tr_B \{ \hat{V}(t)\ V(t')\ \hat{\rho}_B(0) \} \\
&= Tr_B \{ \hat{V}_A(t)\ \hat{V}_A(t')\ \hat{V}_B(t)\ \hat{V}_B(t')\ \hat{\rho}_B(0) \} \\
&= \hat{V}_A(t)\ \hat{V}_A(t')\ C(t,t')
\end{aligned} \tag{8.15}$$

or

$$I = -\frac{1}{\hbar^2}\ Tr_A \{ \hat{V}_A(t)\ \hat{G}_A(t)\ \hat{S}(t)\ \hat{T}_N(\xi) \} \tag{8.16}$$

where the operator $\hat{G}_A(t)$ is given by the integral

$$\hat{G}_A(t) = \int C(t,t')\ \hat{V}_A(t')\ dt' \tag{8.17}$$

If we express the time dependent operators $\hat{V}_A(t)$, $\hat{G}_A(t)$ and $\hat{S}(t)$ in the diagonal representation

$$\begin{aligned}
\hat{S}(t) &= \pi^{-1} \int d^2\alpha\ s(\alpha,t) |\alpha\rangle\langle\alpha| \\
\hat{V}_A(t) &= \pi^{-1} \int d^2\gamma\ v(\gamma,t) |\gamma\rangle\langle\gamma| \\
\hat{G}_A(t) &= \pi^{-1} \int d^2\beta\ g(\beta,t) |\beta\rangle\langle\beta|
\end{aligned} \tag{8.18}$$

then Eq. (8.16) may be written in terms of the three weighting functions $v(\gamma,t)$ $s(\alpha,t)$ $g(\beta,t)$ as

$$I=-\frac{\pi^{-3}}{\hbar^2}\mathrm{Tr}_A\int d^2\alpha d^2\beta d^2\gamma v(\gamma,t)s(\alpha,t)g(\beta,t)|\gamma\rangle\langle\gamma|\beta\rangle\langle\beta|\alpha\rangle\langle\alpha\rangle\langle\alpha|\hat{T}_N(\xi)$$

$$=-\frac{\pi^{-3}}{\hbar^2}\int d^2\alpha\, d^2\beta\, d^2\gamma\; v(\gamma,t)g(\beta,t)s(\alpha,t)\langle\gamma|\beta\rangle\langle\beta|\alpha\rangle\langle\alpha|\hat{T}_N(\xi)|\gamma\rangle \tag{8.19}$$

The coherent state matrix elements for the normally ordered operator $\hat{T}_N(\xi)$ is given by the expression (7.14)

$$\langle\alpha|\hat{T}_N(\xi)|\gamma\rangle = \lim_{s\to1} 2\langle\alpha|\gamma\rangle \frac{e^{2/_{s-1}(\alpha-\xi)^*(\gamma-\xi)}}{1-s}$$

which, upon substitution into Eq. (8.19), yields

$$I=-\lim_{s\to1}\frac{2\pi^{-3}}{\hbar^2}\int d^2\alpha\, d^2\beta\, d^2\gamma\; v(\gamma,t)g(\beta,t)s(\alpha,t)\langle\gamma|\beta\rangle\langle\beta|\alpha\rangle\langle\alpha|\gamma\rangle \frac{e^{2/_{s-1}(\alpha-\xi)^*(\gamma-\xi)}}{1-s} \tag{8.20}$$

$$=-\lim_{s\to1}\frac{2\pi^{-3}}{\hbar^2}\int d^2\alpha d^2\beta\, d^2\gamma\; v(\gamma,t)g(\beta,t)s(\alpha,t)e^{\gamma^*\beta-|\gamma|^2-|\beta|^2+\beta^*\alpha-|\alpha|^2+\alpha^*\gamma} \frac{e^{2/_{s-1}(\alpha^*-s^*)(\gamma-\xi)}}{1-s}$$

Integrating Eq. (8.20) with respect to the variable β and using the identity (3.3)

$$\pi^{-1}\int f(\alpha)\, e^{\alpha^* y} e^{-z|\alpha|^2} d^2\alpha = \tfrac{1}{z} f(y/z);\ \mathrm{Re}\ z>0$$

we obtain

$$I = -\lim_{s\to1}\frac{2\pi^{-2}}{\hbar^2}\int d^2\alpha d^2\gamma \frac{v(\gamma,t)s(\alpha,t)}{1-s}\, e^{-|\gamma|^2-|\alpha|^2+\alpha^*\gamma+\frac{2}{s-1}(\alpha-\xi)^*(\gamma-\xi)}\, g(\alpha,t)\, e^{\gamma^*\alpha} \tag{8.21}$$

which upon integration with respect to the variable γ yields

$$I = -\lim_{s\to 1} \frac{2}{\pi\hbar^2} \int d^2\alpha\; s(\alpha,t)\; v(\alpha,t)\; g(\alpha,t)\; \frac{e^{\frac{2}{s-1}|\alpha-\xi|^2}}{1-s} \tag{8.22}$$

by virtue of (3.3).

The functions $g(\beta,t)\, e^{\gamma^*\beta}$, $v(\gamma,t)\, e^{\alpha^*\gamma + \frac{2}{s-1}\gamma(\alpha-s)^*}$ are analytic functions of β and γ respectively when the functions $g(\beta,t)$, $v(\gamma,t)$ are analytic functions.

A similar procedure may be used for the remaining terms in Eq. (8.11) to yield the equivalent master equation in terms of the phase space distribution $s(\alpha,t)$

$$\frac{\partial S(\xi,t)}{\partial t} = -\lim \frac{2\pi^{-1}}{\hbar^2} \int d^2\alpha\; S(\alpha,t)\; \frac{e^{\frac{2}{s-1}|\alpha-s|^2}}{1-s}$$

$$\times \{ v(\alpha,t)\; g(\alpha,t) - v(\alpha,t)\; g(f(\alpha,\xi),t)$$
$$-v(f(\alpha,\xi),t)\; g(\alpha,t) +$$
$$v(f(\alpha,\xi),t)\; g(f(\alpha,\xi),t)\} ; \tag{8.23}$$

$$f(\alpha,\xi) = \frac{\alpha(1+s) - 2\xi}{s-1}$$

where the individual product vg stems from the commutation terms of Eq. (8.11).

Appendix A

We show that when f is a function of $\hat{\gamma}^\dagger\hat{\gamma}$, where $\hat{\gamma} = \hat{a} + \varepsilon\hat{a}^\dagger$, its normal form $:f(\hat{\gamma}^\dagger\hat{\gamma}):$ is given by the double summations

$$:\hat{f}(\hat{\gamma}^\dagger\hat{\gamma}): = \sum_{r=0}^{\infty} \sum_{s=0}^{r} \frac{(-1)^s\; f[(r-s)(1-\varepsilon^2)]\hat{\gamma}^{\dagger r}\hat{\gamma}^r}{(r-s)!\; s!\; (1-\varepsilon^2)^{r-s}} \tag{A.1}$$

Consider the normally ordered series expansion of $f(\hat{\gamma}^\dagger\hat{\gamma})$.

$$:f(\hat{\gamma}^{\dagger}\hat{\gamma}): \; = \sum_{r=0}^{\infty} C_r \, \hat{\gamma}^{\dagger r} \, \hat{\gamma}^{r} \tag{A.2}$$

To find the expansion coefficients C_r we first show that $\hat{\gamma}^{\dagger r}\hat{\gamma}^{r}$ may be written as the finite product

$$\hat{\gamma}^{\dagger r}\hat{\gamma}^{r} = \prod_{n=1}^{r} [\hat{\gamma}^{\dagger}\hat{\gamma} - (n-1)\, x]; \; x = 1-\varepsilon^2 \tag{A.3}$$

We prove Eq. (A.3) by induction. Let Eq. (A.3) be true for r=r'

$$\hat{\gamma}^{\dagger r}\hat{\gamma}^{r} = \prod_{n=1}^{r'} [\hat{\gamma}^{\dagger}\hat{\gamma} - (n-1)x] \tag{A.4}$$

Multiply both sides of Eq. (A.4) by $(\hat{\gamma}^{\dagger}\hat{\gamma} - r'x)$. The left hand side

$$\hat{\gamma}^{\dagger r'} \, \hat{\gamma}^{r'} \, \hat{\gamma}^{\dagger}\hat{\gamma} - r'x \, \hat{\gamma}^{\dagger r'}\hat{\gamma}^{r'}$$

reduces with the aid of the commutation relation[10],

$$[\hat{\gamma}^{\dagger}, \hat{\gamma}^{r'}] = -(1-\varepsilon^2)\, r'\hat{\gamma}^{r'-1}$$

or

$$\hat{\gamma}^{r'} \, \hat{\gamma}^{\dagger} = \hat{\gamma}^{\dagger} \, \hat{\gamma}^{r'} + (1-\varepsilon^2)\, r' \, \hat{\gamma}^{r'-1},$$

to

$$\hat{\gamma}^{\dagger r'} \{\hat{\gamma}^{\dagger}\hat{\gamma}^{r'} + (1-\varepsilon^2)\, r' \, \hat{\gamma}^{r'-1}\} \, \hat{\gamma} - r' \, x \, \hat{\gamma}^{\dagger r'} \, \hat{\gamma}^{r'} = \hat{\gamma}^{\dagger (r'+1)}\hat{\gamma}^{(r'+1)}$$

Eq. (A.4) now takes the form

$$\hat{\gamma}^{\dagger r'+1}\hat{\gamma}^{r'+1} = \hat{\gamma}^{\dagger}\hat{\gamma}(\hat{\gamma}^{\dagger}\hat{\gamma}-x) \ldots \, (\hat{\gamma}^{\dagger}\hat{\gamma} - r'x) = \prod_{n=1}^{r'+1} [\hat{\gamma}^{\dagger}\hat{\gamma} - (n-1)x]$$

Thus, if Eq. (A.4) is valid for r', it is also valid for r'+1. (A.4) is valid for r'=1 and the statement (A.3) is proven.

We turn now to prove equation (A.1). We may rewrite Eq. (A.2), by virtue of Eq. (A.3), as

$$:\hat{f}\ (\hat{\gamma}^{\dagger}\hat{\gamma}):=\sum_{n=o}^{\infty} C_r(\hat{\gamma}^{\dagger}\hat{\gamma} - x)\ldots(\hat{\gamma}^{\dagger}\hat{\gamma} - (r-1)x)$$

and apply it to the ket $|n>_\gamma$. By virtue of the characteristic equation

$$\hat{\gamma}^{\dagger}\hat{\gamma}|n>_\gamma = nx|n>_\gamma$$

$$\{ :\hat{f}(\hat{\gamma}^{\dagger}\hat{\gamma}):\}|n>_\gamma = f(nx)|n>_\gamma$$

$$= x^n \sum_{r=o}^{n} C_r(n)(n-1)\ldots(n-(n-1))|n>_\gamma$$

Whence

$$\begin{aligned} f(0) &= C_0 \\ f(x) &= (C_0 + C_1)x \\ f(2x) &= (C_1 + 2C_1 + 2C_2)x^2 \\ &\vdots \end{aligned} \tag{A.5}$$

or

$$\begin{aligned} C_0 &= f(0) \\ C_1 &= \frac{f(x)}{x} - f(0) \\ C_2 &= \frac{f(2x)}{2x^2} - \frac{f(x)}{x} + \frac{f(0)}{2} \\ C_r &= \sum_{s=o}^{r} (-1)^s \frac{f[(r-s)x]}{(r-s)!\ s!\ x^{r-s}} \end{aligned} \tag{A.6}$$

in agreement with Eq. (A.1).

As a special case of Eq. (A.1) we consider the normal form of the function

$$f(\hat{\gamma}^\dagger\hat{\gamma}) = e^{y\hat{\gamma}^\dagger\hat{\gamma}}$$

By a direct application of Eq. (A.1) we find the expansion coefficients C_r

$$C_r = \frac{e^{yrx}}{x^r\ r!} \sum_{s=o}^{r} \frac{(-1)^s\ r!\ (e^{-yx})^s}{(r-s)!\ s!\ x^{-s}}$$

$$= \frac{e^{yrx}}{x^r\ r!} \sum_{s=o}^{r} \frac{(-1)^s\ r!\ (xe^{-yx})^s}{(r-s)!\ s!} \qquad (A.7)$$

$$= \frac{e^{yrx}}{x^r\ r!} (1 - xe^{-yx})^r$$

and

$$: e^{y\hat{\gamma}^\dagger\hat{\gamma}} : = \sum_{s=o}^{\infty} \frac{(e^{yx} - x)^r\ \hat{\gamma}^{\dagger r}\ \hat{\gamma}^r}{x^r\ r} . \qquad (A.8)$$

see Eq. (A.2).

References

1. E. P. Wigner, Phys. Rev. *40*, 749 (1932).
2. J. E. Moyal, Proc. Camb. Phil. Soc. *45*, 99 (1948) and *45*, 545 (1949).
3. H. Weyl, *The Theory of Groups and Quantum Mechanics* (Dover, New York, 1931), pp. 274-277.
4. J. C. T. Pool, J. Math. Phys. *7*, 66 (1966).
5. K. E. Cahill and R. J. Glauber, Phys. Rev. *177*, 1857 (1969); *179*, 1882 (1969).
6. I. N. Herstein, *Topics in Algebra* (Blaisdell Publishing Co. Massachusetts, 1964), pp. 48-50.
7. R. J. Glauber, Phys. Rev. Letters *10*, 84 (1963); Phys. Rev. *131*, 2766 (1963); J. R. Klauder and E. C. G. Sudarshan, *Fundamentals of Quantum Optics* (Benjamin, New York, 1968).
8. J. R. Klauder, J. Math. Phys. *4*, 1055 (1963); *5*, 177 (1964).
9. W. H. Louisell, *Radiation and Noise in Quantum Electronics*, (McGraw-Hill Book Co., 1964), Chap. 3.

10. Contents of this section are based on the work of P. P. Betrand, Ph.D. thesis, Polytechnic Institute of Brooklyn, 1969, and K. Moy, M.Sc. thesis, Polytechnic Institute of Brooklyn, 1968. See also Ref. 11, 12.
11. P. P. Betrand and E. A. Mishkin, Bull. Am. Phys. Soc. *15*, 89 Jan (1970).
12. P. P. Betrand, K. Moy and E. A. Mishkin, Phys. Rev. *4*, 1909 (1971).
13. A. E. Glassgold and D. Holliday, Phys. Rev. *139*, A1717 (1965).
14. P. P. Betrand and E. A. Mishkin, Phys. Letters *25A*, 204 (1967).
15. M. M. Miller and E. A. Mishkin, Phys. Letters *24A*, 188 (1967).
16. For an analysis of these constraints see Ref. 5.
17. R. J. Glauber, Phys. Rev. *131*, 2766 (1963).
18. P. Kelly and W. H. Kluner, Phys. Rev. *136*, 316 (1964).
19. R. J. Glauber, Phys. Rev. *130*, 2529 (1963).
20. R. J. Glauber, *Quantum Optics and Electronics* (Les Houches, 1964) edited by C. de Witt, et al., p. 63, (Gordon and Breach, New York, 1965).
21. E. C. B. Sudarshan, Phys. Rev. Letters *10*, 277 (1963).
22. R. Bonifacio, L. M. Narducci, and E. Montaldi, Phys. Rev. Letters *16*, 1125 (1966); Nuovo Cimento *47*, 890 (1967).
23. M. M. Miller and E. A. Mishkin, Phys. Rev. *164*, 1610 (1967).
24. W. H. Louisell, *Quantum Optics*, (Proceedings of the International School of Physics "Enrico Fermi", 1967), edited by R. J. Glauber (Academic Press, New York, 1969).

RADIATION BY MANY EXCITED ATOMS BETWEEN MIRRORS*

C. S. Chang and P. Stehle

University of Pittsburgh, Pittsburgh, Pa.

1. Introduction

Theories of laser generally assume that the interaction between the atoms or molecules of the active laser medium, and the electromagnetic field can be accounted for by considering at most a few modes of the field as determined by the laser mirrors. Examples of this kind are furnished by the work of Lamb[1], Scully and Lamb[2], Fleck[3], Haken[4], and many others. These theories are very successful in describing the properties of real lasers, and they provide the basis for extensions to theories of mode locking[4], harmonic production[5], and a variety of other effects. It has remained somewhat of a puzzle, however, why the restriction to a few modes does work so well, especially when the cavities used in practice are so far from closed and do not come close to providing a complete set of modes. As seen from the center of a typical He-Ne laser used in an instructional laboratory, for example, the mirror occupies only 1% of the total solid angle. There is radiation into free space modes, and its intensity can be used as a measure of population inversion. There is, therefore, finite coupling to all modes of the radiation field.

In one of the first papers (1926) on the quantum theory of interaction between atoms and the electromagnetic field Dirac[7] said following a development of the semiclassical theory:

*Supported in part by Army Research Office, Durham, North Carolina.

"The present theory thus accounts for the absorption and stimulated emission of radiation, and shows that the elements of the matrices representing the total polarisation determine the transition probabilities. One cannot take spontaneous emission into account without a more elaborate theory involving the positions of the various atoms and the interference of their individual emissions, as the effects will depend upon whether the atoms are distributed at random, or arranged in a crystal lattice, or all confined in a volume small compared with a wave-length. The last alternative mentioned, which is of no practical interest, appears to be the simplest theoretically.

"It should be observed that we get the simple Einstein results only because we have averaged over all initial phases of the atoms. The following argument shows, however, that the initial phases are of real physical importance, and that in consequence the Einstein coefficients are inadequate to describe the phenomena except in special cases. If initially all the atoms are in the normal state, then it is easily seen that the expression (29) for ΔN_m holds without the averaging process, so that in this case Einstein coefficients are adequate. If we now consider the case when some of the atoms are initially in an excited state, we may suppose that they were brought into this state by radiation incident on the atoms before the time t = 0. The effect of the subsequent incident radiation must then depend on its phase relationships with the earlier incident radiation, since a correct way of treating the problem would be to resolve both incident radiations into a single Fourier integral. If we do not wish the earlier radiation to appear explicitly in the calculation, we must suppose that it impresses certain phases on the atoms it excites, and that these phases are important for determining the effect of the subsequent radiation. It would thus not be permissible to average over these phases, but one would have to work directly from equation (28)."

The theory of superradiance[7,8] provides a suggestion that mode concentration can occur in an extended atomic system. Dicke[7] showed that such a system excited by a plane wave pulse near resonance tends very strongly to emit in the direction of the exciting pulse. If a field mode provided the pumping mechanism, that mode would be expected to dominate the radiation. Pumping, however, is not done in this way but in a much more random way; by a flash tube or by a gas discharge. Closely related to superradiance is the work of Ernst and Stehle[9] on spontaneous decay of extended systems. In their model the initial excitation is complete so that any spatial variation comes only from the geometry of the system and not from the manner of excitation. They found that the radiation emitted is highly correlated, coming out in the form of a diffraction limited ray. The direction of this ray is determined by the shape of the system, being random for a spherical system and very nearly axial for an elongated system. In the language of Ernst and Stehle, the "forces" producing the ray are much stronger that those determining its direction. This suggests the possibility that the mirrors steer the ray into a mode with little effect on the internal structure of the ray. The axial nature of the ray emitted by a rod then leads to the overwhelming predominance of the axial or nearly axial modes. We show here that this is, indeed, the case.

The method used is that of Weisskopf and Wigner as in Ernst and Stehle[9]. The presence of the mirrors is accounted for by using the Fabry-Perot modes instead of free space modes. This modification has been used by Stehle[10] to discuss the spontaneous radiation of a single atom between mirrors. The mirrors are assumed to be infinite in extent, but to have a reflectivity R less than unity. In view of the results it is clear that off-axial modes are unimportant so that the finiteness of the mirrors does not invalidate their application to realistic models. The most important idealization is the complete initial excitation.

In Section 2 we treat two atoms between mirrors. The calculation parallels that of I[9] exactly, and the comparison of results reveals the effect of the mirrors. In Section 3 the N atom case is treated, again paralleling I. Section 4 summarizes the results and discusses their meaning for laser physics.

2. Two Atoms

The system consisting of two atoms between parallel mirrors is worth investigating in detail because it shows in simple form most of the features of the N atom system. In the absence of mirrors it radiates correlated photons; with mirrors present the Fabry-Perot modes have an effect.

The field labeled with the wave vector $\underline{k}$ consists of the plane wave $\exp[i\underline{k}\cdot\underline{x}]$ together with its reflections, as in II[10].

$$F_{\underline{k}}(\underline{x}) = A_{\underline{k}}\, e^{i\underline{k}\cdot\underline{x}} + B_{\underline{k}}\, e^{i(\underline{k}^{\parallel}-\underline{k}^{\perp})\cdot\underline{x}} \tag{1}$$

$$A_{\underline{k}} = \frac{1-R^4}{(1-R^2)^2 + 4R^2 \sin^2(\underline{k}\cdot\underline{L})} \quad ,$$

(2)

$$B_{\underline{k}} = \frac{2R(1-R^2)\cos(\underline{k}\cdot\underline{L})}{(1-R^2)^2 + 4R^2 \sin^2(\underline{k}\cdot\underline{L})} \quad .$$

Here R is the amplitude reflectivity of the mirrors, $\underline{L}$ is perpendicular to the mirrors whose separation is $L = |\underline{L}|$, and ∥ and ⊥ refer to the mirror surface.

The interaction of an atom at position $\underline{X}$ with a mode of the field is determined by the quantity

$$\tilde{C}^*_{\underline{k}}(\underline{X}) = -ie \int d^3x\, \overline{U}_g(\underline{x}-\underline{X})\, \gamma_\mu e_\mu\, U_e(\underline{x}-\underline{X})\, F^*_{\underline{k}}(\underline{x}) \tag{3}$$

$$\approx F^*_{\underline{k}}(\underline{X})\, D$$

as in (II 9), the atomic size being assumed small compared with the wavelength of the mode. Following (I 3.3) we then define the function

$$\Gamma(\underline{x},\underline{y}) = \frac{\Delta}{2(2\pi)^3} \int d\Omega_{\underline{k}}\, \tilde{C}^*_{\underline{k}}(\underline{x})\, \tilde{C}_{\underline{k}}(\underline{y})$$

(4)

$$\approx \frac{\Delta|D|^2}{2(2\pi)^3} \int d\Omega_{\underline{k}}\, F^*_{\underline{k}}(\underline{x})\, F_{\underline{k}}(\underline{y}) .$$

Γ depends on its arguments separately because of the existence of cavity modes; in free space it depends only on $|\underline{x}-\underline{y}|$. It is symmetric and satisfies Schwarz's inequality in the form

$$\Gamma(\underline{x},\underline{x})\, \Gamma(\underline{y},\underline{y}) \geq [\Gamma(\underline{x},\underline{y})]^2 . \tag{5}$$

When $|\underline{x}-\underline{y}|$ is large the angular integration in (4) will give a small result because the integrand is rapidly oscillating. This is analogous to the free-space case where γ_{12} approaches zero for large atomic separation. To carry out the integration we separate contributions from the various Fabry-Perot modes and the nonmode contribution. The reflectivity is assumed to be near unity so the angular spread of the modes is small and that of mode m is determined by

$$|2(\Delta L\cos\theta - m\pi)| \leq 1-R^2 \ . \tag{6}$$

The contribution of mode m to $\Gamma(\underline{x},\underline{y})$, denoted by Γ^m, is

$$\Gamma^m(\underline{x},\underline{y}) = \frac{|D|^2}{4\pi L(1-R^2)} J_0(\varepsilon_m r_- \sin\theta_-)\left[\cos\left(\frac{m\pi}{L} r_-\cos\theta_-\right) + (-)^m \cos\left(\frac{m\pi}{L} r_+\cos\theta_+\right)\right], \tag{7}$$

where $J_n(z)$ is the Bessel function of order n, and

$$\varepsilon_m = \Delta\left[1-\left(\frac{m\pi}{L\Delta}\right)^2\right]^{1/2} ,$$

$$\underline{x} \pm \underline{y} = r_\pm(\sin\theta_\pm\cos\phi_\pm,\ \sin\theta_\pm\sin\phi_\pm,\ \cos\theta_\pm) \ . \tag{8}$$

The nonmode contribution is

$$\Gamma^{non}(\underline{x},\underline{y}) = \frac{|D|^2\Delta}{2(2\pi)^2} (1-R^2)^2 \frac{\sin\Delta r_-}{\Delta r_-} \ . \tag{9}$$

As might be expected, for high reflectivity the nonmode contribution is negligible.

The Weisskopf-Wigner equations, explicitly given in I and in Chang and Stehle[11] involve both $\Gamma(\underline{x},\underline{x})$ and $\Gamma(\underline{x},\underline{y})$, analogous to Γ_0 and Γ_{12} in those references. $\Gamma(\underline{x},\underline{x})$ is obtained from (7) by summing over all m, with $r_- = 0$. The result depends on the atomic positions only through the last term which contains the position of the center of mass. This term alternates in sign and we neglect it, so we write

$$\Gamma(\underline{x},\underline{x}) = \Gamma_\infty = \frac{|D|^2\Delta}{\pi(1-R^2)} \ . \tag{10}$$

The result for $\Gamma(\underline{x},\underline{y})$ after summing over m in a similar way is

$$\Gamma_{12}^c = \frac{|D|^2\Delta}{\pi(1-R^2)} J_0\left(\frac{\Delta}{Q^{1/2}}|\rho_1-\rho_2|\right)\cos[\Delta(z_1-z_2)], \tag{11}$$

where ρ, z are two of the three cylindrical coordinates of $\underline{x}$. Q is the quality factor of the mirrors:

$$\frac{\Delta}{Q} = \frac{1}{\tau}, \qquad \tau = \frac{L}{1-R^2}. \tag{12}$$

τ is a measure of the lifetime of an excitation of the mirror system.

With expressions (10) and (11) for Γ_∞ and $\Gamma_{12}^c(\underline{x},\underline{y})$, the Weisskopf-Wigner equations can be solved as in I or Reference 11. We do not repeat the deails here. We take $\Gamma_{12}^c \ll \Gamma_\infty$, which is the case for widely separated atoms. Then we find that the correlation between the photons emitted by the two atoms as defined in I is given by

$$\tilde{R}_{\underline{k}_1\underline{k}_2} = 1 + \frac{2|A(\underline{k}_1,\underline{X}_1)A(\underline{k}_2,\underline{X}_2)A(\underline{k}_1,\underline{X}_2)A(\underline{k}_2,\underline{X}_1)|}{|A(\underline{k}_1,\underline{X}_1)A(\underline{k}_2,\underline{X}_2)|^2+|A(\underline{k}_1,\underline{X}_2)A(\underline{k}_2,\underline{X}_1)|^2}$$

$$\times \cos\,[\psi(\underline{k}_1,\underline{X}_1)+\psi(\underline{k}_2,\underline{X}_2)-\psi(\underline{k}_1,\underline{X}_2)-\psi(\underline{k}_2,\underline{X}_1)], \tag{13}$$

where

$$A(\underline{k},\underline{X}) = [1+\cos^2(\underline{k}\cdot\underline{L})+2\cos(\underline{k}\cdot\underline{L})\cos\,(2\underline{k}^{\perp}\cdot\underline{X})]^{1/2},$$

$$\psi(\underline{k},\underline{X}) = \tan^{-1}\left[\frac{\sin(\underline{k}\cdot\underline{X})+\cos(\underline{k}\cdot\underline{L})\sin(\underline{k}'\cdot\underline{X})}{\cos(\underline{k}\cdot\underline{X})+\cos(\underline{k}\cdot\underline{L})\cos(\underline{k}'\cdot\underline{X})}\right], \tag{14}$$

$$\underline{k} = \underline{k}^{\parallel} + \underline{k}^{\perp}, \quad \underline{k}' = \underline{k}^{\parallel} - \underline{k}^{\perp}.$$

This expression is more complicated than the corresponding mirror-less expression of I,

$$R_{\underline{k}_1\underline{k}_2} = 1 + \cos(\underline{k}_1-\underline{k}_2)\cdot(\underline{X}_1-\underline{X}_2), \tag{15}$$

but if $\cos(\underline{k}\cdot\underline{L})=0$, rendering the B_k of (2) zero, then (13) does reduce to (15). Vanishing B_k makes the modes essentially free-space modes, the normalization factor A_k of (2) cancelling out of the correlation function. The more interesting case is that when

$\underline{k}_1, \underline{k}_2$ satisfy the Fabry-Perot conditions

$$\sin(\underline{k}_1 \cdot \underline{L}) = \sin(\underline{k}_2 \cdot L) = 0 \quad . \tag{16}$$

Choosing $\cos(\underline{k}_1 \cdot \underline{L}) = \cos(\underline{k}_2 \cdot \underline{L}) = +1$, the correlation function now takes the form

$$\tilde{R}_{\underline{k}_1\underline{k}_2} = 1 + \frac{2\left|\cos(\underline{k}_1^{\perp}\cdot\underline{X}_1)\cos(\underline{k}_2^{\perp}\cdot\underline{X}_2)\cos(\underline{k}_1^{\perp}\cdot\underline{X}_2)\cos(\underline{k}_2^{\perp}\cdot\underline{X}_1)\right|}{\left|\cos(\underline{k}_1^{\perp}\cdot\underline{X}_1)\cos(\underline{k}_2^{\perp}\cdot\underline{X}_2)\right|^2 + \cos(\underline{k}_1^{\perp}\cdot X_2)\cos(\underline{k}_2^{\perp}\cdot\underline{X}_1)\Big|^2} \tag{17}$$

$$\times \cos[(\underline{k}_1^{\parallel} - \underline{k}_2^{\parallel})\cdot(\underline{X}_1 - \underline{X}_2)] \quad .$$

A similar expression results when $\cos(\underline{k}_1 \cdot \underline{L}) = \cos(\underline{k}_2 \cdot \underline{L}) = -1$. If these two cosines differ in sign, the last factor in (17) is replaced by

$$\cos[(\underline{k}_1^{\perp} - \underline{k}_2^{\parallel})\cdot(\underline{X}_1 - \underline{X}_2)] \text{ or } \cos[(\underline{k}_1^{\parallel} - \underline{k}_2^{\perp})\cdot(\underline{X}_1 - \underline{X}_2)].$$

Because the atoms are alike and are well separated, $|\underline{k}_1| \approx \Delta$, $|\underline{k}_2| \approx \Delta$. For any positions of the atoms we see that the radiation of both photons into the same mode is favored maximally, making $\tilde{R}_{\underline{k}_1\underline{k}_2} = 2$. There are other combinations of $\underline{k}_1$, $\underline{k}_2$, $\underline{X}_1$, and $\underline{X}_2$ which will also give $\tilde{R}_{\underline{k}_1\underline{k}_2}$ this value. Because these do depend on the atomic positions, they will play no role in the many atom case.

3. Many Atoms

The system of N excited atoms between mirrors is treated by the Weisskopf-Wigner method by replacing the plane waves of free space with the mode functions of (1) and (2) appropriate to the presence of mirrors. The characteristic feature of the method, that the number of excited atoms plus the number of photons is constant in time, is retained. The analysis proceeds in parallel with the analysis of I, and here we describe only the most important features of it.

The Weisskopf-Wigner equations couple the amplitudes for states with M deexcited atoms, $j_1 \ldots j_M$, and with M photons

$\underline{k}_1 \ldots \underline{k}_M$ for all values of M from 0 to N. By making certain smoothing assumptions, in particular involving replacing sums over atoms by integrals over the occupied volume, solutions can be obtained which have the form

$$\alpha_{\underline{k}_1 \ldots \underline{k}_M}^{j_1 \ldots j_M}(t) = \varepsilon^{-1}(\underline{k}_1 \ldots \underline{k}_M) \prod_{\ell=1}^{M} \frac{e^{-i(\Delta - k_\ell)t}}{(2Vk_\ell)^{1/2}} \times p_{\underline{k}_1 \ldots \underline{k}_M}^{j_1 \ldots j_M} \beta_{k_1 \ldots k_M}(t). \tag{18}$$

The factors on the right side have the following definitions and meanings as in I:

$$\varepsilon(\underline{k}_1 \ldots \underline{k}_M) = [\pi(n_i!)]^{1/2} , \tag{19}$$

where n_i is the number of times a particular $\underline{k}$ appears among $\underline{k}_1 \ldots \underline{k}_M$. It is unity for distinct photons. $\exp[-i(\Delta - k_\ell)t]$ comes from the use of the interaction picture and gives the time dependence coming from "energy non-conservation" in the emission process.

$$p_{\underline{k}_1 \ldots \underline{k}_M}^{j_1 \ldots j_M} = \underset{\underline{\kappa}_1 \ldots \underline{\kappa}_M}{\overset{\underline{k}_1 \ldots \underline{k}_M}{P}} \prod_{i=i}^{M} F_{\underline{\kappa}_i}(\underline{X}_{ji}). \tag{20}$$

The permutation sum P extends over all M! pairings of the $\underline{\kappa}$'s with $\underline{k}$'s. $F_{\underline{\kappa}}(\underline{X})$ is a field mode.

$$\beta_{k_1 \ldots k_M}(t) = T_M \exp[-\Gamma_\infty t(N-M)(1+\tilde{\gamma}M)] \times \prod_{\ell=1}^{M} \frac{e^{-\Gamma_0 t[1+\tilde{\gamma}(2M-N-1)]} - e^{i(\Delta - k_\ell)t}}{\Delta - k_\ell - i\Gamma_\infty[1+\tilde{\gamma}(2M-N-1)]} \tag{21}$$

This function depends only on the photon energies. The direction dependence appears in the previous factor. The quantity $\tilde{\gamma}$ entering (21) is an average over quantities $\tilde{\gamma}(k_\ell)$ defined by

$$\tilde{\gamma}(k) = \frac{1}{\nu^{*2}} \int_\nu d^3x_1 \int_\nu d^3x_2 \frac{\Gamma^c_{12}(\underline{x}_1,\underline{x}_2)}{\Gamma_\infty} J_0\left(\frac{a^* k}{Q^{1/2}}|\rho_1\ \rho_2|\right) \times \frac{\cos(kz_1)\ \cos(kz_2)}{kL} \quad . \tag{22}$$

This dimensionless quantity is a measure of the degree of cooperation between the atoms in the radiative process, at least as far as this process depends on time. a^* is the radius of the cylindrical (or ellipsoidal) volume containing the atoms, L^* its length, and ν^* its volume.

Information about the angular distribution of the photons is contained in the quantity $p_{\underline{k}_1\ldots\underline{k}_M}{}^{j_1\ldots j_M}$. The probability of finding M photons with these momenta is proportional to

$$P_{\underline{k}_1\ldots\underline{k}_M} = \frac{1}{M!}\left|p_{\underline{k}_1\ldots\underline{k}_M}{}^{j_1\ldots j_M}\right|^2 \quad , \tag{23}$$

where on the left the j's have been suppressed. Inspection of (20) shows that when all the $\underline{k}$'s are alike, the permutation sum consists of M! identical terms. In free space where the $F_{\underline{K}_i}(\underline{X}_{ji})$ are plane waves this makes $P_{\underline{k}_1\ldots\underline{k}_M} = M!$, its maximum value.[1] When mirrors are present there are possible circumstances when P is not maximal with all $\underline{k}$'s equal, but these circumstances are very improbable and we assume equal $\underline{k}$'s to give a maximum value to P, though not M! necessarily. To investigate the dependence of P on the $\underline{k}$'s near a point where they are identical we put

$$P_{\underline{k},\ldots,\underline{k},\underline{k}+\underline{q}} = P_{\underline{k},\ldots\underline{k},\underline{k}}\ \tilde{\eta}^2_M(\underline{q},\underline{k}) \quad . \tag{24}$$

Adapting the results of I to the presence of modes, we obtain

$$\tilde{\eta}_M^2(\underline{q},\underline{k}) = [\sum_{i,j=1}^{M} \cos \underline{k}^{\perp}\cdot\underline{X}_i \cos \underline{k}^{\perp}\cdot\underline{X}_j \cos(\underline{k}^{\perp}+\underline{q}^{\perp})\cdot\underline{X}_i \cos(\underline{k}^{\perp}+\underline{q}^{\perp})\cdot\underline{X}_j \times \cos \underline{q}^{\parallel}\cdot(\underline{X}_i-\underline{X}_j)] \tag{25}$$

$$\times [\sum_{i,j=1}^{M} \cos^2(\underline{k}^{\perp}\cdot\underline{X}_i) \cos^2(\underline{k}^{\perp}\cdot\underline{X}_j)]^{-1} .$$

Setting

$$\tilde{\eta}_M^2(\underline{q},\underline{k}) = \frac{1}{M} + \frac{M-1}{M}\tilde{\eta}^2(\underline{q},\underline{k}), \tag{26}$$

which takes into account the terms with i=j explicitly by the first term, we find

$$\tilde{\eta}(\underline{q},\underline{k}) = \frac{1}{2a^{*2}L^*\pi}\int_0^{2\pi} d\phi \int_0^{a^*} \rho d\rho \int_{-\frac{L^*}{2}}^{\frac{L^*}{2}} dz \cos kz \cos(k+q_z)z \cos(\rho q_\rho \cos\phi)$$

$$= \frac{J_1(a^*q_\rho)}{a^*q_\rho}\left[\frac{\sin(\frac{1}{2}q_z L^*)}{q_z L^*} + \frac{\sin(k+\frac{q_z}{2})L^*}{(2K+q_z)L^*}\right], \tag{27}$$

where $\underline{k}$ is parallel to $\underline{L}$, the most interesting case. For optical frequencies $kL^*>>1$ and the second term in the bracket is negligible. Thus $\tilde{\eta}$ depends only on $\underline{q}$, q_ρ yielding primarily the angular dependence and q_z the frequency dependence of the radiation.

If a second photon momentum deviates from $\underline{k}$ by $\underline{q}'$, another factor $\tilde{\eta}^2(\underline{q}')$ appears, and so on for additional deviation. Near a maximum of $P_{k_1\ldots k_M}$, the dependence on deviations from the central ray momentum $\underline{k}$ factorizes, indicating the mutual independence of the deviations.

As $t\to\infty$ the only amplitude $\beta(t)$ to remain nonzero is $\beta_{k_1\ldots k_N}(t)$ as can be seen from (21), and its absolute square is

$$|\beta_{k_1\ldots k_N}(\infty)|^2 \propto \prod_{\ell=1}^{N} \frac{1}{(\Delta-k_\ell)^2+\Gamma_\infty^2(1+\tilde{s})^2} . \tag{28}$$

with $\tilde{s} = (N-1)\tilde{\gamma} \approx N\tilde{\gamma}$. This is a product of Lorentzian shape functions with width $\Gamma_\infty(1+\tilde{s})$. This width is the width associated with the time dependence; there is another shape factor which comes from the q_z term in (27), and which one dominates depends on the values of L^*, Γ_∞, and $\tilde{s}$. The entire line shape is given by the function

$$\tilde{\mathcal{L}}_{\tilde{s}}(\Delta-k) = \frac{\sin^2\left(\frac{L^*}{2}(\Delta-k)\right)}{[L^*(\Delta-k)]^2} \times \frac{\tilde{C}}{(\Delta-k)^2 + \Gamma_\infty^2(1+\tilde{s})^2} \quad . \tag{29}$$

The normalization constant $\tilde{C}$ makes the integral of $\tilde{\mathcal{L}}$ over k unity. When

$$(1+\tilde{s})\Gamma_\infty L^* >> 1,$$

$$\tilde{C} \approx \frac{(1+\tilde{s})^2\Gamma^2{}_\infty L^*}{\pi} \quad . \tag{30}$$

We now use the fact that $\tilde{\gamma}$ is defined as an average of $\tilde{\gamma}(k)$, with $\tilde{\mathcal{L}}_{\tilde{s}}(\Delta-k)$ as the appropriate weight function. Therefore using (22), (29), and the equation

$$N\tilde{\gamma} = \tilde{s} = N \int_{-\infty}^{\infty} dk\, \tilde{\mathcal{L}}_{\tilde{s}}(\Delta-k)\tilde{\gamma}(k), \tag{31}$$

we obtain an equation for $\tilde{s}$. When (30) is valid this can be solved explicitly to give

$$\tilde{s} = \frac{2N}{1-R^2}\,\frac{\lambda\!\!\!^-}{L}\,\tilde{F}\left(\frac{a^*}{\lambda\!\!\!^- Q^{1/2}}\right) \quad , \tag{32}$$

where

$$\tilde{F}(\xi) = \frac{5}{12}\int_0^1 t_1 dt_1 \int_0^1 t_2 dt_2\, [J_0(\xi|t_1-t_2|)]^2 . \tag{33}$$

For a long rod this reduces to

$$\tilde{s} = \frac{5}{24}\,\frac{N\lambda\!\!\!^-}{(1-R^2)L} \quad . \quad \text{(long rod)} \tag{34}$$

The ratio of the above $\tilde{s}$ to the free space s calculated in the same way for the same system without mirrors is

$$\frac{\tilde{s}}{s} \approx \frac{55L^*}{12\pi(1-R^2)L} \quad . \tag{35}$$

It was pointed out by Rehler and Eberly[8] that their quantity μ is the $s/N=\gamma$ of I. The latter contained no explicit calculation of s but the self-consistent method of calculation given above gives a value agreeing with that of Rehler and Eberly within a numerical factor of order unity, despite the very different approaches in these two works.

For physically interesting values of the parameters, $\tilde{s}$ as given by (34) is a large number so the Lorentzian width in (29) is large and the decay is rapid. The line half width Ω is not determined by the Lorentzian factor under these circumstances, however, but is determined by the condition

$$\frac{\sin^2\left(\frac{L^*}{2}\Omega\right)}{\left(\frac{L^*}{2}\Omega\right)^2} = \frac{1}{2} \quad , \tag{36}$$

which is independent of the rate of radiation. This narrowness does not contradict the energy-time uncertainty relation because this relation concerns the energy of all N photons, not individual photons.

The properties of the ray emitted have been discussed. It remains to point out that the momentum $\underline{k}$ which makes each of the terms in the sum (20) a maximum is one associated with a Fabry-Perot mode, except for very improbable atomic arrangements. It follows that a system of many excited atoms placed between highly reflecting mirrors will radiate in the form of a ray whose central $\underline{k}$ corresponds to a mode of the mirror system.

4. Conclusions

The results of the preceding sections concerning the effect of mirrors on the emission of radiation by a system of many excited atoms show explicitly certain features which were anticipated earlier in I. The existence of a ray of spontaneous radiation from an extended system persists when mirrors are present. The forces which steer this ray, purely geometric in the free space case and strongly influenced by the mode structure in the mirror case, are weaker than those which create it but are strong enough to steer it into a Fabry-Perot mode. Furthermore, the distribution function describing the photon momentum distribution in the ray still factorizes in the neighborhood of a most probable momentum, so the photons can be regarded as having been radiated into this mode independently. As stated in I, the cooperative effect produces a distortion of the radiation pattern for a single photon, but leaves the photons with that distorted distribution independent of each other. This provides the basis for calculations which included coupling to only one mode from the start; the cooperative behavior of many atoms interacting with all modes of the radiation field is almost identical with the behavior of atoms interacting independently with a single mode. This indicates that the single mode approximation will be valid only if enough atoms are present over a large enough volume to define a ray. We know from II that the single mode approximation is invalid for one atom.

Another speculation of I was that mirrors would effectively increase the value of s for the system of atoms, and thus increase the rate of emission of the radiation. This is borne out by (35), which shows that s increases by a factor of about $(1-R^2)^{-1}$. This increased rate is not accompanied by a corresponding line broadening because the line shape is not determined by the width of the Lorentzian factor in (29) but by the diffraction factor.

The question of the role played by "stimulated emission" in this radiation process is difficult to answer. It appears from (34) that $\tilde{s}$ is proportional to N, the number of excited atoms, and hence to the number of photons finally present. $\tilde{s}$ also is proportional to the overall emission rate, so there is a proportionality between the eventual number of photons and the rate of their emission. The detailed dynamics seem more complicated than a simple picture of stimulated emission would yield.

References

1. W. E. Lamb, Jr., Phys. Rev. *134*, A1429 (1964).
2. M. O. Scully and W. E. Lamb, Jr., Phys. Rev. *159*, 208 (1967).
3. J. A. Fleck, Jr., Phys. Rev. *149*, 309 (1966).
4. H. Haken, *Encyclopedia of Physics*, Vol. XXV/2C, S. Flügge, Ed. (Springer, New York 1970).

5. N. Bloembergen, *Nonlinear Optics* (Benjamin, New York 1965).
6. P. A. M. Dirac, Proc. Roy. Soc. (London) *A112*, 661 (1926).
7. R. H. Dicke, Phys. Rev. *93*, 99 (1954).
8. N. Rehler and J. H. Eberly, Phys. Rev. *A3*, 1735 (1971).
9. V. Ernst and P. Stehle, Phys. Rev. *176*, 1456 (1968). Referred to as I.
10. P. Stehle, Phys. Rev. *A2*, 102 (1970). Referred to as II.
11. C. S. Chang and P. Stehle, Phys. Rev. *A5*, 1928 (1972).

TRANSIENT COHERENT EMISSION*

Richard G. Brewer

IBM Research Laboratory, San Jose, California

Recently it was shown that molecules excited by cw laser radiation show coherent transient effects when their transition frequency is suddenly shifted. The use of Stark pulses, which act as a frequency switching mechanism, has led directly to the observation of optical nutation[1] and photon echoes[1]. In addition to reviewing these effects, two other processes which have just been observed are reported here.

1. Optical Free Induction Decay[2]

This effect accompanies the optical nutation signal following a Stark pulse and is the optical analog of free induction decay in NMR. Whereas the nutation signal arises from molecules which are switched into resonance by the Stark field, the emission signal results from those that are switched out. This emission is easily identified because it beats with the laser, producing a damped oscillation whose frequency is the Stark shift.

*Work supported in part by the Office of Naval Research, Contract #N00014-72-C-0153.

2. Two-Photon Superradiance[3]

A transient two-photon process is observed in the infrared which exhibits all the cooperative properties associated with a superradiant two-level system. It arises when a cw laser beam excites a molecular sample whose *level degeneracy* is suddenly removed by a Stark pulse. The resulting emission, which heterodynes with the laser, gives precise ground and excited state Stark splittings, and decays with a homogeneous relaxation time since Doppler dephasing effects are absent in forward scattering.

References

1. R. G. Brewer and R. L. Shoemaker, Phys. Rev. Letters *27*, 631 (1971).
2. R. L. Shoemaker and R. G. Brewer, Bull. Am. Phys. Soc., Series II, *17*, 66 (1972).
3. R. L. Shoemaker and R. G. Brewer, Phys. Rev. Letters *28*, 1430 (1972) and Phys. Rev. (to be published Dec. 1972).

SUPERRADIANCE FROM A SYSTEM OF 3-LEVEL PARTICLES

Y.C. Cho and N.A. Kurnit *

Massachusetts Institute of Technology, Cambridge, Mass.

R. Gilmore

University of Southern Florida, Tampa, Florida

Superradiance, first discussed in 1954 by Dicke [1], has recently re-emerged as one of the central interests in the field of atomic radiation, due in part to experimental developments in the area of short pulse coherent interactions [2],[3].

In his construction of collective states [1], Dicke has implicitly employed a group theoretical method , namely, group SU(2) and permutation group S_N. The cooperation quantum number r can be interpreted as a measure of symmetry of a collective state under permutations of the N constituent particles.

On the other hand, a classical coherent state of particles, which is an eigenstate of macroscopic dipole moment of the system, is a direct product of single particle states. This is known as a Bloch state [4],[5]. A completely phased Bloch state is fully symmetric under permutations on N particles; thus, it can be given as a linear combination of Dicke states of r = N/2. The Dicke state is an energy eigenstate with null macroscopic dipole moment, while the Bloch state can be represented with a mixed density matrix.

A particle with m arbitrarily spaced internal energy levels can be described by means of the group SU(m). [6] However, no simple

* Work supported by National Science Foundation and Army Research Office - Durham.

analog of the ordinary angular momentum formalism exists for $m \geqslant 3$. Thus, Dicke's formalism cannot be immediately applied to a three or more level particle system. A complete description of the system of N identical 3-level particles can be obtained by using $SU(3) \times S_N$ under symmetry assumptions [7]. We will not give a detailed discussion of the group theory here;[6] however, we discuss the physical meaning of quantum numbers and the action of generators of SU(3).

Dicke has pointed out that for the system of identical 2-level particles the decay rate between states of common r is enhanced (for large r) by the existence of unexcited particles prior to decay. We show here that for the system of identical 3-level particles analogous relations exist for each pair of coupled levels.

Throughout the discussion in this paper, we make the following assumptions: a) The system is of such low density that particles have non-overlapping wave functions; b) Internal states of each particle are non-degenerate and well-separated; c) Particle interactions are due to the common radiation field only.

The Hamiltonian of the system of N identical 3-level particles can be written in terms of generators of SU(3), omitting the spatial coordinate dependent part of the unperturbed Hamiltonian (c.f. assumption a) as follows:[8]

$$H = \varepsilon_2 H_1 + \varepsilon_1 H_2 + \hbar c \sum_{\underline{k}\hat{\varepsilon}} k a^{+}_{\underline{k}\hat{\varepsilon}} a_{\underline{k}\hat{\varepsilon}} + V , \tag{1}$$

$$V = - \sum_{\alpha=1}^{3} \sum_{j=1}^{N} \{ \sum_{\underline{k}(\alpha)} g_{\underline{k}(\alpha)} a_{\underline{k}(\alpha)} E_j(\alpha) \exp[i(\underline{k}(\alpha)\underline{x}_j - \omega_\alpha t)] + \text{h.c.}\}$$

with

$$g_{\underline{k}(\alpha)} = [\hbar c/vk(\alpha)]^{1/2} \underline{u}(\alpha)\cdot \hat{\varepsilon}_{\underline{k}(\alpha)} \text{ and } |\underline{u}(\alpha)|^2 = 3\hbar c\gamma_\alpha/4\omega_\alpha .$$

Here $a^{+}_{\underline{k}\hat{\varepsilon}}$ and $a_{\underline{k}\hat{\varepsilon}}$ are creation and annihilation operators of a photon whose propagation and polarization vectors are $\underline{k}$ and $\hat{\varepsilon}$, v is the volume of the system, and $\underline{u}(\alpha)$ are constant vectors determined by the internal structure of a single particle. The three decay modes are labelled by α; ω_α and γ_α are respectively the resonant frequency and the single particle decay rate of the transition mode α (see Fig. 1). The constants ε_1 and ε_2 are respectively the energies of the 1st and 2nd excited states of single particles. $H_1 = \Sigma_j H_{1j}$ and $H_2 = \Sigma_j H_{2j}$; H_{1j}, H_{2j}, $E_j(\pm 1)$, $E_j(\pm 2)$ and $E_j(\pm 3)$

are generators of SU(3) referring to the jth particle. The H_{qj} are mutually commuting hermitian operators and $E_j^{\dagger}(\alpha) = E_j(-\alpha)$.

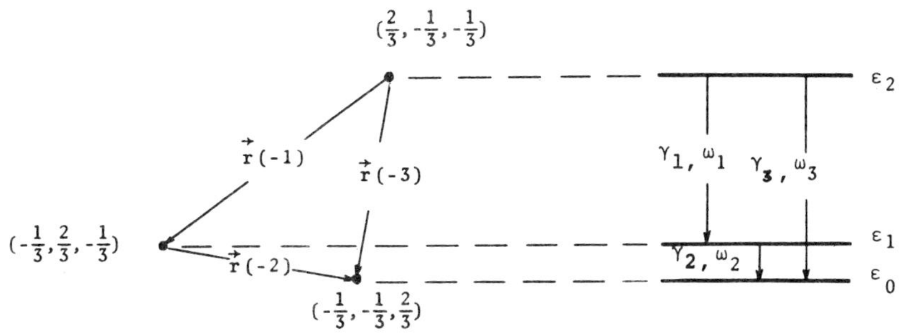

Fig. 1 Single particle energy levels and transition scheme. States correspond to weights (see footnote 12) and transitions correspond to roots of SU(3). Weights and roots are seen from the $\hat{e}_3$ -direction.

These 8 operators satisfy the following commutation relations:[6]

$$[H_{pi}, H_{qj}] = 0 \quad , \quad [H_{qi}, E_j(\alpha)] = \delta_{ij} r_q(\alpha)\, E_j(\alpha),$$

$$[E_i(\alpha), E_j(-\alpha)] = \delta_{ij} \sum_q r_q(\alpha) H_{qj}, \tag{2}$$

$$[E_i(\alpha), E_j(\beta)] = \delta_{ij} N^{\theta}_{\alpha\beta}\, E_j(\theta) \text{ if } \beta \neq -\alpha,$$ where $N^{\theta}_{\alpha\beta}$ is non-zero only if $\underline{r}(\theta)$ exists such that $\underline{r}(\theta) = \underline{r}(\alpha) + \underline{r}(\beta)$.

A vector $\underline{r}(\alpha) = \{r_1(\alpha), r_2(\alpha)\}$ is called a root of SU(3). Roots of SU(3) can be visualized as 3-dimensional vectors by introducing H_3 such that $H_3 = -(H_1+H_2)$. Then, the roots are $\underline{r}(\pm 1) = \pm(\hat{e}_1 - \hat{e}_2)$, $\underline{r}(\pm 2) = \pm(\hat{e}_2 - \hat{e}_3)$ and $\underline{r}(\pm 3) = \pm(\hat{e}_1 - \hat{e}_3)$. where $\hat{e}_1$, $\hat{e}_2$ and $\hat{e}_3$ are mutually orthogonal unit vectors (see Figure 1).[10]

In the absence of any interaction, stationary collective energy eigenstates are given as simultaneous eigenvectors of commuting generators H_1 and H_2.

These energy eigenstates are futher classified by symmetry under permutations of the constituent particles. This is due to the fact that the unperturbed Hamiltonian is invariant under permutations. From the commutation relations, one can show that non-commuting generators $E(\alpha)$ are shift operators between eigenvectors of the commuting generators. The proper selection rules for transitions between collective energy eigenstates are automatically built into V if the single particle transitions are subject to some selection rules, since the $\underline{u}(\alpha)$ are determined from a single particle.

Furthermore, if the interaction is permutationally invariant, the system will remain in the same symmetry representation ($\Delta r=0$ in Ref. 1) during decay. For simplicity, we assume that the system remains in the fully symmetric representation where the cooperative emission is largest.[11]

A typical collective state of the 3-level particle system in the fully symmetric representation is denoted by $|N;n_2,n_1>$ as in the representation of the group SU(3) x S_N. Eigenvalues of H_1 and H_2 to $|N;n_2,n_1>$ are given as follows:[12]

$$H_1|N;n_2,n_1\rangle = n_1|N;n_2,n_1\rangle \text{ and } H_2|N;n_2n_1\rangle = (n_2-n_1)|N;n_2,n_1\rangle. \tag{3}$$

The physical meaning of quantum numbers may be given in terms of the single particle picture as follows: N is the total number of particles and it also refers to the fully symmetric representation, n_1 is the number of particles in the 2nd excited state $|\varepsilon_2>$, (n_2-n_1) is the number of particles in the 1st excited state $|\varepsilon_1>$, and $(N-n_2)$ is the number of particles in the ground state $|\varepsilon_0>$.

We consider first the case $\ell << \lambda$, where ℓ is the linear dimension of the system and λ is a transition wavelength. In this case, $\underline{k}(\alpha) \cdot \underline{x}_j \cong 0$. Then,

$$V = - \sum_{\alpha=1}^{3} \{\sum'_{\underline{k}(\alpha)} g_{\underline{k}(\alpha)} a_{\underline{k}(\alpha)} E(\alpha) e^{-i\omega_\alpha t} + h.c.\}, \tag{4}$$

where $E(\alpha) = \sum_j E_j(\alpha)$ and Σ' stands for a sum over all directions of $\underline{k}(\alpha)$.

Neglecting the secondary interaction of emitted photons with particles, we can calculate matrix elements for the spontaneous coherent decay by using the commutation relations given in Eq.(2). The results are

$$<N;n_2,n_1-1;\underline{k}(1)|V|N;n_2,n_1;\phi> = -g^*_{\underline{k}(1)}\{(n_2-n_1+1)n_1\}^{1/2}$$

$$<N;n_2-1,n_1;\underline{k}(2)|V|N;n_2,n_1;\phi> = -g^*_{\underline{k}(2)}\{(N-n_2+1)(n_2-n_1)\}^{1/2}$$

$$\langle N;n_2-1,n_1-1;\underline{k}(3)|V|N;n_2,n_1;\phi\rangle = -g^*_{\underline{k}(3)}\{(N-n_2+1)n_1\}^{1/2}, \quad (5)$$

where $|\phi\rangle$ is photon-free state and $|\underline{k}(\alpha)\rangle$ is a one-photon state.

These matrix elements reduce to Dicke's result for a two-level particle system in the limit of $\gamma_1=\gamma_2=0$, $\gamma_2=\gamma_3=0$, or $\gamma_1=\gamma_3=0$. For example, if $\gamma_1=\gamma_2=0$, then the occupation number of $|\varepsilon_1\rangle$ would be zero; that is, $n_2-n_1=0$. Consequently, $\langle N;n-1;\underline{k}|V|N;n;\phi\rangle = -g^*_{\underline{k}}\{(N-n+1)n\}^{\frac{1}{2}}$, where we have dropped the subscript α from all quantities because there would exist only one decay mode, γ_3, and we have replaced n_2 and n_1 by n. By recalling that $N/2 = r$, $n=n^+$ and $N-n = n^-$ in Dicke's notation [1]), one can derive Dicke's result from the present one and vice versa.

Coherent spontaneous decay rates for the three-level particle system are calculated as follows:

$$\Gamma(|N;n_2,n_1\rangle \rightarrow |N;n_2,n_1-1\rangle) =$$

$$\frac{2\pi}{\hbar}\int dv \int d\rho_{\underline{k}(1)} |\langle N;n_2,n_1-1;\underline{k}(1)|V|N;n_2,n_1;\phi\rangle|^2 = (n_2-n_1+1)n_1\gamma_1$$

$$\Gamma(|N;n_2,n_1\rangle \rightarrow |N;n_2-1,n_1\rangle) = (N-n_2+1)(n_2-n_1)\gamma_2 \quad (6)$$

$$\Gamma(|N;n_2,n_1\rangle \rightarrow |N;n_2-1,n_1-1\rangle) = (N-n_2+1)n_1\gamma_3 ,$$

where $d\rho_{\underline{k}(\alpha)} = \dfrac{k^2(\alpha)}{(2\pi)^3\hbar c}\, d\Omega_{\underline{k}(\alpha)}$ has been used.

We see that the decay rate is proportional to one plus the occupation number of the single particle state into which decay goes, as well as to the occupation number of the decaying single particle state. This result can be interpreted as arising from the quantum statistical effect in a system of N identical particles. When the system is in the fully symmetric representation it acts as a boson system. [13]

In the case of samples larger than a wavelength [14], the radiation rate from a superradiant state phased for emission in a given direction is similarly enhanced, but only over a small solid angle given by the diffraction aperture of the sample. The dynamics of the coherent decay of a system of two-level particles has been discussed within this context, [15], as well as within the context of a single mode laser model. [16] The latter approach allows propagation effects to be ignored provided the sample is shorter than a "cooperation length", ℓ_c, [4],[16] determined by the condition that the emitted pulse is not shorter than the sample. The system decays coherently provided it is longer than a threshold length which for low mirror reflectivity reduces to an absorption length [17],[18]. We discuss the decay of a system of three level particles within the same model.

Assuming that each transition interacts with only a single mode, we have

$$V = -\sum_{\alpha=1}^{3} \{g_{\underline{k}(\alpha)} a_{\underline{k}(\alpha)} \; E(\underline{k}(\alpha)) + \text{h.c.}\}$$

where

$$E(\underline{k}(\alpha)) = \Sigma_j E_j(\underline{k}(\alpha)) \text{ and } E_j(\underline{k}(\alpha)) = E_j(\alpha) e^{i\underline{k}(\alpha)\cdot\underline{x}_j} . \qquad (7)$$

One can prove that the H_q and phased operators $E(\underline{k}(\alpha))$ in Eq. (7) satisfy the same commutation relations as those for the H_q's and $E(\alpha)$'s (see Eq. [2]) if the $\underline{k}(\alpha)$ are fixed.

This transformation on $E_j(\alpha)$ is accompanied by the following transformations of the single particle states:

$$|\varepsilon_2>_{j,\underline{k}(3)} = e^{i\underline{k}(3)\cdot\underline{x}_j}|\varepsilon_2>_j, \qquad |\varepsilon_1>_{j,\underline{k}(2)} = e^{i\underline{k}(2)\cdot\underline{x}_j}|\varepsilon_1>_j$$

$|\varepsilon_0>_j$ unchanged, where $\underline{k}(1) + \underline{k}(2) \cong \underline{k}(3)$.

The decay rates are then

$$\Gamma(|N;n_2,n_1> \to |N;n_2,n_1-1>) = \frac{2\pi}{\hbar}\int_{\Delta\Omega_{\underline{k}(1)}} d\Omega_{\underline{k}(1)} \frac{k(1)}{(2\pi)^2}|\underline{u}^*(1)|^2 \times (n_2-n_1+1)n_1$$

$$= \gamma_1'(n_2-n_1+1)n_1$$

$$\Gamma(|N;n_2,n_1> \to |N;n_2-1,n_1>) = \gamma_2'\ (N-n_2+1)(n_2-n_1) \qquad (9)$$

$$\Gamma(|N;n_2,n_1> \to |N;n_2-1,n_1-1>) = \gamma_3'(N-n_2+1)n_1 \ ,$$

where

$$\gamma'_\alpha = \frac{3\Delta\Omega_{\underline{k}(\alpha)}}{8\pi}\gamma_\alpha\ ,\ \Delta\Omega_{\underline{k}(\alpha)} \cong \frac{\lambda_\alpha^2}{A}$$ and A is the cross section area of the sample.[19]

When $\ell \ll \ell_c$, damping of the internal energy of the system is balanced by the field energy of the emitted radiation. As $N\to\infty$, n_1 and n_2 can be replaced by continuous functions of time. Then from Eq. (6) and (9) we obtain the following nonlinear differential equations governing the decay:[20]

$$\frac{dX(t)}{dt} = -\Gamma_1 X(t)\{Y(t)+\eta\} - \Gamma_3\{Z(t)+\eta\}X(t),$$

$$\frac{dY(t)}{dt} = -\Gamma_2 Y(t)\{Z(t)+\eta\} + \Gamma_1 X(t)\{Y(t)+\eta\}, \tag{10}$$

$$\frac{dZ(t)}{dt} = \Gamma_3\{Z(t)+\eta\}X(t) + \Gamma_2 Y(t)\{Z(t)+\eta\},$$

where $X = n_1/N$, $Y = (n_2-n_1)/N$, $Z = (N-n_2)/N$, $\eta = 1/N$ and $\Gamma_\alpha = N\tilde{\gamma}_\alpha$ with $\tilde{\gamma}_\alpha = \gamma_\alpha$ for the case of $\ell \ll \lambda$ and $\tilde{\gamma}_\alpha = \gamma'_\alpha$ for a pencil-shaped [14] sample.

We consider the case of $\gamma_2 = 0$, [21] then Eq. (10) reduces to

$$\frac{dY}{dt} = \Gamma_1 X(Y+\eta)$$

$$\frac{dZ}{dt} = \Gamma_3 X(Z+\eta), \tag{11}$$

with $X + Y + Z = 1$.

Solution of Eq. (11) in the physical region (0<[Y and Z]< 1) is

$$(Y+\eta)/(Y_0+\eta) = \{(Z+\eta)/(Z_0+\eta)\}^s, \tag{12}$$

where $s = \tilde{\gamma}_1/\tilde{\gamma}_3$, $Y_0 = Y(t=0)$ and $Z_0 = Z(t=0)$.

From Eqs. (11) and (12) we obtain

$$I(\omega_1)/I(\omega_3) = s(Y_o+\eta)/(Z_o+\eta)\{(Z+\eta)/(Z_o+\eta)\}^{s-1}$$

$$= s(Y_o+\eta)/(Z_o+\eta)\{(Y+\eta)/(Y_o+\eta)\}^{1-1/s} ,$$

where $I(\omega_1)$ and $I(\omega_3)$ are intensities of the emitted photons.

Under most experimental conditions (Y_o and $Z_o >> \eta$), η may be neglected in Eq. (11).[22] Y and Z can be obtained as explicit functions of time for the given numerical values of Y_o, Z_o and s. But, since the decay time is very short for spontaneous coherent decay, the total numbers of the emitted photons, $Q(\omega_1)$ and $Q(\omega_3)$, are more easily measured than $I(\omega_1)$ and $I(\omega_3)$. Then,

$$Q(\omega_1) = N[Y(t=\infty) - Y_o] , \quad Q(\omega_3) = N[Z(t=\infty) - Z_o]$$

with $Y(t=\infty) + Z(t=\infty) = 1$.

In Fig. (2), (3), and (4), $Q(\omega_1)/Q(\omega_3)$ is plotted as a function of Y_o, Z_o and s.[23].

In conclusion, our result provides a direct way of measuring superradiance and affords the possiblity of a superradiant laser [14], in the sense that pumping of the system by the γ_3 transition mode would have negligible effect on coherent decay by the γ_1 transition (c.f. assumption b). Finally, we can experimentally estimate the dependence of the coherent decay on the resonant frequencies and other parameters. This should be relevant to the recent theories on cooperative phenomena of decay [4], [16].

One of us (Y.C.C.) wishes to thank Dr. F. Haake and Dr. H.M. Lee for stimulating discussions.

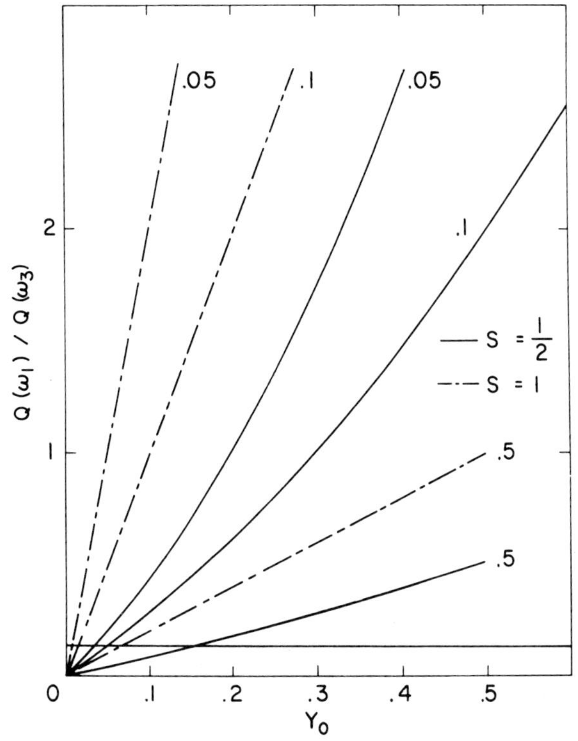

Fig. 2 $Q(\omega_1)/Q(\omega_3)$ vs. Y_0 for s=1/2, 1 and Z_0 = 0.05, 0.1, 0.5. For comparison, the horizontal line shows the ratio of the numbers of photons emitted in incoherent decay for γ_1/γ_3 = 1/8, corresponding to s=1/2 with $\lambda_1 = 2\lambda_3$.

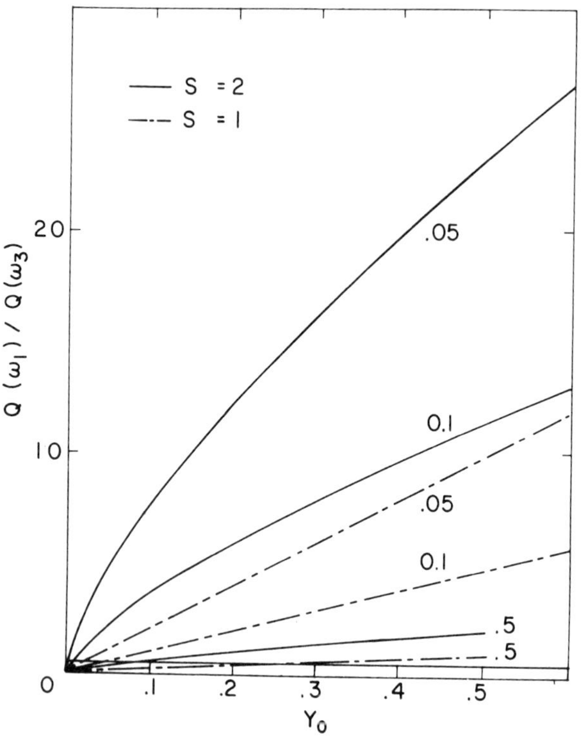

Fig. 3 $Q(\omega_1)/Q(\omega_3)$ vs. Y_0 for s=1,2 and Z_0=0.05, 0.1, 0.5. For comparison, the horizontal line shows the ratio of the numbers of photons emitted in incoherent decay for γ_1/γ_3 = 1/2, corresponding to s=2 with $\lambda_1 = 2\lambda_3$.

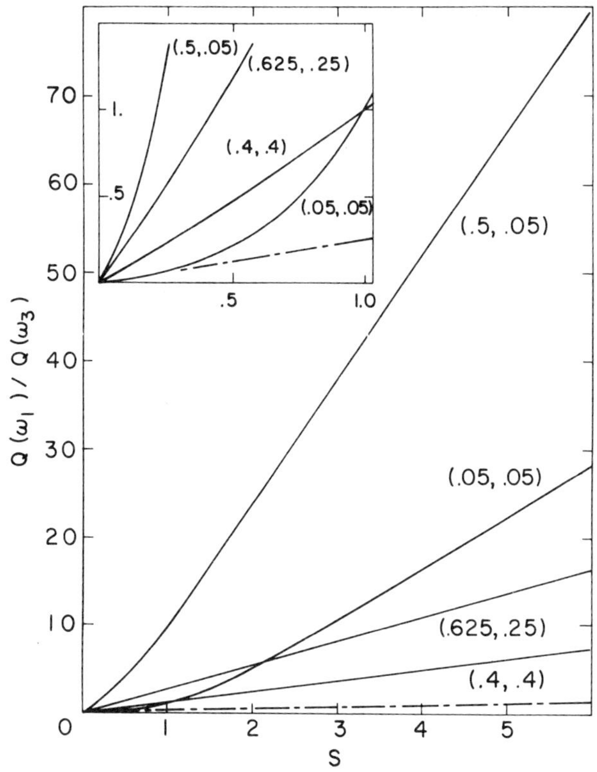

Fig. 4 $Q(\omega_1)/Q(\omega_3)$ vs. s for (Y_o=0.5, Z_o=0.05), (Y_o=0.05, Z_o=0.05), (Y_o=0.625, Z_o=0.25) and (Y_o=0.4, Z_o=0.4). For comparison the broken line shows the ratio of the numbers of photons emitted in incoherent decay for the case $\lambda_1=2\lambda_3$.

References

1. R.H. Dicke, Phys. Rev. *93*, 99 (1954).
2. I.D. Abella, N.A. Kurnit and S.R. Hartmann, Phys. Rev. *141*, 391 (1966).
3. S.L. McCall and E.L. Hahn, Phys. Rev. *183*, 457 (1969).
4. F.T. Arecchi and E. Courtens, Phys. Rev. *A2*, 1730 (1970).
5. F. Bloch, Phys. Rev. *70*, 460 (1946).
6. a. G. Racah, Ergeb. Exakt. Naturw. *37*, 28 (1965).
   b. E.M. Loebl, *Group Theory* (Academic Press, 1968).
   c. M. Hamermesh, *Group Theory* (Addison-Wesley Co., 1962).
   d. H.J. Lipkin, *Lie Groups* (North Holland Pub.Co., Amsterdam, 1966).

e. R. Gilmore, J. of Math. Phys. *11*, 513 and 3420 (1970).

7. This approach has been suggested by F.T. Arecchi and D.M. Kim, who collaborated with one of us (R.G.) on preliminary investigations. See also L.A. Shelepin, Zh. Eksp. Theor. Fiz. *54*, 1463 (1968) (Sov. Phys. JETP *27*, 784 (1968)).
8. Y.C. Cho, Ph.D. Thesis, M.I.T. (1972) (unpublished).
9. We assume here that ω_1, ω_2 and ω_3 are different from one another to the limits of the superradiant line breadth so that interference may be avoided.
10. See p. 187 in Ref. 6(b).
11. The classification into irreducible representations is illustrated for four three-level particles in Fig. 4 of the paper by R. Gilmore in this volume, p. 217.
12. Eigenvalues of H_1 and H_2 are called weights of the representation of SU(3). In this paper, each weight is subtracted by the lowest weight so that the collective ground state is redefined as zero energy level.
13. R. Bonifacio, D.M. Kim and M.O. Scully, Phys. Rev. *187*, 441 (1969).
14. R. H. Dicke, *Quantum Electronics III* (N. Bloembergen and P. Grivet, eds., 1964), vol. **1**, p. 35. See also Refs. 1, 15 and 16.
15. N.E. Rehler and J.H. Eberly, Phys. Rev. *A3* , 1735 (1971).
16. R. Bonifacio and P. Schwendimann, Nuovo Cimento Letters *3*, 509 (1970). See also R. Bonifacio, P. Schwendimann and F. Haake, Phys. Rev. *A4*, 302 and 854 (1971).
17. See also R. Friedberg and S.R. Hartmann, Physics Letters *37A*, 285 (1971).
18. The permutational invariance of the interaction can be maintained even for low mirror reflectivity by assuming, for example, a traveling wave ring laser cavity with distributed losses.
19. See Appendix C in Ref. 2. As discussed here and in Ref. 15, this result pertains to a sample with a large Fresnel number.
20. Similar differential equations, but without the constant η which accounts for quantum fluctuations, have been derived by M.D. Crisp and E.T. Jaynes, Phys. Rev. *179*, 1253 (1969), using a semiclassical treatment.
21. The proceeding discussions can be extended to the case $\gamma_2 \neq 0$.
22. If $(X_0, Y_0) \sim \eta$, the quantum fluctuation would be so large during spontaneous decay that the semiclassical treatment would result in considerable error.
23. It must be emphasized that the interference between the γ_1 transition and the γ_2 transition should have been taken into account for $\lambda_1 = 2\lambda_3$ $(\omega_1 = \omega_2)$ if γ_2 were not assumed to be zero.

A QUANTUM ELECTRODYNAMIC VIEW OF SUPERRADIANCE AS A COMPETITION BETWEEN STIMULATED AND SPONTANEOUS ATOMIC DECAY

K.G. Whitney

Stanford University, Stanford, California

1. Introduction

A number of treatments of cooperative, or "superradiant", spontaneous decay of N similarly excited two-level atoms have been presented lately. [1]-[7] In broad outline all of these treatments give the same result: The peak emitted power is proportional to N^2 and the system's radiative lifetime is shortened by a factor 1/N. In some details, however, marked differences have been noted. In particular, fully quantum electrodynamic and fully semiclassical treatments differ widely in several respects. Consider these two, for example:

(1) *Time to peak emission of a system completely excited.* QED says the time to reach peak emission is finite, while semiclassical theory says that it is infinite.

(2) *Frequency modulation of the emitted field.* QED in general does not account for it, while semiclassical theory says it is present, possibly even large, and describes it in detail.

In this short note we sketch a quantum electrodynamic treatment of the N-atom emission problem which is intended to interpolate between these widely different predictions. This work differs from most previous QED treatments because of our choice of initial atomic state. Instead of a highly correlated many-atom "cooperation number" state of the kind introduced by Dicke, we choose what has been called a Bloch state or a Θ-state. The distinction between these two kinds of state is fairly simple to make, and is well known. A Dicke-type state is generated by the absorption of an integer number m of photons in a completely symmetric way by N ground state atoms. A Bloch-type state is generated by the action on the N ground state atoms, of a pulse of purely classical radiation. Our choice of the Bloch-type states is motivated by the assumption that coherent intense laser-pulse excitation has been used to raise the atoms out of their ground states, leaving them, to a good approximation, initially in a Bloch state. If, on the other hand, a method of excitation were to be used which imparts a well-defined integral number of photons to the atoms, then the Dicke-type state would be a better initial choice.

This "classical" initial condition, which treats all the atoms independently, naturally influences the subsequent evolution of the N-atom system. Although atom-atom correlations build up in the course of the interaction, as dictated by our use of QED, they do not prevent an essentially classical *point of view* being maintained throughout the emission process. Our principle interest here is served merely by showing the extent of the correspondence between the purely semiclassical work of Stroud, Eberly, Lama and Mandel [7] and the restricted quantum electrodynamic Bloch state results of Rehler and Eberly. [3] One consequence of our demonstration of this correspondence is interesting in its own right. We will show that a ready distinction may be made between stimulated and spontaneous aspects of the emission process, even though no external stimulating field is present.

2. Description of the Density Matrix Analysis

Because of our assumption of a classical initial state, we write

$$\rho(t_0) = \rho_A(t_0)\rho_F(t_0) \tag{1}$$

and

$$\rho(t) = \rho_A(t)\rho_F(t) + \rho_{AF}(t) \quad . \tag{2}$$

That is, at the initial time t_0 , immediately after excitation, there are no atom-field correlations, so the density matrix factors into separate atom and field parts. At later times $t > t_0$, the density matrix no longer factors. The existence of $\rho_{AF}(t)$ signifies the build-up of such correlations.

Evolution in time, including the onset of correlations, is governed by the total atom-field Hamiltonian in the usual way. Let us write $\hbar\Omega_T$ for the total Hamiltonian. It is the sum of atomic, free field, and interaction Hamiltonians:

$$\Omega_T = \Omega_A + \Omega_F + \Omega_{AF} \quad . \tag{3}$$

The density matrix time evolution is governed by Schrodinger's equation:

$$i \frac{d}{dt} \rho(t) = [\Omega_T, \rho_A(t)\rho_F(t) + \rho_{AF}(t)] \quad . \tag{4}$$

At any time the density matrix for the atom or field system alone is obtained by tracing over the states of the other system. In this way we obtain

$$i \frac{d}{dt} \rho_A(t) = [\Omega_A + \overline{\Omega}_A , \rho_A(t)] + TR_F[\Omega_{AF} , \rho_{AF}(t)] , \tag{5}$$

$$i \frac{d}{dt} \rho_F(t) = [\Omega_F + \overline{\Omega}_F , \rho_F(t)] + TR_A[\Omega_{AF} , \rho_{AF}(t)] , \tag{6}$$

where the "Hartree" operators, $\overline{\Omega}_A$ and $\overline{\Omega}_F$, are given by traces over the interaction term in the Hamiltonian:

$$\overline{\Omega}_A = TR_F \, \rho_F(t) \, \Omega_{AF} \quad , \tag{7}$$

$$\overline{\Omega}_F = TR_A \, \rho_A(t) \, \Omega_{AF} \quad . \tag{8}$$

When it is needed, an equation for $\rho_{AF}(t)$ can be found from Eqs. (2) and (4)-(6). [8],[9]

Keeping in mind that ρ_{AF} is initially zero, and that Ω_{AF} leads to rates of change that are slow compared to the atomic frequencies in the problem (which we will introduce explicitly shortly), it seems likely that the terms involving products of ρ_{AF} and Ω_{AF} can be neglected in Eqs.(5) and (6), to a first approximation. It is well known that in this form Eqs.(5) and (6) give a completely semiclassical description of the interaction; [10] there are no atom-field correlations initially and none develop in time. The Stroud-Eberly-Lama-Mandel (abbreviated SELM hereafter) treatment of superradiance is the most complete to date at this level of approximation.

At the next level of approximation the contributions of the

$\Omega_{AF}\,\rho_{AF}$ product terms make their influence felt. An incomplete analysis of the problem at this level has been presented by Rehler and Eberly. [3] It is our goal here to obtain the corrections at this level to the SELM analysis, and show how those corrections point to a completion of the Rehler-Eberly (RE hereafter) work.

3. Specification of the Dynamical Problem

To proceed further, we now choose a specific framework of operators and states within which to carry out our calculations. To begin with, it will be assumed that the size of the atomic sample is small (by an order of magnitude) compared to the duration of the atomic-radiative-decay times c. It is still possible, of course, for the sample to be very large compared with an optical wavelength. As the atoms de-excite from their classically prepared initial state, many radiation modes over a broad frequency spectrum and with many directions of propagation are excited. Because of the smallness of the sample size, each atom within the ensemble reacts effectively the same to the radiation modes. Thus, it is appropriate to consider the interaction of the atoms with the radiation field from the mode point of view. The radiation field Hamiltonian $\hbar\Omega_F$ will be decomposed in the conventional way:

$$\Omega_F = \sum_{\kappa} \omega_\kappa\, a^+_\kappa\, a_\kappa \ , \tag{9}$$

and as usual, $[a_\kappa(t),\ a^+_{\kappa'}(t)] = \delta_{\kappa\kappa'}$.

We assume furthermore that all of the N atoms have only two levels of any importance, denoted $|+\rangle$ and $|-\rangle$, and that both the effective transition frequency ω and dipole matrix element, μ, are the same from atom to atom. The atoms may interact with the host lattice, but the atomic decay times, T_1, T_2', and T_2^*, produced by the lattice, are assumed long in comparison to the time it takes the atoms to radiate their energy, so that these effects are also neglected in this paper.

Thus, in a well-known way, [11] the N-atom Hamiltonian, $\hbar(\Omega_A+\Omega_{AF})$, can be written very compactly as a sum of scalar products:

$$\Omega_A + \Omega_{AF} = \sum_{\ell=1}^{N} \hat{\underline{\Omega}}_\ell \cdot \hat{\underline{p}}_\ell \ , \tag{10}$$

where the three components of the operator vector $\hat{\underline{p}}_\ell$, referring to the ℓ-th atom, obey the algebra of spin 1/2 angular momentum:

$$\hat{p}_{\ell i}\,\hat{p}_{\ell j} = \frac{1}{4}\,\delta_{ij} + i/2 \sum_{k=1}^{3} \varepsilon_{ijk}\,\hat{p}_{\ell k} (i,j = 1,2,3) \tag{11}$$

and $\hat{\underline{\Omega}}_\ell$ is the operator triplet,

$$\Omega_{\ell_1} = \sum_\kappa \{f_\kappa^*(r_\ell)\,a_\kappa + \text{c.c.}\} \quad , \tag{12}$$

$$\Omega_{\ell_2} = i \sum_\kappa \{f_\kappa^*(r_\ell)\,a_\kappa - \text{c.c.}\} \quad , \tag{13}$$

$$\Omega_{\ell_3} = \omega \quad . \tag{14}$$

The coupling constant $f_\kappa(r_\ell)$ depends upon the position r_ℓ of the ℓ-th atom as follows (in the limit of a near continuum of modes in a large quantization volume, V):

$$f_\kappa(r_\ell) = e^{-ik\cdot r_\ell}\, f_\kappa \quad , \tag{15}$$

where

$$f_\kappa = i\sqrt{\frac{2\pi|k|c}{\hbar V}}\;\underline{\mu}\cdot\hat{\underline{\varepsilon}}_\lambda \quad , \qquad \kappa = (\underline{k},\lambda), \quad \underline{k}\cdot\hat{\underline{\varepsilon}}_\lambda = 0 \ . \tag{16}$$

Equations (12), (13) and (16) for the radiation field, atom interaction and for the coupling constant result from expressing the electric dipole interaction in rotating wave approximation. In the Heisenberg picture,

$$\frac{d\hat{\underline{p}}_\ell}{dt} = i\,[\Omega_T \ , \ \hat{\underline{p}}_\ell] = \hat{\underline{\Omega}}_\ell \times \hat{\underline{p}}_\ell \tag{17}$$

so that $\hat{\underline{p}}_\ell$ may be called the Bloch vector operator for the ℓth atom.

Concerning the initial state of our system, we assume, for reasons discussed in Section I, that initially, at the time t_0, the radiation field is in its vacuum state $|v\rangle$, and each atom is in a Θ-state, $|\mathring{\theta}_\ell, \mathring{\chi}_\ell\rangle$, such as would be prepared by an intense classical radiation pulse. [11] In terms of $|+\rangle$ and $|-\rangle$,

$$|\overset{\circ}{\Theta}_\ell \overset{\circ}{\chi}_\ell\rangle = e^{\frac{i}{2}\overset{\circ}{\chi}_\ell} \cos\left(\frac{\overset{\circ}{\Theta}_\ell}{2}\right) |-\rangle + e^{-\frac{i}{2}\overset{\circ}{\chi}_\ell} \sin\left(\frac{\overset{\circ}{\Theta}_\ell}{2}\right) |+\rangle .$$

Thus, in Eq. (1), $\rho_F(t_0) = |v\rangle\langle v|$ and $\rho_A(t_0) = \prod_{\ell=1}^{N} |\overset{\circ}{\Theta}_\ell \overset{\circ}{\chi}_\ell\rangle\langle\overset{\circ}{\Theta}_\ell \overset{\circ}{\chi}_\ell|$.

4. QED Corrected Atomic Equations of Motion

We will focus our attention on the corrections to the semi-classical description of atomic motion that result when a lowest-order evaluation of ρ_{AF}, in terms of ρ_A and ρ_F, is inserted in Eq. (5). The motion of each atom is described by the physical variables $\underline{p}_\ell(t) \equiv Tr_A\, \rho_A(t)\hat{\underline{p}}_\ell$, which comprise a numerical Bloch vector. From Eq. (5), we want to obtain equations of motion for p_{ℓ_3}, which relates to the population inversion of the ℓth atom, and $p_\ell \equiv p_{\ell_1} - ip_{\ell 2}$, which describes the current flow of the ℓth atom. This current flow is driven by the radiation field variables $\alpha_\kappa(t) \equiv Tr_F\rho_F(t)\, a_\kappa$. The "Hartree" potential $\bar{\Omega}_A$ of Eq. (5) is expressible in terms of α_κ and supplies the semiclassical terms to the equations of motion for p_{ℓ_3} and p_ℓ that follow from Eq. (5):

$$\frac{dp_{\ell_3}(t)}{dt} = i \sum_\kappa \{f_\kappa(r_\ell)\, \alpha^*_\kappa(t) p_\ell(t) - c.c.\}$$

$$- i\, Tr_A \{\hat{p}_{\ell_3}\, Tr_F\, [\Omega_{AF}\, ,\, \rho_{AF}(t)]\} \quad , \tag{18}$$

$$\frac{dp_\ell(t)}{dt} = - i\omega p_\ell(t) + 2i \sum_\kappa f^*_\kappa(r_\ell)\alpha_\kappa(t) p_{\ell_3}(t)$$

$$- i\, Tr_A \{(\hat{p}_{\ell_1} - i\hat{p}_{\ell_2})\, Tr_F\, [\Omega_{AF}\, ,\, \rho_{AF}(t)]\} . \tag{19}$$

The ρ_{AF} terms in Eqs. (18) and (19), therefore, supply corrections to semiclassical theory.

The expression for ρ_{AF} that we have used to evaluate the traces in Eqs. (18) and (19) is of first order in the interaction Hamiltonain $\hbar\Omega_{AF}$: [9]

$$\rho_{AF}(t) \simeq - i \int_{t_0}^{t} dt' [\Delta\Omega_{AF}(t' - t + t_0 \; ; t'), \rho_A(t)\rho_F(t)], \qquad (20)$$

where $\Delta\Omega_{AF}$ is composed of operators minus their average values:

$$\Delta\Omega_{AF}(t_0; t') = \Omega_{AF} - \overline{\Omega}_A(t') - \overline{\Omega}_F(t') + Tr_{AF}\, \rho_A(t')\rho_F(t')\, \Omega_{AF}.$$

The notation $\Delta\Omega_{AF}(t \; ; t')$ means that the operators in $\Delta\Omega_{AF}$ are evaluated as Heisenberg operators at the time t; whereas, the expectation values are evaluated at the time t'. When Eq. (20) is substituted into Eqs. (18) and (19), one finds that the semiclassical motion of $p_{\ell 3}$ and p_ℓ is modified by the appearance, in these equations of fluctuation correlation functions involving operators less their average values. If the zero-order statistics of the uncorrelated, semiclassical motion (which has the radiation field evolving through α-states while the atoms evolve through Θ-states) are used to evaluate these correlation functions, then most of them are zero. This is reassuring, for it supports the original assumption that the motion remains essentially uncorrelated. However, some correlation functions cannot vanish even when evaluated between α- states or Θ-states. These are the unequal time, mode commutator,

$$< [a_\kappa(t), \; a_\kappa^+(t')] > \simeq e^{-i\omega_\kappa(t-t')} ,$$

as well as the single atom, current-current and current-population correlation functions. They are responsible for the QED corrections to semiclassical theory that we have calculated. The results of the calculation are that Eqs. (18) and (19) become Bloch equations:

$$\frac{dp_{\ell_3}(t)}{dt} = i \sum_\kappa \{f_\kappa(r_\ell)\, \overline{\alpha}_\kappa^*(r_\ell t)\, p_\ell(t) - c.c.\}$$

$$- \frac{1}{\tau_0}\left(p_{\ell_3}(t) + \frac{1}{2}\right) , \qquad (21)$$

$$\left(\frac{d}{dt} + i\omega\right) p_\ell(t) = 2i \sum_\kappa f^*_\kappa(r_\ell)\, \bar{\alpha}_\kappa(r_\ell,t)\, p_{\ell_3}(t)$$

$$- \left(\frac{1}{2\tau_0} + i\Delta\omega\right) p_\ell(t), \tag{22}$$

where

$$\frac{1}{2\tau_0} + i\Delta\omega = \sum_\kappa |f_\kappa(r_\ell)|^2 \{\pi\delta(\omega_\kappa - \omega) + iP\frac{1}{\omega - \omega_\kappa}\} \tag{23}$$

and

$$\bar{\alpha}(r_\ell,t) \equiv -i \sum_{(m\neq\ell)} \int_{t_0}^{t} dt' e^{-i\omega_\kappa(t-t')} f_\kappa(r_m) p_m(t') \tag{24}$$

is the field generated by all atomic currents *except* the current of the ℓth atom. In other words, $\bar{\alpha}_\kappa$ is α_κ with the self-field subtracted out.

The one significant difference between the pure and QED-corrected semiclassical descriptions of N-atom superradiant decay is, therefore, to be found in the way in which generation of a self-field affects the motion of the atomic variables, p_{ℓ_3} and p_ℓ. The total radiation field is still generated by the total of all the atomic currents. However, in this QED-corrected version of semiclassical radiation theory, the current flow is a linearly damped oscillation coupled nonlinearly to other atoms via their field emissions. Thus, at this level of description, the atom interacts semiclassically with the fields generated by other atoms and quantum electrodynamically, so to speak, with its own self-field. The result is a Bloch phenomenological description of the atomic motion in which the phenomenological time constants are due solely to radiation decay and the driving terms represent stimulation by other atoms. Because this latter interaction is a nonlinear one, it can dominate the linear spontaneous decay. In other words, the superradiant emission predicted by Eqs. (21) and (22) should be (and is), for large N, a first-order correction to the semiclassical zero-order prediction (assuming we are away from the metastable semiclassical equilibrium point of complete inversion).

Equations (21) and (22) can be processed further and placed in a form which allows a comparison with the work of RE and SELM. Because of the position dependence contained in Eqs. (21) and (22), each atom will behave somewhat differently from the others. This behavior is in contrast to the RE assumption that all atoms simultaneously time-evolve through the same Θ-states. However, the knowledge of what each atom in the ensemble is doing is much more detailed microscopic knowledge than we need. We are only interested in the macroscopic behavior of the atoms, which is, in effect, the average behavior of the ensemble. The RE assumption is an alternate way of saying that the average motion of the atoms is being studied.

To carry out the averaging of Eqs. (21) and (22), one must isolate the random position dependence of the dynamical variables from the systematic position dependence. One form of systematic dependence is contained in $f_\kappa(r_\ell)$ and has already been mentioned in Eq. (15). A second form can be impressed on the currents p_ℓ by the coherent radiation pulse that excites the atoms to their initial Θ-states:

$$p_\ell(t_0) = e^{+ik_I\cdot r_\ell}\, p'(t_0),$$

where k_1 is the wave vector of the coherent pulse. If it is assumed that the r_ℓ-dependence of $p_{\ell_3}(t)$ and

$$p'_\ell(t) \equiv e^{-ik_1\cdot r_\ell}\, p_\ell(t)$$

is random and uncorrelated with the r_ℓ-dependence of other quantities, then, after substitution of Eq. (24) into Eqs. (21) and (22), Eqs. (21) and (22) can be averaged over the ensemble of atoms to produce the equations,

$$\frac{dp_3(t)}{dt} = -\frac{N\mu}{\tau_0}\left|p(t)\right|^2 - \frac{1}{\tau_0}\left(p_3(t) + \frac{1}{2}\right), \tag{25}$$

$$\left(\frac{d}{dt} + i\omega\right)p(t) = N\left(\frac{\mu}{\tau_0} + i\mu'\Delta\omega\right)p_3(t)p(t) - \left(\frac{1}{2\tau_0} + i\Delta\omega\right)p(t), \tag{26}$$

where the variables p_3 and p, without r_ℓ-dependence, are defined as spacial averages of the atomic variables:

$$p_3(t) \equiv \frac{1}{N}\sum_{\ell=1}^{N} p_{\ell_3}(t), \tag{27}$$

$$p(t) \equiv \frac{1}{N} \sum_{\ell=1}^{N} e^{-ik_1 \cdot r_\ell} p_\ell(t) \ . \tag{28}$$

Here μ/τ_0 is the effective many-atom decay rate introduced earlier [2] by Eberly and Rehler and $\mu'\Delta\omega$ is an effective many-atom shift defined in the same way:

$$(\mu'\Delta\omega + i\,\frac{\mu}{\tau_0}) \equiv 2 \sum_{\kappa} |f_\kappa|^2 \, [\Gamma(k,k_1) - \frac{1}{N}](P\,\frac{1}{\omega-\omega_\kappa} + i\pi\delta(\omega_\kappa-\omega)), \tag{29}$$

where

$$\Gamma(k,k_1) \equiv \left| \langle e^{i(k-k_1)\cdot r_\ell} \rangle_{AV} \right|^2 \tag{30}$$

and f_κ is given in Eq. (16). Equations (25) and (26) are the QED-corrected atomic equations of motion that we were seeking. They have an immediate relationship to the work of RE and SELM.

5. Comparison with Other Work

If one assumes that each atom is in the same Θ-state at time t, then it follows that $|p(t)|^2 + p_3(t)^2 = \frac{1}{4}$. Thus, under the assumption that all atoms evolve in time through the same Θ-states and with the RE definition, $W \equiv Np_3$, it follows that Eq. (25) can be rewritten as

$$-\frac{dW}{dt} = \frac{\mu}{\tau_0}\left(\frac{N}{2} + W\right)\left(\frac{N}{2} - W + \frac{1}{\mu}\right) \ . \tag{31}$$

This is an RE result.

Similarly, if one neglects the linear damping terms in Eqs. (25) and (26), these equations imply that $d/dt\{|p(t)|^2 + p_3(t)^2\} = 0$, or if $|p(t_0)|^2 + p_3(t_0)^2 = \frac{1}{4}$, they imply that $|p(t)|^2 + p_3(t)^2 = \frac{1}{4}$. However, Eqs. (25) and (26), taken together without the neglect of the linear damping terms, imply that the length $p_3^2 + |p|^2$ of the numerical Bloch vector $\underline{p}$ does not remain a constant throughout the decay of the atomic energy. Thus Eqs. (25) and (26) provide correction terms to Eq. (31).

To find these correction terms, one must solve for $|p(t)|^2$ from Eqs. (25) and (26). It follows from these equations that

$$\left(\frac{d}{dt} + \frac{1}{\tau_0}\right)\left(|p|^2 + p_3^2\right) = -\frac{1}{\tau_0}\left(p_3^2 + p_3\right) \quad . \tag{32}$$

If initially, all atoms are in the same Θ- state, then the solution to Eq. (32) is

$$|p(t)|^2 = \frac{1}{4} - p_3(t)^2 + \frac{1}{4}\left(e^{-\frac{1}{\tau_0}(t - t_0)} - 1\right)$$

$$-\frac{1}{\tau_0}\int_{t_0}^{t} dt' e^{-\frac{1}{\tau_0}(t-t')}\left(p_3^2 + p_3\right)(t') . \tag{33}$$

When Eq. (33) is inserted into Eq. (25) and Eq. (25) is multiplied through by N, the result is a corrected version of Eq. (31) which is non-local in time:

$$-\frac{dW}{dt} = \frac{\mu}{\tau_0}\left(\frac{N}{2} + W(t)\right)\left(\frac{N}{2} - W(t) + \frac{1}{\mu}\right)$$

$$+ \frac{\mu}{\tau_0}\frac{N^2}{4}\left(e^{-\frac{1}{\tau_0}(t-t_0)} - 1\right)$$

$$-\frac{\mu}{\tau_0^2}\int_{t_0}^{t} dt' e^{-\frac{1}{\tau_0}(t-t')}\left(NW(t') + W(t')^2\right) \quad . \tag{34}$$

We have solved Eqs. (31) and (34) numerically for W and $dW/d\tau$ as a function of $\tau \equiv t/\tau_0$ in the special case where $N = 200$ and all atoms are situated within a wavelength of each other. In this case, $\Gamma=1$ and, from Eqs. (23) and (29), $\mu=1 - 1/N$. Three sets of curves for $-dW/d\tau$, the emission rate of energy from the atomic system, are shown in Figures 1, 2, and 3. The curve with the highest peak in each figure is computed from Eq. (31). The other curve in each figure is the emission rate predicted by Eq. (34). Each set of curves corresponds to a different initial value for W. *The scales are the same for all figures.*

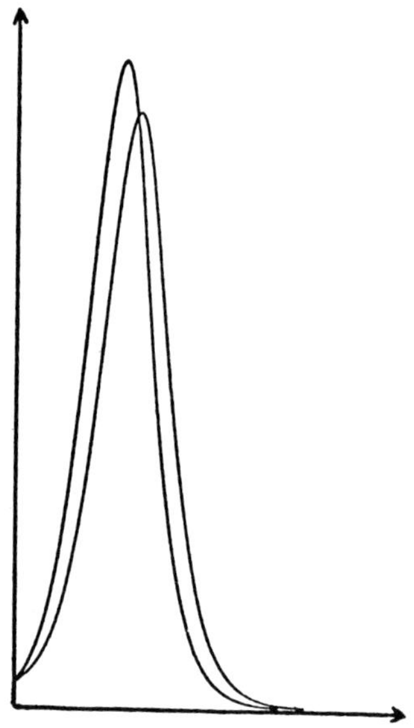

Fig. 1 $-dW/d\tau$ as a function of $\tau \equiv t/\tau_0$. [$W(t_0) = 99$].

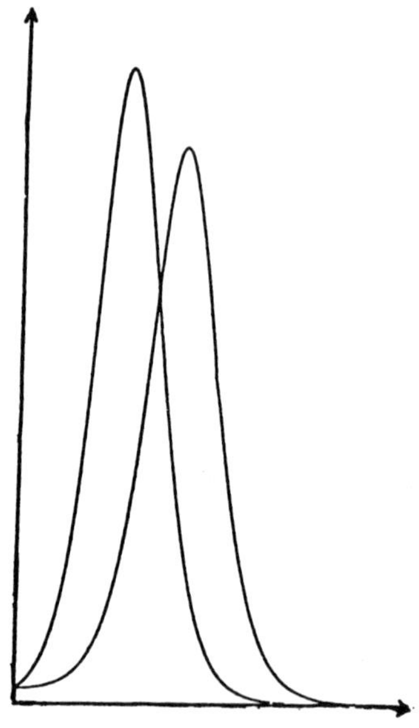

Fig. 2 $-dW/d\tau$ as a function of $\tau \equiv t/\tau_0$. [$W(t_0) = 99.9$].

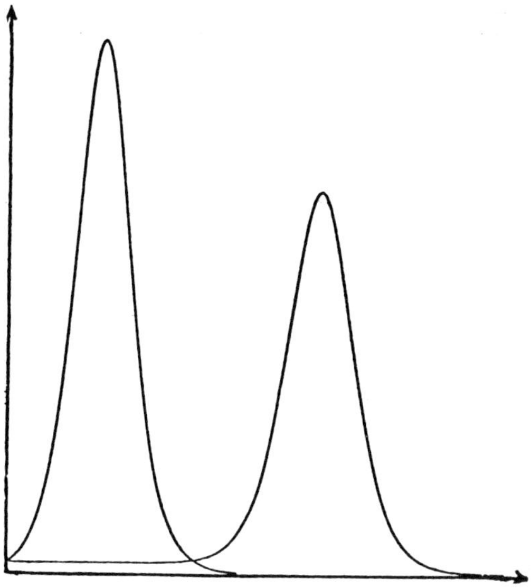

Fig. 3 $-dW/d\tau$ as a function of $\tau \equiv t/\tau_0$. $[W(t_o) = 99.99999]$.

Figures 1-3 show that as W approaches the initial value 100, corresponding to complete inversion of the atomic population, the output pulse is delayed and its peak is diminished. One can understand this behavior with reference to Eqs. (25) and (26). In the limit, W $\rightarrow$ complete inversion, the current magnitude $|p|$ approaches zero. Equations (25) and (26) predict, as in semiclassical theory, that if no current is present initially, none is generated later in time. In semiclassical theory, this means that, at the limit point, the atomic system will never decay, but will remain in the metastable state of total inversion. However, the quantum electrodynamic features of our model lead to the prediction that even when $|p|$ is initially zero, the atoms still decay with an exponential time dependence although no nonlinear N^2 pulse of energy is developed. The limit behavior of Eqs. (25) and (26) as $|p| \rightarrow 0$ initially can thus be understood as follows. The current grows exponentially from its initial value. Therefore, the smaller this value, the longer it takes the current to develop macroscopic values, and the longer it takes for the N^2 pulse to be emitted. If the atoms were not decaying spontaneously at the same time the current is building up, the energy pulse would have the same peak value no matter how long it took to be generated. However, Eqs. (25) and (26) describe a competition between stimulated and spontaneous emission. The longer the N^2 pulse is delayed, the smaller is its energy content. If it is delayed for longer than a natural atomic lifetime, it will not appear at all.

The results obtained from Eq. (34) provide some semiclassical interpretations of some recent results of Bonifacio, Schwendimann, and Haake [5] (BSH) and of Haake and Glauber. [12] Just as in the BSH work, our pulse heights are smaller than those predicted by RE. Moreover, in order to be able to describe the stimulated atomic decay semiclassically by Eq. (34), there is a limit to the inversion level that can be initially attained (in the absence of external, current-stimulating fields). If not enough current is present initially, Eq. (34) predicts that the decay will be entirely spontaneous and there will be no superradiance.

This pulse disappearance is not what BSH found, of course. They used Dicke-state theory to solve exactly for the energy pulse produced by an initially completely inverted ensemble of 200 atoms. They found that the peak emission rate was lower by 23 or 24% from the peak rate predicted by Eq. (31). Correction terms must be added to Eqs. (25) and (26) before they can be used to more accurately describe this limit behavior of the atoms at the point of complete inversion.

Equations (25) and (26) can also be cast into the super-Bloch vector form of SELM through the definitions $X - iY \equiv 2e^{i\omega t}\,Np$ and $Z \equiv 2Np_3$:

$$\frac{dZ}{dt} = -\frac{\mu}{2\tau_0}(X^2 + Y^2) - \frac{1}{\tau_0}(Z + N) \quad , \tag{35}$$

$$\frac{d(X - iY)}{dt} = \frac{1}{2}\left(\frac{\mu}{\tau_0} + i\mu'\Delta\omega\right)Z(X - iY) - \left(\frac{1}{2\tau_0} + i\Delta\omega\right)(X - iY). \tag{36}$$

These equations reduce to those of SELM when the atoms are restricted to lie within a wavelength of each other and terms that are order N are neglected in comparison to the terms, $X^2 + Y^2$ and $Z(X - iY)$, that are order N^2.

The situation regarding zero-order and first-order superradiance theory as viewed through Eqs. (25) and (26) is summarized in the following table:

	Zero-Order Motion	First-Order Motion
Single-atom Radiative Decay	$\frac{dp}{dt} = -\frac{1}{\tau_0}\|p(t)\|^2$	$\frac{dp_3}{dt} = -\frac{1}{\tau_0}(p_3 + \frac{1}{2})$
	$(\frac{d}{dt} + i\omega)p(t) = (\frac{1}{\tau_0} + 2i\Delta\omega)p_3 p$	$[\frac{d}{dt} + i(\omega+\Delta\omega) + \frac{1}{2\tau_0}]\, p = 0$
N-atom Small	$(R \equiv X - iY \quad , \quad R_3 \equiv Z)$	
System (Dicke)	$\frac{dR_3}{dt} = -\frac{1}{2\tau_0}\|R\|^2$	$\frac{dR_3}{dt} = -(\frac{1}{2\tau_0})(1 - \frac{1}{N})\|R\|^2 - \frac{1}{\tau_0}(R_3+N)$
Superradiance	$\frac{dR}{dt} = \frac{1}{2}(\frac{1}{\tau_0} + 2i\Delta\omega)R_3 R$	$\frac{dR}{dt} = \frac{1}{2}(1-\frac{1}{N})(\frac{1}{\tau_0} + 2i\Delta\omega)R_3 R - (\frac{1}{2\tau_0} + i\Delta\omega)R$
N-atom Large	$\frac{dR_3}{dt} = -\beta\|R\|^2$	$\frac{dR_3}{dt} = -\frac{\mu}{2\tau_0}\|R\|^2 - \frac{1}{\tau_0}(R_3 + N)$
System (Rehler-Eberly)		
Superradiance	$\frac{dR}{dt} = (\beta - i\gamma)R_3 R$	$\frac{dR}{dt} = \frac{1}{2}(\frac{\mu}{\tau_0} + i\mu'\Delta\omega)R_3 R - (\frac{1}{2\tau_0} + i\Delta\omega)R$

The nature of the first-order QED correction to the zero-order semiclassical theory is shown in row 1. For "single-atom theory" (N=1), the first-order correction, arising from the commutation relation $[a,a^{+}] = 1$, completely cancels the zero-order, Jaynes-Cummings semiclassical description [13] of the atomic motion. The description of a well-defined charge distribution interacting non-linearly with a classical radiation field is replaced by a totally different statistical (or photon) description of the decay process. That is, instead of following the Jaynes-Cummings hyperbolic secant law, any current that the atom possesses decays exponentially in an atomic natural lifetime (generating a Lorentzian distributed pulse of α-state radiation), and the energy of the atom decays not as a hyperbolic tangent, but exponentially along with the current in the same period.

As N increases, the point of view that the first-order corrections to the semiclassical zero-order theory only slightly modify the results of the zero-order theory seems to be more and more nearly correct for a superradiant system . This feature of Eqs. (25) and (26) is manifested in row 2 of the table for the case of Dicke superradiance. In this case, the zero-order terms are multiplied by the factor 1 - 1/N to obtain the first-order nonlinear terms. In other words, among the N^2 pairwise atomic interactions, the N self-interactions have been subtracted out and replaced by linear damping terms. Finally, in progressing to an extended N-atom system, as shown in row 3 of the table, the form of both the zero-order and first order Dicke theory is changed only by the modification of the coefficients multiplying the nonlinear terms. In this case, one must evaluate the coefficient Γ, which acts to further diminish the effect of these nonlinear terms in determining average atomic motion.

Summary and Conclusions

The derivation given in this paper of the Rehler-Eberly energy decay of a superradiant N-atom system indicates that their results and the phenomena of superradiance can be interpreted in a new way. When the N-atoms are viewed as a single highly-excited quantum system sitting isolated from any external radiation, their radiative decay can be categorized as spontaneous. However, when the motion of each atom is viewed separately, superradiance is seen to be a cooperative phenomena of the whole system, in which each atom, decaying spontaneously in the usual QED fashion, is simultaneously stimulated by the radiation emitted from the other atoms in the ensemble. The collective behavior of the N atoms is predicted when the systematically varying r_k dependences of the atomic currents and coupling constants are factorized from the remaining r_k dependences of the atomic variables, which are taken to be uncorrelated.

A third view of superradiance, midway between the above two views, has an interesting relation to the semiclassical description of single atom spontaneous decay, which views this decay as a radiation reaction process [14]. The N atoms can be thought of as a single system that radiates spontaneously in the midst of pairwise radiation reaction of one atom on another. The zero-order description of this decay is completely analogous to the semiclassical description of single atom decay (especially in a limit $N \to \infty$ for which total charge remains finite in some sense). However, in the first-order theory of this paper, the description of self-reaction (i.e., self-field generation) is altered. The semiclassical description of the motion is lost and is replaced by the usual QED statistical description.

One could imagine that an atomic charge distribution were also divided into N parts and ask to what extent the above first-order density matrix theory could be applied, in principle, to the description of atomic decay. One would have an almost semiclassical description of the decay that conceivably could become perfectly semiclassical as $N \to \infty$ and the total charge remained constant (since self-field radiation reaction goes as N and pair radiation reaction goes as N^2). However, the N currents in the N-atom system were taken to be independent of each other; that is, the atoms were separate noninteracting systems. An atom cannot be similarly decomposed; hence, the description of N-atom superradiance given in this paper is only suggestively extrapolable to the problem of describing single atom radiative decay semiclassically.

Finally, let us point out that the first-order motion described in this paper preserves most of the semiclassical zero-order motion, for large N, somewhat by assumption since $\langle (a_\kappa^+ - \alpha_\kappa^*)(a_\kappa - \alpha_\kappa)\rangle$ field correlation functions were neglected in this work. In effect, we have answered the question: how is the semiclassical description of superradiance altered by the inclusion of field commutators into the theory. The answers have generalized some of the previously obtained results on superradiance. However, a more complete contrast between QED and semiclassical descriptions of superradiance and radiation processes, in general, requires that $\langle (a_\kappa^+ - \alpha_\kappa^*)(a_\kappa - \alpha_\kappa)\rangle$ correlation functions be included in the theory and their effects on the motion understood. The first-order correction to the ρ_F equation of motion describes the build-up of this correlation function in time. $\langle (a^+ - \alpha^*)(a - \alpha)\rangle$ also corrects the ρ_A and $\underline{p}$ equations to first-order and, moreover, is coupled nonlinearly to $\underline{p}$. Its buildup is associated with the quantum corrections provided by Haake and Glauber [12], for example, to the very large fluctuations predicted earlier [15] for emission from a completely inverted state.

Acknowledgments

The author is gratefully indebted to Dr. J.H. Eberly for his participation in the development of this work, and for his proofreading of this manuscript and many improvements of the text. The author would also like to thank Prof. A.L. Schawlow for the use of his computer facilities and Dr. S.M. Curry for his aid in setting up the computer program.

References

1. V. Ernst and P. Stehle, Phys. Rev.*176*, 1456, (1968).
2. J.H. Eberly and N.E. Rehler, Phys. Letters *29A*, 142 (1969).
3. N.E. Rehler and J.H. Eberly, Phys. Rev.*A 3* , 1735 (1971).
4. R. Bonifacio and G. Preparata, Letters Nuovo Cimento *1*, 887 (1969).
5. R. Bonifacio, P. Schwendimann, and F. Haake, Phys. Rev.*A 4*, 302 (1971); Phys. Rev. *A 4*, 854 (1971).
6. G.S. Agarwal, Phys. Rev.*A 4*, 1791 (1971); Phys. Rev.*A 3*, 1783 (1971); Phys. Rev. *A 2*, 2038 (1970).
7. C.R. Stroud, Jr., J.H. Eberly, W.L. Lama, and L. Mandel, Phys. Rev. *A 5*, 1094 (1972).
8. M. Lax, J. Phys. Chem. Solids *25*, 487 (1964).
9. M.O. Scully and K.G. Whitney, in *Progress in Optics*, E. Wolf, editor (North-Holland Publishing Company, Amsterdam), in press.
10. C.R. Willis, J. Mathe. Phys. *5*, 1241 (1964).
11. R.P. Feynman, F.L. Vernon, Jr., and R.W. Hellwarth, J. Appl. Phys. *28*, 49 (1956). We are using the second-quantized version of this theory.
12. F. Haake and R.J. Glauber, Phys. Rev. *A 5*, 1457 (1972).
13. E.T. Jaynes and F.W. Cummings, Proc. IEEE *51*, 89 (1963).
14. M.D. Crisp and E.T. Jaynes, Phys. Rev. *179*, 1253 (1969); and C.R. Stroud, Jr. and E.T. Jaynes, Phys. Rev. *A 1*, 106 (1970) and *A 2*, 1613 (1970).
15. J.H. Eberly and N.E. Rehler, Phys. Rev.*A 2*, 1607 (1970).

THIRD- AND HIGHER-ORDER INTENSITY CORRELATIONS IN LASER LIGHT*

C.D. Cantrell †

Swarthmore College, Swarthmore, Pennsylvania
and
Princeton University, Princeton, New Jersey

M. Lax

Bell Laboratories, Murray Hill, New Jersey
and
City College of New York, New York

Wallace Arden Smith

New York University, Bronx, New York

* Research at Princeton University was supported in part by the U.S. Atomic Energy Commission Division of Biology and Medicine, and by an institutional grant to Swarthmore College from the Alfred P. Sloan Foundation; research at City College was supported in part by A.R.O.D., O.N.R. and by a National Science Foundation Development Grant; research at New York University was supported in part by a National Science Foundation Institutional Grant.

† Part of this work was done while one of us (C.D.C.) was a Visiting Research Fellow in the Department of Physics, Princeton University.

1. Introduction

The statistics of single-mode laser light in the threshold region have been the subject of considerable experimental and theoretical study. The steady-state statistics are found to be well described by essentially the simplest physically reasonable model, the rotating-wave van der Pol (RWVP) model. However, the time-dependent statistical behavior of a laser, as it relaxes towards equilibrium from a fluctuation, involves considerably more details of its dynamics than does the time-independent steady-state behavior. Experimentally, the relaxation processes are revealed in the time dependence of the multi-time correlations of the intensity such as $\langle I(t_1)I(t_2)\rangle$ and $\langle I(t_1)I(t_2)I(t_3)\rangle$. In this paper we shall calculate the third-and higher-order intensity correlations, [1] which are more sensitive to the higher decay rates of the laser than the well-studied second-order intensity correlation. This will provide a more detailed understanding and test of the RWVP model.

A triple-photoelectron-coincidence experiment has been performed by Davidson and Mandel [2] on a He:Ne laser at and below threshold. These measurements are consistent with a simple exponential time decay, with a third-order coherence time which (below threshold) is distincly shorter than the second-order intensity correlation time. According to our calculations the decay slightly below threshold departs significantly from simple exponential behavior, and the times characteristic of this decay are longer than those found by Davidson and Mandel. Above threshold, the decay becomes oscillatory. F. Haake [3] has predicted a damped oscillation time dependence for the intensity correlation functions of lasers where atomic "memory" effects invalidate the Markoffian assumption of the RWVP model. However, we find damped oscillatory behavior even in the RWVP model, that is, on the 10-microsecond time scale, rather than on the nanosecond time scale needed to observe non-Markoffian effects.

The additional complexity of an oscillatory correlation means that the experimental results are sensitive to some of the higher eigenvalues (decay rates) of the Fokker-Planck operator in the RWVP model. Thus in contrast to the second order intensity correlation, it is not possible to describe the expected third-order correlation results in terms of a single correlation time. More precise measurements would be desirable.

2. Calculation

It is convenient to express both experimental and theoretical results for Nth order intensity correlations in terms of quantities which contain only the true Nth-order correlation, with all lower-order effects subtracted away. The quantities which fulfill this criterion are the cumulants, [4]

$$K_{11}(I_1,I_2) = \langle T_N(\Delta I_1 \Delta I_2)\rangle$$

$$K_{111}(I_1,I_2,I_3) = \langle T_N(\Delta I_1 \Delta I_2 \Delta I_3)\rangle$$

$$\begin{aligned} K_{1111}(I_1,I_2,I_3,I_4) = & \langle T_N(\Delta I_1 \Delta I_2 \Delta I_3 \Delta I_4)\rangle \\ & - \langle T_N(\Delta I_1 \Delta I_2)\rangle \langle T_N(\Delta I_3 \Delta I_4)\rangle \\ & - \langle T_N(\Delta I_1 \Delta I_3)\rangle \langle T_N(\Delta I_2 \Delta I_4)\rangle \\ & - \langle T_N(\Delta I_1 \Delta I_4)\rangle \langle T_N(\Delta I_2 \Delta I_3)\rangle \end{aligned} \quad (1)$$

where

$$\Delta I_j = b_j^\dagger b_j - \langle b_j^\dagger b_j\rangle \quad ;$$

$b_j = b(t_j)$ is the positive-frequency (destruction) part of the field operator at time t_j; and T_N places the operators in normal order (creation operators to the left) and in apex time sequence (e.g., $b_1^\dagger b_2^\dagger b_3^\dagger b_3 b_2 b_1$). Experimental results for the intensity cumulants are proportional to the average intensity, solid angles, and other efficiency factors. These can be eliminated by comparing experiment with theory for the dimensionless ratios

$$\begin{aligned} K_n(t_1,\ldots, t_{n-1}) = \langle I\rangle^{-n} k_{11\ldots1} & (I(t_1+t_2+\ldots+t_{n-1}), \\ & I(t_1+t_2+\ldots+t_{n-2}),\ldots,I(t_1), I(0)) \end{aligned} \quad (2)$$

where t_1 is the time delay between the first and second measurements of the intensity, t_2 the delay between the second and third, and so on.

The multi-time correspondence between quantum and classical stochastic systems [5][6]

$$b(t)\to\beta(t), \quad b^\dagger(t)\to\beta^*(t)$$

is such that apex-ordered operators, as in (1), can be averaged by taking the associated c-number average:

$$\langle T_N(\Delta I_1 \Delta I_2 \Delta I_3)\rangle = \langle \Delta\rho_3 \Delta\rho_2 \Delta\rho_1 \rangle_c \tag{3}$$

where

$$\rho_j = \rho(t_j) = |\beta(t_j)|^2 \tag{4}$$

is the intensity and

$$\Delta\rho_j = \rho_j - \langle\rho_j\rangle_c = \rho_j - \overline{\rho} \tag{5}$$

is the intensity fluctuation. The mean intensity can be denoted simply $\overline{\rho}$, since it is independent of the time t_j. Hereafter all averages will be c-number averages, and the subscript will be omitted.

The result, Eq. (3), is valid whether or not the quantum field statistics are Markoffian [7]. Under the Markoffian assumption (and in particular, for the RWVP model) the multi-time probability density needed to carry out the indicated average in (3) factors [5],[8] so that

$$\begin{aligned} &\langle \Delta\rho_3 \Delta\rho_2 \Delta\rho_1 \rangle \\ &= \int_0^\infty d\rho_3 d\rho_2 d\rho_1 (\rho_3 - \overline{\rho}) P(\rho_3, t_3 | \rho_2, t_2) \\ &\times (\rho_2 - \overline{\rho}) P(\rho_2, t_2 | \rho_1, t_1)(\rho_1 - \overline{\rho}) P_0(\rho_1) \end{aligned} \tag{6}$$

where $P_0(\rho)$ is the steady-state intensity probability distribution and $P(\rho, t|\rho_0, t_0)$ is the conditional probability that the intensity will be ρ at time t, given that it was ρ_0 at t_0. In general, the intensity probability distributions obey an equation of motion

$$\frac{\partial P}{\partial t} = - LP \tag{7}$$

where L is a general linear operator. In the Fokker-Planck (or diffusion) approximation, where L is a non-hermitian second order differential operator, the conditional probability (or Green's-function solution of (7)) can be written in the form [9]

$$P(\rho,t|\rho_0,t_0) = \sum_{n=0}^{\infty} e^{-\Lambda_n(t-t_0)} P_n(\rho)\phi_n(\rho_0)^*, \tag{8}$$

where P_n and ϕ_n are the eigenfunctions of L and its Hermitian adjoint L†:

$$LP_n(\rho) = \Lambda_n P_n(\rho)$$

$$L^{\dagger}\phi_n(\rho) = \Lambda_n^* \phi_n(\rho). \tag{9}$$

It has been shown [8],[10] that when time reversal in the form of detailed balance is obeyed,

$$\phi_n(\rho)^* = P_n(\rho)/P_0(\rho) \tag{10}$$

where $P_0(\rho)$ satisfies the equation

$$LP_0(\rho) = 0 \tag{11}$$

corresponding to the eigenvalue $\Lambda_0 = 0$. While L is not in general a hermitian operator it is equivalent to a hermitian operator under a similarity transform given by multiplication by $[P_0(\rho)]^{\frac{1}{2}}$. Thus the eigenvalues of L are real: $\Lambda_n = \Lambda_n^*$. In the RWVP model, $P_n(\rho)$ can be chosen to be real.

The triple intensity cumulant, Eq. (6), reduces to

$$\langle\Delta\rho(t+t_2)\Delta\rho(t)\Delta\rho(t-t_1)\rangle$$

$$= \sum_{m,n=0}^{\infty} (\Delta\rho)_{om}\, e^{-\Lambda_m t_2} (\Delta\rho)_{mn}\, e^{-\Lambda_n t_1} (\Delta\rho)_{no}$$

$$= \sum_{m,n=0}^{\infty} (\rho_{om}-\bar{\rho}\delta_{om})\, e^{-\Lambda_m t_2} (\rho_{mn}-\bar{\rho}\delta mn)$$

$$\times e^{-\Lambda_n t_1} (\rho_{no}-\bar{\rho}\delta_{no}) \tag{12a}$$

$$= \sum_{m,n=1}^{\infty} e^{-\Lambda_m t_1 - \Lambda_n t_2} \rho_{mo}\rho_{no}(\rho_{mn}-\delta_{mn}\bar{\rho}) \tag{12b}$$

where

$$(\rho^k)_{mn} = \int_0^\infty \phi_m(\rho)^* \rho^k P_n(\rho)\, d\rho \tag{13a}$$

and

$$(\Delta\rho^k)_{mn} = \int_0^\infty \phi_m(\rho)^* (\rho-\bar{\rho})^k P_n(\rho)\, d\rho \tag{13b}$$

are the matrix elements of the kth powers of the intensity and intensity fluctuation between the eigenfunctions of L and $L^\dagger$. If k=1 the superscript is omitted, as in (12). The mean intensity is

$$\bar{\rho} = \langle\rho\rangle = \rho_{oo}. \tag{14}$$

The double intensity cumulant can be expressed similarly as

$$\langle\Delta\rho(t)\Delta\rho(0)\rangle = \sum_{n=0}^{\infty} (\rho_{on} - \bar{\rho}\,\delta_{on})\, e^{-\Lambda_n t} \times(\rho_{no} - \bar{\rho}\,\delta_{no}) \tag{15a}$$

$$= \sum_{n=1}^{\infty} (\rho_{no})^2\, e^{-\Lambda_n t} \tag{15b}$$

Note that using $\Delta\rho$ instead of ρ results in the elimination of the n=0 term in (15b).

The normalized intensity moments about the mean are

$$H_2(t) = \langle\Delta\rho(t)\Delta\rho(0)\rangle/\bar{\rho}^{\,2}, \tag{16a}$$

$$H_3(t_1,t_2) = \langle\Delta\rho(t_1+t_2)\Delta\rho(t_1)\Delta\rho(0)\rangle/\bar{\rho}^{\,3}, \tag{16b}$$

$$H_4(t_1,t_2,t_3) = \langle\Delta\rho(t_1+t_2+t_3)\Delta\rho(t_1+t_2)\Delta\rho(t_1)\Delta\rho(0)\rangle/\bar{\rho}^{\,4}. \tag{16c}$$

The associated normalized *cumulants* are

$$K_2(t) = H_2(t) \tag{17a}$$

$$K_3(t_1,t_2) = H_3(t_1,t_2) \tag{17b}$$

$$K_4(t_1,t_2,t_3) = H_4(t_1,t_2,t_3) - H_2(t_1)H_2(t_3)$$

$$- H_2(t_2)H_2(t_1+t_2+t_3)$$

$$- H_2(t_1+t_2)H_2(t_2+t_3). \tag{17c}$$

When all delay times except one are zero,

$$K_3(t,0) = (\bar{\rho})^{-3} \sum_{n=1}^{\infty} (\Delta\rho^2)_{0n}\, e^{-\Lambda_n t} (\Delta\rho)_{n0} \tag{18a}$$

$$K_4(t,0,0) = (\bar{\rho})^{-4} \sum_{n=1}^{\infty} (\Delta\rho)_{n0}\{(\Delta\rho^3)_{0n} - 3(\Delta\rho^2)_{00}(\Delta\rho)_{0n}\}e^{-\Lambda_n t} \tag{18b}$$

$$K_5(t,0,0,0) = (\bar{\rho})^{-5} \sum_{n=1}^{\infty} (\Delta\rho)_{n0}\{(\Delta\rho^4)_{0n} - 4(\Delta\rho^3)_{00}(\Delta\rho)_{0n}$$

$$- 6(\Delta\rho^2)_{00}(\Delta\rho^2)_{0n}\}e^{-\Lambda_n t} \quad . \tag{18c}$$

3. Results

To evaluate Eqs. (16) we need the eigenvalues and eigenfunctions of the Fokker-Planck operator L. For the RWVP model,

$$- L = \frac{\partial}{\partial\rho} (2\rho^2 - 2p\rho - 4) + \frac{\partial^2}{\partial\rho^2} (4\rho) \tag{19}$$

where p is the dimensionless net pump rate (equal to zero at threshold, positive above threshold, and negative below). Some of the numerical techniques for integrating (7) and (19) and calculating ρ_{mn} have been described elsewhere [11][12]. The first 10 non-zero eigenvalues Λ_m for assorted integral values of p in the range $-10 \leq p \leq 10$ have been published, [12] together with values of the coefficients,

$$p_{on} = (\rho_{on})^2 / \sum_{j=1}^{\infty} (\rho_{oj})^2. \tag{20}$$

The values of ρ_{mn} for this range of p, and values of Λ_m and $(\rho^k)_{mn}$ for $1 \leq k \leq 4$, $0 \leq m, n \leq 20$, are available.[10][13]

Our numerical results for the time dependence of $H_3=K_3$ at $p = 3,2,1,0,-1$ are displayed in Figures 1-5. Our results for p=-2 and -4 are not displayed. In these graphs we have measured time in units of the second order coherence time, T_c, at the given intensity,

$$s = t_1\bar{\Lambda} \quad , \quad s' = t_2\bar{\Lambda} \tag{21}$$

where

$$T_c \equiv (\bar{\Lambda})^{-1} = \sum_{n=1}^{\infty} p_{on}(\Lambda_n)^{-1} \tag{22}$$

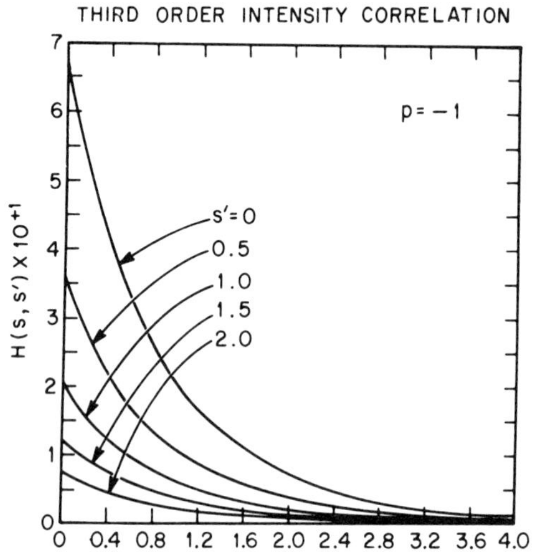

Fig. 1 The normalized third-order intensity cumulant $H(s,s') \equiv K_3(s,s')$(Eq. (18a)) is shown as a function of the normalized time delays s and s' for pump parameter p=-1. The times s and s' are measured in units of the laser second-order coherence time (Eqs. (21) and (22)), and H(s,s') is measured in units of $\bar{\rho}^3$, where $\bar{\rho}$ is the mean intensity at this pump parameter.

Evidently, below threshold

$$H(s,s') \equiv H_3(s,s') \tag{23}$$

is positive for all (normalized times s,s' but somewhat above threshold H_3 becomes negative for non-zero times. The third-order cumulant is a measure of the asymmetry of a probability distribution: for example, if

$$K_3(s,0) = H_3(s,0)$$
$$= \langle \Delta\rho(s)\ (\Delta\rho(0))^2 \rangle \qquad (24)$$

is positive, then the joint probability distribution of $\Delta\rho(s)$ and $(\Delta\rho(0))^2$ is asymmetric towards positive fluctuations

$$\Delta\rho(s) = \rho(s) - \bar{\rho} > 0.$$

If $K_3(s,0)<0$, then the distribution is asymmetric towards negative fluctuations.

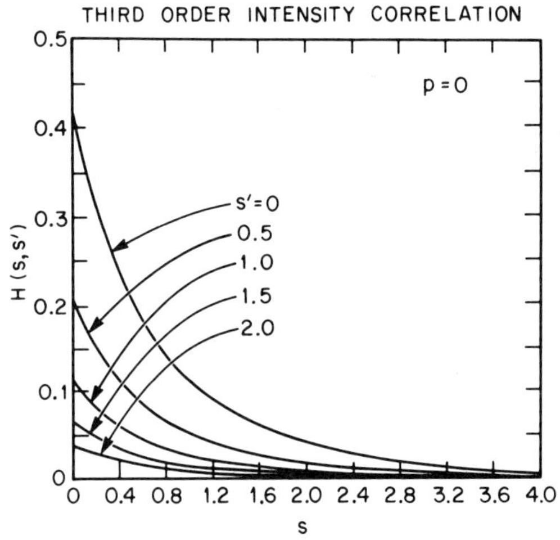

Fig. 2. Plot of H(s,s') for p=0, with units as described for Fig. 1.

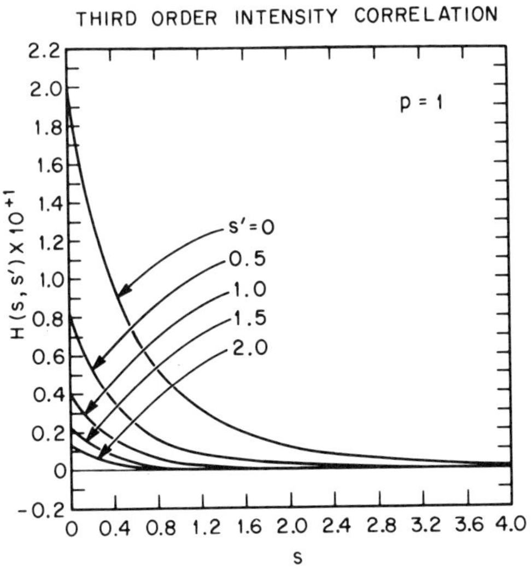

Fig. 3. Plot of H(s,s') for p=1, with units as described for Fig. 1.

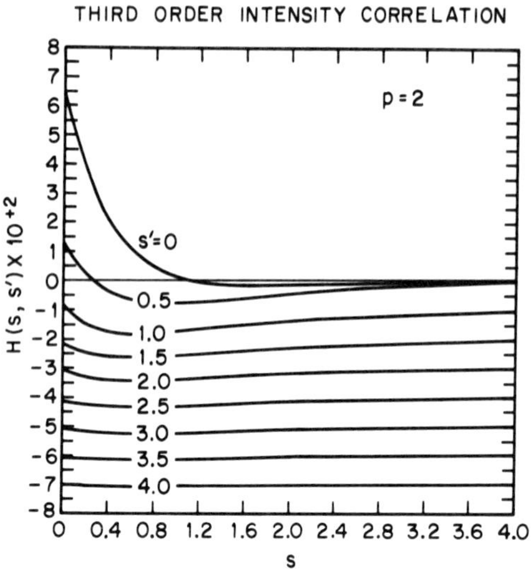

Fig. 4. Plot of H(s,s') for p=2, with units as described for Fig. 1. For clarity, the curves for s' > 0.5 have been displaced downwards successively by one unit per curve. In each case, the actual asymptote of H(s,s') for large values of s is the s-axis.

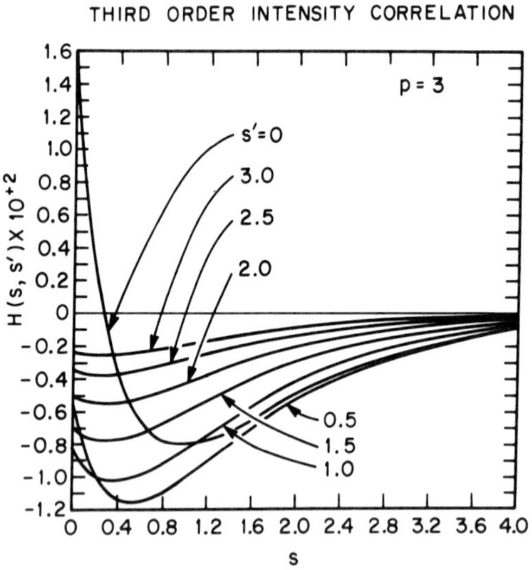

Fig. 5. Plot of H(s,s') for p = 3, with units as described for Fig. 1.

Figure 6 shows that somewhat above threshold the asymmetry of the joint probability distribution depends on the normalized delay time s. For short delay times, the distribution is asymmetric towards positive fluctuations (as is always true below threshold); but for longer delay times the asymmetry favors negative fluctuations. If one can intuitively separate the appearance of a fluctuation from its subsequent decay, $<(\Delta\rho(0))^2 \Delta\rho(s)>$ can be viewed as an average of the value of a fluctuation s-seconds after it appears weighted with the square of its original magnitude. Thus while fluctuations above the mean tend to be larger ($K_3(s,0)>0$ for $s=0$), they also decay more rapidly leaving the negative fluctuations to predominate for longer delay times ($K_3(s,0) < 0$ for s not near zero).

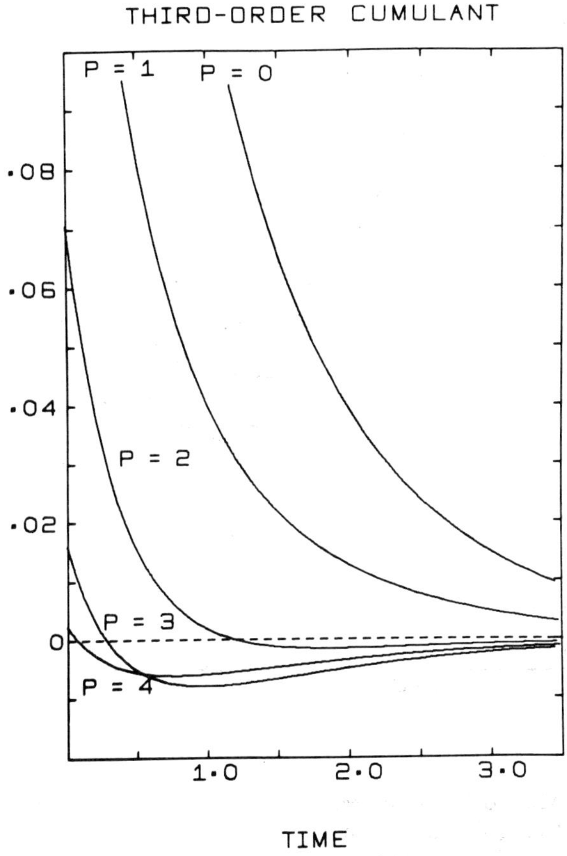

Fig. 6. Comparison plot of H(s,0) for several values of the pump parameter p in the threshold region. The oscillatory behavior of H(s,0) for p>1.5 indicates a change in the asymmetry of the joint (two-time) intensity probability distribution. The units are the same as in Fig. 1.

This behavior is physically reasonable, in view of saturation of the atomic population difference by the laser light. The same saturation effects which stabilize the intensity in a laser above threshold, evidently damp the large intensity fluctuations above the mean more rapidly than those fluctuations below the mean.

Our results for the fourth-and fifth-order cumulants calculated from (18) are presented in Figures 7 and 8.

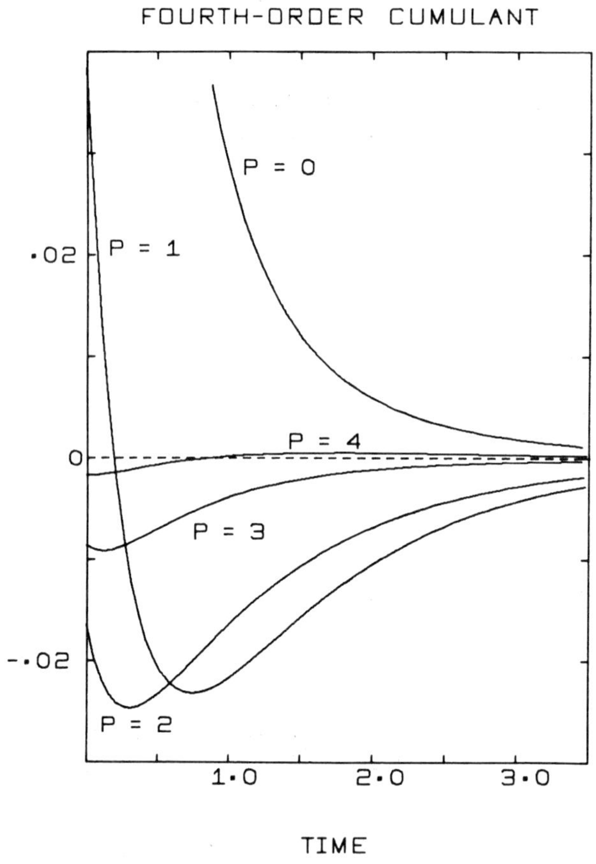

Fig. 7 Comparison plot of the normalized fourth-order intensity cumulant $K_4(s,0,0)$ (Eq. (18b)) as a function of normalized time delays s, for several values of the pump paramter p in the threshold region. The units of time are the same as those used in Fig. 1. For each curve, K_4 is measured in units of $\bar{\rho}^{\,4}$, where $\bar{\rho}$ is the mean intensity of the laser at that value of p.

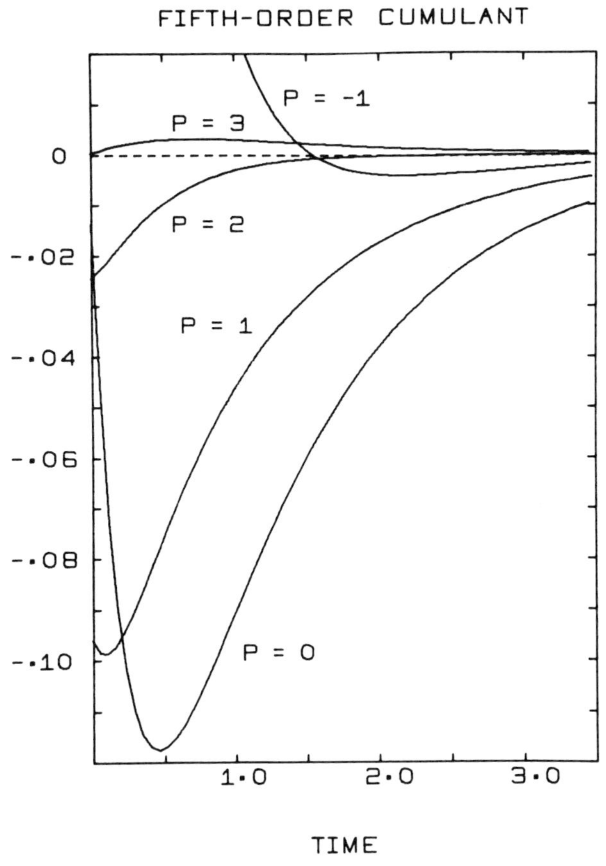

Fig. 8 Comparison plot of the normalized fifth-order intensity cumulant $K_5(s,0,0,0)$ (Eq(18c)) as a function of normalized time delay s, for several values of the pump parameter p in the threshold region. The units of time are the same as those used in Fig. 1. For each curve, K_5 is measured in units of $\bar{\rho}^5$ where $\bar{\rho}$ is the mean intensity of the laser at that value of p.

In the thresold region (p = -1 to p = 2) K_5 becomes negative for non-zero delay times, thus supporting the conclusions drawn from K_3.

To present our results for H_3 in a compact form, we have fit a five-parameter expression of the form

$$\begin{aligned} H(s,s') = {} & c_1 \exp\,[-\,\lambda_1(s+s')/\bar{\Lambda}] \\ & + c_2 \exp\,[-\,\lambda_2(s+s')/\bar{\Lambda}] \\ & + c_{12}[\exp[-\,(\lambda_1 s+\lambda_2 s')/\bar{\Lambda}] \\ & \quad + \exp[-\,(\lambda_1 s'+\lambda_2 s)/\bar{\Lambda}\,] \end{aligned} \tag{25}$$

to our exact results for $H_3(s,s')$. The constants $c_1, c_2, c_{12}, \lambda_1$, and λ_2 have been chosen so that the values of

$$H_3(0,0),\ \left.\frac{\partial H_3(s,0)}{\partial s}\right|_{s=0},\ \int_0^\infty H_3(s,0)\,ds\ \int_0^\infty sH_3(s,0)\,ds,$$

and $\int_0^\infty H_3(s,s)ds$ are given exactly. While the form of (25) has no special significance, it is simple and gives a reasonably good fit to the values of $H_3(s,s')$ calculated from the more cumbersome expression (12) . For the indicated values of p, Table 1 contains: the constants $c_1, c_2, c_{12}, \lambda_1$, and λ_2; the reciprocal second-order coherence time $\bar{\Lambda}$; ΔH, the maximum difference between the exact $H_3(s,s')$ and (25); and $\Delta H/H_{max}$, the ratio of ΔH to the maximum value of $|H_3(s,s')|$, expressed as a percent. Figure 9 contains a comparison plot of $H_3(s,s')$ and the five paramter fit (25).

To compare our calculations to the experiments of Davidson and Mandel, we determined the experimental threshold intensity I_0 by fitting the measurements of $H_3(0,0)$ versus intensity to the theoretical curves, thus finding the values of p corresponding to the measured intensities. For each value of the laser intensity the experimental delay times at which the third order cumulant was measured, were divided by the measured second-order coherence time to obtain the normalized times s,s'. Davidson and Mandel report that $H_3(s,s')$ decays essentially exponentially to zero and can be characterized by a third-order coherence time. For the measured intensities (which were all below threshold) and the range of decrease H_3(roughly a factor of five) the report of essentially exponential decay is consistent with these calculations.

TABLE 1. Parameters to be used in the approximate expression, eq. (25), for the third-order intensity cumulant $H_3(s,s')$

p	$\overline{\Lambda}$	λ_1	λ_2	C_1	C_2	C_{12}	$\Delta H \times 10^3$	$(\Delta H/\lvert H_{max}\rvert) \times 10^2$
-10	21.4794	21.4694	45.0143	1.6938952	.0032445	.0539338	.0055	.0003
-8	17.7829	17.7667	37.9667	1.5751517	.0059024	.0710724	.014	.0008
-6	14.2229	14.1955	31.2890	1.3875394	.0111778	.0940294	.041	.0026
-4	10.8965	10.8474	25.1834	1.0891405	.0211136	.1204209	.13	.010
-3	9.3735	9.3070	22.4388	.8856358	.0280608	.1313229	.24	.020
-2	7.9889	7.8987	19.9689	.6477631	.0352551	.1364293	.43	.043
-1	6.7927	6.6721	17.8441	.3935962	.0401263	.1314991	.69	.10
0	5.8539	5.7027	16.1628	.1620673	.0386370	.1142497	1.1	.26
1	5.2688	5.1770	15.0958	.0000313	.0252275	.0904824	1.5	.73
2	5.1750	3.3616	14.5447	-.0325333	.0490249	.0270513	1.1	1.6
3	5.7508	4.4155	15.6461	-.0442604	.0174704	.0213626	.43	2.7
4	7.1122	5.2632	18.1747	-.0258039	.0077065	.0101828	.22	2.9

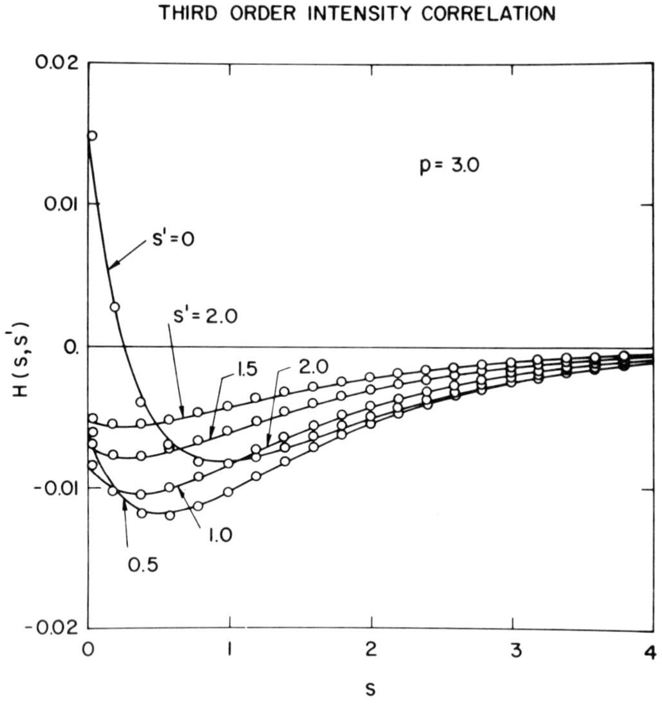

Fig. 9 Comparison plot of the exact third-order intensity cumulant H(s,s') and the approximate five-parameter fit (Eq.(25) and Table I).

However, the measured third-order coherence times (shown in Fig. 10) are distinctly smaller than the calculated times which characterize the decay of H_3. The small amount of freedom in our determination of the time and intensity scales is inadequate to produce a substantially better agreement than is shown in Fig. 10. This situation contrasts sharply with the generally excellent agreement between experiment and the RWVP model for the mean reciprocal coherence time, the normalized second-order cumulant at zero time delay (Fig. 11), and the normalized higher-order cumulants at zero time delay. [4c] These results suggest that more detailed measurements of the third-order intensity cumulant should be made, particularly above threshold, in order to look for the predicted oscillatory decay of $H_3(s,s')$ and provide more exacting tests of the RWVP model.

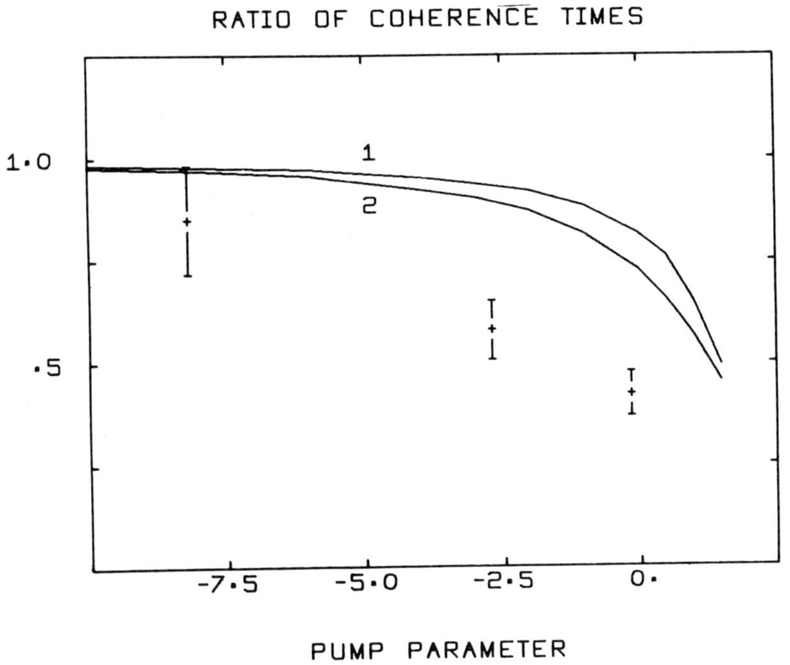

Fig. 10 Comparison between the calculated time required for H(0,s) to decay to e^{-1} of its initial value (curve 1), one-half the time required to decay to e^{-2} of its initial value (curve 2), and the experimental results of Davidson and Mandel (Ref. 2). Far below threshold (p large and negative), curves 1 and 2 asymptotically approach 1.0, indicating that the second-order and third-order coherence times agree. Nearer threshold, the difference between curves 1 and 2 indicates the non-exponential time dependence of H(0,s).

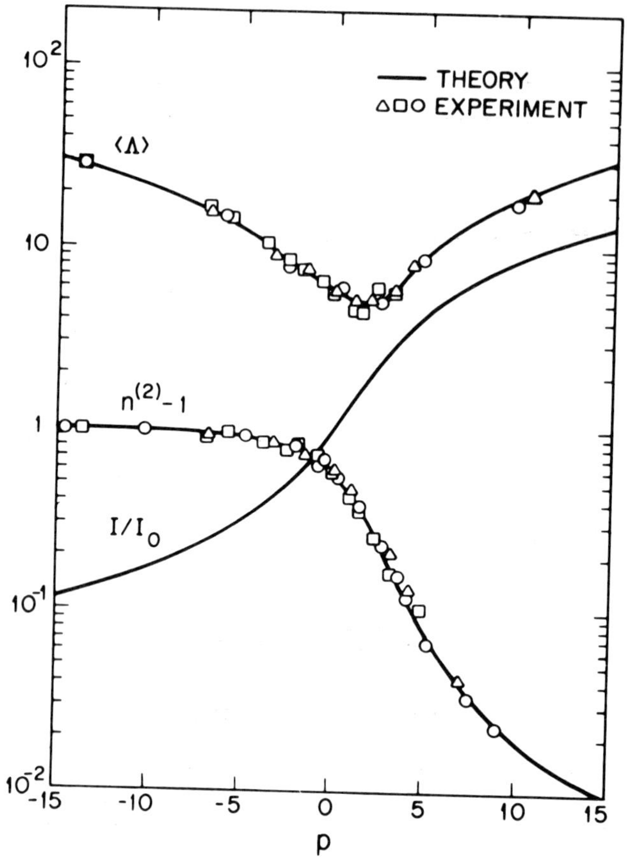

Fig. 11 Summary of experimental measurements of the reciprocal of the second-order coherence time, $<\Lambda>$, Eq. (22), and the second-order intensity cumulant at zero time delay, $n^{(2)}-1$ (Eq. (15b), with t=0), as functions of the pump parameter p.

Acknowledgements

One of us (C.D.C.) wishes to thank Prof. G.T. Reynolds of Princeton for his support of portions of this work, and the Aspen Center for Physics for their hospitality during preparation of an initial version of this paper. Another of us (W.A.S.) would like to thank his wife, Beverly Ann, for her valuable assistance with the computer programming. We also thank Prof. Frederic Davidson for useful conversations, and Ms. B.C. Chambers for assistance in computer programming.

References

1. Some of the results reported in this paper were reported earlier by C.D. Cantrell and W.A. Smith, Physics Letters *37A*, 167(1971); M. Lax arrived at the same results independently. We have decided to publish them together.
2. F. Davidson and L. Mandel, Physics Letters *25A*, 700 (1967); more details are contained in F. Davidson, Phys. Rev. *185*, 446 (1969).
3. F. Haake, Zeits. Physik *227*, 179 (1969); Phys. Rev. *A3*, 1723 (1971).
4. A statistician's formulation of cumulants is contained in (a) M.G. Kendall and A. Stuart, *The Advanced Theory of Statistics* (New York, Hafner Publishing Co., 1958), Vol. I, Chapter 3. A general formulation for many times and for non-commuting operators was given by (b) R. Kubo, J. Phys. Soc. Japan *17*, 1100 (1962). The use of cumulants in quantum optics has been emphasized by (c) R.F. Chang, V. Korenman, C.O. Alley, and R.W. Detenbeck, Phys. Rev. *178*, 612 (1969), and (d) C.D. Cantrell, Phys. Rev. *A 1*, 672 (1970).
5. M. Lax, Phys. Rev. *172*, 350 (1968).
6. H. Haken and W. Weidlich, Zeits. Physik *205*, 96 (1967).
7. The Markoffian property was not used in Ref. 5 in setting up the multi-time correspondence, but only in factoring the resulting probability distributions.
8. M. Lax, 1966 Brandeis University Summer Institute in Theoretical Physics, *Fluctuation and Coherence Phenomena in Classical and Quantum Physics, in Statistical Physics, Phase Transitions, and Superfluidity*, Vol. 2, edited by M. Chretien, E.P. Gross, and S. Deser (Gordon and Breach, Science Publishers, Inc., New York, 1968).
9. P.M. Morse and H. Feshback *Methods of Theoretical Physics* (New York, McGraw-Hill Book Co., 1953), pp. 884-886.
10. M. Lax and M. Zwanziger, *Exact Photocount Statistics: Lasers Near Threshold*, Table 4 (unpublished).
11. H. Risken and H.D. Vollmer, Zeits. Physik *201*, 323 (1967).
12. R.D. Hempstead and M. Lax, Phys. Rev. *161* , 350 (1967).
13. C.D. Cantrell and W.A. Smith (unpublished).

CORRELATION FUNCTION OF A LASER BEAM NEAR THRESHOLD*

S. CHOPRA and L. MANDEL

University of Rochester, Rochester, N.Y.

1. Introduction

As is now well known, the rotating-wave van der Pol oscillator theories of the laser [1-4] lead to a precise prediction for the form of the intensity correlation function of the light. In the steady state, we may define the normalized intensity correlation function $\lambda(\tau)$ by the ratio

$$\lambda(\tau) \equiv \langle\Delta I(t)\Delta I(t+\tau)\rangle/\langle I\rangle^2 \quad , \tag{1}$$

where I(t) is the light intensity at time t and $\Delta I(t)$ is the difference between I(t) and its expectation value. The intensity may be treated as a classical random process, when the average in (1) is taken over the ensemble of realizations, or the intensity may be regarded as a Hilbert space operator, in which case the average denotes the quantum expectation, and it is understood that all operators are written in normal order.

The theoretically predicted form of $\lambda(\tau)$ for a laser beam may be written[1-4]

$$\lambda(a,\tau) = C(a) \sum_{r=1}^{\infty} M_r(a) \exp[-\lambda_r(a)\tau] \quad , \tag{2}$$

*This work was supported by the Air Force Office of Scientific Research and by the National Science Foundation.

where a is a parameter, the so-called pump parameter, that characterizes the level of excitation of the laser. At threshold a = 0, whereas a is negative or positive below or above threshold. It has become customary to express all times in dimensionless form, in terms of the correlation time at threshold $T_c'(a=0)$. If we write

$$\tau = 0.171\tau'/T_c'(a=0) \quad , \tag{3}$$

where τ' and T_c' are experimentally measured times, then $C(a), M_r(a), \lambda_r(a)$ in Eq.(2) become universal functions that should hold for any laser, with $\sum_r M_r(a) = 1$. According to Eq.(2), $\lambda(a,\tau)$ has the form of a series of exponentially decaying functions of τ, with each term having a different time constant in general. $\lambda(a,\tau)$ is therefore not at all exponential. However, a quantitative analysis shows that the expected departure from exponential form is very slight. The first term in Eq.(2) dominates the behavior, unless the laser is working appreciably above threshold. But appreciably above threshold, the light fluctuations, the function C(a), and $\lambda(a,\tau)$, all become very small. The departure of $\lambda(a,\tau)$ from exponential form has therefore proved to be difficult to observe, although evidence for such departure has recently been obtained[5].

By making use of a new electronic correlation technique[6], that allows far more data to be accumulated in a given time than before, and therefore leads to greater statistical accuracy, we have been able to measure the intensity correlation function with sufficient accuracy to demonstrate the non-exponential form. The results are all in very good agreement with the theory.

2. Principle of the Method

The light source for these experiments was a single-mode He-Ne laser operating at 6328 Å, that was stabilized by a feedback arrangement, so that the laser could be operated for long periods of time at any intensity either below, at, or above threshold[7] (see Fig. 1). The beam from the laser was divided by a half-silvered mirror and the two beams fell on two photodetectors whose outputs, after amplification, were passed to two discriminators.

The outputs of the discriminators were fed to inputs 1 and 2 of the electronic correlator shown in Fig. 1. The pulse at input 1 is the start pulse, that initiates a 100 μsec gate pulse, and allows 1 μsec clock pulses to be applied to each of 6 B.C.D. counters. When a pulse is received at input 2, it is passed to a shift register, that effectively stops B.C.D. counter 1. The number stored

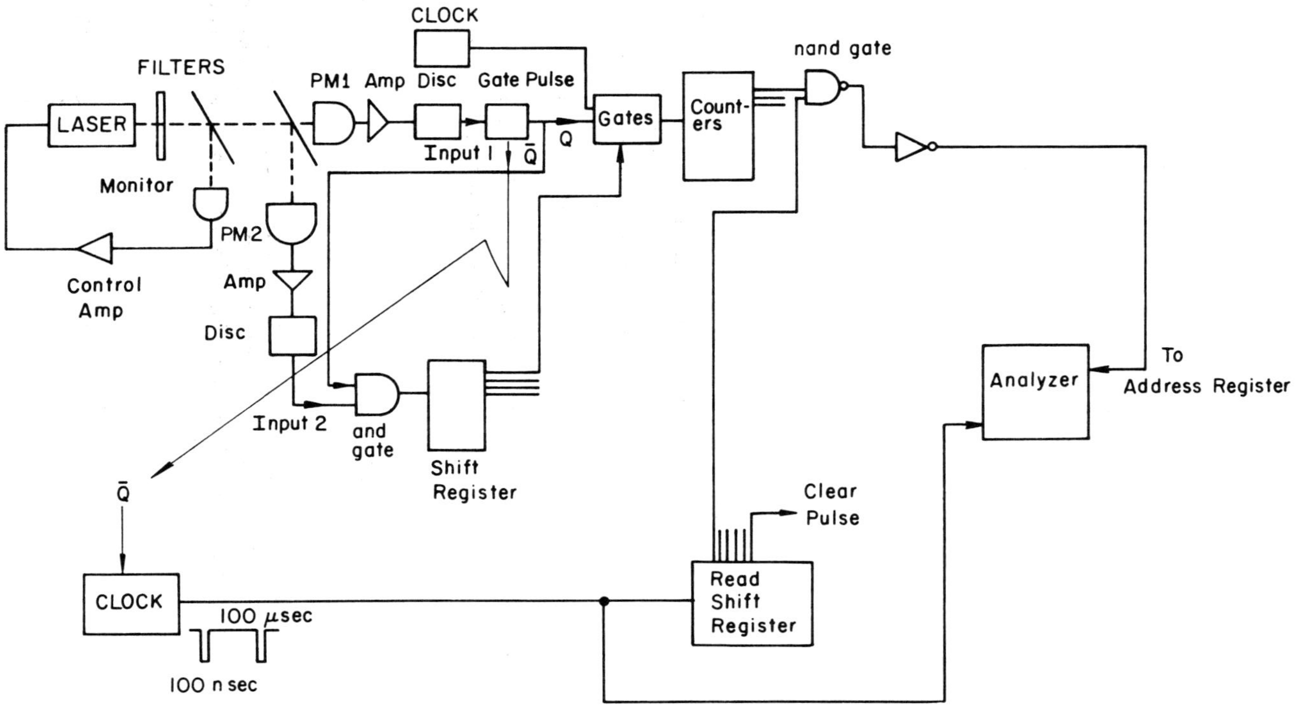

Fig. 1 Outline of the apparatus used for making correlation measurements.

on this counter is therefore a measure of the time interval in microseconds between this pulse and the start pulse. A subsequent pulse at input 2 stops B.C.D. counter number 2, and causes the shift register to shift to counter number 3, and so on. Up to six arrival times can be stored in this way within the same 100 μsec measurement time interval initiated by the start pulse. At the end of the measurement time, the stored numbers are read into the memory of a 100-channel analyzer with the help of a Read shift register, such that each stored number is used as an address in the analyzer. The counters are then reset to zero, and the correlator is ready for another counting sequence initiated by the next start pulse.

In the steady state, the conditional probability that a start pulse at time t from detector 1 will be followed by a pulse from detector 2 at a later time $t+\tau$, within $\Delta\tau$ is given by[8]

$$P_c(\tau)\Delta\tau = \alpha_2 c S_2 \langle I_2\rangle \Delta\tau[1+\lambda(\tau)] \quad , \tag{4}$$

provided each photodetector is illuminated normally and is effectively exposed to the same light beam. Here $\langle I_2\rangle$ is the mean light intensity at detector 2 in photons per unit volume, S_2 is the exposed surface area of the photocathode, and α_2 is the quantum efficiency. The histogram of the distribution of time intervals τ in the analyzer memory is therefore a direct measure of the correlation function $\lambda(\tau)$. In practice, a slight complication arises from the dark counting rates r_1 and r_2 of the two photodetectors. When this is taken into account, we find that the expected number of counts stored in a channel of the analyzer corresponding to a delay τ is given by[6]

$$\langle n(\tau)\rangle = n_s(R_2\Delta\tau/\theta_2)[1+\theta_1\theta_2\lambda(\tau)] \quad . \tag{5}$$

Here n_s is the total number of times that the start sequence is initiated, $\Delta\tau$ is the effective width in time of one analyzer channel, R_1, R_2 are mean counting rates of the two photodetectors due to the light alone, and

$$\left.\begin{aligned} \theta_1 &\equiv R_1/(R_1+r_1) \\ \theta_2 &\equiv R_2/(R_2+r_2) \quad . \end{aligned}\right\} \tag{6}$$

Because our correlator contained only six counters, it was necessary to limit the counting rate of detector 2, in order to ensure that the probability of getting more than six pulses from detector 2 in any one measurement interval T (T was 100 μsec) was sufficiently small. It can be shown that this probability is less than 10^{-3} as long as $R_2T \lesssim 1$. This represents a much less severe limitation on the counting rate than is encountered in measurements with a time-to-amplitude converter with the same maximum interval T[9].

In order to determine the time scaling parameter in Eq.(3), it was necessary first to identify the threshold (a=0) of the laser. This was done by adjusting the operating point of the laser until the zero time correlation satisfied

$$\lambda(a=0,\tau=0) = C(a=0) = \pi/2-1 \approx 0.571 \tag{7}$$

as required by the theory. The correlation time $T_c'(a=0)$ was then determined from the slope of $\lambda(0,\tau)$.

The mean light intensity $\langle I(a)\rangle$ produced by the laser is expected to vary with pump parameter a according to the relation[1-4]

$$\langle I(a)\rangle/\langle I(0)\rangle = \sqrt{\pi}a/2 + [\exp(-\tfrac{1}{4}a^2)]/[1+\Phi(\tfrac{1}{2}a)] \quad , \tag{8}$$

where $\Phi(x)$ is the Gaussian error integral.

The adjustment of the laser to the other predetermined values of the pump parameter a was therefore accomplished by ensuring that the counting rates R_1-r_1 and R_2-r_2 of the two detectors, which were proportional to $\langle I_1(a)\rangle$ and $\langle I_2(a)\rangle$, satisfied Eq.(8) for any selected a. Our measurements covered the range a = 1, 2, and 3, where the departures of $\lambda(a,\tau)$ from exponential form are expected to be noticeable.

3. Results

The results of three measurements of the correlation function $\lambda(a,\tau)$, for the three different working points of the laser, are shown in Figs. 2, 3, and 4, together with the theoretically predicted curves, according to Eq.(2). Typical correlation times were in the neighborhood of 40 μsec. It will be seen that there is good agreement between experiment and theory, and that the predicted small departure of the correlation function from exponential form is confirmed.

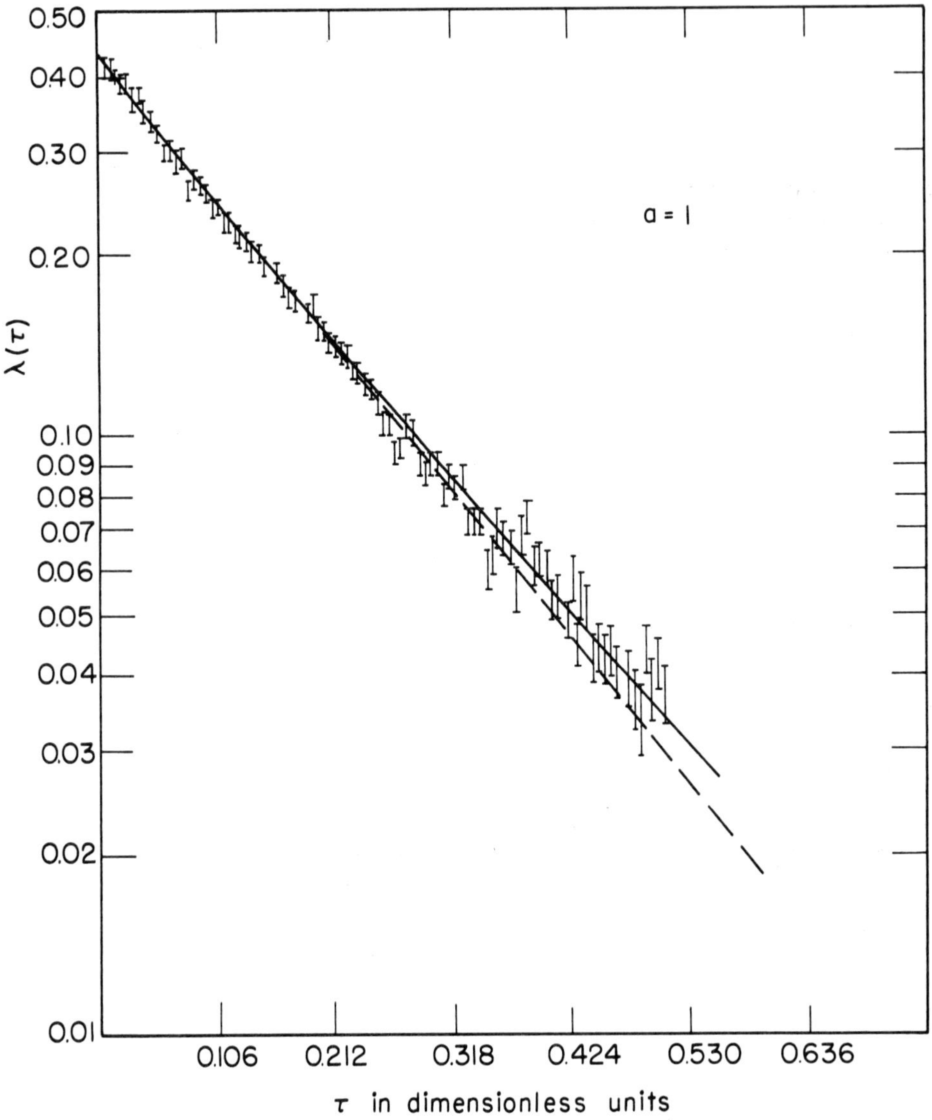

Fig. 2 Experimental results of the measurement of $\lambda(a,\tau)$ for $a = 1$, together with the standard deviations. The full curve is the theoretically predicted curve. The broken curve is the extrapolated exponential form.

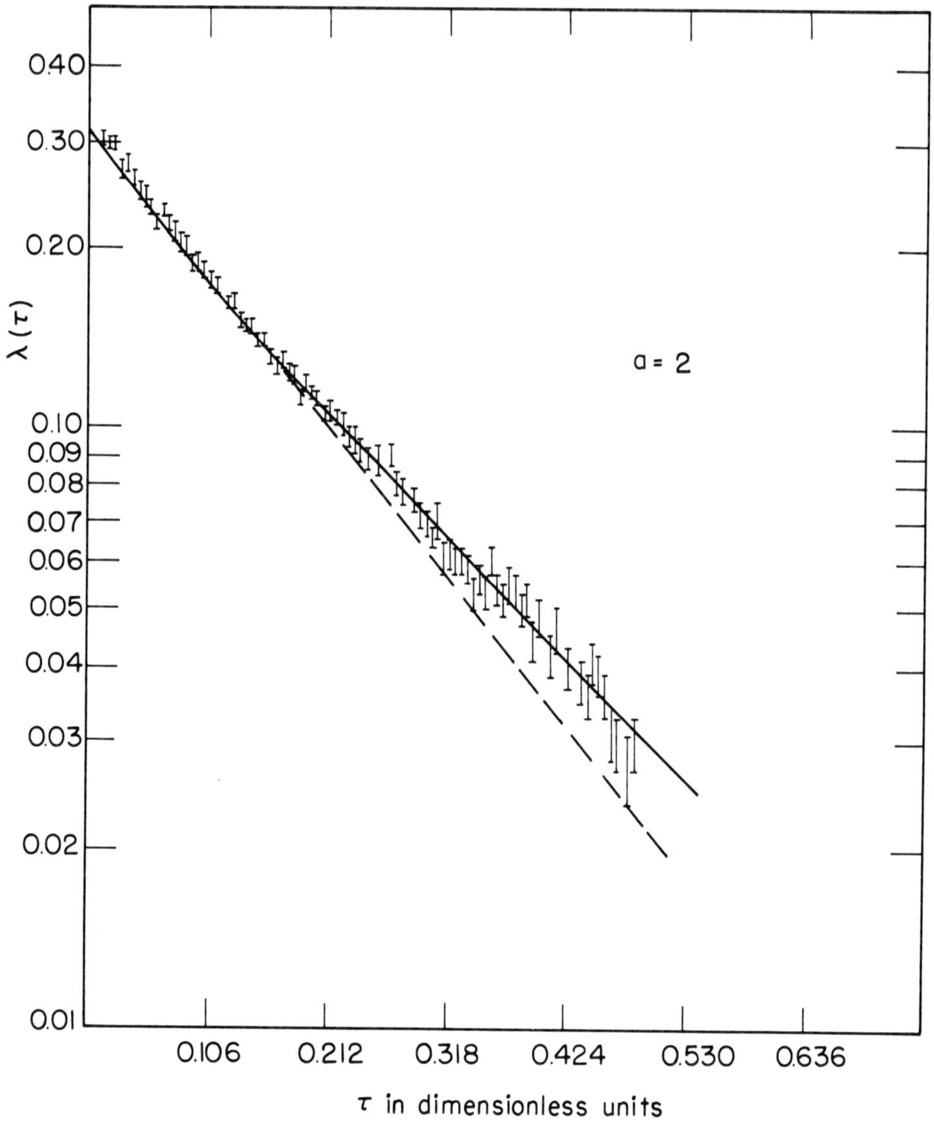

Fig. 3 Experimental results of the measurements of $\lambda(a,\tau)$ for $a = 2$, together with the standard deviations. The full curve is the theoretically predicted curve. The broken curve is the extrapolated exponential form.

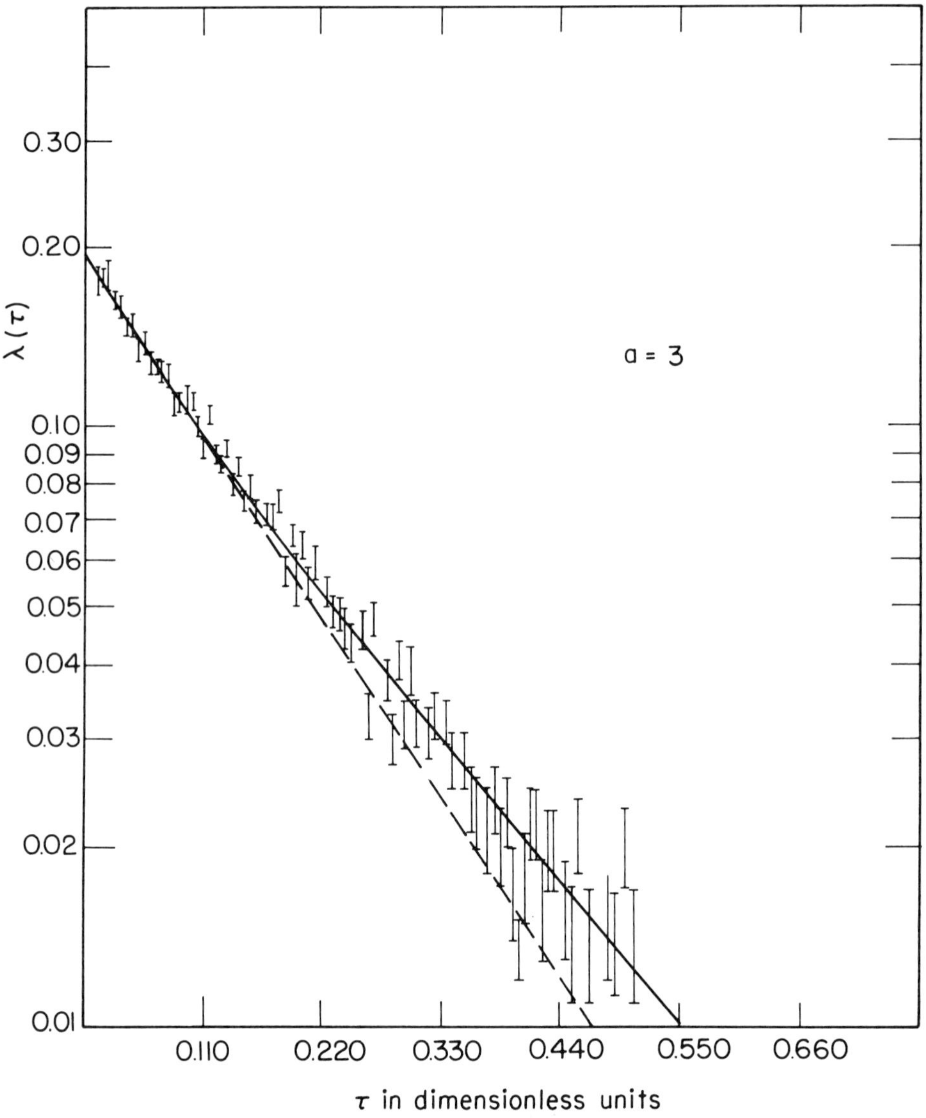

Fig. 4 Experimental results of the measurements of $\lambda(a,\tau)$ for $a = 3$, together with the standard deviations. The full curve is the theoretically predicted curve. The broken curve is the extrapolated exponential form.

References

1. H. Risken, Z. Phys. *186*, 85 (1965) and *191*, 302 (1966); *Progress in Optics*, Vol. 8, ed. E. Wolf (North-Holland Publishing Co., Amsterdam, 1970) p. 239.

2. R.D. Hempstead and M. Lax, Phys. Rev. *161*, 350 (1967).

3. M. Lax and W.H. Louisell, IEEE J. Quantum Electronics *QE-3*, 47 (1967).

4. H. Risken and H.D. Vollmer, Z. Phys. *201*, 323 (1967).

5. S. Chopra and L. Mandel, IEEE J. Quantum Electronics *QE-8*, 324 (1972).

6. For a description of the instrument and its operation, see S. Chopra and L. Mandel, Rev. Sci. Instru. *43*, 1489 (1972).

7. For details see D. Meltzer and L. Mandel, IEEE J. Quantum Electronics *QE-6*, 661 (1970) and D. Meltzer and L. Mandel, Phys. Rev. A *3*, 1763 (1971).

8. See for example L. Mandel and E. Wolf, Rev. Mod. Phys. *37*, 231 (1965).

9. For a comparison see F. Davidson and L. Mandel, J. Appl. Phys. *39*, 62 (1968).

LOCALIZATION OF PHOTONS

John F. Clauser

University of California, Berkeley, California

In a popular quantum mechanics textbook one reads..."If we have some indications that classical wave theory is macroscopically correct, it is nevertheless clear that on the microscopic level only the corpuscular theory of light is able to account for typical absorption and scattering phenomena such as the photoelectric effect and the Compton effect, respectively. One must still ascertain how the photon hypothesis may be reconciled with the essentially wave-like phenomena of interference and diffraction...[1], and in another "...We have, on the one hand, the phenomena of interference and diffraction, which can be explained only on the basis of a wave theory; on the other, phenomena such as photoelectric emission and scattering by free electrons, which show that light is composed of small particles."[2]

Recent considerations, however, have called into question whether or not actual experiments have unambiguously established a particle nature for photons. Thus, the observations usually thought to do so--those of the photoelectric effect [3], the Compton effect [4], spontaneous emission [5], and the Lamb shift [5]--can be predicted semiclassically with surprising accuracy. A particle nature for photons is apparently not required for the description of these experiments; they may be described with photons acting solely as waves. These results are indeed surprising, since photons are the simplest and presumably the best understood elementary particles. Naturally, it is highly desirable to experimentally demonstrate unambiguously their particle-like behavior.

What is the simplest and most conspicuous difference between particles and waves? It is evidently the fact that only particles are localizable to arbitrarily small volumes. A quantum mechanical description of electromagnetic radiation predicts photon localization through the "collapse" of a photon's wave function, which occurs as the result of position measurement. This collapse is foreign, however, to classical waves, and neither semiclassical theories nor any other linear classical-wave picture predict a localization of photons to dimensions smaller than the size of their classical interference patterns.

It must be recognized that any theory in which the radiation field is to be described specifically by the classical Maxwell's equations will yield predictions which violate the observed polarization correlation of atomic-cascade photons [6]. It may be possible, perhaps, that there are classical-wave theories not describable by Maxwell's equations, to which the arguments of Ref.6 do not apply. The following discussion applies to any linear classical-wave theory of electromagnetic radiation.

It is the purpose of this paper to first review various experiments which suggest a localization of photons, and show that they are not in conflict with a simple wave description. Included are the Compton effect, and the photon angular correlations in π° and/or positronium $\rightarrow 2\,\gamma$ decays. Next we discuss Lamb and Scully's semiclassical treatment of the photoelectric effect, and describe a situation in which its predictions are in conflict with those of a usual quantum mechanical treatment of the electromagnetic field. The difference is found in the localizability of electromagnetic emissions. Finally the requirements for a conclusive experiment are derived, and existing experimental tests are reviewed but found inconclusive. A distinguishing experiment to actually demonstrate this particle-like localization is currently being performed at this laboratory.

Correlation Experiments

When one is asked to think of processes in which photons act as localized particles, those of positronium annihilation, $\pi^{\circ} \rightarrow 2\gamma$, and Compton scattering immediately come to mind. In these, the detection of a γ-ray (or an electron in the case of Compton scattering) localizes the remaining γ-ray. Since the γ-rays may be emitted with a spherically symmetric distribution, it seems that these experiments locate them to a volume much smaller than the size their classical interference patterns.

The following simple consideration shows that this is not the case. Consider a *Gedankenexperiment* in which a positronium atom is confined in the x direction to a dimension Δx, perhaps by a system of slits as is shown in Figure 1. The momentum of the atom in the x direction is thus rendered uncertain by an amount $\Delta p_x \gtrsim h/\Delta x$.

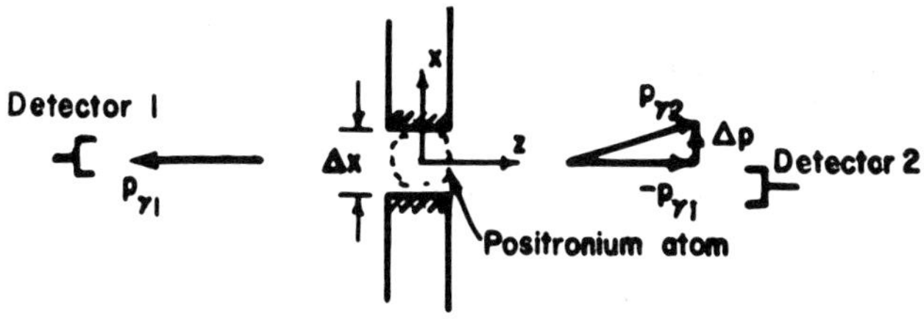

Figure 1. Scheme for attempting to localize a γ-ray to a volume smaller than its classical interference pattern. A positronium atom is confined in the x direction by slit system to a dimension Δx, and the annihilation quanta are detected by detectors 1 and 2.

Suppose now that the atom decays into two γ-rays, and one of these is detected by a detector subtending an infinitesimal solid angle, and located on the -z axis. The sum of the x components of the γ momenta must be uncertain by the same amount Δp_x, thus the γ-rays will not be exactly collinear. If the momentum of the first γ is denoted by p_γ the momentum of the second γ will have a distribution of angles with respect to the z axis. The beam width will thus be,

$$\Delta\Theta_{QM} \simeq \Delta p_x/p_\gamma \gtrsim h/(p_\gamma \Delta x).$$

Next let us consider the above process viewing the γ-rays as waves. The positronium atom with transverse dimension d = Δx coherently radiates the second γ-ray. The classical-wave picture suggests that the radiation may be sent out in a beam with width $\Delta\Theta_{SCT} \gtrsim \lambda/d$ where $\lambda = h/p_\gamma$ is the γ-ray's wavelength. This "diffraction limit" is characteristic of any linear wave theory. The semiclassical beam width is thus given by $\Delta\Theta_{SCT} \gtrsim h/(p_\gamma \Delta x)$. Comparing this with the previous result, we find equal minimum beam widths in both descriptions. Thus a particle picture and a classical wave picture both predict that the 2γ's will be found collinear only to the same angular precision. The predictions for this experiment are then consistent with a semiclassical theory in which the atom sends out thin diffraction limited beams of classical waves - not particles!

A similar analysis applied to Compton scattering achieves the same result. Such an analysis has in fact been carried out in detail by Schrödinger and by Gordon [4] who present a semiclassical theory which predicts identically the results of the usual quantum mechanical treatment of the radiation field for the γ-ray's wavelength shift and electron recoil direction, and to a close approximation the γ-ray intensity dependence. It is conceivable that the residual differences may be accounted for by higher order effects not included in these calculations, and/or the breakdown of Maxwell's theory implied by Ref.6.

Photoelectric Effect, à la Lamb and Scully

Lamb and Scully have shown for the simple case of radiation propagating from a source to a detector that both the wave and particle pictures can predict the same results for the photoelectric effect [3]. In the particle view a source atom may emit a particle which then strikes an atom in the photocathode and ionizes it. (See Figure 2.) In the wave view a source atom may emit a

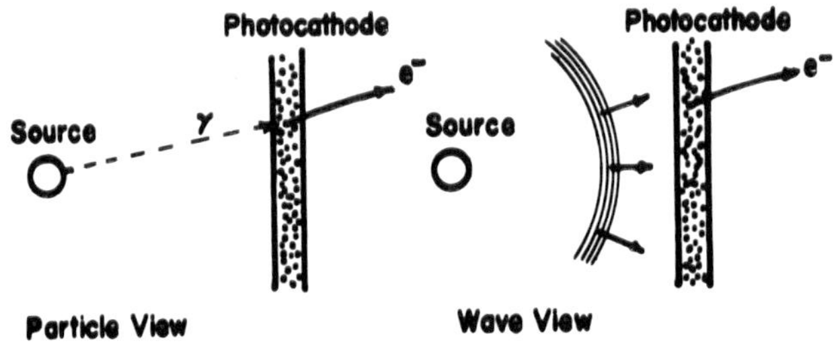

Figure 2. Comparison of wave and particle views of the photoelectric effect. In the particle view a particle-like γ has a certain probability for striking any of the atoms in the photocathode. In the collision there is a certain probability for the ejection of a photoelectron. In the wave view, a wave impinges upon all of the atoms in the photocathode, and the resultant oscillating electric field has a certain probability for photoionizing any of them.

spherically expanding wave which will have a certain probability for photoionizing any of the atoms in the screen. Lamb and Scully show that in a semiclassical description the probability of a photoionization is proportional to the classical field intensity, and becomes applicable without the time lag necessary for an accumulation of the classical field energy. Thus experiments of this type may only establish the localization of photoionizations, not photons themselves.

When one views the source with two detectors preceded by different wavelength filters resonant to opposite wavelengths of a two-photon cascade, coincidences are observed. Again both models apply. From the particle view, two particles are emitted in sequence. In the wave view the observation of coincidences implies that for a single photon the wave must manifest itself as a short pulse (perhaps similar to the usual wave packet). Thus in a cascade two pulses are successively emitted, each with the appropriate wavelength. Indeed, the semiclassical theory of Jaynes, Crisp and Stroud exhibits exactly this model [5].

Suppose now that one places two detectors within the interference pattern of a single photon pulse and employs two wavelength filters both resonant to the same transition, as is shown in Figure 3. Here the similarity between the two viewpoints ends. A

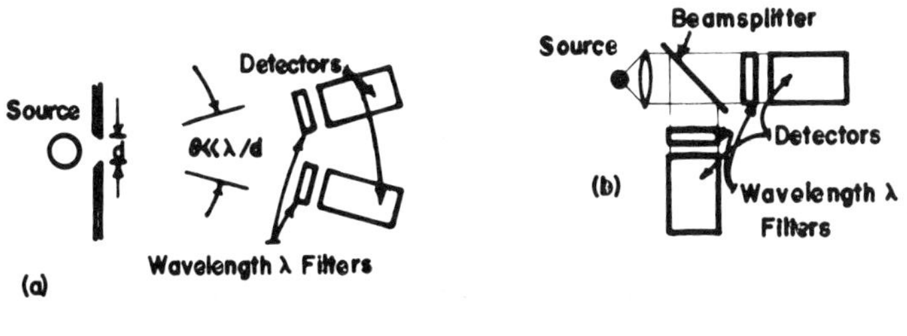

Figure 3. Experiments to distinguish between semiclassical and quantum mechanical predictions of photoelectric effect. (a) Two detectors are placed within the interference pattern of a single photon and coincidences are sought. (b) Alternative scheme which assures equal illumination of both detectors. Scheme (a) localizes photons in the θ and ϕ coordinates, while (b) localizes them in the radial coordinate.

particle model predicts that for each pulse only one photoelectron will be liberated at one of the detector photocathodes. Indeed this is the prediction by a quantum mechanical description of the radiation field. Van Neumann's reduction postulate requires the photon wave function to "collapse" when one of the detectors responds. The probability of a second response at the other detector immediately vanishes. In this way, energy conservation is assured.

The collapse does not occur in a simple wave model, however, since the wave-packet reduction is unique to quantum mechanical systems. Indeed, no classical process can be responsible for the collapse when the arrivals of a given pulse at the two detectors have a space-like separation [7]. Thus a classical pulse will be present simultaneously at both detectors, and there is a certain probability that photoelectrons will be simultaneously liberated at the photocathodes of both detectors! This will be true even though only one photoelectron is liberated per pulse on the average. Given an ensemble of identical pulses, for some of these, more than one electron will be liberated, and for others, none will be liberated.

Thus there is a difference for this *Gedankenexperiment* between the predictions of the classical-wave and quantum mechanical descriptions of the electromagnetic field. The former predicts an "excess" coincidence rate for the two detectors. We shall consider the conditions necessary for an actual experimental test of this difference, and find that no distinguishing experiments have been performed so far. Before we do this, however, a digression is warranted concerning energy conservation in the semiclassical scheme.

Energy Conservation in Semiclassical Theories

A frequently voiced objection to Lamb and Scully's description of the photoelectric effect is that superficially it appears to violate energy conservation. Before condemning the theory on this ground, however, one should carefully re-examine what energy conservation actually means. In descriptions of the electromagnetic field two physically different measures of the field energy arise for a single photon. First there is the total classical energy as calculated from Maxwell's equations, thus

$$E_C = \int dV\ (|\underline{E}|^2 + |\underline{H}|^2)/8\pi \qquad (1)$$

Second there is the energy-frequency relation given by

$$E_Q = h\nu$$

To be sure, in a quantum field theory these are equal, but in a semiclassical theory this restriction does not hold. Thus if ever $E_C \neq E_Q$ applies, the conservation of at least one of these is violated. Indeed the nonconservation of E_C occurs in a semiclassical description of the photoelectric effect. It is most dramatically demonstrated by the process in which two photoelectrons are liberated following a single atomic decay. E_Q is, however, conserved for each individual process.

But to dismiss semiclassical theories for this reason alone is prejudicial. Physics is an experimental science, and one may argue plausibility only on experimental grounds. What then is the experimental evidence for the equality of E_C and E_Q and for their simultaneous conservation? Reasonably accurate comparisons of E_C and E_Q have only been made for average values of E_C and E_Q, (e.g. bolometric measurements) which we have seen, can easily be accounted for by an appropriate statistical balance between processes in which several photoelectrons are emitted, and others in which none are emitted. Accurate wavelength comparisons for atomic systems (the Ritz combination principle) again can only test conservation of E_Q for individual radiative transitions. A demonstration of point-wise conservation of E_C must come from an analysis of experiments of the type currently being discussed.

Indeed the notion of statistical energy conservation was considered earlier by Bohr, Kramers, and Slater [8], in response to Einstein's discussion of thermodynamic equilibrium [9]. They theorized that energy is conserved only statistically in all radiative processes, but were forced to abandon this idea when Bothe and Geiger observed electron-γ momentum correlations in Compton scattering [10]. In the present light we see that this dismissal may have been premature. The straightforward prediction by a semiclassical theory permits a classical-wave picture for both Compton and photoelectric effects, and employs statistical conservation of the classical field energy only for the latter process.

Experimental Requirements

We now discuss the necessary experimental conditions for a realization of our *Gedankenexperiment* to distinguish the semiclassical from the quantum mechanical prediction. If E pulses per second are emitted per unit time by a source, and if p is the average probability per pulse that a photomultiplier will yield a count, then the count rate at that detector is $S = Ep$. In either theory we will have $p = Q \times L \times \Omega/4\pi$ where Ω is the solid angle subtended by the detector, Q is the photocathode quantum efficiency, and L represents other losses, either in the optics, electron-multiplier or electronics.

In this experiment it is necessary to assure that both detectors are within the interference pattern of a given pulse, and are equally illuminated by it. The easiest way to do this is to use a beam splitter as is shown in Figure 3b. That this will occur is evidenced by the fact that transmitted and reflected components of a single photon can be made to interfere. (e.g. in a Michelson interferometer.) All photons will then have approximately the same probability for generating a count at either detector. Thus the expected excess coincidence rate predicted by a classical-wave theory is given approximately by

$$C \simeq p^2E. \tag{3}$$

Assuming negligible detector dark rates, the accidental coincidence background rate from which C must be distinguished is

$$A \simeq p^2E^2 2\tau \quad , \tag{4}$$

where τ is the resolving time of the system. One can now calculate the time required to measure to a precison of N standard deviations the difference between the excess coincidence rate given by (3) and the zero excess rate predicted by quantum mechanics. Doing this we obtain

$$T \approx (1 + 4E\tau)N^2/(p^2E) \tag{5}$$

which in the limit of high source rates takes the form

$$T \simeq 4N^2\tau/p^2 \quad . \tag{6}$$

Measured detector efficiences in cascade experiments employing fast optics, and the most modern photomultiplier tubes and electronics typically yield values [12] $p \simeq 10^{-3}$. For equation (3) to apply, τ may not be shortened to less than the length of a given pulse, which is presumably the order of the atomic state lifetime (~ 5nsec. for typical allowed atomic transitions). Taking N = 5, we see from equation (6) with the above parameters that a total integration time of T = 1 second suffices.

Experiment of Ádám, Jánossy and Varga

In 1954 Ádám, Jánossy and Varga performed an experiment to search for an effect similar to the one discussed above [13]. Their experiment is frequently referenced in discussions of the wave-particle paradox [14]. As the only existing test of this aspect of photon localization, it is worthwhile to examine it carefully.

Figure 4 reproduces a diagram of their experiment. In it they selected the light of a single spectral line with a monochrometer, and focused it through a beam splitter onto two photomultipliers operating in coincidence. They assumed their detector efficiency to be p = 1/300. With a resolving time $\tau = 2.3\mu$sec (good by 1954 standards) one calculates from (6) T = 20.7 sec for N = 5. They claimed thus to be able to easily detect the expected excess coincidence rate, if present.

However, their efficiency p = 1/300 is the efficiency for detection of photons in a beam, not that for wave-like pulses emitted spherically by the source. Their use of this value ignores the serious loss in efficiency suffered because of the narrow acceptance solid-angle of their monochrometer. Conservatively estimating from their diagram this additional loss of efficiency to be 1/400, their actual detector efficiency for wave like pulses was undoubtedly less than 8.3×10^{-6}, in which case the required integration time for even N = 1 becomes $T \simeq 1.3 \times 10^{5}$sec. This is an order of magnitude longer than the duration of their experiment. Thus the experiment of Ádám, Jánossy and Varga appears to be considerably less conclusive than has usually been assumed.

It is noteworthy that similar experiments--those measuring the Brown-Twiss effect--accept light within only very small solid angles from the source, and thus are inapplicable for the same reason. Moreover the excess coincidence rate predicted by a semiclassical theory should be easily distinguishable from that of the Brown-Twiss effect. The latter occurs only for small detector solid angles with the excess coincidence rate varying with the

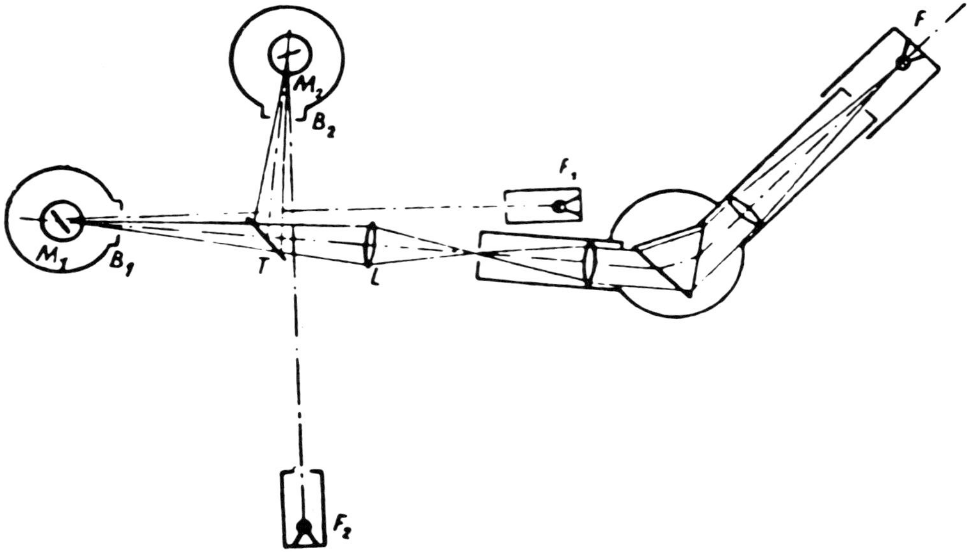

Figure 4. Optical system of Ádám, Jánossy and Varga [13]. Light from source F is forced through a monochrometer on photomultipliers M_1 and M_2 via beam splitter T. (Figure after Ádám, Jánossy and Varga).

square of the excitation rate. The excess coincidence rate predicted by a semiclassical theory, on the other hand, will occur only at large detector solid angles, and will vary linearly with excitation rate.

Conclusion

The most conspicuous difference between particles and waves is that only particles may be localized. In the foregoing discussion we have indicated that there is apparently no existing experimental result which requires photons be viewed as particles. Any linear classical-wave description of the photoelectric effect, though, does lead to an experimentally observable distinction between its predictions in which photons are not localized, and those by a quantum mechanical treatment in which they are. An experimental test is currently is progress at this laboratory.

These experimental results, in addition to their relevance to the foundations of quantum mechanics and to a consideration of semiclassical radiation theories, will be significant in another respect. They are related to experiments which seek to determine whether or not nature may be viewed objectively. It seems reasonable to assume that photons objectively exist, propagate, and in so doing carry information independently of external observers. However, extensions of Bell's theorem have shown that any objective model of nature must be in conflict with the quantum mechanical predictions for suitably devised polarization correlation experiments [15]. Since fully conclusive experiments are presently technologically difficult (due to low available polarizer and/or photodetector efficiencies), conclusions drawn from present experimental results have had to rely upon additional assumptions concerning the behavior of photons. One of these assumptions is that photons may be described as localized particles. The above experimental results may thus lend additional support for the experimental evidence found by Freedman and Clauser [16] against such models.

References

1. A. Messiah, *Quantum Mechanics* (John Wiley and Sons, Inc., New York, 1961), Vol. I, p.18.
2. P. A. M. Dirac, *Principles of Quantum Mechanics*, 4th ed. (Oxford University Press, Oxford, 1958), p.2.
3. W. E. Lamb and M.O. Scully in *Polarization: Matière et Rayonnement, Jubilee volume in honor of A. Kastler*, edited by

Société Francaise de Physique (Presses Universitaires de France, Paris, 1969); M.O. Scully and M. Sargent III, Phys. Today *25*, 38 (1972).

4. E. Schrödinger, Ann. Physik *82*, 257 (1927) [English translation in E. Schrödinger, *Collected Papers on Wave Mechanics* (Blackie and Sons., Ltd., London, 1928), p.124]; Nuovo Cimento *9*, 162 (1958); W. Gordon, Z. Physik *40*, 117 (1927); For a comparison with experiment see A. Bernstein and A. K. Mann, Amer. J. Phys. *24*, 445 (1956).
5. M. D. Crisp, and E. T. Jaynes, Phys. Rev. *179*, 1253 (1969); *185*, 2046 (1969); C. R. Stroud Jr., and E. T. Jaynes, Phys. Rev. *A1*, 106 (1970); E. T. Jaynes, ibid *2*, 260 (1970).
6. J. F. Clauser "Experimental Limitations to the Validity of Semiclassical Radiation Theories", this volume, p. 111; Phys. Rev. *A6*, 49 (1972).
7. The above argument resembles an objection to quantum mechanics by A. Einstein concerning alpha decay which he presented at the 1927 Solvay Conference.
8. N. Bohr, H. A. Kramers, and J. C. Slater, Phil. Mag. *47*, 785 (1924).
9. A. Einstein, Phys. Z. *18*, 121 (1917).
10. For a discussion of this point see article by N. Bohr in *A. Einstein: Philospher-Scientist*, edited by P. Schilpp (Library of the Living Philosphers, Evanston, Illinois 1949), p.207.
11. Similar situations will be encountered in a consideration of point wise momentum conservation. For example, a curious phenomenon arises when $\lambda >> d$ applies, as it does in the case of optical emissions from thermal atoms. The classical interference pattern of the emitted radiation then fills all space. Evidently a single photon should here be described classically as a spherically expanding wave. Symmetry implies that a spontaneously radiating atom will then experience no radiation recoil. It is unfortunate that the only experimental attempt to directly observe the recoil following spontaneous emission was not conclusive. [R. Frisch, Z. Physik, *86*, 42 (1935)]. With only statistical momentum and energy conservation, Einstein's thermodynamic argument against classical-wave models (Ref. 9) does not apply.
An observation of recoil, however, would not alone invalidate semiclassical radiation theories, since it is possible to classically generate the asymmetric radiations patterns necessary to achieve a recoil. The usual semiclassical radiation patterns do not, however, exhibit such asymmetries.
12. The quantity p may be directly measured in cascade experiments from the ratio of coincidence rate to singles rate for the second photon of a cascade. Atomic cascade experiments have been reviewed by C. Camhy-Val and A. M. Dumont, Astron. and Astrophys. *6*, 27 (1970). See also Ref.16.

13. A. Ádám, L. Jánossy and P. Varga, Acta Phys. Hung. *4*, 301 (1955), and Ann. Physik *16*, 408 (1955); L. Jánossy and Zs. Náray, Acta Phys. Hung. *7*, 403 (1957).
14. See for example J. M. Jauch in *Foundations of Quantum Mechanics, Proceedings of the International School of Physics 'Enrico Fermi', Course IL*, (Academic Press, New York, to be published) and J. M. Jauch, *Dialogue on the Question "Are Quanta Real"*, (Section de Physique, Univ. of Geneva, Geneva, 1971).
15. J. F. Clauser, M. A. Horne, A. Shimony, R. A. Holt, Phys. Rev. Letters *23*, 880 (1969).
16. S. J. Freedman and J. F. Clauser, Phys. Rev. Letters *28*, 938 (1972).

ADIABATIC FOLLOWING AND THE SELF-DEFOCUSING OF LIGHT IN RUBIDIUM VAPOR

D. Grischkowsky and J. A. Armstrong*

IBM Thomas J. Watson Research Center, Yorktown Heights, N. Y.

The narrow-line output of a dye laser on the low-frequency side of $^2P_{1/2}$ resonance line (7948 Å) of rubidium was self-defocused by passage through dilute rubidium vapor. The defocusing was caused by the electronic nonlinearity associated with adiabatic following of the laser field by the pseudomoment of the resonant atoms. Using the corresponding nonlinear dielectric constant, the wave equation was solved numerically and gave excellent quantitative agreement with experiment.

The term adiabatic following describes the situation in which the pseudomoment $\underline{p}$ of the near-resonant transition follows (remains parallel to) the effective field $\mathcal{E}_e$ of the laser pulse[1]. Adiabatic following occurs when two conditions are satisfied. First, in the rotating coordinate frame the direction of $\mathcal{E}_e$ must change slowly compared to the precession frequency Δ of $\underline{p}$ about $\mathcal{E}_e$ ($\mathcal{E}_e$ must change adiabatically); second the pulse width must be short compared to T_1 and T_2 of the atomic system. The response time of $\underline{p}$ and of the corresponding resonant electronic nonlinearity to changes in $\mathcal{E}_e$ is of the order of Δ^{-1} (for the work reported here the response time was less than 100 psec). Both adiabatic following and the steady-state model of Javan and Kelley[2] give constitutive relations which are known analytically for all intensities.

---

*Work of this author partially supported by ONR contract N00014-70-C-0187

Also, the resonant electronic nonlinearity of each model causes self-focusing when $\nu_0<\nu$ and self-defocusing when $\nu<\nu_0$. However, in contrast to adiabatic following the steady-state model is insensitive to how the pulse is applied and requires the pulse duration to be long compared to both T_1 and T_2. The response time for the steady-state model is approximately T_1.

The experiments were done with the rubidium vapor in a magnetic field in order to remove all degeneracies. The incident light with frequency ν was circularly polarized and propagated along the field. The precession frequency of $\underline{p}$ about $\mathcal{E}_e$ (in the rotating frame) is $\Delta = [(\nu-\nu_0)^2+2p_{12}^2E^2/h^2]^{1/2}$, where p_{12} is the absolute value of the matrix element of the electric dipole moment for the σ transition of frequency ν_0. $\mathcal{E}_e$ changes adiabatically when $\delta\nu \ll |\nu-\nu_0|$, where $\delta\nu$ is the linewidth of the incident light.

The analytic expression for ε resulting from adiabatic following is given by $\varepsilon = 1 - 4\pi N_e p_{12}^2/[h(\nu-\nu_0)(1 + 2E^2/E_s^2)^{1/2}]$. N_e is the effective number density of atoms, and $E_s = |h(\nu-\nu_0)/p_{12}|$. When $E^2 \ll E_s^2$, $\varepsilon \simeq \varepsilon_0 + \varepsilon_2 E^2$, with $\varepsilon_0 = 1 - 4\pi N_e p_{12}^2/[h(\nu-\nu_0)]$ and $\varepsilon_2 = 4\pi N_e p_{12}^4/[h(\nu-\nu_0)]^3$.

Since the experiments were done with the Rb vapor in an 8 kG magnetic field, we were able to study the interaction of σ^+ and σ^- circularly polarized light separately and hence verify the strongly resonant nature of the nonlinearity. For our experiments the cell temperature was 124°C; $\delta\nu < 0.005$ cm^{-1}; $(\nu_0-\nu)=0.45$ cm^{-1} for σ^- light; $(\nu_0-\nu) = 1.45$ cm^{-1} for σ^+ light. $\varepsilon_2 = -2.5 \times 10^{-7}$ esu for σ^- light which was 33.4 times larger than ε_2 for σ^+ light. For comparison $\varepsilon_2 = 4.2 \times 10^{-11}$ esu for CS_2, one of the most nonlinear Kerr liquids.

The experiments were performed as follows. For each 10 nsec pulse of the dye laser a Fabry-Perot interferogram and an oscillogram were taken before the beam was circularly polarized (the helicity could be changed from σ^+ to σ^- by rotating a quarter-wave plate 90°) and encountered a centered 0.5 mm diameter aperture (the peak power through the aperture was approximately 200 watts). The aperture was 50 cm from the entrance window of the 100 cm heated rubidium vapor cell (the number density of Rb atoms could be changed by changing the cell temperature), so the transverse intensity distribution of the beam entering the cell was the Fraunhofer diffraction pattern of the aperture. After passage through the cell near-field and far-field photographs were taken of the beam, and the output pulse was monitored.

After passage through the cell, the beams of the two polarizations were qualitatively different. The distribution of σ^+

light deviated only slightly from a Fraunhofer diffraction pattern, corresponding to the input pulse having had an accurate Fraunhofer pattern. The distribution of σ^- light (which also had a Fraunhofer distribution at the cell input) had been strongly modified. In the near-field the peak was flattened and broadened, and two rings appeared where there would have been one in the Fraunhofer pattern. Also, in the far-field the central peak broadened and rings appeared.

A numerical integration of the nonlinear wave equation was performed using the analytic expression for ε resulting from adiabatic following. The results agreed extremely well with the near-field data for both σ^+ and σ^- light. As the computer computation used the experimentally measured parameters, the good agreement established the validity of the adiabatic following model.

The adiabatic following model will be discussed in detail and the limits for the application of the model will be given. The results for both theory and experiment will be presented.

References

1. D. Grischkowsky, Phys. Rev. Letters *24*, 866 (1970).
2. A. Javan and P.L. Kelley, IEEE J. Quantum Electronics *QE-2*, 470 (1966).

EXPERIMENTAL EVIDENCE OF AN X-RAY LASER

John G. Kepros, Edward M. Eyring, and

F. William Cagle, Jr.

University of Utah, Salt Lake City, Utah

A three-stage neodymium:glass laser has been used to produce collimated x-radiation from a copper containing target. The 1.06 μm Q-switched pulse from the neodymium laser was focused by a cylindrical lens to a horizontal line traversing the thin target. The emission from the target was detected as a 0.2-mm spot on shielded, standard x-ray photographic film at distances that varied from 30 to 110 cm from the sandwich. Spot size did not increase with distance, suggesting coherence of the x-rays.

QUANTUM THEORY OF PHOTODETECTION AND COHERENCE PROPERTIES OF ELECTRONS

F. Rocca

University of Nice, Nice, France

We are interested in a completely quantum mechanical description of the photodetection mechanism, in order to determine the exact relations between electron counting statistics and the incident light.

We propose the following simple model:
We only consider the elementary processes:

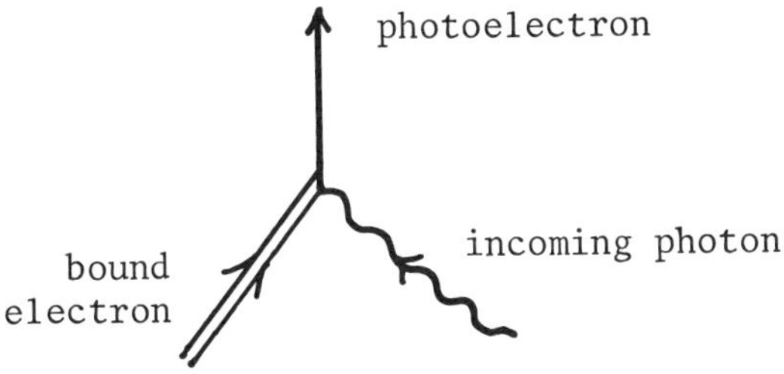

such that, when n photoelectrons are counted in the time interval $[t_o,t]$, the final state is obtained from the initial state at time t_o by n repetitions of this event.

By the standard procedure, the counting probability is written, in the general quantum theory of interacting fields:

$$P_n(t_o,t) = \mathrm{Tr}\ \{\ \rho_D(t_o)\ \rho_\gamma(t_o)\ U^{(n)}(t,t_o)^* U^{(n)}(t,t_o)\ \}$$

ρ_D and ρ_γ are the density operators, in the Dirac picture, of the detector and of the variation field, and

$$U^{(n)}(t,t_o) = \frac{(-i)^n}{n!} \int_D .. \int_D .. \int_{t_o}^{t} .. \int_{t_o}^{t} N[H(x_1)....H(x_n)]d^4x_1...d^4x_n ,$$

where N is the normal product operator and $H(x) = J^e_\mu(x)A^+_\mu(x)$ is the effective interaction Hamiltonian of the previous process. Using the P representation for $\rho_\gamma(t_o)$[1], it is sufficient to calculate $P_n(t_o,t)$ when $\rho_\gamma(t_o) = |\Phi \text{ coh}><\Phi \text{ coh}|$, with $A^+_\mu(x)|\Phi \text{ coh}> = \phi_\mu(x)|\Phi \text{ coh}>$; we have

$$P_n(t_o,t) = Tr \{\rho_D(t_o)\ U_c^{(n)}(t,t_o)^* U_c^{(n)}(t,t_o)\} , \tag{1}$$

where $U_c^{(n)}$ is obtained from $U^{(n)}$ by the substitution $A^+_\mu(n) \to \phi_\mu(n)$. Our problem is now in complete analogy with an external field problem.

Then, the total system is assumed to be in thermal equilibrium at time t_o, and it is a known result of Statistical Mechanics that the equilibrium states of a fermions system in an external field are "quasi-free" states , that is states such that[2]:

$$Tr \{\rho A_1....A_{2n}\} = \sum_{i=1}^{2n-1} (-1)^{i+1} Tr\{\rho\ A_i A_{2n}\} Tr\{\rho \prod_{\substack{j=1 \\ j\neq i}}^{2n-1} Aj\} , \tag{2}$$

where the A_i are fermion field operators.

We apply this general formula to the relation (1). In fact, it is convenient to consider the ratio P_{n+1}/P_n . We obtain after calculations:

$$\frac{P_{n+1}(t_o,t)}{P_n(t_o,t)} = \frac{1}{n+1} Tr\{\rho_D(t_o) \left| \int_D \int_{t_o}^{t} J^e_\mu(x)\phi_\mu(x)d^4x \right|^2\} + \text{other terms.}$$

If we neglect the "other terms", which arise from electron correlations effects in the detector, the solution is immediate. We have after normalization[3]:

$$P_n(t_o,t) = \frac{\lambda(t,t_o)^n}{n!} e^{-\lambda(t,t_o)} ,$$

with

$$\lambda(t,t_o) = \mathrm{Tr}\{\rho_D(t_o)\left|\int_D \int_{t_o}^{t} J^e_\mu(x)\phi_\mu(x)d^4x\right|^2\} . \qquad (3)$$

So, in this case, the "apparent statistics" given by the detection mechanism is a Poisson distribution, that is really the "true statistics" of the coherent incident field.

Nevertheless, mathematically speaking, it is possible to choose a state of the detector system such that the "other terms" cannot be ignored, and, in this case, the "apparent statistics" become quite different from the "true statistics".

From this remark, we propose a definition of coherence properties of electrons connected with the importance of the correlations effects described by the "other terms" when the fermion field interacts with an electromagnetic field. An interesting extremal situation is obtained when the fermion field is in a state such that the "other terms" are exactly zero. This state, which corresponds to a classical behavior of the detector, would be a "coherent" state of the field, by analogy with the quasi-classical behavior of a photon field in a coherent state. It is such that:

$$\langle J^{e*}_{\mu_1}(x_1)\ldots J^{e*}_{\mu_n}(x_n)J^{e}_{\mu_{n+1}}(x_{n+1})\ldots J^{e}_{\mu_{2n}}(x_{2n})\rangle =$$

$$\sum_{\sigma\varepsilon G_n} \prod_{i=1}^{n} \langle J^{e*}_{\mu}(x_i)\ J^{e}_{\mu_{\sigma i+n}}(x_{\sigma i+n})\rangle$$

where <...> stands for $\mathrm{Tr}\{\rho_D(t_0)\ldots\}$.

We can note that, for this state, the effective current operator of the fermion field plays a similar role as the field operators in a gaussian boson field.

References

1. R. J. Glauber, Phys. Rev. *6*, 2766(1963).
2. A. A. Abrikosov, L. P. Gorkov and I. E. Dzyaloshinsky, *Methods of Quantum Field Theory in Statistical Physics* (Prentice Hall, 1963).
For this writing:
F. Rocca, M. Sirugue and D.Testard, Comm. Math. Phys. *13*, 317 (1969).
3. For a previous and different derivation see R. H. Lehmberg, Phys. Rev. *167*, 1152 (1968).

STATISTICS OF LIGHT SCATTERED BY NON-GAUSSIAN FLUCTUATIONS

Dale W. Schaefer and P.N. Pusey

IBM T.J. Watson Research Center, Yorktown Heights, N.Y.

Introduction

In most scattering experiments the scattered light field represents a chaotic superposition of fields scattered from a large number of individual scatterers and therefore obeys Gaussian statistics [1]. That is, the amplitude distribution of the fluctuating scattered field is Gaussian. It is the purpose of this paper to investigate the single time-interval statistics of the scattered light when the number of scatterers is not large. Under these conditions the scattered field is not Gaussian.

The non-Gaussian nature of the field scattered from a finite number of scatterers was pointed out by Schaefer and Berne [2]. They showed that for a small number of scatterers not only is the scattered field non-Gaussian, but also the scattered intensity fluctuates on two widely different time scales. The rapid or so-called interference fluctuations arise because of the random superposition of the phases of the field scattered from each individual scatterer in the scattering volume. The slow or so-called occupation number fluctuations arise because of changes in the total number of scatterers in the scattering volume. The characteristic time of the interference fluctuations is roughly the time required for a scatterer to move (diffuse, swim, flow etc.) a wavelength of light (typically milliseconds for macromolecular solutions) whereas the characteristic time of occupation number fluctuations is the time required for a scatterer to move the dimensions of the scattering volume (typically seconds). The relative contribution of occupation number fluctuations scales as $1/\langle N\rangle$, where $\langle N\rangle$ is the average number of particles in the illuminated volume. Occupation number fluctuations are therefore significant only when the density of scatterers is very low and/or the scattering volume is very small.

Because of the wide disparity in the characteristic time of occupation number and interference fluctuations, two measurements are suggested. In the first, the sampling interval is short compared to both characteristic times, and in the second the sampling interval is short compared to the characteristic time of occupation number fluctuations but long compared to that of interference fluctuations, thus effectively averaging over the latter. We have investigated both regimes theoretically and experimentally.

Theory

Insight into the question of non-Gaussian statistics can be gained by consideration of the intensity distribution for scattering from a uniformly illuminated scattering volume. The problems of non-uniform illumination, spatial coherence, and the relation of the intensity distribution to the photocount distribution will be treated later.

If one samples the intensity of light scattered from a macromolecular solution using a sampling interval long compared to the characteristic time of interference fluctuations, but short compared to occupation number fluctuations, then the observed integrated intensity $\bar{I}(t)$ is just proportional to $N(t)$, the instantaneous number of particles in the scattering volume:

$$\bar{I}(t) = \alpha\, N(t) \quad .$$

The bar indicates an average over the rapid fluctuations. $P(\bar{I})$, the probability of observing an intensity $\bar{I}$, will directly mirror $P(N)$, the probability of finding N particles in the scattering volume. Since for independent particles $P(N)$ is Poisson [3],

$$P(\bar{I}) = \{\langle\bar{I}\rangle^{\bar{I}}/\Gamma(\bar{I} - 1)\} \;\exp(-\langle\bar{I}\rangle) \qquad (2)$$

where $<\overline{I}> = <N> = \int d\overline{I}\ \overline{I}\ P(\overline{I})$ is the average scattered intensity and Γ is the gamma function. The proportionality factor α will be taken as unity. As $<N>$ increases, $P(\overline{I})$ specializes to a Gaussian distribution and finally to a delta function distribution:

$$P(\overline{I}) \longrightarrow (<N> \longrightarrow \infty) \longrightarrow \delta(\overline{I} - <\overline{I}>) \quad . \qquad (3)$$

In Fig.1, the envelope of Eq.2 is plotted for $<\overline{I}> = 2$ ($\overline{I}$ is actually discontinous). This curve will be compared with other distributions derived below.

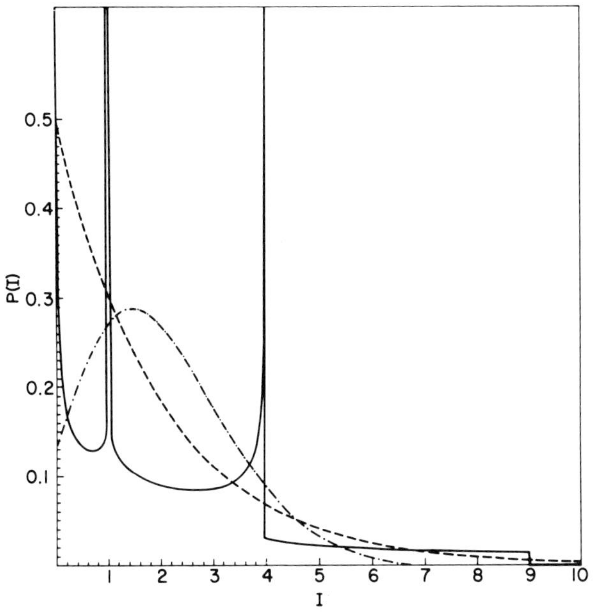

Figure 1. Intensity distributions for uniform illumination, $<I> = <N> = 2$: •—•—•—• Poisson (Eq. 2), ——— Random walk (Eq. 8), ------- Exponential (Eq. 6).

If the interference fluctuations are not averaged (either spatially or temporally), then it is necessary to take explicit account of the phase of the field scattered from each individual particle in the scattering volume. In fact, for a uniformly illuminated scattering volume, the total scattered field E is just proportional to the sum of the phase factors introduced by the individual scatterers situated at positions r_j in the scattering volume,

$$E = \sum_{j}^{N} \exp(i\underline{K}\cdot\underline{r}_j) \tag{4}$$

where K is the scattering vector [2]. If the scattering volume is large compared to 1/K, then the sum in Eq.4 is described by a two-dimensional random walk of unit step length in the complex plane, with random angles between steps [4]. For large N (many step random walk) Eq.4 leads to a Gaussian field distribution [1],

$$P(E) = (1/\pi\langle|E|^2\rangle) \exp(-|E|^2/\langle|E|^2\rangle) \quad , \tag{5}$$

or an exponential intensity distribution,

$$P(I) = (1/\langle I\rangle) \exp(-I/\langle I\rangle) \quad , \tag{6}$$

where $\langle I\rangle = \langle|E^2|\rangle = \langle N\rangle$.

If $\langle N\rangle$ is finite, deviations from Gaussian statistics for E are expected. In fact, the finite step two dimensional random walk has been solved by Kluyver [5] in terms of J_o, Bessel function of order zero. It follows from Kluyver's work that

$$P_{\langle N\rangle}(I) = (1/2) \int_o^\infty du\, u\, J_o\,(u\sqrt{I})\ \{J_o(u)\}^N \quad . \tag{7}$$

If this result is then averaged over the Poisson occupation number distribution, one obtains

$$P_{<N>}(I) = (1/2) \int_0^\infty du\, u\, J_0(u\sqrt{I})\, \exp\{<N>[J_0(u) - 1]\} \tag{8}$$

for the desired intensity distribution.

The deviation of Eq.8 from Gaussian form is very striking for small $<N>$. In Fig. 1, Eq.8 is plotted for $<I> = 2$. For comparison an exponential of the same mean is also plotted. Fig.1 not only graphically illustrates the deviation of I from exponential form for small occupation numbers, but also illustrates the change in the distribution if interference effects are averaged ($I \to \bar{I}$).

Unfortunately, sharp discontinuities shown in Fig.1 have not yet been observed. Several factors tend to preclude observation of discontinuous curves like Fig. 1. First, one measures P(n), the

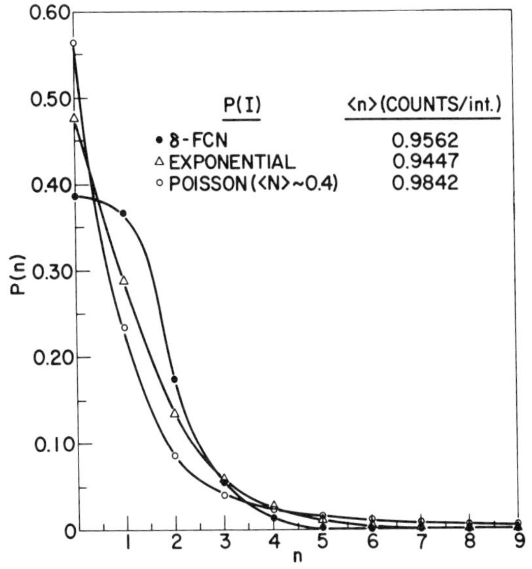

Figure 2. Measured photocount distributions: • Occupation number fluctuations, △ Gaussian concentration fluctuations, • Constant intensity illumination.

probability of observing n photocounts in the sampling interval, rather than the intensity distribution P(I)[6]. The stochastic nature of the photoemission process tends to mask the form of P(I). In fact, P(n) approaches P(I) only for $n \gg 1$. In addition, a uniformly illuminated scattering volume is difficult to achieve in practice. The more common scattering volume would be defined by a Gaussian laser beam profile clipped by a narrow slit. In this case, the illumination profile is expected to be approximately Gaussian in all three dimensions [7]. Finally, some averaging of the interference fluctuations is expected due to the finite area of any real detector. P(n) for uniform illumination can be obtained by folding Eq.8 with the Poisson photoemission probability [6]. Treatment of the Gaussian beam and finite detector, however, requires reformulation of the problem from Eq.4. Although we have not obtained closed form expressions for P(n) for non-uniform illumination and finite detector area, we have obtained the lower order factorial moments of P(n).

The factorial moments of the P(n) can be calculated from the known relation between the photocount factorial moments and the normal moments of the intensity [8]:

$$F_M = \langle n(n-1) \cdots (n-M+1)\rangle / \langle n\rangle^M = \langle I^M\rangle / \langle I\rangle^M \; . \tag{9}$$

F_M is the Mth normalized factorial moment of P(n). The moments of the intensity can be calculated from a properly weighted sum of the fields scattered from each individual scatterer. If $e(\underline{r})$ is the amplitude of the field scattered by a particle at position $\underline{r}$,

$$\langle I^M\rangle = \langle (|E|^2)^M\rangle$$

$$\langle |E|^2\rangle = \langle \sum e\{\underline{r}_i(t)\}\, e\{\underline{r}_j(t)\} \exp\{i\underline{K}\cdot(\underline{r}_i(t) - \underline{r}_j(t))\}\rangle \tag{10}$$

The observed intensity would represent a temporal average of Eq.10 over the sampling interval T and a spatial average over the range of K within the detector. Realizing that the probability of observing a particle at position $\underline{r}_j$ is proportional to the density ρ, the normalized second moment follows directly:

$$\langle I^2\rangle/\langle I\rangle^2 = (\Xi_4/\rho\Xi_2^{\,2}) + 1 + \beta \tag{11}$$

where

$$\Xi_c \equiv \int d\underline{r}\; e^c(\underline{r}) \;. \tag{12}$$

The effects of spatial and temporal averaging are contained in the factor β. The conditions under which $\beta \neq 1$ have been discussed elsewhere [9], [10]. β can be obtained from the second moment of the intensity distribution in the exponential limit (ρ large)[11]. Coherence factors which occur in moments higher than the second, however, cannot be obtained from the exponential limit. For this reason interpretation of experimental data for the higher moments is suspended. Interpretation of higher moments is possible if the sampling interval is sufficiently long or the detector sufficiently large to effectively average interference fluctuations ($\beta \to 0$). In this case, the third and fourth normalized moments become

$$\langle I^3\rangle/\langle I\rangle^3 \simeq \{\Xi_6/\rho^2\Xi_2^{\,3}\} + \{\Xi_4(3 + 6\beta)/\rho\Xi_2^{\,2}\} + 1 + 3\beta + \cdots \tag{13}$$

$$\langle I^4\rangle/\langle I\rangle^4 \simeq \{\Xi_8/\rho^3\Xi_2^{\,4}\} + \{(2+12\beta)\Xi_4^{\,2}/\rho^2\Xi_2^{\,4}\} + \{(5+12\beta)\Xi_6/\rho^2\Xi_2^{\,3}\}$$
$$+ \{\Xi_4(6+30\beta)/\rho\Xi_2^{\,2}\} + 1 + 6\beta \;+\cdots \tag{14}$$

Note that for uniform illumination Ξ_c is proportional to the scattering volume and Eqs. 11, 13 and 14 specialize ($\beta = 0$) to the moments of the Poisson distribution, Eq. 2.

For the data reported here, illumination was approximately Gaussian, i.e.

$$e(\underline{r}) = \exp\{-(x^2 + y^2)/\sigma^2\} \exp\{-z^2/\sigma'^2\}, \tag{15}$$

where σ is the width of the field profile of the Gaussian beam and σ' is the effective width of the collection slit. In this case,

$$\Xi = (V \equiv \pi^{3/2}\sigma^2 \sigma')/c^{3/2} \tag{16}$$

Experimental

Experimental work consisted of measurement of the single interval photocount statistics for a series of water solutions of polystyrene spheres. First the photocount distribution for light scattered from occupation number fluctuations was compared with an exponential source (many scatterers) and a constant intensity source. Then the photocount distribution was measured on the same sample with sampling intervals such that $\beta \to 0$ and $\beta \to 1$. These data were interpreted in terms of factorial moments.

Apparatus used here was by Pusey et. al. [12], [13] and consisted of a Krypton ion laser (λ = 5682) focused on the center of a 1cm x 1cm square cuvette by a 3cm focal length lens. Collection optics consisted of a imaging lens, aperture and slit. An ITT FW 130 was used as the detector. The photocount probability was analyzed by a modified version of the digital correlator described by Pusey et. al. [13]. Details of this device will be published elsewhere [4].

The illumination profile was measured by studying the dynamics of occupation number fluctuations for a very dilute sample ($\langle N\rangle$=.03) of .481μ polystyrene spheres under uniform translational motion. Under these conditions it can be shown [7] that the intensity autocorrelation function is homologous with a one dimensional convolution of the illumination profile with itself. Both the laser beam profile and the slit transmission profile were found to be Gaussian so that Eqs. 15 and 16 are applicable with $\sigma^2 = 280 \pm 30\mu^2$ and $\sigma'^2 = 130 \pm 20\mu^2$.

Fig. 2 shows the photocount distribution for light scattered from occupation number fluctuations. These data were taken on a very dilute sample of .481μ diameter polystyrene spheres (manufactured by Dow Chemical Co.) using a sampling interval T of 65 msec. Since the time constant for interference fluctuations is 1 msec for these particles, the rapid fluctuations are time averaged (β = .025) so that P(I) is expected to be Poisson. P(n) is then the "Poisson transform" of P(I) [6]. For comparison, P(n) is also shown for a concentrated sample of .109μ polystyrene spheres with short counting interval (β = .75). Here the intensity distribution is nearly exponential (Eq.4) and P(n) is expected to be Bose-Einstein [1]. Finally, a constant intensity source (broad band thermal) is also shown such that P(I) = δ(I - <I>) and P(n) is Poisson.

As stated above, data for P(n) can be analyzed through F_M the factorial moments of P(n). The normalized factorial moments of the curves in Fig.2 are plotted in Fig. 3. The solid line is

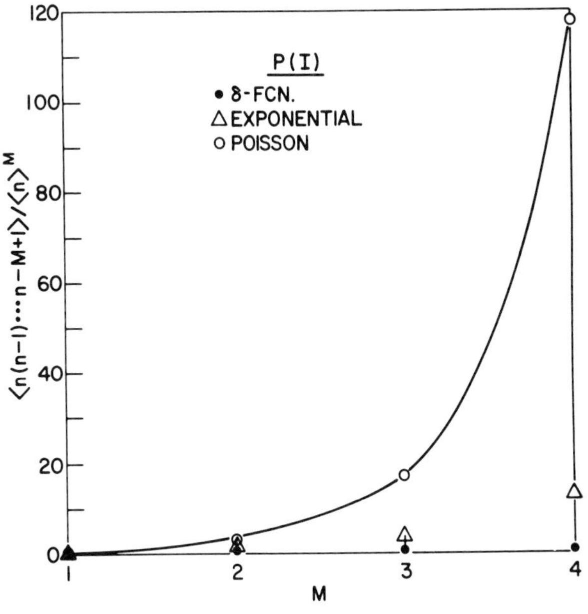

Figure 3. Factorial moments of photocount distributions in Fig. 2. Solid curve is calculated from Eqs. 11, 13 and 14 with β = .025 and <N> = ρV = .41.

calculated from Eqs. 11, 13 and 14 with $\langle N\rangle = \rho V = .41$ particles. In the case of the constant intensity source, F_M are all unity as expected. For the exponential source, the actual F_M are somewhat less that the expected M! due to incomplete spatial coherence [11].

A second set of experiments was performed on a solution of .234μ polystyrene spheres with concentration such that $\langle N\rangle \simeq 3$. Sampling intervals of 5×10^{-5} and 5×10^{-2} sec were used. Interference fluctuations are effectively averaged for the longer interval. In the short interval experiment, the sample cell was translated transverse to the laser beam at a rate of 500 μ/sec in order to reduce the time constant of the occupation number fluctuations and thereby reduce the run time required to obtain a good ensemble average. The normalized factorial moments of the measured P(n) are tabulated in Tables 1 and 2. The lowest order coherence factor β as well as higher order coherence factors were

TABLE 1

Order	F_M	Coherence Factor	Calculated $\langle N\rangle$
2	2.05	.754(β)	3.37
3	6.62	.65	-
4	29	.56	-

Table 1. Normalized factorial moments of P(n): .234μ dia spheres, $T = 5 \times 10^{-5}$ sec, $\langle n\rangle = .41$ cts/int, Run time 8180 sec.

obtained from the factorial moments of P(n) in the exponential limit ($\langle N\rangle$ large) as suggested by Cantrell and Fields [11].

As indicated above, interpretation of the higher moments for $\beta \not\to 0$ is beyond the scope of this paper. Therefore the second factorial moment of the short interval data was used to calculate $\langle N\rangle$ for this sample. The resulting value of $\langle N\rangle = 3.37$ was then used to calculate F_M for the long interval experiment. These moments as well as the measured moments are tabulated in Table 2 and the agreement is good.

TABLE 2

Order	F_M	Coherence Factor	Calculated F_M
2	1.322	0.027 (β)	1.324
3	2.18	0.00	2.16
4	4.28	0.00	4.18

Table 2. Normalized factorial moments of P(n) for occupation number fluctuations: .234μ dia spheres. $T = 5 \times 10^{-2}$ sec., $\langle n \rangle = 2.38$ cts/int, Run time = 3514 sec.

The conclusion of this study is that the field scattered from a small number of scatterers is not Gaussian. The factorial moments of photocount distribution have been shown not only to be an effective measure of the deviation from Gaussian behavior, but also to be a simple measure of the number density of macromolecules in solutions.

Acknowledgement

We are grateful to D. E. Koppel who designed the probability analyzer used here and participated in many helpful discussions. We also thank M. C. Gutzwiller who pointed out the random walk description of Eq.4. Finally we thank S. H. Koenig for critical reading of this manuscript.

References

1. F. T. Arecchi, M. Giglio and U. Tartari, Phys. Rev. *163*, 187 (1967).
2. D. W. Schaefer and B. J. Berne, Phys. Rev. Let. *28*, 475 (1972).
3. S. Chandrasekhar, Rev. Mod. Phys. *15*, 1 (1943).
4. P. N. Pusey and D.W. Schaefer, manuscript in preparation.

5. J.C. Kluyver, Konink. Akad. Wetenshap. (Amsterdam) *14*, 325 (1905).
6. E. Wolf and C.L. Mehta, Phys. Rev. Letters *13*, 705 (1964).
7. D.E. Koppel and D.W. Schaefer, Appl. Phys. Letters (to be published).
8. C.D. Cantrell, Phys. Rev. *A1*, 672 (1970).
9. D.E. Koppel, J. Appl. Phys. *42*, 3216 (1971).
10. E. Jakeman, C.J. Oliver and E.R. Pike, J. Phys. *A3*, L45 (1970).
11. C.D. Cantrell and J.R. Fields, unpublished Princeton Technical Report.
12. P.N. Pusey, D.W. Schaefer, D.E. Koppel, R.C. Camerini-Otero and R.M. Franklin, *Proceedings of the International Conference on the Scattering of Light by Fluids*, Paris, 1971 (to be published).
13. P.N. Pusey, D.E. Koppel, D.W. Schaefer and R.C. Camerini-Otero, to be published.

UNIFIED THERMODYNAMICS OF DISSIPATIVE STRUCTURES AND COHERENCE IN NONLINEAR OPTICS

R. Graham†

New York University, New York, N. Y. and

Universität Stuttgart, Stuttgart, Germany

1. Introduction

Threshold phenomena in lasers and nonlinear optics have been studied extensively by theoreticians within the last decade, mainly by means of microscopic, quantum mechanical approaches[1,2]. However, in recent times, the macroscopic nature of many of the important results like probability densities and linewidths was made apparent by showing their pronounced analogy to results of the macroscopic Landau theory of phase transitions[3]. Therefore, it became interesting to look at these threshold phenomena also from a macroscopic point of view.

It turned out that many optical systems satisfy a detailed balance principle in the stationary state. It was shown[4,5] that in these cases a generalized thermodynamic potential can be defined by the relation*

$$\Phi = \ln P, \tag{1.1}$$

(P = probability density), which has all the properties of an

†Supported in part by the National Science Foundation Grant, NSF GU-3186.

*Note a change of sign compared with Ref. [4,5].

entropy; in particular it is always increasing in time and it is decreasing for fluctuations from the stable stationary state.

By using the properties of this potential it was possible to develop a phenomenological theory of fluctuations near optical instabilities in complete analogy to the Landau theory of phase transitions[6]. According to this analogy coherence is introduced into optical fields in the same way as coherence (or order) is introduced into other systems like superconductors or ferromagnets, namely by means of a change in the symmetry of the state of the system. In the optical case, the symmetry change usually alters the symmetry of the light field amplitude or the amplitude of the atomic polarization with respect to their phase, thereby giving rise to coherence. Both the change of symmetry (instability), and the fluctuations could be described by the same potential Φ, Eq. (1.1). However, the relation of Φ to non-equilibrium thermodynamics remained still unclear.

2. Dissipative Structures

Symmetry changing instabilities far from equilibrium occur also in many other systems, e.g., in hydrodynamics and chemistry[7]. In purely dissipative systems, like in chemistry, they are particularly surprising. The order or structure which is introduced by the symmetry change has been called "dissipative structure"[8]. We will show later that it is not important whether the system is purely dissipative or not; therefore we will use the term "dissipative structures" more generally to indicate structure in a stationary non-equilibrium state.

A particularly nice example is the Bénard instability of heat conduction across a fluid layer heated from below[9] (cf. Fig. 1). It displays all typical features of a dissipative structure[10] (cf. Fig. 2). Starting from the thermal equilibrium state one generates a sequence (branch) of stationary states by applying some external force of increasing strength. In the domain where the system responds linearly to the external force, one can show[11] (cf. below) that the stationary states are always stable, if the equilibrium is stable. However, in the nonlinear domain instabilities may occur and do occur in many systems. The system then follows a new branch, which will have, in general, a different symmetry (i.e., a dissipative structure). The structure, which appears beyond the Benard instability is shown in Fig. 3.

All the above mentioned general features can also be found in optics[6]. Indeed, it is immediately clear that from a

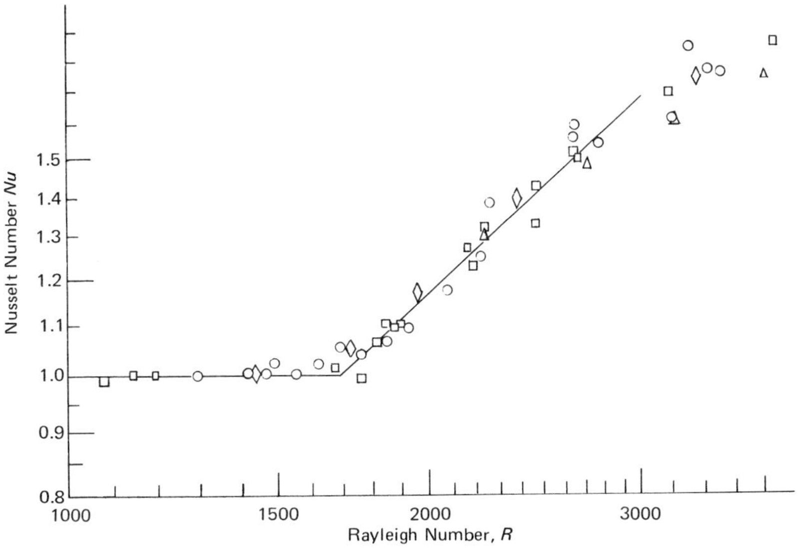

Fig. 1. Bénard instability in a liquid layer heated from below (from Chandrasekhar, Ref. 9). The Nusselt number, Nu (defined as the ratio of total heat flux and conductive heat flux) is plotted as a function of the Rayleigh number, R (which is proportional to the temperature gradient). (Reproduced by permission of Clarendon Press)

phenomenological point of view the structure appearing in the Benard problem and the coherence appearing in nonlinear optics, e.g., lasers, have many things in common (cf. Table 1). It is the aim of this paper to propose a thermodynamic theory, valid far from equilibrium, which is general enough to span the gap between fluctuations and coherence in nonlinear optics, and instabilities in hydrodynamic or chemical systems.

3. Stability and Fluctuations in the Theory of Glansdorff and Prigogine

A general thermodynamic theory for macroscopic systems far from equilibrium was developed in recent years by Glansdorff and Prigogine[7]. Their theory has three main aspects: the general time evolution of the system[10], the stability of stationary states[11], and the fluctuations from the stationary states[12].

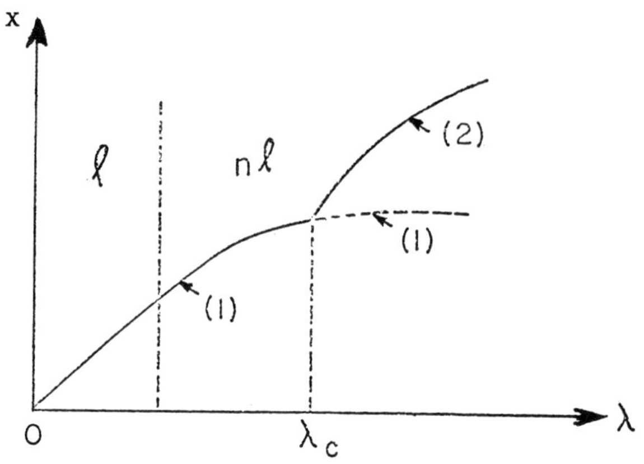

Fig. 2. Instability far from equilibrium. A variable x, describing the stationary state of a system, is plotted as a function of the strength of externally applied forces (λ). Near equilibrium (point 0) the response is linear (ℓ) and the branch (1) is stable. Far from equilibrium the response is nonlinear (nℓ), the branch (1) becomes unstable (dashed line), and a new branch (2) with a new symmetry is stable.

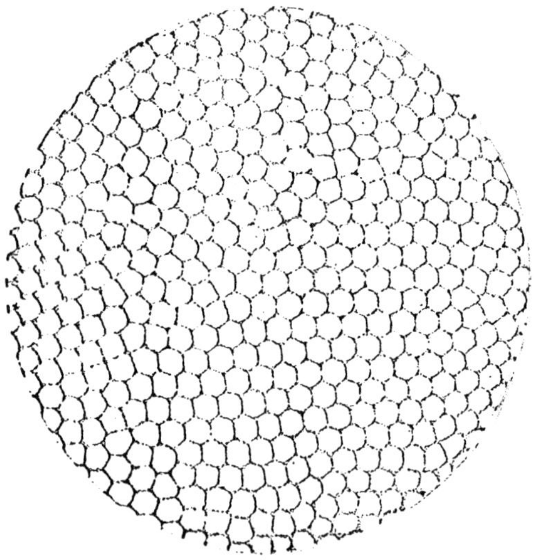

Fig. 3. Structure of a liquid layer beyond the Bénard instability (from Chandrasekhar, Ref. 9).(Reproduced by permission of Clarendon Press).

TABLE 1

Common features of lasers and dissipative structures (e.g., Bénard instability)

Stationary non-equilibrium states
Various branches of states each branch with different symmetry
Thermodynamic branch has highest symmetry (least structure)
Thermodynamic branch is stable in linear vicinity of equilibrium
Instability (threshold) of symmetric branch occurs far from equilibrium; non-linearity in deviation from equilibrium induces the instability
The state of the system changes continuously at threshold
The new branch of states has lower symmetry (more structure)
The transition is driven by fluctuations

It is, of course, tempting to apply this theory to optics as well. In fact, it has turned out to be possible to obtain a general evolution criterion and a stability criterion for non-linear optics by applying the Glansdorff-Prigogine theory. However, the application of the theory to fluctuations from the stationary state leads to difficulties. In order to explain this in more detail we have to give a brief review first.

In thermodynamic equilibrium the entropy s, the extensive variables x_i, and the intensive variables p_i are all related by Gibbs' relation*

$$ds = \sum_i p_i dx_i \tag{3.1}$$

*We use discrete variables through this paper. A continuum of variables (fields) is then a limiting case.

In linear non-equilibrium thermodynamics, valid near equilibrium, Eq. (3.1) is used to define s by taking for p_i the same thermodynamic variables as in equilibrium (e.g., the intensive variable 1/T is still associated with the extensive variable U, where T is the temperature and U is the internal energy). Glansdorff and Prigogine went one step further and extended the same definition to the domain far from thermodynamic equilibrium. Furthermore, they assume that the matrix

$$g_{ik} = \frac{\partial^2 s}{\partial x_i \partial x_k} = \frac{\partial p_i}{\partial x_k}, \text{ (negative definite)}, \tag{3.2}$$

which is negative definite in stable thermodynamic equilibrium, retains this property also far from equilibrium.*

From Eqs. (3.1), (3.2) the stability criterion of Glansdorff and Prigogine[11] immediately follows. The excess entropy $\Delta^{(2)}s$ due to a fluctuation Δx_i from the stationary state is given by

$$\Delta^{(2)}s = 1/2 \sum_{i,k} g_{ik} \Delta x_i \Delta x_k < 0. \tag{3.3}$$

$\Delta^{(2)}s$ has a well-defined sign because of (3.2). It can be used, therefore, as a Lyapunoff function in the (sufficient) stability criterion

$$\frac{\partial \Delta^{(2)}s}{\partial t} = \sum_{i,k} g_{ik} \Delta x_i \Delta \dot{x}_k \geqslant 0. \tag{3.4}$$

The quantity $\partial \Delta^{(2)}s/\partial t$ is called the excess entropy production.

The criterion (3.4) has been used to prove[11] that purely dissipative systems are stable in the linear vicintiy of thermodynamic equilibrium. In the Appendix 1 we extend this proof to systems with combined dissipative and conservative motion, like optical or hydrodynamical systems. It is then clear that structures in purely dissipative systems and coherent fields in nonlinear optics are analogous, and can only appear if the system is driven sufficiently far away from equilibrium. The important feature of Eqs. (3.3), (3.4) is that the role of a thermodynamic

*In the case of space dependent extensive and intensive variables, Eq. (3.2) is the assumption of stable local equilibrium.

potential is played by the excess entropy production rather than the excess entropy itself. This point is a direct consequence of the definition (3.1) and the assumption (3.2).

In order to incorporate the theory of fluctuations from a stationary state into their theory, Glansdorff and Prigogine[12] introduced the additional assumption that the formula

$$P \sim \exp\ (\Delta^{(2)} s) \tag{3.5}$$

which is known to hold in equilibrium, still holds far from equilibrium. However, they emphasized that (3.5) need not hold necessarily, and that their stability analysis as well as their evolution criterion remains valid regardless of the validity of (3.5).

Eqs. (3.4), (3.5) make clear why an application to optical fluctuations leads to difficulties: the fluctuations are determined as usual by the excess entropy in Eq. (3.5), whereas the excess entropy *production* plays the role of a thermodynamic potential in stability conditions (3.4). On the other hand it is known for many optical examples that both the stability and the fluctuations are governed by the *same* generalized thermodynamic potential Φ(1.1)[6]. We mention also that (3.5) with the definition (3.3) differs from some recent results for the fluctuations near the Bénard instability[13].

In Appendix 2 we show under some general assumptions that the use of Eq. (3.5) is restricted to the linear vicinity of thermodynamic equilibrium in the frame of the Glansdorff-Prigogine theory.

We conclude that a unified treatment of thermodynamics and fluctuations far from equilibrium is still missing. An attempt to give such a theory is made in the next section.

4. Unified Thermodynamics and Fluctuation Theory Far From Equilibrium:

The basic idea of our approach is the following: we use the Gibbs' relation (3.1)

$$ds' = \sum_i p_i' \, dx_i \ ; \ p_i' = \frac{\partial s'}{\partial x_i} \tag{4.1}$$

to define the entropy s' also far from equilibrium; however, we do *not* associate with x_i the *same* intensive variable p_i as in equilibrium; we rather allow the p_i' to change their meaning as we move away from equilibrium. Eq. (4.1) can, therefore, be viewed as a definition of the intensive variables p_i', as well, if s' is determined by an additional relation. Let us assume now that the (nonlinear) equations of motion for the extensive variables are known

$$\dot{x}_i = K_i(x) \tag{4.2}$$

with nonlinear functions $K_i(x)$. We, furthermore, assume that we know the matrix of Onsager coefficients K_{ik} which, *near equilibrium*, relate the irreversible part of the fluxes $\dot{x}_i$ to the forces $\partial s'/\partial x_k$:

$$\left(\frac{\partial x_i}{\partial t}\right)_{irr} = \sum_k K_{ik} \frac{\partial s'}{\partial x_k} \qquad \text{(near eq.)} \tag{4.3}$$

From the second law of thermodynamics follows that the matrix K_{ik} is non-negative. Furthermore, it satisfies Onsager's symmetry relations.

Eq. (4.3) will not hold, in general, *far from equilibrium*; however it suggests to split K_i into the two parts

$$K_i = \sum_k K_{ik} \, p_k' + J_i \quad , \tag{4.4}$$

where J_i is that part of K_i which does not satisfy (4.3). We can use now Eqs. (4.1), (4.2), (4.4) to calculate the entropy production

$$\dot{s}' = \sum_i K_i p_i' = \sum_{i,k} K_{ik} \, p_i' \, p_k' + \sum_i J_i \, p_i' \ . \tag{4.5}$$

The first term on the right-hand side is always non-negative, since K_{ik} is non-negative. The sign of the second term is unknown and will, apparently, depend on the detailed properties of the system. However, we are still free to define the entropy s' by one equation. The equation we propose is

$$\sum_i J_i \, p_i' = 0 \quad . \tag{4.6}$$

In Eq. (4.6) we postulate that the entropy s' will form a surface in phase space which contains the fluxes J_i.

Eqs. (4.1), (4.4), (4.6) in principle allow the calculation of s', J_i and p_i' if the quantities K_i and K_{ik} are given.

The choice of Eq. (4.6) is certainly not unique. Therefore, it has to be considered as our *definition* of entropy and has to be judged by its usefulness. We list now some features of this definition.*

(i) Due to (4.6) the entropy production (4.5) is always non-negative

$$\dot{s}' = \sum_{i,k} K_{ik}\, p_i'\, p_k' \geq 0 \tag{4.7}$$

This is an extension of the second law of thermodynamics. We feel that inequality (4.7) should be satisfied by any physical definition of entropy.

(ii) Due to Eq. (4.7) s' can be used as Lyapunoff function even for large fluctuations from the stationary state. The stability condition, which is now necessary and sufficient, reads

$$s' - s_0' < 0, \tag{4.8}$$

where s_0' is the entropy in the stationary state. Note that Eq. (4.8) applies even in the case of large fluctuations, where no normal mode stability analysis is possible.

(iii) If we specialize our definition to the quasi-linear vicinity of a *stationary state*, we recover a definition of entropy, which was already proposed by Lax[14]. If we specialize further to the linear vicinity of *equilibrium* we recover linear irreversible thermodynamics. In the latter region our theory coincides with the theory of Glansdorff and Prigogine. This is shown in Appendix 3.

(iv) In the case of large fluctuations and in the domain far from equilibrium our entropy s', if specialized to cases where detailed balance is present, reduces to the potential Φ, Eq. (1.1).

---

*A more detailed account will be published elsewhere.

(v) The entropy s' gives the probability of a fluctuation by the formula

$$P \sim \exp(s' - s'_0) \tag{4.9}$$

if the fluxes J_i, defined in Eq. (4.4) fulfill the relation

$$\sum_i \frac{\partial J_i}{\partial x_i} = 0 \ . \tag{4.10}$$

In this case, Eq. (4.9) can be shown to satisfy the time independent Fokker-Planck equation

$$\sum_i \frac{\partial}{\partial x_i} (-K_i P + \sum_k K_{ik} \frac{\partial P}{\partial x_k}) = 0 \ . \tag{4.11}$$

It can be shown that (4.10) is always fulfilled in the quasi-linear vicinity of a stationary state. Furthermore, (4.10) is fulfilled if detailed balance holds. Eq. (4.10) is not necessarily restricted to these conditions, however.

(vi) Whenever Eq. (4.10) is satisfied, our definition (4.6) determines s', p_i' uniquely,* since a solution of Eq. (4.11) was shown to be unique under rather general conditions[15].

For all these reasons we propose Eq. (4.6) as the definition of an entropy function which unifies thermodynamics and fluctuation theory far from equilibrium.

We note that the entropy production (4.7) has the same form as in equilibrium. Therefore, our theory becomes formally similar to the Glansdorff-Prigogine theory if we exchange

$$\Delta^{(2)}s \longrightarrow \frac{\partial s'}{\partial t}$$

$$\frac{\partial \Delta^{(2)}s}{\partial t} \longrightarrow s'$$

---

*s' is determined apart from a constant.

$$\begin{aligned} x_i &\longrightarrow p_i' \\ 1/2g_{ik} &\longrightarrow K_{ik} \end{aligned} \tag{4.12}$$

(cf. Table 2).

The scheme (4.12) and Table 2 make clear that in our approach, the entropy recovers its role as a thermodynamic potential which is played by the entropy production in the theory of Glansdorff and Prigogine. This fact makes our unification of thermodynamics and fluctuation theory possible.

5. Symmetry Restrictions on s' in the Vicinity of an Instability

The fact that s' is unique, at least if Eq. (4.10) is satisfied, has important consequences. First of all, s' must possess all symmetries in phase space, which are present in the system. Therefore, s' has a maximum or a minimum in all points in phase space which are left invariant by symmetry operations.

From (4.8) follows that only maxima of s' are stable. This simple fact explains the appearance of symmetry changing transitions (cf. Fig. 4): If the strength of the external forces acting on the system is changed without changing the symmetry in phase space, then a symmetric stable stationary state remains stable as long as s' retains it maximum property. However, for some well-defined strength of the external forces, the maximum of s' can change into a minimum. Because of the boundary conditions for s' (overall stability of the system) a *new* maximum must form in a new point (or a set of new points) of phase space. These new points will have a new symmetry, in general. Therefore, a symmetry changing transition takes place and a dissipative structure appears or disappears.

The same phenomenological picture occurs in the Landau theory of phase transitions. Therefore, expressions for s' in the vicinity of a transition may be obtained by applying the Landau theory. This procedure is very general and can be used to study instabilities and fluctuations in hydrodynamics, optics or other systems. It has been applied to fluctuations in optics in [6].

VICINITY OF STABLE EQUILIBRIUM	$\Delta^{(2)}s = 1/2 \sum_{i,k} g_{ik}\, \Delta x_i\, \Delta x_k < 0$ (stability) $\frac{\partial s}{\partial t} = \sum_{i,k} K_{ik}\, p_i\, p_k \geq 0$ (2nd law)	
Far From Equilibrium	Glansdorff and Prigogine	Approach of Section 4
Assumption Taken From Equilibrium Theory	$\Delta^{(2)}s = 1/2 \sum_{i,k} g_{ik}\, \Delta x_i\, \Delta x_k < 0$	$\frac{\partial s'}{\partial t} = \sum_{i,k} K_{ik}\, p_i'\, p_k' \geq 0$
Evolution Criterion	$\psi = 1/2 \sum_{i,k} g_{ik}\, \Delta\dot{x}_i\, \Delta\dot{x}_k < 0$	$\psi' = \sum_{i,k} K_{ik}\, \dot{p}_i'\, \dot{p}_k' \geq 0$
Stability Condition	$\frac{\partial \Delta^{(2)}s}{\partial t} \geq 0$ (sufficient)	$(s' - s_0') < 0$ (necessary and sufficient)
Fluctuations	$P \sim e^{\Delta^{(2)}s}$ (near equilibrium, quasi-linear)	$P \sim e^{(s' - s_0')}$ (also far from equilibrium, non-linear)

TABLE 2

Comparison of thermodynamic theories far from equilibrium

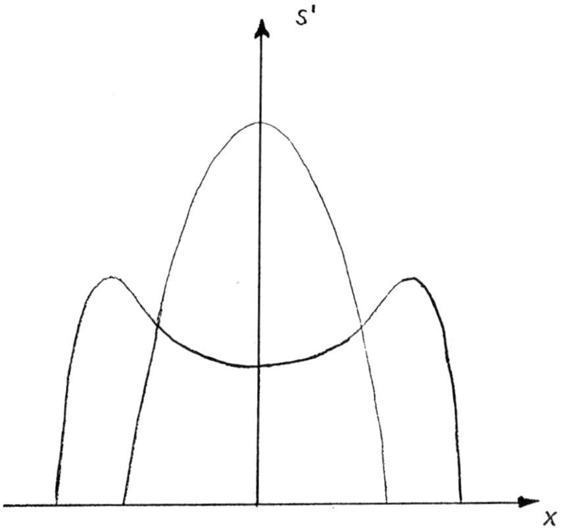

Fig. 4. The entropy s' in the vicinity of a symmetry changing instability in a one dimensional phase space (x) with inversion symmetry ($x \to -x$). The absolute value of s' is determined by the normalization of $P \sim e^{s'}$ and decreases if the system goes into a state of higher order (less symmetry).

6. Application of the Thermodynamic Theory to Nonlinear Optical Scattering

The example, considered here, has been treated from a slightly different point of view in [6]. We assume that a coherent pump field at frequency ω_p is shined into a medium where new optical fields can be generated by nonlinear scattering. For simplicity we consider a case where the fields are in a cavity and the spatial structure of the optical fields is unimportant. Furthermore we assume linearly polarized fields. The internal energy U of the optical medium can then be written in the form (cf. [16])

$$U = \hbar\gamma \left(\beta_p \Omega^* (\beta_i^*) + \beta_p^* \Omega(\beta_i)\right) \quad , \tag{6.1}$$

where γ is a coupling constant proportional to a nonlinear susceptibility, β_p is the complex amplitude of the field at the pump frequency ω_p and β_i are the complex amplitudes of the fields generated by nonlinear scattering. We consider scattering processes where the function Ω depends on the amplitudes β_i only, whereas the function Ω^*, in the same way, depends on the β_i^* only. For usual scattering processes Ω is a simple product of field amplitudes. For parametric scattering, e.g., we have to put $\Omega=\beta_1\beta_2$, for subharmonic scattering we have $\Omega=\beta_1^2$, etc. The coupled mode equations for the field amplitudes take the form:

$$\dot{\beta}_i = -\kappa_i\beta_i - i\gamma\,\beta_p \frac{\partial\Omega^*}{\partial\beta_i^*} \tag{6.2}$$

and

$$\dot{\beta}_p = -\kappa_p\beta_p - i\gamma\,\Omega + F_p \quad . \tag{6.3}$$

In derivatives we treat β_i, β_i^* formally as independent variables. κ_i are the damping constants of the various modes. We take $\kappa_i = \kappa$. Eq. (6.3) is the equation for the amplitude at the pump-frequency ω_p. F_p is an externally applied pump field, κ_p is the damping constant. We assume that the amplitude β_p adjusts itself quasi-instantaneously ($\kappa_p \gg \kappa$) to the external pump field F_p and to the field amplitudes of the generated modes (saturation) and put

$$\beta_p = \frac{F_p}{\kappa_p} - \frac{i\gamma}{\kappa_p}\Omega \ . \tag{6.4}$$

Therefore, the scattered field amplitudes obey the equations

$$\dot{\beta}_i = -\kappa\beta_i - i\frac{\gamma F_p}{\kappa_p}\frac{\partial\Omega^*}{\partial\beta_i^*} - \frac{\gamma^2}{\kappa_p}\frac{\partial|\Omega|^2}{\partial\beta_i^*} \ , \tag{6.5}$$

where we used the fact that Ω does not depend on β_i^*.

The field amplitudes are now adopted as our "extensive" variables* x_i. The functions K_i in Eq. (4.2) are then given by the right-hand side of Eq. (6.5). In order to apply our theory we still have to determine the Onsager coefficients K_{ik}. We obtain them by specializing Eq. (6.5) to equilibrium and comparing it with the form

*The term "extensive" here loses its original meaning. The variables x_i of section 3 and 4 need not be proportional to the volume of the system. They only have to satisfy "balance equations" like Eq. (4.2).

$$\dot{\beta}_i = \sum_k K_{ik} \frac{\partial s'_{eq}}{\partial \beta_k^*} \quad . \tag{6.6}$$

The equilibrium entropy s'_{eq} of the set of harmonic field amplitudes is obtained from $s'_{eq}=\ln P_{eq}$ and is given by the quadratic expression

$$s'_{eq} = - \sum_i \frac{|\beta_i|^2}{\langle|\beta_i|^2\rangle_{eq}} + \text{const.}, \tag{6.7}$$

where $\langle|\beta_i|^2\rangle_{eq}$ are the average field intensities in equilibrium. If we are only interested in quantum fluctuations we may put, according to semiclassical theory,

$$\langle|\beta_i|^2\rangle_{eq} = 1/2 \quad . \tag{6.8}$$

Inserting (6.7) with (6.8) in (6.6) and comparing with (6.5) specialized to equilibrium (and to small amplitudes) we obtain

$$K_{ik} = \frac{\kappa}{2} \delta_{ik} \quad . \tag{6.9}$$

We can now apply Eq. (4.6) which takes the form

$$\sum_i (J_i \frac{\partial s'}{\partial \beta_i} + J_i^* \frac{\partial s'}{\partial \beta_i^*}) = 0 \quad . \tag{6.10}$$

According to Eq. (4.4), J_i is given by

$$J_i = -\kappa\beta_i - i \frac{\gamma F_p}{\kappa_p} \frac{\partial \Omega^*}{\partial \beta_i^*} - \frac{\gamma^2}{\kappa\kappa_p} \frac{\partial |\Omega|^2}{\partial \beta_i^*} - \frac{\kappa}{2} \frac{\partial s'}{\partial \beta_i^*} \quad . \tag{6.11}$$

For our present example, Eq. (6.10) can be solved by putting

$$J_i = 0 \quad . \tag{6.12}$$

This is a special feature of our example, introduced by putting

$\kappa_i = \kappa$. This condition is known to introduce detailed balance into the stationary state in the present example[6]. From (6.11) and (6.12) we obtain

$$s' = -2\sum_i |\beta_i|^2 + 4\mathrm{Im}\left(\frac{\gamma F_p}{\kappa\kappa_p}\Omega^*\right) - \frac{2\gamma^2}{\kappa\kappa_p}|\Omega|^2 + \mathrm{const.} \qquad (6.13)$$

Because of (6.12), the condition (4.10) is satisfied by the present example. The probability distribution of fluctuations is, therefore, obtained from Eq. (4.9). This result is in agreement with an earlier result obtained by the solution of a Fokker-Planck equation [6].

The entropy (6.13) governs now the fluctuations and the stability in the stationary state. If special expressions for Ω are inserted in Eq. (6.13), one can easily check that, in the vicinity of a symmetry changing transition, s' behaves like expected from the symmetry arguments in section 5 (cf. Figs 5, 6). This was discussed in some detail in [6].

The above example has been carried through so easily because of the special feature (6.12). The more general case $J_i \neq 0$ can still be treated explicitly without difficulty if either detailed balance holds or if the system is quasi-linear. However, we have not yet treated an example where neither detailed balance nor

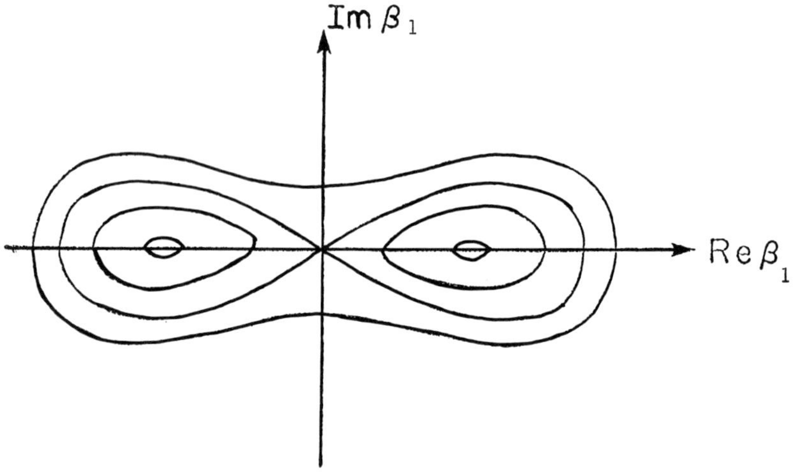

Fig. 5. Contour line plot of the entropy s' for subharmonic generation above threshold (schematic). The stationary states break the symmetry $\beta_1 \to -\beta_1$.

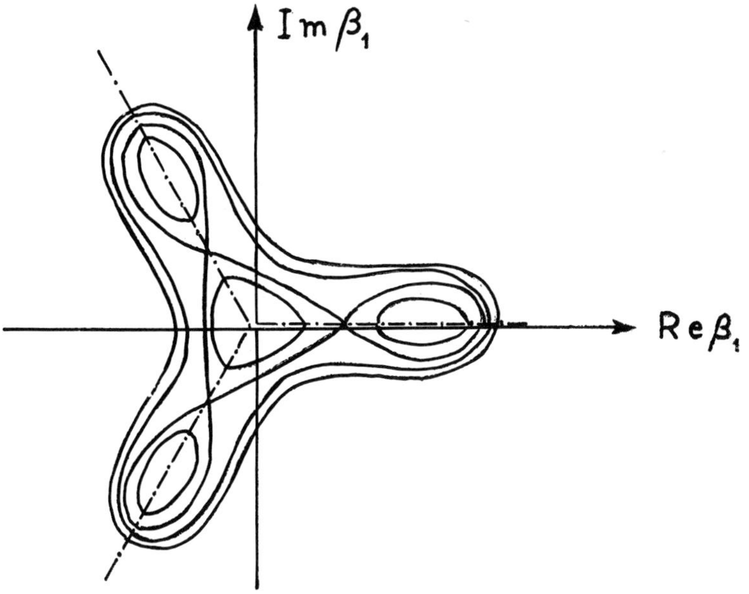

Fig. 6. Contour line plot of the entropy s' for third order subharmonic generation above threshold (schematic). The entropy has four maxima, three of which break the symmetry $\beta_1 \to \beta_1 \exp(2\pi i/3) \to \beta_1 \exp(4\pi i/3)$ The state $\beta_1 = 0$ is stable against small fluctuations, but unstable against large fluctuations.

quasi-linearity holds. The explicit treatment of such an example would be very helpful for judging the final usefulness of Eq. (4.6).

Acknowledgement

The author wants to thank K. Kaufmann, Universität Stuttgart, for very fruitful discussions, and is grateful to the Physics Department of New York University for its hospitality.

Appendix I: Stability of stationary states near thermodynamic equilibrium.

In the linear vicinity of equilibrium one may write the equations of motion for the quantities Δx_i (cf. (3.3)) in the form

$$\Delta\dot{x}_i = \sum_k K_{ik} \frac{\partial \Delta^{(2)} s}{\partial \Delta x_k} + \left(\frac{\partial \Delta x_i}{\partial t}\right)_{rev} . \tag{I.1}$$

K_{ik} are Onsager's coefficients. Their matrix is non-negative. The second term on the right-hand side describes a reversible motion of Δx_i only. Therefore, this part has to satisfy

$$\left(\frac{\partial \Delta^{(2)} s}{\partial t}\right)_{rev} = \sum_i \frac{\partial \Delta^{(2)} s}{\partial \Delta x_i} \left(\frac{\partial \Delta x_i}{\partial t}\right)_{rev} = 0 . \tag{I.2}$$

As a result we obtain for the excess entropy production

$$\frac{\partial \Delta^{(2)} s}{\partial t} = \sum_{i,k} K_{ik} \frac{\partial \Delta^{(2)} s}{\partial \Delta x_i} \frac{\partial \Delta^{(2)} s}{\partial \Delta x_k} \geq 0 . \tag{I.3}$$

Hence, the stability condition (3.4) is satisfied.

Appendix II: Restriction of Eq. (3.5) to near equilibrium.

We assume that in the quasi-linear vicinity of a stationary state, the equations of motion for the fluctuations Δx_i may be written in the form

$$\Delta\dot{x}_i = \sum_k B_{ik} \Delta x_k + F_i(t) , \tag{II.1}$$

where B_{ik} is a constant matrix and $F_i(t)$ are rapidly fluctuating forces with the properties

$$\langle F_i \rangle = 0$$

$$\langle F_i(t) F_k(t+\tau) \rangle = 2K_{ik} \delta(\tau) . \tag{II.2}$$

We assume furthermore that the forces F_i are Gaussian. Near equilibrium the coefficients K_{ik} reduce to Onsager's coefficients due to the fluctuation dissipation theorem[5]. The matrices B_{ik} and K_{ik} are in this region connected through Einstein's relation

$$B_{ij} = (Kg)_{ij} \quad , \tag{II.3}$$

where g_{ij} is defined in (3.2). This may be seen easily by comparing Eqs. (II.1) and (I.1). Far from equilibrium Eq. (II.3) does not hold in general. The probability distribution obtained from (II.1) and (II.2) is well-known[14,17]. It takes the form

$$P \sim e^{\Phi} \quad , \tag{II.4}$$

with

$$\Phi = 1/2 \sum_{i,k} \Delta x_i \; U_{ik} \; \Delta x_k \quad , \tag{II.5}$$

where the matrix U is given by the relation

$$B\,U^{-1} + U^{-1}\,B^{T} = 2K \quad . \tag{II.6}$$

In the linear vicinity of equilibrium Eq. (II.3) holds and we obtain from (II.6):

$$U = g \quad ; \quad \Phi = \Delta^{(2)}s \quad . \tag{II.7}$$

Therefore, $P \sim \exp(\Delta^{(2)}s)$ holds.

Far from equilibrium Eq. (II.3) does not hold, in general, and we obtain

$$U \neq g \quad ; \; \Phi \neq \Delta^{(2)}s \quad \text{(in general)} \quad . \tag{II.8}$$

Therefore, in this region P is given by $P \sim \exp(\Phi)$ and not by $P \sim \exp(\Delta^{(2)}s)$.

Appendix III: The entropy s' in the vicinity of a stationary state.

By quasi-linearizing Eq. (4.2) around the stationary state Eq. (4.2) takes the form

$$\Delta\dot{x}_i = \sum_k B_{ik}\,\Delta x_k \quad ; \quad B_{ik} = \left(\frac{\partial K_i}{\partial x_k}\right)_{st} \quad . \tag{III.1}$$

We expand the entropy around the stationary state

$$\Delta^{(2)} s' = 1/2 \sum_{i,k} g'_{ik}\,\Delta x_i\,\Delta x_k \quad , \tag{III.2}$$

which yields

$$\Delta p'_i = \sum_k g'_{ik}\,\Delta x_k \quad . \tag{III.3}$$

Using the definition (4.4) for J_i,

$$\Delta J_i = \sum_k (B_{ik} - \sum_j K_{ij}\,g'_{jk})\,\Delta x_k \quad , \tag{III.4}$$

the condition (4.6) takes the form

$$\sum_i \sum_k (B_{ik} - \sum_j K_{ij}\,g'_{jk}) \sum_\ell g'_{i\ell} \Delta x_k\,\Delta x_\ell = 0 \quad . \tag{III.5}$$

Comparing coefficients of $(\Delta x_k\,\Delta x_\ell + \Delta x_\ell\,\Delta x_k)$ we get

$$B^T g' + g' B = 2 g'\,K\,g' \tag{III.6}$$

or

$$B\,g'^{-1} + g'^{-1}\,B^T = 2K \quad . \tag{III.7}$$

By comparison with Eq. (II.6) we find

$$g' = U \tag{III.8}$$

$$\Delta^{(2)} s' = \Phi \quad . \tag{III.9}$$

Therefore, Eq. (4.9) is always valid in the quasi-linear vicinity of a stationary state. The definition of entropy (III.9) was already proposed by Lax[14].

In the vicinity of equilibrium $\Delta^{(2)}s' = \Phi$ reduces to $\Delta^{(2)}s$

$$\Delta^{(2)}s = \Delta^{(2)}s' \quad \text{(near equilibrium)} \qquad \text{(III.10)}$$

This is shown in Appendix 2.

References

1. See, e.g., H. Haken, *Encyclopedia of Physics*, Vol. 25/2c, (Springer, New York, 1970); H. Risken, *Progress in Optics*, Vol. 8, ed. E. Wolf (North Holland, Amsterdam, 1970); M. Lax, *Statistical Physics, Phase Transitions and Superfluidity*, ed. H. Chretian et al., (Gordon and Breach, New York 1968); M. Scully and W. E. Lamb, Jr., Phys. Rev. *159*, 208 (1967); *166*, 246 (1968).
2. R. Graham, Z. Physik *210*, 319 (1968); *211*, 469 (1968); R. Graham and H. Haken, Z. Physik *210*, 276 (1968); R. Graham, Phys. Lett. *32A*, 373 (1970); D. R. White and W. H. Louisell, Phys. Rev. A *1*, 1347 (1970).
3. R. Graham and H. Haken, Z. Physik *213*, 420 (1968); *237*, 31 (1970); V. DeGiorgio and M. O. Scully, Phys. Rev. A *2*, 1170 (1970); R. Graham, *Proceedings of the International Symposium on Synergetics*, ed. H. Haken, (Teubner, Stuttgart, 1972).
4. R. Graham and H. Haken, Z. Physik *243*, 289 (1971).
5. R. Graham and H. Haken, Z. Physik *245*, 141 (1971).
6. R. Graham, *Springer Tracts in Modern Physics*, to be published (Springer, New York 1972).
7. P. Glansdorff and I. Prigogine, *Thermodynamic Theory of Structure, Stability, and Fluctuations*, (Wiley Interscience, New York, 1971).
8. I. Prigogine and G. Nicolis, J. Chem. Phys. *46*, 3542 (1967); I Prigogine and R. Lefever, J. Chem. Phys. *48*, 1695 (1968); R. Lefever, J. Chem. Phys. *49*, 4977 (1968).
9. S. Chandrasekhar, *Hydrodynamic and Hydromagnetic Stability*, (Clarendon Press, Oxford 1961).
10. P. Glansdorff and I. Prigogine, Physica *30*, 351 (1964).
11. P. Glansdorff and I. Prigogine, Physica *46*, 344 (1970).
12. I. Prigogine and P. Glansdorff, Physica *31*, 1242 (1965).
13. V. M. Zaitsev and M. I. Shliomis, Sov. Phys. JETP *32*, 866 (1971).
14. M. Lax, Rev. Mod. Phys. *32*, 25 (1960).

15. J. L. Lebowitz and P. G. Bergmann, Annals of Physics *1*, 1 (1957).
16. N. Bloembergen, *Nonlinear Optics*, (Benjamin, New York 1965); P. S. Pershan, Phys. Rev. *130*, 919 (1963).
17. M. C. Wang and G. E. Uhlenbeck, Rev. Mod. Phys. *17*, 323 (1945).

THE APPROACH TO THERMAL EQUILIBRIUM IN SYSTEMS OF COUPLED QUANTUM OSCILLATORS INITIALLY IN A GENERALIZED GAUSSIAN STATE

Martine Rousseau

Université de Paris, Centre d'Orsay, France

The class of generalized Gaussian fields was recently exposed by B. Picinbono and myself[1]. We have shown that the chaotic field is no more than a particular case among other Gaussian fields called generalized Gaussian fields. Among them, the real Gaussian field, or Gaussian field with real amplitude, is especially interesting for many reasons and it has been realized by Perrot and Bendjaballah[5] in our laboratory.

It can be defined by this classical relation:

$$I(t) = x^2(t) \tag{1}$$

where x(t) is a real Gaussian random function. From this relation it follows in particular that the bunching effect is 1.5 times greater for the real Gaussian field than for the chaotic one. The authors cited before[5] have verified many properties related to the real Gaussian field: counting distributions, time intervals, and so on.

Now we shall study some interactions of this real Gaussian field with matter by considering field and matter as a system of quantum oscillators.

The temporal evolution of systems of coupled quantum oscillators has been treated by many authors, especially Louisell, Mollow[4], Robl and Raitford[2].

For solving such a problem we need two sets of parameters:

1. The Hamiltonian

2. The initial conditions.

As it was done by others for reasons of facilities, we limit our problem to quadratic Hamiltonians. That is to say, if a_j^+ and a_k are the creation and annihilation operators for the oscillators j and k, respectively, we shall consider Hamiltonians such as,

$$H = \sum_{i,j} a_i \, a_j \, \alpha_{ij}(t) + \sum_{i,k} a_i^+ \, a_k \, \gamma_{ik}(t) + \sum_i \delta_i(t) a_i + \text{H.C.} \qquad (2)$$

Such Hamiltonians, in fact, cover almost all the cases encountered, even in nonlinear optics, nowadays.

The P-representation of each quantum oscillator can be determined from the characteristic functions $\chi_j(\xi,\eta,t)$

$$\chi_j(\xi,\eta,t) = \mathrm{Tr}_j\{\rho(t) \, e^{-i(\xi Q_j+\eta P_j)}\} = \mathrm{Tr}\{\rho(o) \, e^{-i(\xi Q_j(t) + \eta P_j(t))}\} \qquad (3)$$

with

$$Q = (a+a^+)\frac{1}{\sqrt{2}} , \qquad P = i(a-a^+)\frac{1}{\sqrt{2}} ,$$

since it is only the Fourier transform of $\chi_j(\frac{\xi}{\sqrt{2}}, \frac{\eta}{\sqrt{2}}, t)$.

It has been shown by Mollow[4] and Robl[2] that, for such Hamiltonians with $k \neq j$, initially coherent states remain coherent, and initially chaotic oscillators remain in a chaotic state.

The initial state in our case is a real Gaussian one. More precisely, we shall concentrate our attention upon one oscillator in this system, whose initial P-representation is a real Gaussian function of α:

$$P(\alpha,o) = \frac{1}{\sqrt{2\pi}\sigma_x} \exp(-\alpha_x^2/2\sigma_x^2) \, \delta(\alpha_y) . \qquad (4)$$

We suppose that the other oscillators in the system are initially independent and Gaussian (chaotic or real Gaussian):

$$\left.\begin{array}{l} P(\{\alpha_j\},0) = \pi_j\ P_j(\alpha_j,0) \\ \\ P_j(\alpha_j,0) \quad \text{Gaussian functions} \ . \end{array}\right\} \qquad (4')$$

We can present two conclusions about this problem:

1) *The Gaussian character is maintained.*

2) *The interactions thermalize the system.*

The first one follows immediately from the calculus since the P-representation P_j for each oscillator is only the Fourier transform of

$$\chi_j\left(\frac{\xi}{\sqrt{2}}, \frac{\eta}{\sqrt{2}}, t\right) \quad .$$

The second one is not evident at all and perhaps depends upon the model chosen for interactions, but it holds true for the two following examples we have treated.

In particular, we apply the general calculus to the following problems:

The first example concerns the forced oscillation of a mode ω_1 damped by a reservoir and excited by a pump ω. The Schroedinger Hamiltonian of this system is[2]

$$H_S(t) = \hbar\omega_1(a_1^+ a_1, + \frac{1}{2}) + \hbar \sum_\lambda \omega_\lambda(a_{\lambda\mu}^+ a_{\lambda\mu} + \frac{1}{2})$$
$$- \hbar g \sum_{\lambda\mu} (a_1^+ a_{\lambda\mu} + a_1 a_{\lambda\mu}^+) - \hbar G(a_1^+ e^{-i\omega_0 t} + a_1 e^{i\omega_0 t})$$

where g (resp. G) is the coupling constant between the oscillator ω_1 and the reservoir (resp. the harmonic force).

Let us consider a harmonic oscillator ω_1, which has initially a real Gaussian amplitude. When it is damped by a set of harmonic oscillators, the oscillator ω_1 becomes chaotic whether the damping oscillators are initially real Gaussian or chaotic ones.

The forced oscillation of a Gaussian mode damped by a reservoir is a microscopic explanation of thermalization phenomenon that obeys the law of entropy growth. In particular, this effect may represent the change in the statistical properties of a radiation field after it propagates through different media. Thermalization of solar beams, after they go across the atmosphere, is an analogous phenomenon, although the initial states are not real Gaussian.

The second example concerns the parametric amplification. Generally, it can be described by the following Hamiltonian:

$$H(t) = \frac{\hbar}{2}\omega_a\, a^+(t)\, a(t) + \frac{\hbar}{2}\omega_b\, b^+(t)\, b(t) - \hbar K(a^+(t) b^+(t) e^{-i\omega t})$$

$$+ \hbar \sum_{\lambda\mu} \omega_\lambda\, a^+_{\lambda\mu}\, a_{\lambda\mu} - \hbar \sum_{\lambda\mu} a^+_{\lambda\mu}\, a_{\lambda\mu} \quad .$$

Two oscillators ω_a and ω_b are coupled by a classical driving force with the sum frequency $\omega = \omega_a + \omega_b$, which exchanges one photon with each oscillator, and two photons with both.

We have shown that when some conditions are fulfilled concerning the mean number of photons in the modes A and B, the mode ω_a becomes chaotic whether it is damped by a reservoir or not.

In particular, if the modes are initially empty, that is to say, if there is no photon, each one becomes chaotic. This is the parametric fluorescence phenomenon.

A particular case of such a parametric amplification with two modes is the stimulated Raman effect: an initially real Gaussian field becomes a scattered chaotic light. In the perspective of an experimental study of the statistical properties of the coherent Raman effect, it should be interesting to introduce a resonant, real Gaussian field in the atomic medium and to study the statistical properties of the stimulated light scattered by the medium.

It is also possible to have parametric amplification when one oscillator ω_a makes two-photon transitions thanks to a pump $\omega = 2\omega_a$. This is a degenerate case[2] versus the preceding one because there is only one oscillator. Here the field becomes real Gaussian except for a phase factor; we call it a pseudo-real Gaussian field. This result seems to be due to two-photon interactions between ω_a and the pump. In all other cases each oscillator interacts with others with only one photon.

Thus thermalization appears to be due to one-photon interactions. Recently, Christine Benard[3] showed that, if interactions in the source are not taken into account, a chaotic source cannot be described as an assembly of atoms emitting independent wave packets. It would be interesting to know if interactions similar to those discussed here, would "thermalize" the electromagnetic field described by Benard.

References

1. B. Picinbono and M. Rousseau, Phys. Rev. *A1*, 635 (1970).

2. H.R. Robl, Phys. Letters *165*, 1426 (1968); M.T. Raitford, Phys. Rev. *A2*, 1541 (1970).

3. C. Benard, Thèse de Doctorat d'Etat, Université de Paris-Sud (1972), and this volume, page no. 879.

4. B.R. Mollow, Phys. Rev. *162*, 1256 (1967).

5. C. Bendjaballah and F. Perrot, Optics Commun. *3*, 21 (1971).

COMPARISON BETWEEN INCOHERENT AND CHAOTIC SOURCES : CONTRIBUTION TO THE STUDY OF THE ORIGIN OF BUNCHING EFFECT

Christine Bénard

Université de Paris, Orsay, France

We propose a theoretical model of an atomic source, which we call incoherent source. Such a source consists of a great number of independent two-level atoms, excited to their upper state by broadline excitation. It emits spontaneously an incoherent optical field. We compare this source with chaotic sources and show that those two types of sources cannot be identified. The differences in the properties of the two emitted fields are important enough to be observed experimentally.

Wave Packet Formalism

For performing a comparison between incoherent and chaotic sources, we use the wave packet formalism [1][2][3]. This is the reason why we shall first give a quick review of this formalism. It consists in associating with every particle of a system of bosons or fermions a state $|\phi_i\rangle$ of the Hilbert space H, and building by symmetrization or antisymmetrization the state $|n(\{r_i\})\rangle$ of the Fock space, describing n particles of the system. If the states $|\phi_i\rangle$ were orthogonal, this construction would be nothing but the usual construction of the Fock space from the one-particle-state-space H. But, the states $|\phi_i\rangle$ are not assumed to be orthogonal. In fact they are defined in the following way: every $|\phi_i\rangle$ is deduced from a given state $|\phi\rangle$ by a translation T_i (characterized by a vector r_i) in the volume V of the space where the considered system is confined. We shall say that the states $|\phi_i\rangle$ are one-particle-wave-packets: the impulsion of a particle in such a state is not defined completely.

We show the following result, which is of great importance in our present study: the states as well as the operators of the Fock space are expressed in a *unique* way in terms of the state $|n(\{r_j\})>$, provided that certain symmetry conditions are prescribed.

Theoretical Model of an Incoherent Source

Let us consider the electromagnetic field emitted spontaneously by an assembly of independent two-level atoms in a *given direction* (angular momenta of the emitted photons are not taken into account and angular correlations between the emitted photons are not considered either). Moreover, let us assume that the atoms are excited by very short pulses, stochastically independent (their width in time is very small compared with the width of the upper atomic state). Provided that only a given emission direction is considered, the different excitation time-instants θ_j are easily related to the vectors r_j defining the translations T_j (in a one dimension space L),

$$r_j = X_j - c\theta_j ,$$

where X_j is the position of the atom j, and c the light velocity. The photon emitted by the atom j can be described by a one-particle wave-packet $|\phi_j>$. Taking into account the indiscernability and symmetry of photons, the total emitted field is described by the density matrix ρ_I, the expression of which is unique,

$$\text{(I)} \quad \rho_I = \sum_{n=0}^{+\infty} \frac{J_n}{n!} \int_{L^n} |n(\{r_j\})\rangle\langle n(\{r_j\})| \prod_{j=1}^{n} dr_j$$

where J_n characterizes a Poisson process in L,

$$J_n = e^{-\Omega L}\,\Omega^n ,$$

Ω being the linear density of the process.

Chaotic Sources

By definition, we shall call a chaotic source, a source emitting a field described by the density matrix

$$\rho_c = \sum_{\{n_k\}} \prod_k \frac{\langle n_k\rangle^{n_k}}{(1+\langle n_k\rangle)^{1+n_k}} \,|\{n_k\}\rangle\langle\{n_k\}|.$$

The vector $|\{n_k\}\rangle$ describes a state where there are n_k particles in mode k, and $\langle n_k\rangle$ is the mean number of particles in this mode.

By rewritting ρ_c in the wave packet formalism [3], we obtain the unique expression

$$\text{(II)}\quad \rho_c = \sum_{n=0}^{+\infty} c_n \int_{L^n} W_n[\{r_j\}]\, |n(\{r_j\})\rangle\langle n(\{r_j\})| \prod_{j=1}^{n} \frac{dr_j}{L}$$

where

$$c_n = \left[\prod_k (1+\langle n_k\rangle)^{-1}\right] \frac{\sum_k \langle n_k\rangle(1+\langle n_k\rangle)^{-1}}{n!}$$

and

$$W_n[\{r_j\}] = \sum P_\alpha \prod_{j=1}^{n} \langle\phi_j|\phi_{\alpha j}\rangle \quad .$$

The symbol $\sum P_\alpha$ indicates a sum over all the permutation $\{\alpha\}$ of n elements.

By comparing (I) and (II) we see that the field emitted by a chaotic source is different from the one emitted by an incoherent source.

More precisely, we can compare the bunching effect [4] obtained in these two cases, assuming that the emitted field is weak (the degeneracy parameter is small compared to one, as it is the case for natural sources such as Hg-lamps for instance). The spatial second-order coincidence probabilities (probabilities for detecting a photon in a given point and another one, at the same time, in a point at the distance a) $P^{CB}{}_2(a)$ and $P^{IB}{}_2(a)$ corresponding to the chaotic and incoherent cases respectively are drawn on figure 1, in terms of a/ℓ_c, where ℓ_c is the coherence length of the considered fields. (The spectrum of the field is assumed to be lorentzian).

From considering this figure, several conclusions can be drawn:

1) In the incoherent case, a bunching effect is obtained, the origin of which *is uniquely due to the properties of symmetry of photons*. We must emphazise the importance of this result: in fact the origin of the bunching effect is still not very clear as shown

by many contradictory publications on the subject. Some of them consider the bunching effect, for chaotic sources, as a consequence of correlations in the emitting source [3][5], and some other as a consequence of the symmetry of photons [2][6].

Fig. 1

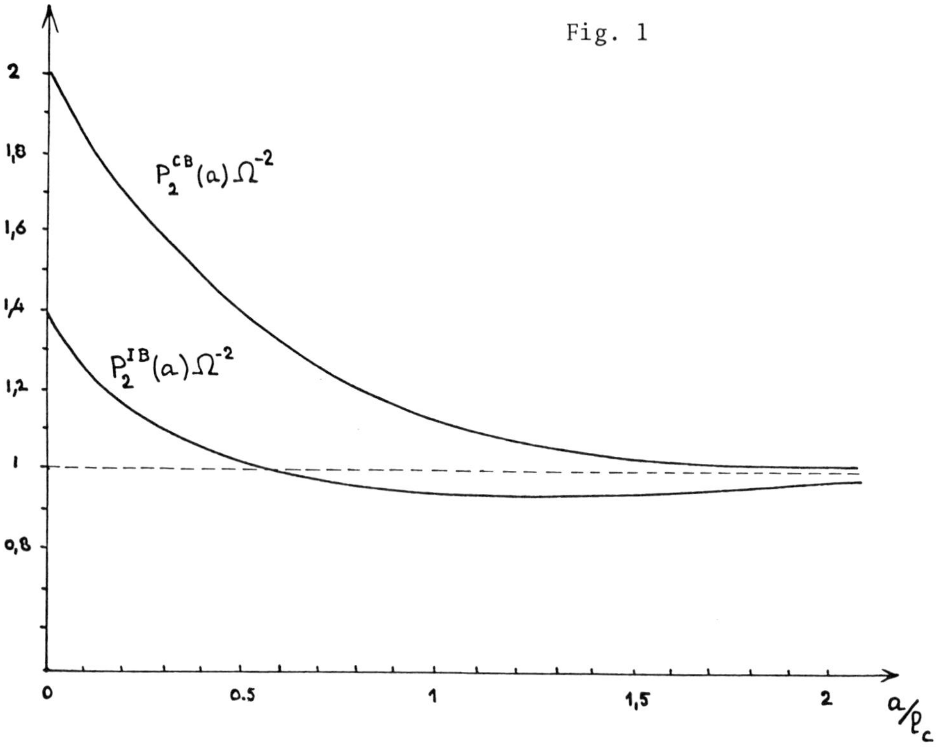

2) The incoherent bunching effect is smaller than the chaotic one.

3) The difference between chaotic and coherent bunching effect can be observed experimentally: in fact the coincidence experiments performed by D.B. Scarl [6] show that a precision high enough to distinguish between these two cases can be attained.

Moreover, it seems possible to build an incoherent source by exciting an assembly of independent atoms (a low pressure gas) by short and low stochastic light pulses.

Thus our theoretical results might be experimentally checked.

References

1. M.L. Goldberger and K.M. Watson, Phys. Rev. *137*,B 1396,(1965).

2. C. Bénard, C.R.A.S. *268*, p. 1504, (1969).
C. Bénard, in "Quantum Optics" edited by S.M. Kay and A. Maitland (Academic Press, London, New York ,1970).
C. Bénard, Phys. Rev. *A-2*, p. 2140, (1970).

3. R.J. Glauber, in "Quantum Optics" edited by S.M. Kay and A. Maitland (Academic Press, London, New York, 1970).

4. L. Mandel and E. Wolf, Rev. Mod. Phys.*37* , p. 231, (1965).

5. R.H. Dicke, Phys. Rev. *93*, p. 99, (1954).

6. D.B. Scarl, Phys. Rev. *175*, p. 1661, (1968).

SELF-PULSING IN LASER AMPLIFICATION OF BROADBAND NOISE

M.M. Miller* and A. Szöke

Tel-Aviv University, Ramat-Aviv, Israel

1. Introduction

The development of high gain gas laser amplifiers[1] has renewed interest in the problem of non-linear amplification of broadband noise. Such signals may be externally applied or intrinsic to the amplifying medium itself, i.e, amplified spontaneous emission. In the linear regime, where the population inversion of the atomic medium is unsaturated, it is well-known that the amplified noise is spectrally narrowed; much less is known about the behavior of the field in the non-linear regime. In previous work either a random phase approximation of the field has been evoked to make the problem tractable analytically [2],[3],[4], or the possibility of phase coupling has been included in time domain computer analysis [5]. In the present study we have considered the possibility of the occurence of coherence effects in the non-linear regime from the following points of view:

(a) We make the assumption that in the non-linear regime the envelope of the amplifier field can be decomposed into a slowly varying, steady state solution plus a small perturbation. This ansatz, similar to that used in laser theory [6], has recently been applied by Graham and Haken [7] to the problem of the occurence of instabilities in an infinitely extended, homogeneously broadened laser active medium, and also by Risken and Nummedal [8] in a study of self-pulsing in a homogeneously broadened ring laser. In Sec. 2 we show that this model is valid for the noise-driven laser

* Permanent Address: Purdue University, Lafayette, Indiana

amplifier if the field bandwidth $\Delta\omega$ is less than γ_1, the relaxation rate of the atomic population inversion. The instability threshold is calculated for both homogeneously broadened and weakly inhomogeneously broadened systems.

(b) We apply the pulse area:pulse energy description [9],[10] of coherent propagation in a resonant atomic medium to the noise-driven amplifier, and show in Sec. 3 that these considerations can be used to predict instabilities in situations where the ansatz of (a) is not applicable, i.e., when $\Delta\omega > \gamma_1$.

A more complete account of this work , including a discussion of the effects of self-focusing and non-resonant feedback, will be published elsewhere.

2. Small Perturbation Model

The coupled Maxwell-Bloch equations which describe the interaction of a classical electromagnetic field with a system of quantum mechanical two-level atoms are by now standard in the literature (see e.g., reference 11), and therefore we do not repeat their derivation here. In the slowly varying envelope and phase approximation for the linearly polarized plane wave field

$$\underline{E}(\underline{r},t) = \varepsilon(z,t)\cos[(kz - \nu t) + Q(z,t)]\,\hat{a}_x,\quad k = \frac{\nu}{c}, \tag{1}$$

these equations are

$$\frac{\partial\varepsilon(z,t)}{\partial z} + \frac{1}{c}\frac{\partial\varepsilon(z,t)}{\partial t} + \frac{\kappa}{2}\varepsilon(z,t) = -\frac{\nu}{2c\varepsilon_0}\int P(\Delta,z,t)g(\Delta)d\Delta \tag{2}$$

$$\frac{\partial P}{\partial t} = -\gamma_2 P - \Delta Q - \frac{\mu^2}{\hbar}\varepsilon W \tag{3}$$

$$\frac{\partial Q}{\partial t} = -\gamma_2 Q + \Delta P \tag{4}$$

$$\frac{\partial W}{\partial t} = -\gamma_1(W - \overline{W}) + \frac{\varepsilon P}{\hbar}\;, \tag{5}$$

where Q and P are the in-phase and out-of-phase components of the macroscopic polarization $\underline{P}$ as a function of the difference $\Delta = (\omega - \nu)$ between the atomic resonance and central field frequencies

$$P(\Delta,z,t) = Q(\Delta,z,t)\cos(kz - \nu t) + P(\Delta,z,t)\sin(kz - \nu t)$$

W is the population inversion density which relaxes to $\overline{W}$ in the absence of a field, $g(\Delta)$ is the normalized inhomogeneous lineshape function, and $\mu, \kappa, \gamma_1, \gamma_2$ are the transition dipole moment, non-resonant linear loss per unit length, and population and polarization relaxation rates, respectively. With the assumption that $g(\Delta)$ is symmetrical around $\Delta = 0$, we have set $Q(z,t)$ equal to zero [11]. It is convenient to make the change of variable $\varepsilon \rightarrow -\mu\varepsilon/\hbar$ and $P,Q \rightarrow P/\mu, Q/\mu$. In this way the new variables P,Q as well as W are dimensionless and ε has the dimensions of inverse time. The transformed equations are:

$$\frac{\partial\varepsilon(z,t)}{\partial z} + \frac{1}{c}\frac{\partial\varepsilon(z,t)}{\partial t} + \frac{\kappa}{2}\varepsilon(z,t) = \frac{\alpha'}{2}\int P(\Delta,z,t)g(\Delta)\,d\Delta \tag{6}$$

$$\frac{\partial P}{\partial t} = -\gamma_2 P - \Delta Q + \varepsilon W \tag{7}$$

$$\frac{\partial Q}{\partial t} = -\gamma_2 Q + \Delta P \tag{8}$$

$$\frac{\partial W}{\partial t} = -\gamma_1(W - \overline{W}) - \varepsilon P \tag{9}$$

where $\alpha' = \mu^2\nu/\varepsilon_o\hbar c$.

The standard method of testing the system of equations (6)-(9) for the occurence of instabilities is to assume that the envelope $\varepsilon(z,t)$ can be representated as the sum of a large, slowly varying, steady state solution plus a small perturbation. If, under certain conditions, the perturbation starts to grow exponentially in space or time we have an instability. There are two basic questions involved in applying this technique:

(1) In what regime is the steady state solution plus small perturbation ansatz a valid representation of the field envelope? For example, Risken and Nummedal [8] have assumed that the steady state solution for the ring laser is strictly constant in space and time. Because of the stability of the laser oscillator above threshold

this should be a very good approximation; however, the situation is more subtle in the case of a noise driven amplifier. In the absence of phase coupling, both externally applied noise signals and amplified spontaneous emission can be represented as narrowband Gaussian noise with an envelope which changes slowly over time intervals small compared with the inverse of the field bandwidth $\Delta\omega(z)$ [12]. If $\Delta\omega(z)$ narrows sufficiently during linear amplification so that the inequality

$$\Delta\omega < \gamma_1 \tag{10}$$

is satisfied, the atomic inversion will approach its steady state (asymptotic) value before the envelope has changed appreciably, and will thereafter adiabatically maintain this value. Thus the system can be described by equations characteristic of an atomic system interacting with a monochromatic field with the constant field envelope replaced by a slowly varying function. The proof of this result is identical to that given by Icsevgi and Lamb for the propagation of long (duration $\Delta\tau > 1/\gamma_1$) coherent pulses in a laser amplifier [13]. From a physical point of view Eq. (10) implies that the coherence time of the amplified noise is longer than the lifetime of the atomic transition so that the atoms "remember" an approximately constant field. In particular, note that Eq. (10) will be satisfied in the case of amplified spontaneous emission in a homogeneously broadened amplifier where $\Delta\omega(0) = \gamma_1$. As will be shown in Sec. 3, the field in the non-linear regime grows in an approximately linear fashion for moderate saturation, and tends asymptotically to a constant when the saturated gain is balanced by the non-resonant loss. We will derive the instability threshold for a constant steady state envelope; the case of a perturbation to the linear growth solution can be treated in a similar fashion.

(2) The second question concerns the nature of the instability. As discussed above, if Eq. (10) is satisfied the perturbed field can be represented as

$$\varepsilon(z,t) = \varepsilon_0(z,t) + \{\varepsilon' \exp i(kz - \nu' t) + c.c.\} \tag{11}$$

where $\varepsilon_0(z,t)$ is the steady state solution. Two types of instability, absolute and convective, can occur depending on the physics of the situation. The former means exponential growth in time at a fixed point in space (complex ν' with $\mathrm{Im}\nu' > 0$ for real k), while the latter gives exponential growth in space (complex k with $\mathrm{Im}k < 0$ for real ν'). In both the uniformly pumped ring laser and high gain noise-driven amplifier the instability is convective. However, because of the feedback in the ring, the propagating instability is returned to the system and becomes an absolute one. [14]

The proviso that the pumping be uniform means that in the amplifier the instability at a point in the saturation region is driven by the input noise and not by the local spontaneous emission. [15] If the pumping is inhomogeneous, absolute instabilities can also occur.

It is convenient to rewrite the ansatz, Eq. (11), for the amplifier as

$$\varepsilon(z,t) = \varepsilon_o + \{\varepsilon e^{Gz} e^{i(kz-\nu' t)} + c.c.\}. \tag{12}$$

The instability threshold is defined as the condition which gives G=0 for real ν' and k. Substituting Eq. (12) and similar expressions for the atomic variables into Eqs. (6) - (9), and linearizing the resulting equations in the small perturbations, we obtain in a straight-forward manner the following equation for the instability threshold

$$[i(k - \frac{\nu'}{c}) + \kappa] = \alpha' \int d\Delta g(\Delta) \frac{[W_o(\Delta) - \frac{\varepsilon_o P_o(\Delta)}{(\gamma_1 - i\nu')}]}{[(\gamma_2 - i\nu') + \frac{\Delta^2}{(\gamma_2 - i\nu')} + \frac{\varepsilon_o^2}{(\gamma_1 - i\nu')}]} \tag{13}$$

where ε_o, $P_o(\Delta)$, and $W_o(\Delta)$, the steady state values of ε, P, and W are given by

$$\kappa\varepsilon_o = \alpha' \int d\Delta g(\Delta) P_o(\Delta), \tag{14}$$

$$P_o(\Delta) = \frac{\varepsilon_o \overline{W} \gamma_2}{[\Delta^2 + \gamma_2^2 + (\gamma_2/\gamma_1)\varepsilon_o^2]}, \tag{15}$$

$$W_o(\Delta) = \frac{(\Delta^2 + \gamma_2^2)}{[\Delta^2 + \gamma_2^2 + (\gamma_2/\gamma_1)\varepsilon_o^2]}. \tag{16}$$

In general, Eq. (13) must be evaluated by numerical techniques. However, for the homogeneously broadened amplifier it is possible to derive an analytic expression for the instability threshold. [7],[8]. Substituting $g(\Delta) = \delta(\Delta)$ we obtain the following quartic equation for ν'

$$[\nu'^2 - \gamma_1\gamma_2(1+\lambda)]^2 + \nu'^2(\gamma_1 + \gamma_2)\gamma_1$$

$$-\gamma_1\gamma_2(1-\lambda)[\gamma_1\gamma_2(1+\lambda) - \nu'] = 0 \tag{17}$$

where $\lambda = \frac{(\overline{W}-W_o)}{W_o} = \frac{\varepsilon_o^2}{\gamma_1\gamma_2}$.

From Eq. (18) it follows directly that ν' real implies

$$\lambda \geqslant \lambda_{th}$$

$$= \{3(\gamma_1/\gamma_2) + 4\} + 2\{2(\gamma_1/\gamma_2)^2 + 6(\gamma_1/\gamma_2) + 4\}^{\frac{1}{2}} . \tag{18}$$

The corresponding spiking frequency is

$$\nu'^2_{th} = \{\frac{3\gamma_1\gamma_2\lambda_{th} - \gamma_1^2}{2}\} = \{\frac{3}{2}\ \varepsilon^2_{th} - \frac{\gamma_1^2}{2}\} , \tag{19}$$

where $\varepsilon^2_{th} = \gamma_1\gamma_2\lambda_{th}$.

In the limits $(\gamma_1/\gamma_2) << 1$ and $(\gamma_1/\gamma_2) \simeq 1$, Eqs. (18) and (19) reduce to [16]

$$\lambda_{th} \simeq 8; \quad \nu'^2_{th} \simeq 12\gamma_1\gamma_2, \quad (\gamma_1/\gamma_2) << 1 \tag{20}$$

$$\lambda_{th} \simeq 14; \quad \nu'^2_{th} \simeq 20\gamma_1\gamma_2 , \quad (\gamma_1/\gamma_2) \simeq 1. \tag{21}$$

Turning now to the inhomogeneously broadened amplifier, the instability threshold is determined by first equating the real parts of Eq. (13), and then eliminating $P_o(\Delta)$ and $W_o(\Delta)$ using Eqs. (15) and (16). We obtain

$$\frac{1}{W}\ \frac{\kappa}{\alpha'} = \gamma_2 \int_{-\infty}^{+\infty} \frac{d\Delta g(\Delta)}{\{\Delta^2+\gamma^2+ (\gamma_2/\gamma_1)\varepsilon_o^2\}} \equiv F(\varepsilon_o)$$

$$= \mathrm{Re} \int_{-\infty}^{+\infty} d\Delta g(\Delta) \frac{\left[\dfrac{(\Delta^2+\gamma_2^2)}{(\Delta^2+\gamma_2^2+(\gamma_2/\gamma_1)\varepsilon_o^2)} - \dfrac{\varepsilon_o^2\gamma_2}{(\gamma_1-i\nu')[\Delta^2+\gamma_2^2+(\gamma_2/\gamma_1)\varepsilon_o^2]}\right]}{\left[(\gamma_2 - i\nu') + \dfrac{\Delta^2}{(\gamma_2-i\nu')} + \dfrac{\varepsilon_o^2}{(\gamma_1-i\nu')}\right]}$$

$$\equiv R_e G(\varepsilon_o,\nu') . \tag{22}$$

We have evaluated $F(\varepsilon_o)$ and $R_e G(\varepsilon_o,\nu')$ on a computer assuming a Gaussian form for $g(\Delta)$, i.e.,

$$g(\Delta) = \frac{1}{\gamma_2^* \sqrt{\pi}} \exp(-\Delta^2/\gamma_2^*) .$$

The procedure is as follows: choosing values for γ_1,γ_2 and ε_o^2, with γ_2^* set equal to one to fix the scale, we compute $F(\varepsilon_o)$ and $R_e G(\varepsilon_o,\nu')$ for a range of values of ν'. The process is repeated varying ε_o^2 until a value of ε_o^2 is found for which Eq. (22) is satisfied. Representative results are given in table 1, part a. For comparison we also give in part b similar results for the homogeneously broadened amplifier computed using Eqs. (19) and (20). It is seen that the instability threshold is lower and the spiking frequency is about the same for weak inhomogeneous broadening vis a vis homogeneous broadening. However, it should be borne in mind that the ansatz Eq. (11) is only valid when the inequality Eq. (10) is satisfied. In the absence of phase locking, the field bandwidth, which is initially equal to the inhomogeneous width γ_2^* for amplified spontaneous emission noise, narrows during linear amplification, and rebroadens in the non-linear regime due to saturation. Maximum gain narrowing occurs approximately at the beginning of the saturation regime $z=z_s$, where the unsaturated gain α_o is reduced by a factor of 1/2. The field bandwidth at this point is [3]

$$\frac{\Delta\omega(z_s)}{\gamma_2^*} \simeq \sqrt{\frac{2}{\alpha_0 z_s}} \quad . \tag{23}$$

Combining Eqs. (10) and (23) we obtain

$$\gamma_2^* < \sqrt{\frac{\alpha_0 z_s}{2}} \quad \gamma_1 \tag{24}$$

as the criterion for the validity of the ansatz Eq. (11) in the vicinity of z_s. Experimental evidence bearing on Eq. (24) is only available for the Xenon amplifier [2],[4] where it has been found that

$$\gamma_2^* \simeq 10^8\text{Hz} \quad ; \gamma_1 \simeq 10^6\text{Hz}, \sqrt{\frac{\alpha_0 z_s}{2}} \simeq 5,$$

so that Eq. (24) does not hold in this case. The situation is worse further into the saturated regime since the field bandwidth eventually rebroadens to the inhomogeneous width. The model would have greater validity in an amplifier where the distributed losses are just sufficient to keep the gain from ever saturating. Calculations by Casperson and Yariv [4] indicate that the Doppler line in a distributed loss Xenon amplifier would be narrowed by a factor of about 4×10^7. We also note that Eq. (24) may be well-satisfied for an applied narrowband noise signal.

3. Pulse Area: Pulse Energy Model

We have seen in Sec. 2 that the usual linearization ansatz can be applied to the problem of instabilities in the noise-driven amplifier only if the field bandwidth $\Delta\omega$ is less than the relaxation rate of the atomic population inversion, γ_1. Previous work on the propagation of intense coherent light pulses in a resonant atomic medium has demonstrated the efficacy of the pulse area: pulse energy description [9],[10] of these phenomena. In this section we outline an approach to the question of instability in an inhomogeneously broadened noise-driven amplifier, for which $\gamma_2^* > \Delta\omega > \gamma_1$ from this point of view. We start from the observation that the envelope of the unlocked narrowband Gaussian noise propagating in the amplifier can be considered as a train of pulses of average duration $\tau(z) = 2/\Delta\omega(z)$ and average amplitude equal to the noise intensity $I(z,\omega)$. The fluctuations in the duration and amplitude of these noise pulses are a consequence of the randomness in the phases of the

Table 1

Spiking Threshold and Spiking Frequency

(a) Inhomogeneous Broadening ($\gamma_2^* = 1$)

γ_1	γ_2	ε_{th}^2	$(\varepsilon_{th}^2/\gamma_1\gamma_2)$	ν'^2_{th}	$\nu_{th}^2/\gamma_1\gamma_2$
10	10	1350	13.5	2000	20
3	3	122	13.5	175	19.5
2	2	54	13	78	19.5
1	1	8.7	8.7	13.2	13.2
1	10	82	8.2	125	12.5
2	20	350	8.7	525	13
0.1	10	7	7	10	10
0.1	1	0.2	2	0.5	5
0.01	1	0.018	1.8	0.05	5

(b) Homogeneous Broadening

10	10	1400	14	2000	20
3	3	126	14	180	20
2	2	56	14	80	20
1	1	14	14	20	20
1	10	80	8	120	12
0.1	1	0.8	8	1.2	12

Fourier components of the field. Using the Wiener-Khinchine theorem we can characterize the individual noise pulses by their average area and energy. The instability manifests itself in the following manner. In the saturation region of the amplifier some of the noise pulses will grow in area more than an average pulse, while others will grow less. We assume that the spatial evolution of these large area pulses can be followed using the area:energy description. The effect of the small area pulses is incorporated in the gain constant in the area: energy equations, i.e, the gain is a function of the background noise intensity averaged over a time interval T_1 preceding the large area pulses. If $\Delta\omega \gg 1/T$, then many pulses will occur in a time T_1 and the "T_1" gain will not differ much from the long time average gain of the noise train. However, the gain immediately following a large area pulse will be suppressed, and hence pulses in the wake of such a pulse will grow less or may even be attenuated if the area of the large pulse is greater than $\pi/2$. Since the gain will recover to its average value in a time $(P+1/T_1)$ where P is the pump rate, this should be the average period of the large area pulses. It should be noted that in this approach the "frequency" of the instability is prescribed by the width of the pulses, in contrast to the small perturbation model where it was a derived result. Moreover, the latter is approximately equal to the Rabi flopping frequency (see Eq. (19)), which suggests that the nature of the instability is connected to the growth of coherent polarization in the medium, while the instability considered here is non-sinusoidal and statistical in nature. We develop below the quantiative aspects of the model ; first, however, the basic ideas of the area: energy description of coherent pulse propagation will be reviewed.

The area theorem of McCall and Hahn [18] was originally derived in order to explain the phenomenon of self-induced transparency of short (compared with T_2) intense, coherent pulses propagating in an inhomogeneously broadened attentuating medium. The area theorem can also be applied to an amplifying medium including the effect of nonresonant loss which is necessary for steady state propagation in the asymptotic regime. More recently, McCall and Hahn [9] and Courtens [10] have shown that the "spirit" of the area theorem can be extended to give a more complete description of coherent pulse propagation in terms of pulse area, energy, and duration. In the following we use the notation of reference 10.

From Eqs. (2)-(5) with $\gamma_1 \gamma_2 = 0$ it follows directly [18] that the spatial evolution of the pulse area A(z) defined by

$$A(z) = \frac{\mu}{\hbar} \int_{-\infty}^{+\infty} \varepsilon(z,t)\, dt \tag{25}$$

is determined by the area theorem

$$\frac{dA(z)}{dz} = \frac{\alpha}{2} \sin A(z) - \frac{\kappa}{2} A(z) \quad , \tag{26}$$

where $\alpha = \frac{\pi\mu^2 W}{\varepsilon_o \hbar} g(0)$

is the gain and W the inversion at line center before the arrival of the pulse. One would expect that the pulse energy

$$S(z) = \frac{c}{8\pi} \int_{-\infty}^{+\infty} \varepsilon^2(z,t)\, dt \tag{27}$$

the pulse width $\tau(z)$ and the area $A(z)$ are related by $\tau \propto A^2/s$ where the proportionality constant is determined by the pulse shape. This notion can be made precise if we define the pulse duration $\tau(z)$ in the following manner [19]

$$\tau(z) = \frac{2}{\pi^2} \quad \frac{[\int_{-\infty}^{+\infty} \varepsilon(z,t)dt]^2}{[\int_{-\infty}^{+\infty} \varepsilon^2(z,t)dt]} \quad . \tag{28}$$

It follows from Eqs. (25), (27), and (28) that

$$S(z) = \frac{\hbar^2 c}{4\pi^3 \mu^2} \quad \frac{A^2(z)}{\tau(z)} \quad . \tag{29}$$

In references 9 and 10 it is shown that S(z) approximately obeys the following equation of motion

$$\frac{1}{S(z)} \quad \frac{dS(z)}{dz} = 2\alpha \left[\frac{1 - \cos A}{A^2}\right] - \kappa \quad . \tag{30}$$

This equation is exact for small area pulses, i.e., when $(1-\cos A) \simeq A^2/2$ Eq. (30) becomes

$$\frac{1}{S(z)} \frac{dS(z)}{dz} \simeq (\alpha-\kappa), \tag{31}$$

which correctly describes the Beer's Law exponential growth of weak signals. We now turn to the area:energy description of unlocked noise pulses. The average energy of a single noise pulse can be related to the intensity of the entire noise train using the Wiener-Khinchin Theorem in the form

$$\frac{c}{8\pi} \overline{\varepsilon^2(z)} = \lim_{T\to\infty} \frac{1}{T} \int_{-T/2}^{+T/2} \frac{c}{8\pi} \varepsilon^2(z,t)\, dt$$

$$= \frac{1}{2\pi} \int_{-\infty}^{+\infty} \lim_{T\to\infty} \frac{1}{T} \frac{c}{8\pi} \overline{|\tilde{\varepsilon}_T(z,\omega)|^2} d\omega$$

$$= \frac{1}{2\pi} \int_{-\infty}^{+\infty} I(z,\omega)\, d\omega , \tag{32}$$

where we have assumed that $\varepsilon(z,t)$ is an ergodic random process. $\tilde{\varepsilon}_T(z,\omega)$ is the Fourier spectrum of $\varepsilon(z,t)$ over a long time interval T.

$$\varepsilon_T(z,\omega) = \int_{-T/2}^{+T/2} \varepsilon(z,t) e^{-i\omega t}\, dt,$$

and

$$I(z,\omega) = \lim_{T\to\infty} \frac{1}{T} \frac{c}{8\pi} \overline{|\varepsilon_T(z,\omega)|^2}$$

is the power spectral density per unit cross-sectional area of the amplifier. (The bar over $|\varepsilon_T(z,\omega)|^2$ indicates that an ensemble average must be taken before the limit). From Eqs. (27), (29), and (32) we obtain the following approximate expressions for the average energy $S_{UL}(z)$ and area $A_{UL}(z)$ of a single unlocked noise pulse

$$S_{U.L}(z) = \frac{c}{8\pi} \overline{\varepsilon^2(z)} \cdot \tau_{U.L}(z)$$

$$= \frac{c}{8\pi}\overline{\varepsilon^2(z)} \quad \frac{2}{\Delta\omega(z)} = \frac{1}{\pi\Delta\omega(z)} \int_{-\infty}^{+\infty} I(z,\omega)\, d\omega$$

$$\simeq \frac{1}{\pi\Delta\omega(z)} \; I(z,\omega{=}0) \quad \Delta\omega(z)$$

$$= \frac{1}{\pi} \; I(z,\omega{=}0) \; \equiv \frac{1}{\pi} \; I(z) \; . \tag{33}$$

$$A^2_{UL}(z) = \frac{4\pi^3\mu^2}{\hbar^2 c} \quad S_{U.L}(z) \; \tau_{U.L}(z)$$

$$= \frac{8\pi^2\mu^2}{\hbar^2 c} \quad \frac{I(z)}{\Delta\omega(z)} \; . \tag{34}$$

As previously explained, our viewpoint towards the instability problem is that some of the noise pulses will grow more than the average and that their growth can be specified by using the area theorem. Mathematically, this is equivalent to the statement that there are noise pulses for which the inequality

$$\frac{1}{A_{UL}(z)} \quad \frac{dA_{UL}(z)}{dz} = \frac{1}{2I(z)}\frac{dI(z)}{dz} - \frac{1}{2}\frac{1}{\Delta\omega(z)} \quad \frac{d\Delta\omega(z)}{dz}$$

$$= \frac{(\bar{\alpha} - \kappa)}{2} - \frac{1}{2} \quad \frac{1}{\Delta\omega(z)} \quad \frac{d\Delta\omega(z)}{dz}$$

$$< \frac{1}{A(z)} \quad \frac{dA(z)}{dz} \simeq \frac{(\alpha - \kappa)}{2} - \frac{\alpha A^2}{12} \tag{35}$$

is satisfied. Here $\bar{\alpha}$ is the long time average gain as a function of the intensity $I(z)$ and α is the "T_1" gain seen by a large area pulse. Setting $\alpha \simeq \bar{\alpha}$ Eq. (35) tells us that a large area pulse will grow until area $A(z\ )$ attains the value

$$A_m(z) \simeq \frac{6}{\bar{\alpha}} \quad \frac{1}{\Delta\omega(z)} \quad \frac{d\Delta\omega(z)}{dz} \; . \tag{36}$$

To obtain a quantitative estimate of $A_m(z)$ we calculate $\bar{\alpha}$ and $\Delta\omega(z)$ in the inhomogeneously (Doppler) amplifier assuming there is no phase coupling. The basic equation governing the growth of the intensity is

$$\frac{dI(z,\omega)}{dz} = [\bar{\alpha}(\omega) - \kappa] I(z,\omega)$$

$$= \left\{ \frac{\alpha_0(\omega)}{[1 + \frac{I(z,\omega)}{I_S}]} - \kappa \right\} I(z,\omega), \tag{37}$$

where $\alpha_0(\omega) = \alpha_0 \exp-(\omega^2/\Delta^2\omega_D), \Delta\omega_D$ is the Doppler width and the saturation intensity I_S is given by

$$I_S = [\frac{4\pi^2\mu^2(\gamma_a^{-1} + \gamma_b^{-1})}{\hbar^2 c}]^{-1} \tag{38}$$

(γ_a, γ_b are the relaxation rates of the upper and lower laser levels, respectively).

If we define the asymptotic intensity $I(\infty,\omega) \equiv I_\infty$ by

$$\frac{dI}{dz} = 0 = \left\{ \frac{\alpha_0(\omega)}{(1 + \frac{I_\infty}{I_S})} - \kappa \right\} I_\infty \tag{39}$$

and denote by z_S the point in the amplifier at which saturation effects become important, i.e., $I(z_S, \omega=0) = I_S$, then the exact solution of Eq. (37) is

$$[\alpha_0(\omega) - \kappa]\ (z - z_S)$$

$$= \frac{\alpha_0(\omega)}{\kappa} \ln [\frac{I_\infty - I(z_S,\omega)}{I_\infty - I(z,\omega)}] + \ln \frac{I(z,\omega)}{I(z_S,\omega)} \cdot \tag{40}$$

Parametrizing $I(z,\omega) = I(z)\exp(-\omega^2/\Delta^2\omega(z))$ and equating the leading terms in a power series expansion in ω^2 we obtain from Eq. (29) the following solutions for $I(z)$ and $\Delta\omega(z)$ valid for $I(z) < I_\infty$, i.e., in the region where the intensity is still growing.

$$\frac{\alpha_o/\kappa}{(\alpha_o/\kappa-2)}\ \frac{I(z)}{I_S} + \ln\frac{I(z)}{I_S} = [(\alpha_o-\kappa)(z-z_S)+1], \tag{41}$$

$$\begin{aligned}
&\left[\frac{\alpha_o^2}{\kappa}\left(\frac{1}{\Delta\omega_S^2} - \frac{2}{\Delta\omega_D^2}\right) - \alpha_o\left(\frac{2}{\Delta\omega_S^2} - \frac{3}{\Delta\omega_D^2}\right) + \frac{\kappa}{\Delta\omega_S^2}\right](z-z_S)\\
&= -\frac{\alpha_o}{\kappa}\frac{1}{\Delta\omega_D^2}\left[\frac{I(z)}{I_S} - 1\right] - \frac{\alpha_o}{\kappa}\frac{I(z)}{I_S}\left(\frac{1}{\Delta\omega^2} - \frac{1}{\Delta\omega_S^2}\right)\\
&- \left(\frac{\alpha_o}{\kappa} - 2\right)\left(\frac{1}{\Delta\omega^2} - \frac{1}{\Delta\omega_S^2}\right) + \ln\frac{I(z)}{I_S}\left[\left(\frac{\alpha_o}{\kappa} - 1\right)\frac{1}{\Delta\omega_S^2} - \frac{\alpha_o}{\kappa}\frac{1}{\Delta\omega_D^2}\right]
\end{aligned} \tag{42}$$

where $\Delta\omega_S = \Delta\omega(z_S)$.

Assuming that $(\kappa/\alpha_o) \ll 1$ the presence of non-resonant loss adds only a small correction outside the asymptotic region. Setting $K = 0$ we obtain via Eqs. (23), (41) and (42) the physically more transparent results.

$$\frac{I(z)}{I_S} + \ln\frac{I(z)}{I_S} = [\alpha_o(z-z_S) + 1] \tag{43}$$

and

$$\left(\frac{\Delta\omega}{\Delta\omega_D}\right)^2 = \frac{1}{\alpha_o z}\left(\frac{I(z)}{I_S} + 1\right). \tag{44}$$

Solving Eq. (43) for $z \simeq z_S$ and $z >> z_S$ and substituting the result into Eqs. (36) and (44) we obtain

(a) $z \simeq z_S$

$$\frac{I(z)}{I_S} \simeq [\frac{\alpha_o}{2}(z-z_S) + 1] \tag{45a}$$

$$\bar{\alpha} = \frac{\alpha_o}{(1 + \frac{I(z)}{I_S})} \simeq \frac{\alpha_o}{[\frac{\alpha_o}{2}(z-z_S) + 2]} \tag{45b}$$

$$(\frac{\Delta\omega}{\Delta\omega_D})^2 \simeq \frac{1}{\alpha_o z}[\frac{\alpha_o}{2}(z-z_S) + 2] \tag{45c}$$

(b) $z >> z_S$

$$\frac{I(z)}{I_S} \simeq [\alpha_o(z-z_S) + 1] - \ln[\alpha_o(z-z_S) + 1] \tag{46b}$$

$$\bar{\alpha} = \frac{\alpha_o}{[\alpha_o(z-z_S) + 2] - \ln[\alpha_o(z-z_S) + 1]} \tag{46b}$$

$$(\frac{\Delta\omega}{\Delta\omega_D})^2 = \frac{1}{\alpha_o z}\{[\alpha_o(z-z_S) + 2] - \ln[\alpha_o(z-z_S) + 1]\} . \tag{46c}$$

Differentiating Eqs. (44c) and (45c) and substituting the resulting expressions and Eqs. (44b) and (45b) into Eq. (35), we obtain the following estimates of $A_m^2(z)$

$$A_m^2(z) \simeq \frac{3}{\alpha_o z}(\frac{\alpha_o}{2} z_S - 2) , \qquad z \simeq z_S$$

$$\simeq 1.5, \qquad \alpha_o z_S >> 1 \tag{46}$$

$$A_m^2(z) \simeq \frac{3}{\alpha_o z}(\alpha_o z_S - 2) \quad , \qquad z >> z_S$$

$$\simeq 0.3, \ z = 10z_S \quad , \quad \alpha_o z_S >> 1. \tag{47}$$

These values should be compared with the average area of an unlocked noise pulse at $z = z_S$. From Eqs. (33), (37), and (23)

$$A^2_{UL}(z_S) = \frac{8\pi^2\mu^2}{\hbar^2 c} \quad \frac{I_S}{\Delta\omega(z_S)}$$

$$= \frac{2}{(\gamma_a^{-1}+\gamma_b^{-1})} \quad \frac{1}{\Delta\omega_D} \sqrt{\frac{\alpha_o z_S}{2}} \ . \tag{48}$$

We estimate $A^2_{UL}(z_S)$ using paramters characteristic of the Xenon amplifier [2], i.e., for

$$\gamma_a^{-1} = 1.35 \ \mu\text{sec},$$

$$\gamma_b^{-1} = 44 \ \text{n sec},$$

$$\Delta\omega_D = 117 \ \text{MHz}.$$

$$\sqrt{\frac{\alpha_o z_S}{2}} \simeq 5,$$

we obtain $A^2_{UL}(z_S) \simeq 4 \times 10^{-3}$.

It is seen that a large area pulse can grow to an area orders of magnitude larger than an unlocked pulse. This is especially true near the onset of the saturation region where the linewidth starts to rebroaden after considerable gain narrowing ($\alpha_o z_S$>>1). The rebroadening is a necessary condition for the instability to develop. The rate of growth of the area of the unstable pulse can be estimated by integrating the area theorem in the small pulse approximation sin A≃A. Substituting Eq. (36) with $I(z)/I_S$ given by Eq. (44a) into Eq. (26) with sin A ≃ A we get

$$A(z) \simeq A(z_S) \ [\alpha_o(z-z_S)]^{\frac{1}{2}} \tag{49}$$

in contrast to the unlocked pulse which grows slower

$$A_{UL}(z) \simeq A_{UL}(z_S)[\alpha_o(z-z_S)]^{1/4} . \qquad (50)$$

In order to close the argument we must show that the gain in the wake of a large area pulse is considerably depressed, thus enhancing the instability. By integrating the Bloch equations, neglecting relaxation terms, it follows that the fractional change in the gain following a large area pulse $\Delta\alpha/\alpha$ is given by

$$\Delta\alpha/\alpha = \Delta W/W = (\cos A - 1) \simeq -A^2/2 . \qquad (51)$$

Our argument eventually breaks down for large area pulses.

This paper is a preliminary discussion of the onset of instability in a noise driven amplifier. The medium consists of a system of continuously pumped, inhomogenoeusly broadened two-level atoms interactions, with a plane wave electromagnetic field treated in the slowly varying amplitude approximation with a constant phase. Four important effects are neglected: the feedback through Rayleigh scattering and through the neglected second order terms in the field equation, the transverse structure of the field, and the effects of spontaneous emission throughout the amplifier volume. The latter adds to the noise locally, but more importantly, it gives rise to a wave travelling in the backward direction thus saturating the medium in a manner different from that discussed here. Within the approximations of our model we have obtained results for the following regimes

(a) $\Delta\omega < \gamma_1$ (Sec. 2)

(b) $\gamma_1 < \mu E/\hbar < \Delta\omega$ (Sec. 3)

In addition to (a) and (b) there are two other important regimes

(c) $\gamma_1 < \Delta\omega < \mu E/\hbar$

(d) $\gamma_1 < \Delta\omega < \gamma_2$.

Case (c) means that there are large area (π) pulses propagating in the amplifier. From the work of e.g., Hopf and Scully, [11] we expect these large area pulses will break up into smaller pulses, and in this sense the amplifier is also unstable in this regime. We hope to discuss case (d) and more refined considerations concerning the statistics of the gain in the area:energy model in a subsequent publication.

References

1. GaAs: A. Yariv and R. Leite, J. Appl. Phys. *34*, 3410 (1963).
   Pb: W.T. Silfvast and J.S. Deech, Appl. Phys. Letters *11*, 97 (1967).
   Th: A.A. Isaev, P.I.Ishchenko and G.G. Petrash, JETP Letters *6* , 118 (1967)
   Ne: D.A. Leonard, W.R. Zinky, Appl Phys. Letters *12*, 113 (1968).
   HF,DF: J. Goldhar, R.M. Osgood, Jr., A. Javan, Appl. Phys. Letters *18*, 167 (1971).
   Xe: H. Gamo, S. Chuang, to be published.
2. J.H. Parks, A. Javan and A. Szöke, Phys. Rev., to be published.
3. M.M. Litvak, Phys. Rev. *A2*, 2107 (1970).
4. L.W. Casperson and A. Yariv, IEEE J. Qu. Elec. *8* , 80 (1972).
5. F.A. Hopf and J.H. Parks, J. Opt. Soc. Am. *61*, 659 (1971).
6. See, e.g., H. Haken, Z. Physik *181*, 96 (1964). This ansatz is also useful in other situations where instabilities occur; for an application to the theory of turbulence, see L.D. Landau and E.M. Lifshitz, *Fluid Mechanics* (Pergamon Press, Oxford, 1959) Chapter III.
7. R. Graham and H. Haken, Z. Physik *213* , 420 (1968).
8. H. Risken and K. Nummedal, J. Appl. Phys. *39*, 4662 (1968).
9. S.L. McCall and E.L. Hahn, The University of Arizona, Technical Report (45, Vol. 1. p 323. (1969)
10. E. Courtens, *Lectures on Coherent Resonant Propagation in Absorbers*, Chania Conference on Short Laser Pulses and Coherent Interactions, (1969).
11. F.A. Hopf and M.O. Scully, Phys. Rev. *179*, 399 (1969).
12. The field bandwidth is defined in the conventional manner; see e.g., reference 2. A more detailed discussion is given in Section III.
13. A. Icsevgi and W.E. Lamb, Jr., Phys. Rev. *185*, 517 (1969).
14. A.I. Akhiezer and R.V. Palovin, Soviet Physics Uspekhi, *14*, 278 (1971).
15. We thank F.A. Hopf for stimulating correspondence concerning this point.
16. These results have also been found previously by F.A. Hopf (private communication.
17. For a comprehensive review, see E. Courtens, Coherent Pulse Propagation in *Laser Handbook*, edited by F.T. Arrechi and E.O. Schulz-DuBois (North-Holland Publishing Company, 1972).
18. S.L. McCall and E.L. Hahn, Phys. Rev. Letters, *18*, 908 (1967).
19. The coefficient is appropriate for hyperbolic secant pulses: other pulse shapes give results of the same order of magnitude.

AUTHOR INDEX

SUBJECT INDEX